Developments in Environmental Modelling
Volume 26

Ecological Modelling and Engineering of Lakes and Wetlands

Developments in Environmental Modelling

1. ENERGY AND ECOLOGICAL MODELLING edited by W.J. Mitsch, R.W. Bossermann and J.M. Klopatek, 1981
2. WATER MANAGEMENT MODELS IN PRACTICE: A CASE STUDY OF THE ASWAN HIGH DAM by D. Whittington and G. Guariso, 1983
3. NUMERICAL ECOLOGY by L. Legendre and P. Legendre, 1983
4A. APPLICATION OF ECOLOGICAL MODELLING IN ENVIRONMENTAL MANAGEMENT PART A edited by S.E. Jørgensen, 1983
4B. APPLICATION OF ECOLOGICAL MODELLING IN ENVIRONMENTAL MANAGEMENT PART B edited by S.E. Jørgensen and W.J. Mitsch, 1983
5. ANALYSIS OF ECOLOGICAL SYSTEMS: STATE-OF-THE-ART IN ECOLOGICAL MODELLING edited by W.K. Lauenroth, G.V. Skogerboe and M. Flug, 1983
6. MODELLING THE FATE AND EFFECT OF TOXIC SUBSTANCES IN THE ENVIRONMENT edited by S.E. Jørgensen, 1984
7. MATHEMATICAL MODELS IN BIOLOGICAL WASTE WATER TREATMENT edited by S.E. Jørgensen and M.J. Gromiec, 1985
8. FRESHWATER ECOSYSTEMS: MODELLING AND SIMULATION by M. Straškraba and A.H. Gnauck, 1985
9. FUNDAMENTALS OF ECOLOGICAL MODELLING by S.E. Jørgensen, 1986
10. AGRICULTURAL NONPOINT SOURCE POLLUTION: MODEL SELECTION AND APPLICATION edited by A. Giorgini and F. Zingales, 1986
11. MATHEMATICAL MODELLING OF ENVIRONMENTAL AND ECOLOGICAL SYSTEMS edited by J.B. Shukia, T.G. Hallam and V. Capasso, 1987
12. WETLAND MODELLING edited by W.J. Mitsch, M. Straškraba and S.E. Jørgensen, 1988
13. ADVANCES IN ENVIRONMENTAL MODELLING edited by A. Marani, 1988
14. MATHEMATICAL SUBMODELS IN WATER QUALITY SYSTEMS edited by S.E. Jørgensen and M.J. Gromiec, 1989
15. ENVIRONMENTAL MODELS: EMISSIONS AND CONSEQUENCES edited by J. Fenhann, H. Larsen, G.A. Mackenzie and B. Rasmussen, 1990
16. MODELLING IN ECOTOXICOLOGY edited by S.E. Jørgensen, 1990
17. MODELLING IN ENVIRONMENTAL CHEMISTRY edited by S.E. Jørgensen, 1991
18. INTRODUCTION TO ENVIRONMENTAL MANAGEMENT edited by P.E. Hansen and S.E. Jørgensen, 1991
19. FUNDAMENTALS OF ECOLOGICAL MODELLING by S.E. Jørgensen, 1994
20. NUMERICAL ECOLOGY 2nd English edition by Pierre Legendre and Louis Legendre
21. FUNDAMENTALS OF ECOLOGICAL MODELLING, Third Edition by G. Bendoricchio and S.E. Jørgensen
22. ENVIRONMENTAL FORESIGHT AND MODELS A MANIFESTO edited by M.B. Beck
23. ENTROPY PRINCIPLE FOR THE DEVELOPMENT OF COMPLEX BIOTIC SYSTEMS: ORGANISMS, ECOSYSTEMS, THE EARTH by I. Aoki
24. NUMERICAL ECOLOGY 3rd English Edition by Pierre Legendre and Louis Legendre
25. MODELS OF THE ECOLOGICAL HIERARCHY: FROM MOLECULES TO THE ECOSPHERE edited by Ferenc Jordán and Sven Erik Jørgensen

Developments in Environmental Modelling
Volume 26

Ecological Modelling and Engineering of Lakes and Wetlands

Edited by

Sven Erik Jørgensen
Professor Emeritus, Environmental Chemistry,
University of Copenhagen,
Copenhagen, Denmark

Ni-Bin Chang
Department of Civil,
Environmental, and Construction Engineering,
University of Central Florida, Orlando, Florida, USA

Fu-Liu Xu
College of Urban and Environmental Sciences,
Peking University, Beijing, China

ELSEVIER AMSTERDAM • BOSTON • HEIDELBERG • LONDON • NEW YORK • OXFORD
PARIS • SAN DIEGO • SAN FRANCISCO • SINGAPORE • SYDNEY • TOKYO

Elsevier
Radarweg 29, PO Box 211, 1000 AE Amsterdam, The Netherlands
The Boulevard, Langford Lane, Kidlington, Oxford OX5 1GB, UK

British Library Cataloguing in Publication Data
A catalogue record for this book is available from the British Library

Library of Congress Cataloging-in-Publication Data
A catalog record for this book is available from the Library of Congress

For information on all **Elsevier** publications
visit our web site at store.elsevier.com

ISBN: 978-0-444-63249-4
ISSN: 0167-8892

Front cover image courtesy of the photographer, Mette Vejlgaard Jørgensen.
All rights reserved.

Contents

Contributors

Carles Alcaraz
IRTA Aquatic Ecosystems, E-43540 Sant Carles de la Ràpita, Catalonia, Spain

Bruce C. Anderson
Department of Civil Engineering, Queen's University, Kingston, Ontario, Canada

Gordon C. Balch
Centre for Alternative Wastewater Treatment, Fleming College, Lindsay, Ontario, Canada

Simone Bastianoni
Department of Earth, Environmental and Physical Sciences, Ecodynamics Group, University of Siena, Siena, Italy

Nuno Caiola
IRTA Aquatic Ecosystems, E-43540 Sant Carles de la Ràpita, Catalonia, Spain

Antonio Camacho
Cavanilles Institute of Biodiversity and Evolutionary Biology and Department of Microbiology and Ecology, Edificio de Investigacion, Campus de Burjassot, Universitat de Valencia, E-46100 Burjassot, Spain

María-Francisca Carreño
Department of Ecology and Hydrology, University of Murcia, Murcia, Spain

Ni-Bin Chang
Department of Civil, Environmental, and Construction, University of Central Florida, Orlando, FL, USA

Bin Chen
State Key Laboratory of Water Environment Simulation, School of Environment, Beijing Normal University, Beijing, China

G.Q. Chen
State Key Laboratory for Turbulence and Complex Systems, College of Engineering, Peking University, Beijing, China, and NAAM Group, King Abdulaziz University, Jeddah, Saudi Arabia

Shaoqing Chen
State Key Laboratory of Water Environment Simulation, School of Environment, Beijing Normal University, Beijing, China

Z.M. Chen
School of Economics, Renmin University of China, Beijing, China

Annie Chouinard
Department of Civil Engineering, Queen's University, Kingston, Ontario, Canada

Miguel-Angel Esteve-Selma
Department of Ecology and Hydrology, University of Murcia, Murcia, Spain

Pablo Farinós
Department of Ecology and Hydrology, University of Murcia, Murcia, Spain

Feng Han
Center for Water Research, College of Engineering, Peking University, Beijing, China

Wei He
MOE Laboratory for Earth Surface Processes, College of Urban & Environmental Sciences, Peking University, Beijing, China

Hongxing Hu
School of Resource and Environmental Science, Wuhan University, Wuhan, China

Carles Ibáñez
IRTA Aquatic Ecosystems, E-43540 Sant Carles de la Ràpita, Catalonia, Spain

T.A. Irene
Department of Environmental Studies, The Open University of Tanzania, Dar es Salaam, Tanzania

Jamie Jones
Department of Civil, Environmental, and Construction, University of Central Florida, Orlando, FL, USA

Sven Erik Jørgensen
Professor Emeritus, Environmental Chemistry, University of Copenhagen, Copenhagen, Denmark

Xiang-Zhen Kong
MOE Laboratory for Earth Surface Processes, College of Urban & Environmental Sciences, Peking University, Beijing, China

Yuzhao Li
Key Laboratory of Water and Sediment Sciences (MOE), College of Environmental Science and Engineering, Peking University, Beijing, China

Zhongrong Lin
Center for Water Research, College of Engineering, Peking University, Beijing, China

Wen-Xiu Liu
MOE Laboratory for Earth Surface Processes, College of Urban & Environmental Sciences, Peking University, Beijing, China

Yong Liu
Key Laboratory of Water and Sediment Sciences (MOE), College of Environmental Science and Engineering, Peking University, Beijing, China

Michela Marchi
Department of Earth, Environmental and Physical Sciences, Ecodynamics Group, University of Siena, Siena, Italy

Zachary A. Marimon
Department of Civil, Environmental, and Construction, University of Central Florida, Orlando, FL, USA

Julia Martínez-Fernández
Department of Ecology and Hydrology, University of Murcia, Murcia, Spain

Javier Martínez-López
Department of Ecology and Hydrology, University of Murcia, Murcia, Spain

Jose-Miguel Martínez-Paz
Department of Applied Economics, University of Murcia, Murcia, Spain

M. Mbogo
Department of Environmental Studies, The Open University of Tanzania, Dar es Salaam, Tanzania

T.S.A. Mbwette
Department of Environmental Studies, The Open University of Tanzania, Dar es Salaam, Tanzania

Stephen D. Murphy
Faculty of Environment, Waterloo, Ontario, Canada

Federico Maria Pulselli
Department of Earth, Environmental and Physical Sciences, Ecodynamics Group, University of Siena, Siena, Italy

Ning Qin
MOE Laboratory for Earth Surface Processes, College of Urban & Environmental Sciences, Peking University, Beijing, China

Enrique Reyes
Department of Biology, East Carolina University, Greenville, North Carolina, USA

Francisco Robledano
Department of Ecology and Hydrology, University of Murcia, Murcia, Spain

Carlos Rochera
Cavanilles Institute of Biodiversity and Evolutionary Biology and Department of Microbiology and Ecology, Edificio de Investigacion, Campus de Burjassot, Universitat de Valencia, E-46100 Burjassot, Spain

M. Senzia
Civil Engineering, Arusha Technical College, Arusha, Tanzania

Ling Shao
State Key Laboratory for Turbulence and Complex Systems, College of Engineering, Peking University, Beijing, China

Rong Tian
Institute of Remote Sensing Applications, Chinese Academy of Sciences, Beijing, China

Yong Tian
Center for Water Research, College of Engineering, Peking University, Beijing, China

Benjamin Vannah
Department of Civil, Environmental, and Construction, University of Central Florida, Orlando, FL, USA

Juan Antonio Villaescusa
Cavanilles Institute of Biodiversity and Evolutionary Biology and Department of Microbiology and Ecology, Edificio de Investigacion, Campus de Burjassot, Universitat de Valencia, E-46100 Burjassot, Spain

Qing-Mei Wang
MOE Laboratory for Earth Surface Processes, College of Urban & Environmental Sciences, Peking University, Beijing, China

Martin P. Wanielista
Department of Civil, Environmental, and Construction, University of Central Florida, Orlando, FL, USA

Brent C. Wootton
Centre for Alternative Wastewater Treatment, Fleming College, Lindsay, Ontario, Canada

Bin Wu
Center for Water Research, College of Engineering, Peking University, Beijing, China

Zi Wu
State Key Laboratory for Turbulence and Complex Systems, College of Engineering, Peking University, Beijing, China

Fu-Liu Xu
MOE Laboratory for Earth Surface Processes, College of Urban & Environmental Sciences, Peking University, Beijing, China

Zhemin Xuan
Department of Civil, Environmental, and Construction, University of Central Florida, Orlando, FL, USA

Colin N. Yates
Faculty of Environment, Waterloo, Ontario, Canada

L. Yohana
Department of Environmental Studies, The Open University of Tanzania, Dar es Salaam, Tanzania

L. Zeng
State Key Laboratory of Simulation and Regulation of Water Cycle in River Basin, China Institute of Water Resources and Hydropower Research, Beijing, China

Yanyan Zhang
MOE Laboratory for Earth Surface Process, College of Urban and Environmental Sciences, Peking University, Beijing, China

Lei Zhao
Yunnan Key Laboratory of Pollution Process and Management of Plateau Lake-Watershed, Kunming, China

Yi Zheng
Center for Water Research, College of Engineering, Peking University, Beijing, China

Rui Zou
Tetra Tech, Inc., Fairfax, Virginia, USA

Introduction

Sven Erik Jørgensen[a,*], Ni-Bin Chang[b], Fu-Liu Xu[c]

[a]*Professor Emeritus, Environmental Chemistry, University of Copenhagen, Copenhagen, Denmark*
[b]*Department of Civil, Environmental, and Construction, University of Central Florida, Orlando, FL, USA*
[c]*MOE Laboratory for Earth Surface Processes, College of Urban & Environmental Sciences, Peking University, Beijing, China*
Corresponding author: e-mail address: msijapan@hotmail.com

1.1 MODELS OF LAKES AND WETLANDS

Ecological modeling has developed rapidly in the last decades in which modeling practices for the restoration of lakes and wetlands were very much a focus. Looking at the statistics to date, lakes and wetlands have been the most well-studied ecosystems via a myriad of modeling analyses. It is noticeable that biogeochemical models are the most applicable type of ecological model. The types of ecosystems that have been well investigated using biogeochemical models from 1975 until 2010 are summarized in Table 1.1. Two-thirds of all ecological system models are classified as biogeochemical models in the 1970s and 1980s, whereas one-third of all ecological system models with salient popularity belong to biogeochemical models in the 2000s. A numerical scale from 0 to 5 illustrates the modeling effort from 1975 until 2010. In this context, 5 means that a very intense modeling effort is identified; more than 100 different modeling approaches can be classified in this category in the literature. Similarly, 4 stands for an intense modeling effort, and 25–99 different modeling approaches in the literature can be classified under this category. Cases where there is some modeling effort are represented by 3, and 10–24 different modeling approaches in the literature can be classified under this category; 2 indicates that few modeling efforts have been adopted, and 4–9 different models in the literature can be classified under this category albeit with fairly good performance. The scale number 1 indicates a few not sufficiently well calibrated and validated ecological systems models, and only 1–3 good studies in the literature can be classified under this category. Finally, 0 is associated with the situation where almost no modeling effort can be identified. Note that the classification is based on the number of different models rather than the number of case studies where these models were applied. On most occasions the same models could be used in several cases studies to address

Table 1.1 Biogeochemical Models of Ecosystems

Ecosystem	Modeling Effort
Rivers	5
Lakes, reservoirs, ponds	5
Estuaries	5
Coastal zone	4
Open sea	3
Wetlands	5
Grassland	4
Desert	1
Forests	5
Agriculture land	5
Savanna	2
Mountain lands (above timberline)	1
Arctic ecosystems	2
Coral reef	3
Wastewater systems	5

Partially reproduced from Jørgensen and Fath (2011).

different types of issues. More relevant statistics can be found in the peer-reviewed journal, *Ecological Modelling*, which has published 42% of modeling papers within all international journals during the 10-year time span from 1997 to 2006.

It is clear that lakes and wetlands are the most well-modeled ecosystems (Table 1.1). Although many of these models were developed in the 1970s and 1980s, several of them are still in use. However, it would be interesting to examine recent progress in our effort to better understand these two important ecosystems. Are there at present some new results and approaches that could be used to develop better models than we could achieve 10 or 20 years ago? In addition, are there new ecological engineering approaches that can lay down the foundation for ecological system modeling to signify and magnify the synergistic effect between both scientific regimes and help deepen our understanding of the essence of ecosystems? This book attempts to answer these questions, and the editors are of the opinion that these questions can be answered by a "yes." Hopefully, the readers will agree with us when they have read the contents.

The advances in ecological modeling of lakes and wetlands presented in the book can be summarized in the following points:

(1) Structurally dynamic models of lakes and wetlands can be developed on the basis of many successful case studies. From our experience in the use of this model type, it is possible to conclude that we are able to capture the adaptation and the shifts in species composition within a structurally dynamic modeling framework. This experience has been gained mainly during the last 10–15 years.

Chapter 2 presents the structurally dynamic model classification associated with its development and application on lakes (and wetlands).

Chapters 12 and 14 present two new interesting and illustrative case studies. Both case studies are insightful because they show the reaction of lakes to invasive species and in recovering macrophyte vegetation, which gives new insight into the possibilities of recovering lakes and wetlands. Structurally dynamic modeling analysis is not a new type of model, given the fact that the first case study was published in 1986. Yet, the progress reported in this book in Chapters 2, 12 and 14 is important because this type of modeling practice can be applied more widely when structural changes of ecosystem are of concern, as demonstrated by these case studies.

(2) Our modeling experience has recently been expanded to the Arctic and Antarctic climate-zone. Chapter 9 presents a model of an Antarctic lake where it is remarkable that the main energy driver is organic matter through the drainage water that is transported to the lake, unlike the counterparts in temperate or tropical lakes where solar radiation is the dominant factor. It implies that the state variables of the lake model are different from what we had learned and used in most other lake studies. Development of wetland models for the Arctic climate is presented in Chapters 19 and 22, both of which illustrate the minor but important changes that are needed for Arctic wetland models compared with the models developed for a tropical climate (see Chapter 24 for more insight).

(3) Only a few ecotoxicological models have been developed for lakes, but Chapters 3 and 4 give rise to two useful examples. Chapter 3 uses a so-called fugacity approach, whereas the model presented in Chapter 4 represents a unique type of biogeochemical model. The two chapters show that both types of model are applicable for environmental management of toxic substances in aquatic environments. The model developed in Chapter 7 looks into the effect of heavy metal removal by wetlands—a very important ecotoxicological question that is very relevant for the application of ecotechnological solutions to the heavy metal pollution problem in wetlands. It can be concluded that wetlands are able to remove heavy metals with a relatively good efficiency.

(4) Watershed models to control nonpoint sources are another important application of models in environmental management of lakes and wetlands, as illustrated in Chapter 5. Generally, gaining more experience with watershed models is very much needed, but it has been the general trend that more spatial models consisting of several ecosystems covering entire landscapes have been developed in the last 10 years.

(5) Some new types of model can be found in Chapters 8 and 10, in which the biophysical processes are emphasized, although they are deemed as variants of existing models. Similarly, the use of an extended network analysis in Chapter 6 illustrates a new variant of the application of ecological network modeling analysis.

(6) Modeling the effect of climate change is another new challenge. Several papers published during the last 10 years in the journal *Ecological Modelling* have covered this topic. A model covering the effect of climate change on zebra mussel dynamics in a reservoir is presented in Chapter 18, showing an additional dimension of lake/reservoir management. It has been widely discussed how climate change will affect lakes, reservoirs, and wetlands, and therefore it is important to obtain more experience in regard to the impact of climate change on these ecosystems. The content presented in Chapter 18 is very useful and should inspire the development of future models dealing with this environmental problem.

(7) Several 3D models of lakes and wetlands have been developed, although 3D models are not generally applied to lakes and wetlands due to the limited data available in most cases for the model development. However, Chapter 15 gives a 3D example that focuses on a catastrophic shift of a shallow lake. The chapter illustrates the possibilities and the data requirements for developing a 3D lake model.

(8) The use of stormwater ponds has been increasing in urban regions during the last decade due to more concentrated rainfalls as a consequent of climate change. The need for the development of good ecological models for addressing stormwater pond issues is therefore of great urgency to facilitate the design of these ponds. Chapter 17 shows how to develop a simple but very useful stormwater pond model that is able to answer the core questions about the capacity and the water quality.

(9) A model for an upflow subsurface constructed wetland has been developed by use of tracers; the resulting model is presented in Chapter 25. The model is very useful for the design of this type of subsurface wetland, which should be considered as a very applicable constructed wetland for both stormwater and wastewater treatment, particularly where effective nutrient removal is important.

The proper collection and integration of the nine types of advancements mentioned in this summary represent some of the very important progress that we have been able to realize in the context of ecosystem modeling of lakes and wetlands during the last 5 years or so. Although these ecosystems have been modeled extensively from 1970 to 2000, the new experience gained is valuable for further development of lake and wetland models, particularly under extreme climatic conditions, leading to the assessment of impacts caused by climate change and the construction of effective stormwater ponds and upflow constructed subsurface wetland for effective removal of nutrients. The progress being made in modeling for such a wealth of applications for lakes and wetlands include structurally dynamic models, 3D models, biophysical models, entire watershed models, and ecotoxicological models.

1.2 ECOLOGICAL ENGINEERING APPLIED TO LAKES AND WETLANDS

Ecological engineering is sometimes denoted as ecotechnology that is extensively applied for the management of lakes and wetlands. Ecological engineering can be classified as (see Mitsch and Jørgensen, 2003; Jørgensen, 2009 for more information):

A. Methods applied for restoration of ecosystems
B. Application of natural ecosystems to solve environmental problems
C. Construction of ecosystems to solve environmental problems
D. Ecological planning of our application and changes of nature

Ecological engineering has been widely used for the environmental management of lakes and wetlands, especially for the classes A, B, and C as listed above. In the last 5–10 years, we have expanded our experience in the application of the ecological engineering approach to gain better environmental management of lakes and wetlands. This book covers several recent advancements and precious experience, which generally is very valuable for broadening the use of the ecological engineering approach for pollution abatement. The important new experience that is covered in this book may be summarized as follows:

(1) Use of constructed wetlands for solving wastewater problems. Although some experience was available before 2003, we needed to obtain extended experience particularly under different climatic conditions and different types of wastewater characteristics. Chapter 7 presents information about the possibilities for removing and reducing phosphorus and heavy metal concentrations of wastewater by wetlands.

A software package named SubWet 2.0 was developed based on data collected from Tanzania and Canada. Because the wetland processes are different in the tropical and arctic climates, the softwarepackage has two versions—one is for warm regions and the other for cold regions. The software package is able to raise the reliability of the design of wetlands that are able to treat wastewater streams with known influent characteristics of BOD5, total-N, organic-N, ammonium-N, nitrate-N, and total P. The application of SubWet 2.0 is clearly advantageous in expanding our experience as presented in Chapter 19 for the cold climate version of the software package. In addition, Chapters 20 and 21 present additional interesting case studies and result in more applications of wetlands for the treatment of wastewater in the arctic zone. Chapter 24 demonstrates useful data for a subsurface wetland for treatment of wastewater in Tanzania, which is located in the tropical zone. Although the chapter does not present a full test of SubWet 2.0, a full test has now been performed with acceptable results, based on the presented data.

(2) Two chapters focus on the development of biophysical models of constructed wetlands, and the presented models can be considered a supplement to the to-date frequently applied biogeochemical models by using SubWet 2.0 to answer the core question regarding whether the constructed (or natural) wetland is able to reduce the waste load sufficiently. These biophysical models are important to assess the physical capacity of wetlands expressed as m/24 h, which may sometimes be the limiting factor for the amount of water that can be treated by a constructed wetland per day regardless of the required water quality. Due to a low hydraulic conductivity, the capacity of wetlands expressed as volume per area per time (m/24 h) is often a crucial property for the applicability of wetlands for treatment of wastewater and drainage water.

(3) Wetlands are widely used in Europe to treat agricultural drainage water. Such wetlands are often ecologically poor in the sense that they tend to have a low biodiversity. It is therefore important to try to find a trade-off between biodiversity and removal of nutrients for wetlands applied to treat agricultural drainage water. Chapter 11 discusses how it is possible to find a trade-off between biodiversity conservation and nutrient removal. The results presented in Chapter 11 are therefore important to consider when wetlands are used for treatment of drainage water treatment.

(4) Seeking sustainable lake restoration strategies has received wide attention, particularly in Europe, to solve eutrophication problems. These strategies can sometimes be extended to other types of aquatic ecosystems. Various lake restoration methods are discussed in the chapters dealing with structurally dynamic models because the application of structurally dynamic models is almost a "must" when restorating a lake. Chapter 2 presents cases where biomanipulation is applied as a restoration method, and Chapter 12 discusses the restoration of a lake that has been invaded by crayfish, resulting in an increased eutrophication. Chapters 13 and 14 focus on the restoration of a shallow lake by reestablishment of macrophytes.

(5) With the increasing use of stormwater ponds, it is important to gain experience in the maintenance of good ecological conditions and thereby a good water quality in stormwater ponds via the use of an ecological engineering approach. Chapters 16 and 17 present useful results for the solution of this problem.

(6) Chapter 23 presents the results of using three ecological engineering methods in series, namely a waste stabilization pond, followed by a constructed wetland and a fishpond. This solution associated with the three methods in series seems very attractive in many tropical regions where wastewater treatment can be combined with fishery production. The results in Chapter 23 present a very attractive method for wastewater treatment in tropical Africa.

(7) A new constructed subsurface wetland using an upflow hydraulic pattern has been developed and tested fully in Florida. The results are summarized in Chapter 24. The design of this type of constructed wetland is reliable as a model for this wetland type and has been developed with the aid of tracers. The model, which was formulated by a straightforward approach for the design of such a wetland, is presented in Chapter 25.

The seven ecological engineering topics mentioned in this summary emphasize the advancements being made in the application of an ecological engineering regime for better environmental management of lakes and wetlands. Important insights have been gained in the application of constructed wetlands for the treatment of wastewater, both in tropical and arctic regions. Furthermore, some progress has been achieved in the application of ecological models for the design of constructed wetlands. A readily available software package—SubWet 2.0—can be applied in the design of wetlands in various climatic conditions across tropical and arctic regions. The results obtained by the use of the software package seem promising. Furthermore, biophysical models have been developed for wetlands and a model that is well supported by tracer data is also ready for use in a new type of constructed wetland—a subsurface upflow wetland, which is able to offer effective nutrient removal. With this collection, new insight can be gained in the application of stormwater ponds, their design, and their environmental management too. Further progress has been made in ecological engineering applied to lake restoration through biomanipulation and restoration of macrophytes in shallow lakes, which deserves further attention in this context.

References

Jørgensen, S.E. (Ed.), 2009. Applications in Ecological Engineering. Elsevier, Amsterdam, p. 380.

Jørgensen, S.E., Fath, B., 2011. Fundamentals of Ecological Modelling, fourth ed. Elsevier, Amsterdam, p. 340.

Mitsch, W.J., Jørgensen, S.E., 2003. Ecological Engineering and Ecosystem Restoration. John Wiley and Sons, New York, NY, p. 412.

Structurally Dynamic Models of Lakes

Sven Erik Jørgensen*

Professor Emeritus, Environmental Chemistry, University of Copenhagen, Copenhagen, Denmark
*Corresponding author: e-mail address: msijapan@hotmail.com

2.1 INTRODUCTION

Ecological models attempt to capture the characteristics of ecosystems. However, ecosystems differ from most other systems by being extremely adaptive, having the ability of self-organization, and having a large number of feedback mechanisms. Even a *shift in species composition* can take place. The real challenge of modeling is therefore: How can we construct models that are able to reflect these very dynamic characteristics? Some recent development have attempted to answer this question by applying what is denoted structurally dynamic models or variable parameter models—sometimes also called the fifth generation of models. The thermodynamic variable eco-exergy (free energy or work capacity defined for ecosystems; the definition and presentation are given below) has been applied to develop structurally dynamic models (SDMs) in 25 cases; see Zhang et al. (2010), Jørgensen (2008, 2009), and Jørgensen and Fath (2011). Fourteen of these models have been lake models:

(1–8) Eight eutrophication models of six different lakes,
(9) A model to explain the success and failure of biomanipulation based on removal of planktivorous fish,
(10) A model to explain under which circumstances submerged vegetation and phytoplankton are dominant in shallow lakes,
(11) A model of Lake Balaton which was used to support the intermediate disturbance hypothesis,
(12) The structurally dynamic model included in Pamolare 1 devloped by UNEP has been applied on Lake Fure in Denmark.
(13) A model to assess the effect of a restoration of Lake Chaohu, China. See Chapter 14.
(14) A model for Lake Chozas, which has been invaded bu crayfish, see Chapter 12.

This paper will summarize the experience gained by using SDMs on lakes and present a few characteristic illustrative examples to demonstrate the benefits that SDMs can offer the ecological modelers of lakes: better prognoses and improved calibration. Before presenting these benefits, however, it is necessary to understand how to construct structurally dynamic models and define eco-exergy that is used as goal function in this model type.

2.2 HOW TO CONSTRUCT STRUCTURALLY DYNAMIC MODELS AND DEFINITIONS OF ECO-EXERGY

Species are continuously tested against the prevailing conditions (external as well as internal factors). The best fitted are selected and are able to maintain and even increase their biomass (see Figure 2.1). The property of fitness must be heritable to have any effect on the species composition and the ecological structure of the ecosystem in the long run. How can we account for these dynamic changes of the properties in modeling?

If we follow the generally applied modeling procedure presented in most textbooks on ecological modeling (see Jørgensen, 2009), we will develop a model that describes the processes in the focal ecosystem, but the parameters will represent the properties of the state variables as they are found in the ecosystem during the examination period represented by the observations. They are not necessarily valid for another period because an ecosystem can regulate, modify, and change the properties of

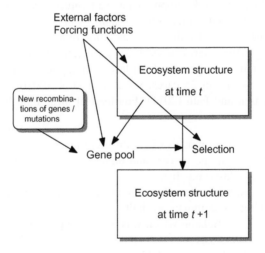

FIGURE 2.1

Conceptualization of how the external factors steadily change the species composition. The possible shifts in species composition are determined by the gene pool, which is steadily changed due to mutations and new sexual recombinations of genes. The development is, however, more complex. This is indicated by (1) arrows from "external factors" to "structure" and "selection" to account for the possibility that the species can modify their own environment (see below) and thereby their own selection pressure; (2) an arrow from "structure" to "gene pool" to account for the possibilities that the species can to a certain extent change their own gene pool.

the species if needed as a response to changes in the prevailing ecosystem conditions (see Figure 2.1). The prevailing conditions are determined by the forcing functions and the interrelations between the state variables, meaning the other species present in the ecosystem. Our present models have rigid structures and a fixed set of parameters, meaning that no changes or replacements of the components are possible. We need to introduce parameters (properties) that can change according to changing forcing functions and general conditions for the state variables (components) to optimize continuously the ability of the system to move away from thermodynamic equilibrium. The idea is to test as the state variables change, if a change of the most crucial parameters produces a higher value of a well-selected and -defined goal function of the system. An appropriate goal function would be the work energy of the ecosystem including the *information* (=eco-exergy), as discussed in system ecology (see Jørgensen et al., 2007; Jørgensen, 2002, 2012). If a higher eco-exergy = work energy including information, is obtained, the corresponding parameter set represents the best fitted properties of the species.

The model type that can account for the change in species composition as well as for the ability of the species, that is, the biological components of our models, to change their properties, that is, to adapt to the existing conditions imposed on the species, is, as mentioned above, called a structurally dynamic model, to indicate that species are able to capture structural changes. These models may also be called the next or fifth generation of ecological models to underscore that they are radically different from previous modeling approaches and can do more, namely describe adaptation and changes in species composition.

It could be argued that the ability of ecosystems to replace present species with other better fitted species can be illustrated by constructing models that encompass all actual and possible species for the entire period that the model attempts to cover. However, this approach has two essential disadvantages. The model becomes first of all very complex, because it will contain many state variables for each trophic level. Therefore, the model will contain many more parameters that have to be calibrated, thus introducing a high uncertainty to the model and rendering the application of the model very case specific (Nielsen, 1992a,b). In addition, the model will be rigid and not possess continuously changing parameters as a result of adaptation.

Straskraba (1979) and (1980) uses a maximization of biomass of the key species as the governing principle. The model computes the biomass and adjusts one or more selected parameters to achieve the maximum biomass at every instance. The model has a routine that computes the biomass for all possible combinations of parameters within a given realistic range. The combination that gives the maximum biomass is selected for the next time step, and the process continues. Biomass can only be used when only one state variable is adapting or shifted to other species.

Eco-exergy (work capacity or free energy for a far-from-thermodynamic equilibrium system, including the work energy of the information) calculated for ecosystems by the use of a special reference system has been used widely as a goal function in ecological models, and some of the most illustrative case studies of lake models will be presented and discussed below. Eco-exergy has two pronounced advantages as goal function for development of structurally dynamic models. It is defined as far-from-thermodynamic equilibrium and it is related to the state variables, which are easily determined or measured, in contrast to instance maximum power that is related to the flows. Furthermore, eco-exergy can be applied also when two or more species are adapting and shifted to other species, which is often the case, as all the species in an ecosystem are coevolving and codeveloping. For instance, phytoplankton and zooplankton are often changed simultaneously when the forcing function for lakes is changed. Because eco-exergy is not a generally used thermodynamic function, we need to present this concept. Eco-exergy expresses energy with a built-in measure of quality. It is defined as the ecosystem content of free energy (work energy), including the work energy of the information embodied in the ecosystem (see Figure 2.3), with the same ecosystem at thermodynamic equilibrium as reference state.

Let us try to translate Darwin's theory into thermodynamics, applying eco-exergy as the basic concept. Survival implies biomass maintenance, and growth means biomass increase. It costs free energy that can do work to construct biomass, and biomass therefore possesses eco-exergy, which is transferable to support other exergetic (energetic) processes. Survival and growth can therefore be measured by use of the thermodynamic concept eco-exergy, which may be understood as the free energy relative to a reference state for a far-from-thermodynamic equilibrium system. Darwin's theory can therefore be reformulated in thermodynamic terms as follows: *The prevailing conditions of an ecosystem steadily change and the system will continuously select the species and thereby the processes that can contribute most to the maintenance or even growth of the eco-exergy of the system. It means moving further away from thermodynamic equilibrium.*

Jørgensen and Mejer (1979) have shown by the use of thermodynamics that the following equation is valid for the eco-exergy density of an ecosystem:

$$\text{Eco-ex} = RT \sum_{i=1}^{i=n} \left(C_i^* \ln \left(\frac{C_i}{C_{\text{eq},i}} \right) - (C_i - C_{\text{eq},i}) \right),$$

where R is the gas constant, T the temperature of the environment (Kelvin), while C_i represents the ith component expressed in a suitable unit, for example, for phytoplankton in lake C_i units could be milligrams of a focal nutrient in the phytoplankton per liter of lake water; $C_{\text{eq},i}$ is the concentration of the ith component at thermodynamic equilibrium, which is used as reference state.

The idea of SDMs is to find continuously a new set of parameters (limited for practical reasons to the most crucial, i.e., sensitive, parameters) that are better fitted for the prevailing conditions of the ecosystem. "Fitted" is defined in the Darwinian sense by the ability of the species to survive and grow, which may be measured by the use of eco-exergy as mentioned above (see Jørgensen and Mejer, 1977, 1979; Jørgensen, 1982, 1986, 1988, 1990, 1992a,b). Figure 2.2 shows the proposed modeling procedure, which has been applied in the two cases presented below.

As reference for an ecosystem, we have proposed the same system at thermodynamic equilibrium, meaning that all the components are (1) inorganic, (2) at the

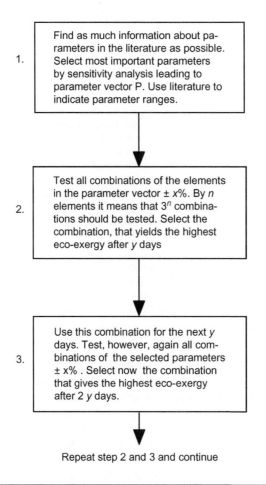

1. Find as much information about parameters in the literature as possible. Select most important parameters by sensitivity analysis leading to parameter vector P. Use literature to indicate parameter ranges.

2. Test all combinations of the elements in the parameter vector $\pm$ $x\%$. By n elements it means that 3^n combinations should be tested. Select the combination, that yields the highest eco-exergy after y days

3. Use this combination for the next y days. Test, however, again all combinations of the selected parameters $\pm$ $x\%$. Select now the combination that gives the highest eco-exergy after 2 y days.

Repeat step 2 and 3 and continue

FIGURE 2.2

The procedure used for the development of structurally dynamic models.

highest possible oxidation state signifying that all free energy has been utilized to do work, and (3) homogeneously distributed in the system, meaning no gradients (see Figure 2.3). Temperature and pressure differences between systems and their reference environments make only small contributions to overall eco-exergy and for present purposes can be ignored. We will compute the eco-exergy based entirely on biochemical energy: $\Sigma_i(\mu_c - \mu_{c,o})N_i$, where i is the number of exergy-contributing compounds; c and μ_c are the chemical potential relative to that at a reference inorganic state, $\mu_{c,o}$. Our (chemical) exergy index for a system will be taken with reference to the same system at the same temperature and pressure, but in the form of a prebiotic environment without life, biological structure, information, or organic molecules—an inorganic soup.

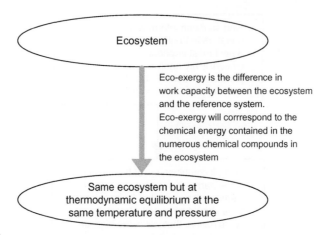

FIGURE 2.3

The definition of eco-exergy is shown. The work capacity in the ecosystem in the form of the chemical energy of the many different and complex chemical compounds relative to the reference system is the eco-exergy. The reference system is the same ecosystem but at thermodynamic equilibrium, that is, a homogeneous system without life. All the chemical compounds are inorganic and there are no gradients.

By using this particular exergy, denoted eco-exergy, based on the same system at thermodynamic equilibrium as the reference and at the same temperature and pressure, the eco-exergy becomes dependent only on the chemical potential of the numerous biochemical components controlling the life processes. These components include the amino bases in the DNA carried by the genome and determining the amino acid sequence, controlling the life processes. In other words, eco-exergy includes the work energy of the information embodied in the species. The work energy of the information was already presented by Boltzmann (1905).

In accordance with Jørgensen and Svirezhev (2005), it is possible to show that eco-exergy density for a model can be found as:

$$\text{Eco-exergy density} = \sum_{i=1}^{i=n} \beta_i c_i$$

The equation uses as unit g organic matter energy equivalents per unit of area or per unit of volume. The eco-exergy due to the "fuel" value of organic matter (chemical energy) is about 18.7 kJ/g (compared with coal: about 30 kJ/g and crude oil: 42 kJ/g). The above presented equation should therefore be multiplied by 18.7 to obtain the eco-exergy density in kJ per unit of area or per unit of volume.. The information eco-exergy $= (\beta -) \times$ biomass or density of information eco-exergy $= (\beta - 1) \times$ concentration. The information eco-exergy controls the function of the many biochemical processes. The ability of a living system to do work is contingent upon its functioning as a living dissipative system. Without the information eco-exergy, the organic matter could only be used as fuel similar to fossil fuel. But due to the information eco-exergy, organisms are able to make a network of the sophisticated biochemical processes that characterize life. The eco-exergy (of which the major part is embodied in the information) is a measure of the organization (Jørgensen and Svirezhev, 2005). This is the intimate relationship between energy and organization that Schrødinger (1944) was struggling to find. The β-values can be by various methods. They express the information embodied in the genomes of various species or the information in the amino acid sequence. Table 2.1 lists the found β-values in accordance with Jørgensen et al. (2005).

The application of eco-exergy to develop SDMs is based on what may be considered thermodynamic translation of survival of the fittest, as already discussed above. Biological systems have many possibilities for moving away from thermodynamic equilibrium, and it is important to know along which pathways among the possible ones a system will develop.

In the next sections, SDMs of lakes will be presented as illustrative examples. Three examples of structural changes that SDMs have been able to capture are presented. In the example of development of an SDM to describe the competition phytoplankton and submerged vegetation, it will furthermore be shown that application of an SDM is also able to improve the calibration.

Table 2.1 ß-Values = Exergy Content Relatively to the Exergy of Detritus (Jørgensen et al., 2005)

Early Organisms	Plants		Animals
Detritus		1.00	
Virus		1.01	
Minimal cell		5.8	
Bacteria		8.5	
Archaea		13.8	
Protists	Algae	20	
Yeast		17.8	
		33	Mesozoa, Placozoa
		39	Protozoa, amoebe
		43	Phasmida (stick insects)
Fungi, molds		61	
		76	Nemertina
		91	Cnidaria (corals, sea anemones, jelly fish)
	Rhodophyta	92	
		97	Gatroticha

Continued

Table 2.1 ß-Values = Exergy Content Relatively to the Exergy of Detritus (Jørgensen et al., 2005)—cont'd

Early Organisms	Plants		Animals
Porifera, sponges		98	
		109	Brachiopoda
		120	Platyhelminthes (flatworms)
		133	Nematoda (round worms)
		133	Annelida (leeches)
		143	Gnathostomulida
	Mustard weed	143	
		165	Kinorhyncha
	Seedless vascula plants	158	
		163	Rotifera (wheel animals)
		164	Entoprocta
	Moss	174	
		167	Insecta (beetles, fruit flies, bees, wasps, bugs, ants)
		191	Coleodiea (Sea squirt)
		221	Lepidoptera (buffer flies)
		232	Crustaceans, Mollusca, bivalvia, gastropodea
		246	Chordata
	Rice	275	
	Gynosperms (incl. pinus)	314	
		322	Mosquito
	Flowering plants	393	
		499	Fish
		688	Amphibia
		833	Reptilia
		980	Aves (Birds)
		2127	Mammalia
		2138	Monkeys
		2145	Anthropoid apes
		2173	*Homo sapiens*

2.3 BIOMANIPULATION

This example of the use of SDMs to understand observed reactions caused by the use of biomanipulation is presented in more details in Jørgensen and Fath (2011). The eutrophication and remediation of a lacustrine environment do not proceed according to a linear relationship between nutrient load and vegetative biomass, but display

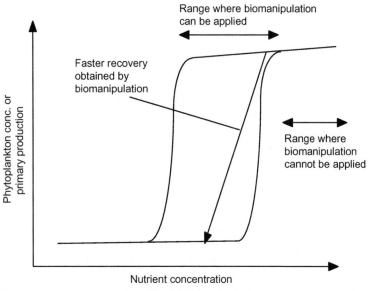

FIGURE 2.4

The hysteresis relation between nutrient level and eutrophication measured by the phytoplankton concentration is shown. The possible effect of biomanipulation is shown. An effect of biomanipulation can hardly be expected above a certain concentration of nutrients, as indicated on the diagram. The biomanipulation can only give the expected results in the range where two different structures are possible.

rather a sigmoid trend with delay, as shown in Figure 2.4. The hysteresis reaction is completely in accordance with observations (Hosper, 1989; Van Donk et al., 1989), and it can be explained by structural changes (de Bernardi, 1989; Hosper, 1989; Sas, 1989; de Bernardi and Giussani, 1995). A lake ecosystem shows a marked buffering capacity to increasing nutrient level, which can be explained by a current increasing removal rate of phytoplankton by grazing and settling. Zooplankton and fish abundance are maintained at relatively high levels under these circumstances. At a certain level of eutrophication, it is not possible for zooplankton to increase the grazing rate further, and the phytoplankton concentration will increase very rapidly by slightly increasing concentrations of nutrients. When the nutrient input is decreased under these conditions, a similar buffering capacity to variation is observed. The structure has now changed to a high concentration of phytoplankton and planktivorous fish, which causes a resistance and delay to a change where the second and fourth trophic levels become dominant again.

Willemsen (1980) distinguishes two possible conditions:

(1) A bream state characterized by turbid water, high eutrophication, low zooplankton concentration, absent of submerged vegetation, large amount of breams, while pike are hardly found at all.

(2) A pike state, characterized by clear water, low eutrophication. Pike and zooplankton are abundant and there are significant fewer breams.

The presence of two possible states in a certain range of nutrient concentrations may explain why biomanipulation has not always been used successfully. According to the observations referred to in the literature, success is associated with a total phosphorus concentration below 50 µg/l (Lammens, 1988) or at least below 100–200 µg/l (Jeppesen et al., 1990), while disappointing results are often associated with phosphorus concentration above the level of more than approximately 120 µg/l (Benndorf, 1990), with difficulty controlling the standing stocks of planktivorous fish (Shapiro, 1990; Koschel et al., 1993).

Scheffer (1990) has used a mathematical model based on catastrophe theory to describe these shifts in structure. However, this model does not consider the shifts in species composition, which is of particular importance for biomanipulation. The zooplankton population undergoes a structural change when we increase the concentration of nutrients, for example, from a dominance of calanoid copepods to small caldocera and rotifers according to the following references: de Bernardi and Giussani (1995) and Giussani and Galanti (1995). Hence, a test of structurally dynamic models could be used to give a better understanding of the relationship between concentrations of nutrients and the vegetative biomass and to explain possible results of biomanipulation. This section refers to the results achieved by a structural dynamic model with the aim of understanding the above described changes in structure and species composition (Jørgensen and de Bernardi, 1998). The applied model has six state variables: (1) dissolved inorganic phosphorus; (2) phytoplankton, phyt.; (3) zooplankton, zoopl.; (4) planktivorous fish, fish 1; (5) predatory fish, fish 2; and (6) detritus. The forcing functions are the input of phosphorus, in P, and the throughflow of water determining the retention time. The latter forcing function determines also the outflow of detritus and phytoplankton.

Simulations have been carried out for phosphorus concentrations in the inflowing water of 0.02, 0.04, 0.08, 0.12, 0.16, 0.20, 0.30, 0.40, 0.60, and 0.80 mg/l. For each of these cases the model was run for any combination of a phosphorus uptake rate of 0.06, 0.05, 0.04, 0.03, 0.02, 0.01 1/24 h and a grazing rate of 0.125, 0.15, 0.2, 0.3, 0.4, 0.5, 0.6, 0.8, and 1.0 1/24 h. When these two parameters were changed, simultaneous changes of phytoplankton and zooplankton mortalities were made according to allometric principles (see Peters, 1983). The parameters that are made variable to account for the structural dynamics are for phytoplankton growth rate (uptake rate of phosphorus) and mortality and for zooplankton growth rate and mortality.

The settling rate of phytoplankton was made proportional to the $(length)^2$. Half of the additional sedimentation when the size of phytoplankton increases corresponding to a decrease in the uptake rate was allocated to detritus to account for resuspension or faster release from the sediment. A sensitivity analysis has revealed that exergy is most sensitive to changes in these five selected parameters, which also represent the parameters that change significantly by size. The 6 respectively 9 levels selected above represent approximately the range in size for phytoplankton and zooplankton.

For each phosphorus concentration, 54 simulations were carried out to account for all combinations of the two key parameters. Simulations over 3 years, 1100 days, were applied to ensure that steady state, limit cycles, or chaotic behavior would be attained. This structural dynamic modeling approach presumed that the combination with the highest exergy should be selected as representing the process rates in the ecosystem. If eco-exergy oscillates even during the last 200 days of the simulation, then the average value for the last 200 days was used to decide which parameter combination would give the highest eco-exergy. The combinations of the two parameters, the uptake rate of phosphorus for phytoplankton and the grazing rate of zooplankton giving the highest exergy at different levels of phosphorus inputs, are plotted in Figures 2.5 and 2.6. The uptake rate of phosphorus for phytoplankton is gradually decreasing when the phosphorus concentration increases. As seen, the zooplankton grazing rate changes at the phosphorus concentration 0.12 mg/l from 0.4 1/24 h to 1.0 1/24 h, that is, from larger species to smaller species, which is according to expectations.

Figure 2.7 shows the eco-exergy named on the diagram information with an uptake rate according to the obtained. The phytoplankton concentration increases for both parameter sets with increasing phosphorus input, as shown Figure 2.8, while the planktivorous fish show a significantly higher level by a grazing rate of 1.0 1/24 h, when the phosphorus concentration is ≥ 0.12 mg/l (=valid for the high exergy level). Below this concentration the difference is minor. The concentration of fish 2 is higher for case 2, corresponding to a grazing rate of 0.4 1/24 h for phosphorus concentrations below 0.12 mg/l. Above this value the differences are minor, but at a phosphorus concentration of 0.12 mg/l, the level is significantly higher for a grazing

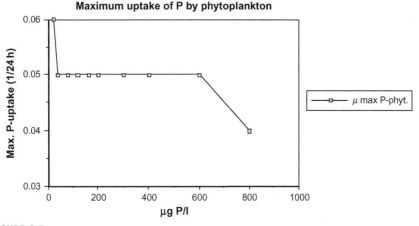

FIGURE 2.5

The maximum growth rate of phytoplankton obtained by the structural dynamic modeling approach is plotted versus the phosphorus concentration.

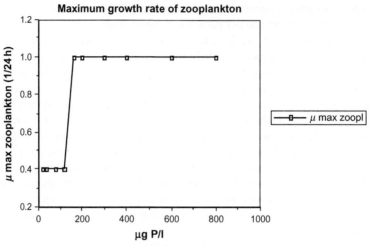

FIGURE 2.6

The maximum growth rate of zooplankton obtained by the structural dynamic modeling approach is plotted versus the zooplankton concentration.

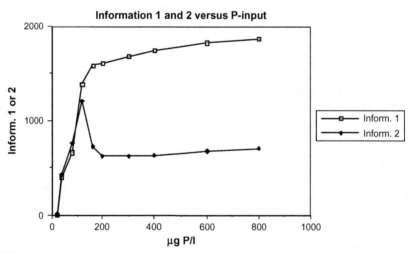

FIGURE 2.7

The exergy is plotted versus the phosphorus concentration. Information 1 corresponds to a maximum zooplankton growth rate of 1 (1/24 h) and information 2 corresponds to a maximum zooplankton growth rate of 0.4 (1/24 h). The other parameters are the same for the two plots, including the maximum phytoplankton growth rate as a function of the phosphorus concentration. (For color version of this figure, the reader is referred to the online version of this chapter.)

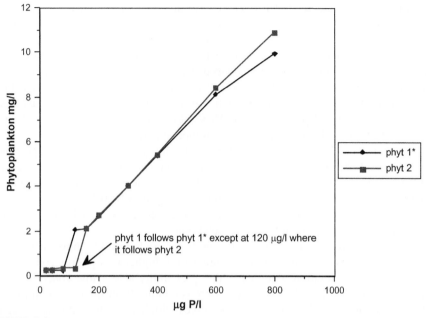

FIGURE 2.8

The phytoplankton concentration as function of the phosphorus concentration for parameters corresponding to "information 1" and "information 2"; see Figure 2.6. The plot named "phyt 1*" coincides with "phyt 1," except for a phosphorus concentration of 0.12 mg/l, where the model shows limit cycles. At this concentration, information 1* represents the higher phytoplankton concentration, while information 1 represents the lower phytoplankton concentration. Notice that the structural dynamic approach can explain the hysteresis reactions. (For color version of this figure, the reader is referred to the online version of this chapter.)

rate of 1.0 1/24 h, particularly for the lower exergy level, where also the zooplankton level is highest.

If it is presumed that eco-exergy can be used as a goal function in ecological modeling, then the results seem to be able to explain why we observe a shift in grazing rate of zooplankton at a phosphorus concentration in the range of 0.1–0.15 mg/l. The ecosystem selects the smaller species of zooplankton above this level of phosphorus because it means a higher level of the eco-exergy, which can be translated to a higher rate of survival and growth. It is interesting that this shift in grazing rate only gives a slightly higher level of zooplankton, while the eco-exergy index level gets significantly higher by this shift, which may be translated as survival and growth for the entire ecosystem. Simultaneously, a shift from a zooplankton, predatory fish dominated system to a system dominated by phytoplankton and particularly by planktivorous fish takes place.

It is interesting that the levels of eco-exergy and the four biological components of the model for phosphorus concentrations at or below 0.12 mg/l parameter combinations are only slightly different for the two parameter combinations. It can explain why biomanipulation is more successful in this concentration range. Above 0.12 mg/l the differences are much more pronounced, and the exergy index level is clearly higher for a grazing rate of 1.0 1/24 h. It should therefore be expected that the ecosystem after the use of biomanipulation easily falls back to the dominance of planktivorous fish and phytoplankton. These observations are consistent with the general experience of success and failure of biomanipulation; see above.

An interpretation of the results points toward a shift at 0.12 mg/l, where a grazing rate of 1.0 1/24 h yields limit cycles. It indicates an instability and probably an easy shift to a grazing rate of 0.4 1/24, although the exergy level is on average highest for the higher grazing rate. A preference for a grazing rate of 1.0 1/24 h at this phosphorus concentration should therefore be expected, but a lower or higher level of zooplankton is dependent on the initial conditions.

If the concentrations of zooplankton and fish 2 are low and high for fish 1 and phytoplankton, that is, the system is coming from higher phosphorus concentrations, then the simulation gives with high probability also a low concentration of zooplankton and fish 2. When the system is coming from high concentrations of zooplankton and fish 2, the simulation gives with high probability also a high concentration of zooplankton and fish 2, which corresponds to an eco-exergy index level slightly lower than obtained by a grazing rate of 0.4 1/24 h. This grazing rate will therefore still persist. Because it also takes time to recover the population of zooplankton and particularly of fish 2, these observations explain the presence of hysteresis reactions.

The model is considered to have general applicability and has been used to discuss the general relationship between nutrient level and vegetative biomass and the general experiences of the application of biomanipulation. The model could probably be improved by introducing size preference for grazing and the two predation processes, which is in accordance with numerous observations. In spite of these shortcomings of the applied model, it has been possible to give a semiquantitative description of the reaction to changed nutrient level and biomanipulation, and even to indicate an approximately correct phosphorus concentration where the structural changes may occur. This may be due to an increased robustness by the structural dynamic modeling approach. It is possible to model competition between a few species with quite different properties, but the structural dynamic modeling approach makes it feasible to include more species even with only slightly different properties, which is impossible by the usual modeling approach; see also the unsuccessful attempt to do so by Nielsen (1992a,b). The rigid parameters of the various species make it difficult for the species to survive under changing circumstances. After some time only a few species will still be present in the model, opposite what is the case in reality, where more species survive because they are able to adapt to changing circumstances. It is important to capture this feature in our models. The structural dynamic models seem promising to apply in lake management, because this type of model is applicable to explain our experience in the use of biomanipulation. It has the advantage compared with the use of catastrophe models, which can also be used to explain success and failure of biomanipulation that it is able also to describe the observed shifts in species composition.

2.4 DEVELOPMENT OF A SDM TO DESCRIBE THE COMPETITION BETWEEN PHYTOPLANKTON AND SUBMERGED VEGETATION

This illustration of the use of SDM has been presented in more details in Jørgensen (2009). Zhang et al. (2003a,b) have developed a structurally dynamic model by using STELLA. The conceptual diagram of the model is shown in Figure 2.9. The model

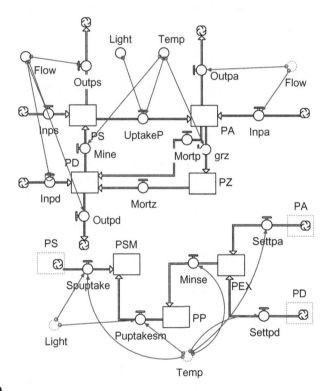

FIGURE 2.9

The conceptual diagram of the Lake Mogan eutrophication model focusing on the cycling of phosphorus. The model has seven state variables: soluble P, denoted PS; phosphorus in phytoplankton, PA; phosphorus in zooplankton, PZ; phosphorus in detritus, PD; phosphorus in submerged plants, denoted PSM; exchangeable phosphorus in the sediment, PEX; and phosphorus in pore water, PP. (For color version of this figure, the reader is referred to the online version of this chapter.)

was developed by using data from Lake Mogan, which is close to Ankara, Turkey. Phosphorus is the liming factor for eutrophication in the lake, which is interesting because it is a shallow lake with competition between phytoplankton and submerged vegetation. The model has seven state variables: soluble P, denoted PS; phosphorus in phytoplankton, PA; phosphorus in zooplankton, PZ; phosphorus in detritus, PD; phosphorus in submerged plants, denoted PSM; exchangeable phosphorus in the sediment, PEX; and phosphorus in pore water, PP. The processes are inflows and outflows of phosphorus, phosphorus in phytoplankton, and phosphorus in detritus. Soluble phosphorus is taken up by phytoplankton—the process is named uptakeP. Zooplankton grazes on phytoplankton, indicated as grz on the diagram. The settling of detritus and phytoplankton is covered by a first-order reaction. A part of the settled material is lost as nonexchangeable phosphorus, while the exchangeable fraction goes to the state variable exchangeable phosphorus, PEX. A mineralization of the exchangeable phosphorus takes place in the sediment—the process is named minse in the diagram.

Mineralization in the water phase of detritus phosphorus is a process called mine. Temperature influences all the process rates. Light is, of course, considered a climatic forcing function influencing the growth of both phytoplankton and submerged plants. The submerged plants take most phosphorus up from the sediment, but the model considers both uptake of phosphorus from water and sediment by the submerged plants.

The SDM approach was used in the presented eutrophication model to examine:

(1) The possibilities for improving calibration. Usually eutrophication models use one set of parameters for the entire annual cycle, but almost all lakes have different species of phytoplankton and zooplankton in the spring, in the summer, and in the fall, because the conditions are different from season to season. The question is, if we could include these changes of the parameters corresponding to the shifts in species composition from season to season, would we then obtain a better calibration and validation? In other words, could we improve calibration and validation if we would use the SDM approach not only for development of prognoses but also for the calibration and validation?

(2) Whether the catastrophic changes from submerged vegetation dominance to phytoplankton dominance that is described by Scheffer et al. (2001) could be covered by the SDM.

The use of the SDM approach in the calibration phase was carried out using the following stepwise method:

I. To reduce the number of parameter combinations, the allometric relationships between parameters of phytoplankton and zooplankton and their sizes were applied: In accordance to the procedure for development of SDMs (see Figure 2.2), the model should be tested for all combinations of at least three possible values of the variable parameters. It means that if seven parameters are made structurally dynamic as decided for the eutrophication model in this case, it is required to run the model 3^7 times $= 2187$, but if allometric principles are used, it is possible to reduce the parameters to two—namely, the size of phytoplankton and the size of zooplankton—and the model only needs to run 3×3 or 9 times.

II. First, the model was calibrated using the usual trial-and-error method to find the combinations of parameters that would give the smallest discrepancy between model results and observations. These parameters were maintained for the subsequent application of the structurally dynamic approach, except for the seven parameters listed above, which would be determined by a current change of the size according to the SDM approach, described as point III.

III. Nine runs (three phytoplankton sizes—the same, 10% increase and 10% decrease and three zooplankton sizes—the same, 10% increase and 10% decrease) are performed from day 0 to day x ($x = 10$ days was chosen for the Lake Mogan model). The combination of size for phytoplankton and zooplankton that gave the highest eco-exergy value was selected for the 10 days. Eco-exergy was calculated for the model as Eco-exergy $= 21 \times$ phytoplankton $+ 135 \times$ zooplankton $+ 100 \times$ submerged vegetation $+$ detritus. This procedure was repeated every 10 days. After an annual model run, it was possible to make a graph or table of the phytoplankton and zooplankton sizes that would currently—every 10 days—optimize the eco-exergy of the model. The sizes were "translated" to the seven parameters that were selected as structurally dynamic (see the list above).

IV. The size of phytoplankton and zooplankton that was found as function of time was introduced to the model and the other model parameters, that is, all the parameter minus the seven phytoplankton and zooplankton parameters, were now calibrated again to account for the influence of our knowledge about the phytoplankton and zooplankton size as function of time. In principle, III and IV should be repeated until no further changes of the parameters are obtained.

This stepwise procedure (see steps 1–3 in Figure 2.10) is of course more cumbersome than using an automatic SDM programmed in C++, but still it takes only a few days to obtain this SDM calibration, while the conventional calibration by trial and error may even require much more time. The procedure I–IV is represented schematically in Figure 2.10. For the Lake Mogan model, it was found that the standard deviation expressing the difference between the modeled value and observed value of phytoplankton phosphorus for the SDM calibration was 10.9% versus 18% for the usual calibration procedure. For phytoplankton, zooplankton, and soluble phosphorus, which were considered the most important state variables, the standard deviation was 8.2% for the SDM calibration and 10.7% for the usual calibration. The graphs for the observations and the model outputs of the two different calibration procedures are shown in Figure 2.11.

The structurally dynamic model was used after calibration to answer the following question: Is it possible to describe the catastrophic changes from submerged vegetation dominance to phytoplankton dominance and back again that is described by Scheffer et al. (2001) by application of the SDM of Lake Mogan? In accordance with

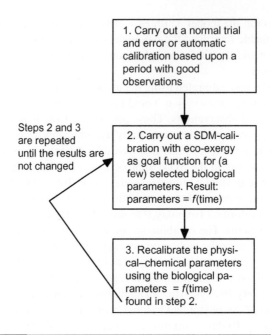

FIGURE 2.10

The diagram shows how the structurally dynamic approach is applied for an improved calibration procedure. The steps 2 and 3 are applied iterative until the results of the calibration are not changed further.

Scheffer et al. we should expect that submerged vegetation is replaced by phytoplankton if we increase the phosphorus concentration to 250 µg/l and that the submerged vegetation is not returned when we decrease the phosphorus concentration before at about 100 µg/l. In other words, can we simulate the described hysteresis behavior by use of the developed SDM for Lake Mogan? The phosphorus concentration in the lake is about 80–85 µg/l, according to the observations, and it was the concentration applied for the calibration described above.

To answer the question, the phosphorus concentration in the water was increased a factor 5×, which implies that we should expect a phosphorus concentration after several years corresponding to the retention time of about 400 µg/l. Afterwards, the phosphorus concentration was reduced to the present level of 80–85 µg/l. Figure 2.12 illustrates the results of these changes in the phosphorus concentration from 80–85 to 400 µg/l and from 400 back to 80–85 µg/l

Figure 2.13 shows the phytoplankton-P concentrations as function of time for five different phosphorus concentrations of the water flowing into the lake: (1) 0.5 × the present level, (2) the present level, (3) 2 × the present level, (4) 5 × the present level, and (5) 10 × the present level. As indicated above the present level is 80–85 µgP/l. Figure 2.14 shows the submerged plant as g/m^2 as function of time for the same five phosphorus concentrations of the water flowing into the lake. The shift in dominance

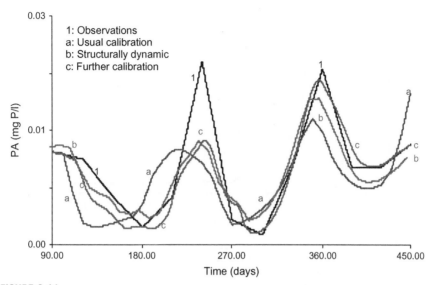

FIGURE 2.11

The phytoplankton-P as mg/l is shown for (1) observations, (a) the calibration obtained by the conventional trial and error calibration, (b) the calibration of the SDM (the method see the text), and (c) the calibration obtained after the non-structurally dynamic parameters have been calibrated by use of the structurally dynamic parameters obtained by procedure b. The time is from 1 October (day 90) to 1 October (day 450). The results of the final calibration, SDM calibration followed by the normal calibration of the non-structurally dynamic parameters (c), are very well in accordance with the observation, except for the first peak late April. The deviation between the modeled and the observed phytoplankton concentration at the second peak in July is only 8% relatively, while the trial and error here gives a deviation of almost 40%. (For color version of this figure, the reader is referred to the online version of this chapter.)

from submerged vegetation to phytoplankton is very clear for the phosphorus concentrations $5\times$ and $10\times$ the present value.

It is possible to conclude from the presented application of SDM that

(1) Use of the SDM approach for the calibration can improve the calibration results and
(2) That it is possible to answer the raised questions about the catastrophic changes by use of SDM.

2.5 SDM DEVELOPED FOR LAKE FURE

Lake Fure is the deepest lakes in Denmark. It has a surface area of 941 ha and an average depth of 13.5 m. It consists of two ecologically different parts that are connected: the main basin, which is the large and deep part with an average depth

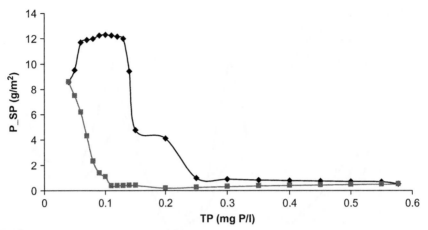

FIGURE 2.12

The graphs show the reaction of submerged plant phosphorus to an increase of phosphorus to about 400 μg/l followed by return to the original about 80–85 μg/l. When the phosphorus concentration is increased, P-SP is increasing, but at about 250 μg/l the submerged vegetation disappears and is replaced by phytoplankton-P. At the return to the 80–85 μg/l, the submerged vegetation emerged at 100 μg/l. The hysteresis behavior is completely in accordance with Scheffer et al. (2001). (For color version of this figure, the reader is referred to the online version of this chapter.)

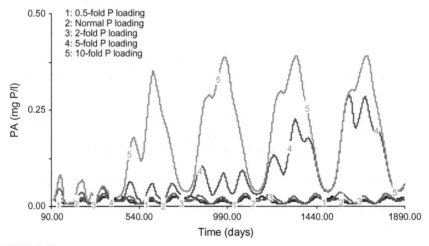

FIGURE 2.13

The phytoplankton-P concentration as function of time is shown when the phosphorus concentration of the water flowing into the lake is increased by (1) a factor of 0.5, (2) normal (factor of 1.0), (3) a factor of 2, (4) a factor of 5, and (5) a factor of 10. The present concentration of phosphorus in the lake is 80–85 μg/l. (1)–(3) Give no changes, while (4) and (5) give a significant increase of phytoplankton, which becomes dominant. (For color version of this figure, the reader is referred to the online version of this chapter.)

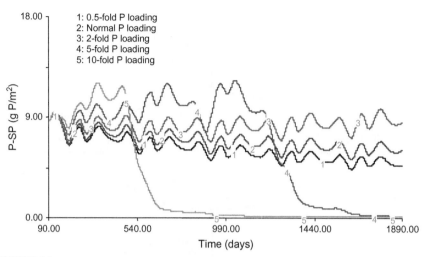

FIGURE 2.14

The submerged plant-concentration as gP/m^2 as a function of time is shown when the phosphorus concentration of the water flowing into the lake is increased by (1) a factor of 0.5, (2) normal (factor of 1.0), (3) a factor of 2, (4) a factor of 5, and (5) a factor of 10. The present concentration of phosphorus in the lake is 80–85 μg/l. (1)–(3) Give no changes, while (4) and (5) first give a minor increase of the concentration followed by a significant decrease, when phytoplankton becomes dominant (compare with Figure 2.13). (For color version of this figure, the reader is referred to the online version of this chapter.)

of 16.5 m and a maximum depth of 37.7 m, and the shallow small part, which is called Store Kalv and has a mean and a maximum depth of 2.5 and 4.5 m, respectively. The lake is situated 12–17 km from the center of Copenhagen and has therefore great recreational value.

The eutrophication of the lake increased significantly in the 1950s and 1960s due to growing populations in the suburbs north of Copenhagen. The wastewater was treated mechanically and biologically, but there was no removal of nutrients. It was therefore decided in the late 1960s to take measures to reduce the eutrophication. In 1970, it was decided that the three municipalities discharging waste water to the lake could either choose to treat the wastewater effectively by removal of nutrients or pump the wastewater to the sea. A maximum phosphorus concentration of 0.2 mg/l and a maximum nitrogen concentration of 8 mg/l was required for the treated waste-water if the municipality selected the treatment solution. One of the three municipalities selected the treatment solution, while the two other municipalities preferred to pump the mechanically, biologically treated water to the sea. A model developed at that time showed that the treatment solution was the best solution for the lake due to a faster recovery of the lake. The lake had a water retention of approximately 16 years and the pumping of round 2 million m^3 to the sea prolonged the retention to about 21 years, which implied of course a slower reduction of the eutrophication.

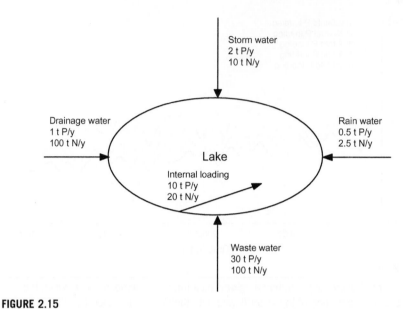

FIGURE 2.15

The nutrient balances in 1972 before the loading of nutrients from wastewater was reduced.

The nutrient balances of the lake in 1972 just before the implementation of the above-mentioned solution are shown in Figure 2.15. The measures taken reduced the phosphorus and nitrogen discharge to the lake by wastewater to less than 1 t and about 3 t, respectively. The three other sources of nutrients, drainage water, storm water overflow, and precipitation, were unchanged. During the period from 1972 to 2000 the municipalities were able to reduce the other sources slightly as can be seen in Figure 2.16 showing the nutrient balances for year 2000. The nutrient input from storm water was reduced by enlarging the storm water capacity. Notice, however, that the internal loading was not reduced, which is at least partly due to long retention time. Part of the story is that the two municipalities that preferred to pump the wastewater to the sea in the early 1970s and did not accept the model result that the treatment solution would give a faster recovery were forced in 1986 to include nitrogen and phosphorus removal in the wastewater treatment due to introduction of a maximum standard for discharge of all wastewater, even wastewater discharged directly to the sea.

Shortly after year 2000 it was decided to use restoration methods to recover the lake faster. Two restoration methods were proposed:

(1) Aeration of the hypolimnion by oxygen from late April to late October when the lake had a thermocline. The release of phosphorus from the sediment would thereby be reduced significantly.

(2) Biomanipulation by massive removal of planktivorous fish.

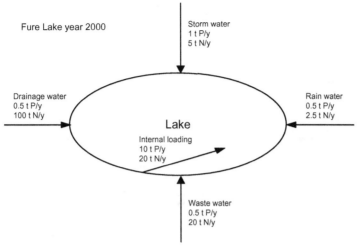

FIGURE 2.16

The nutrient balances in year 2000.

The restoration of the lake started in 2003. An SDM was developed in 2005 (see the paper by Gurkan et al., 2006). An SDM was natural to apply in this case as structural changes were expected due to reduced internal loading, the removal of a significant part of the planktivorous fish, and further increase of the storm water capacity. The model was used for prognosis about the water quality in year 2010. Figure 2.17 shows the application of the model results on the nutrient balance, and Table 2.2 shows the changes in the important state variables from 2004 to 2010 according to the prognosis made by the SDM of Pamolare.

2.6 SUMMARY AND CONCLUSIONS

Because ecosystems are adaptive and are able to change species compositions, there is a need for models that are able to consider this characteristic dynamic. Structurally dynamic models are able to account for these ecological, dynamic properties. By the use of eco-exergy as goal function, it is possible to develop models that are able to describe adaptation and shifts in species composition. Twelve structurally dynamic lake models have been developed, and three illustrative case studies among these 12 case studies are presented in this paper. The three presented case studies show that SDMs are able to describe the observed structural changes and thereby are able to improve the prognoses developed by the models. In addition, it is also possible to improve the calibration of ecological lake models by the use of SDMs because there are in lakes clearly seasonal changes of the dominant species, which makes it difficult to calibrate the models with one parameter to cover the entire year. In contrast, the application of SDMs makes it possible to capture the seasonal changes of the

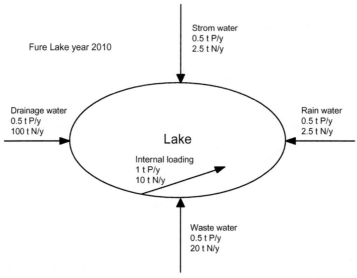

FIGURE 2.17

The nutrient balances according to the result of the SDM. The internal loading is reduced due mainly to the aeration of the hypolimnion: The phosphorus and nitrogen loading from storm water is reduced due to increase of the storm water capacity.

Table 2.2 Changes in Important State Variables from 2000 to 2010 According to the Results of the SDM Applied Due to the Restoration Project (Aeration of Hypolimnion, Biomanipulation, and Increased Storm Water Capacity)

State Variable	Year 2000	Year 2010
Total P mg/l	0.18–0.25	0.02–0.07
Chl.a maximum (early Aug.) mg/l	0.044	0.022
Transparency minimum (early Aug.) m	2.2	3.6
Zooplankton mg d.w./l	0.8	1.8

properties of phytoplankton and zooplankton, for example, and thereby improve significantly the calibration.

All the 12 case studies of SDMs of lakes show the same picture (see Jørgensen, 2009; Zhang et al., 2010; Jørgnesen and Fath, 2011): It is in most cases possible to improve the calibration and the accuracy of the prognoses increases, and it is possible with an acceptable standard deviation to capture the dynamic changes.

It can be recommended to apply SDMs whenever adaptation and structural changes are presumed to take place. It makes SDMs particularly attractive when models are used for prognoses in environmental management, because prognoses often consider radical changes of the forcing functions; therefore, structural changes could most probably take place.

References

Benndorf, J., 1990. Conditions for effective biomanipulation. Conclusions derived from whole-lake experiments in Europe. Hydrobiologia 200/201, 187–203.

Boltzmann, L., 1905. The Second Law of Thermodynamics, (Populare Schriften. Essay No.3 (Address to Imperial Academy of Science in 1886)). Reprinted in English in: Theoretical Physics and Philosophical Problems, Selected Writings of L. Boltzmann. D. Riedel, Dordrecht.

de Bernardi, R., 1989. Biomanipulation of aquatic food chains to improve water quality in eutrophic lakes. In: Ravera, O., Ravera, O. (Eds.), Ecological Assessment of Environmental Degradation, Pollution and Recovery. Elsevier Sci. Publ., Amsterdam, pp. 195–215, 356 pp.

de Bernardi, R., Giussani, G., 1995. Biomanipulation: bases for a top-down control. In: De Bernardi, R., Giussani, G. (Eds.), Guidelines of Lake Management, Volume 7. Biomanipulation in Lakes and Reservoirs. ILEC and UNEP, pp. 1–14, 211 pp.

Giussani, G., Galanti, G., 1995. Case study: Lake Candia (Northern Italy). In: De Bernardi, R., Giussani, G. (Eds.), Guidelines of Lake Management, Volume 7. Biomanipulation in Lakes and Reservoirs. ILEC and UNEP, pp. 135–146, 211 pp.

Gurkan, Z., Zhang, J., Jørgensen, S.E., 2006. Development of a structurally dynamic model for Forecasting the effects of restoration of lakes. Ecol. Model. 197, 89–103.

Hosper, S.H., 1989. Biomanipulation, new perspective for restoring shallow, eutrophic lakes in The Netherlands. Hydrobiol. Bull. 73, 11–18.

Jeppesen, E.J., et al. 1990. Fish manipulation as a lake restoration tool in shallow, eutrophic temperate lakes. Cross-analysis of three Danish Case Studies. Hydrobiologia 200/201, 205–218.

Jørgensen, S.E., 1982. A holistic approach to ecological modelling by application of thermo-Dynamics. In: Mitsch, W., et al. (Eds.), Systems and Energy. Ann Arbor Science, Ann Arbor, MI.

Jørgensen, S.E., 1986. Structural dynamic model. Ecol. Modell. 31, 1–9.

Jørgensen, S.E., 1988. Use of models as experimental tools to show that structural changes are accompanied by increased exergy. Ecol. Modell. 41, 117–126.

Jørgensen, S.E., 1990. Ecosystem theory, ecological buffer capacity, uncertainty and complexity. Ecol. Modell. 52, 125–133.

Jørgensen, S.E., 1992a. Parameters, Ecological constraints and exergy. Ecol. Modell. 62, 163–170.

Jørgensen, S.E., 1992b. Development of models able to account for changes in species composition. Ecol. Modell. 62, 195–208.

Jørgensen, S.E., 2002. Integration of Ecosystem Theories: A Pattern. Kluwer, Dordrecht, p. 386.

Jørgensen, S.E., 2008. An overview of the model types available for development of ecological models. Ecol. Model. 215, 3–9.

Jørgensen, S.E., 2009. Ecological Modelling—An Introduction. WIT, Southampton, p. 190.

Jørgensen, S.E., Fath, B., 2011. Fundamentals of Ecological Modelling, fourth ed. Elsevier, Amsterdam, 400 p.

Jørgensen, S.E., 2012. Introduction to Systems Ecology. CRC, Boca Raton, FL, p. 330.

Jørgensen, S.E., de Bernardi, R., 1998. The use of structural dynamic models to explain the success and failure of biomanipulation. Hydrobiologia 379, 147–158.

Jørgensen, S.E., Mejer, J.F., 1977. Ecological buffer capacity. Ecol. Modell. 3, 39–61.

Jørgensen, S.E., Mejer, H.F., 1979. A holistic approach to ecological modelling. Ecol. Modell. 7, 169–189.

Jørgensen, S.E., Svirezhev, Y., 2005. Toward a Thermodynamic Theory for Ecological Systems. Elsevier, Amsterdam, p. 366.

Jørgensen, S.E., Ladegaard, N., Debeljak, M., Marques, J.C., 2005. Calculations of exergy for organisms. Ecol. Model. 185, 165–175.

Jørgensen, S.E., Fath, B., Bastiononi, S., Marques, J.C., Mueller, F., Nielsen, S.N., Patten, B.C., Tiezzi, E., Ulanowicz, R., 2007. A New Ecology. Elsevier, Amsterdam, p. 276.

Koschel, R., Kasprzak, P., Krienitz, L., Ronneberger, D., 1993. Long term effects of reduced nutrient loading and food-web manipulation on plankton in a stratified Baltic hard water lake. Verh. Int. Ver. Limnol. 25, 647–651.

Lammens, E.H.R.R., 1988. Trophic interactions in the hypertrophic Lake Tjeukemeer: top-down and bottom-up effects in relation to hydrology, predation and bioturbation, during the period 1974-1988. Limnologica (Berl.) 19, 81–85.

Nielsen, S.N., 1992a. Application of Maximmn Exergy in Structmal Dynamic Models. Ph.D. Thesis, National Environmental Research Institute, Denmark, p. 51.

Nielsen, S.N., 1992b. Strategies for structural-dynamical modelling. Ecol. Modell. 63, 91–101.

Peters, R.H., 1983. The Ecological Implications of Body Size. Cambridge University Press, Cambridge, 329 p.

Sas, H., 1989. Lake restoration by reduction of nutrient loading. Expectations, experiences, extrapolations. Academia Verl. Richarz, St. Augustin, (Coordination) 497 pp.

Scheffer, M., 1990. Simple Models as Useful Tools for Ecologists. Elsevier, Amsterdam, p. 192.

Scheffer, M., Carpenter, S., Foley, J.A., Folke, C., Walker, B., 2001. Castrophic changes Ecosystems. Nature 413, 591–596.

Schrødinger, E., 1944. What is Life? Cambridge University Press, Cambridge, p. 186.

Shapiro, J., 1990. Biomanipulation: the next phase-making it stable. Hydrobiologia 200/210, 13–27.

Straskraba, M., 1979. Natural control mechanisms in models of aquatic ecosystems. Ecol. Modell. 6, 305–322.

Straskraba, M., 1980. Cybernetic-categories of ecosystem dynamics. ISEM J. 2, 81–96.

Van Donk, E., Gulati, R.D., Grimm, M.P., 1989. Food web manipulation in lake Zwemlust: positive and negative effects during the first two years. Hydrobiol. Bull. 23, 19–35.

Willemsen, J., 1980. Fishery aspects of eutrophication. Hydrobiol. Bull. 14, 12–21.

Zhang, J., Jørgensen, S.E., Tan, C.O., Beklioglu, M., 2003a. A structurally dynamic modelling - Lake Mogan, Turkey as a case study. Ecol. Model. 164, 103–120.

Zhang, J., Jørgensen, S.E., Tan, C.O., Beklioglu, M., 2003b. Hysteresis in vegetation shift— Lake Mogan Prognoses. Ecol. Model. 164, 227–238.

Zhang, J., Gurkan, Z., Jørgensen, S.E., 2010. Application of eco-energy for assessment of ecosystem health and development of structurally dynamic models. Ecol. Model. 221, 693–702.

Development of Level-IV Fugacity-Based QWASI Model for Dynamic Multimedia Fate and Transport Processes of HCHs in Lake Chaohu, China

Xiang-Zhen Kong, Fu-Liu Xu*, Wei He, Ning Qin

MOE Laboratory for Earth Surface Processes, College of Urban & Environmental Sciences,
Peking University, Beijing, China
*Corresponding author: e-mail address: xufl@urban.pku.edu.cn

3.1 INTRODUCTION

3.1.1 Hexachlorocyclohexanes and the isomers

Hexachlorocyclohexanes (HCHs) are compounds of major concern among the organochlorine pesticides (OCPs), which have been heavily used in the past around the world. There are eight geometric isomers of HCHs, which differ in the axial and equatorial positions of the chlorine atoms (Walker et al., 1999). HCH isomers (α-, β-, and γ-HCH) were recently added to the list of persistent organic pollutants (POPs) at the Stockholm Convention (Vijgen et al., 2011). The isomers of α-HCH and γ-HCH are the main ingredients for the commercial product of technical HCHs (containing 60–70% α-HCH and 10–15% γ-HCH), and γ-HCH is the absolute ingredient of insecticide "lindane" (containing more than 99.9% γ-HCH). The ratio of α-/γ-HCH is frequently used as the indicator to identify the source of the HCHs (Iwata et al., 1993, 1995; Walker et al., 1999; Law et al., 2001; Liu et al., 2012, 2013; Ouyang et al., 2012, 2013; Xu et al., 2013b). Among the isomers of HCHs, α-HCH can cause human neurological disorders and gastrointestinal discomfort, resulting in liver and kidney damage, human endocrine system disorders and immune system abnormalities (SCPOPs, 2009). In addition, the study showed that pesticides like lindane have been posing pressures on the freshwater ecosystem (Qu et al., 2011; Matozzo et al., 2012). As the only isomer with specific insecticidal properties (Walker et al., 1999), γ-HCH was found to be a tumor promoter (Dich et al., 1997). Kalantzi et al. (2004) indicated that subtle alterations in breast and prostate cells could be induced by environmental concentrations of lindane.

3.1.2 HCHs usage and residue level in the study site

In China, there was extensive use of technical HCHs from 1953 to 1983 (Li et al., 2001), when lindane began to be applied instead. HCHs were totally prohibited by the government in 1992 (Tao et al., 2006). Anhui Province was among the regions with the highest usage of pesticides in China (Cai, 2010). The application rate of technical HCHs in Anhui was close to Jiangxi Province, with a total usage of nearly 2×10^5 ton between 1952 and 1984 (Li et al., 2001). High levels of HCH residues could persist in the environment under such intensive application. However, the residues have been decreasing rapidly in the environment since the prohibition. It has been reported that the residue level of HCHs in the topsoil in Anhui has dropped from 0.349 (Xia et al., 1987) to 0.150 mg/kg (Yue et al., 1990) and 0.0286 mg/kg (Wang et al., 2011). Lake Chaohu is located in Anhui Province. It is known as the fifth largest shallow freshwater lake in China with a population of nearly 8 million in 1987 living in the lake basin (Tu et al., 1990). According to historical data, the OCPs emissions in Chaohu water bodies amounted to 1.16 ton in 1984 (Zhang and Lu, 1986). The lake has been severely polluted by HCH pesticides through surface runoff, undercurrent, osmosis, and leaching from the soil into surface water. The concentrations of HCHs in the soil and sediment of Lake Chaohu basin were 1.4 µg/kg (Gao and Zhao, 2012) and 0.58 µg/kg (Wang et al., 2012), respectively. Although the residue level of HCHs has decreased significantly, the HCHs will still remain in the environment for a long time due to their persistence. In addition, the level of γ-HCH in fish was 11.42 ± 29.98 µg/kg dw (Wang, 2012), which might be consumed by human beings and cause negative impacts on health. Therefore, understanding the environmental behaviors of α-HCH and γ-HCH in the lake environment are of great concern.

3.1.3 Level-IV fugacity-based QWASI model

The environmental multimedia model is a mathematical model based on the concept that the physical and chemical properties of environmental systems and pollutants synergistically determine the concentration distribution and migration of contaminants throughout the transformation process between environmental compartments (Mackay, 2001). They were usually applied in assessment of the fate and transport of organic pollutants in the environment (von Waldow et al., 2008).

The multimedia model based on fugacity was proposed by Mackay (1979) and Mackay and Paterson (1981, 1982). Rather than formulating the model in terms of concentration, fugacity is used so that the kinetic and equilibrium parameters can adopt forms that facilitate interpretation of the model output and assist the identification of the dominant processes (Mackay et al., 1983). It has been widely used to describe the environmental behavior of organic pollutants in global, regional and local scales (Wania and Mackay, 1995; Tao et al., 2003; Liu et al., 2007; Xu et al., 2013a; Xia et al., 2011). In order to apply this approach to lakes or sections of rivers, a fugacity-based model termed the Quantitative Water Air Sediment Interaction (QWASI) model was developed (Mackay et al., 1983) so that the contaminant behavior in an air, water, and sediment system can be described. This model has been applied in several studies (Mackay and Diamond, 1989; Diamond et al., 2005; Xu et al., 2013a), all of which were modeling the contaminant under the steady state, that is, the level-III fugacity model. In fact, there are four levels in the fugacity-based multimedia model (Mackay and Paterson, 1982). A level-IV fugacity model should be utilized when continuous changes in the concentrations of particular pollutants are studied over a period of time (Cao et al., 2007; Lang et al., 2007; Liu et al., 2007; Ao et al., 2009). Liu et al. (2007), Cao et al. (2007) and Ao et al. (2009) have applied the level-IV fugacity model to simulate the long-term fate of OCPs contaminates in certain areas in China. On the other hand, Lang et al. (2007) simulated the seasonal variations of polycyclic aromatic hydrocarbons (PAHs) in Pearl Delta region based on a detailed PAHs emission inventory. To date, however, few studies have focused on seasonal variations in pollutants in subtropical aquatic systems using the level-IV fugacity-based QWASI model, probably because of the lack of the seasonal emission data as the source. For the insecticides like HCHs, which had been prohibited for more than twenty years, the emission is supposed to be ignored. In addition, for an aquatic environment such as Lake Chaohu, the source of HCHs originated from the air and water input advection, which were relatively easy to obtain by field sampling. Therefore, it is feasible to simulate the seasonal variation of HCHs in Lake Chaohu using the level-IV fugacity-based QWASI model.

Sensitivity and uncertainty analysis are both crucial in model development and verification. A normal Morris method (Morris, 1991) was used in this study for sensitivity analysis. On the other hand, in studies that focused on the uncertainty of the fugacity model (Tao et al., 2006; Ao et al., 2009), the basic Monte Carlo simulation was commonly used, which relies on user-defined probability distributions for the model input parameters (Saloranta et al., 2008). Due to data limitations, the distribution of certain parameters was difficult to determine, which might lead to an unsatisfactory fit between the model output and the observed data and overestimated uncertainty of the model results in the Monte Carlo simulation. A Bayesian Markov chain Monte Carlo (MCMC) method was then proposed to combine the uncertainty analysis and the calibration of the fugacity model (Saloranta et al., 2008). MCMC updates the parameter distributions during the simulation according to the comparison between the model outcomes and observations so that improbable parameter combinations are discarded. However, Saloranta et al. (2008) illustrated their model based on the long-term simulation for yearly observations. An improved fit of the model output to the observations or a better estimate of model uncertainty was not clearly shown; both of these are considered as the two advantages of the MCMC method. With a higher time-resolution observations, such as the seasonal observations in this study, the two advantages of the MCMC over basic Monte Carlo can be further verified.

3.2 DEVELOPMENT OF LEVEL IV FUGACITY-BASED QWASI MODEL

3.2.1 Model framework

The framework of the fugacity-based QWASI model is illustrated in Figure 3.1, while the characteristics of the phases and subphases in the model are shown in Table 3.1. This model includes three main compartments: atmosphere, water, and sediment, which were represented by the subscripts "1," "2," and "4," respectively. The number "3" was skipped because it was assigned to the soil compartment in many studies (Tao et al., 2006; Cao et al., 2007; Lang et al., 2007), which were not included in the QWASI model. The atmospheric phase was comprised of two subphases: gaseous and particulate matter. The aqueous phase consisted of three subphases: water, suspended solids, and fish. The sediment phase consisted of pore water and solids. Two major adaptations in this model in comparison to that in Mackay et al. (1983) were as follows: first, a biota subphase of fish was included in the water, while the corresponding process of fish production (T_{2f}) was also added in the model framework due to availability of HCHs data in fish. As a result, the effect of bioaccumulation can be evaluated, which is important for the assessment of the negative effect of contaminants on lake ecosystem. Second, atmospheric advection input and output of the system were both incorporated into the model. As the sampling data of HCHs both on the shore and in the lake as well as the wind data were available, the

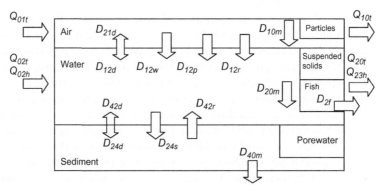

FIGURE 3.1

Transport fluxes of HCHs in and out of the Lake Chaohu area and between the adjacent compartments. D_{12d}, D_{21d}, D_{24d}, and D_{42d} represent the diffusion processes between air/water and water/sediment. D_{12p} and D_{12w} represent the dry and wet deposition from air to water, respectively. D_{12r} represents scavenging by precipitation. Q_{01t}, Q_{02t}, and Q_{02h} represent the input from air advection, water inflows and wastewater discharge, respectively. Q_{10t}, Q_{20t}, and Q_{23h} represent the output from air advection, water outflows, and water reuse by industry and agriculture, respectively. D_{10m}, D_{20m}, and D_{40m} represent the degradation occurring in the air, water, and sediment, respectively. D_{2f} represents losses from water by fish production.

Table 3.1 Volume and Properties of the Phases and Subphases

Main Phase	Area (m²)	Depth (m)	Organic Carbon (%)	Subphases and Volume Fraction (X_{ij})		
				Air (1)	Water (2)	Solid Particles (3)
Air (1)	7.58E+08	1.00E+02	–	1.00E+00	–	8.31E−11[a]
Water (2)	7.58E+08	2.69E+00[a]	8.77E−02	–	1.00E+00	1.65E−05[a]
Sediment (3)	7.58E+08	1.00E−01	5.27E−02	–	7.00E−01	3.00E−01

[a]Seasonal data were available for simulation.

atmospheric advection can help to interpret the seasonal variations of HCHs in the air of the lake. Details can be found in Section 3.2.2.

Processes included in the framework refer to Kong et al. (2012) and Kong et al. (2014; in press). Table 3.2 provides a detailed description of all the processes in the model. Advections include the input and output in air and water; transport processes include two types, namely solute diffusion transfer processes, which are driven by fugacity differences, and material transfer processes, which are driven by material physically movements from phase to phase (such as deposition and resuspension). Transformation processes such as biolysis, oxidation, photolysis, and hydrolysis are included as a first-order rate, namely degradation. Note that all transport and transformation processes are expressed as the product of a D (mol/h Pa) term and a fugacity f (Mackay et al., 1983), and advection processes are expressed as the product of volume Q (m³/d) and concentration C (mol/m³).

Table 3.2 Definitions of the Transfer and Transformation Processes

Symbol	Formula	Explanation
System Input		
T_{01t}	$Q_{01t} \times C_{01t}$	Air advection flows into the area
T_{02t}	$Q_{02t} \times C_{02t}$	Water advection flows into the area
T_{02h}	$Q_{02h} \times C_{02h}$	Locative wastewater discharge
System Output		
T_{10t}	$D_{10t} \times f_1$	Air advection flows out of the area
T_{20t}	$D_{20t} \times f_2$	Water advection flows out of the area
T_{10m}	$D_{10m} \times f_1$	Degradation in air
T_{20m}	$D_{20m} \times f_2$	Degradation in water
T_{40m}	$D_{40m} \times f_4$	Degradation in sediment
T_{23h}	$D_{23h} \times f_2$	Industry and agriculture water usage
T_{2f}	$D_{2f} \times f_2$	Fish production
Air–Water Transfer		
T_{12d}	$D_{12d} \times f_1$	Diffusion from air to water
T_{21d}	$D_{21d} \times f_2$	Diffusion from water to air
T_{12p}	$D_{12p} \times f_1$	Dry deposition from air to water
T_{12w}	$D_{12w} \times f_1$	Wet deposition from air to water
T_{12r}	$D_{12r} \times f_1$	Rain scavenging
Water–Sediment Transfer		
T_{24d}	$D_{24d} \times f_2$	Diffusion from water to sediment
T_{42d}	$D_{42d} \times f_4$	Diffusion from sediment to water
T_{24s}	$D_{24s} \times f_2$	Sedimentation from water to sediment
T_{42r}	$D_{42r} \times f_4$	Resuspension from sediment to water

In the fugacity-based model, a fugacity capacity Z (mol/m^3 Pa) is defined relating the fugacity f (Pa) to the concentration C (mol/m^3) as $C = f \cdot Z$. Details for calculation of Z (mol/m^3 Pa) and the D (mol/h Pa) term can be found in Xu et al. (2013a). In addition, it was assumed that the fugacities for all of the subphases in the air, water column, and bottom sediments (f_i, $i = 1$, 2, and 4) were the same (Mackay and Paterson, 1991), that is, thermodynamic equilibrium exists within all three main compartments in these phases, because each main phase is individually well mixed (Mackay et al., 1983). The differential mass balance equations with fugacities as variables are shown in Equation (3.1):

$$
\begin{cases}
\dfrac{V_1 Z_1 \mathrm{d}f_1}{\mathrm{d}t} = T_{01t} - \left(D_{12d} + D_{12p} + D_{12r} + D_{12w} + D_{10m} - D_{10t}\right) \cdot f_1 + D_{21d} \cdot f_2 \\[2mm]
\dfrac{V_2 Z_2 \mathrm{d}f_2}{\mathrm{d}t} = T_{02t} + \left(D_{12d} + D_{12p} + D_{12g} + D_{12w}\right) \cdot f_1 \\[2mm]
\qquad\qquad - \left(D_{21d} + D_{24d} + D_{24s} + D_{20m} + D_{20t} + D_{23h} + D_{2f}\right) \cdot f_2 \\[2mm]
\qquad\qquad + \left(D_{42d} + D_{42r}\right) \cdot f_4 \\[2mm]
\dfrac{V_4 Z_4 \mathrm{d}f_4}{\mathrm{d}t} = \left(D_{24d} + D_{24s}\right) \cdot f_2 - \left(D_{42d} + D_{42r} + D_{40m}\right) \cdot f_4
\end{cases}
\tag{3.1}
$$

3.2.2 Model simulation and validation

Simulation started from May 2010 to February 2011. The fourth-order Runge–Kutta method was applied to solve differential equations with a simulation time step of 1 h. The available monthly observations of both α-HCH and γ-HCH from May 2010 to February 2011 for air and air particles (Ouyang et al., 2012), water and suspended solids (He et al., 2012; Liu et al., 2012), as well as fish of 2010 (Wang, 2012) and sediment of August 2009 (Wang et al., 2012), were all used for model validation. Details of the sampling and measuring of the data can be found in the relevant references.

3.2.3 Parameter determination

Model parameter symbols, units, values, and data sources are provided in Tables 3.3 and 3.4. Fifty parameters were included, including 24 environmental parameters, 16 mass transfer parameters, and 10 physicochemical parameters of the pollutant. The environmental parameters, including temperature, lake area, height, and subphase volume fraction, were determined by the literature or laboratory measurements. The environmental kinetics of the process parameters, including the rate of degradation, the rate of diffusion, migration constant, molecular diffusion path length, atmospheric wet and dry deposition rates, deposition rate, and cleaning coefficients, were obtained from the relevant literature or calculated based on the conditions in Lake Chaohu. The physicochemical parameters, such as the gas constant, Henry's constant, and saturated vapor pressure, were obtained from the literature. In addition, Henry's constant, the saturated vapor pressure, and the fugacity rate of the pollutant

Table 3.3 Environmental Parameters for the Model

Symbol	Unit	Parameters	Reference	Data Numbers	Allocate mean	Allocate SD	Geometric mean	Geometric SD
A_2	m^2	Interface areas of air/water	Tu et al. (1990) and Yin (2011)	2	7.58E+08	2.69E+06	7.58E+08	1.00E+00
h_1	m	Thickness of air	Tu et al. (1990)	1	1.00E+03	–[a]	1.00E+03	6.12E+00
$h_2{}^b$	m	Depth of water	AHTIS (http://61.191.22.154/yc_web/yc_index_frame.aspx)	8760	2.69E+00	–	2.69E+00	1.30E+00
h_4	m	Thickness of sediment	Tu et al. (1990)	1	1.00E−01	–[a]	1.00E−01	1.73E+00
$X_{13}{}^b$	v/v	Volume fractions of solids in air	Mackay (2001) and Note A	12	8.36E−11	3.88E−11	7.55E−11	1.56E+00
$X_{23}{}^b$	v/v	Volume fractions of solids in water	Mackay (2001) and Note A	12	1.60E−05	1.29E−05	1.26E−05	2.02E+00
X_{43}	v/v	Volume fractions of solids in sediment	Mackay (2001)	1	3.00E−01	–[a]	3.00E−01	1.08E+00
X_{42}	v/v	Volume fractions of water in sediment	Mackay (2001)	1	7.00E−01	–[a]	7.00E−01	1.02E+00
X_{2f}	v/v	Volume fractions of fish in water	Tu et al. (1990), Zhang and Lu (1986), and Guo (2005)	3	8.93E−06	1.39E−06	7.60E−06	2.07E+00
$O_{23}{}^b$	%	Contents of organic carbon in solids in water	Mackay (2001), Zhou et al. (2007), and Note A	12	8.77E−02	7.31E+00	7.83E−02	1.90E+00
O_{43}	%	Contents of organic carbon in solids in sediment	Tu et al. (1990), Yin (2011), Zhang (2009), and Note A	61	5.27E−02	2.92E−02	7.92E−02	1.72E+00
ρ_{23}	t/m^3	Densities of solids in water	Tu et al. (1990)	1	2.50E+00	–[a]	2.50E+00	1.16E+00
ρ_{43}	t/m^3	Densities of solids in sediment	Tu et al. (1990)	1	2.76E+00	–[a]	2.76E+00	1.18E+00

Q_{O1t} ($=Q_{1ot}$)[b]	m³/h	Air advection flow in and out of the area	Calculated	12	2.17E+11	–	2.17E+11	1.47E+00
Q_{O2t}[b]	m3/h	Water advection flow into the area	Tu et al. (1990)	12	5.35E+05	–	5.35E+05	3.22E+00
Q_{2ot}[b]	m3/h	Water advection flow out of the area	Tu et al. (1990)	12	6.46E+05	–	6.46E+05	3.07E+00
Q_{O2h}[b]	m³/h	Rate of local wastewater discharge	Tu et al. (1990)	12	4.67E+04	–	4.67E+04	1.45E+00
Q_{23h}[b]	m³/h	Irrigation rates	Tu et al. (1990)	12	1.19E+05	9.27E+04	9.26E+04	2.07E+00
T[b]	K	Local average temperature	Tu et al. (1990), CMDSSS (http://www.cma.gov.cn/2011qxfw/2011qsjgx/index.htm)	1	2.88E+02	9.75E+00	2.88E+02	–[c]
C_{O1t}[b]	mol/m³	γ-HCH concentration in air	Note A	12	2.20E−14	1.47E−14	1.71E−14	1.42E+00
C_{O2t}	mol/m³	γ-HCH concentration in water advection flow	Zhang (2009)	1	6.75E−10	–[a]	6.75E−10	1.71E+00
C_{O2h}	mol/m³	γ-HCH concentration in wastewater	Tu et al. (1990)	1	4.83E−08	–[a]	4.83E−08	1.56E+00
C_{O1t}[b]	mol/m³	α-HCH concentration in air	Note A	12	3.81E−14	2.32E−14	2.31E−14	1.43E+00
C_{O2t}	mol/m³	α-HCH concentration in water advection flow	Tu et al. (1990) and Zhang (2009)	1	1.34E−08	–[a]	1.34E−08	1.63E+00
C_{O2h}	mol/m³	α-HCH concentration in wastewater	Tu et al. (1990)	1	1.56E−08	–[a]	1.56E−08	1.82E+00
Y_f	t/a	Fish production rate	Zhang and Lu (1986)	1	5.71E−01	–[a]	5.71E−01	1.18E+00

Note A: Measured in Laboratory.
[a]One value only; Geometric SD assigned manually (log-normal distribution assumed).
[b]Dynamic parameters for simulation, temporal values were shown in Kong et al. (2012).
[c]Geometric SD assigned manually (normal distribution assumed).

Table 3.4 Mass Transfer Kinetic and Physical–Chemical Parameters for the Model

Physical–Chemical Parameters for α-HCH

Symbol	Unit	Parameters	Reference	Data Numbers	Allocate mean	Allocate SD	Geometric mean	Geometric SD
P_S	Pa	Local vapor pressure	Cao et al. (2007), Mackay (2001), Cao et al. (2004), Cao et al. (2005), Ao et al. (2009), and Dong et al. (2009)	6	4.92E−02	1.08E−01	9.01E−03	1.50E+00
F_{25}	–	Fugacity ratio at 25 °C	Paasivirta et al. (1999)	1	1.21E−02	–[a]	1.21E−02	1.00E+00
H_{25}	Pa m³/mol	Henry's constant	Cao et al. (2007), Cao et al. (2004), Ao et al. (2009), and Dong et al. (2009)	4	6.12E−01	4.18E−02	6.11E−01	1.07E+00
B_F	–	Fugacity ratio temperature correction factor	Paasivirta et al. (1999)	1	1.62E+03	–[a]	1.62E+03	1.00E+00
B_H	–	Henry's law constant temperature correction factor	Paasivirta et al. (1999)	1	1.71E+03	–[a]	1.71E+03	1.00E+00
BP_S	–	Saturation vapor pressure temperature correction factor	Paasivirta et al. (1999)	1	4.95E+03	–[a]	4.95E+03	1.00E+00
K_{OC}	m³/t, 1/h	Adsorption coefficient	Cao et al. (2007), Ao et al. (2004), Ao et al. (2009), and Mackay (2001)	3	1.69E+03	1.86E+02	1.68E+03	1.11E+00
K_{m1}	–	Degradation rate of α-HCH in air	Cao et al. (2007), Mackay (2001), Ao et al. (2009), Prinn et al. (2001), and Brubaker and Hites (1998)	4	2.43E−03	3.70E−03	1.09E−03	3.86E+00

Symbol	Unit	Description	References	N			
K_{m2}	—	Degradation rate of α-HCH in water	Cao et al. (2007), Mackay (2001), Ao et al. (2009), and Breivik and Wania (2002)	4	3.17E−04	3.33E−04	2.18E−04	2.63E+00
K_{m4}	—	Degradation rate of α-HCH in sediment	Cao et al. (2007), Mackay (2001), Cao et al. (2004), Ao et al. (2009), and Breivik and Wania (2002)	5	4.87E−04	7.37E−04	1.22E−04	5.52E−01

Physical-Chemical Parameters for γ-HCH

Symbol	Unit	Description	References	N			
P_S	Pa	Local vapor pressure	Cao et al. (2007), Mackay (2001), Cao et al. (2004), Cao et al. (2005), Ao et al. (2009), and Dong et al. (2009)	7	1.13E−02	1.93E−02	1.24E−03	3.20E+00
F_{25}	—	Fugacity ratio at 25 °C	Paasivirta et al. (1999)	1	8.84E−02	−[a]	8.84E−02	1.00E+00
H_{25}	Pa m³/mol	Henry's constant	Cao et al. (2007), Cao et al. (2004), Ao et al. (2009), and Dong et al. (2009)	5	2.98E−01	1.40E−01	2.61E−01	1.90E+00
B_F	—	Fugacity ratio temperature correction factor	Paasivirta et al. (1999)	1	1.24E+03	−[a]	1.24E+03	1.00E+00
B_H	—	Henry's law constant temperature correction factor	Paasivirta et al. (1999)	1	3.09E+03	−[a]	3.09E+03	1.00E+00
BP_S	—	Saturation vapor pressure temperature correction factor	Paasivirta et al. (1999)	1	5.57E+03	−[a]	5.57E+03	1.00E+00
K_{OC}	m³/t, 1/h	Adsorption coefficient	Cao et al. (2007), Cao et al. (2004), and Ao et al. (2009)	4	1.64E+03	2.79E+02	1.62E+03	1.19E+00

Continued

Table 3.4 Mass Transfer Kinetic and Physical–Chemical Parameters for the Model—cont'd

Symbol	Unit	Parameters	Reference	Data Numbers	Allocate mean	Allocate SD	Geometric mean	Geometric SD
K_{OW}	–	Octanol/water partition coefficient	Cao et al. (2007), Cao et al. (2004), Ao et al. (2009), and Mackay (2001)	4	4.82E+03	2.30E+03	4.60E+03	2.32E+00
BCF_f	–	Bioconcentration factors for fish	Cao et al. (2007)	1	8.77E+02	–[a]	8.77E+02	1.66E+00
K_{m1}	–	Degradation rate of γ-HCH in air	Cao et al. (2007), Mackay (2001), Ao et al. (2009), Prinn et al. (2001), and Brubaker and Hites (1998)	5	9.69E−04	6.65E−04	6.66E−04	1.73E+00
K_{m2}	–	Degradation rate of γ-HCH in water	Cao et al. (2007), Mackay (2001), Ao et al. (2009), and Breivik and Wania (2002)	6	1.11E−04	5.55E−05	4.08E−05	1.74E+00
K_{m4}	–	Degradation rate of γ-HCH in sediment	Cao et al. (2007), Mackay (2001), Cao et al. (2004), Ao et al. (2009), and Breivik and Wania (2002)	7	2.38E−04	2.79E−04	1.26E−05	3.98E+00
Mass Transfer Kinetic Parameters								
R	$Pa\,m^3/mol\,K$	The gas constant	Cao et al. (2007)	1	8.31E+00	0.00E+00	8.31E+00	1.00E+00
B_1	m^2/h	Molecular diffusivities in air	Mackay (2001), Breivik and Wania (2002), Mackay and Paterson (1991), and Cao et al. (2004)	4	2.49E−02	1.03E−02	2.36E−02	1.44E+00
B_2	m^2/h	Molecular diffusivities in water	Mackay (2001), Breivik and Wania (2002), Mackay and Paterson (1991), and Cao et al. (2004)	4	2.86E−06	1.17E−06	2.68E−06	1.54E+00

Symbol	Units	Description	Reference					
B_4	m^2/h	Molecular diffusivities in sediment	Mackay (2001) and Cao et al. (2004)	2	1.43E−05	1.09E−05	1.20E−05	2.34E+00
K_{12}^{b}	m/h	Air-side molecular transfer coefficient over water	Mackay (2001), Calculated	1	7.11E+00	−[a]	6.68E+00	1.43E+00
K_{21}^{b}	m/h	Water-side molecular transfer coefficient over air	Mackay (2001), Calculated	1	2.08E−03	−[a]	2.02E−03	1.24E+00
K_{24}	m/h	Water-side molecular transfer coefficient over sediment	Mackay (2001)	1	1.00E−02	−[a]	1.00E−02	1.59E+00
K_{42}	m/h	Sediment-side molecular transfer coefficient over water	Tu et al. (1990)	1	5.39E−06	−[a]	5.39E−06	1.52E+00
K_{42r}^{b}	m/h	Sediment resuspension rate	Mackay (2001), Calculated	1	2.92E−05	−[a]	2.56E−05	1.70E+00
L_4	m	Diffusion path lengths in sediment	Mackay (2001)	2	3.50E−02	2.12E−02	3.16E−02	1.91E+00
K_P	m/h	Dry deposition velocity	Mackay (2001) and Breivik and Wania (2002)	2	1.04E+01	5.66E−01	1.04E+01	1.06E+00
K_S	m/h	Water sedimentation rates	Tu et al. (1990) and Gu (2005)	5	3.62E−06	4.34E−06	1.66E−06	4.32E+00
K_w^{b}	m/h	Wet deposition velocity	Tu et al. (1990)	1	1.33E−04	−[a]	1.33E−04	5.96E+00
S_c	−	Scavenging Ratio	Mackay (2001) and Breivik and Wania (2002)	2	1.34E+05	9.33E+04	1.17E+05	2.14E+00

[a]One value only; Geometric SD assigned manually (log-normal distribution assumed).
[b]Dynamic parameters for simulation, temporal values were shown in Kong et al. (2012).

were primarily obtained using a temperature of 25 °C. The temperature correction equation required for these parameters is shown as follows (Paasivirta et al., 1999):

$$\log_{10} P_T = \log_{10} P_{25} + A \times \left(\frac{1}{298} - \frac{1}{T+273} \right) \tag{3.2}$$

For each Henry's constant, the saturation vapor pressure or fugacity rate, P_T, is the physical and chemical parameter values at T (°C); P_{25} is the physical and chemical parameters at 25 °C; A is the temperature correction coefficient.

Sixteen parameters had time-varying values, including the environmental parameters (h_2, X_{13}, Q_{01t} Q_{10t}, Q_{02t}, Q_{20t}, Q_{23h}, Q_{02h}, T, C_1, O_{23}, and X_{23}) and the mass transfer parameters (K_{12}, K_{21}, K_{42r}, and K_w). Among them, the h_2 has hourly data; T, K_{12}, K_{21}, K_{42r}, and K_w had daily data; and X_{13}, Q_{01t} (Q_{10t}), Q_{02t}, Q_{20t}, Q_{23h}, Q_{02h}, C_1, O_{23}, and X_{23} had monthly data. Other parameters used annual average values and remained constant during the simulation.

To obtain the total river inflows of Lake Chaohu, monthly data from May 1987 to April 1988 were collected (Tu et al., 1990) along with the corresponding daily precipitation data from the China Meteorological Data Sharing Service System (http://www.cma.gov.cn/2011qxfw/2011qsjgx/index.htm). There was a significant linear relationship between the river inflow and the precipitation data. Using this linear relationship and the monthly precipitation data from May 2010 to February 2011 for Lake Chaohu, the river inflow (Q_{02t}) for the simulation period was easily calculated. The average monthly river outflow (Q_{20t}) of Lake Chaohu was based on the water balance calculation of inflow and water level in addition to the rates of industrial and agricultural water consumption (Q_{23h}).

Atmospheric HCHs input originated from atmospheric advection. The HCH concentrations in the advection within the study area (C_{01t}) were determined according to the sampled values on the lake side (four samples in total). The daily average wind speed and direction during the simulation period in Lake Chaohu area were obtained from the China Meteorological Data Sharing Service System. The volumes of atmospheric advections (Q_{01t} and Q_{10t}) were calculated according to the corresponding atmospheric height, the length of Lake Chaohu area, and the wind speed. HCH inputs originated from water inflows (C_{02t}) and were determined by the summation of the input amount from all the rivers around the lake.

The mass transfer coefficients of both sides of the gas–water interface (K_{12} and K_{21}) were calculated according to the method proposed by Southworth (1979). The resuspension coefficient (K_{42r}) was calculated according to the formula from Tu et al. (1990). The specific equations are as follows:

$$K_{12} = 11.375 \cdot (\text{WS} + \text{RS}) \cdot \left(\frac{18}{\text{MW}} \right)^{0.5} \tag{3.3}$$

$$K_{21} = 0.2351 \cdot \text{RS}^{0.969} \cdot \frac{(32/\text{MW})^{0.5} \cdot a}{h_{\text{W}}^{0.673}} \tag{3.4}$$

$$(\text{if WS} \leq 1.9 \,\text{m/s}, a = 1; \text{else } a = \exp[0.529 \cdot (\text{WS} - 1.9)])$$

$$K_{42r} = 3 \times 10^{-8} \cdot \frac{WS}{h_W} \tag{3.5}$$

where WS is the average wind speed (m/s); RS is the surface flow velocity (m/s); MW is the molecular weight (g/mol); and h_W is the water depth (m).

3.2.4 Sensitivity analysis

The accuracy of the parameters, particularly regarding some sensitive parameters, is one of the most important factors in model study when a system error of the model cannot be eliminated. All 50 parameters were included in the sensitivity analysis except the gas constant (R). For constant parameters, a local sensitivity analysis was applied that implemented a "perturbation" near the best estimate value of a parameter, and the variation of model outputs was studied under the condition that other parameters remained unchanged. The Morris classification screening method, a widely applied local sensitivity analysis method, was used (Morris, 1991). A variable was selected, and the value changed to the fixed step size, while the other parameters remained the same. The sensitivity index of the parameter was the average of the multiple disturbance calculated Morris coefficient:

$$S = \frac{\sum_{i=0}^{n-1} \frac{(Y_{i+1} - Y_i)/Y_0}{(P_{i+1} - P_i)/100}}{(n-1)} \tag{3.6}$$

where S is the Morris coefficient; Y_i is the model output value in the ith simulation; Y_0 is the model calculation result when the parameter is set at the initial value; P_i is the percentage change of the parameter value the for the ith simulation; and n is the number of runs.

Cao et al. (2004) proposed that when the step size is small enough, the nonlinear effects of the parameters of the model output are negligible. In this study, it was assumed that the parameters increased and decreased by 10% on the basis of the original value. $Y_{0.9}$, Y_0, and $Y_{1.1}$ are the output results when the parameter was multiplied by 0.9, 1, and 1.1, respectively. The sensitivity coefficient (C_s) is calculated by Equation (3.7):

$$C_s = \text{Abs}\left(\frac{Y_{1.1} - Y_{0.9}}{0.2 \times Y_0}\right) \tag{3.7}$$

The effect of the parameters on the model output was not only associated with corresponding C_s values of the parameters but was also related to the fluctuation range of the parameters in the environment. With the same C_s value, those parameters with higher variability have greater impacts on the model than those with lower variability. In this study, the sensitivity coefficient after the correction of the coefficient of variation (C_n) for the parameters (Tao et al. 2006) was also calculated such that $C_n = C_s \times CV$, where CV is the coefficient of variation of the parameter.

Moreover, for the 16 dynamic parameters in the model, the variability-based sensitivity coefficient (SCV; Lang et al., 2007) were calculated to assess their influence on the variation of the model outputs. The dynamic sensitivity coefficient (SCV) is calculated by Equation (3.8):

$$SCV_i = \frac{\Delta CV_i^Y / CV_i^Y}{\Delta CV_i^X / CV_i^X} \qquad (3.8)$$

where CV_i^X and CV_i^Y indicate the corresponding coefficients of variation of the ith input parameter and the output parameter, respectively, and ΔCV_i^X and ΔCV_i^Y represent the variations of the corresponding coefficients of variation of the ith input parameter and the output parameter, respectively.

3.2.5 Uncertainty analysis

Basic Monte Carlo simulation was applied to assess the uncertainty of the model by studying the impact of the simultaneous changes in the parameters on the model results. Both static and dynamic parameters with higher sensitivity coefficients were selected, and the original values were retained for the remaining parameters in the simulation process. All of the selected static parameters, except for temperature (T; normal distribution), were assumed to obey the lognormal distribution. The distributions of the parameters were obtained from prior knowledge (Tao et al., 2006; Lang et al., 2007) and the analysis of collected data in this study. The geometric mean and standard deviation could be calculated for static parameters with multiple values. On the other hand, if one value was obtained only, the corresponding coefficients of variation for the parameters were assigned using values based on the literature (for more details see Kong et al., 2012). For dynamic parameters, the monthly geometric mean and standard deviation were calculated from hourly or daily data. When only monthly data were available, the coefficients of variation were manually assigned. A total of 2200 Monte Carlo simulation runs were conducted. Each run was implemented with values for each parameter, which were randomly selected in the range of the mean $\pm$ standard deviation. Semi-interquartile ranges for the model output in different phases were then obtained for the uncertainty analysis.

Furthermore, the MCMC simulation method associated with the adaptive Metropolis–Hastings algorithm (Saloranta et al., 2008) based on Bayesian inference was also conducted for uncertainty analysis. This algorithm updates the parameter set values in each run, and then calculates the likelihood measure $l(\varphi)$ for each of the parameter set φ by the following Equation (3.9):

$$l(\varphi) = \frac{1}{\text{sigma0}} \exp\left(-\sum_{i=1}^{M} \frac{1}{2 \cdot \text{sigma0}^2} \left(C_i^{\text{obs}} - C_i^{\text{cal}} \right)^2 \right) \qquad (3.9)$$

where C_i^{obs} and C_i^{cal} are the observed and calculated concentrations (log transformed) for one phase in month i; sigma0 is the standard deviation of the normal distribution of observations in log-scale. *M* is the observation number. Then an acceptance probability α is calculated to decide whether the new parameter set is accepted or discarded, and a new set is subsequently obtained. This process is repeated at least for 10^5 so that a Markov Chain of the parameter is obtained and the prior distributions are updated to the posterior distributions. More details can be found in Saloranta et al. (2008). The density of observations in time and various media in this study provided sufficient information for the MCMC simulation, which can be used to update the prior distributions of the parameters to obtain the posterior distributions. Fifteen static parameters without seasonal variations were selected (listed in Table 3.5) primarily based on the results of the sensitivity analysis, together with sigma0. In addition, the statistical characteristics of the 16 dynamic parameters with seasonal variations were not updated but were also randomly selecting values during the simulation, which was the same as the simulation by Basic Monte Carlo method. The remaining parameters were fixed to their original distributions. The length of the Markov chain was set as 10^5. The convergence of the chain was judged by visual inspection and the first 2×10^4 runs were discarded (considered as the burn-in period). The uncertainty results obtained by random sampling (2200 runs) from the rest of the chain were compared with the former results derived from the basic Monte Carlo simulation.

3.3 RESULTS AND DISCUSSION
3.3.1 Simulation of seasonal variations

The simulated annually average concentrations of α-HCH and γ-HCH are shown in Figure 3.2 and were found to be in agreement with the measured data. For α-HCH, the differences in the main phases were 0.21, 0.6, 0.06, 1.69, and 0.07 logarithmic units for air, air particles, water, suspended solids, and sediment solids, respectively; while for γ-HCH, the deviations were 0.36, 0.25, 0.12, 1.36, 0.66, and 0.13 logarithmic units for air, air particles, water, suspended solids, fish, and sediment solids,

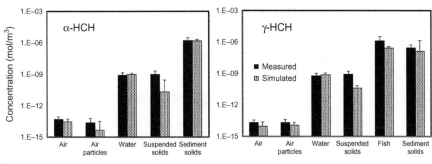

FIGURE 3.2

Comparison between the annual simulated and measured α-HCH and γ-HCH concentrations in Lake Chaohu. The error bars included in this figure represent the standard deviations.

respectively. These deviations were all within one logarithmic unit, except for suspended solids in water, which had residual errors more than one logarithmic unit for both α-HCH and γ-HCH. This underestimation might be attributed to sample collection. The suspended solids were filtered from the water samples with a 0.45-μm glass fiber filter (He et al., 2013), which included the phytoplankton in the water. As a heavily eutrophic shallow lake, the phytoplankton was dominant in the lake, especially in the summer when algae were blooming. In addition, aquatic organisms, especially plankton, can substantially affect the fate of POPs in aquatic environments by strong uptaking (Dachs et al., 1999). However, in this study, only absorption by the organic matter in the suspended solids was considered in the model, resulting in the large underestimation of HCH concentrations. Further studies are needed to ascertain the relative uptake coefficient constants for HCHs in phytoplankton. For other phases, the relatively high level of γ-HCH concentration in fish should be the result of the bioconcentration effect (Hargrave et al., 2000). Furthermore, the HCH concentration in the sediment solids was much higher than that in the atmosphere or in the water column. It was concluded that sediment acts as a sink of HCHs (Walker et al., 1999).

Seasonal variations in the modeled α-HCH and γ-HCH concentrations in various media compared with the observations are indicated from Figures 3.3–3.6. The model output of the α-HCH and γ-HCH concentrations in the atmosphere was relatively consistent with the measured values (Figure 3.3). The gaseous α-HCH concentrations in the summer and winter, notably in August and December, were higher than in other seasons, while the gaseous γ-HCH concentration was

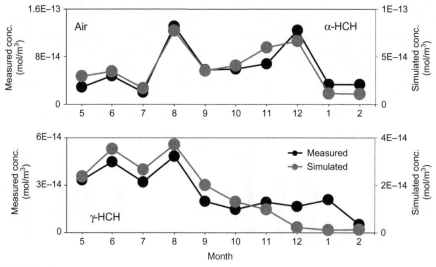

FIGURE 3.3

Seasonal variations of α-HCH and γ-HCH concentrations in the air. Both measured and simulated concentrations are presented for model validation.

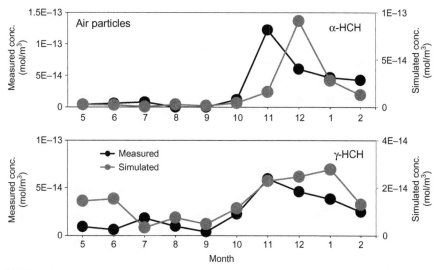

FIGURE 3.4

Seasonal variations of α-HCH and γ-HCH concentrations in the air particles. Both measured and simulated concentrations are presented for model validation.

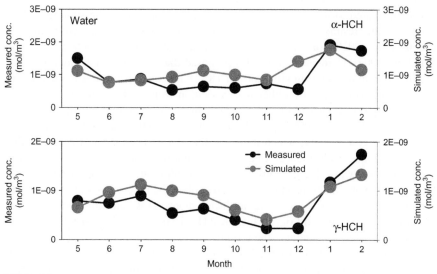

FIGURE 3.5

Seasonal variations of α-HCH and γ-HCH concentrations in the water. Both measured and simulated concentrations are presented for model validation.

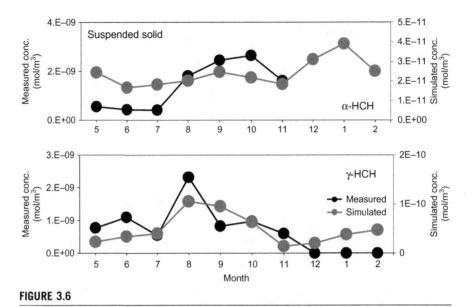

FIGURE 3.6

Seasonal variations of α-HCH and γ-HCH concentrations in the suspended solid. Both measured and simulated concentrations are presented for model validation.

significantly higher in the summer (May–August) than in other seasons. Ridal et al. (1996) observed a similar pattern as gaseous HCHs concentrations were relatively high in summer in Lake Ontario. In this study, the high values in summer were primarily attributed to increased input from air advection during the summer, which could be the results of higher temperature that favors volatilization in land (Zheng et al., 2010), or remote inputs from the long-range atmospheric transport (Li et al., 2002). Haugen et al. (1998) proposed that the regression slope between $\ln P$ (air partial pressure; in atm) and $1/T$ (K^{-1}) indicated that atmospheric concentrations were controlled by either re-evaporation from surfaces in the vicinity of the sampling site (steep) or air advection (shallow). It was found that the regression coefficients (R^2) were 0.0007 (p>0.1) and 0.4317 (p<0.01) for α-HCH and γ-HCH (Figure 3.7), respectively, which indicated that the gaseous α-HCH concentration in Lake Chaohu was influenced to a greater extent by remote input than by volatilization in lake, while the gaseous γ-HCH concentration was strongly affected by local surface volatilization. This could also explain the higher gaseous α-HCH concentration in winter, when a strong northwest monsoon is controlling the lake area (Kong et al., 2013). In addition, for technical HCHs, the ratio of α-/γ-HCH is approximately 4–7; while for lindane, the ratio of α-/γ-HCH is less than 0.1 (Iwata et al., 1993, 1995; Walker et al., 1999). It was found that the α/γ ratio was much lower in both air and water in the summer (Figure 3.8), indicating possible recent illegal applications of lindane around Lake Chaohu. The agricultural land accounted for 61.1% of the total land area in the Lake Chaohu basin (Tu et al., 1990), resulting in high levels of γ-HCH residues in the soil. The soil could be converted from a major

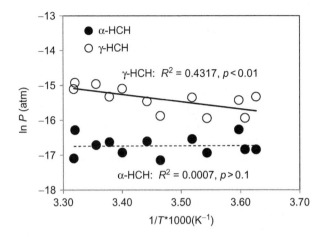

FIGURE 3.7

Logarithm of the partial pressure of α-and γ-HCH in air plotted against the reciprocal of the ambient temperature.

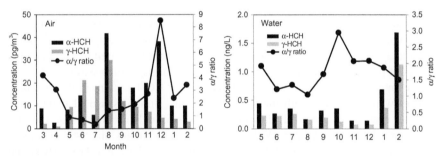

FIGURE 3.8

α-HCH and γ-HCH concentration and the α/γ-HCH ratio in air and water in Lake Chaohu.

sink to an important emissions source of OCPs after the ban (Tao et al., 2008). Higher precipitation levels and optimized plant growth in the summer also resulted in more rapid release of HCHs from soil (Waliszewski, 1993).

For the atmospheric particles, the α-HCH and γ-HCH concentration were both lower in the summer and higher in the winter (Figure 3.4). The primary reason could be that since the temperature rises in the summer, the atmospheric vapor pressure (P_s) for α-HCH and γ-HCH was higher, resulting in a lower fugacity capacity of the air particles so that the gas–solid balance of HCHs in the air shifts toward the gaseous phase. The situation was opposite in the winter (Carlson et al., 2004). Therefore, temperature was the key factor controlling the partition of HCHs between the air vapor and particulate phases (Carlson et al., 2004). In addition, α-HCH in the atmospheric particulate peaked in November according to the measured values, while the calculated value peaked in December, which corresponded to the peak of gaseous α-HCH concentration but failed to capture the November peak. This discrepancy may due to higher concentrations of α-HCH in the remote input of atmospheric particulate matter in November. The specific mechanisms underlying this difference require further study.

The measured and simulated values of α-HCH and γ-HCH in the water were also in good agreement (Figure 3.5). The model captured the high value in the winter and the variation in the other seasons, which was also consistent with the data in Lake Ontario (Ridal et al. 1996). Although the winter temperatures are lower, leading to reduced water fugacity capacity (Carlson et al., 2004), the precipitation and water inflow are also lower in the winter, resulting in a significant decrease in water levels, which may cause a concentration effect. In the contrast, lower concentrations were simulated in the summer and the autumn. A lower peak value was also observed in the summer (July). The significantly higher gas–water interface processes in summer (Figure 3.10), especially wet deposition (T_{12w}), might lead to the relatively higher concentration than those in late spring or autumn. The HCHs concentrations in water began to be overestimated from August to December, which coincides with an observed increase in the seasonal distribution of phytoplankton in Lake Chaohu (Xu et al., 1999). It can be speculated that disregarding aquatic organisms, particularly the phytoplankton phase, can lead to a significant deviation between the measured data and simulation results.

The simulated results for suspended solids in water were consistent with the seasonal pattern of the observations but were underestimated by nearly one and half orders of magnitude (Figure 3.6). Further research was still required to introduce the process of phytoplankton uptake and ascertain the corresponding parameters.

The annual averages of the sampled values of α-HCH content in the sediment particles were consistent with the simulated results. Similar seasonal variation in the water bodies was obtained, showing the trends of higher values in the summer and lower values in the winter. With smaller seasonal changes, the α-HCH content in the sediment was relatively stable compared to that in the water. For γ-HCH, the average of the model results for fish and sediment agreed with the observations. The concentration in fish had relatively small seasonal variations. Similar seasonal variations in water bodies were observed for sediment solids. Although the sedimentation (T_{24s}) had an elevated level in summer rather than winter, the much lower Henry's law function might have had a higher influence, leading to the peak value of the concentration in sediment solids in winter time.

3.3.2 Transfer fluxes

The annual and seasonality of the fluxes had been estimated by the model and presented in Figures 3.9 and 3.10. As shown in Figure 3.9, the net input of α-HCH into the Lake Chaohu environment is approximately 0.294 t/a, while the net output is approximately 0.412 t/a. It can be observed that the α-HCH content in the Lake Chaohu watershed is diminishing by 0.118 t/a. The net input of γ-HCH into Lake Chaohu was estimated at approximately 0.160 t/a, while the net output was approximately 0.196 t/a. Therefore, the total quantity of γ-HCH in Lake Chaohu had decreased by 0.036 ton, which was about one-third of α-HCH.

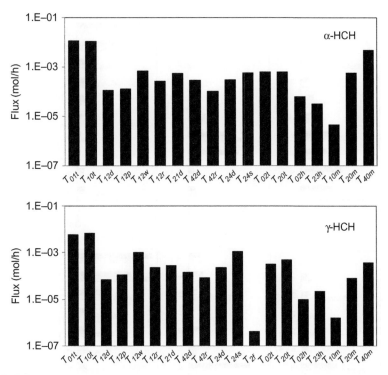

FIGURE 3.9

α-HCH and γ-HCH fluxes in and out of the Lake Chaohu area and between the adjacent compartments. T_{01t}, air advection flows into the area; T_{10t}, air advection flows out of the area; T_{12d}, diffusion from air to water; T_{12p}, dry deposition from air to water; T_{12w}, wet deposition from air to water; T_{12r}, rain scavenging; T_{21d}, diffusion from water to air; T_{42d}, diffusion from sediment to water; T_{42r}, resuspension from sediment to water; T_{24d}, diffusion from water to sediment; T_{24s}, sedimentation from water to sediment; T_{02t}, water advection flows into the area; T_{20t}, water advection flows out of the area; T_{02h}, locative wastewater discharge (zeros); T_{23h}, industry and agriculture water usage; T_{10m}, degradation in air; T_{20m}, degradation in water; T_{40m}, degradation in sediment; T_{2f}, fish production.

For α-HCH, the atmospheric advection input was found to be the main source (T_{01t}) (0.278 t/a), which corresponded to the atmospheric advection output (T_{10t}) (0.277 t/a). By contrast, the α-HCH input from water inflows was very small (0.016 t/a). An important output was the degradation in the sediments (0.119 t/a), which accounted for 89.05% of the total degradation in the environment, while the degradation in the water was 0.015 t/a, which accounted for 10.86% of the total degradation. For γ-HCH, the main source of input was atmospheric advection of 0.151 t/a, which corresponded to an advection output of 0.171 t/a. Similar to α-HCH, the input from water inflow was relatively small (0.008 t/a). Degradation processes were significantly lower than α-HCH, as the value were estimated as

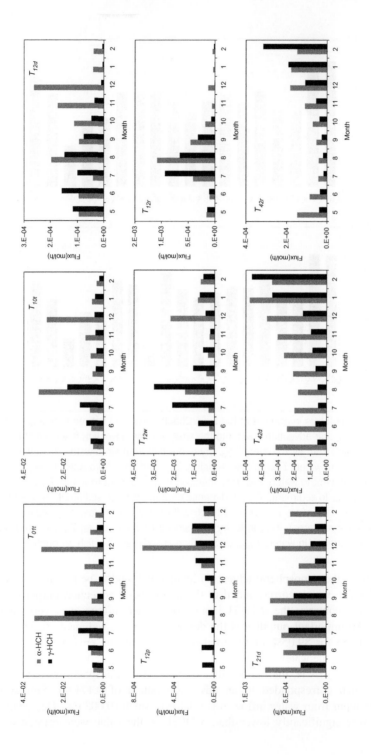

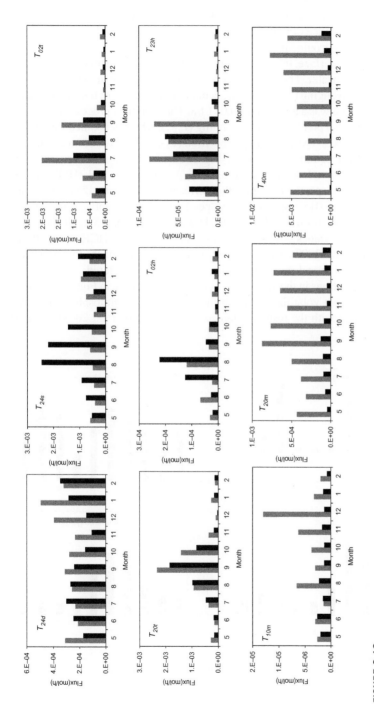

FIGURE 3.10

Seasonal variations of the γ-HCH fluxes in and out of the Lake Chaohu area and between the adjacent compartments. T_{01t}, air advection flows into the area; T_{10t}, air advection flows out of the area; T_{12d}, diffusion from air to water; T_{12p}, dry deposition from air to water; T_{12w}, wet deposition from air to water; T_{12n}, rain scavenging; T_{21d}, diffusion from water to air; T_{42n}, resuspension from sediment to water; T_{24d}, diffusion from water to sediment; T_{24s}, sedimentation from water to sediment; T_{02h}, water advection flows into the area; T_{20t}, water advection flows out of the area; T_{02h}, locative wastewater discharge (zeros); T_{23h}, industry and agriculture water usage; T_{10m}, degradation in air; T_{20m}, degradation in water; T_{40m}, degradation in sediment.

0.001, 0.002, and 0.009 t/a for air, water, and sediment, respectively, which could be attributed to the persistence of γ-HCH (Cao et al., 2004) and lower levels of γ-HCH than α-HCH in the environment.

On the air–water interface, there was an annual α-HCH net input of 0.016 t/a from the atmosphere to the water. The dominant process of atmospheric input to the water was precipitation scavenging (T_{12w}), which accounted for 57.80% of the gas-to-water flux. For γ-HCH, there was a net input of 0.031 t/a. The dominant process on the air–water interface was also precipitation scavenging (T_{12w}), which accounted for 72.3% of the air-to-water fluxes. Except for dry deposition (T_{12p}), all fluxes between air and water had a seasonal pattern of relatively higher in summer and lower in winter (Figure 3.10). Diffusion from air to water (T_{12d}) was 0.003 and 0.002 t/a for α-HCH and γ-HCH, respectively, while the corresponding process from water to air (T_{21d}) amounted to 0.014 and 0.007 t/a, respectively. Consequently, annual average of net volatilization for α-HCH and γ-HCH was about 11 and 5 kg. This result was comparable to the values in Lake Taihu (49 and 4.1 kg for α-HCH and γ-HCH, respectively; Qiu et al., 2008). However, net input from atmosphere to water was ascertained if deposition and rain scavenging fluxes were considered.

The seasonal variations in the air–water exchange were shown in Figure 3.11. For α-HCH, there was a net volatilization from the water into the atmosphere in May, which was consistent with the results obtained by Taihu (Qiu et al., 2008). During

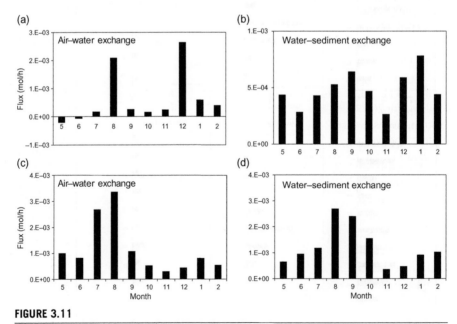

FIGURE 3.11

Seasonal α-HCH (a and b) and γ-HCH (c and d) fluxes over the air–water and the water–sediment interface. Positive values indicate net inputs from air to water or from water to sediment.

the other seasons, however, there is a net input from the atmosphere to the water, which is the converse of the results observed in Lake Taihu. The main cause of this difference may be that the research in Lake Taihu did not include deposition from air to water. It is also worth noting that the α-HCH concentrations in the Lake Taihu atmosphere and water are 32 ± 28 pg/m^3 and 1887 ± 1372 pg/L, respectively, while in Lake Chaohu, the corresponding concentrations are 16 ± 11 pg/m^3 and 423 ± 395 pg/L, which are 50.0% and 22.4% of the values of Lake Taihu, respectively. The lower α-HCH concentration in the water of Lake Chaohu may be due to historically lower HCH pesticide usage. Ridal et al. (1996) proposed that, due to a reduction in the atmospheric concentration, the α-HCH flux in Lake Ontario has shifted from net settlement to net volatilization when compared with the years prior to 1990. For Chaohu, however, due to a reduction in the water α-HCH concentration, the air–water interface may have still been net settlement. Therefore, despite the net volatile flux in the summer, the annual net flux is from the gas to the water. For γ-HCH, a net deposition happened throughout the simulation duration. It was estimated that approximately 0.031 ton of γ-HCH were added to Lake Chaohu by air–water interface fluxes in one year. The large quantities of γ-HCH cycling through air–water interface fluxes indicated that this pathway might be a major factor in determining HCH concentrations in large lakes (Ridal et al., 1996). In addition, it is worth noting that wet deposition (T_{12w}) was higher in the summer, particularly in August, and lower during the other seasons. In contrast, dry deposition (T_{12p}) was higher in the winter, notably in December, and lower during the other seasons.

The flux from the water to the sediment was 0.022 t/a for α-HCH, and sedimentation (T_{24s}) accounted for 65.49% of this flux. In addition, the flux from the sediment to the water was 0.010 t/a, and diffusion flux (T_{42d}) accounted for 73.28% of this flux. There was a net input of 0.012 t/a from the water to the sediment (Figure 3.11). Although the sediment resuspension flux was 0.003 t/a, which accounted for 26.72% of the flux from the sediment to the water, this flux still reflects the strong resuspension process in Lake Chaohu (Tu et al., 1990). For γ-HCH, the interface between water and sediment provided a net input of 0.013 t/a (Figure 3.11). Sedimentation accounted for 82.8% of the processes from water to sediment, while resuspension dominated the fluxes from sediment to water (85.4%).

In the sensitivity analysis, those parameters related to relatively important processes will always be observed with higher sensitivity (see Section 3.3.3).

3.3.3 Sensitivity analysis of the model parameters

Results of a sensitivity analysis were similar for α-HCH and γ-HCH. Therefore, only the results for α-HCH and the comparison with γ-HCH are illustrated here in order to be concise, while the corresponding details for γ-HCH can be found in Kong et al. (2014; in press). The CV corrected sensitivity coefficients (C_n) for the 15 most sensitive parameters are displayed in Figure 3.12.

For some of the static parameters, the sensitivity coefficients changed significantly after correction with the coefficients of variation. Thus, despite the high

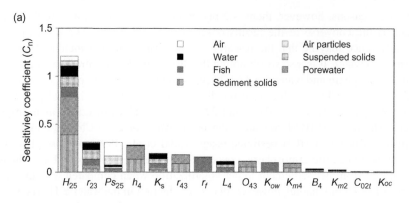

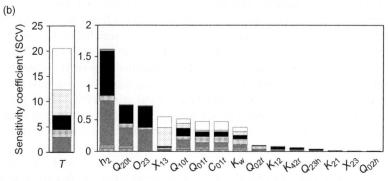

FIGURE 3.12

Coefficient of variability normalized sensitivity coefficients (C_n) of static parameters (a) and variability-based sensitivity coefficient (SCV) of the dynamic parameters (b) to the model outputs in various environmental media. (For the color version of this figure, the reader is referred to the online version of this chapter.)

sensitivities regarding K_{oc}, B_{ps}, B_H, and Sc, the corrected sensitivity coefficients for those parameters with lower variability were significantly reduced. Thus, these parameters were considered to be insensitive. The sensitivity reduction in K_{oc} was also observed by Cao et al. (2004). In contrast, due to higher variability, the sensitivity coefficients of h_4, k_{m4}, k_{m2}, and L_4 increased after correction, and they were found to be important parameters. h_4 is related to sediment volume, and sediment is found as the sink for α-HCH in lakes; k_{m4} is directly related to the degradation of α-HCH in the sediments, which has been found to be the most important degradation process in the environment. Thus, the two static parameters exerted considerable influence on the model results. L_4 and k_{m2} become more important parameters due to their high variability. However, it was worth noting that parameters associated with degradation, such as k_{m4} and k_{m2}, had relatively lower sensitivity for γ-HCH than α-HCH, because degradation fluxes of γ-HCH in all media were significantly lower than those of α-HCH.

Other parameters, including H_{25}, Ps_{25}, C_{02t}, O_{43}, and K_s, had relatively similar high sensitivity coefficients before and after correction. H_{25} (Henry's law constant at 25 °C) was determined to be the most sensitive parameter, which mainly resulted from large variation in H_{25} (Table 3.4). In addition, all of the calculations of the fugacity capacity in the water and sediment phases were related to this parameter (Mackay and Paterson, 1991), which played a fundamental role in predicting the environmental behavior of POPs (Odabasi et al., 2008). Ps_{25} determines the fugacity capacity of the atmospheric particulates (Mackay et al., 1986). Both Ps_{25} and H_{25} play a decisive role on the fate of POPs in the environment. Although the variability of these two parameters is negligible, the collected values in this study are based on the results from different time periods using different methods. Therefore, the sensitivities of these two parameters remain high after the correction. C_{02t} strongly affects the α-HCH content in the water and suspended matter. O_{43} determines the adsorption capacity of the particles in the sediments so that it exerts a great influence on the model output. K_s had a relatively stronger influence on the concentration of γ-HCH than α-HCH, probably due to the addition of fish subphase. This addition also resulted in the high sensitivity of K_{ow} and r_f, which were important parameters for fish.

Each of the parameters has a different influence on the various environmental compartments. For example, k_{m4} has a higher sensitivity coefficient for the sediment than for the water or atmosphere, while k_{m2} has the highest sensitivity coefficient for the water. Overall, for α-HCH, the average values of C_n for the air, water, and sediment were 1.17%, 2.78%, and 3.42%, respectively, while these values were 1.42%, 4.59%, and 6.23% for γ-HCH. Although the water compartment was the center of the system and had the largest number of parameters in the model among the three main phases (Tao et al., 2006), sediment serves as an important sink and was under the greater influence of the parameters than either the air or the water phases.

Similar variability-based sensitivity coefficients (SCV) for α-HCH and γ-HCH for the 16 dynamic parameters were found and the results for γ-HCH were indicated in Figure 3.12. Temperature (T) was found to have a much greater influence on the variation of the model output than the other parameters, probably because temperature had very strong effects on Ps and H, the two important parameters in the model. Consequently, temperature played a decisive role in the distribution of HCHs between the gaseous and particulate phases as well as between the air and water (Walker et al., 1999). h_2 strongly affected the variations in γ-HCH concentrations in water and fish, which might be associated with the apparent seasonality of high water in the summer and the drought period in the winter. The SCV for Q_{20t} was much higher than for Q_{02t}, which might be attributed to the nearly one and half times higher quantity of water outflow (T_{20t}) than inflow (T_{02t}) (Figure 3.9). Lower concentration in the water inflow resulted in the higher influence of Q_{20t} on the variation than Q_{02t}. X_{13} had a relatively strong influence on the seasonal changes in the concentration in the atmosphere and the water bodies as well as the particulate and suspended matter content, which is in agreement with the conclusion of the Pearl River Delta study (Lang et al., 2007). O_{23} was relatively influential on the subphases in water with lower SCV on the suspended solids, which might be due to the significant underestimation of this phase. Q_{01t}, Q_{10t}, and C_{01t} were associated with the atmospheric advection, which was the main source of the γ-HCH in Lake Chaohu. Thus, the seasonal variations in these three parameters also had significant impacts. K_{12}, K_{21}, and K_w were the main parameters influencing the air–water interface flux due to their direct impacts and significant seasonal variations. However, in contrast to α-HCH, K_{12} and K_{21} were both considered to be unimportant to seasonality of γ-HCH, probably because of lower diffusion fluxes for γ-HCH than α-HCH between air and water. Without considering the biological phase, the importance of X_{23} was also reduced. The low sensitivity coefficient of K_{42r} was due to the corresponding low resuspension flux.

3.3.4 Uncertainty analysis of the model simulation

Results of uncertainty analysis by basic Monte Carlo method for α-HCH were similar to those for γ-HCH. Therefore only γ-HCH is discussed here. The model uncertainty for each phase is illustrated in terms of the semi-interquartile range (i.e., the range between 25th and 75th percentiles) by dashed lines in Figure 3.13. It was found that the uncertainty of the model was relatively small from May to September and began to increase significantly in October. This increase was attributed to our finding that from October, the coefficients of variation in the gas–water diffusion rate (K_{12} and K_{21}) significantly increased, leading to an increase of variation in the air–water diffusion flux. This also contributed to a significant increase in the uncertainty of the other phases. Lang et al. (2007) similarly found that the coefficient of variability of diffusion was associated with wide variability in the gaseous PAHs concentrations. The rates of diffusion across the gas–water interface (K_{12} and K_{21}) were related to wind speed and water depth, and the coefficient of variation of water depth (h_2) did not increase during October–December. It can be speculated that elevated variation in the wind speed in this period causes the increasing uncertainty.

In the MCMC simulation, 15 parameters determined to be influential to model outputs in the sensitivity analysis were selected to be estimated and updated in the MCMC simulation. After the first 2×10^4 runs, the chains for all of the parameters converged, which were judged by visual inspection (Figure 3.14). Subsequently, 2200 runs of the model were performed, and the 15 parameters (excluding sigma0) were randomly resampled from the last 8×10^4 parameter chains. The results of the uncertainty analysis for all media are illustrated by the solid line in Figure 3.13. The dispersion of the model output, shown as semi-interquartile ranges, was significantly reduced when compared to the results by basic Monte Carlo method. It was revealed that the uncertainty of the model included both inherent variability and true uncertainty of the model estimates (McKone, 1996). Tao et al. (2006) proposed that D_{Rp}, which was defined as the difference between the concentration ranges in which a given percentage (p) of the simulated and the measured results fall, could be considered the indicator for the true uncertainties. Because the multi-site monthly samples were only available for water and suspended solids in

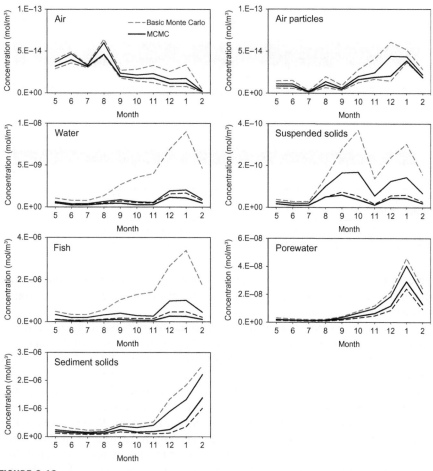

FIGURE 3.13

The dynamics of the simulated dispersions of γ-HCH concentrations in various media from 2010.5 to 2011.2, which are presented as the 0.25 (dashed line at the bottom) and 0.75 (dashed line on the top) percentiles, determined by the basic Monte Carlo simulation and the 0.25 (solid line at the bottom) and 0.75 (solid line on the top) percentiles, determined by MCMC simulation.

this study, while the simulated data did not provide rational results for the suspended solids, the results in water were selected for uncertainty reduction estimation. For p, 95% was used as the criterion. The annual averages of D_{R95} for the water phase in the basic Monte Carlo and MCMC simulations were 3.208 and 0.741, respectively. The true uncertainty provided by the basic Monte Carlo was unreasonably high, giving a false impression of the reliability of the model (Saloranta et al., 2008). The MCMC simulation decreased the ascertained true uncertainty to 23% of the original,

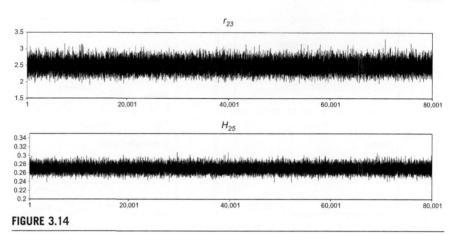

FIGURE 3.14

Examples (H_{25} and r_{23}) of the corresponding converged chain from MCMC simulation. The last 80,000 data from the total 10^5 runs were taken for presentation.

indicating that 77% of the true uncertainty of model results in water column had been eliminated. A significant decrease of the true uncertainty of model prediction in other media could be speculated.

This decrease in model uncertainty was primarily attributed to the reduced dispersion of the 15 parameters after the MCMC simulation. Details of the statistical characteristics for these parameters before and after the MCMC are listed in Table 3.5. The prior and posterior probability distributions of r_{23} and H_{25} are also shown in Figure 3.15. The estimations of these parameters based on their posterior distributions were relatively close to the original values, while the semi-interquartiles ranges were all remarkably reduced. However, K_s was estimated to be approximately two orders of magnitude lower compared to its original value. During the simulation, the model could be trying to fit the largely underestimated simulated data of suspended solids to the corresponding observations by decreasing the output of this subphase, such as sedimentation, which was strongly affected by K_s (Figure 3.12). However, the model error was too high to be reduced by simply updating the parameters. Moreover, this simulation also led to an incorrect continuous decrease in this parameter, resulting in an unreasonably low value for K_s after the simulation. Therefore, K_s retrieved its original distribution during the reproduction of the model uncertainty after the MCMC simulation. Except for K_s, most of the parameters centered on their prior distributions with a significantly narrowed dispersion range, such as h_4, L_4, and K_{ow}. Several other parameters had been slightly displaced compared to their prior distributions. O_{43} was updated to a higher value, which could be attributed to the underestimation of the concentration in sediment solids in the original simulation. The estimated K_{m2} was lower by a factor of two, the primary reason for which should be the underestimation of the suspended solids. The MCMC simulation was attempting to increase the levels in water so that the estimated concentration in suspended solids would be simultaneously elevated to fit the

Table 3.5 Statistical Characteristics of Prior and Posterior Distribution for the 16 Parameters in the MCMC Simulation

Parameters	Prior			Posterior		
	25%	50%	75%	25%	50%	75%
h_4	6.907E−02	1.000E−01	1.448E−01	1.255E−01	1.350E−01	1.452E−01
L_4	2.042E−02	3.162E−02	4.896E−02	2.850E−02	3.393E−02	4.039E−02
K_{ow}	2.605E+03	4.600E+03	8.124E+03	4.561E+03	4.601E+03	4.641E+03
K_{oc}	1.441E+03	1.619E+03	1.818E+03	1.475E+03	1.588E+03	1.710E+03
B_4	6.785E−06	1.205E−05	2.140E−05	1.096E−05	1.372E−05	1.718E−05
sigma0	1.141E+00	1.500E+00	1.972E+00	5.265E−01	5.476E−01	5.696E−01
r_{23}	2.264E+00	2.500E+00	2.760E+00	2.545E+00	2.552E+00	2.559E+00
r_{43}	2.474E+00	2.760E+00	3.079E+00	2.514E+00	2.555E+00	2.596E+00
r_f	1.436E+00	1.500E+00	1.567E+00	1.636E+00	1.706E+00	1.779E+00
Ps_{25}	5.655E−04	1.240E−03	2.719E−03	1.219E−03	1.340E−03	1.473E−03
K_s	6.161E−07	1.655E−06	4.446E−06	2.992E−08	8.542E−08	2.438E−07
O_{43}	5.494E−02	7.920E−02	1.142E−01	1.083E−01	1.187E−01	1.301E−01
K_{m4}	4.961E−06	1.260E−05	3.200E−05	1.086E−05	1.115E−05	1.145E−05
K_{m2}	2.811E−05	4.077E−05	5.912E−05	2.193E−05	2.407E−05	2.642E−05
C_{O2t}	4.693E−10	6.750E−10	9.708E−10	7.245E−10	7.707E−10	8.199E−10
H_{25}	1.692E−01	2.610E−01	4.025E−01	3.704E−01	4.439E−01	5.320E−01

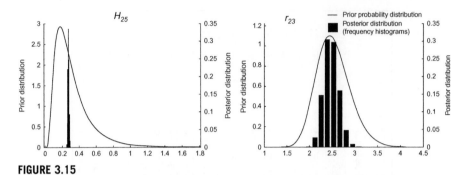

FIGURE 3.15

Examples (H_{25} and r_{23}) of the prior distributions (shown by the black line) and corresponding posterior distributions derived from the converged parameter chain form MCMC simulation (black histograms).

observations. Posterior estimated H_{25} was nearly $2 \times$ higher than the prior values. All of the calculations of the fugacity capacity in the water and sediment phases were related to this parameter. Therefore, this update might have been the combined effect of the model fitting for both compartments in water and sediment.

The model simulation results with the updated parameter estimations for various media were not indicated because the output did not significantly fit the observation better. For the multimedia fugacity model, it was always acceptable if the differences between the simulated and observed data were less than 0.5 or 0.7 orders of magnitude (Cao et al., 2004). The patterns of variation were more valuable than the prediction accuracy for seasonal modeling by the level-IV fugacity model (Lang et al., 2007). Thus, model calibration might not be important for the fugacity model. The calibration in the MCMC simulation aimed at avoiding improbable parameter combinations, which led to a poor fit between the model results and observed data and the subsequent overestimation of model uncertainty, instead of fitting the model results as close as possible to the observations. As a result, it was suggested that the main functions of the MCMC in this study should be to conduct a proper uncertainty analysis, disregard improbable parameter combinations, and avoid the overestimation of the confidence bands in model prediction (Saloranta et al., 2008), rather than calibrating the model with the observations.

3.4 CONCLUSION

A fugacity-based level-IV QWASI model was developed to simulate the seasonal variation of α-HCH and γ-HCH in the air, water, and sediment, as well as the various environmental fluxes in Lake Chaohu, China. The simulated results were consistent with the measured data with deviations of generally less than one order of magnitude. However, disregarding the effects of aquatic organisms resulted in large deviations between the simulated and measured values of HCHs in suspended solids in water.

The factors leading to the seasonal variations of α-HCH and γ-HCH in various compartments were revealed. The major source of α-HCH and γ-HCH in Lake Chaohu was input from atmospheric advection inflows, while the major environmental outputs were atmospheric advection outflows. It was estimated that total quantity of α-HCH and γ-HCH in Lake Chaohu decreased by 118 and 36 kg each year, respectively. Approximately 16 kg α-HCH of 31 kg of γ-HCH was added to Lake Chaohu by air–water interface fluxes and 12 kg α-HCH of 13 kg of γ-HCH was added to the sediment by the water–sediment interface fluxes. Thus, sediment is an important sink for HCHs in this lake. The sensitivity analysis found that Henry's law constant at 25 °C was the most sensitive static parameter. The sediment was influenced more by the combined effects of the various parameters than air and water were. In addition, temperature variation had a much greater impact on the dynamics of the model output than other dynamic parameters. Uncertainty analysis based on the basic Monte Carlo method showed that the model uncertainty was relatively low, especially from May to August during the simulation. From October, due to the increase in the gas–water diffusion flux variability, uncertainty of the model significantly increased in all of the compartments. The MCMC simulation eliminated 77% of the true model uncertainty in water column ascertained by the basic Monte Carlo method. It was suggested that rather than calibrating the model, the main function of the MCMC for the fugacity model should be to conduct a proper uncertainty analysis and avoid overestimating the uncertainty in the model prediction.

References

Ao, J.T., Chen, J.W., Tian, F.L., Cai, X.Y., 2009. Application of a level IV fugacity model to simulate the long-term fate of hexachlorocyclohexane isomers in the lower reach of Yellow River basin, China. Chemosphere 74, 370–376.

Breivik, K., Wania, F., 2002. Evaluating a model of the historical behavior of two hexachlorocyclohexanes in the Baltic sea environment. Environ. Sci. Technol. 36, 1014–1023.

Brubaker, W.W., Hites, R.A., 1998. OH reaction kinetics of gas-phase alpha- and gamma-hexachlorocyclohexane and hexachlorobenzene. Environ. Sci. Technol. 32, 766–769.

Cai, R., 2010. Agri-chemicals inputs and its impact on environment. China Popul. Res. Environ. 20, 107–110 (in Chinese).

Cao, H.Y., Tao, S., Xu, F.L., Coveney, R.M., Cao, J., Li, B.G., Liu, W.X., Wang, X.J., Hu, J.Y., Shen, W.R., Qin, B.P., Sun, R., 2004. Multimedia fate model for hexachlorocyclohexane in Tianjin, China. Environ. Sci. Technol. 38, 2126–2132.

Cao, H.Y., Liang, T., Tao, S., 2005. Dynamic simulation and prediction of BHC transfer and residues in Beijing during 50 years. Sci. China Ser. D Earth Sci. 35, 980–988.

Cao, H.Y., Liang, T., Tao, S., Zhang, C.S., 2007. Simulating the temporal changes of OCP pollution in Hangzhou, China. Chemosphere 67, 1335–1345.

Carlson, D.L., Basu, I., Hites, R.A., 2004. Annual variations of pesticide concentrations in great lakes precipitation. Environ. Sci. Technol. 38, 5290–5296.

Dachs, J., Eisenreich, S.J., Baker, J.E., Ko, F.C., Jeremiason, J.D., 1999. Coupling of phytoplankton uptake and air-water exchange of persistent organic pollutants. Environ. Sci. Technol. 33, 3653–3660.

Diamond, M.L., Bhavsar, S.P., Helm, P.A., Stern, G.A., Alaee, M., 2005. Fate of organochlorine contaminants in arctic and subarctic lakes estimated by mass balance modelling. Sci. Total Environ. 342, 245–259.

Dich, J., Zahm, S.H., Hanberg, A., Adami, H.O., 1997. Pesticides and cancer. Cancer Causes Control 8, 420–443.

Dong, J.Y., Gao, H., Wang, S.G., Yao, H.J., Ma, M.Q., 2009. Simulation of the transfer and fate of HCHs since the 1950s in Lanzhou, China. Ecotoxicol. Environ. Saf. 72, 1950–1956.

Gao, J.J., Zhao, H.Q., 2012. Residues of organic chlorine pesticide and their distribution characteristics in arable land soil of Chaohu Lake region. J. Anhui Agric. Univ. 39, 613–618 (in Chinese).

Gu, C.J., 2005. Historical Sedimentary Records and Environmental Changes in Chaohu lake. East China Normal University, Shanghai, Master dissertation.

Guo, L.G., 2005. Studies on Fisheries Ecology in a Large Eutrophic Shallow Lake, Lake Chaohu. Institude of Hydrobiology, Chinese Acadamy of Sciences, Wuhan, Doctoral dissertation.

Hargrave, B.T., Phillips, G.A., Vass, W.P., Bruecker, P., Welch, H.E., Siferd, T.D., 2000. Seasonality in bioaccumulation of organochlorines in lower trophic level arctic marine biota. Environ. Sci. Technol. 34, 980–987.

Haugen, J.E., Wania, F., Ritter, N., Schlabach, M., 1998. Hexachlorocyclohexanes in air in southern Norway. Temporal variation, source allocation, and temperature dependence. Environ. Sci. Technol. 32, 217–224.

He, W., Qin, N., He, Q., Wang, Y., Kong, X., Xu, F., 2012. Characterization, ecological and health risks of DDTs and HCHs in water from a large shallow Chinese lake. Ecol. Inform. 12, 77–84.

He, W., Qin, N., Kong, X.Z., Liu, W.X., He, Q.S., Ouyang, H.L., Yang, C., Jiang, Y.J., Wang, Q.M., Yang, B., Xu, F.L., 2013. Spatio-temporal distributions and the ecological and health risks of phthalate esters (PAEs) in the surface water of a large, shallow Chinese lake. Sci. Total Environ. 461–462, 672–680.

Iwata, H., Tanabe, S., Tatsukawa, R., 1993. A new view on the divergence of HCH isomer compositions in oceanic air. Mar. Pollut. Bull. 26, 302–305.

Iwata, H., Tanabe, S., Ueda, K., Tatsukawa, R., 1995. Persistent organochlorine residues in air, water, sediments, and soils from the Lake Baikal region, Russia. Environ. Sci. Technol. 29, 792–801.

Kalantzi, O.I., Hewitt, R., Ford, K.J., Cooper, L., Alcock, R.E., Thomas, G.O., Morris, J.A., McMillan, T.J., Jones, K.C., Martin, F.L., 2004. Low dose induction of micronuclei by lindane. Carcinogenesis 25, 613–622.

Kong, X.Z., He, W., Qin, N., He, Q.S., Yang, B., Ouyang, H.L., Wang, Q.M., Yang, C., Jiang, Y.J., Xu, F.L., 2012. Simulation of the fate and seasonal variations of alpha-hexachlorocyclohexane in Lake Chaohu using a dynamic fugacity model. Sci. World J. 2012, 691539. http://dx.doi.org/10.1100/2012/691539.

Kong, X.Z., He, W., Qin, N., He, Q.S., Yang, B., Ouyang, H.L., Wang, Q.M., Xu, F.L., 2013. Comparison of transport pathways and potential sources of PM10 in two cities around a large Chinese lake using the modified trajectory analysis. Atmos. Res. 122, 284–297.

Kong, X.Z., He,W., Qin, N., He, Q.S., Yang, B., Ouyang, H.L.,Wang, Q.M., Yang, C., Jiang, Y.J., Xu, F.L. 2014. Modeling the multimedia fate dynamics of γ-hexachlorocyclohexane in a large Chinese lake. Ecol. Indic. in press.

Lang, C., Tao, S., Wang, X.J., Zhang, G., Li, J., Fu, J.M., 2007. Seasonal variation of polycyclic aromatic hydrocarbons (PAHs) in Pearl River Delta region, China. Atmos. Environ. 41, 8370–8379.

Law, S.A., Diamond, M.L., Helm, P.A., Jantunen, L.M., Alaee, M., 2001. Factors affecting the occurrence and enantiomeric degradation of hexachlorocyclohexane isomers in northern and temperate aquatic systems. Environ. Toxicol. Chem. 20, 2690–2698.

Li, Y.F., Cai, D.J., Shan, Z.J., Zhu, Z.L., 2001. Gridded usage inventories of technical hexachlorocyclohexane and lindane for china with 1/6 degrees latitude by 1/4 degrees longitude resolution. Arch. Environ. Contam. Toxicol. 41, 261–266.

Li, Y.F., et al., 2002. The transport of beta-hexachlorocyclohexane to the western Arctic Ocean: a contrast to alpha-HCH. Sci. Total Environ. 291, 229–246.

Liu, Z.Y., Quan, X., Yang, F.L., 2007. Long-term fate of three hexachlorocyclohexanes in the lower reach of Liao River basin: dynamic mass budgets and pathways. Chemosphere 69, 1159–1165.

Liu, W.X., He, W., Qin, N., Kong, X.Z., He, Q.S., Ouyang, H.L., Yang, B., Wang, Q.M., Yang, C., Jiang, Y.J., Wu, W.J., Xu, F.L., 2012. Residues, distributions, sources and ecological risks of OCPs in the water from Lake Chao, China. Sci. World J. 2012, 16. Article ID 897697. http://dx.doi.org/10.1100/2012/897697.

Liu, W.X., He, W., Qin, N., Kong, X.Z., He, Q.S., Ouyang, H.L., Xu, F.L., 2013. The residues, distribution, and partition of organochlorine pesticides in the water suspended solids, and sediments from a large Chinese lake (Lake Chaohu) during the high water level period. Environ. Sci. Pollut. Res. 20, 2033–2045.

Mackay, D., 1979. Finding fugacity feasible. Environ. Sci. Technol. 13 (10), 1218–1223.

Mackay, D., 2001. Multimedia Environmental Models: The Fugacity Approach, second ed. Lewis Publishers, New York, NY.

Mackay, D., Diamond, M., 1989. Application of the QWASI (quantitative water air sediment interaction) fugacity model to the dynamics of organic and inorganic chemicals in lakes. Chemosphere 18, 1343–1365.

Mackay, D., Paterson, S., 1981. Calculating fugacity. Environ. Sci. Technol. 15 (9), 1006–1014.

Mackay, D., Paterson, S., 1982. Fugacity revisited—the fugacity approach to environmental transport. Environ. Sci. Technol. 16, A654–A660.

Mackay, D., Paterson, S., 1991. Evaluating the multimedia fate of organic-chemicals—a level-III fugacity model. Environ. Sci. Technol. 25, 427–436.

Mackay, D., Joy, M., Paterson, S., 1983. A Quantitative Water Air Sediment Interaction (QWASI) fugacity model for describing the fate of chemicals in lakes. Chemosphere 12, 981–997.

Mackay, D., Paterson, S., Schroeder, W.H., 1986. Model describing the rates of transfer processes of organic chemicals between atmosphere and water. Environ. Sci. Technol. 20, 810–816.

Matozzo, V., Binelli, A., Parolini, M., Previato, M., Masiero, L., Finos, L., Bressan, M., Marin, M.G., 2012. Biomarker responses in the clam Ruditapes philippinarum and contamination levels in sediments from seaward and landward sites in the Lagoon of Venice. Ecol. Indic. 19, 191–205.

McKone, T.E., 1996. Alternative modeling approaches for contaminant fate in soils: uncertainty, variability, and reliability. Reliab. Eng. Syst. Safety 54, 165–181.

Morris, M.D., 1991. Factorial sampling plans for preliminary computational experiments. Technometrics 33, 161–174.

Odabasi, M., Cetin, B., Demircioglu, E., Sofuoglu, A., 2008. Air-water exchange of polychlorinated biphenyls (PCBs) and organochlorine pesticides (OCPs) at a coastal site in Izmir Bay, Turkey. Mar. Chem. 109, 115–129.

Ouyang, H.L., He, W., Qin, N., Kong, X.Z., Liu, W.X., He, Q.S., Jiang, Y.J., Wang, Q.M., Yang, C., Yang, B., Xu, F.L., 2012. Levels, temporal-spatial variations and sources of organochlorine pesticides in ambient air of Lake Chaohu, China. Sci. World J. 2012, 12. Article ID 504576. http://dx.doi.org/10.1100/2012/504576.

Ouyang, H.L., He, W., Qin, N., Kong, X.Z., Liu, W.X., He, Q.S., Jiang, Y.J., Wang, Q.M., Yang, C., Yang, B., Xu, F.L., 2013. Water–gas exchange of organochlorine pesticides at Lake Chaohu, a large Chinese lake. Environ. Sci. Pollut. Res. 20, 2020–2032.

Paasivirta, J., Sinkkonen, S., Mikkelson, P., Rantio, T., Wania, F., 1999. Estimation of vapor pressures, solubilities and Henry's law constants of selected persistent organic pollutants as functions of temperature. Chemosphere 39, 811–832.

Prinn, R.G., Huang, J., Weiss, R.F., Cunnold, D.M., Fraser, P.J., Simmonds, P.G., et al., 2001. Evidence for substantial variations of atmospheric hydroxyl radicals in the past two decades. Science 292, 1882–1888.

Qiu, X.H., Zhu, T., Wang, F., Hu, J.X., 2008. Air-water gas exchange of organochlorine pesticides in Taihu Lake. China Environ. Sci. Technol. 42, 1928–1932.

Qu, C.S., Chen, W., Bi, J., Huang, L., Li, F.Y., 2011. Ecological risk assessment of pesticide residues in Taihu Lake wetland, China. Ecol. Model. 222, 287–292.

Ridal, J.J., Kerman, B., Durham, L., Fox, M.E., 1996. Seasonality of air-water fluxes of hexachlorocyclohexanes in Lake Ontario. Environ. Sci. Technol. 30, 852–858.

Saloranta, T.M., Armitage, J.M., Haario, H., Naes, K., Cousins, I.T., Barton, D.N., 2008. Modeling the effects and uncertainties of contaminated sediment remediation scenarios in a Norwegian Fjord by Markov chain Monte Carlo simulation. Environ. Sci. Technol. 42, 200–206.

Southworth, G.R., 1979. Role of volatilization in removing polycyclic aromatic hydrocarbons from aquatic environments. Bull. Environ. Contam. Toxicol. 21, 507–514.

Stockholm Convention on persistent organic pollutants (SCPOPs), 2009. The Nine New POPs: an introduction to the nine chemicals added to the Stockholm Convention by the Conference of the Parties at its fourth meeting. http://www.pops.int/.

Tao, S., Cao, H.Y., Liu, W.X., Li, B.G., Cao, J., Xu, F.L., Wang, X.J., Coveney, R.M., Shen, W.R., Qin, B.P., Sun, R., 2003. Fate modeling of phenanthrene with regional variation in Tianjin. China Environ. Sci. Technol. 37, 2453–2459.

Tao, S., Yang, Y., Cao, H.Y., Liu, W.X., Coveney, R.M., Xu, F.L., Cao, J., Li, B.G., Wang, X.J., Hua, J.Y., Fang, J.Y., 2006. Modeling the dynamic changes in concentrations of γ-hexachlorocyclohexane (γ-HCH) in Tianjin region from 1953 to 2020. Environ. Pollut. 139, 183–193.

Tao, S., Liu, W.X., Li, Y., Yang, Y., Zuo, Q., Li, B.G., Cao, J., 2008. Organochlorine pesticides contaminated surface soil as reemission source in the Haihe Plain. China Environ. Sci. Technol. 42, 8395–8400.

Tu, Q.Y., Gu, D.X., Yi, C.Q., Xu, Z.R., Han, G.Z., 1990. The Researches on the Lake Chaohu Eutrophication. Publisher of University of Science and Technology of China, Hefei (in Chinese).

Vijgen, J., Abhilash, P.C., Li, Y.F., Lal, R., Forter, M., Torres, J., Singh, N., Yunus, M., Tian, C.G., Schaffer, A., Weber, R., 2011. Hexachlorocyclohexane (HCH) as new Stockholm Convention POPs-a global perspective on the management of Lindane and its waste isomers. Environ. Sci. Pollut. Res. 18, 152–162.

von Waldow, H., Scheringer, M., Hungerbuhler, K., 2008. Modelled environmental exposure to persistent organic chemicals is independent of the time course of emissions: proof and significance for chemical exposure assessments. Ecol. Model. 219, 256–259.

Waliszewski, S.M., 1993. Residues of lindane, HCH isomers and HCB in the soil after lindane application. Environ. Pollut. 82, 289–293.

Walker, K., Vallero, D.A., Lewis, R.G., 1999. Factors influencing the distribution of lindane and other hexachlorocyclohexanes in the environment. Environ. Sci. Technol. 33, 4373–4378.

Wang, Y., 2012. Residual levels characteristics and ecological risks assessment of organochlorine pesticides in Lake Chaohu ecosystem. Master thesis (in Chinese).

Wang, X.Q., Hua, R.M., Pan, J.Y., Gao, Q., Li, X.D., Cao, H.Q., Wu, X.W., Tang, J., 2011. Distribution and composition of organochlorine pesticides in farmland top soils of Anhui Province. China J. Appl. Ecol. 22, 3285–3292 (in Chinese).

Wang, Y., He, W., Qin, N., He, Q.S., 2012. Residual levels and ecological risks of organochlorine pesticides in surface sediments from Lake Chaohu. Acta Sci. Circumst. 32, 308–316 (in Chinese).

Wania, F., Mackay, D., 1995. A global distribution model for persistent organic-chemicals. Sci. Total Environ. 160–61, 211–232.

Xia, H.S., Yue, Y.D., Du, X.Y., 1987. Residues of HCHs in soil of Anhui tea plant regions. J. Tea Business 3, 30–33 (in Chinese).

Xia, X., Hopke, P.K., Holsen, T.M., Crimmins, B.S., 2011. Modeling toxaphene behavior in the Great Lakes. Sci. Total Environ. 409, 792–799.

Xu, F.L., Jørgensen, S.E., Tao, S., Li, B.G., 1999. Modeling the effects of ecological engineering on ecosystem health of a shallow eutrophic Chinese lake (Lake Chao). Ecol. Model. 117, 239–260.

Xu, F.L., Qin, N., Zhu, Y., He, W., Kong, X.Z., Barbour, M.T., He, Q.S., Wang, Y., Ou-Yang, H.L., Tao, S., 2013a. Multimedia fate modeling of polycyclic aromatic hydrocarbons (PAHs) in Lake Small Baiyangdian, Northern China. Ecol. Model. 252, 246–257.

Xu, F.L., Kong, X.Z., He, W., Qin, N., Zhu, Y., Tao, S., 2013b. Distributions, sources, and ecological risks of hexachlorocyclohexanes in the sediments from Haihe Plain, Northern China. Environ. Sci. Pollut. Res. 20, 2009–2019.

Yin, F.C., 2011. A Study on Evaluation and Control Instruments of Chao Lake Eutrophication. China Environmental Science Press, Beijing.

Yue, Y.D., Hua, R.M., Zhu, L.Z., Zhu, S.W., Chen, D.D., 1990. Residual form and level of BHC and DDT in argoenvironment of Anhui. J. Anhui Agric. Coll. 17, 194–197 (in Chinese).

Zhang, M., 2009. Distribution Characteristic and Assessment of Typical Persistent Organic Pollutions—Organochlorine Pesticides in Water of Chaohu Lake Watershed. Anhui Agriculture University, Hefei, Master dissertation.

Zhang, T.F., Lu, X.P., 1986. Investigation and evaluation of water environmental quality of Lake Chaohu. In: Lake Chaohu Water Environment, Ecological Evaluation and Countermeasures-Special Report (No.11). Hefei (in Chinese).

Zheng, X.Y., et al., 2010. Spatial and seasonal variations of organochlorine compounds in air on an urban-rural transect across Tianjin, China. Chemosphere 78, 92–98.

Zhou, Z.H., Liu, C.Q., Li, J., Zhu, Z.Z., 2007. Record of ecosystem evolvement processes provided by $\delta13Corg$ and $\delta15N$ values in Chaohu Lake sediments. Environ. Sci. 28, 1338–1343 (in Chinese).

Eco-Risk Assessments for Toxic Contaminants Based on Species Sensitivity Distribution Models in Lake Chaohu, China

Fu-Liu Xu*, Xiang-Zhen Kong, Ning Qin, Wei He, Wen-Xiu Liu

MOE Laboratory for Earth Surface Processes, College of Urban & Environmental Sciences,
Peking University, Beijing, China
*Corresponding author: e-mail address: xufl@urban.pku.edu.cn

4.1 INTRODUCTION

4.1.1 Ecological risk assessments

An ecological risk assessment has been defined as the process of estimating the likelihood that a particular event will occur with a given set of circumstances (Maltby et al., 2005; Domene et al., 2008). During recent decades, some indicators and methods of different complexities have been proposed for the ecological risk assessment of toxic chemicals in water. In the early stages of a risk assessment, the Hazard Quotient, which is the quotient of the measured or estimated environmental concentration divided by the toxicant reference value, was proposed for the individual-value estimate (Solomon et al., 2000). The species sensitivity distribution (SSD) approach is one frequently used method for ecological risk assessment (Solomon et al., 1996; Steen et al., 1999). An SSD model is a statistical distribution describing, among a set of species, the variation in toxicity of a certain compound or mixture (van Straalen, 2002). To assess the eco-risk of toxic pollutants using the SSD model, some indicators, such as the maximum permissible concentration, negligible concentration, potential affected fraction (PAF), hazardous concentration at which $p\%$ of the selected species will be affected (HC_p), and margin of safety (MOS10) can be calculated for both the ecological risk of an individual chemical and the combined ecological risk of multiple substances (Solomon et al. 1996; Steen et al. 1999). The SSD method has been proven as a useful site-to-site estimate both for the eco-risk of individual chemicals and for the joint eco-risk of multiple substances (Solomon et al., 1996; Steen et al., 1999). Although significant progress and improvements have been made for the SSD methods, there are still some flaws (e.g., the lack of uncertainty analysis) (Solomon et al., 2000; Forbes and Calow, 2002). To address this issue, a probabilistic risk assessment (PRA) was proposed (Solomon and Sibley, 2002). The PRA method

Developments in Environmental Modelling, Volume 26, ISSN 0167-8892, http://dx.doi.org/10.1016/B978-0-444-63249-4.00005-1

considers the estimate of uncertainty and the stochastic properties of exposure and effects, and it allows the variability of exposure concentrations and the distributions of species sensitivity in the risk assessment process. It can better describe the likelihood exceeding the effect thresholds and the risk of adverse effects (Solomon and Sibley, 2002; Yang et al., 2006). The indicators, including the overlap area between the exposure and effect curves, and the joint probability are calculated by the PRA method to assess the ecological risks (Wang et al., 2002; Shi et al., 2004). The PRA method has proven very useful in estimating both the exposure of a population or community to potentially hazardous pollutants and their responses to the chemicals in the research area (Wang et al., 2002). However, the PRA method requires as many measured data as possible to construct the probabilistic distribution of the exposure levels. The combinations of multiple risk indicators based on the SSD and PRA models for the ecological risk assessment are necessary to provide the more general information on the spatial variations and on the probabilities of the potential ecological risks of the individual and multiple pollutants.

4.1.2 **Organochlorine pesticides**

As typical persistent organic pollutants (POPs), organochlorine pesticides (OCPs) were widely used due to the need for pest control and have threatened the ecosystem and human health. There are 13 OCPs in the list produced by the Stockholm Convention on Persistent Organic Pollutants, which forbids the production and use of 21 types of chemical substances, including DDT, chlordane, mirex, aldrin, dieldrin, endrin, heptachlor, toxaphene, α-HCH, β-HCH, lindane (γ-HCH), chlordecone (Kepone), and pentachlorobenzene; in addition, endosulfan was also investigated (UNEP, 2001, 2007). Although these OCPs have been banned (especially DDT) and the residue levels have gradually decreased since the 1980s, OCPs can still be detected in various environmental and biological media (Furst et al., 1991; Gao and Jiang, 2005; Frenich et al., 2006).

OCPs can enter water, one of the environmental media that is most vulnerable to OCP contaminants through a variety of routes, such as surface runoff and atmospheric wet and dry deposition. At present, there are residues of OCPs in the surface water including rivers, lakes, and oceans, such as the Kucuk Menderes River in Turkey (Turgut, 2003), Ebro River in Spain (Fernandez et al., 1999), Gomti River in India (Singh et al., 2005), and the section from the Sea of Japan to the Bering Sea (Chernyak and McConnell, 1995). Much research has also been done on the distribution of OCPs in the environment, such as in the Huaihe River (Feng et al., 2003), Pearl River (Yang et al., 2004), Guanting Reservoir in Beijing (Wang et al., 2003), and Lake Small Baiyangdian (Wang et al., 2012). The residue concentrations in these regions were different.

4.1.3 **Polycyclic aromatic hydrocarbons**

Polycyclic aromatic hydrocarbons (PAHs) are a group of ubiquitous POPs that are generally formed by the incomplete combustion of fossil fuels and biomass fuels

(Rogge et al., 1993; Tao et al., 2005; Xu et al., 2006). PAHs are a major concern because of their potentially toxic, mutagenic, and carcinogenic properties (Khalili et al., 1995; Fernandes et al., 1997; Larsen and Baker, 2003; Li et al., 2009). The U.S. Environmental Protection Agency (USEPA) has established 16 PAHs as priority control pollutants, and seven of them are potentially carcinogenic to humans, according to the International Agency for Research on Cancer. Furthermore, PAHs enter a water body through wastewater discharge, surface runoff, atmospheric deposition, and other means, such as crude oil leaks (Heemken et al., 2000). PAHs can adversely affect not only human health (through drinking water and skin contact) but also aquatic ecosystems. The ecological health risks of PAHs are being increasingly studied by environmental researchers. In China, PAH emissions in excess of 27,000 ton/year have resulted in the contamination of various environmental media (Zhang et al., 2007).

4.1.4 The study site of Lake Chaohu

Lake Chaohu, the fifth largest freshwater lake in China, is located near the Yangtze River delta region (Figure 4.1), one of the most developed regions in China. With the rapid urbanization of the surrounding area, Lake Chaohu is becoming increasingly polluted by OCPs and PAHs from human activities, such as the application of pesticides, as well as the burning of fossil fuels and agricultural and industrial practices. These types of pollution will damage the lake ecosystem and compromise the safe use of the lake water as a source for drinking, industrial production, and agricultural

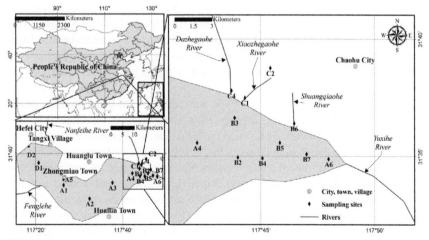

FIGURE 4.1

Location of Lake Chaohu and water sampling sites. Fifteen sites for PAHs belong to river sites (C1, C2, C3, C4), West Lake (D1), East Lake (A1, A2, A3, A4, A5), and water source area (A6, B2, B3, B4, B5 and B7), and four sites for OCPs as TX (D2), MS (A1), ZM (A5), and JC (B5).

irrigation. However, there is little information on the residual levels and ecological risks of OCPs and PAHs in the water from Lake Chaohu.

There are three primary objectives of this study: (1) to investigate the residual levels and distributions of 18 detected OCPs and 16 priority PAHs in the water of Lake Chaohu; (2) to estimate the potential ecological risk of the individual and multiple congeners of OCPs and PAHs, based on both the SSD and PRA methods; and (3) to discuss the uncertainty of the ecological risks of the studied PAH components. A platform, named the Bayesian Matbugs Calculator (BMC), was developed to perform the best fittings of the distribution model, the ecological risk index calculations, and the uncertainty analysis.

4.2 MATERIALS AND METHODS

4.2.1 Measurements of OCPs and PAHs

4.2.1.1 Measurement of OCPs in the water

The water samples were collected from May 2010 to February 2011 monthly, and the distribution of the sample sites is shown in Figure 4.1. The MS (A1) and ZM (A5) are located at 200 m south of the Zhongmiao Temple and 200 m east of Mushan Island, respectively; the JC (B5) and TX (D2) represent the city water intake near the Chaohu automatic monitoring station and western TangXi, 150 m south of the intake of an original waterworks, respectively.

Surface, middle, and bottom water samples were collected separately and then mixed together. In the sampling sites having depths of more than 1 m, the water samples were collected from the surface water (0–0.15 m below the surface), mid-water (0.5–0.65 m below surface), and bottom water (0–0.15 m above the sediment) and mixed. In the sites having depths of less than 1 m, the surface water and bottom water were collected and mixed. The water samples were stored in brown glass jars that were washed with deionized water and the water samples before use. From each site, 20 l of water was collected.

As a recovery indicator, 100 ng PCNB was added to the water samples, which were then filtered through a 0.45–μm glass fiber filter (ashed at 450 °C for 4 h) using a peristaltic pump (80EL005, Millipore Co., USA) and a filter plate with a 142-mm diameter to remove the suspended particles. A solid-phase extraction (SPE) system was used to extract the filtered water samples. Before extraction, the octadecylsilane SPE cartridges (SPE, C18, 6 ml, 500 mg, Supelco, Co.,) were first washed with 6 ml dichloromethane (DCM) and conditioned with 6 ml methanol and 6 ml ultrapure water, and the cartridges were not be dried before loading the samples. After the activation, the water samples were loaded using a large volume sampler (Supelco Co., USA) that was connected to the SPE vacuum manifold (Supelco Co., USA), and the cartridges were dried by vacuum pump after the extraction step. The SPE cartridges were sealed and delivered back to laboratory prior to the elution and purification.

Each cartridge was connected to an anhydrous sodium sulfate (5 g) cartridge and eluted using DCM (3×, 6 ml/elution). The extracts were concentrated to approximately 1 ml with a vacuum rotary evaporator (Eyela N-1100, Tokyo Rikakikai Co., Japan). The solvent was changed to hexane, and then the samples were again concentrated to approximately 1 ml. TCMX (2, 4, 5, 6-tetrachloro-*m*-xylene) was added to the samples as an internal standard. The samples were transferred to vials and sealed until analysis.

The samples were analyzed using an Agilent 6890-5976C gas chromatography and mass spectrometer detector and a HP-5MS fused silica capillary column (30 m × 0.25 mm × 0.25 mm, Agilent Co., USA). Helium was used as the carrier gas at a flow rate of 1 ml/min. Samples (1 μl) were injected by the autosampler under a splitless mode at a temperature of 220 °C. The column temperature program was the following: 50 °C for 2 min, 10 °C/min to 150 °C, 3 °C/min for 240 °C, 240 °C for 5 min, 10 °C/min for 300 °C, and 300 °C for 5 min. The ion source temperature of the mass spectrometer was 200 °C, the temperature of the transfer line was 250 °C, and the temperature of the quadrupole was 150 °C. The compounds were quantified in the selected ion mode, and the calibration curve was quantified with the internal standard. There were two parallel samples in each sampling site. The samples, method blanks, and procedure blanks were prepared in the same manner. The average recovery ranged from 27.4% (p,p′-DDE) to 107.4% (Endrin). The detection limits were in the range of 0.01 ng/l (HCB) to 1.2 ng/l (o,p′-DDT).

4.2.1.2 Measurement of PAHs in the water

Water samples from 15 sites (Figure 4.1) were collected in August 2009. An emphasis was placed on the eastern drinking-water source area with six sites (A6, B2, B3, B4, B5, and B7) and the inflow rives with four sites (C1, C2, C4, and B6). Twenty liters of water was collected from each sampling site. After shaking and mixing, a 1-l aliquot of each collected water sample was filtered through a 0.45-μm glass fiber filter (burned at 450 °C for 4 h) using a filtration device consisting of a peristaltic pump (80EL005, Millipore Co., USA) and a filter plate with a diameter of 142 mm. Surrogate standards of 2-fluoro-1, 1′-biphenyl and p-terphenyl-d14 (J&K Chemical, USA, 2.0 mg/ml) were added to the water samples to indicate the recovery before extraction.

The water samples were extracted using a SPE system (Supelco). C18 cartridges (500 mg, 6 ml, Supelco) were prewashed with DCM and conditioned with methanol and deionized water. A 1-l water sample passed through the SPE system and was extracted. The cartridges were eluted with 10 ml of DCM. The volume of the extracts was reduced by a vacuum rotary evaporator (R-201, Shanghai Shen Sheng Technology Co., Ltd., Shanghai, China) in a water bath and was adjusted to a volume of 1 ml with hexane. Internal standards (Nap-d8, Ace-d10, Ant-d10, Chr-d12, and Perylene-d12) were added for the GC analysis.

All samples were analyzed on a gas chromatograph with a mass spectrometer detector (Agilent 6890GC/5973MSD). A 30 m × 0.25 mm i.d. with a 0.25-μm film thickness HP-5MS capillary column (Agilent Technology) was used. The column temperature was programmed to increase from 60 °C to 280 °C at 5 °C/min and then was held isothermal for 20 min. The MSD was operated in the electron impact mode at 70 eV, and the ion source temperature was 230 °C. The mass spectra were recorded using the selected ion monitoring mode. The concentrations of 16 PAHs were determined: naphthalene (Nap), acenaphthene (Ace), acenaphthylene (Acy), fluorine (Flo), phenanthrene (Phe), anthracene (Ant), fluoranthene (Fla), pyrene (Pyr), benz(a)anthracene (BaA), chrysene (Chr), benzo(b)fluoranthene (BbF), benzo(k)fluoranthene (BkF), benzo(a)pyrene (BaP), dibenz(a, h) anthracene (DahA), indeno(1,2,3-cd)pyrene (IcdP), and benzo(g, h, i) perylene (BghiP).

The quantification was performed by the internal standard method using Nap-d8, Ace-d10, Ant-d10, Chr-d12, and Perylene-d12 (J&K Chemical, Beijing, China). All of the solvents used were HPLC-grade pure (J&K Chemical, Beijing, China). All of the glassware was cleaned using an ultrasonic cleaner (KQ-500B, Kunshan, China) and heated to 400 °C for 6 h. In the sampling process, three parallel samples were collected from each sample site. The laboratory blanks were analyzed with the true samples. The average recovery for Nap, Ace, Acy, Flo, Phe, Ant, Fla, and Pyr ranged from 75% to 117%, and for BaA, Chr, BbF, BkF, BaP, DahA, IcdP, and BghiP the ranges were 68%, 67%, 54%, 51%, 77%, 45%, 34%, and 35%, respectively. The detection limits were in the range of 0.54–4.22 ng/l.

4.2.2 Ecological risk assessments by SSD and PRA

In this study, the SSD model was applied to evaluate the separate and combining ecological risks of typical OCPs, while the multiple risk indicators based on both the SSD and PRA models were calculated to obtain a comprehensive picture of the potential ecological risks of the PAHs in the water from Lake Chaohu. The SSD method (Wheeler et al., 2002; Liu et al., 2009; Wang et al., 2009a,b) was used for the site-specific assessment of the ecological risk for both the individual and multiple congeners of OCPs and PAHs, while the PRA method was used for the probabilistic assessment of the ecological risk of individual PAH congeners based on all of the sampling sites.

4.2.2.1 General procedures of the SSD and PRA methods

The basic assumption of the SSD method is that the sensitivity of a group of organisms can be described by a distribution and that the available toxicological data are considered to be a sample of this distribution. Thus, the SSD is estimated from the sample of toxicity data and visualized as a cumulative distribution function. To assess ecological risk using the SSD method, there are usually four steps: (1) obtain the toxicity data of the pollutants; (2) fit the SSD curves; (3) calculate the PAFs of the individual pollutants for the ecological risk assessment of an individual pollutant; and (4) calculate the accumulated multi-substance potentially affected fractions (msPAFs) for the joint ecological risk assessment of multiple pollutants (Liu et al., 2009).

In contrast to the SSD method and focusing on the potential risk of an individual sampling site, the PRA method determines the ecological risk of pollutants through the analysis of the probability distribution curves of the toxicity and exposure levels in the entire lake. In the same manner as the SSD method, the toxicity data for all species are combined to produce an effect distribution curve; in addition, the exposure levels are plotted on the same axes as the toxicity effects data. The extent of overlap between the toxicity and exposure curves indicates the probability of exceeding an exposure concentration associated with a particular effect probability of the substance of concern. The use of the distribution curves for the exposure and toxicity data allows for the application of a joint probability curve (JPC) to describe the nature of the risks posed by the environmental concentrations measured. To assess the ecological risk using the PRA method, there are generally four steps: (1) obtain the toxicity data and exposure data of the pollutants; (2) fit the distributions of the toxicity data and exposure levels; (3) calculate the overlap area of the toxicity and exposure distribution curves for the ecological risk assessment of an individual pollutant; and (4) produce the JPCs of the pollutants for assessing the ecological risk probability of exceeding an exposure concentration associated with a particular effect probability. Therefore, it is concluded that the SSD and PRA methods have the same steps: collecting the toxicity data, fitting the distribution curves of the toxicity data, calculating the risk index, and estimating the ecological risk.

4.2.2.2 Collecting toxicity data

Based on the availability of the toxicity data, the studied OCPs compounds included p,p′-DDT, γ-HCH, heptachlor, aldrin, and endrin, and the PAH compounds included naphthalene (Nap), acenaphthylene (Ace), fluorine (Flo), phenanthrene (Phe), anthracene (Ant), pyrene (Pyr), fluoranthene (Fla), and benzo[a]pyrene (BaP). The toxicity data were collected from the database ECOTOX provided by the U.S. EPA (www.epa.gov/ecotox). The 24- to 96-h acute toxicity data (LC50 or EC50) to multiple aquatic species were collected. For OCPs, in order to understand the ecological risks to different types of freshwater organisms comprehensively, the toxicity data for the OCPs were classified into three patterns: (1) all species were not subdivided; (2) all species were subdivided into vertebrates and invertebrates; and (3) three subcategories for which the toxicity data were rich were selected, which included fish, insects and spiders, and crustaceans. For PAHs, the species include green algae (*Selenastrum capricornutum*), a diatom (*Skeletonema costatum*), the southern house mosquito (*Culex quinquefasciatus*), the yellow fever mosquito (*Aedes aegypti*), a water flea (*Daphnia magna*), the sheepshead minnow (*Cyprinodon variegatus*), the channel catfish (*ctalurus punctatus*), the brown trout (*Oncorhynchus mykiss*), the fathead minnow (*Pimephales promelas*), a scud (*Gammarus annulatus*), a freshwater prawn (*Palaemonetes*), and a pond snail (*Physa heterostropha*). Because of the differences between the personnel and laboratory environment, there are many toxicity data on the same pollutant for the same species. In this study, the data point was the geometric mean of the toxicity data for the same species (Newman et al., 2000). The toxicity data of the five OCPs compounds and eight PAH compounds are presented in Tables 4.1 and 4.2, respectively.

Both the acute data (e.g., LC50 and EC50) and chronic data (e.g., NOEC) of ecotoxicity can be used to build the SSD. However, in the present study, only the acute ecotoxicity data was used to build the SSD for the following two reasons. First, there

Table 4.1 The Statistical Characteristics of the Toxicity Data for OCPs (After Logarithmic Transformation) (µg/l)

	p,p'-DDT			γ-HCH			Heptachlor		
	Numbers	Mean	SD	Numbers	Mean	SD	Numbers	Mean	SD
All species	151	1.782	1.148	122	2.323	1.068	48	2.08	1.11
Vertebrates	62	1.802	1.083	60	2.475	0.896	32	2.11	0.65
Invertebrates	89	1.769	1.196	62	2.175	1.201	16	1.79	0.93
Fishes	57	1.678	1.022	54	2.352	0.854	31	2.09	0.65
Crustaceans	28	1.496	1.127	20	2.048	1.151	8	1.67	0.48
Insects and spiders	50	1.516	0.939	28	1.509	0.663	6	1.48	1.14

	Aldrin			Endrin		
	Numbers	Mean	SD	Numbers	Mean	SD
All species	55	2.08	1.11	83	83	83
Vertebrates	31	1.72	0.66	46	46	46
Invertebrates	24	2.54	1.39	37	37	37
Fishes	29	1.64	0.58	40	40	40
Crustaceans	13	2.59	1.71	10	10	10
Insects and spiders	6	1.71	0.55	21	21	21

Table 4.2 The Statistical Characteristics of the Toxicity Data for PAHs (After Logarithmic Transformation) (µg/l)

	Nap	Ace	Flo	Phe	Ant	Fla	Pyr	Bap
Minimum	40.70	240.00	212.00	212.13	1.93	1.20	2.63	4.00
Maximum	175,927	3499	5800	195,697	17,822	41,590	894,000	17,660
Arithmetic mean	5020	660	1591	870	56	48	60	13
Geometric mean	4818	847	1340	1992	76	67	167	76
Standard deviation	37,796	987	1967	1992	4278	7985	269,027	6273
Total data number	24	11	12	17	17	27	11	12

were very limited chronic ecotoxicity data available for the studied OCPs and PAHs. The amount of data is too small to easily generate a large deviation. The least amount of ecotoxicity data for the development of SSD model was five (Hose and Van den Brink, 2004) or eight (Wheeler et al., 2002). Second, some studies developed the chronic SSD model using the chronic ecotoxicity data that converted from the acute data based on the empirical formula (e.g., Heger et al., 1995; Lange et al., 1998). However, such an approach on the basis of the empirical formula caused high uncertainty of the SSD curve, which led further to the high uncertainty of assessment results as well.

4.2.2.3 *Fitting the distribution of toxicity and exposure data*

The PRA method requires a fit of the distributions of both the toxicity and exposure data, while the SSD method requires only a fit of the distribution of the toxicity data. In most cases, the exposure data are of a log-normal distribution. However, for the toxicity data, because no certain distribution has been proved theoretically (Wheeler et al., 2002), different distribution models have been selected, including log-normal (Wagner and Lokke, 1991), log-logistic (Aldenberg and Slob, 1993), and Burr Type III (Shao, 2000). For OCPs, the BurrliOZ software, which was designed by Australia's Commonwealth Scientific and Industrial Research Organization (CSIRO, 2008), was employed to fit the SSD curves and calculate the relevant parameters. SSD curves for five OCPs for vertebrates, invertebrates, fish, crustaceans, and insects and spiders are shown in Figure 4.2 along with the fitting parameters of SSD curves calculated by BurrliOZ for the five OCPs, that is, p,p'-DDT, γ-HCH, heptachlor, aldrin, and endrin, are given in Table 4.3.

In addition, to select the best SSD model for PAHs, a platform named the BMC (He et al., 2014) was developed based on the Bayesian Inference, WinBUGS Software (Lunn et al., 2000; Ntzoufras, 2009), and Matlab graphical user interface. Five frequently used distribution models, including the log-normal, log-logistic, BurrIII, Reweibull, and Weibull models (van Straalen, 2002), were compared. The best fitting distribution mode was determined by the index of deviance information criterion (DIC) using Equation (4.1):

$$DIC = \bar{D} + p_D \tag{4.1}$$

$\bar{D}$ and p_D were used to measure the quality of the fit and the complexity of the model, respectively. The smaller the value of the DIC, the better the curve fit the toxicity data. In the winBUGs software, the DIC value can be obtained directly by a Markov Chain Monte Carlo simulation. As shown in Table 4.4, the BurrIII model has the smallest values for Phe, Ant, Fla, and Pyr; however, no significant differences in the DIC values for the different models were found by a one-way ANOVA test. The Kolmogorov–Smirnov tests revealed that both the toxicity and exposure data followed a log-normal distribution. Thus, a log-normal model was chosen to simulate the SSD curve of the PAH toxicity data.

SSD curves for vertebrates and invertebrates to ACP, FLA, and NAP are shown in Figure 4.3, while SSD curves for ACP, FLA, and NAP to vertebrates and for eight PAH congeners to invertebrates are illustrated in Figure 4.4. SSD curves for all the species to eight PAH congeners are given in Figure 4.5. The parameters of the log-normal SSD distributions of the eight PAHs were listed in Table 4.5.

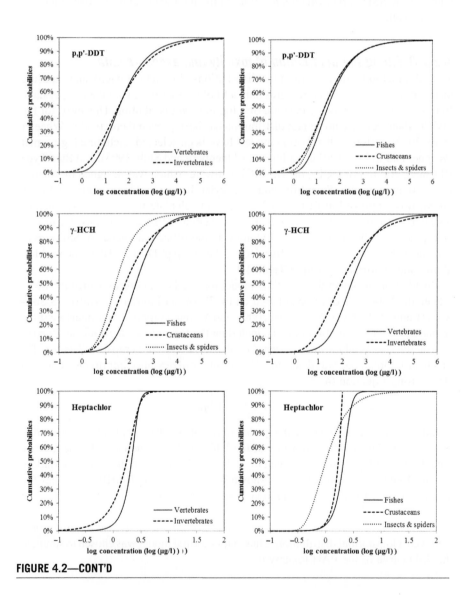

FIGURE 4.2—CONT'D

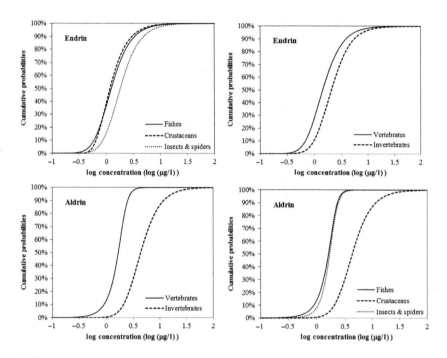

FIGURE 4.2

Typical OCPs SSD curves for different species.

4.2.2.4 Ecological risk assessment based on multiple risk indices

For the SSD method, the PAF of the single pollutant can be calculated by the following Burr III equation:

$$F(x) = \frac{1}{\left[1 + \left(\frac{b}{x}\right)^c\right]^k} \tag{4.2}$$

where x is the concentration of the pollutant (μg/l) in the environment, and b, c, and k are the three parameters of the model (the same as below). When k tends to infinity, the Burr III distribution model transforms into a ReWeibull distribution model:

$$F(x) = \exp\left(-\frac{b}{x^c}\right) \tag{4.3}$$

When c tends to infinity, it transforms into a RePareto distribution:

$$F(x) = \left(\frac{x}{x_0}\right)^\theta, \ I\{x \le x_0\}(x_0, \theta > 0) \tag{4.4}$$

The parameters are calculated by the BurrliOZ program. When k is greater than 100 or c is greater than 80, the software will use ReWeibull or RePareto to calculate the relevant parameters automatically.

Table 4.3 The Parameters of SSD Curves Calculated by BurrliOZ

	p,p'-DDT				Lindane (γ-HCH)			
	Fitted Curve	Parameters and Values			Fitted Curve	Parameters and Values		
All species	BurrIII	0.082(b)	0.489(c)	14.626(k)	BurrIII	2.519(b)	0.515(c)	6.043(k)
Vertebrates	ReWeibull	5.146(b)	0.541(c)		BurrIII	58.638(b)	0.708(c)	2.259(k)
Invertebrates	BurrIII	0.146(b)	0.468(c)	9.786(k)	ReWeibull	5.450(b)	0.456(c)	
Fishes	ReWeibull	5.365(b)	0.593(c)		BurrIII	57.899(b)	0.784(c)	2.085(k)
Crustaceans	BurrIII	1.960(b)	0.577(c)	3.214(k)	ReWeibull	6.430(b)	0.526(c)	
Insects and spiders	ReWeibull	3.906(a)	0.551(b)		BurrIII	1.560(b)	0.780(c)	6.655(k)

	Heptachlor				Aldrin			
	Fitted Curve	Parameters and Values			Fitted Curve	Parameters and Values		
All species	BurrIII	2.704(b)	8.188(c)	0.280(k)	BurrIII	1.860(b)	2.000(c)	3.000(k)
Vertebrates	BurrIII	2.614(a)	8.839(b)	0.357(k)	BurrIII	2.086(b)	6.654(c)	0.413(k)
Invertebrates	BurrIII	2.586(b)	5.919(c)	0.284(k)	BurrIII	2.230(b)	2.000(c)	3.000(k)
Fishes	BurrIII	2.490(b)	7.902(c)	0.429(k)	BurrIII	2.042(b)	8.036(c)	0.343(k)
Crustaceans	RePareto	2.000(x_o)	4.093(θ)		BurrIII	2.180(b)	2.000(c)	3.000(k)
Insects and spiders	ReWeibull	0.699(b)	1.636(c)		BurrIII	1.956(b)	7.174(c)	0.521(k)

	Endrin			
	Fitted Curve	Parameters and Values		
All species	BurrIII	0.987(b)	2.000(c)	3.000(k)
Vertebrates	BurrIII	0.724(b)	2.000(c)	3.000(k)
Invertebrates	BurrIII	1.041(b)	2.000(c)	3.000(k)
Fishes	BurrIII	0.634(b)	2.000(c)	3.000(k)
Crustaceans	ReWeibull	0.957(b)	2.011(c)	
Insects and spiders	BurrIII	0.924(b)	2.000(c)	3.000(k)

Note: The letter in parentheses mean the parameters b, c, k, x_o, and θ.

Table 4.4 Deviance Information Criterion (DIC) Values of the Five Different Models for the SSD Fitting

	Nap	Ace	Flo	Phe	Ant	Fla	Pyr	BaP
Log-normal	484.1	178.4	212.0	335.0	228.9	339.9	175.3	163.4
Log-logistic	482.2	179.0	212.9	335.0	229.4	333.6	172.3	169.8
Burr III	489.8	180.3	213.9	329.4	223.4	332.8	168.6	163.4
ReWeibull	492.3	188.6	219.3	331.6	227.7	340.4	170.3	169.8
Weibull	482.9	163.9	199.8	339.7	232.7	355.2	179.9	163.4

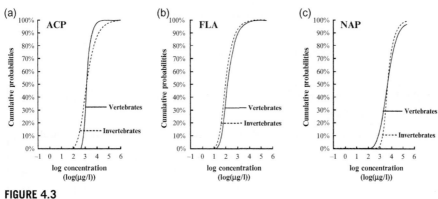

FIGURE 4.3

SSD curves for vertebrates and invertebrates to ACP (a), FLA (b), and NAP(c).

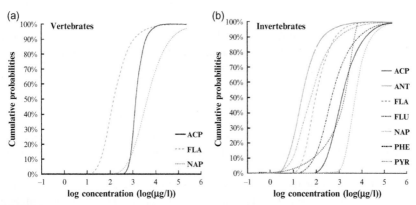

FIGURE 4.4

SSD curves for ACP, FLA, NAP to vertebrates (a) and for eight PAH congeners to invertebrates (b).

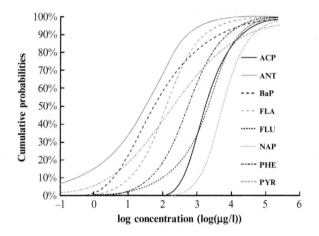

FIGURE 4.5

SSD curves for all the species to eight PAH congeners.

Table 4.5 SSD Parameters of the Toxicity Data and Exposure Levels of Eight Individual PAHs (μg/l)

	Nap	**Ace**	**Flo**	**Phe**	**Ant**	**Fla**	**Pyr**	**BaP**
μ_0	3.683	2.928	3.127	3.299	1.882	1.825	2.222	1.879
σ_0	0.758	0.356	0.471	0.915	1.056	0.802	1.568	1.458
μ_1	−1.163	−2.110	−1.692	−1.370	−2.551	−2.041	−2.041	−4.026
σ_1	0.159	0.222	0.155	0.153	0.274	0.255	0.280	0.333

Note: μ_0 and σ_0 as well as μ_1 and σ_1 with μg/l as unit are the mean values and standard deviations for the log-transformed toxicity data and for the log-transformed exposure data, respectively.

The potentially affected fraction (PAF) of the individual PAH compound and the accumulated PAFs of eight PAH compounds (msPAFs) are used for the ecological risk assessments. The PAF was calculated by Equation (4.5):

$$\text{PAF} = \frac{1}{\sqrt{2\pi}\sigma_0} e^{-\frac{[\lg(C-\mu_0)]^2}{2\sigma_0^2}} \tag{4.5}$$

where C is the log-transformed exposure concentration of eight specific PAHs, μ_0 is the mean value of the toxicity data, and σ_0 is the standard deviation of the toxicity data in Table 4.5.

The msPAFs are calculated based on the toxic mode of action (TMoA) of the pollutants. In cases where the pollutants had the same or similar TMoA, a concentration-addition method could be used to calculate the msPAF; otherwise, a response-addition approach will be used (Posthuma et al., 2002; Traas et al., 2002). Because the TMoA of PAHs is not currently clear, both the concentration-addition and response-addition approaches were used to calculate the msPAF in this study. The msPAFs of eight PAHs were calculated by Equations (4.6), (4.7), and (4.8) for the concentration-addition approach (Traas et al., 2002):

$$\text{HU}_i = \frac{C_i}{\bar{C}_i} \tag{4.6}$$

$$\text{HU}_{\text{TMoA}} = \sum_i^n \text{HU}_x \tag{4.7}$$

$$\text{msPAF} = \frac{1}{\sqrt{2\pi}\sigma} e^{-\frac{\left[\lg\left(\sum \text{HU}_{\text{TMoA}}\right)\right]^2}{2\sigma^2}} \tag{4.8}$$

where HU_i and HU_{TMoA} are the dimensionless hazard units (HUs) of the ith PAH component and of the addition of all HUs for the studied individual PAH components; C_i and $\bar{C}_i$ are the concentrations of the measured data and the geomean value of the toxic data of the ith PAH component ($\mu g/l$), respectively; and σ is the average standard deviation of the toxicity data of all studied individual PAH components ($\mu g/l$).

The advantage of the SSD is that the msPAF can be calculated and consequently the combining ecological risks of multiple pollutants can be evaluated. According to the TMoA by different pollutants, the msPAF was calculated using concentration addition or response addition (Traas et al., 2002). In this study, the TmoAs of the five OCPs and eight PAHs were different, and thus the response addition was adopted. The equation is as follows (Traas et al., 2002):

$$\text{msPAF} = 1 - \prod_i (1 - \text{PAF}) \tag{4.9}$$

For $i = 1$ to n substances, msPAF represents the multi-substance potential affected fractions of the various compounds calculated by the response addition.

For the PRA method, the potential ecological risks were assessed by calculating the overlap area between the exposure and effect curves (Wang et al., 2002) and by calculating the JPCs (Shi et al., 2004). The JPCs can be obtained by plotting the cumulative probability distributions of the exposure and toxicity data for each chemical on the same axes. Each point on the curve represents both the probability that the chosen proportion of a species will be affected and the frequency with which that level of effect would be exceeded (Solomon et al., 2000; Wang et al., 2002).

4.2.2.5 Uncertainty analysis

For the selected specific method for the assessment ecological risk, such as the SSD and PRA methods, the sources of uncertainty in the assessment results are mainly from the exposure level and toxicity data. The uncertainty in the exposure levels originates from the sampling and laboratory analysis errors. The toxicity data for the limited species cultured and tested in the laboratory extrapolated to the responses of natural taxa may lead to uncertainty because the laboratory LC50 values may overestimate the field effects at the population level. The lack of toxicity data also causes obvious uncertainty for the assessment results of the ecological risks.

A Monte Carlo simulation was used to demonstrate the uncertainties of the exposure and toxicity data. Both the exposure and toxicity data were represented as a probability density function that defined both the range of values and the likelihood of the data having that value. All of the data were assumed to follow the log-normal distribution. The simulation was performed $5000\times$, with new values randomly selected for the data within the range of the mean $\pm$ standard deviation. The winBUGs developed in this study were used to randomly select the values for the data. The uncertainty was ascertained by the statistical analysis of the output result. To quantify the differences, the coefficients of variation (CVs) were calculated based on the log-transformed data.

4.3 ECO-RISK ASSESSMENTS FOR OCPs IN LAKE CHAOHU
4.3.1 The residues of OCPs in the water (Liu et al., 2012)

Eighteen OCPs were found in the water from Lake Chaohu (Table 4.6): HCH isomers (α-, β-, γ-, and δ-HCH), DDT and its metabolites (o,p'-, p,p'-DDE, DDT, and DDD), heptachlor, hexachlorobenzene (HCB), aldrin, isodrin, endosulfan isomers (endosulfan I, endosulfan-II), γ-chlordane, and endrin. The annual mean concentration of the region's total OCPs was 6.99 ng/l, and the arithmetic mean was 7.14 ± 4.19 ng/l. The detection rates of aldrin, HCB, α-HCH, β-HCH, and γ-HCH were 100%, while the rates of γ-chlordane and endrin were less than 50%; the rates of the other pollutants ranged from 64.86% to 97.3%. The residual level of aldrin (2.83 ± 2.87 ng/l) was the highest, followed by the DDTs (1.91 ± 1.92 ng/l) and HCHs (1.76 ± 1.54 ng/l); together, these residual levels accounted for 91% of the total OCPs. The residual levels of the pollutants are illustrated in Figure 4.6.

Compared with other studies, the level of aldrin in Lake Chaohu was lower than that in the Pearl River artery estuary during the low flow season (4.17 ± 3.07 ng/l) (Yang et al., 2004), the karst subterranean river in Liuzhou (9.22 ± 1.90 ng/l) (Wei et al., 2011), and the Kucuk Menderes River in Turkey (17–1790 ng/l) (Turgut, 2003), higher than that in the Changsha section of the Xiangjiang River (0.22–0.51 ng/l) (Tian et al., 2010) and the Wuhan section of the Yangtze River (1.88 ng/l) (Zhi et al., 2008), and comparable with that in the Huaxi River in Guizhou (2.079 ng/l) (Ban and Li, 2009) and the Guanting Reservoir in Beijing (2.26 ± 2.84 ng/l) (Kang et al., 2003). The levels of HCHs were similar to those in Lake Baiyangdian (2.1 ± 0.8 ng/l) (Hu et al., 2010), considerably lower than those in the Qiantang River in Zhejiang (33.07 ± 14.64 ng/l) (Zhou et al., 2008), the Chiu-lung River in Fujian (71.1 ± 85.5 ng/l) (Maskaoui et al., 2005), and the Kucuk Menderes River in Turkey (187–337 ng/l) (Turgut, 2003), and higher than those in

Table 4.6 The Residues of OCPs in Water from Lake Chaohu (ng/l)

	SD	Maximum	Minimum	Arithmetic Mean	Geometric Mean	Detection Rate (%)
α-HCH	0.53	2.40	0.11	0.47	0.33	100.00
β-HCH	0.51	2.19	0.36	0.92	0.80	100.00
γ-HCH	0.38	1.77	0.06	0.29	0.19	100.00
δ-HCH	0.13	0.60	N.D.	0.08	0.06	83.78
HCHs	1.45	6.92	0.55	1.76	1.41	100.00
o,p'-DDE	1.93	7.03	N.D.	1.42	1.01	62.16
p,p'-DDE	0.04	0.16	N.D.	0.02	0.03	32.43
o,p'-DDD	0.08	0.38	N.D.	0.03	0.18	16.22
p,p'-DDD	0.22	1.06	N.D.	0.06	0.07	24.32
o,p'-DDT	0.46	2.32	N.D.	0.16	0.15	35.14
p,p'-DDT	0.30	1.15	N.D.	0.23	0.30	62.16
DDTs	1.92	7.03	N.D.	1.91	1.10	97.30
HCB	0.08	0.35	0.06	0.17	0.15	100.00
Heptachlor	0.25	1.09	N.D.	0.17	0.15	64.86
Aldrin	2.87	12.22	0.15	2.83	1.76	100.00
Isodrin	0.17	0.63	N.D.	0.16	0.12	91.89
γ-Chlordane	0.02	0.14	N.D.	0.01	0.01	32.43
Endosulfan I	0.03	0.15	N.D.	0.02	0.03	62.16
Endosulfan II	0.45	2.70	N.D.	0.09	0.01	48.65
Endosulfan	0.46	2.80	N.D.	0.10	0.30	86.26
Endrin	0.08	0.37	N.D.	0.03	0.08	27.03

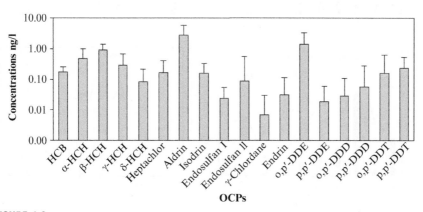

FIGURE 4.6

Annual mean concentrations of 18 OCPs in the water from Lake Chaohu (ng/l).

Meiliang Bay in Lake Taihu (>0.4 ng/l) (Ta et al., 2006), Lake Co Ngoin in Tibet (0.3 ng/l) (Zhang et al., 2003), and Lake Baikal in Russia (0.056–0.96 ng/l) (Iwata et al., 1995). The concentrations of DDTs were also at low levels, which were roughly equal to those in the Nanjing section of the Yangtze River (1.57–1.79 ng/l) (Jiang et al., 2000) and lower than those in the Guanting Reservoir (3.71–16.03 ng/l) (Wan et al., 2009), the Huangpu River (3.83–20.90 ng/l) (Xia et al., 2006), the Pearl River artery estuary during the low flow season (5.85–9.53 ng/l) (Yang et al., 2004), the Kucuk Menderes River in Turkey (ND–120 ng/l) (Turgut, 2003), and Lake Baikal in Russia (ND–0.015 μg/l) (Iwata et al., 1995).

4.3.2 The spatial and temporal distribution of OCPs in the water

The changes in the concentrations of the total OCPs and the three main pollutants (HCHs, DDTs, and aldrin) in Lake Chaohu and the three subregions from May 2010 to February 2011 are shown in Figure 4.7. There were similar trends for the OCPs over time both in the entire lake and in the central lake. The OCP levels increased jaggedly from May to September, and the peak was in September. Then, the residues declined rapidly, reached the bottom in November, and rose again from December to February. The trend in the western lake from September to February was the same, but the trend in the eastern lake was different. One of the main causes

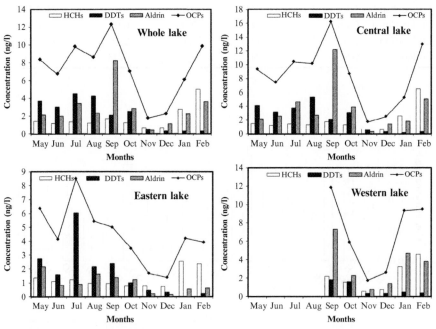

FIGURE 4.7

The spatial and temporal variation of OCPs in water from Lake Chaohu.

was that the concentrations of DDT in July were excessive, resulting in the higher OCPs from the eastern lake in July than in the other months. There was presumably a temporary point source pollution in July. Moreover, the high values of aldrin both in the western and central lake in September, which were not observed in the eastern lake, made the overall trends of the eastern lake different from the other subregionals.

Ten months were divided into four seasons, with spring just using the data of May as a reference. The concentrations of HCHs in the four seasons were 1.44, 1.25, 1.19, and 2.81 ng/l, and the concentrations of DDTs were 3.61, 3.75, 1.53, and 0.24 ng/l. The variable trends of the HCHs and DDTs were similar except during winter, and the concentrations were higher in the spring and summer than in autumn. The levels of HCHs in the winter were greater than those in any other season, but the levels of DDTs were the opposite and an order of magnitude lower in the winter than in the other seasons. The possible reasons for this phenomenon included water changes and the use of related pesticides. Beginning in June, the input amount of water from Lake Chaohu was higher than the output amount, reaching the highest level in July and August. After September, the output amount of water was greater than the input, and the water of Lake Chaohu was gradually reduced. On the one hand, the increase in water diluted the pollutants in the lake, and on the other hand, new pollutants were added to the lake from the area along the river. Furthermore, the use of OCPs around the lake would result in an increase in the OCP residues in the spring and summer, when there are more agricultural activities. Additionally, other technical products that include HCHs or DDTs may result in this irregular seasonal variation.

Seasonal differences in the remaining pollutants were analyzed as follows: the seasonal trends of HCB and heptachlor, which were similar to those of HCHs, were the highest residues in the winter; the residue of aldrin was at a high concentration, but the seasonal variation was inconspicuous; the pollution of isodrin and γ-chlordane were severe in the summer while the concentrations of endosulfan and endrin had high values in the spring. These values may have certain relationships with the application characteristics of these pollutants in general without uniform trends.

Based on the spatial distribution, the sampling site JC represented the eastern lake and its water source areas, MS and ZM represented the central lake and the lakeside area of Zhongmiao Temple, and TX represented the western lake region, which was near the region of the water intake. The data in TX just included September 2010 to February 2011. To ensure the comparability among the sampling sites, the monitoring data of the other three sites were also selected from this period (Table 4.7). The concentration of the OCPs was 3.33 ng/l from the eastern lake, 7.56 ng/l from the central lake, and 6.83 ng/l from the western lake. The pollution levels, from heavy to light, followed central lake > western lake > eastern lake and the water source area, and the levels of OCPs in the western and central lakes were more than twice those in the eastern lake and the water source area. The main pollutants in each region of the lake were different. The main pollutants were HCHs and DDTs in the eastern lake and the water source area and aldrin in the western lake and central lake in addition to HCHs and DDTs. Because of fewer sampling sites, the spatial differences they reflected may be influenced by the environment around the sites. There was an

Table 4.7 The Residues of OCPs at Each Site from September 2010 to February 2011 (ng/l)

Pollutants	MS	ZM	JC	TX
HCB	0.16	0.17	0.15	0.20
HCHs	2.30	2.13	1.36	2.14
DDTs	0.87	1.01	0.78	0.78
Heptachlor	0.29	0.23	0.22	0.14
Aldrin	3.90	3.84	0.68	3.36
Isodrin	0.08	0.07	0.11	0.09
γ-Chlordane	0.00	0.00	0.00	0.00
Endosulfan	0.03	0.02	0.03	0.10
Endrin	0.01	0.01	0.01	0.03
OCPs	7.65	7.48	3.33	6.83

unpopulated region near the site of JC, whereas the relatively dense residential areas were located near the sites of ZM and MS. The life or industrial emissions were also one of the factors that led to the high pollution levels of the lake.

4.3.3 Eco-risk assessments for OCPs

The SSD model was employed to assess the ecological risks for all species at four sampling sites. The average and maximum ecological risks are given in Tables 4.8 and 4.9, respectively. By comparing the mean values, the ecological risk of site MS, where the pollution of p,p'-DDT and aldrin was heavy, was slightly higher than those of the other sites. The potential risk of γ-HCH at site TX was relatively higher, while at sites JC and ZM, the risks of heptachlor and isodrin were higher. In five OCPs, the ecological risk of heptachlor was the highest, followed by γ-HCH, p,p'-DDT, aldrin, and endrin. However, Tables 4.8 and 4.9 indicate that the potential risks of the OCPs for all species at the four sites were very low, ranging from 7.885×10^{-28} to 1.639×10^{-8}. The maximum risk probability of a single pollutant was less than 10^{-7}. Comparing by species, the risks of p,p'-DDT and heptachlor for vertebrates were less than those for invertebrates, and the risks of the other three pollutants for vertebrates were higher. For further classification of the three subcategories, the risk of p,p'-DDT for crustaceans was 10^{-7}, which was the highest, whereas the risk of p,p'-DDT was mostly harmless for fish and insects and spiders. The risk of γ-HCH was highest for fish (10^{-8}) and was up to 10^{-16} for insects and spiders and less for crustaceans. Heptachlor had no risk for insects and spiders, but its risk for fish was two orders of magnitude higher than those for crustaceans, at 10^{-12} and 10^{-14}, respectively. The risk of aldrin and endrin was ranked as followed: fish > insects and spiders ≫ crustaceans. The risk of aldrin for fish was up to 10^{-7}, whereas endrin generally had a low risk.

Table 4.8 The Spatial Variation of the Mean Ecological Risk of Typical OCPs (PAF)

Pollutant	Site	Mean Value (μg/L)	PAF					
			All Species	Vertebrates	Invertebrates	Fishes	Crustaceans	Insects and Spiders
p,p′-DDT	MS	3.556E−4	4.692E−18	7.440E−165	6.083E−13	2.494E−259	1.128E−07	1.329E−135
	ZM	2.897E−4	1.187E−18	4.218E−184	2.503E−13	9.462E−293	7.730E−08	9.952E−152
	JC	2.237E−4	2.072E−19	1.247E−211	8.114E−14	0.000E+00	4.799E−08	7.572E−175
	TX	3.470E−4	3.983E−18	4.823E−167	5.472E−13	4.123E−263	1.078E−07	1.957E−137
γ-HCH	MS	1.940E−4	1.508E−13	1.717E−09	2.159E−117	1.123E−09	4.103E−251	5.312E−21
	ZM	1.911E−4	1.440E−13	1.676E−09	3.391E−118	1.096E−09	4.184E−253	4.913E−21
	JC	1.555E−4	7.615E−14	1.205E−09	8.951E−130	7.825E−10	5.199E−282	1.686E−21
	TX	2.282E−4	2.490E−13	2.226E−09	4.571E−109	1.465E−09	1.286E−230	1.233E−20
Heptachlor	MS	1.436E−4	1.582E−10	3.606E−14	7.024E−08	4.264E−15	1.094E−17	0.000E+00
	ZM	1.603E−4	2.036E−10	5.102E−14	8.450E−08	6.191E−15	1.717E−17	0.000E+00
	JC	2.039E−4	3.535E−10	1.090E−13	1.266E−07	1.400E−14	4.595E−17	0.000E+00
	TX	1.176E−4	1.001E−10	1.920E−14	5.020E−08	2.166E−15	4.831E−18	0.000E+00
Aldrin	MS	2.446E−3	5.172E−18	8.825E−09	1.741E−18	8.852E−09	1.995E−18	1.412E−11
	ZM	2.412E−3	4.755E−18	8.492E−09	1.601E−18	8.517E−09	1.835E−18	1.340E−11
	JC	7.304E−4	3.667E−21	3.186E−10	1.235E−21	3.164E−10	1.415E−21	1.542E−13
	TX	2.580E−4	7.123E−24	1.825E−11	2.398E−24	1.797E−11	2.748E−24	3.154E−15
Endrin	MS	8.664E−5	4.575E−25	2.937E−24	3.324E−25	6.513E−24	0.000E+00	6.796E−25
	ZM	1.177E−4	2.876E−24	1.846E−23	2.089E−24	4.094E−23	0.000E+00	4.272E−24
	JC	3.000E−5	7.885E−28	5.062E−27	5.728E−28	1.123E−26	0.000E+00	1.171E−27
	TX	7.348E−5	1.703E−25	1.093E−24	1.237E−25	2.424E−24	0.000E+00	2.529E−25

Table 4.9 The Spatial Variation of the Maximum Ecological Risk of Typical OCPs (PAF)

Pollutant	Site	Max Value (μg/l)	Month	PAF					
				All Species	Vertebrates	Invertebrates	Fishes	Crustaceans	Insects and Spiders
p,p'-DDT	MS	1.145E−3	8	9.797E−15	6.536E−88	8.683E−11	5.427E−130	9.655E−07	1.537E−71
	ZM	1.067E−3	10	6.248E−15	2.650E−91	6.476E−11	1.628E−135	8.485E−07	2.397E−74
	JC	3.700E−4	10	6.117E−18	2.285E−161	7.220E−13	2.590E−253	1.213E−07	1.100E−132
	TX	4.967E−4	10	4.326E−17	1.040E−137	2.561E−12	7.684E−213	2.086E−07	6.401E−113
γ-HCH	MS	1.770E−3	2	1.333E−10	5.887E−08	2.691E−43	4.167'	5.447E−79	4.983E−16
	ZM	1.467E−3	2	7.530E−11	4.361E−08	4.212E−47	3.066E−08	4.093E−87	1.889E−16
	JC	2.833E−4	6	4.853E−13	3.145E−09	6.840E−99	2.086E−09	6.763E−206	3.784E−20
	TX	1.053E−3	2	2.738E−11	2.567E−08	1.134E−54	1.783E−08	1.417E−103	3.400E−17
Heptachlor	MS	1.087E−3	2	1.639E−08	2.143E−11	2.110E−06	4.072E−12	4.337E−14	0.000E+00
	ZM	4.867E−4	2	2.598E−09	1.697E−12	5.466E−07	2.672E−13	1.618E−15	0.000E+00
	JC	6.300E−4	1	4.695E−09	3.832E−12	8.435E−07	6.409E−13	4.651E−15	0.000E+00
	TX	4.400E−4	1	2.062E−09	1.235E−12	4.613E−07	1.898E−13	1.070E−15	0.000E+00
Aldrin	MS	1.201E−2	9	7.246E−14	6.996E−07	2.440E−14	7.111E−07	2.796E−14	5.407E−09
	ZM	1.222E−2	9	8.041E−14	7.338E−07	2.707E−14	7.459E−07	3.102E−14	5.769E−09
	JC	2.140E−3	5	2.320E−18	6.112E−09	7.810E−19	6.124E−09	8.948E−19	8.570E−12
	TX	7.313E−3	9	3.694E−15	1.790E−07	1.244E−15	1.812E−07	1.425E−15	8.466E−10
Endrin	MS	3.225E−4	7	1.217E−21	7.812E−21	8.840E−22	1.732E−20	0.000E+00	1.808E−21
	ZM	3.667E−4	5	2.630E−21	1.688E−20	1.911E−21	3.744E−20	0.000E+00	3.907E−21
	JC	3.000E−5	10	7.885E−28	5.062E−27	5.728E−28	1.123E−26	0.000E+00	1.171E−27
	TX	9.000E−5	9	5.748E−25	3.690E−24	4.176E−25	8.183E−24	0.000E+00	8.539E−25

The results of the combining ecological risk of each site are shown in Table 4.10. The mean combining ecological risk probability of each site for all species was approximately 10^{-10}, following the order of MS > JC > ZM > TX. The site of the highest combining risk was MS in February (1.652×10^{-10}). A species-by-species comparison revealed that the potential combining ecological risk probability for invertebrates was 10^{-6} at the MS site in February, which was higher than that for vertebrates. Among the three subcategories, the probability of the combining ecological risks was ranked as crustaceans > fish > insects and spiders, with the maximum probability being close to 10^{-6} at the MS and ZM sites. Nevertheless, the risk was actually very low because of its order of magnitude, and the pollutants had little influence on aquatic organisms. Overall, the ecological risk of OCPs for aquatic organisms in Lake Chaohu was very low.

4.4 ECO-RISK ASSESSMENTS FOR PAHs IN LAKE CHAOHU
4.4.1 The residues of PAHs in the water (Qin et al., 2013)

The total PAH concentrations (PAH_{16}) (sum of the 16 EPA priority pollutants) in the water from Lake Chaohu are provided in Figure 4.8. The total PAH concentrations in the 15 sample sites ranged from 95.2 to 370.1 ng/l, with a mean value 181.5 ± 70.8 ng/l. The highest level of total PAHs was detected at site B6, the Shuangqiao River, and the second highest level was at site C2, the Xiaozhigao River. With a mean value of 267.3 ± 80.0 ng/l, the total PAH concentration in four of the river sites was much greater than that in the lake, which had a mean concentration of 150.3 ± 32.9 ng/l. Inside the lake, the maximum PAH level was found in the western lake (183.1 ng/l), followed by the eastern lake (excluding the water source area) (163.8 ± 23.3 ng/l), and then the eastern water source area (135.8 ± 30.4 ng/l). Of the 16 priority PAHs, Nap had the highest concentration (68.8 ng/l), followed by Phe, Flo, Fla, Pyr, Ant, and Acy, with concentrations of 42.7, 20.3, 9.1, 9.1, 2.8, and 1.2 ng/l, respectively. The concentrations for the rest of the 16 priority PAHs were less than 1 ng/l. Figure 4.9 shows that the content of the low-molecular-weight PAHs were much greater than that of the high-molecular-weight PAHs.

The PAH_{16} in the inflow rivers to Lake Chaohu (267.3 ± 80.0 ng/l) was close to that found in the Tianjin rivers (281.6 ± 336.9 ng/l) (Shi et al., 2004), was less than that reported in the Yangtze River (0.242–6.235 μg/l) (Feng et al., 2007), and greater than that found in the Yellow River (121.3 ng/l) (Wang et al., 2009a,b) and in the Luan River (99.4 ng/l) (Cao et al., 2010). The PAH_{16} in Lake Chaohu (150.3 ± 31.4 ng/l) was close to that reported in Lake Taihu (134.5 ± 54.8 ng/l) and greater than that reported in the Pearl River Estuary (Luo et al., 2006). Compared with reports from abroad, the PAH_{16} level in Lake Chaohu was lower than the levels in Victoria Lake (Kwach and Lalah, 2009), Great Bitter Lake, and El Temsah Lake (Said and El Agroudy, 2006).

Table 4.10 The Spatial and Temporary Variation of Combining Ecological Risks (msPAF)

Site	Month	All Species	Vertebrates	Invertebrates	Fishes	Crustaceans	Insects and Spiders
MS	2010.5	1.926E−13	1.011E−08	0.000	9.464E−09	0.000	1.270E−11
	2010.6	1.414E−13	1.263E−08	0.000	1.210E−08	0.000	1.899E−11
	2010.7	1.449E−11	2.659E−08	1.192E−08	2.576E−08	1.954E−08	5.418E−11
	2010.8	5.499E−09	3.986E−09	9.473E−07	3.632E−09	9.655E−07	3.243E−12
	2010.9	3.067E−11	7.012E−07	2.099E−08	7.121E−07	2.798E−14	5.407E−09
	2010.10	5.895E−11	1.259E−08	3.408E−08	1.239E−08	6.597E−07	2.120E−11
	2010.11	5.894E−11	4.801E−10	3.404E−08	3.270E−10	1.316E−07	1.732E−14
	2010.12	1.728E−12	3.372E−09	2.555E−09	3.274E−09	4.271E−08	3.417E−12
	2011.1	1.689E−09	9.302E−09	3.986E−07	8.156E−09	0.000	8.082E−12
	2011.2	1.652E−08	1.056E−07	2.109E−06	8.880E−08	2.892E−08	1.364E−10
	Mean	**1.584E−10**	**1.054E−08**	**7.024E−08**	**9.976E−09**	**1.128E−07**	**1.412E−11**
ZM	2010.5	4.681E−13	6.548E−09	0.000	5.509E−09	0.000	3.955E−12
	2010.6	7.053E−12	9.245E−09	7.078E−09	8.706E−09	0.000	1.159E−11
	2010.7	3.411E−13	9.264E−08	0.000	9.266E−08	8.477E−09	3.325E−10
	2010.8	1.130E−13	2.610E−08	0.000	2.574E−08	0.000	5.703E−11
	2010.9	3.008E−10	7.356E−07	1.125E−07	7.473E−07	2.788E−08	5.775E−09
	2010.10	4.374E−14	6.094E−08	6.467E−11	6.118E−08	8.481E−07	1.920E−10
	2010.11	3.754E−11	2.460E−10	2.446E−08	1.563E−10	1.884E−07	0.000E+00
	2010.12	2.436E−10	1.099E−09	9.637E−08	9.890E−10	5.187E−08	5.504E−13
	2011.1	1.444E−09	4.838E−09	3.553E−07	3.832E−09	0.000E+00	1.708E−12
	2011.2	2.673E−09	1.249E−07	5.465E−07	1.127E−07	1.076E−07	2.893E−10
	Mean	**2.038E−10**	**1.017E−08**	**8.451E−08**	**9.613E−09**	**7.730E−08**	**1.340E−11**
JC	2010.5	1.744E−13	7.963E−09	0.000E+00	7.338E−09	0.000E+00	8.570E−12
	2010.6	4.855E−13	3.589E−09	0.000E+00	2.527E−09	0.000E+00	2.413E−13

2010.7	3.836E−13	3.308E−09	0.000E+00	2.362E−09	8.930E−09	3.029E−13
2010.8	7.927E−14	4.147E−09	0.000E+00	3.716E−09	0.000E+00	3.134E−12
2010.9	4.730E−14	2.677E−09	0.000E+00	2.340E−09	0.000E+00	1.546E−12
2010.10	2.770E−10	1.692E−09	1.059E−07	1.536E−09	1.213E−07	1.017E−12
2010.11	1.116E−10	3.722E−10	5.439E−08	2.388E−10	7.582E−08	0.000E+00
2010.12	4.545E−11	4.205E−10	2.814E−08	2.681E−10	6.636E−08	0.000E+00
2011.1	4.695E−09	2.838E−09	8.435E−07	1.923E−09	4.663E−15	5.462E−14
2011.2	8.353E−10	2.243E−09	2.378E−07	1.538E−09	4.654E−08	7.860E−14
Mean	**3.536E−10**	**1.524E−09**	**1.266E−07**	**1.099E−09**	**4.799E−08**	**1.542E−13**
TX 2010.9	2.189E−11	1.822E−07	1.620E−08	1.833E−07	0.000E+00	8.468E−10
2010.10	9.437E−14	8.476E−09	2.560E−12	8.024E−09	2.086E−07	1.057E−11
2010.11	4.774E−15	6.127E−10	1.243E−13	5.042E−10	5.747E−08	1.592E−13
2010.12	2.461E−12	2.107E−09	3.311E−09	1.980E−09	7.582E−08	1.589E−12
2011.1	2.068E−09	6.356E−08	4.614E−07	6.032E−08	1.692E−07	1.575E−10
2011.2	9.511E−10	5.529E−08	2.561E−07	4.766E−08	9.465E−08	7.329E−11
Mean	**1.003E−10**	**2.244E−09**	**5.021E−08**	**1.483E−09**	**1.078E−07**	**3.109E−15**

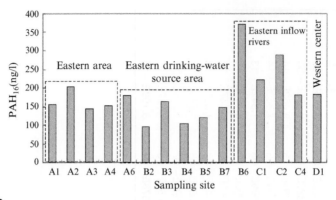

FIGURE 4.8

Distributions of total PAHs (PAH_{16}) in the water from Lake Chaohu.

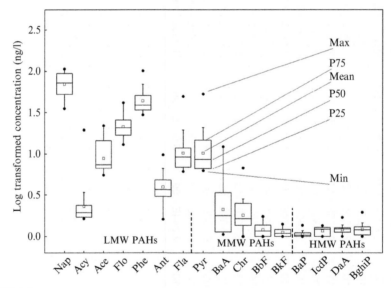

FIGURE 4.9

Contents of 16 PAHs in the water from Lake Chaohu. LMW PAHs include Nap, Acy, Ace, Flo, Phe, Ant, and Fla; MMW PAHs includes Pyr, BaA, Chr, BbF, and BkF; HMW PAHs includes BaP, IcdP, DahA, and BghiP. The vertical axis is log-transformed total PAH concentrations $log(C_{PAHs} + 1)$.

4.4.2 Site-specific ecological risk of PAHs based on the SSD method

Exposure concentrations of eight individual PAHs for each sampling site were translated into the ecological risk values based on the SSD curves. The risks of concentration-addition and response-addition msPAFs were also calculated to represent the combined ecological risk of the eight PAHs. The results are listed in Table 4.11.

HMW PAHs have greater risks than the MMW PAHs: Pyr had the greatest ecological risk with PAFs ranging from 2.85×10^{-3} (B3) to 1.56×10^{-2} (B6). The average risk of Pyr is two orders of magnitude greater than that of Ant, three orders of magnitude greater than that of Fla, and four orders of magnitude greater than those of Phe and BaP. The LMW PAHs have the least ecological risk; Nap risks were 12 orders of magnitude greater than Flo and 27 orders of magnitude greater than the risks of Ace, which had the least ecological risk 1.61×10^{-46} (B2) to 2.91×10^{-36} (B6) with a mean of 5.57×10^{-37}. The results can be attributed to the major toxicity of the HMW PAHs and the significant concentrations of the MMW PAHs.

The msPAFs of the response addition ranged from 2.86×10^{-3} (B4) to 1.58×10^{-2} (B6), and the msPAF of the concentration addition varied from 2.43×10^{-5} (B4) to 8.51×10^{-4} (B6). The msPAFs based on concentration were even lower than the PAFs of some individual PAHs, such as Pyr, which indicates that the risks based on concentration are underestimated. The concentration addition can be applied based on the hypothesis that the chemicals that have the same toxicity mode so that the exposure data can be scaled into dimensionless HUs. In other words, the chemicals have similar distribution modes and the same variance (Traas et al., 2002). However, the PAH toxicity data have quite different standard deviations (Table 4.5). There may be two reasons for this: First, the PAHs may have different toxicity modes in the water system, so the concentration-addition method should not be used in the water ecological risk assessment, and second, there are insufficient data to form the complete distribution curves. Before the toxicity mode was clear, the msPAF based on the response addition may be a credible result.

The results of the response-addition msPAF are used to analyze the geographical risk distribution of the PAHs in Lake Chaohu. The msPAF in Lake Chaohu ranged from 0.29% (B3) to 1.58% (B6). The ecological risks in rivers, with mean value 0.93%, were greater than those in the lake, where the mean PAF value was 0.35%. Site B6 had the largest PAF value (1.58%), which is $2.0\times$ greater than the ecological risk in the upstream of the Xiaozhigao River, $2.2\times$ greater than that in the estuary of the Xiaozhigao River, and $2.6\times$ greater than that in Dazhigao River. In the lake, the mean value of the PAF in East Lake (0.34%) was less than the PAF in West Lake (0.42%) and was at the same level as that in the water source area (0.34%). The result indicated that rivers may be the main sources of the PAHs pollution in Lake Chaohu. Controlling the PAH discharge into the river inflows may be important to control the PAH pollution in Lake Chaohu.

4.4.3 Probability of ecological risk of PAHs based on the PRA method

The toxicity and exposure distribution curves of seven PAHs were produced. The BaP level was lower than the detection limit in seven sites; thus, BaP was not included due to a lack of exposure data. The ecological risks of the exposure PAHs

Table 4.11 Potentially Affected Factions (PAFs) for Eight Individual PAHs at 15 Sample Sites

Sampling Site	PAFs for Eight Individual PAHs (%)								msPAF (%)	
	Nap (10^{-9})	Ace (10^{-43})	Flo (10^{-23})	Phe (10^{-5})	Ant (10^{-4})	Fla (10^{-5})	Pyr (10^{-1})	BaP (10^{-5})	Con-add (10^{-3})	Res-add (10^{-1})
A1	9.02	49.06	321.95	3.04	27.35	8.73	4.20	–	6.12	4.23
A2	72.83	135.88	662.05	3.46	32.57	6.30	3.47	9.04	5.88	3.60
A3	16.60	0.41	28.57	1.83	1.13	6.02	3.03	–	3.25	3.03
A4	31.91	7.55	22.16	1.29	10.72	2.45	2.93	–	2.85	2.95
A6	64.68	68.27	124.10	1.70	11.35	3.07	3.11	5.26	3.50	3.18
B2	1.16	0.16	1.23	1.06	6.54	3.18	2.97	–	2.45	2.97
B3	42.56	52.98	45.29	1.48	11.11	2.43	2.85	–	2.95	2.86
B4	2.25	17.63	3.53	1.17	10.92	2.02	3.09	–	2.43	3.10
B5	2.26	105.36	17.05	2.09	15.69	7.40	3.84	20.30	4.86	4.06
B6	27.86	2.91×10^{9}	6.43×10^{4}	20.57	127.59	555.19	15.61	1.05	85.10	15.80
B7	5.40	1.86	123.25	3.29	29.28	7.73	4.41	2.01	5.96	4.46
C1	18.22	6.81×10^{-6}	6.38×10^{3}	6.90	62.36	32.40	7.23	4.69	16.22	7.35
C2	46.47	5.42×10^{-9}	2.88×10^{4}	8.67	58.22	59.72	7.99	–	22.19	8.06
C4	14.26	2.52×10^{7}	964.31	4.03	32.42	15.49	6.04	4.69	10.02	6.12
D1	68.56	9.78	10.79	1.65	17.90	10.32	4.18	2.52	6.03	4.23
Mean	28.27	5.57×10^{8}	6783.03	4.15	30.34	48.16	5.00	6.20	11.99	5.07
GM	15.48	1007.07	156.78	2.77	18.70	9.38	4.37	4.28	6.35	4.43

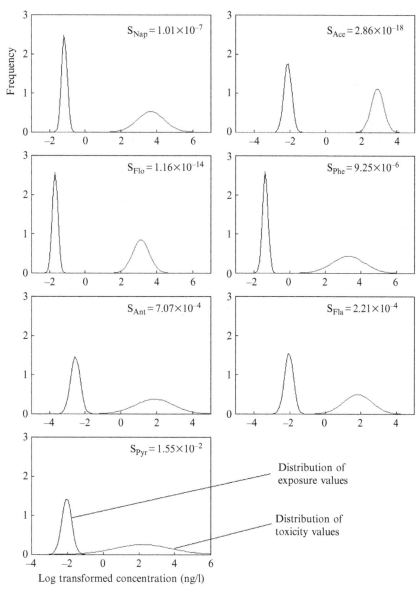

FIGURE 4.10

Toxicity and exposure distribution curves of seven PAHs. (For the color version of this figure, the reader is referred to the online version of this chapter.)

in Lake Chaohu were qualified by the calculation of the overlap area S (Figure 4.10). The black line is the distribution of the log-transformed exposure values, and the red line is the distribution of the log-transformed toxicity values. The risks based on the PRA method ranged from 2.86×10^{-16}% to 1.55%. The greatest ecological risk probability was found for Pyr (1.55%), followed by Ant (7.07×10^{-2}%), Fla (2.21×10^{-2}%), Phe (9.25×10^{-6}%), Nap (1.01×10^{-5}%), Flo (1.16×10^{-14}%), and Ace (2.86×10^{-16}%).

A JPC for each chemical was generated (Figure 4.11). The position of the joint risk probability curve reflects the risk of the PAHs; the closer the curve is to the axis, the less risk the chemical poses. The JPC of the seven PAHs demonstrates that the risk gaps between the PAHs are quite large. Fla and Ant were plotted on the same scale, Pyr was plotted on a large-scale axis, and the other four PAHs were plotted on a smaller-scale axis. As the results from the overlapping area calculation, Pyr caused the greatest risk among the compounds studied, followed by Ant, Fla, Phe,

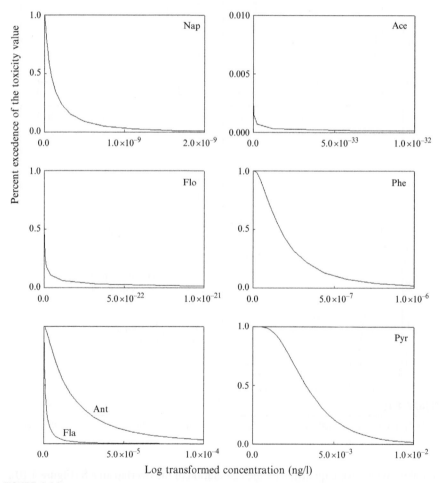

FIGURE 4.11

Joint probabilistic curves of seven PAHs.

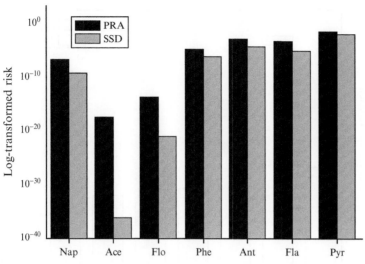

FIGURE 4.12

Comparisons of ecological risk assessment results of seven PAHs based on the SSD and PRA methods.

Nap, Flo, and Ace. The order was the same as that calculated by the overlap area and the average risk values by SSD.

The comparisons of the ecological risk assessment results of the seven PAHs based on the SSD and PRA methods are presented in Figure 4.12. The ecological risks calculated by the SSD and PRA methods were not of the same order of magnitude, and the risks estimated by the PRA method were greater than those estimated by the SSD method. However, the ecological risk orders of seven PAHs were the same. A significant difference was identified in the Ace ecological risks produced by the two methods. This might be caused by the uncertainty that exists with the Ace data.

4.4.4 Uncertainty analysis

Monte Carlo simulation was used to demonstrate the uncertainties of the exposure and toxicity data. The more simulation times performed, the closer the results obtained to the real situation. The simulation times 1000, 3000, 5000, and 10000 iterations commonly used in the literatures. In the present study, the results based on the simulation times 1000, 3000, 5000, and 10,000 iterations were compared and showed that $5000\times$ iterations were sufficient enough to ensure the stability of results. For the simulation results, there were no statistically significant differences between 5000 and $10,000\times$ iterations.

The comparisons of the CV values of the toxicity and exposure data for eight PAHs based on 5000 Monte Carlo simulations are presented in Figure 4.13.

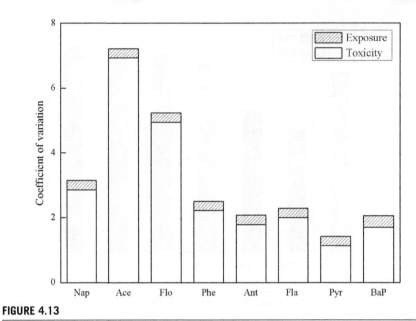

FIGURE 4.13

Comparisons of CV values of the exposure data and toxicity data for eight PAHs.

The larger CV values were found for the toxicity data of eight PAHs, indicating the uncertainties were larger for the toxicity data than for those of the measured exposure data (Figure 4.13). The CVs varied from 1.14 to 6.92 for the toxicity data, while the CVs were from 0.28 to 0.36 for the measured exposure data. For the toxicity data, Ace had the largest CV value (6.92), followed by Flo (4.94), Nap (4.85), Phe (2.22), Fla (2.01), Ant (1.78), BaP (1.70), and Pyr (1.14). For the measured exposure data, BaP had the largest CV (0.36) because there were eight measured exposure values. Figure 4.13 illustrates that the uncertainties of the exposure data of eight PAHs were very similar, as indicated by their CV values. This result suggested that the uncertainties of the toxicity data might be the primary source of the ecological risk uncertainties.

4.4.5 Discussion

The combination of the SSD and PRA methods can provide the more general information on the spatial variations and on the probabilities of the potential ecological risks of the individual and multiple pollutants. The combination of the SSD and PRA methods can join the advantages and overcome the disadvantages of the two methods used alone.

The SSD model can provide a site-specific risk assessment and make it possible for the comparison of spatial distribution of ecological risks. Furthermore, the SSD model provides a method to calculate msPAFs of different contaminates. In contrast, the PRA method, introducing the concept of probability into the risk evaluation, considers the variability of exposure concentrations and provides a PRA for the whole area. In addition, the results from the PRA method can be illustrated in the joint probabilistic curves to reflect the relationship of exposure concentration and ecological risks that cannot be obtained by the SSD method.

The SSD and PRA methods have their disadvantages. First, both of the models are established on the probabilistic distribution of toxicity data and depend on the adequacy of the toxicity data. As mentioned in the uncertainty analysis, the uncertainties were larger for the toxicity data than for the measured exposure data. Therefore, the toxicity data determine the reliability of the assessment results. Furthermore, the PRA method depends on the adequacy of the exposure data for the establishment of exposure distribution curve. Second, both methods have shortcomings in the estimation of msPAFs. The PRA method cannot be used in the evaluation of multi-risk assessment of different compounds. In contrast, although the msPAFs can be calculated based on two toxicological modes of action, it is still difficult to obtain the joint risk before the toxicological modes of action are clear. The response-addition msPAFs of the studied PAHs in Lake Chaohu, for example, was two orders of magnitude greater than the concentration-addition msPAFs. Clarifying the toxicological modes of action for different contaminants is necessary to calculate the msPAF of multi-contaminants.

References

Aldenberg, T., Slob, W., 1993. Confidence-limits for hazardous concentrations based on logistically distributed NOEC toxicity data. Ecotoxicol. Environ. Saf. 25 (1), 48–63.

Ban, R., Li, Y.M., 2009. Contamination of OCPs in Huaxi River in Guiyang, China. In: Persistent organic pollutants Forum 2009 and the fourth persistent organic pollutants in National Academic Symposium [C], pp. 1–43.

Cao, Z.G., Liu, J.L., Luan, Y., Li, Y.L., Ma, M.Y., Xu, J., Han, S.L., 2010. Distribution and ecosystem risk assessment of polycyclic aromatic hydrocarbons in the Luan River, China. Ecotoxicology 19 (5), 827–837.

Chernyak, S.M., McConnell, L.L., 1995. Fate of some chlorinated hydrocarbons in arctic and far eastern ecosystems in the Russian Federation. Sci. Total Environ. 160–161, 75–85.

CSIRO (Australia's Commonwealth Scientific and Industrial Research Organisation), 2008. A Flexible Approach to Species Protection from: http://www.cmis.csiro.au/envir/burrlioz.

Domene, X., Ramirez, W., Mattana, S., Alcaniz, J.M., Andres, P., 2008. Ecological risk assessment of organic waste amendments using the species sensitivity distribution from a soil organisms test battery. Environ. Pollut. 155 (2), 227–236.

Feng, K., Yu, B.Y., Ge, D.M., Wong, M.H., Wang, X.C., Cao, Z.H., 2003. Organochlorine pesticide (DDT and HCH) residues in the Taihu Lake Region and its movement in soil-water system I. Field survey of DDT and HCH residues in ecosystem of the region. Chemosphere 50, 683–687.

Feng, C.L., Xia, X.H., Shen, Z.Y., Zhou, Z., 2007. Distribution and sources of polycyclic aromatic hydrocarbons in Wuhan section of the Yangtze River, China. Environ. Monit. Assess. 133 (1–3), 447–458.

Fernandes, M.B., Sicre, M.A., Boireau, A., Tronczynski, J., 1997. Polyaromatic hydrocarbon (PAH) distributions in the Seine River and its estuary. Mar. Pollut. Bull. 34 (11), 857–867.

Fernandez, M.A., Alonso, C., Gonzalez, M.I., Hernandez, L.M., 1999. Occurrence of organochlorine insecticides, PCBs and PCB congeners in water and sediments of the Ebro River (Spain). Chemosphere 38, 33–43.

Forbes, V.E., Calow, P., 2002. Species sensitivity distributions revisited: a critical appraisal. Hum. Ecol. Risk. Assess. 8 (3), 473–492.

Frenich, A.G., Vidal, J.L.M., Sicilia, A.D.C., 2006. Multiresidue analysis of organochlorine and organophosphorus pesticides in muscle of chicken, pork and lamb by gas chromatography-triple quadrupole mass spectrometry. Anal. Chim. Acta. 558, 42–52.

Furst, P., Furst, C., Wilmers, K., 1991. PCDDs and PCDFs in human milk statistical evaluation of a 6-years survey. Chemosphere 25 (7–10), 1029–1038.

Gao, H.J., Jiang, X., 2005. Bioaccumulation of organochlorine pesticides and quality safety in vegetables from Nanjing suburb. Acta Sci. circumstantiae 25 (1), 90–93.

He, W., Qin, N., Kong, X.Z., Liu, W.X., He, Q.S., Ouyang, H.L., Yang, C., Jiang, Y.J., Wang, Q.M., Yang, B., Xu, F.L., 2014. Ecological risk assessment and priority setting for typical toxic pollutants in the water from Beijing-Tianjin-Bohai area using Bayesian matbugs calculator (BMC). Ecol. Indic. (in press).

Heemken, O.P., Stachel, B., Theobald, N., Wenclawiak, B.W., 2000. Temporal variability of organic micropollutants in suspended particulate matter of the River Elbe at Hamburg and the River Mulde at Dessau, Germany. Arch. Environ. Contam. Toxicol. 38 (1), 11–31.

Heger, W., Jung, S.J., Martin, S., Peter, H., 1995. Acute and prolonged toxicity to aquatic organisms of new and existing chemicals and pesticides. Chemosphere 31 (2), 2707–2726.

Hose, G.C., Van den Brink, P.J., 2004. Confirming the species sensitivity distribution concept for endosulfan using laboratory, mesocosm, and field data. Arch. Environ. Contam. Toxicol. 47 (4), 511–520.

Hu, G.C., Luo, X.J., Li, F.C., Dai, J.Y., Guo, J.Y., Chen, S.J., Hong, C., Mai, B.X., Xu, M.Q., 2010. Organochlorine compounds and polycyclic aromatic hydrocarbons in surface sediment from Baiyangdian Lake, North China: concentrations, sources profiles and potential risk. J. Environ. Sci. 22, 176–183.

Iwata, H., Tanabe, S., Ueda, K., Tatsukawa, R., 1995. Persistent organochlorine residues in air, water, sediments, and soils from the Lake Baikal region, Russia. Environ. Sci. Technol. 29, 792–801.

Jiang, X., Xu, S.F., Martens, D., Wang, L.S., 2000. Polychlorinated organic contaminants in waters, suspended solids and sediments of the Nanjing section, Yangtze River. China Environ. Sci. 20 (3), 193–197.

Kang, Y.H., Liu, P.B., Wang, Z.J., Lv, Y.B., Li, Q.J., 2003. Persistent organochlorine pesticides in water from Guanting Reservoir and Yongdinghe River, Beijing. J. Lake Sci. 15 (2), 125–132.

Khalili, N.R., Scheff, P.A., Holsen, T.M., 1995. PAH source fingerprints for coke ovens, diesel and gasoline-engines, highway tunnels, and wood combustion emmisions. Atmos. Environ. 29 (4), 533–542.

Kwach, B.O., Lalah, J.O., 2009. High concentrations of polycyclic aromatic hydrocarbons found in water and sediments of car wash and Kisat areas of Winam Gulf, Lake Victoria-Kenya. Bull. Environ. Contam. Toxicol. 83 (5), 727–733.

Lange, R., Hutchinson, T.H., Scholz, N., Solbe, J., 1998. Analysis of the ECETOC aquatic toxicity (EAT) database II—comparison of acute to chronic ratios for various aquatic organisms and chemical substances. Chemosphere 36 (1), 115–127.

Larsen, R.K., Baker, J.E., 2003. Source apportionment of polycyclic aromatic hydrocarbons in the urban atmosphere: a comparison of three methods. Environ. Sci. Technol. 37 (9), 1873–1881.

Li, X.R., Zhao, T.K., Yu, Y.X., Zhang, C.J., Li, P., Li, S.J., 2009. Population exposure to PAHs and the health risk assessment in Beijing area. J. Agro-Environ. Sci. 28 (8), 1758–1765 (in Chinese).

Liu, L., Yan, X.P., Wang, Y., Xu, F.L., 2009. Assessing ecological risks of polycyclic aromatic hydrocarbons (PAHs) to freshwater organisms by species sensitivity distributions. Asian J. Ecotoxicol. 4 (5), 647–654 (in Chinese).

Liu, W.X., He, W., Qin, N., Kong, X.Z., He, Q.S., Ouyang, H.L., Yang, B., Wang, Q.M., Yang, C., Jiang, Y.J., Wu, W.J., Xu, F.L., 2012. Residues, distributions, sources and ecological risks of OCPs in the water from Lake Chao, China. ScientificWorldJournal 2012, Article ID 897697, 16 pages, http://dx.doi.org/10.1100/2012/897697.

Lunn, D.J., Thomas, A., Best, N., Spiegelhalter, D., 2000. WinBUGS—a Bayesian modeling framework: concepts, structure, and extensibility. Stat. Comput. 10, 325–337.

Luo, X.J., Chen, S.J., Mai, B.X., Yang, Q.S., Sheng, G.Y., Fu, J.M., 2006. Polycyclic aromatic hydrocarbons in suspended particulate matter and sediments from the Pearl River Estuary and adjacent coastal areas, China. Environ. Pollut. 139 (1), 9–20.

Maltby, L., Blake, N., Brock, T.C.M., Van Den Brink, P.J., 2005. Insecticide species sensitivity distributions: importance of test species selection and relevance to aquatic ecosystems. Environ. Toxicol. Chem. 24 (2), 379–388.

Maskaoui, K., Zhou, J.L., Zheng, T.L., Hong, H., Yu, Z., 2005. Organochlorine micropollutants in the Jiulong River Estuary and Western Xiamen Sea, China. Mar. Pollut. Bull. 51, 950–959.

Newman, M.C., Ownby, D.R., Mezin, L.C.A., Powell, D.C., Christensen, T.R.L., Lerberg, S.B., Anderson, B.A., 2000. Applying species-sensitivity distributions in ecological risk assessment: assumptions of distribution type and sufficient numbers of species. Environ. Toxicol. Chem. 19 (2), 508–515.

Ntzoufras, I., 2009. Bayesian Modeling Using WinBUGS. A John Wiley & Sons, Inc., Hoboken, NJ.

Posthuma, L., Traas, T.P., Suter, G.W., 2002. Species Sensitivity Distributions in Ecotoxicology [C]. Lewis Publishers, Boca Raton, FL, pp. 3–9.

Qin, N., He, W., Kong, X.Z., Liu, W.X., He, Q.S., Yang, B., Ouyang, H.L., Wang, Q.M., Xu, F.L., 2013. Ecological risk assessment of polycyclic aromatic hydrocarbons (PAHs) in the water from a large Chinese lake based on multiple indicators. Ecol. Indic. 24, 599–608.

Rogge, W.F., Hildemann, L.M., Mazurek, M.A., Cass, G.R., Simoneit, B.R.T., 1993. Sources of fine organic aerosol. 3. Road dust, tire debris, and organometallic brake lining dust—roads as sources and sinks. Environ. Sci. Technol. 27 (9), 1892–1904.

Said, T.O., El Agroudy, N.A., 2006. Assessment of PAHs in water and fish tissues from Great Bitter and El Temsah lakes, Suez Canal, as chemical markers of pollution sources. Chem. Ecol. 22 (2), 159–173.

Shao, Q.X., 2000. Estimation for hazardous concentrations based on NOEC toxity data: an alternative approach. Environmetrics 11 (5), 583–595.

Shi, X., Yang, Y., Xu, F.L., Liu, W.X., Tao, S., 2004. Ecological risk assessment of polycyclic aromatic hydrocarbons in surface water from Tianjin. Acta Sci. Circumst. 24 (04), 619–624.

Singh, K.P., Malik, A., Mohan, D., Takroo, R., 2005. Distribution of persistent organochlorine pesticide residues in Gomti River, India. Bull. Environ. Contam. Toxicol. 74, 146–154.

Solomon, K.R., Sibley, P., 2002. New concepts in ecological risk assessment: where do we go from here? Mar. Pollut. Bull. 44 (4), 279–285.

Solomon, K.R., Baker, D.B., Richards, R.P., Dixon, D.R., Klaine, S.J., LaPoint, T.W., Kendall, R.J., Weisskopf, C.P., Giddings, J.M., Giesy, J.P., Hall, L.W., Williams, W.M., 1996. Ecological risk assessment of atrazine in North American surface waters. Environ. Toxicol. Chem. 15 (1), 31–74.

Solomon, K., Giesy, J., Jones, P., 2000. Probabilistic risk assessment of agrochemicals in the environment. Crop. Prot. 19 (8–10), 649–655.

Steen, R., Leonards, P.E.G., Brinkman, U.A.T., Barcelo, D., Tronczynski, J., Albanis, T.A., Cofino, W.P., 1999. Ecological risk assessment of agrochemicals in European estuaries. Environ. Toxicol. Chem. 18 (7), 1574–1581.

Ta, N., Zhou, F., Gao, Z.Q., Zhong, M., Sun, C., 2006. The status of pesticide residues in the drinking water sources in Meiliangwan Bay, Taihu Lake of China. Environ. Monit. Assess. 123, 351–370.

Tao, S., Xu, F.L., Wang, X.J., Liu, W.X., Gong, Z.M., Fang, J.Y., Zhu, L.Z., Luo, Y.M., 2005. Organochlorine pesticides in agricultural soil and vegetables from Tianjin, China. Environ. Sci. Technol. 39 (8), 2494–2499.

Tian, G., Chen, Y.Q., Wan, X.Z., Wang, S.C., 2010. Investigations and control measures on seven persistent organic pollutants in protected region of drinking water sources of Xiangjiang in Changsha. Environ. Monit. China 26 (1), 58–62.

Traas, T.P., Van de Meent, D., Posthuma, L., 2002. The potentially affected fraction as ameasure of ecological risk. In: Posthuma, L., Traas, T.P., Suter, G.W. (Eds.), Species Sensitivity Distributions in Ecotoxicology [M]. Lewis, Boca Raton, FL, USA, pp. 315–343.

Turgut, C., 2003. The contamination with organochlorine pesticides and heavy metals in surface water in Kucuk Menderes River in Turkey, 2000–2002. Environ. Int. 29 (1), 29–32.

UNEP, 2001. Final Act of the Plenipotentiaries on the Stockholm Convention on Persistent Organic Pollutants. United Nations Environment Program Chemicals, Geneva.

UNEP, 2007. Proposal for Listing Endosulfan in the Stockholm Convention on Persistent Organic Pollutants, Stockholm Convention on Persistent Organic Pollutants Persistent Organic Pollutants Review Committee, Third meeting, Geneva.

van Straalen, N.M., 2002. Threshold models for species sensitivity distributions applied to aquatic risk assessment for zinc. Environ. Toxicol. Pharmacol. 11 (3–4), 167–172.

Wagner, C., Lokke, H., 1991. Estimation of ecotoxicological protection levels from NOEC toxicity data. Water Res. 25 (10), 1237–1242.

Wan, Y.W., Kang, T.F., Zhou, Z.L., Li, P.N., Zhang, Y., 2009. Health risk assessment of volatile organic compounds in water of Beijing Guanting reservoir. Res. Environ. Sci. 22 (2), 150–154.

Wang, X.L., Tao, S., Dawson, R.W., Xu, F.L., 2002. Characterizing and comparing risks of polycyclic aromatic hydrocarbons in a Tianjin wastewater-irrigated area. Environ. Res. 90 (3), 201–206.

Wang, X.T., Chu, S.G., Xu, X.B., 2003. Organochlorine pesticide residues in water from Guanting Reservoir and Yongding River, China. Bull. Environ. Contam. Toxicol. 70, 351–358.

Wang, L.L., Yang, Z.F., Niu, J.F., Wang, J.Y., 2009a. Characterization, ecological risk assessment and source diagnostics of polycyclic aromatic hydrocarbons in water column of the

Yellow River Delta, one of the most plenty biodiversity zones in the world. J. Hazard. Mater. 169 (1–3), 460–465.

Wang, Y., Wang, J.J., Qin, N., Wu, W.J., Zhu, Y., Xu, F.L., 2009b. Assessing ecological risks of DDT and lindane to freshwa ter organisms by species sensitivity distributions. Acta Sci. Circumst. 29 (11), 2407–2414 (in Chinese).

Wang, Y., Wu, W.J., He, W., Qin, N., He, Q.S., Xu, F.L., 2012. Residues and ecological risks of organochlorine pesticides in Lake Small Baiyangdian, North China. Environ. Monit. Assess. 185, 917–929.

Wei, L.L., Guo, F., Wang, J.Z., Kang, C.X., 2011. Distribution characteristics of organochlorine pesticides in karst subterranean river in Liuzhou. Carsolog. Sin. 30 (1), 16–21.

Wheeler, J.R., Grist, E.P.M., Leung, K.M.Y., Morritt, D., Crane, M., 2002. Species sensitivity distributions: data and model choice. Mar. Pollut. Bull. 45 (1–12), 192–202.

Xia, F., Hu, X.X., Han, Z.H., Wang, W.H., 2006. Distribution characteristics of organochlorine pesticides in surface water from the Huangpu River. Res. Environ. Sci. 19 (2), 11–15.

Xu, S.S., Liu, W.X., Tao, S., 2006. Emission of polycyclic aromatic hydrocarbons in China. Environ. Sci. Technol. 40 (3), 702–708.

Yang, Q.S., Mai, B.X., Fu, J.M., Sheng, G.Y., Wang, J.X., 2004. Spatial and temporal distribution of organochlorine pesticides (OCPs) in surface water from the Pearl River artery estuary. Environ. Sci. 25 (2), 150–156.

Yang, Y., Shi, X., Wong, P.K., Dawson, R., Xu, F.L., Liu, W.X., Tao, S., 2006. An approach to assess ecological risk for polycyclic aromatic hydrocarbons (PAHs) in surface water from Tianjin. J. Environ. Sci. Health A Tox. Hazard. Subst. Environ. Eng. 41 (8), 1463–1482.

Zhang, W.L., Zhang, G., Qi, S.H., Peng, P.A., 2003. A preliminary study of organochlorine pesticides in water and sediments from two Tibetan lakes. Geochimica 32 (4), 363–367.

Zhang, Y.X., Tao, S., Cao, J., Coveney, R.M., 2007. Emission of polycyclic aromatic hydrocarbons in China by county. Environ. Sci. Technol. 41 (3), 683–687.

Zhi, X., Niu, J.F., Tang, Z.W., 2008. Ecological risk assessment of typical organochlorine pesticides in water from the Wuhan reaches of the Yangtze River. Acta Sci. Circumst. 28 (1), 168–173.

Zhou, R.B., Zhu, L.Z., Chen, Y.Y., 2008. Levels and source of organochlorine pesticides in surface waters of Qiantang River, China. Environ. Monit. Assess. 136, 277–287.

Addressing the Uncertainty in Modeling Watershed Nonpoint Source Pollution

5

Yi Zheng*, Feng Han, Yong Tian, Bin Wu, Zhongrong Lin

Center for Water Research, College of Engineering, Peking University, Beijing, China
*Corresponding author: e-mail address: yizheng@pku.edu.cn

5.1 INTRODUCTION TO THE ISSUE

5.1.1 Current status of nonpoint source pollution

Within the context of water quality protection, nonpoint source (NPS) pollution refers to the diffuse pollution primarily driven by the rainfall-runoff process. Typical nonpoint sources include fertilizers and pesticides from agricultural lands and residential areas, oil, grease, and toxic chemicals from urban stormwater, bacteria and nutrients from livestock, pet wastes, and faulty septic systems, sediment from improperly managed lands and eroding streambanks, acid drainage from abandoned mines, wet and dry atmospheric deposition, etc. Through complex watershed processes, pollutants from NPS would eventually be deposited into lakes, rivers, wetlands, coastal waters, and groundwaters, and pose risk to the aquatic system.

In developed countries where point sources are now strictly regulated, the NPS pollution has become the major threaten to the aquatic environment. In the United States, for example, NPS pollution has been the primary cause of impairments in over 33,000 waters, roughly three-quarters of the impaired waters for which total maximum daily loads (TMDLs) have been calculated (USEPA, 2011). For the rivers and lakes (Figure 5.1), the top impairment causes are mostly of NPS nature, and municipal discharges/sewage (i.e., the typical point sources) only ranks 6th and 7th, respectively. In Europe, almost 50–80% of the total nitrogen load and roughly half of the total phosphorus load are contributed by nonpoint sources (ETC/ICM, 2012). Figure 5.2 demonstrates the percentages of assessed water bodies that are affected by point and nonpoint sources.

Nationwide assessments of NPS pollution in developing countries have been rare, and the management focus is still on point sources. Yet, in China, it has been well recognized that the nation's NPS pollution has become increasingly prominent. For example, the NPS load of COD increased from 8.31 million tons in 2000 to 11.86 million tons in 2011 (Wang, 2006; SEPA, 2012a). Figure 5.3 illustrates the contribution of NPS to COD and NH_3–N loads in several key basins of China

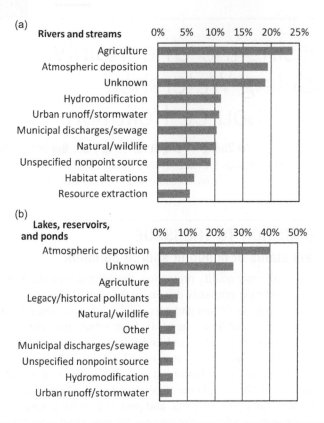

FIGURE 5.1

Top 10 causes of impairment for the assessed waters in the United States. The *x*-axis is the percentage of impairments attributed to a specific cause. Rivers are counted in length, and lakes are counted in area. (For the color version of this figure, the reader is referred to the online version of this chapter.)

Data source: Water Quality Assessment and Total Maximum Daily Loads Information (ATTAINS), http://www.epa.gov/waters/ir/ (last accessed on 26 May 2013).

(SEPA, 2012b). Although not as dominant as in developed countries, the COD contribution of nonpoint sources has already exceeded 50% in some China's basins, such as Songhuajiang, Huaihe, Haihe, and Liaohe Basins.

5.1.2 Management efforts

The pace of managing NPS pollution varies across the world. With its Water Framework Directive (WFD) initiated from 2000, European Union (EU) now leads in practicing integrated river basin management for water quality protection. The Nitrates Directive (1991) forms an integral part of the WFD and is one of the key instruments for controlling agricultural nonpoint sources. The implementation of the Directive consists of the following key elements: identification of water polluted (or at risk

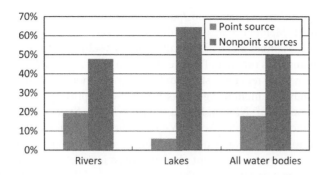

FIGURE 5.2

Contribution of point and nonpoint sources to impairments of water bodies in European Union (EU). (For the color version of this figure, the reader is referred to the online version of this chapter.)

Data source: WISE WFD database, http://www.eea.europa.eu/data-and-maps/data/wise_wfd.

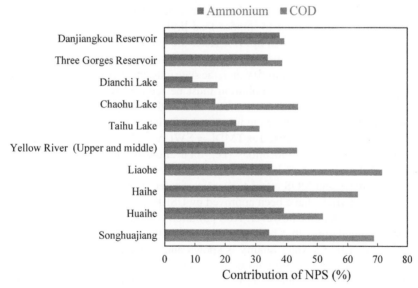

FIGURE 5.3

Contribution of nonpoint sources in key basins of China, as of year 2010. (For the color version of this figure, the reader is referred to the online version of this chapter.)

of pollution), designation as "Nitrate Vulnerable Zones" (NVZs), establishment of Codes of Good Agricultural Practice to be implemented by farmers on a voluntary basis, establishment of action programs to be implemented by farmers within NVZs on a compulsory basis, and national monitoring and reporting. The Nitrates Directive is aggressive and appears to be effective so far. As indicated by official statistics (http://ec.europa.eu/environment/pubs/pdf/factsheets/nitrates.pdf, last accessed on 26 May 2013), all Member States have drawn up more than 300 action programs. Across the 27 EU Member States, 39.6% of territory is subject to the implementation of action programs. It is proving effective: Between 2004 and 2007, nitrate concentrations in surface water remained stable or fell at 70% of monitored sites, and quality at 66% of groundwater monitoring points is stable or improving.

The management of NPS pollution in the United States is articulated by Section 319 of the Clean Water Act (CWA). The Section requires states to identify waters that fail to achieve water quality standards due to nonpoint sources and to develop comprehensive management plans that must include a process for identifying "best management practices (BMPs)" and other controls to reduce NPS pollution. Unfortunately, the Section offers no penalties and very weak financial incentives to motivate state action, and the U.S. Environmental Protection Agency (USEPA) has no authority to prepare and implement NPS controls if the state's program is inadequate or a state refuses to develop a plan.

On the other hand, the TMDL program, originated from Section 303(d) of the 1972 CWA, is now considered to be a more promising policy measure for NPS management. While the TMDL provision does not directly mandate regulation of nonpoint sources, it does require state planning to impose load limitations on those sources. Under TMDL regulations promulgated in 1992, states are required to include waters within their boundaries that are not meeting water quality standards into a "Section 303(d) list." For each listed (i.e., impaired) water, the state must identify the amount by which point and nonpoint sources of pollution must be reduced to achieve compliance through a rigorous planning process (i.e., the "TMDL process"). A TMDL (i.e., the maximum pollutant load a receiving water can assimilate and still meet water quality objectives) can be conceptualized as $TMDL = \Sigma WLA + \Sigma LA + MOS$, where ΣWLA is the sum of waste load allocations (i.e., total allowable load allocated to point sources), ΣLA is the sum of load allocations (i.e., total allowable load allocated to nonpoint sources), and MOS refers to margin of safety (MOS). From 1996 to 2013, around 50,000 TMDLs have been approved by USEPA.

In developing nations, however, NPS pollution is largely overlooked by the environmental regulation. In China, for instance, although the threat of nonpoint sources to the nation's water quality is known and regulation of nonpoint sources is required by laws, there exist no enforceable legal measures for the regulation. In China's "Total Load Control (TLC)" program, a counterpart of the TMDL program in the United States, the total allowed pollutant load is only allocated to point sources. The paucity of data and significant uncertainty in estimating watershed NPS loads appear to be the major impediments, among many others, for China to formulate effective strategies for regulating nonpoint sources.

5.1.3 Existing models

A great number of watershed hydrology and NPS pollution models have been developed (Singh, 1995; Singh and Frevert, 2002a,b; Borah and Bera, 2003). Groundbreaking work of the model development mostly took place in the 1970s and 1980s.

The existing models can be categorized in different dimensions, including whether physical processes are explicitly considered (i.e., empirical or process-based models), spatial representation of watershed (i.e., lumped, semi-distributed, or distributed models), temporal scale of modeling (i.e., storm event-based or long-term continuous models), and whether a stochastic or deterministic approach is used for model input or parameter specification.

Empirical models usually consist of empirical functions for fitting available data, such as regression models, artificial neural networks (Lek et al., 1999; Raghuwanshi et al., 2006), fuzzy logic algorithms (Tayfur and Singh, 2006; Barreto-Neto and Filho, 2008), or export coefficient models of various types (Johnes, 1996; Bowes et al., 2008; Ding et al., 2010). Process-based models, on the contrary, explicitly consider watershed hydrological processes and fate and transport of pollutants. Mass and heat balances are the underlying principles for such models. Figure 5.4 illustrates the modeling hierarchy of a typical process-based watershed NPS model.

Lumped models treat a watershed as a single homogeneous unit where the parameters and variables are averaged. Semidistributed and distributed models account for the spatial variability of processes and data by delineating the watershed into a number of sub-basins (i.e., semidistributed) or grids/cells (i.e., distributed) with different characteristics. With respect to temporal scale, storm event-based models simulate individual rainfall-runoff events with a focus on loading or event mean concentration (EMC) of target pollutants. Long-term continuous models simulate pollutant loading and water quality parameters as long as data allow, with either fixed or variable time steps.

This chapter is focused on models that are process-based, continuous, and spatially distributed, since these models offer great flexibility to evaluating and comparing pollution management scenarios. Soil and Water Assessment Tool (SWAT) (Arnold et al., 1998), Hydrological Simulation Program - FORTRAN (HSPF)

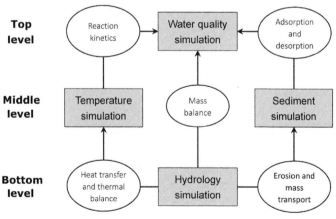

FIGURE 5.4

The modeling hierarchy of a typical process-based watershed NPS pollution model.

(Bicknell et al., 2001), MIKE SHE (Refsgaard and Storm, 1995), and Watershed Analysis Risk Management Framework (WARMF) (Chen et al., 1999) are the representative models, widely used in both academic research and management. These models are able to capture both temporal and spatial characteristics (e.g., long-term trends, critical timing, critical zones) of the pollution, and therefore effectively support management decisions. Nevertheless, these complex models have significant uncertainty in their simulations and are often questioned with regards to their prediction accuracy.

Figure 5.5 shows the magnitude of the simulation/prediction accuracy of SWAT, based on the SWAT literature (Gassman et al., 2007; Parajul et al., 2009; Wu and Chen, 2009; Shen et al., 2010; Gong et al., 2011; Yang et al., 2011; Niraula et al., 2012). Nash-Sutcliffe Efficiency Coefficient (NSE) was used as the evaluation criterion. The figure shows that, although the model performance is generally acceptable (i.e., all the medians of NSE are above 0.6), the chance of having poor simulation/prediction results is still very high (i.e., all the 25th percentiles of NSE are below 0.5, and even negative values of NSE are present). Other competing models such as HSPF and AnnAGNPS have the similar level of accuracy (Luo et al., 2006; Jeon et al., 2007; Das et al., 2008; Shamshad et al., 2008; Parajul et al., 2009; Liu and Tong, 2011). Note that Figure 5.5 in fact represents an optimistic situation (i.e., poor modeling results have rarely been published in good peer-reviewed journals), and the uncertainty issue should be much more serious in reality. Thus, systematic uncertainty assessment for such complex models would be

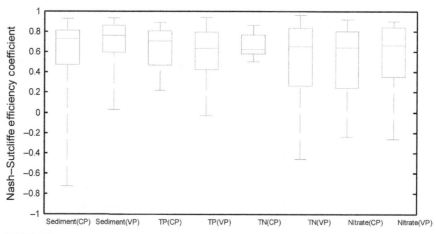

FIGURE 5.5

Performance of SWAT regarding different water quality parameters. In each box, the central mark is the median, and the upper and lower edges are the 25th and 75th percentiles, respectively. The upper and lower whiskers are 5th and 95th percentiles, respectively. CP and VP indicate calibration period and validation period, respectively. The number of samples for drawing each box-and-whisker plot ranges from 9 to 36. (For the color version of this figure, the reader is referred to the online version of this chapter.)

extremely useful for those who implement these models and/or depend on the output of these models to make decisions.

5.1.4 **Modeling for decision support**

The watershed-scale water quality models have been playing a critical role in developing management strategies for NPS pollution control. For example, models of varying complexity have been implemented in developing TMDLs. However, given the current status of data availability and model accuracy, the model simulations are subject to significant uncertainty, which often induces the criticism on the scientific basis of the TMDL process. It has been well recognized (NRC, 2001) that uncertainty must be explicitly acknowledged both in the models selected to develop TMDLs and in the results generated by those models, and prediction uncertainty must be estimated in a rigorous way, and systematically accounted for by an MOS. Based on a real TMDL case, Zheng and Keller (2008) demonstrated that watershed water quality modeling, with its uncertainty appropriately accounted for, would provide a strong support to NPS pollution management. This case study is briefly introduced below, as it is relevant to the discussion of this chapter.

The Newport Bay watershed (NBW) (Figure 5.6) is located in Orange County, California, with an area of 399 km^2. It has been highly urbanized, especially its

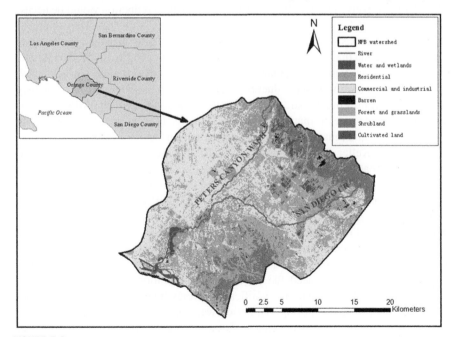

FIGURE 5.6

Newport Bay watershed. (For the color version of this figure, the reader is referred to the online version of this chapter.)

middle and lower portions. The climate is characterized by short, mild winters, and dry summers, typical of coastal southern California. San Diego Creek, the major drainage channel of NBW, contributes about 95% of the flow volume into the upper Newport Bay. Diazinon (an organophosphate pesticide once widely used in California for both agriculture and urban landscapes) pollution in San Diego Creek was included in the 1998 and 2002 Section 303(d) lists. In 2002, a TMDL was established for diazinon, based on a limited number of surface water samples taken from 1996 to 2000. Two water quality targets were considered: Criterion continuous concentration (CCC) (i.e., the highest 4-day average concentration that should not cause unacceptable effects during a long-term exposure) equal to 50 ng/l as a criterion for chronic effects, and criterion maximum concentration (CMC) (i.e., the highest one-hour average concentration that should not result in unacceptable effects on aquatic organisms and their uses) equal to 80 ng/l as a criterion for acute effects. The regulatory objective was that neither the CCC nor the CMC should be exceeded more than once every 3 years on average.

In developing this TMDL, no specific model was used to analyze the linkage between the source loading and the instream concentration. The maximum single storm average concentration (848 ng/l) at the outlet of San Diego Creek was used as a substitute for the level of diazinon loading. To account for the MOS required by the TMDL program, both the CCC and CMC were simply reduced by 10% to 45 and 72 ng/l, respectively. The conclusion was that a 95% load reduction would be necessary to meet the CCC and a 92% load reduction to meet the CMC. There were a number of problems with this simple approach. First of all, the relation between the diazinon source loading and the instream concentration was not necessarily linear. Second, the important seasonal variation in diazinon loading and concentration was neglected. Third, the concentration-based reduction targets are difficult to implement since they do not articulate how much source load reduction is required or which BMPs should be implemented to reduce the management goals. Finally, the arbitrary 10% MOS has no scientific basis, and thus inappropriately accounted for the uncertainty.

Zheng and Keller (2008) revisited this TMDL case and adopted a watershed modeling approach instead. The WARMF model was used to simulate the response of instream diazinon concentration to the loading from nonpoint sources. WARMF is a physically based watershed model recommend by USEPA for TMDL development (http://www.epa.gov/athens/wwqtsc/html/warmf.html, last accessed on 26 May 2013). The uncertainty associated with the modeling was systematically assessed using a stochastic simulation approach called management objectives constrained analysis of uncertainty (MOCAU) (Zheng and Keller, 2007b). The stochastic simulations adequately reproduced the historical concentration for the period before 2003. From 2003, the simulations significantly overestimated the concentration, because the effect of BMPs during that period was not adequately taken into consideration by the model. It was concluded that a 95% load reduction by phasing out diazinon uses alone would provide no chance for meeting the CCC. Given the significant uncertainty, to achieve the compliance with acceptable confidence, the diazinon usage on all land uses would have to be reduced to almost zero if no effective

BMPs were taken. The study results disproved the calculation in the original TMDL, but validated the effectiveness of the BMPs scheduled for 2003–2005.

This case study demonstrated the advantages and feasibility of conducting watershed water quality modeling to support NPS pollution management. It also reinforced the importance of systematic uncertainty analysis (UA). Nevertheless, UA itself remains a great challenge for complex watershed water quality models. This chapter summarizes the state-of-the-art understandings of the uncertainty issue and introduces some advances in addressing this issue.

5.2 UNCERTAINTY IN MODELING NPS POLLUTION: STATE OF THE ART

5.2.1 A framework for analysis

When a complex watershed water quality model is used for supporting real-life management decisions, uncertainty in the model simulations or predictions usually has very complicated origins. Zheng and Keller (2007a) proposed a framework for analyzing the uncertainty within the context of management-oriented water quality modeling (Figure 5.7). This framework discerns different sources of uncertainty and conceptualizes their interaction. The entire process of the management-oriented modeling is divided into a linking/calibration stage and a planning/prediction stage. Here "linking" refers to connecting cause and effect of water quality based on the complex model. In the first stage, stochastic simulations are generated assimilating the uncertainty from model parameter values, inputs and model structure and then constrained by observational data (involving observational error) to derive a linkage with quantifiable uncertainty. Using this framework, two types of model inputs are discerned. One is dynamic driving forces including meteorological data (e.g., daily precipitation, temperature), pollutant loads from various sources (e.g., wastewater treatment plant, land application, atmospheric deposition), and human activities (e.g., irrigation, water diversion). The other is distributed geographic data including topography, land use, and river network data. The linkage may take the form of an optimal parameter set along with explicitly assumed error term(s), as in calibration-based methods, or multiple parameter sets with respective probability values, as in formal and informal Bayesian approaches.

In the second stage, the established linkage is used to generate stochastic predictions that lead to probabilistic management judgment. Two new types of uncertainty are involved in this stage. The first one is change in linkage (hereafter referred as "linkage uncertainty"), which means that the established relationship between concerned water quality parameters and driving forces might not be applicable to future conditions. Such variability may be due to internal changes of the watershed system not captured by the model (e.g., surface roughness could be substantially altered due to habitat restoration). It can also be caused by external changes (e.g., different loading scenarios may result in different chemical or biochemical reactions if redox

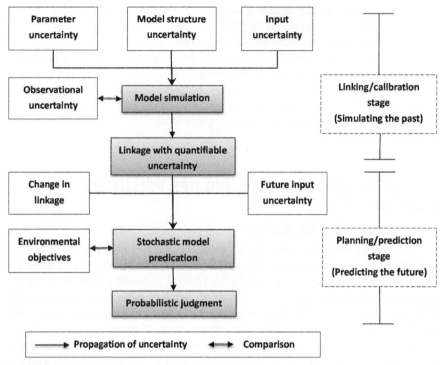

FIGURE 5.7

A framework for uncertainty analysis within the context of management-oriented water quality modeling.

This figure was adapted from Zheng and Keller (2007a), Fig. 1, p. 3.

conditions are modified). Another uncertainty is future input uncertainty, such as that associated with future (not occurred yet) weather or pollutant loads. Considering the uncertainty in the planning stage is crucial for evaluating management options, but this has been largely overlooked in both academic research and management practice. Note that the "environmental objectives" in Figure 5.7 may also have uncertainty, because the objectives may be derived based on uncertain data (Reckhow et al., 2005).

In this framework, the measured (i.e., observed) water quality in the linking stage can be decomposed as:

$$Z(\mathbf{X}^*, \boldsymbol{\theta}^*, t) = y(\mathbf{X}, \boldsymbol{\theta}, t) + \delta(\mathbf{X}, \boldsymbol{\theta}, t) + \varepsilon(\mathbf{X}^*, \boldsymbol{\theta}^*, t) \tag{5.1}$$

where $\mathbf{X}$ and $\boldsymbol{\theta}$ represent the input matrix and parameter vector required by the watershed model $y(\cdot)$; $\mathbf{X}^*$ and $\boldsymbol{\theta}^*$ represent *true* inputs and *true* parameters, both of which are *unknown* system attributes in reality; t is the time index; and $Z(\cdot, t)$, $y(\cdot, t)$, $\delta(\cdot, t)$, and $\varepsilon(\cdot, t)$ stand for observation, model response, model structure error, and observational error at time t, respectively. Differentiating $\mathbf{X}^*$ and $\boldsymbol{\theta}^*$ from $\mathbf{X}$ and

θ reflects the assumption that observational error ε (the vector form of $\varepsilon(\cdot, t)$) is independent of model simulation y and structure error δ (the vector form of $\delta(\cdot, t)$). Additionally, the *true* watershed response $Y(\cdot, t)$ can be expressed as

$$Y(\mathbf{X}^*, \boldsymbol{\theta}^*, t) = Z(\mathbf{X}^*, \boldsymbol{\theta}^*, t) - \varepsilon(\mathbf{X}^*, \boldsymbol{\theta}^*, t) \qquad (5.2)$$

In the planning stage, what is of real importance for management is the future *true* watershed response $\mathbf{Y}$ (the vector form of $Y(\cdot, t)$), not the potential observation $\mathbf{Z}$ (the vector form of $Z(\cdot, t)$). Thus this stage can be modeled as:

$$Y_f\left(\mathbf{X}_f^*, \boldsymbol{\theta}_f^*, t\right) = y\left(\mathbf{X}_f, \boldsymbol{\theta}_f, t\right) + \delta_f\left(\mathbf{X}_f, \boldsymbol{\theta}_f, t\right) \qquad (5.3)$$

where the subscript f is the index for the planning stage. The equation assumes that the same watershed model $y(\cdot)$ is used in both stages. With respect to Equations (5.1) and (5.3), parameter uncertainty and input uncertainty are associated with $\boldsymbol{\theta}$ and $\mathbf{X}$, respectively; model structure uncertainty and observational uncertainty are correspondingly represented by δ and ε; future input uncertainty is reflected by the change from $\mathbf{X}$ to $\mathbf{X}_f$ (or $\mathbf{X}^*$ to $\mathbf{X}_f^*$); and linkage uncertainty is reflected by the change from $\boldsymbol{\theta}$ to $\boldsymbol{\theta}_f$ (or $\boldsymbol{\theta}^*$ to $\boldsymbol{\theta}_f^*$), as well as from δ to δ_f.

Since parameter uncertainty significantly contributes to simulation inaccuracy and can be easily addressed in a Monte Carlo simulation (MCS) framework, it is often treated explicitly by assigning prior probabilistic distributions to uncertain parameters $\boldsymbol{\theta}$. In many cases, water quality measurements are point measurements in time and in space, while the modeling and management involve coarser temporal (e.g., daily) and spatial (e.g., reach-average) resolution. Hence, observational error ε generally consists of temporal variability and commensurability error (depending on the watershed system's properties $\mathbf{X}^*$ and $\boldsymbol{\theta}^*$), as well as errors associated with the measurement techniques. Directly considering the uncertainty in model inputs ($\mathbf{X}$) may be difficult since $\mathbf{X}$ contains both dynamic and spatially distributed variables with largely unknown covariance matrices. Finally, the model structure error δ remains to be the most intractable item in the UA. Models are assemblies of assumptions and simplifications of the complex real world, and the "perfect" model structure is generally never known.

Given the diversified sources of uncertainty and their complicated interactions (Figure 5.7), how to adequately quantify the overall simulation/prediction uncertainty for complex NPS pollution models and attribute the uncertainty to individual sources is still a challenging scientific question. Proper stochastic techniques as well as case-specific strategies are desired to address the issue in different modeling cases. Some advances in this field are introduced in Section 5.3.

5.2.2 **Parameter uncertainty**

Process-based models for NPS pollution (e.g., SWAT, HSPF, WARMF) usually have hundreds of model parameters. Preparing any one of these models for simulation involves significant parameter uncertainty resulting from natural variability and/or imperfect knowledge. Sensitivity analysis (SA) is a typical measure to quantify the impact of parameter uncertainty on overall simulation/prediction uncertainty, and a variety of SA techniques have been developed (Helton, 1993; Saltelli et al., 2000). Nevertheless, parameter uncertainty has not been sufficiently studied with regard to the NPS pollution models, mainly due to the large number of parameters and high computational cost of the models.

Zheng and Keller (2006) investigated the parameter uncertainty of a WARMF model of diazinon pollution using generalized sensitivity analysis (GSA), a global SA technique based on "behavioral separation" of model simulations (Hornberger and Spear, 1981). A total number of 121 model parameters were investigated. For flow, sediment, and diazinon simulations, the number of involved parameters are $N_Q = 59, N_S = 97$, and $N_P = 121$, respectively. Eight management concerns were proposed, and each of them was formulated into a behavioral separation criterion. Three behavioral criteria were for flow (BQ-1, BQ-2, and BQ-3), three for sediment (BS-1, BS-2, and BS-3), and two for diazinon (BP-1 and BP-2). Despite the large number of uncertain parameters, only a few parameters, identified as the "sensitive parameters," significantly impact the behavioral separations (Figure 5.8). Simulation at a higher hierarchy level (e.g., diazinon simulation) may even involve less sensitive parameters (i.e., less complicated parameter uncertainty) than one at a lower hierarchy level (e.g., sediment simulation). An important conclusion from this study is that the parameter uncertainty is largely determined by the most influential processes for concerned output variables, and, in a management application context, is conditioned on specific management concerns as well.

Zheng et al. (2011) further studied dynamics of parameter uncertainty in modeling NPS pollution, using a probabilistic collocation method (PCM) based approach of SA, named PCM-VD. A WARMF model was built for the diazinon pollution in the NBW as introduced in Section 5.1.4. Figure 5.9 illustrates the temporal variation of first-order sensitivity index for three parameters. P3 is a precipitation weighting factor, ICB stands for the initial concentration of diazinon in river bed sediment, and BMP represents the street sweeping efficiency. As Figure 5.9 demonstrates, the parameter BMP took effect after the rain began, and its effect lasted for a while, since the sweeping could effectively reduce the pesticide load accumulated on street surface before the rain. ICB was the dominant source of parameter uncertainty before the rain, indicating that contaminated bed sediment was an important source of diazinon until a runoff event occurs. The spike of P3 in the first rain day reflects the first-flush effect of the small rainfall event. In the study area, small rainfall events usually led to high diazinon concentrations due to the synergy of significant washout and insufficient dilution. The SA not only revealed the complicated dynamics of parameter uncertainty, but provided a unique angle to scrutinize the model structure.

In an SA, probabilistic distribution or ranges of uncertain parameters should be defined *a priori*, based on best available knowledge on the parameters. The distributions or ranges may impact the uncertainty estimation. Uniform distribution is the most commonly adopted assumption. When observational data are available,

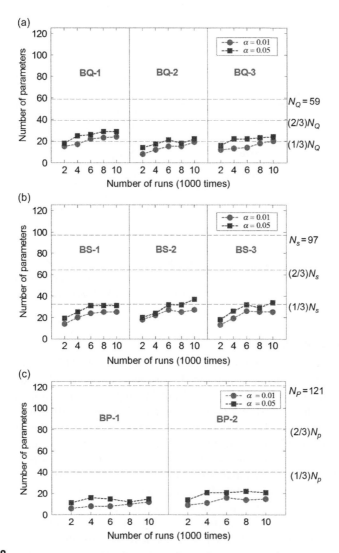

FIGURE 5.8

Number of sensitive parameters with regard to different behavioral separation criteria. α is the confidence level for the GSA analysis. (For the color version of this figure, the reader is referred to the online version of this chapter.)

This figure was adapted from Zheng and Keller (2006), Fig. 3, p. 7.

the distributions could be updated by formal or informal Bayesian approaches, such as Markov chain Monte Carlo (MCMC) (formal) and Generalized Likelihood Uncertainty Estimation (GLUE; informal). Advances in the Bayesian approaches are introduced in Section 5.3.

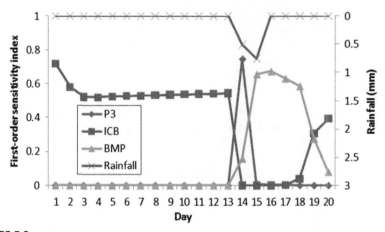

FIGURE 5.9

Temporal variation of parameter sensitivity indices. The secondary *Y* axis is for rainfall. (For the color version of this figure, the reader is referred to the online version of this chapter.)

This figure was adapted from Zheng et al. (2011), Fig. 9, p. 896.

5.2.3 Input uncertainty

Since inputs (per the definition in Section 5.2.1) for watershed NPS pollution models are dynamic and/or spatially distributed, it is much more difficult to quantify their uncertainty and its impact on the model simulation/prediction. Major types of input uncertainty in the context of NPS pollution modeling are briefly discussed below.

As NPS pollution is largely controlled by hydrology processes, the quality of meteorological forcing data like precipitation, wind speed, temperature are crucial to the model accuracy. In modeling watershed hydrology, precipitation is probably the most important source of input uncertainty (Maskey et al., 2004; Aronica et al., 2005; Ekström and Jones, 2009). Its impact on hydrology simulations could be further propagated into water quality simulations. For example, Shen et al. (2012) evaluated the impact of spatial rainfall variability on SWAT simulations in a large watershed. The spatial variability was characterized by different interpolation methods including Centroid, Thiessen Polygon, Inverse Distance Weighted (IDW), Dis-Kriging, and Co-Kriging. The choice of interpolation methods led to significant uncertainty in the rainfall inputs. The uncertainty was magnified by and propagated through the SWAT model, eventually resulting in significantly simulation uncertainty for water quality parameters. To explicitly consider the rainfall uncertainty, especially in a Bayesian framework, one common strategy is to introduce rainfall depth multipliers (Kavetski et al., 2003; Ajami et al., 2007). For example, if $\tilde{r}_j$ represents the true rainfall depth, and r_j is the observed rainfall depth, their input error model has the following form (Ajami et al., 2007):

$$r_j = \phi_j \tilde{r}_j; \ \phi_j \sim N\left(m, \sigma_m^2\right) \tag{5.4}$$

where j indicates the storms within the rainfall series and ϕ_j is a rainfall depth multiplier that follows the identical normal distribution with unknown mean m and variance σ_m^2. ϕ_j corrupts the true rainfall depth and yields the observed rainfall depth. m and σ_m^2 are the so-called latent variables or hyper-parameters added to the model system. By this mean, the input uncertainty is translated into the parameter uncertainty.

Pollutant source loading is another model input with substantial uncertainty. For example, NPS pollution models usually require timeseries data of fertilizer and pesticide application as forcing functions. Unfortunately, it is extremely difficult to achieve accurate information about the land application. Even if available, the data may be of very coarse time and spatial resolutions (Zheng and Keller, 2008). In modeling NPS pollution, the input uncertainty in source loading data would be almost linearly transferred into the model simulation/prediction uncertainty. How to characterize this input uncertainty has not been sufficiently discussed. In Zheng and Keller (2008), load adjustment multipliers for different land use types were proposed as the "latent variable," similar as ϕ_j in Equation (5.4). Uniform distributions were assumed for the random multipliers. In this way, the input uncertainty was translated into parameter uncertainty externally. Some models, like SWAT, have submodels for simulating land application based on the plant growth process. In such cases, the input uncertainty is converted to parameter uncertainty internally.

There are additional types of inputs that may considerably contribute to the overall simulation/prediction uncertainty. For example, Chaplot (2005) studied how the spatial resolution of DEM and soil map would affect SWAT simulations of flow, sediment, and nitrate loads at the outlet of an agricultural watershed. It found that when the Digital elevation model (DEM) was coarser than 50 m, the prediction errors of nitrogen and sediment yields were significant. A detailed soil map had to be used in order to guarantee the accuracy of the load calculation. Unfortunately, no systematical approaches have been developed to address such input uncertainty yet. To reduce its impact on water quality modeling, using the best available data is the general strategy.

5.2.4 Observational uncertainty

In watershed-scale studies, observational uncertainty resulting from measurement techniques is relatively small, if not negligible. Temporal variability and commensurability error of flow and water quality parameters are the dominant sources of observational error. For example, modeling in the TMDL context usually addresses daily average concentrations of pollutants for specific segments of water bodies. However, observations available for model calibration are most likely instantaneous measurements at discrete points. Continuous or composite sampling, temporally and/ or spatially, is rare in practice, since it is too resource-demanding even if technically feasible.

The concentrations of NPS pollutants are subject to significant diurnal fluctuation during storm events, which causes large observational error (e.g., Robinson et al., 2005). On base-flow days, due to the stable flow and reduced NPS loading, the observational error could be much smaller. As Table 5.1 suggests, the diurnal

Table 5.1 Diurnal Fluctuation of Pollutant Concentrations

Pollutant	Water Body	Range in a Typical Day	References
Methyl-mercury (ng/l)	Wetland	0.30–0.65	Naftz et al. (2011)
Nitrate (mg N/l)	River	0.8–1.1	Scholefield et al. (2005)
Phosphate (µg P/l)	River	160–250	Scholefield et al. (2005)
Ammonium (mg N/l)	River	0.00–0.10	Scholefield et al. (2005)
DO (mmol/l)	River	0.30–0.34	Guasch et al. (1998)
CO_2 (mmol/l)	River	0.03–0.05	Guasch et al. (1998)
Zn (µg/l)	Lake	0.25–1.25	Pokrovsky and Shirokova (2013)
Mn (µg/l)	Lake	10–30	Pokrovsky and Shirokova (2013)

fluctuation of concentration varies among different pollutants (Guasch et al., 1998; Scholefield et al., 2005; Naftz et al., 2011; Pokrovsky and Shirokova, 2013), and low-frequency (e.g., once per day) sampling could introduce substantial errors in estimating their daily average concentrations. On the other hand, many factors (e.g., source location, flow regime) could lead to the variability of water quality parameters over space. Measurements at one spot (or several unrepresentative spots) may introduce substantial observational uncertainty as well. The assumption of complete mixing is often invalid in reality, even at a very small scale. For instance, Rovira et al. (2012) investigated dynamics of suspended sediment (SS) in a semi-meandering cross-section under different hydrological conditions and found that the SS concentration varied significantly in the cross-section in high-flow periods.

In order to understand the *true* watershed response **Y**, the observational uncertainty ε has to be explicitly addressed. Nevertheless, as implied by Table 5.1, the mathematical form of ε could be highly case-dependent, and substantial monitoring efforts are required to characterize the uncertainty for various pollutants. Few studies have investigated this issue. Zheng and Keller (2007b) developed a simple error model for instream diazinon concentration, which is briefly introduced below.

From March 1999 to August 2002, California Department of Pesticide Regulation monitored stream water for diazinon every month within Orange County, California. Several storm events were monitored at different sites with temporally intensified sampling. Up to six measurements were conducted through each storm (with a duration less than 24 h). Two storm events with relatively complete data were selected for analyzing the observational error under high-flow conditions. It was assumed that the sample means approximate the average daily concentrations, which are the management objective, and the sample ranges approximate the actual ranges. Data from seven monitoring sites were used to build a regression model (Figure 5.10). Apparently, in this case the error $\varepsilon(\cdot, t)$ is heteroscedastic and its variance (reflected by the range) increases with its mean. Additional assumptions include: (1) temporal variation was the dominant source of $\varepsilon(\cdot, t)$; (2) $\varepsilon(\cdot, t)$ was uniformly distributed in each day, and has no autocorrelation. Given the assumptions

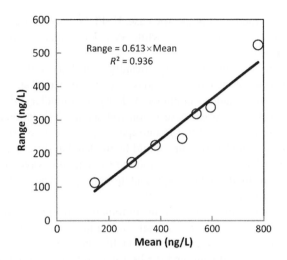

FIGURE 5.10

The regression model for the range and mean of diazinon concentration.

This figure was adapted from Zheng and Keller (2007b), Fig. 2, p. 6.

and the statistics in Figure 5.10, the error model for high-flow conditions can be written as

$$\varepsilon(t) = v \cdot Y(t) \tag{5.5}$$

where $\varepsilon(t)$ and $Y(t)$ represent observational error and *true* average daily concentration on day t, respectively; and v is a variable uniformly distributed within the interval $[-0.613/2, 0.613/2]$. Because $Z(t)$ (instantaneous observation on day t) is the sum of $Y(t)$ and $\varepsilon(t)$, it can then be derived that

$$Y(t) = \frac{Z(t)}{1+v} \tag{5.6}$$

It is worth emphasizing that observational uncertainty could be highly case-dependent, and the above error model only represents a specific case.

5.2.5 Model structure uncertainty

For a specific watershed case, the performance of different models may significantly vary, which embodies the model structure uncertainty. For example, Boomer et al. (2013) used six models to simulate water, nitrogen, and phosphorus discharges to the Patuxent Estuary. The results showed that no model was consistently superior to others, and the simulation results of the models could differ as much as 150%. In addition, models that agreed best with the observations in one basin (or for one pollutant) were often among the worst in another basin (or for another pollutant). Saleh and Du (2004) compared the performance of SWAT and HSPF at the Upper North

Bosque River watershed. In this specific case, HSPF better reproduced the temporal variations of daily flow and sediment, while SWAT provided better predictions for daily and monthly nutrient loads.

Model structure uncertainty can significantly contribute to the overall accuracy of model simulations (Montanari and Baldassarre, 2013). Model structure includes a whole range of choices and assumptions made by the modeler either explicitly or implicitly, among which process description and spatial delineation are two major ones. Process description can differ in multiple ways: A process could be included or excluded from the model structure, could be modeled at different degrees of approximation (e.g., kinematic wave flow routing vs. dynamic wave flow routing), or could have different mathematical representations (e.g., linear reservoir vs. Darcy flow equation in groundwater simulation).

Butts et al. (2004) compared different combinations of process descriptions and spatial delineation strategies within a single modeling platform. For the study area (the Blue River Basin), the variation in modeling performance among the tested model structures was significant, and the best performance was achieved by using dynamic wave routing and a grid-based delineation scheme. It was found that an increase in complexity of model structure did not necessarily result in an improved performance. For example, the dynamic wave routing did not outperform the Muskingum-Cunge routing if MIKE11 NAM was used for rainfall-runoff simulation. Also, there appeared to be no straightforward relationship between spatial resolution of watershed delineation and accuracy of model simulations. Cho et al. (2008) studied the impact of spatial delineation regime on the flow simulation by the Agricultural Nonpoint Source (AGNPS) model, using two small agricultural watersheds in Korea as the case studies. By using irregular cells instead of the uniform cells currently in AGNPS, the number of cells necessary for the modeling was significantly reduced and better agreement was achieved for surface runoff and peak flow rate.

At present, fully accounting for model structure uncertainty in modeling watershed NPS pollution is not feasible even in theory, simply because a "perfect" model structure does not exist. To mitigate the influence of model structure uncertainty, using multiple model structures to compensate each other has become a popular strategy (Duan et al., 2007; Ramin et al., 2012).

5.2.6 Uncertainty about the future

Change in linkage (i.e., linkage uncertainty) and future input uncertainty (Figure 5.7) should be taken into consideration if the modeling is for the purpose of planning. Linkage uncertainty implies that the watershed may behave differently in the future with regard to the calibration period, even if all driving forces are kept the same. It essentially reflects either the model structure deficit (i.e., the model's failure to represent certain processes that take effect under the future conditions) or parameter uncertainty (i.e., physical characteristics of the watershed system significantly changed in the future). To address this type of future uncertainty, innovation of model structure would be the ultimate solution. For example, additional processes may be introduced into the model structure, and/or submodels may be developed to replace originally time-invariant parameters. Such innovation has to be based on improved understanding about the pollution process.

Future input uncertainty is much harder to characterize than the input uncertainty in calibration period, since the former involves extrapolation, rather than interpolation. Forecasting the input data may require other complex models, like a weather prediction model or a land use changing model, and the uncertainty of such forecasts itself would be substantial and difficult to quantify. One alternative to account for the future input uncertainty is to do scenario analysis, that is, designing multiple future scenarios and running the model, respectively. The scenario design would be an *ad hoc* activity. For management-related inputs such as land application of fertilizers and pesticides, it is relatively easy to design scenarios. But for natural driving forces like precipitation, the scenario design needs a scientific basis. The modeler has to first identify the most relevant features of the concerned inputs (e.g., annual rainfall depth, storm intensity and frequency, impervious area), with regard to specific model outputs and then propose future scenarios based on these features. For instance, in our previous studies on the diazinon pollution in NPB, it has been found that non-attainment frequencies of different water quality targets were significantly correlated with number of storms (NS). In this case, NS would be a good index for designing future scenarios of rainfall. Moreover, if long enough historical rainfall data are available, the approach of hydrologic statistics can be applied to derive the probability distribution of NS, which gives a quantitative description about the future input uncertainty. Nevertheless, studies in this direction have been very rare.

5.3 UNCERTAINTY ANALYSIS FOR COMPLEX NPS POLLUTION MODELS

A number of general methods for UA are available, among which MCS-based approaches are considered to be the most applicable for complex environmental models including watershed NPS pollution models (Helton, 1993; Zheng and Keller, 2007a). The feasibility of other methods is seriously limited by the nonlinearity, discontinuity, analytical intractability, and complicated parameter interactions within complex models. In addition, MCS-based approaches are easy to apply since they treat a model as a black box and require no modifications of the model itself. MCS coupled with Bayesian inference is now the mainstream of UA for watershed models. Nevertheless, many challenges remain to be addressed.

The first challenge is to explicitly consider all sources of uncertainty. MCS can be straightforwardly implemented for the parameter uncertainty, but not for the uncertainty associated with model inputs, model structure, observational data, and future conditions. Moreover, how to separate the effect of individual uncertainty sources remains an open question. The second challenge is to develop appropriate error models in a formal Bayesian analysis. For formal Bayesian approaches like MCMC, determining the mathematical forms of the lumped error (i.e., $\boldsymbol{\delta} + \boldsymbol{\varepsilon}$ per Equation 5.1) or the individual ones (i.e., $\boldsymbol{\delta}$ and $\boldsymbol{\varepsilon}$) is critical to the uncertainty estimation, but is still very arbitrary. The third challenge is to reduce the computational cost of MCS. For large watershed models, one single run may take minutes or even hours (e.g., for those with detailed groundwater simulation) to finish. The computational burden prevents a rigorous UA from such models, especially when the models are used for decision support purpose, which unfortunately is often the case. The fourth challenge is data limitations. Compared to flow data, water quality monitoring data are more difficult and expensive to achieve and usually of lower quality in terms of frequency, accuracy, and consistency. Without sufficient high-quality data, the advantage of Bayesian inference would be significantly diminished. The advances introduced below addressed one or more challenges mentioned above.

5.3.1 MCMC approaches

MCMC is a specific type of MCS, which constructs a Markov chain that has, at least in an asymptotic sense, the desired posterior distribution as its stationary distribution (Gilks et al., 1996; Gelman et al., 2004; Robert and Casella, 2004). MCMC approaches incorporate Bayes' rule and have been a popular choice for uncertainty quantification. A general form of Bayes' rule can be written as

$$p(\theta|Z) \propto p(Z|\theta)p(\theta) \tag{5.7}$$

where θ represents one realization of random variable(s) (e.g., uncertain model parameters); Z stands for information (e.g., observed model outputs); $p(\theta|Z)$ and $p(\theta)$ are the posterior and prior distributions of θ with and without the information Z, respectively; and $p(Z|\theta)$ is a likelihood function.

The Metropolis–Hastings (MH) algorithm (Hastings, 1970) is the basic element of many existing MCMC approaches. The transition kernel of the MH algorithm, denoted as $K(\theta, \theta')$, can be expressed as the multiplication of a proposal distribution $q(\theta, \theta')$ and an acceptance probability $\alpha(\theta, \theta')$. Here, θ represents the current state of the Markov chain, and θ' is a proposed candidate state. $q(\theta, \theta')$ represents the proposal density of θ' from θ, and $\alpha(\theta, \theta')$ is the probability for accepting the candidate. The typical procedure of an MH-based MCMC is demonstrated in Figure 5.11. The Bayesian inference step in the procedure is illustrated by Figure 5.12. e represents a lumped error term, that is, $e = \boldsymbol{Z} - \boldsymbol{y}$ per the notation of Equation (5.1), and φ denotes the hyper-parameter(s) (e.g., mean, variance) for the error model.

Proposal distribution $q(\theta, \theta')$ is the critical component of an MCMC approach. One common proposal scheme considers a multivariate normal distribution (Kuczera and Parent, 1998; Gelman et al., 2004). There are no general rules to specifying the covariance matrix of $q(\theta, \theta')$. Different strategies have been developed to determine the covariance matrix, such as the adaptive Metropolis (Haario et al., 2001), the delayed rejection adaptive Metropolis (DRAM) (Haario et al., 2006), the shuffled complex evolution Metropolis algorithm (SCEM-UA) (Vrugt et al., 2003). Strategy of tuning the covariance matrix using a limited-memory multiblock

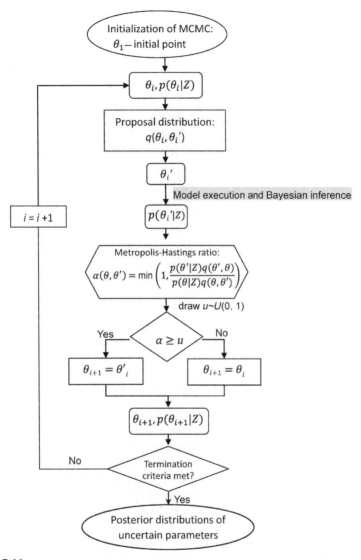

FIGURE 5.11

Schematic diagram of a typical Metropolis-Hastings-based MCMC approach. (For the color version of this figure, the reader is referred to the online version of this chapter.)

presampling step was also proposed (Kuczera et al., 2010). Another proposal scheme adopts the Differential Evolution algorithm. ter Braak (2006) developed the Differential Evolution Markov Chain (DE-MC). DE-MC runs multiple chains simultaneously in parallel and the states of all chains at one stage define a population. Candidates are generated as follows:

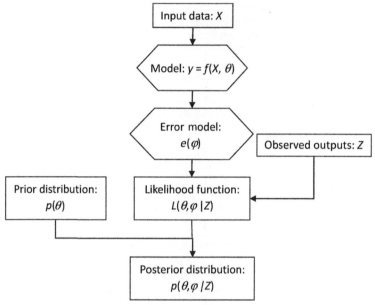

FIGURE 5.12

The Bayesian inference step in a Metropolis–Hastings-based MCMC approach.

$$\mathbf{x}_p = \mathbf{x}_i + \gamma(\mathbf{x}_{R1} - \mathbf{x}_{R2}) + \mathbf{E} \tag{5.8}$$

where $\mathbf{x}_i$ and $\mathbf{x}_p$ are the current and proposal points of the ith chain, respectively, $\mathbf{x}_{R1}$ and $\mathbf{x}_{R2}$ are two random points from the current population, $\mathbf{E}$ is drawn from a symmetric distribution with a relatively small variance, and γ is a scale factor. Metropolis ratio is used to determine whether a candidate point is accepted. To enhance the searching efficiency, many variants of the DE-MC algorithm have been proposed, including the differential evolution adaptive Metropolis (DREAM) (Vrugt et al., 2008, 2009), DREAM$_{(ZS)}$ (ter Braak and Vrugt, 2008), and MT-DREAM$_{(ZS)}$ (Laloy and Vrugt, 2012).

Determining the error model is another key step of using MCMC for uncertainty qualification. In hydrology studies, mutually uncorrelated Gaussian errors are a common assumption for the lumped residual errors e in Figure 5.12 (e.g., Campbell et al., 1999; Marshall et al., 2004; Vrugt et al., 2006; Ajami et al., 2007; Duan et al., 2007; Samanta et al., 2008; Hsu et al., 2009; Smith and Marshall, 2010), in which case the likelihood function $L(\theta, \varphi|Z)$ is $\prod_{i=1}^{n} \frac{1}{\sqrt{2\pi\sigma_e^2}} \exp\left(-\frac{(Z_i - y_i)^2}{2\sigma_e^2}\right)$, where σ_e^2 is the variance of e (i.e., φ per the definition of Figure 5.12) and n is the number of observations. Nevertheless, residual errors are often non-Gaussian, nonstationary, and autocorrelated (e.g., Kuczera, 1983; Bates and Campbell, 2001; Schaefli et al., 2007; Schoups and Vrugt, 2010). For example, Schoups and Vrugt (2010) introduced a heteroscedastic, autocorrelated error model with a skewed exponential power distribution to describe residuals in hydrologic modeling. In this error model, the heteroscedasticity and autocorrelation of errors are accounted for by a linear model and a pth-order autoregressive model, respectively. The likelihood function derived from the error model was in a logarithmic form, with up to 11 parameters, and hereafter referred to as the "Generalized Likelihood". Recently, some algorithms have been developed to explicitly consider different sources of uncertainty in a formal Bayesian inference, such as the integrated Bayesian uncertainty estimator (Ajami et al., 2007), the Bayesian total error analysis (Kavetski et al., 2006; Kuczera et al., 2006; Thyer et al., 2009; Renard et al., 2010, 2011), the Framework for Understanding Structural Errors (Clark et al., 2008; Clark and Kavetski, 2010), and some others (e.g., Vrugt et al., 2008; He et al., 2011; Li et al., 2012).

Although MCMC approaches have been widely used in hydrological modeling, few studies have applied the approaches to NPS pollution models. Raat et al. (2004) applied the SCEM-UA algorithm to a semi-distributed nutrient model (INCA) for a small synthetic catchment. The catchment was assumed to be homogeneous and only four model parameters were considered random. The study also assumed a Gaussian error model. Overall, the performance of MCMC approaches for complex NPS pollution models has not been systematically evaluated. This section introduces a case study in which the DREAM algorithm (Vrugt et al., 2008, 2009) was tested on a SWAT model of the nitrate pollution in the NBW (discussed in Section 5.1.4). DREAM was developed out of the DE-MC algorithm and has four main modifications that improve the searching efficiency, including: generating candidate points using more than two parent points, generating candidate points from a randomized subspace with respect to a vector of probability CR, updating "outlier chains" to accelerate convergence, and adaptively choosing the CR values. More technical details about DREAM can be found in Vrugt et al. (2008, 2009). DREAM has been shown to be a reliable and efficient technique in hydrologic modeling. In this case study, both the autocorrelated, heteroscedastic, and non-Gaussian error model (Schoups and Vrugt, 2010) and the Gaussian error model were tested.

In the SWAT model, the watershed was delineated into 11 sub-basins with a total number of 59 hydrological response units (HRUs). The simulation period is from 01 July 1997 to 30 June 2005. The first 3 years were treated as the warm-up period. Data of DEM, river network, soil, land use, meteorology, and fertilizer usage were collected and fed into the model. Observations of instream nitrate concentration at the outlet of the watershed were also obtained. From July 2000 to June 2004, a total number of 217 observations were available, roughly one observation per week. To better test the performance of DREAM, synthetic observations were also generated by corrupting a "true" model prediction (generated by assuming a "true" parameter set) with an artificial noise (autocorrelated, heteroscedastic, and non-Gaussian). A preliminary SA using Morris screening (Morris, 1991; Campolongo et al., 2007) was conducted to rank the model parameters. Up to 19 most sensitive parameters were considered as uncertain (i.e., random). Uniform prior distributions were assumed for the uncertain parameters. The parameter ranges were determined based

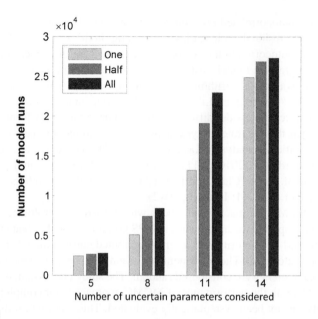

FIGURE 5.13

The number of model runs before the Markov chains converge. "One," "half," and "all" indicate the first one, a half, and all of the uncertain parameters meeting the convergence criterion, respectively. In this experiment, the synthetic observations and the non-Gaussian error model were used.

on the SWAT manual and literature. Relatively wide *a priori* ranges were considered in the numerical experiments. Some results are presented below.

Figure 5.13 shows the convergence speed of Markov Chains given different number of uncertain parameters. $\hat{R}$ statistic by Gelman and Rubin (1992) was used for convergence diagnosis. Markov chains are deemed to reach convergence if $\hat{R} < 1.2$ for all random variables. It shows that the convergence speed dramatically decreased as the number of uncertain parameters increased. This was not only due to the increased dimension of the parameter space, but also because the interaction and equifinality among parameters became more prominent. Figure 5.14 demonstrates that in the case of 19 uncertain parameters, none of the parameters reached convergence even the number of model evaluations was approaching 100,000. The implication here is that, although MCMC approaches like DREAM could be efficiently implemented for hydrological models, their application to NPS pollution models may encounter the problem of computational cost. In general, NPS pollution models consider more processes and have more parameters than hydrological models, and their computational cost is higher. For example, the SWAT model used in this case study is quite simple, but still requires 13 s per model run with a high-performance quad-core processor. To finish a 100,000-run DREAM analysis, it requires around 15 days if no parallel computing is used. When the modeling and UA are aimed to

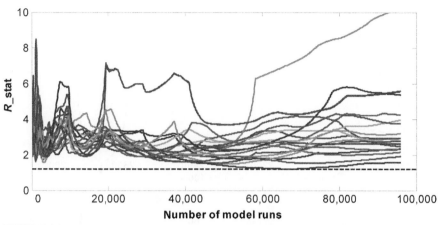

FIGURE 5.14

Dynamics of the $\hat{R}$ (R_stat) values of 19 uncertain parameters (each parameter is represented by a colored line in the figure). The dashed line represents the convergence criterion $\left(\hat{R} = 1.2\right)$. (For the color version of this figure, the reader is referred to the online version of this chapter.)

support management decisions, such a computational burden is rarely affordable in practice. Thus, the existing MCMC approaches need to be improved and tailored for NPS pollution models. Future studies are highly desired in this direction.

Figure 5.15 illustrates the parameter uncertainty band (the darker and narrower one) and the predictive uncertainty band (the lighter and wider one, which takes into account the lumped error e). For both bands, the upper and lower limits represent 97.5% and 2.5% confidence levels, respectively. As the two subplots on the bottom show, for both base flow and peak flow periods, the stochastic simulation with predictive uncertainty well embraces the observations (synthetic), and the dynamics of nitrate concentration is also reproduced by the stochastic simulation. Figure 5.16 presents the uncertainty results with a Gaussian error model instead. First of all, the predictive uncertainty band turns narrower and fails to embrace most of the observations during the peak flow period. Second, the parameter uncertainty becomes more significant. This comparison tells that it is critical to have an appropriate error model in an MCMC analysis.

Figure 5.17 demonstrates the results of the DREAM analysis considering real nitrate observations. Although the predictive uncertainty band well embraces the real observations in general, the dynamics of the nitrate concentration is not well reproduced either by the predictive uncertainty band or the parameter uncertainty band. This mostly reflects the model structure uncertainty and input uncertainty that were not well represented by the error model. Thus, to apply MCMC approaches to complex NPS pollution models, appropriate error models need to be developed. The mathematical form of the error model could be highly pollutant and/or watershed dependent though. Further studies are needed in this direction.

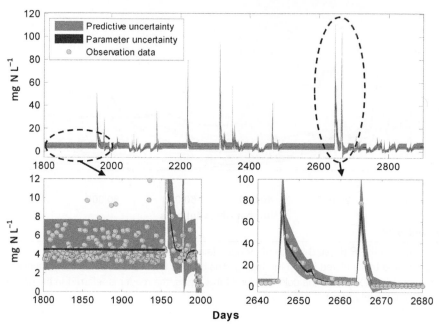

FIGURE 5.15

Parameter and overall predictive uncertainty bands generated by DREAM, assuming an autocorrelated, heteroscedastic, and non-Gaussian error model.

5.3.2 Informal Bayesian approaches

MCMC approaches represent a formal Bayesian inference, since the likelihood function is statistically rigorous given the error model $e(\varphi)$. Nevertheless, the approaches relies strongly on the statistical assumptions of $e(\varphi)$, which are inherently difficult to satisfy in practical applications (Beven et al., 2008). Generalized Likelihood Uncertainty Estimation (GLUE) (Beven and Binley, 1992) represents a different category of UA approaches that adopts informal Bayesian inference instead. The GLUE approach is based on the concept of equifinality (Beven and Freer, 2001; Beven, 2006) of parameter sets and/or model structure in providing acceptable (i.e., behavioral) fits to observational data. Like other MCS-based UA approaches, GLUE deals with parameter uncertainty explicitly. Basically, observational error ε and model structure uncertainty δ are implicitly considered in GLUE. Although GLUE allows evaluation of multiple model structures, in practice at most a few structures can be considered concurrently, and therefore only a small proportion of δ can be treated directly.

In a GLUE analysis, a number of candidate parameter sets are first identified through MCS. Next, the Bayes' rule (Equation 5.7) is applied to update the prior probability of each parameter set. All parameter sets meeting the predefined behavioral criterion (i.e., threshold) are then retained as behavioral sets. Finally, posterior

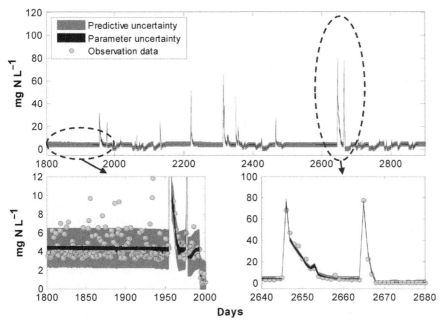

FIGURE 5.16

Parameter and overall predictive uncertainty bands generated by DREAM, assuming a Gaussian error model.

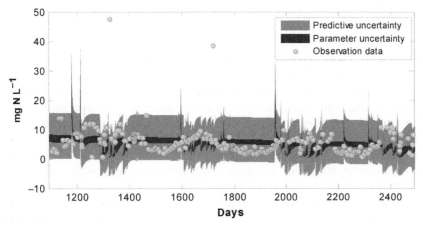

FIGURE 5.17

The DREAM analysis considering the real nitrate observations.

probability weighted simulations generated with the behavioral parameter sets can be used to calculate percentile-based uncertainty limits at each time step, and dynamic uncertainty bounds can be consequently constructed. In a typical GLUE application, the error terms are not explicitly modeled, and the likelihood function is in a form of goodness-of-fit measure that actually measures the *degree* (not probability) to which a model simulation *y approximates* observation Z. This is the main reason why GLUE has been deemed as an informal Bayesian inference.

GLUE has been applied to a variety of models in different research fields. The main advantages of this method include the following: First, it accounts for all sources of uncertainty, either explicitly or implicitly; second, it is conceptually simple and requires no restricted error assumptions if a goodness-of-fit measure is used as its likelihood function. However, GLUE has been frequently criticized for its subjective choices (e.g., likelihood measures, behavioral criteria), because these choices may significantly affect the uncertainty estimation (Montanari, 2005). The effect may be amplified in watershed NPS pollution modeling, since the uncertainty associated with model structure, inputs, and observational data is more substantial and complicated. Recently, a limited number of studies reported applications of GLUE to NPS pollution models such as SWAT (Muleta and Nicklow, 2005; Pohlert et al., 2007; Gong et al., 2011; Kemanian et al., 2011), WARMF (Zheng and Keller, 2007a), and INCA (Rankinen et al., 2006; Dean et al., 2009). It has been found that GLUE's performance could depend on available data and its subjective choices of likelihood measures and behavioral criteria (Zheng and Keller, 2007a; Dean et al., 2009; Gong et al., 2011). Zheng and Keller (2007a) also showed that potential management decisions would be significantly biased if inadequate choices were made and/or unrepresentative observations were used in a GLUE analysis.

Zheng and Keller (2007b) developed an informal Bayesian approach specifically for management-oriented watershed NPS pollution modeling. The approach, named MOCAU, inherits GLUE's equifinality ideology, while explicitly considering management objectives and observational uncertainty. Figure 5.18 shows the procedure of MOCAU.

In MOCAU, model simulations are constrained by a set of management objectives, rather than directly by the original observations **Z**. The objectives are mathematically represented by *management variables* (a management variable is denoted as *M*), which are metrics like nonattainment frequency (NAF) (the frequency or percentage time with which the concentration timeseries exceeds a given target) or severity of exceedance. Management variables can be evaluated for either **y**, **Y**, or **Z** (i.e., $M(\mathbf{y})$, $M(\mathbf{Y})$, and $M(\mathbf{Z})$), per the notation in Equation (5.1). Defining management variables is problem-dependent and inevitably involves subjectivity. However, unlike traditional goodness-of-fit metrics, management variables in MOCAU have clear physical and managerial meanings. Thus, the uncertainty estimates generated by MOCAU are conditioned on explicit management interests instead of on goodness-of-fit measures that are much less interpretable. MOCAU requires that the probability distribution of $M(\mathbf{Y})$ be determined separately, with explicit assumptions about the observational error $\boldsymbol{\varepsilon}$. Then, in an MCS regime, $M(\mathbf{y}(\boldsymbol{\theta}_i))$

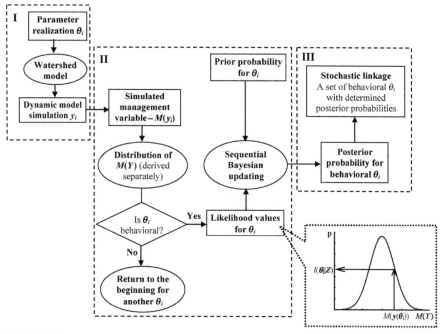

FIGURE 5.18

MOCAU procedure. The graph in the lower right callout illustrates the determination of the likelihood values. Stage I: Monte Carlo simulation of the watershed model; Stage II: Bayesian analysis; Stage III: uncertainty characterization. (For the color version of this figure, the reader is referred to the online version of this chapter.)

This figure was reproduced from Zheng and Keller (2007b), Fig. 1, p. 3.

for each θ_i is compared against the distribution of $M(\mathbf{Y})$ to derive the likelihood value. More details about MOCAU can be found in Zheng and Keller (2007b).

Based on the same modeling case in Zheng and Keller (2007a), the performance of MOCAU was examined and compared against that of GLUE. Figure 5.19 displays the calculated uncertainty intervals for the management variable NAF_1 (i.e., NAF in a 3-year period). It shows that the GLUE analysis resulted in a much wider interval for $NAF_1(\mathbf{y})$, and the central tendency of $NAF_1(\mathbf{y})$ using the specific GLUE analysis does not match the central tendency of either $NAF_1(\mathbf{Y})$ or $NAF_1(\mathbf{Z})$. This highlights a desirable feature of MOCAU: internally (which means no significant trial-and-error work is necessary for identifying appropriate likelihood measures and behavioral criteria) reducing both bias and magnitude of the uncertainty with regards to specific management concerns. In addition, MOCAU enables comparison of uncertainties with different origins. Let W represent the width of NAF_1's uncertainty interval, and subscripts Y, M, G indicate "$NAF_1(\mathbf{Y})$," "MOCAU," and "GLUE," respectively. The ratios W_Y/W_M and W_Y/W_G are metrics of the weight of the observational uncertainty in the global uncertainty. For example, in Figure 5.19a W_Y/W_M and W_Y/W_G are

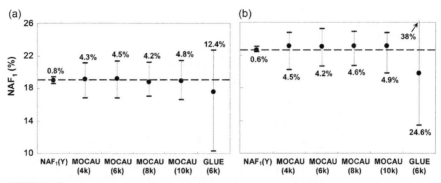

FIGURE 5.19

MOCAU uncertainty intervals (between 5th and 95th percentile) for overall nonattainment frequency (NAF_1) for (a) CMC and (b) CCC. The circles are medians, the numbers above or below the bars are the interval widths (W), and the dashed lines indicate the NAF_1 evaluated for observation Z (i.e., $NAF_1(Z)$). The first point in each x-axis is the uncertainty interval for the simulated $NAF_1(Y)$, which solely reflects the magnitude of ε. Other points along the x-axis are the intervals for $NAF_1(y)$, the magnitude of the global uncertainty for different numbers of Monte Carlo runs.

This figure was reproduced from Zheng and Keller (2007b), Fig. 5, p. 10.

about 0.18 and 0.06, respectively. W_G is significantly larger than W_M due to the greater parameter uncertainty estimated by the GLUE analysis.

MOCAU has unique features specifically desirable for management-oriented watershed water quality modeling, including: (1) it separates observational error ε and treats the error explicitly, and "equifinality" is extended to ε; (2) it focuses on the true watershed response (instead of an error-contaminated observation), which is of real management interest; (3) it uses probability-based likelihood measures that assimilate physical knowledge about ε; and (4) it directly factors management concerns into likelihood estimation and behavioral separation. Hence, it has an internal mechanism to ensure uncertainty results are adequate from the management point of view, as long as the model settings are appropriate. One major implication of the MOCAU study is that the modeling process should not be independent of management concerns; rather, management concerns should be an integrated part of the modeling and its uncertainty assessment.

5.3.3 Probabilistic collocation method (PCM)

For complicated water quality issues, the modeling process usually involves iterations of trial and error. In addition, if data for modeling (e.g., observed water quality, land applications) are sparse, further data collection efforts would be necessary. It is therefore critical to streamline the whole modeling process such that adequate simulation results could be achieved with affordable costs. Zheng et al. (2011) proposed a framework for watershed water quality modeling under data-scarce conditions

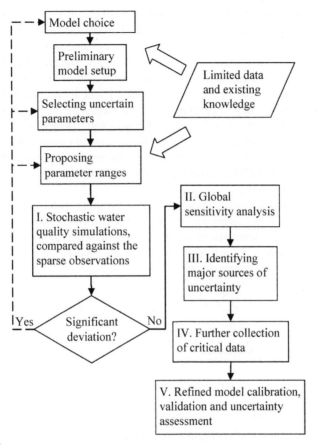

FIGURE 5.20

An integrated framework for watershed water quality modeling under data-scarce conditions.

This figure was reproduced from Zheng et al. (2011), Fig. 1, p. 888.

(Figure 5.20). "Data-scarce conditions" here specifically refers to the situations in which real observations for water quality are sparse. Given the data limitation, full-blown model calibration, validation and uncertainty assessment (Step V in Figure 5.20) may not be feasible at the beginning. Instead, stochastic simulations (Step I) can be practiced after preliminary model setup is finished and parameter uncertainty is considered. If the spread of stochastic simulations does not adequately embrace sparse observations, the modeler should check whether the problem is due to inappropriate parameter ranges, important parameters missed, poor preliminary model setup, or even bad model choice; otherwise, the preliminary model setup and uncertainty assumptions could be trusted, and a global sensitivity analysis (Step II) should follow, through which major sources of uncertainty may be identified

(Step III). Further data collection should then be strategically performed (Step IV) to support refined model calibration, validation, and uncertainty assessment (Step V).

As a pivot of the entire procedure, Step I need be repeated as necessary, and therefore requires a computationally efficient approach of UA. The Step II is critical to the reduction of modeling error, and a quantitative approach of affordable computing expense is highly desired. In addition, as Steps I and II are finished under the data-scarce conditions, Bayesian inference could hardly be applied in the UA. As for the Step V, approaches like Ensemble Kalman filter (EnKF) (Xie and Zhang, 2010) and SCEM-UA (Vrugt et al., 2003), which unify data assimilation and uncertainty assessment, would be appropriate choices. The research introduced in this section demonstrates that PCM is an option to efficiently streamline UA and SA under the data-scarce conditions.

PCM is accomplished by approximating a model output with a Polynomial Chaos Expansion (PCE) in terms of random inputs. The unknown coefficients contained in the expansion can be determined based on model simulations at selected collocation points (each collocation point is a realization of the random inputs). With PCE, a random output y can be expressed as

$$y = a_0 + \sum_{i_1=1}^{k} a_{i_1} \Gamma_1\left(\xi_{i_1}\right) + \sum_{i_1=1}^{k} \sum_{i_2=1}^{i_1} a_{i_1 i_2} \Gamma_2\left(\xi_{i_1}, \xi_{i_2}\right)$$
$$+ \sum_{i_1=1}^{k} \sum_{i_2=1}^{i_1} \sum_{i_3=1}^{i_2} a_{i_1 i_2 i_3} \Gamma_3\left(\xi_{i_1}, \xi_{i_2}, \xi_{i_3}\right) + \cdots \tag{5.9}$$

where $\Gamma_p\left(\xi_{i_1}, \cdots, \xi_{i_p}\right) = (-1)^p e^{(1/2)\boldsymbol{\xi}^T \boldsymbol{\xi}} \frac{\partial^p}{\partial \xi_{i_1} \partial \xi_{i_2} \cdots \partial \xi_{i_p}} e^{-(1/2)\boldsymbol{\xi}^T s}$ denotes p-order Hermite polynomial in terms of standard normal variables (SNVs) $\left\{\xi_{i_1}, \ldots, \xi_{i_p}\right\}$ ($\boldsymbol{\xi}$ is the vector form); k is the number of **different** SNVs; all the a's are expansion coefficients to be determined. Equation (5.9) is often rewritten into Equation (5.10), in which the subscript i is a sequence index.

$$y = a_0 + \sum_{i=1}^{\infty} a_i \Gamma_i(\boldsymbol{\xi}) \tag{5.10}$$

In many cases, a truncated PCE with only low-order terms could produce good approximations. The degree of a truncated PCE (denoted as d) is defined as the highest order of the Hermite polynomials it includes. For example, second-degree PCEs in two dimensions can be written as $\hat{y} = a_0 + a_1 \xi_1 + a_2 \xi_2 + a_3 \left(\xi_1^2 - 1\right) + a_4 \left(\xi_2^2 - 1\right) + a_5 \xi_1 \xi_2$, where $\hat{y}$ indicates the approximation due to truncation.

In fact, each $\Gamma_i(\boldsymbol{\xi})$ in Equation (5.10) can be deemed as a multiplication of univariate Hermite polynomials (UHPs) denoted as $H_p(\xi)$, that is, $\Gamma_i(\boldsymbol{\xi}) = \prod_{j=1}^{M_i} H_{p_{ij}}\left(\xi_{t_{ij}}\right)$, where M_i ($M_i \geq 1$) is the total number of UHPs in $\Gamma_i(\boldsymbol{\xi})$, $\xi_{t_{ij}}$ is the random input variable of the jth UHP in $\Gamma_i(\xi)$ ($t_{ij} \in \{1, 2, \ldots, k\}$), and p_{ij} is the order of the jth UHP in $\Gamma_i(\boldsymbol{\xi})$. For example, $\Gamma_5(\boldsymbol{\xi})$ in Equation (5.10) equals to $H_1(\xi_1) \cdot H_1(\xi_2)$. It can be derived that (Zheng et al., 2011) $E(y) = a_0$ and

$V(y) = \sum_{i=1}^{\infty} a_i^2 \prod_{j=1}^{M_i} p_{ij}!$. This implies that a random variable's expected value and variance can be simply treated as a function of its PCE coefficients and orders of the UHPs in its PCE.

Usually input variables X (i.e., uncertainty model parameters in the context of this chapter) are not SNVs (i.e., ξ), and sometime the variables may be correlated. In such cases, appropriate transformations are required. For example, a uniformly distributed variable $x \sim U(a, b)$ can be represented as $x = a + (b-a)(\frac{1}{2} + \frac{1}{2}erf(\frac{\sqrt{2}}{2}\xi))$ here ξ is an SNV. For a d-degree PCE in k dimensions, the total number of its expansion coefficients is equal to C_{k+d}^d. If cross terms (e.g., $\xi_1\xi_2$) are all neglected, the number of expansion coefficients can be reduced to $(kd+1)$. The selection of collocation points should follow a standard procedure (Tatang et al., 1997) in which points are selected so that each SNV takes the values of either zero or one of the roots of the higher order Hermite polynomial. Based on this procedure, the number of possible collocation points is $(d+1)^k$. An improved selection strategy, named efficient collocation method (ECM) (Isukapalli et al., 1998), gives priority to points that are closer to the origin, as they fall in regions of higher probability. If the number of selected points equals the number of expansion coefficients, a well-determined linear equation system can be obtained and solved for the coefficients. To improve the results for complex models, Isukapalli et al. (1998) suggested obtaining extra collocation points to form an over-determined equation system, and solve it using regression for the dependent variable expansion coefficients.

A new global SA approach integrating PCM with Sobol' indices was also developed by Zheng et al. (2011), and named PCM-VD. Based on high-dimension model representation proposed by Sobol', a variance decomposition for random variable y can be written as $V(y) = \sum_i V_i + \sum_i \sum_{j>i} V_{ij} + \cdots + V_{12\ldots k}$ (Saltelli et al., 2008), where V_i is the variance attributed to the single effect (i.e., first-order effect) of input x_i, and $V_i = V(E(y|x_i))$; V_{ij} is the variance attributed to the interaction effect (i.e., second-order effect) of x_i and x_j, and $V_{ij} = V(E(y|x_i, x_j)) - V(E(y|x_i)) - V(E(y|x_j))$; and higher-order variances are defined in a similar fashion. Dividing both sides of the decomposition equation by the unconditional variance $V(y)$, it can be obtained that $\sum_i S_i + \sum_i \sum_{j>i} S_{ij} + \cdots + S_{12\ldots k} = 1$, where each S on the left side is a sensitivity index. The first- and second-order indices can be expressed respectively as $S_i = V(E(y|x_i))/V(y)$ and $S_{ij} = [V(E(y|x_i, x_j)) - V(E(y|x_i)) - V(E(y|x_j))]/V(y)$, with higher-order sensitivity indices derived in a similar fashion. For x_i, its total-effect index can be written as $S_{T_i} = S_i + \sum_{j \neq i} S_{ji} + \sum_{j \neq i} \sum_{l>j} S_{ijl} + \cdots + S_{12\ldots i\ldots k}$. These indices $(S_{T_i}, S_i, S_{ji}, \ldots)$ are often referred as Sobol' sensitivity indices.

Suppose that a truncated PCE $\hat{y} = g(\xi)$ has been determined, and is a good approximation to the original model $y = f(X)$. We can then write

$$y|X^F \cong \hat{y}|\xi^F = a_0 + \sum_{i=1}^{n} a_i \Gamma_i(\xi)|_{\xi^F} = a_0 + \sum_{l=1}^{n_1} a_l \Gamma_l(\xi^F) + \sum_{m=1}^{n_2} a_m \Gamma_m(\xi^{NF}) \quad (5.11)$$

where the superscripts F and NF identify the input variables to be fixed and those not to be fixed, respectively. It can be further derived that

$$E(y|X^F) \cong E(\hat{y}|\boldsymbol{\xi}^F) = a_0 + \sum_{l=1}^{n_1} a_l \Gamma_l (\boldsymbol{\xi}^F) \tag{5.12}$$

$$V\left[E(y|X^F)\right] \cong V\left(E(\hat{y}|\boldsymbol{\xi}^F)\right) = \sum_{l=1}^{n_1} a_l^2 \prod_{q=1}^{M_l} p_{lq}! \tag{5.13}$$

$$V(y) \cong V(\hat{y}) = \sum_{i=1}^{n} a_i^2 \prod_{j=1}^{M_i} p_{ij}! \tag{5.14}$$

With Equations (5.13) and (5.14), all the Sobol' sensitivity indices $\left(S_{T_i}, S_i, S_{ji}, \ldots\right)$ can be directly estimated via algebraic expressions of the coefficients defining parametric uncertainty. Take the second-degree PCEs in two dimensions for example. We have

$$S_1 \cong \frac{V(E(\hat{y})|\xi_1)}{V(\hat{y})} = \frac{a_1^2 + 2a_3^2}{a_1^2 + a_2^2 + 2a_3^2 + 2a_4^2 + a_5^2} \tag{5.15}$$

$$S_2 \cong \frac{V(E(\hat{y})|\xi_2)}{V(\hat{y})} = \frac{a_2^2 + 2a_4^2}{a_1^2 + a_2^2 + 2a_3^2 + 2a_4^2 + a_5^2} \tag{5.16}$$

$$S_{12} \cong \frac{V(E(\hat{y}|\xi_1,\xi_2)) - V(E(\hat{y}|\xi_1)) - V(E(\hat{y}|\xi_2))}{V(\hat{y})}$$
$$= \frac{a_5^2}{a_1^2 + a_2^2 + 2a_3^2 + 2a_4^2 + a_5^2} \tag{5.17}$$

Also, the total-effect indices can be calculated as $S_{T_1} = S_1 + S_{12}$, and $S_{T_2} = S_2 + S_{12}$.

The developed approaches were applied to the WARMF model of diazinon pollution at the NBW (see Section 5.1.4). Some of the results are presented below. Figure 5.21 illustrates the cumulative distribution functions (cdfs) derived in different numerical experiments. In MCS-0, 10,000 sets of the 19 uncertain parameters were randomly sampled and the original WARMF model was run with all these sets. In MCS-2, a response surface was first constructed, using a PCM analysis with 300 collocation points, for the output variable d_{80} (i.e., number of days with diazinon concentration exceeding 80 ng/l in the simulation period). Ten thousand sets of the 19 parameters were randomly sampled and transformed into the standard normal space (i.e., the space of $\boldsymbol{\xi}$), and the response surface was then evaluated with the transformed parameter sets. MCS-3 is similar as MCS-2, except that the output variable is c_m (i.e., average diazinon concentration in the simulation period). As Figure 5.21 shows, the probabilistic features of the two output variables are well captured by the stochastic response surfaces, especially the central tendency. Small discrepancies exist in the tails of the distributions, which may be due to the experimental design of PCM. The collocation points in PCM define the high probability region in input parameter space, and thus PCM is typically good within the most probable range of values for the input parameters.

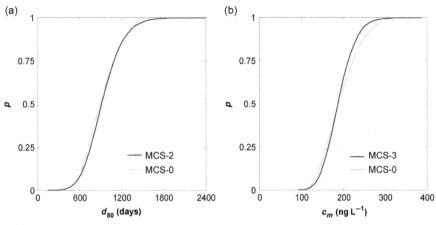

FIGURE 5.21

Comparison of the cumulative distribution functions (cdfs) derived in different numerical experiments. (a) is for d_{80} and (b) is for c_m. (For the color version of this figure, the reader is referred to the online version of this chapter.)

This figure was reproduced from Zheng et al. (2011), Fig. 5, p. 894.

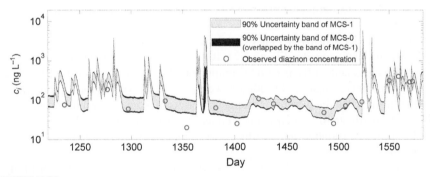

FIGURE 5.22

Comparison of the 90% uncertainty bands (between the 5th and the 95th percentiles) derived in different Monte Carlo simulations. The band of MCS-0 is slightly wider than that of MCS-1, as indicated by the dark color. (For the color version of this figure, the reader is referred to the online version of this chapter.)

This figure was reproduced from Zheng et al. (2011), Fig. 6, p. 895.

Figure 5.22 illustrates the stochastic simulations of c_i (daily instream diazinon concentration) by both experiments MCS-1 and MCS-0. MCS-1 is similar to MCS-2 and MCS-3, except that it constructs a response surface for each of the 1583 simulation days, and thus has 1583 stochastic response surfaces in total. For better illustration, the results are shown for only 1 year. The PCM-based simulation (MCS-1) approximates the original WARMF simulation (MCS-0) quite well.

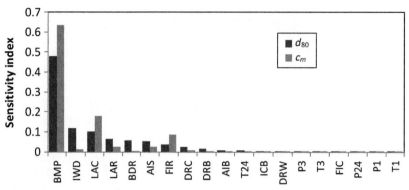

FIGURE 5.23

First-order sensitivity index values calculated by PCM-VD. (For the color version of this figure, the reader is referred to the online version of this chapter.)

This figure was reproduced from Zheng et al. (2011), Fig. 7, p. 895.

Figure 5.22 also includes a limited number of real concentration observations (the circles) for comparison. As we can see, the stochastic simulations adequately embrace the observations. This provides a preliminary validation of the model choice, model setup, uncertain parameter selection, and parameter ranges (Step I in Figure 5.20).

The first-order sensitivity indices of the 19 uncertain parameters calculated by PCM-VD are summarized in Figure 5.23. It shows that only a limited number of parameters have notable influence on the output variables, and d_{80} is sensitive to more uncertain parameters than c_m. In addition, we have $\sum_{i=1}^{19} S_i = 0.969$ for d_{80} and for c_m, which indicates that the parameter interaction effect (e.g., cross terms) does not significantly contribute to the output uncertainty in this modeling case. The sensitivity results provide insights into further data collection and model calibration (Steps III–V in Figure 5.20).

Figure 5.24 compares PCM-VD and Morris screening with respect to sensitivity ranking. Morris screening (Morris, 1991) is a classic SA technique. For both d_{80} and c_m, the ranking results by these two methods are very similar, as very few markers are more than two positions away from the 45° line in Figure 5.24. Thus, the feasibility of PCM-VD was well validated. Moreover, PCM-VD has advantages over Morris screening, including the following: (1) As a variance-decomposition based approach, PCM-VD captures the influence of the full range of variation of each input variable; (2) PCM-VD is able to calculate percentage contribution of individual input variables, as well as of interactions among multiple input variables, to the output uncertainty; (3) SA and UA can be fully integrated in an application of PCM-VD, as they share the same response surface. On the contrary, Morris screening is a stand-alone SA, which does not quantify the uncertainty of output variables; (4) PCM-VD requires no extra runs of the original complex model for SA, while Morris screening usually demands tens or hundreds of model runs.

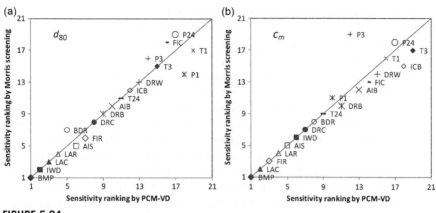

FIGURE 5.24

Sensitivity ranking by PCM-VD and Morris screening. (For the color version of this figure, the reader is referred to the online version of this chapter.)

This figure was reproduced from Zheng et al. (2011), Fig. 8, p. 896.

5.4 IMPROVING DATA AND MODEL STRUCTURE: FUTURE DIRECTIONS

Section 5.3 introduced advanced techniques for quantifying the uncertainty in modeling watershed NPS pollution. However, UA alone does not guarantee the reduction of uncertainty. Improving data and model structure would be the ultimate solution. This section briefly discusses the needs for better data and models, and points out some directions for future research.

5.4.1 Strategic data collection

For water quality issues, data collection (i.e., monitoring, field experiments, survey) is in general an expensive and time-consuming task. Modelers rarely have enough resources to collect data as desired. Thus, collecting data in a strategic way becomes critical. To identify best locations for monitoring/sampling, two strategies are often adopted. One is to do multiobjective optimization. Per this strategy, multiple criteria and their associated weights are defined first, and a certain searching algorithm is then applied to identify the best location(s) per the predefined criteria. For example, Park et al. (2006) used a genetic algorithm (GA) to design a network of water quality monitoring. The criteria included representativeness of a river basin, compliance with water quality standards, supervision of water use, surveillance of pollution sources, and examination of water quality changes/estimation of pollution loads. Each criterion was a simple evaluation function. Different weights were assigned to the criteria and the weighted sum was deemed as the final objective function. Ning and Chang (2004) performed a fuzzy multiobjective programming analysis to design water quality networks. The objectives in this study included percentage of nonattainment time period, mode of water utilization, distance required for a decay of half concentration of pollutant, population, distance between candidate monitoring stations, and water intake.

Another strategy is to maximize the uncertainty reduction resulted from adding new monitoring locations, which, however, has been mainly adopted by groundwater studies. This strategy requires stochastic simulation coupled with a searching algorithm. Herrera and Pinder (2005) used EnKF as the stochastic simulation approach to optimize a spatiotemporal design for groundwater sampling. The objective was to minimize the variance of a groundwater model's outputs. The searching strategy was to go through all possible locations. EnKF is a strictly nonintrusive and straightforward method that avoids the handling of vast autocovariance matrices. Zhang et al. (2005) also used EnKF to quantify the uncertainty of pollutant concentration simulated by a groundwater model, but a GA was adopted for the searching. Leube et al. (2012) applied the bootstrap filter (BF) method coupled with greedy search to find the optimal sampling design. The study showed that BF outperformed EnKF in their case, as the system was highly nonlinear. All these studies used number of monitoring wells as the constraint.

Timing of data collection has been much less discussed in the context of NPS pollution. Zheng and Keller (2007b) conducted numerical experiments to reveal the importance of strategic sampling design. Within the context of the diazinon TMDL in the NBW, nonstrategic sampling plans were compared against strategic sampling plans. In nonstrategic plans, water samples were taken twice a month on fixed dates, reflecting the routine monitoring behavior. In developing the strategic plans, the potential sampling days were first divided into groups based on the prior knowledge about the pollution dynamics. It has been shown that the strategic plans would substantially reduce the simulation uncertainty, and therefore lead to more reliable management decisions.

Optimization with regard to data types received few studies in the field of watershed water quality modeling. Value of Information (VOI), a well-established theory in economic and management sciences, has been recently introduced to addressing this optimization issue (Borisova et al., 2005; Wu and Zheng, 2013). Wu and Zheng (2013) proposed a framework for evaluating and comparing the VOI in watershed NPS pollution modeling and management. This study addressed a management scenario in which a water body does not meet its ambient-based water quality standard(s) or pollutant load limit(s), and therefore abatement is required for both point and nonpoint sources of pollution within its drainage area. This is typical of both U.S. TMDL and China's TLC regulations.

Overall, strategic data collection in the context of NPS modeling and management deserves further research. Critical issues to be systematically addressed include, but are not limited to, the following. First of all, to address real management issues, more efficient UA techniques are yet to be developed to make the stochastic optimization for data collection computationally feasible. Second, data collection cost, in addition to pollution abatement cost, needs to be factored into a cost-minimization analysis, as the cost is not trivial in NPS modeling and management. Third, an analysis framework integrating uncertainty quantification, data collection optimization and data assimilation for complex watershed-scale water quality models is highly desired but has never been attempted.

5.4.2 Process understanding and representation

NPS pollution models developed in 1970s and 1980s have been tremendously improved in the past 20 years, with regard to user friendliness, numerical efficiency, integration with GIS, and coupling with other environmental models. However, relatively slow progress has been achieved in gaining new understandings on the pollution process and representing them appropriately in the models. Some imperative research needs are briefly discussed below.

While many new categories of pollutants, such as Persistent Organic Pollutants, have been identified as critical threatens to human health, their NPS behaviors have been poorly understood and not adequately represented by existing models. The NPS modeling for such pollutants inevitably involves significant uncertainty. For example, Zheng et al. (2012) and Luo et al. (2013) revealed that the traditional enrichment theory for the transport of soil-bound chemicals by runoff, which has been widely adopted by current NPS pollution models, interprets the enrichment phenomenon from an inappropriate perspective. The theory is not valid for chemicals with strong hydrophobicity such as polycyclic aromatic hydrocarbons (PAHs), if the original soil has significant amount of anthropogenic carbon materials like black carbon (BC). These studies indicate that, as new pollution issues keep emerging, even classic theories and methods may need renovation. Great efforts are needed to further improve the existing NPS pollution models.

Atmospheric processes have been largely overlooked by current NPS pollution models. In the models, atmospheric pollutant loading to watershed is usually treated as external inputs for driving the model simulation, rather than internally simulated variables, and meteorological parameters only have a direct impact on hydrology. For pollutants of a significant atmospheric origin (e.g., urban road dusts, PAHs, mercury), this would introduce substantial uncertainty in the model simulation, since monitoring data for pollutant atmospheric loading are often of poor quality. Research efforts to address this issue have been rare. As one of the few attempts, Zhang et al. (2009) developed a model to simulate the buildup and washoff processes of dusts and PAHs on urban roads, which explicitly accounts for the dependence of the pollution level on weather conditions. More studies in this direction are highly desired.

Many semi-distributed models (e.g., SWAT) use HRUs as the basic computational units. Within an HRU, the model parameters have constant values. HRUs are defined based on land use conditions and hydrologic features of soil and mainly reflect the spatial variation of flow-generation capacity. Nevertheless, for pollutants whose fate and transport are significantly impacted by additional factors, using HRUs as the basic units would notably distort the spatial features of NPS pollution. For example, Luo et al. (2013) revealed that anthropogenic carbonaceous materials

(CMs) like BC components, if abundant in the soil, largely control the transport of soil-bound PAHs under rainfall runoff conditions. An important implication is that to appropriately model the NPS pollution of PAHs (possibly other hydrophobic chemicals as well), the content, mobility, and adsorption capacity of major CMs in soil have to be well understood. Thus, in modeling NPS pollution, watershed delineation based on water quality response units (WQRUs), which factor in the spatial variation of pollutant fate and transport, would be preferred over the traditional HRUs. Nevertheless, how to define a WQRU remains an open question.

In summary, the next generation of watershed NPS pollution models must incorporate recent and future advances in understanding the pollution process and reduce the model structure uncertainty by better representing the process in the models. Some important aspects have been discussed above, and more are to be further explored.

References

Ajami, N.K., Duan, Q.Y., Sorooshian, S., 2007. An integrated hydrologic Bayesian multimodel combination framework: confronting input, parameter, and model structural uncertainty in hydrologic prediction. Water Resour. Res. 43 (1), W01403. http://dx.doi.org/10.1029/2005WR004745.

Arnold, J.G., Srinivasan, R., Muttiah, R.S., Williams, J.R., 1998. Large-area hydrologic modeling and assessment: part I. Model development. J. Am. Water Resour. Assoc. 34 (1), 73–89.

Aronica, G., Freni, G., Oliveri, E., 2005. Uncertainty analysis of the influence of rainfall time resolution in the modelling of urban drainage systems. Hydrol. Process. 19 (5), 1055–1071.

Barreto-Neto, A.A., Filho, C.R.D., 2008. Application of fuzzy logic to the evaluation of runoff in a tropical watershed. Environ. Model. Softw. 23 (2), 244–253.

Bates, B.C., Campbell, E.P., 2001. A Markov chain Monte Carlo scheme for parameter estimation and inference in conceptual rainfall runoff modeling. Water Resour. Res. 37 (4), 937–947.

Beven, K., 2006. A manifesto for the equifinality thesis. J. Hydrol. 320, 18–36.

Beven, K.J., Binley, A., 1992. The future of distributed models-model calibration and uncertainty prediction. Hydrol. Process. 6 (3), 279–298.

Beven, K.J., Freer, J., 2001. Equifinality, data assimilation, and uncertainty estimation in mechanistic modeling of complex environmental systems using the GLUE methodology. J. Hydrol. 249, 11–29.

Beven, K., Smith, P.J., Freer, J.E., 2008. So just why would a modeler choose to be incoherent. J. Hydrol. 354, 15–32.

Bicknell, B.R., Imhoff, J.C., Kittle, J.L., Jobes, T.H., Donigian, A.S., 2001. Hydrological Simulation Program-FORTRAN: HSPF Version 12, User's Manual. National Exposure Research Laboratory, Office of Research and Development, U.S. EPA, Athens, GA.

Boomer, K.M.B., Weller, D.E., Jordan, T.E., Linker, L., Liu, Z.J., Reilly, J., Shenk, G., Voinov, A.A., 2013. Using multiple watershed models to predict water, nitrogen, and phosphorus discharges to the Patuxent Estuary. J. Am. Water Resour. Assoc. 49 (1), 15–39.

Borah, D.K., Bera, M., 2003. Watershed-scale hydrologic and nonpoint source pollution models: review of mathematical bases. Trans. ASAE 46 (6), 1553–1566.

Borisova, T., Shortle, J., Horan, R.D., Abler, D., 2005. Value of information for water quality management. Water Resour. Res. 41, W06004. http://dx.doi.org/10.1029/2004WR003576.

Bowes, M.J., Smith, J.T., Jarvie, H.P., Neal, C., 2008. Modelling of phosphorus inputs to rivers from diffuse and point sources. Sci. Total Environ. 395, 125–138.

Butts, M.B., Payne, J.T., Kristensen, M., Madsen, H., 2004. An evaluation of the impact of model structure on hydrological modelling uncertainty for streamflow simulation. J. Hydrol. 298 (1–4), 242–266.

Campbell, E.P., Fox, D.R., Bates, B.C., 1999. A Bayesian approach to parameter estimation and pooling in nonlinear flood event models. Water Resour. Res. 35 (1), 211–220.

Campolongo, F., Cariboni, J., Saltelli, A., 2007. An effective screening design for sensitivity analysis of large models. Environ. Model. Softw. 22 (10), 1509–1518.

Chaplot, V., 2005. Impact of DEM mesh size and soil map scale on SWAT runoff, sediment, and NO_3–N loads predictions. J. Hydrol. 312 (1–4), 207–222.

Chen, C.W., Herr, J., Ziemelis, L., Goldstein, R.A., Olmsted, L., 1999. Decision support system for total maximum daily load. J. Environ. Eng. 125 (7), 653–659.

Cho, J., Park, S., Im, S., 2008. Evaluation of Agricultural Nonpoint Source (AGNPS) model for small watersheds in Korea applying irregular cell delineation. Agric. Water Manage. 95 (4), 400–408.

Clark, M.P., Kavetski, D., 2010. Ancient numerical daemons of conceptual hydrological modeling: 1. Fidelity and efficiency of time stepping schemes. Water Resour. Res. 46, W10510. http://dx.doi.org/10.1029/2009WR008894.

Clark, M.P., Slater, A.G., Rupp, D.E., Woods, R.A., Vrugt, J.A., Gupta, H.V., Wagener, T., Hay, L.E., 2008. Framework for understanding structural errors (FUSE): a modular framework to diagnose differences between hydrological models. Water Resour. Res. 44, W00B02. http://dx.doi.org/10.1029/2007WR006735.

Das, S., Rudra, R.P., Gharabaghi, B., Gebremeskel, S., Goel, P.K., Dickinson, W.T., 2008. Applicability of AnnAGNPS for Ontario conditions. Can. Biosyst. Eng. 50 (1), 1–11.

Dean, S., Freer, J., Beven, K., Wade, A.J., Butterfield, D., 2009. Uncertainty assessment of a process-based integrated catchment model of phosphorus. Stoch. Env. Res. Risk Assess. 23, 991–1010.

Ding, X., Shen, Z., Hong, Q., Yang, Z., Wu, X., Liu, R., 2010. Development and test of the export coefficient model in the Upper Reach of the Yangtze River. J. Hydrol. 383 (3–4), 233–244.

Duan, Q., Ajami, N.K., Gao, X., Sorooshian, S., 2007. Multi-model ensemble hydrologic prediction using Bayesian model averaging. Adv. Water Resour. 30 (5), 1371–1386.

Ekström, M., Jones, P.D., 2009. Impact of rainfall estimation uncertainty on streamflow estimations for catchments Wye and Tyne in the United Kingdom. Int. J. Climatol. 29 (1), 79–86.

European Topic Centre. Inland, Coastal, Marine waters (ETC/ICM), 2012. Ecological and chemical status and pressures in European waters: Thematic assessment for EEA water 2012 Report: ETC/ICM technical report. Available from: http://icm.eionet.europa.eu/ETC_Reports/EcoChemStatusPressInEurWaters_201211.

Gassman, P.W., Reyes, M.R., Green, C.H., Arnold, J.G., 2007. The soil and water assessment tool: historical development, applications, and future research directions. Trans. ASAE 50 (4), 1211–1250.

Gelman, A., Rubin, D.B., 1992. Inference from iterative simulation using multiple sequences. Stat. Sci. 7, 457–472.

Gelman, A., Carlin, J.B., Stern, H.S., Rubin, D.B., 2004. Bayesian Data Analysis. Chapman & Hall/CRC, London, New York/Washington, D.C.

Gilks, W.R., Richardson, S., Spiegelhalter, D.J., 1996. Markov Chain Monte Carlo in Practice. Chapman & Hall, London.

Gong, Y.W., Shen, Z.Y., Hong, Q., Liu, R.M., Liao, Q., 2011. Parameter uncertainty analysis in watershed total phosphorus modeling using the GLUE methodology. Agric. Ecosyst. Environ. 142 (3–4), 246–255.

Guasch, H., Armengol, J., Eugènia, M., Sergi, S., 1998. Diurnal variation in dissolved oxygen and carbon dioxide in two low-order streams. Water Res. 32, 1067–1074.

Haario, H., Saksman, E., Tamminen, J., 2001. An adaptive Metropolis algorithm. Bernoulli 7, 223–242.

Haario, H., Laine, M., Mira, A., Saksman, E., 2006. DRAM: efficient adaptive MCMC. Stat. Comput. 16, 339–354.

Hastings, W.K., 1970. Monte Carlo methods using Markov chains and their applications. Biometrika 57, 97–109.

He, M., Hogue, T.S., Franz, K.J., Margulis, S.A., Vrugt, J.A., 2011. Corruption of parameter behavior and regionalization by model and forcing data errors: a Bayesian example using the SNOW17 model. Water Resour. Res. 47. http://dx.doi.org/10.1029/2010WR009753, W07546.

Helton, J.C., 1993. Uncertainty and sensitivity analysis techniques for use in performance assessment for radioactive-waste disposal. Reliab. Eng. Syst. Safe. 42 (2–3), 327–367.

Herrera, G.S., Pinder, G.F., 2005. Space-time optimization of groundwater quality sampling networks. Water Resour. Res. 41, W12407. http://dx.doi.org/10.1029/2004WR 003626.

Hornberger, G.M., Spear, R.C., 1981. An approach to the preliminary-analysis of environmental systems. J. Environ. Manage. 12 (1), 7–18.

Hsu, K.l, Moradkhani, H., Sorooshian, S., 2009. A sequential Bayesian approach for hydrologic model selection and prediction. Water Resour. Res. 45, W00B12. http://dx.doi.org/10.1029/2008WR006824.

Isukapalli, S.S., Roy, A., Georgopoulos, P.G., 1998. Stochastic response surface methods (SRSMs) for uncertainty propagation: application to environmental and biological systems. Risk Anal. 18 (3), 351–363.

Jeon, J.H., Yoon, C.G., Donigian, Jr, Jung, K.W., 2007. Development of the HSPF-Paddy model to estimate watershed pollutant loads in paddy farming regions. Agric. Water Manage. 90 (1–2), 75–86.

Johnes, P.J., 1996. Evaluation and management of the impact of land use change on the nitrogen and phosphorus load delivered to surface waters: the export coefficient modelling approach. J. Hydrol. 183, 323–349.

Kavetski, D., Franks, S.W., Kuczera, G., 2003. Confronting input uncertainty in environmental modeling, in calibration of Watershed Models. Water Sci. Appl. 6, 49–68.

Kavetski, D., Kuczera, G., Franks, S.W., 2006. Bayesian analysis of input uncertainty in hydrological modeling: 1. Theory. Water Resour. Res. 42, W03407. http://dx.doi.org/10.1029/2005WR004368.

Kemanian, A.R., Julich, S., Manoranjan, V.S., Arnold, J.R., 2011. Integrating soil carbon cycling with that of nitrogen and phosphorus in the watershed model SWAT: theory and model testing. Ecol. Model. 222, 1913–1921.

Kuczera, G., 1983. Improved parameter inference in catchment models. 1. Evaluating parameter uncertainty. Water Resour. Res. 19 (5), 1151–1162.

Kuczera, G., Parent, E., 1998. Monte Carlo assessment of parameter uncertainty in conceptual catchment models: the metropolis algorithm. J. Hydrol. 211 (1–4), 69–85.

Kuczera, G., Kavetski, D., Franks, S., Thyer, M., 2006. Towards a Bayesian total error analysis of conceptual rainfall-runoff models: characterising model error using storm-dependent parameters. J. Hydrol. 331 (1–2), 161–177. http://dx.doi.org/10.1016/j.jhydrol.2006.05.010.

Kuczera, G., Kavetski, D., Renard, B., Thyer, M., 2010. A limited memory acceleration strategy for MCMC sampling in hierarchical Bayesian calibration of hydrological models. Water Resour. Res. 46, W07602. http://dx.doi.org/10.1029/2009WR008985.

Laloy, E., Vrugt, J.A., 2012. High-dimensional posterior exploration of hydrologic models using multiple-try DREAM (ZS) and high-performance computing. Water Resour. Res. 48, W01526. http://dx.doi.org/10.1029/2011WR010608.

Lek, S., Guiresse, M., Giraudel, J.L., 1999. Predicting stream nitrogen concentration from watershed features using neural networks. Water Res. 33 (16), 3469–3478.

Leube, P.C., Geiges, A., Nowak, W., 2012. Bayesian assessment of the expected data impact on prediction confidence in optimal sampling design. Water Resour. Res. 48, W02501. http://dx.doi.org/10.1029/2010WR010137.

Li, M., Yang, D., Chen, J., Hubbard, S.S., 2012. Calibration of a distributed flood forecasting model with input uncertainty using a Bayesian framework. Water Resour. Res. 48, W08510. http://dx.doi.org/10.1029/2010WR010062.

Liu, Z., Tong, S.T.Y., 2011. Using HSPF to model the hydrologic and water quality impacts of Riparian land-use change in a small watershed. J. Environ. Inform. 17 (1), 1–14.

Luo, B., Li, J.B., Huang, G.H., Li, H.L., 2006. A simulation-based interval two-stage stochastic model for agricultural nonpoint source pollution control through land retirement. Sci. Total Environ. 361 (1–3), 38–56.

Luo, X.L., Zheng, Y., Wu, B., Lin, Z.R., Han, F., Zhang, W., Wang, X.J., 2013. Impact of carbonaceous materials in soil on the transport of soil-bound PAHs during rainfall-runoff events. Environ. Pollut. 182, 233–241.

Marshall, L., Nott, D., Sharma, A., 2004. A comparative study of Markov chain Monte Carlo methods for conceptual rainfall-runoff modeling. Water Resour. Res. 40, W02501. http://dx.doi.org/10.1029/2003WR002378.

Maskey, S., Guinot, V., Price, R.K., 2004. Treatment of precipitation uncertainty in rainfall-runoff modelling: a fuzzy set approach. Adv. Water Resour. 27 (9), 889–898.

Montanari, A., 2005. Large sample behaviors of the generalized likelihood uncertainty estimation (GLUE) in assessing the uncertainty of rainfall-runoff simulations. Water Resour. Res. 41, W08406. http://dx.doi.org/10.1029/2004WR003826.

Montanari, A., Baldassarre, G., 2013. Data errors and hydrological modelling: the role of model structure to propagate observation uncertainty. Adv. Water Resour. 51, 498–504.

Morris, M.D., 1991. Factorial sampling plans for preliminary computational experiments. Technometrics 33, 161–174.

Muleta, M.K., Nicklow, J.W., 2005. Sensitivity and uncertainty analysis coupled with automatic calibration for a distributed watershed model. J. Hydrol. 306, 127–145.

Naftz, D.L., Cederberg, J.R., Krabbenhoft, D.P., Beisner, K.R., Whitehead, J., Gardberg, J., 2011. Diurnal trends in methylmercury concentration in a wetland adjacent to Great Salt Lake, Utah, USA. Chem. Geol. 283, 78–86.

National Research Council (NRC), 2001. Assessing the TMDL Approach to Water Quality Management. National Academy Press, Washington, D.C.

Ning, S.K., Chang, N.B., 2004. Optimal expansion of water quality monitoring network by Fuzzy optimization approach. Environ. Monit. Assess. 91 (1), 145–170.

Niraula, R., Kailin, L., Wang, R., Srivastava, P., 2012. Determining nutrient and sediment critical source areas with SWAT: effect of lumped calibration. Trans. ASAE 55 (1), 137–147.

Parajul, P.B., Nelson, N.O., Frees, L.D., Mankin, K.R., 2009. Comparison of AnnAGNPS and SWAT model simulation results in USDA-CEAP agricultural watersheds in south-central Kansas. Hydrol. Process. 23 (5), 748–763.

Park, S.Y., Choi, J.H., Wang, S., Park, S.S., 2006. Design of a water quality monitoring network in a large river system using the genetic algorithm. Ecol. Model. 199 (3), 289–297.

Pohlert, T., Huisman, J.A., Breuer, L., Frede, H.G., 2007. Integration of a detailed biogeochemical model into SWAT for improved nitrogen predictions—model development, sensitivity, and GLUE analysis. Ecol. Model. 203, 215–228.

Pokrovsky, O.S., Shirokova, L.S., 2013. Diurnal variations of dissolved and colloidal organic carbonand trace metals in a boreal lake during summer bloom. Water Res. 47, 922–932.

Raat, K.J., Vrugt, J.A., Bouten, W., Tietema, A., 2004. Towards reduced uncertainty in catchment nitrogen modelling: quantifying the effect of field observation uncertainty on model calibration. Hydrol. Earth Syst. Sci. 8 (4), 751–763.

Raghuwanshi, N.S., Singh, R., Reddy, L.S., 2006. Runoff and sediment yield modeling using artificial neural networks: Upper Siwane River, India. J. Hydrol. Eng. 11 (1), 71–79.

Ramin, M., Labencki, T., Boyd, D., Trolle, D., Arhonditsis, G.B., 2012. A Bayesian synthesis of predictions from different models for setting water quality criteria. Ecol. Model. 242, 127–145.

Rankinen, K., Karvonen, T., Butterfield, D., 2006. An application of the GLUE methodology for estimating the parameters of the INCA-N model. Sci. Total Environ. 365, 123–139.

Reckhow, K.H., Arhonditsis, G.B., Kenney, M.A., Hauser, L., Tribo, J., Wu, C., Elcock, K.J., Steinberg, L.J., Stow, C.A., Mcbride, S.J., 2005. A predictive approach to nutrient criteria. Environ. Sci. Technol. 39 (9), 2913–2919.

Refsgaard, J.C., Storm, B., 1995. MIKE SHE. In: Singh, V.P., Yadava, R.N. (Eds.), Computer Models of Watershed Hydrology. Water Resources Publications, Highlands Ranch, Colorado, USA, pp. 809–846 (Chapter 23).

Renard, B., Kavetski, D., Kuczera, G., Thyer, M., Franks, S.W., 2010. Understanding predictive uncertainty in hydrologic modeling: the challenge of identifying input and structural errors. Water Resour. Res. 46, W05521. http://dx.doi.org/10.1029/2009WR008328.

Renard, B., Kavetski, D., Leblois, E., Thyer, M., Kuczera, G., Franks, S.W., 2011. Toward a reliable decomposition of predictive uncertainty in hydrological modeling: characterizing rainfall errors using conditional simulation. Water Resour. Res. 47 (11), W11516. http://dx.doi.org/10.1029/2011WR010643.

Robert, C.P., Casella, G., 2004. Monte Carlo Statistical Methods. Springer-Verlag New York, Inc, Secaucus, NJ, USA.

Robinson, T.H., Leydecker, A., Keller, A.A., Melack, J.M., 2005. Steps towards modeling nutrient export in Coastal Californian streams in a Mediterranean Climate. Agric. Water Manage. 77 (1–3), 144–158.

Rovira, A., Alcaraz, C., Ibanez, C., 2012. Spatial and temporal dynamics of suspended load at-a-cross-section: the lowermost Ebro River, Catalonia, Spain. Water Res. 46, 3671–3681.

Saleh, A., Du, B., 2004. Evaluation of SWAT and HSPF within BASINS program for the upper north Bosque river watershed in central Texas. Trans. ASAE 47 (4), 1039–1049.

Saltelli, A., Chan, K., Scott, E.M., 2000. Sensitivity Analysis. John Wiley & Sons Ltd., West Sussex, England.

Saltelli, A., Ratto, M., Andres, T., Campolongo, F., Cariboni, J., Gatelli, D., 2008. Global Sensitivity Analysis: The Primer. John Wiley & Sons Ltd, Chichester, England.

Samanta, S., Clayton, M.K., Mackay, D.S., Kruger, E.L., Ewers, B.E., 2008. Quantitative comparison of canopy conductance models using a Bayesian approach. Water Resour. Res. 44, W09431. http://dx.doi.org/10.1029/2007WR006761.

Schaefli, B., Talamba, D.B., Musy, A., 2007. Quantifying hydrological modeling errors through a mixture of normal distributions. J. Hydrol. 332 (3–4), 303–315.

Scholefield, D., Goff, T.L., Braven, J., Ebdon, L., Long, T., Butler, M., 2005. Concerted diurnal patterns in riverine nutrient concentrations and physical conditions. Sci. Total Environ. 344, 201–210.

Schoups, G., Vrugt, J.A., 2010. A formal likelihood function for parameter and predictive inference of hydrologic models with correlated, heteroscedastic and non-Gaussian errors. Water Resour. Res. 46, http://dx.doi.org/10.1029/2009WR008933, W10531.

Shamshad, A., Leow, C.S., Ramlah, A., Wan Hussin, W.M.A., Mohd. Sanusi, S.A., 2008. Applications of AnnAGNPS model for soil loss estimation and nutrient loading for Malaysian conditions. Int. J. Appl. Earth Obs. 10 (3), 239–252.

Shen, Z.Y., Hong, Q., Yu, H., Niu, J.F., 2010. Parameter uncertainty analysis of non-point source pollution from different land use types. Sci. Total Environ. 408 (8), 1971–1978.

Shen, Z., Chen, L., Liao, Q., Liu, R., Hong, Q., 2012. Impact of spatial rainfall variability on hydrology and nonpoint source pollution modeling. J. Hydrol. 472–473, 205–215.

Singh, V.P. (Ed.), 1995. Computer Models of Watershed Hydrology. Water Resources Publications, Highlands Ranch, Colorado.

Singh, V.P., Frevert, D.K. (Eds.), 2002a. Mathematical Models of Large Watershed Hydrology. Water Resources Publications, Highlands Ranch, Colorado.

Singh, V.P., Frevert, D.K. (Eds.), 2002b. Mathematical Models of Small Watershed Hydrology and Applications. Water Resources Publications, Highlands Ranch, Colorado.

Smith, T.J., Marshall, L.A., 2010. Exploring uncertainty and model predictive performance concepts via a modular snowmelt-runoff modeling framework. Environ. Model. Softw. 25 (6), 691–701.

State Environmental Protection Administration of China (SEPA), 2012a. Report on the state of the environment in China in 2011. Available from: http://jcs.mep.gov.cn/hjzl/zkgb/2011zkgb/201206/t20120606_231039.htm, in Chinese.

State Environmental Protection Administration of China (SEPA), 2012b. Key river basin water pollution prevention plan. Available from: http://www.mep.gov.cn/gkml/hbb/bwj/201206/W020120601534091604205.pdf, in Chinese.

Tatang, M.A., Pan, W.W., Prinn, R.G., McRae, G.J., 1997. An efficient method for parametric uncertainty analysis of numerical geophysical model. J. Geophys. Res. Atmos. 102 (D18), 21925–21932.

Tayfur, G., Singh, V.P., 2006. ANN and Fuzzy Logic models for simulating event-based rainfall-runoff. J. Hydrol. Eng. 132 (12), 1321–1330.

ter Braak, C.J.F., 2006. A Markov chain Monte Carlo version of the genetic algorithm differential evolution: easy Bayesian computing for real parameter space. Stat. Comput. 16 (3), 239–249.

ter Braak, C.J.F., Vrugt, J.A., 2008. Differential evolution Markov chain with snooker updater and fewer chains. Stat. Comput. 18 (4), 435–446.

Thyer, M., Renard, B., Kavetski, D., Kuczera, G., Franks, S.W., Srikanthan, S., 2009. Critical evaluation of parameter consistency and predictive uncertainty in hydrological modeling: a case study using Bayesian total error analysis. Water Resour. Res. 45, W00B14. http://dx.doi.org/10.1029/2008WR006825.

U.S. Environmental Protection Agency, USEPA. Office of Wetlands, Oceans, & Watersheds. Assessment & Watershed Protection Division. Nonpoint Source Control Branch. A National Evaluation of the Clean Water Act: Section 319 Program. Assessment & Watershed Protection Division. Nonpoint Source Control Branch, November 2011.

Vrugt, J.A., Gupta, H.V., Bouten, W., Sorooshian, S., 2003. A shuffled complex evolution Metropolis algorithm for optimization and uncertainty assessment of hydrologic model parameters. Water Resour. Res. 39 (8), 1201. http://dx.doi.org/10.1029/2002WR001642.

Vrugt, J.A., Gupta, H.V., Dekker, S.C., Sorooshian, S., Wagener, T., Bouten, W., 2006. Application of stochastic parameter optimization to the Sacramento soil moisture accounting model. J. Hydrol. 325 (1–4), 288–307.

Vrugt, J.A., ter Braak, C.J.F., Clark, M.P., Hyman, J.M., Robinson, B.A., 2008. Treatment of input uncertainty in hydrologic modeling: doing hydrology backward with Markov chain Monte Carlo simulation. Water Resour. Res. 44, W00B09. http://dx.doi.org/10.1029/2007WR006720.

Vrugt, J.A., ter Braak, C.J.F., Diks, C.G.H., Higdon, D., Robinson, B.A., Hyman, J.M., 2009. Accelerating Markov chain Monte Carlo simulation by differential evolution with self-adaptive randomized subspace sampling. Int. J. Nonlinear Sci. Numer. Simul. 10 (3), 273–290.

Wang, X.Y., 2006. Management of agricultural nonpoint source pollution in China: current status and challenges. Water Sci. Technol. 53 (2), 1–9.

Wu, Y.P., Chen, J., 2009. Simulation of nitrogen and phosphorus loads in the Dongjiang River basin in South China using SWAT. Front. Mater. Sci. 3 (3), 273–278.

Wu, B., Zheng, Y., 2013. Assessing the value of information for water quality management: a watershed perspective from China. Environ. Monit. Assess. 185 (4), 3023–3035.

Xie, X., Zhang, D., 2010. Data assimilation for distributed hydrological catchment modeling via ensemble Kalman filter. Adv. Water Resour. 33, 678–690.

Yang, S.T., Dong, G.T., Zheng, D.H., Xiao, H.L., Gao, Y.F., Lang, Y., 2011. Coupling Xinanjiang model and SWAT to simulate agricultural non-point source pollution in Songtao watershed of Hainan, China. Ecol. Model. 222 (20–22), 3701–3717.

Zhang, W., Keller, A.A., Wang, X.J., 2009. Analytical modeling of polycyclic aromatic hydrocarbon loading and transport via road runoff in an urban region of Beijing, China. Water Resour. Res. 45. http://dx.doi.org/10.1029/2008WR007004.

Zhang, Y., Pinder, G.F., Herrera, G.S., 2005. Least cost design of groundwater quality monitoring networks. Water Resour. Res. 41 (8), W08412. http://dx.doi.org/10.1029/2005WR003936.

Zheng, Y., Keller, A.A., 2006. Understanding parameter sensitivity and its management implications in watershed-scale water quality modeling. Water Resour. Res. 42, W05402. http://dx.doi.org/10.1029/2005WR004539.

Zheng, Y., Keller, A.A., 2007a. Uncertainty assessment in watershed-scale water quality modeling and management: 1. Framework and application of generalized likelihood uncertainty estimation (GLUE) approach. Water Resour. Res. 43, W08407. http://dx.doi.org/10.1029/2006WR005345.

Zheng, Y., Keller, A.A., 2007b. Uncertainty assessment in watershed-scale water quality modeling and management: 2. Management objectives constrained analysis of uncertainty, MOCAU. Water Resour. Res. 43, W08408. http://dx.doi.org/10.1029/2006WR005346.

Zheng, Y., Keller, A.A., 2008. Stochastic watershed water quality simulation for TMDL development—a case study in the Newport Bay watershed. J. Am. Water Resour. Assoc. 44 (6), 1397–1410.

Zheng, Y., Wang, W., Han, F., Ping, J., 2011. Uncertainty assessment for watershed water quality modeling: a probabilistic collocation method based approach. Adv. Water Resour. 34 (4), 887–898.

Zheng, Y., Luo, X., Zhang, W., Wu, B., Han, F., Lin, Z., Wang, X., 2012. Enrichment behavior and transport mechanism of soil-bound PAHs during rainfall-runoff events. Environ. Pollut. 171, 85–92.

Extending the Application of Network Analysis to Ecological Risk Assessment for Aquatic Ecosystems

Shaoqing Chen, Bin Chen*

State Key Laboratory of Water Environment Simulation, School of Environment, Beijing Normal University, Beijing, China
*Corresponding author: e-mail address: chenb@bnu.edu.cn

Nature does not dictate the outcome of any process or experiment, even in the simplest of situations...
Hawking and Mlodinow, The Grand Design

6.1 INTRODUCTION

This is a world of change, and when change occurs abruptly, whether it is induced by intensive human activities or by natural disturbances, living organisms in the perturbed environment will have to struggle for survival, frequently the outcome of this struggle is determined at a more complex and self-regulated organizational level (e.g., an ecosystem) mediated by the competition among individuals and species. Projecting the effect of a certain perturbation on an organism alone is no easy task, and ecologists must consider a whole ecosystem of highly interrelated interactions where probability has to be introduced to enable meaningful predictions (Novak et al., 2011).

When it comes to the probability of potential harm, "risk" is a well-suited concept for this type of prediction (Kaplan and Garrick, 1981). The awareness of risk has long been embedded in human genetic memory, but risk analysis did not evolve into a systemic research field until the late 1960s, when rocket scientists realized the importance of appraising the probability of mission completion and the occurrence of injury after a severe accident occurred with Apollo 1 (Bedford and Cooke, 2001). Since then, the conception of probabilistic risk analysis has been so deeply embedded into various disciplines that it plays a crucial role in almost all safety evaluations and managements of human society involving economics (Kahneman and Tversky, 1979), politics (Slovic, 1999), engineering (Kumamoto and Henley, 1996), environmental health (Hallenbeck, 1986), and many others. But the application of risk assessment for ecological theory is, comparatively, quite a recent interest. By definition ecological risk assessment (ERA) is focused on the rational appraisal of the possible damages or potential diverse effects by computing the risk values associated with possible eco-environmental hazards under uncertainty (USEPA, 1992; Freedman, 1998; Suter, 2007). The goal of ERA is to provide information about the statistical distribution of possible ecological effects arising from exposure to one or more stressors (USEPA, 1998; Findlay and Zheng, 1999). In order to achieve this goal, three elements complete the basic profile of ERA: (1) the scenario, (2) the likelihood, and (3) the consequence, or rather—what can happen; how likely things are to happen; and what are the end points from sets of occurrences (i.e., "sets of triplets" in Helton, 1993). However, the random and nonlinear characteristics inherent in ecosystems often make it difficult to predict the precise ecological fate. Furthermore, multi-process scenarios (multi-source, multi-factor, and multi-receptor scenarios) are inevitably encountered by environmental managers when performing risk analyses, all of which urge them to resort to more powerful models, preferably mechanistic ones. So far, for example, mathematical models employed in ERA includes holographic neural networks (Findlay and Zheng, 1999), Bayesian networks (Lee and Lee, 2006; Pollino et al., 2007), comprehensive aquatic systems models (DeAngelis et al., 1989; Bartell et al., 1999), and environmental contaminant dispersion models (Chen et al., 2010). Helpful and instructive as they are for guiding risk-based decision making, most of them are either restricted to the evaluation of species biology and population on the microcosm scale under circumstances of single stressors, or developed on the profile capable of limited risk receptors. Therefore they have relatively poor capacity for fitting into iterative and adaptive management (Yeardley and Roger, 2000). Furthermore, the instant cause-effect type of computation might neglect the information of the indirect effect carried by the interactive components within communities or ecosystems.

Alternatively, network environ analysis (NEA), an important branch of network analysis first developed by Patten (1978a,b, 1982), is a system-oriented modeling technique for examining the structure and flow of materials in ecosystems (Leontief, 1951, 1966; Hannon, 1973). NEA places great emphasis on the interactions between components rather than the characteristics of individuals, and the dynamic attributes within the system are identified and quantified via network structural and functional analytic methods (e.g., storage analysis, throughflow analysis, utility analysis, control analysis) (Fath, 1998, 2004a,b; Fath and Patten, 1998, 1999; Ulanowicz, 2004; Fath and Borrett, 2006; Schramski et al., 2006; Kazanci, 2007; Schramski et al., 2011). Fundamentally, the underlying strength of these concepts and methods is the incorporation of direct and indirect effects that construct the whole regime of the interacted network, and that system wholeness is arguably more critical in determining the system's behavior than the direct effects alone, or further, the holistic picture of the concerned system can only be delineated when interactions of all lengths are clarified.

In view of these important insights, one of the most promising applications of NEA is identified as a methodology platform for modeling the integrated eco-environmental impact of natural systems under human interference (Fath,

2004a). The implication is that NEA is conceivably promising for indexing the holistic ecological risk of perturbed ecosystems. In fact, ecological network analysis (a more general version of NEA) has been proven useful as a complementary tool for assessing disturbed ecosystems in the context of system-based management. The most recent cases concerned the determination of possible ecosystem impacts of fishing on estuarine ecosystem (Manickchand-Heileman et al., 2004), the evaluation of environmental stress due to soil contamination on terrestrial ecosystem (Tobor-Kapłon et al., 2007), and the functional assessment of an estuary ecosystem exposed to eutrophication (Christian et al., 2009). Unfortunately, the fact that the model may encounter flow incompatibility in a material- or energy-oriented NEA remains impeditive when evaluating the adverse impact or managing the transitive risk on a system scale. In other words, energy and material, the conventionally used mediates for network synthesis, are not essentially suitable for a system-wide ERA. Herein, we present a novel network approach for holistic ERA by shifting from energy/material flow to information flow. The information network analysis (INA) is developed to address ecological risk in which both direct and indirect effects are together considered and various risk factors and receptors are technically compatible in the same model.

6.2 GENERAL FRAMEWORK OF APPLYING NETWORK ANALYSIS TO ERA

The INA model has three main aims in the face of risk assessment for aquatic ecosystems:

- To assess the potential impacts of various risk factors (or stressors) via direct and indirect paths after human disturbance.
- To illustrate the effectiveness of adding NEA methodology to the existing ecosystem risk assessment framework.
- To provide a comprehensive tool for regulatory ecosystem management based on the network indicators elicited.

In order to achieve these ends, a technical framework for picturing the holistic ERA based on INA is presented (Figure 6.1). A number of stressors (signified as Sr_1, Sr_2, Sr_i, $\cdots$, Sr_m) impact the biotic system (the aggregation of all living organisms) after a specific event we call risk trigger takes place. The food web of the disturbed ecosystem is explicitly examined through field investigation, whereby all the compartments are identified and all the energy or material flows are traced and quantified in an ecological network. Based on this, the control allocation (CA) of the established network is derived, and ecological risks are distributed among different components within the ecosystem after the evaluation of different stressors. Thereafter, with the consideration of the stressor's sensitivities to different components, the risk operating scenario from risk generation to risk distribution is achieved. Additionally, the energy/

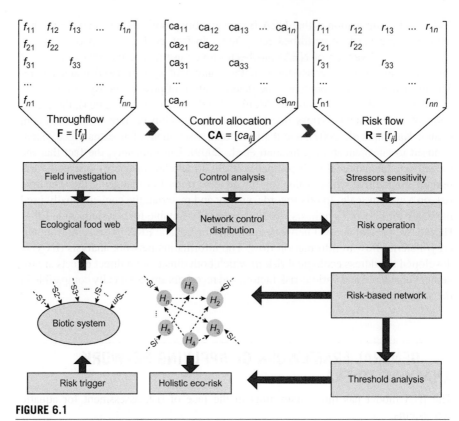

FIGURE 6.1

Comprehensive framework of modeling the holistic risk assessment based on INA (Chen et al., 2011a).

material throughflow, control flow, and integral risk flow produced along the process are displayed in matrices. Hereto, the holistic picture of the perturbed ecosystem showing both the system-wide risk condition and the component-scale microdynamic is derived. Ultimately, with the introduction of threshold theory, the holistic ecological risk of the disturbed ecosystems is applied for ecosystem management.

6.3 INA FOR ERA: METHODOLOGY AND RATIONALE

6.3.1 Food web investigation

There are a number of techniques that can be used for identifying compartments of the food web, such as taxonomy, stable isotope data, or expert recommendation. They entail information from field data, literatures, isotope data, and knowledge of experts familiar with the area. Essentially, stable isotope is assumed to be more precise than other manners in defining the trophic roles of species (Luczkovich et al., 2002), but it is very difficult using such analysis to handle large number of samples at a big geographical scale. Usually, the combination of taxonomy and local expert recommendation are combined for compartmentation. The food chain data of organisms in biomass unit (which was then transformed to energy unit) is derived based on system-specific local investigation and system and nonsystem-specific literatures. A quantitative food web is derived after accounting for the throughflows to detritus, respiration, and harvest of the functional guilds. The food web is then balanced to obtain an equilibrium energy flow model for network analysis (Fath et al., 2007).

6.3.2 **Network CA**

Modern perspectives have shown that there are no absolute controllers in an ecosystem or other interconnected systems (Fath, 2004b; Patten, 2006; Schramski et al., 2006). Instead, each element contributes to the complexity of system organization through its interactions with the other elements. In this sense, control is distributed among the system elements, characterized by the combination of these input and output environs.

As to the methodology layer, distributed control analysis was promoted by Patten (1978b) as a NEA-based measure of control or dominance one component over another. In Patten's control analysis (which was then improved by Patten and Auble, 1981; Fath, 2004b), network control is characterized by the ratio of pair-wise integral flows through network flow and storage analysis, representing the control each component exerts in the overall system configuration. An approach recently developed by Schramski et al. (2006, 2007) employed metrics termed *control difference* and *control ratio* to implicate the absolute open-loop control relationships between components. Based on the food web investigation, a modified version of distributed control index between compartments termed as CA is further developed to formulate the control strength between components via which domination-related information is allocated and dispersed (Chen and Chen, 2012a).

$$\underbrace{\mathbf{N}}_{\text{integrl}} = [n_{ij}] = \underbrace{\mathbf{G}^0}_{\text{initial input}} + \underbrace{\mathbf{G}}_{\text{direct flow}} + \underbrace{\mathbf{G}^2 + \mathbf{G}^3 + \cdots + \mathbf{G}^m}_{\text{indirect flow}} = (\mathbf{I} - \mathbf{G})^{-1} \quad (6.1)$$

$$\underbrace{\mathbf{N}'}_{\text{integrl}} = [n'_{ij}] = \underbrace{\mathbf{G}'^0}_{\text{initial input}} + \underbrace{\mathbf{G}'}_{\text{direct flow}} + \underbrace{\mathbf{G}'^2 + \mathbf{G}'^3 + \cdots + \mathbf{G}'^m}_{\text{indirect flow}} = (\mathbf{I} - \mathbf{G}')^{-1} \quad (6.2)$$

$$\mathbf{CA} = (ca_{ij}) = \begin{cases} \text{if } n_{ij} - n'_{ji} > 0, ca_{ij} = \dfrac{n_{ij} - n'_{ji}}{\sum\limits_{i=1}^{n}\left(n_{ij} - n'_{ji}\right)} \\ \text{if } n_{ij} - n'_{ji} \leq 0, ca_{ij} = 0 \end{cases} \quad (6.3)$$

where, $\mathbf{G} = (g_{ij}), g_{ij} = f_{ij}/T_j$, $\mathbf{G}' = (g'_{ij}), g'_{ij} = f_{ij}/T_i$. f_{ij} denotes energy or material flow from j to i. T_j is the sum of flows into or out of the j-th compartment, T_i is the sum of flows into or out of the i-th compartment, and ca_{ij} signifies the control strength j exerts on i. The integral flows are unfolded in Equations (6.1) and (6.2), as the integration of initial input, direct flow, and indirect flow.

By definition, CA is the difference of two pair-wise integral flows that normalized by the output environ of the dominator (the component that controls the other component). In this case, the transitive control originated from a component is determined by the aggregated configuration of all the ecological flows it involved. CA is able to trace the inner control regime of the ecosystem, which is significant information extracted from the differences in relative energy flows between compartments. CA is adapted as the metrics for determining the dispersion fate of information (in this case, ecological risk) in a perturbed ecosystem, in which the normalized magnitudes of control that each component has over the others are calculated. In the INA, the initial risk values only served as inputs to a network-characterized ecosystem rather than the final responses because the interactions between species will alter those values given a certain period of time for ecosystem reorganization. The ecological risk is assumed to propagate between compartments based on $\mathbf{CA} = (ca_{ij})$ once the ecosystem is exposed to a perturbation, in this case dam construction.

6.3.3 A conversion of flow currency

In order to adapt network analysis to ERA and derive a holistic picture of the ecological risk in a dynamic way, the primary challenge is to establish a suitable transitive medium (or flow currency) for the risk network model. The flow incompatibility inherent in a material- or energy-oriented network makes it impossible to address system-wide properties in the ERA context. An alternative solution into the proper medium, naturally, comes to a conceptual conversion from the material/energy-based network to the information-based one (Figure 6.2).

In light of network theory, an object receives materials and energy from other available compartments through its input environ, and simultaneously generates and transfers materials and energy outside via its output environ, thus completing the storage of useful energy with these continual processes. In this sense, we can define these compartments as energy entities (marked as E). In a bounded ecosystem, the existence of f_{ij} (the direct flow from compartment j to compartment i) indicates energy generated by E_j (could be species or other aggregated functional groups) will be transferred to E_i partially to support its survival or sustainability, while the opposite flow is the feedback from E_i to E_j for the regulation of the energy flow, which is an active and probably indirect feedback generally in that it completes the energy cycle and contributes to the overall network system functioning of different lengths. But if we look into the ecosystem from an information perspective, components within the interactive network should contain the transitive information carried by materials or energy flows for deciding how to behave within their ecological niches and adapt themselves to the changing environment. Similar to an energy-based network, each component should serve as an information entity (marked as H) constantly receiving and generating all kinds of information critical for their survival in the network.

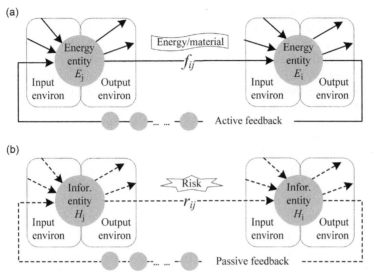

FIGURE 6.2

The conceptual conversion of NEA: From energy-based flow (solid lines) (a) to risk-based flow (dashed lines) (b) (Chen et al., 2011a).

An ecosystem is information-rich, and the operational risk information we are interested in accounts for the vulnerability of the ecosystem. It is assumed that when exposed to a specific hazard, some sensitive components will suffer from it instantly and be forced to change their energy storage and dynamic attributes and generate quantitative risks in their output environs that then exert ensuing intervention on other components they control in the ecosystem through their input environs. In this sense, all the components within the interactive system will inevitably suffer from the influence of the hazard to some extent, either as a direct risk receptor or an indirect risk receptor. Here, we marked the risk flow from information entity H_j to H_i as r_{ij}, indicating the risk which H_j exerts on H_i. Feedback also operates in the INA from H_i to H_j. Different from energy-based networks, it expresses itself in a passive manner in that it retransfers a certain risk back to the first component from where it receives the risk originally and even exacerbates the information state of that component.

6.3.4 Development of risk flow

Risk flow, as has been introduced, is not an energy- or mass-based interaction but an information one, theoretically negative and basically undesirable for nature. The existence of a risk flow indicates that the donor discharge a risk it generated previously, while the receptor suffers the risk from the donor with which it naturally linked. Ecological risk is generally defined as the undesired probability event with the possible occurrence of certain potential harms or damages to ecosystems it exposes.

According to the fundamental risk measurement (Nath et al., 1993; Bolger, 1997), risk flow is generally formulated by combining the change magnitude of an involved factor with its probability of its occurrence. Here we added the component-specific sensitivity toward a concrete exposure to the metric in that the sensitivity disparity induces instant effects on some hazard-susceptive components to a different degree before transferring them to other components through indirect pathways (Bartell et al., 1992, 1999; Knoben et al., 1998). Hereafter, the development risk-based flow is parsed into two different situations, that is, risk flow from the external environment and risk flow within the system.

In terms of input risk value to compartment i from the environment, three parameters—the risk intensity, its corresponding probability and the compartment-specific sensitivity together determine the input risk value (R_i), which is formulated as follows:

$$r_{i0} = \Delta I_x P_x V_{ix}, \ \ 0 \le r_{i0} \le 1 \tag{6.4}$$

in which, ΔI_x refers to the risk intensity resulting from a certain change in risk factor x, P_x refers to its probability of occurrence, while V_{ix} represents the sensitivity of compartment i to stressor x. Specifically, the risk intensity is computed as:

$$\Delta I_x = \frac{|I_{tx} - I_{0x}|}{\max(I_{tx}, I_{0x})} \tag{6.5}$$

where, I_{tx} represents the value of a measurable environmental indicator at a certain time point t of interest after the occurrence of the hazard (e.g., a month, a year, or a century following the hazard), and I_{0x} represents the background value of the indicator, which could be the value of an adjacent ecosystem with almost a same habitat but free from effects of the hazard or an previous state exactly before the risk.

Hereto, the input risk is defined in a dimensionless form by using the ratio of the change magnitude associated with the hazard to the background value (on account that probability and sensitivity per se are dimensionless). Further, the risk flow that travelled through system components should be formulated. Bartell et al. (1999) pointed out that risk analysis should be processed by expressing the toxic effects factors as statistical distributions. As has been elaborated in their risk assessment system, each modeled population has a specific distribution of toxic effects for each chemical contaminant assigned to it. The implication of this decentralized effect for system organization, we contend, also holds for other risk factors. On the other hand, with respect to the environmental flow distribution, Patten and Witkamp (1967) discovered that the distribution of a radionuclide within an ecosystem is significantly influenced by its flow configuration in which way components are coupling or interacting. It has been recently recognized that these distributed control mechanisms are actually formed by compartmental interconnectivity over the system's indirect pathways, as a mechanism for self-organization in an input/output environ characterized ecosystem (Gattie et al., 2006; Schramski et al., 2006). This distributed control between entities indicates the extent or degree to which elements influence each other and thus contribute to the system's overall flow pattern, which possibly determines the functional fate of ecosystems (Straskraba, 2001; Fath, 2004b; Gattie et al., 2006; Schramski et al., 2007). As has been noted, the strength of control information is quantified based on CA. The assumption is that ecological risks received from the environment are transferred and distributed throughout the system components (i.e., the risk receptors) based on the message of CA and impact the whole disturbed ecosystem. Conforming to this, the risk flow within the ecosystem is therefore defined (see Equation 6.6). Assumed that compartment i first receives an input risk (the magnitude of which is r_{i0}) from the external environment, and then it is transferred to compartment j through their control relationship, and subsequently, the resultant risk flow reaches compartment k. Herein, r_{ji} is formulated by multiplying the input risk value of compartment i by the control H_i exerts on H_j, and r_{kj} is formulated by multiplying r_{ji} and the control H_j exerts on H_k. Likewise, all risk flows travelled through the disturbed ecosystem can be explicitly quantified. These internal risk flows compose the elements of direct risk network (matrix $\mathbf{R}$). In order to explore the integral risk flow scenario between different functional guilds with the consideration of both direct flows (pathway length $= 1$) and indirect interactions (pathway length > 1), the integral risk network (matrix $\vec{\mathbf{R}}$) is further defined to aggregate possible risk pathways of all lengths within the disturbed ecosystem (see Equation 6.7). Technically, the values of input risk value (r_{i0}) and CA (ca_{ji}) are both within the interval $[0, 1]$. R_i denotes the risks of component i received from its input environ in direct risk network ($\mathbf{R}$), $\vec{R}_i$ denotes the risks of component i received from its input environ in integral risk network ($\vec{\mathbf{R}}$).

$$\mathbf{R} = \left(r_{ji}\right), \begin{cases} r_{ji} = r_{i0}\,ca_{ji} \\ r_{kj} = r_{ji}\,ca_{kj} \end{cases}, \quad 0 \le r_{ji}, r_{kj} \le 1 \tag{6.6}$$

$$\vec{\mathbf{R}} = \left(\vec{r}_{ji}\right) = \sum_{m=0}^{\infty} \mathbf{R}^m = \left(\mathbf{I} - \mathbf{R}\right)^{-1}, \quad \vec{r}_{ji} \ge 0, \tag{6.7}$$

where, r_{ji} signifies certain risk allocates from i to j, r_{kj} signifies certain risk allocates from j to k, ca_{ji} refers to the control i exerts on j, and ca_{kj} refers to the control j exerts on k.

6.4 A CASE STUDY OF THE APPLICATION OF INA: ERA OF A RIVER ECOSYSTEM INTERCEPTED BY DAMMING

Natural ecosystems, once exposed to a certain stressor or even a set of specific risk factors triggered by an abrupt alteration of their habitats, will inevitably suffer conceivable risks that compel them to corresponding risk processes. Herein, we took the reservoir river ecosystem intercepted by the Manwan dam of Lancang River (N24 25′–24°40′, E100°05′–100°25′) as an example of such ecosystems subjected to human interference. The unique climatic and geographic condition of Lancang River and Manwan Reservoir gives birth to an ecosystem of rich biodiversity, containing the most abundant freshwater fish species and aquatic invertebrates in the world (many are under state first-class protection). It has been well documented that dam construction results in influences on hydrology, river flow pattern, and habitats, therefore disturbing the organisms that reside in the changed environment (Tiffan et al., 2002; Wu et al., 2003; He et al., 2006; Tomsica et al., 2007; Chen et al., 2011b,c). Among these changes, downstream water quality, especially the heavy metal levels, show a tendency to increase (Zhang et al., 2005; Zhai et al., 2010). An investigation into the environmental changes was conducted in Manwan in 1994 (a year after dam construction), serving as the basic data source of this case study. Four system components, or so called functional guilds (i.e., piscivorous fish, detritus, phytoplankon, and zooplankton) are selected to analyze their conditions after exposure to an increased heavy metal contamination. Piscivorous fish, phytoplankon, and zooplankton are regarded as the main components directly sensitive to heavy metal contamination, and detritus is a necessary node connecting other susceptible organisms concerned.

Based on the field investigation and relevant literatures, we structured the energy flows between all components of the Manwan reservoir ecosystem quantitatively based on ecological food web analysis. The natural river bank within the reservoir area serves as the system's boundary. All the input flow, output flow, and throughflow of the four selected compartments are examined a year after dam construction. The energy-based network model of the four functional guilds (i.e., four energy entities: E_1 (piscivorous fish), E_2 (detritus), E_3 (phytoplankon), and E_4 (zooplankton)) extracted from Manwan reservoir ecosystem after dam construction is shown in the form of flow digraph and matrix (F) (Figure 6.3). The **CA** matrix among these interactive compartments is derived to reveal the influence one compartment exerts on another within this biosystem (Figure 6.4).

Taking the potential contamination of a heavy metal Cr in the downstream ecosystem as a case study, we then quantified the input risk scenario within the information network (shown in Table 6.1). Average downstream Cr content was 0.017 mg/L before water impoundment (serving as the background value (I_{0Cr}),

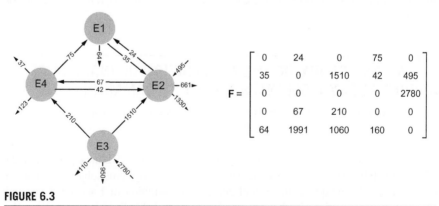

FIGURE 6.3

The energy flow digraph and its flow matrix (**F**) (kJ m^{-2} y^{-1}).

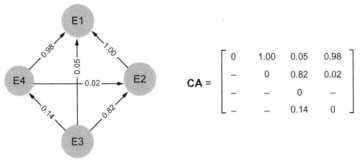

$$CA = \begin{bmatrix} 0 & 1.00 & 0.05 & 0.98 \\ - & 0 & 0.82 & 0.02 \\ - & - & 0 & - \\ - & - & 0.14 & 0 \end{bmatrix}$$

FIGURE 6.4

The control allocation digraph and its matrix (**CA**).

Table 6.1 Input Risk Values of Four Components in Manwan Downstream Ecosystem

Component i	V_{iCr}	P_{Cr}	ΔI_{Cr}	r_{i0}
Piscivorous fish (E_1)	0.20	0.66	$\frac{|I_{tCr}-I_{0Cr}|}{\max(I_{0Cr},\,I_{tCr})}=0.41$	0.0546
Detritus (E_2)	–	–	0	0
Phytoplankton (E_3)	0.50	0.66	$\frac{|I_{tCr}-I_{0Cr}|}{\max(I_{0Cr},\,I_{tCr})}=0.41$	0.1366
Zooplankton (E_4)	0.20	0.66	$\frac{|I_{tCr}-I_{0Cr}|}{\max(I_{0Cr},\,I_{tCr})}=0.41$	0.0546

while that value increased to 0.029 mg/L a year after (serving as the background value I_{tCr}), based on which the risk intensity (RI_{Cr}) is quantified. This increment over the background value is found in 8 months in 1994. Based on our knowledge of the biology, E_1 (piscivorous fish), E_3 (phytoplankton), and E_4 (zooplankton) are regarded as instantly susceptible to the risk exposure of Cr by different sensitivities (i.e., V_{1Cr}, V_{3Cr}, V_{4Cr}) (Knoben et al., 1998; Baird and Brink, 2007). On this basis, the input risk values (r_{i0}) to the corresponding components (functional guilds) from the external environment are calculated.

Provided that the CA and the input risk values of pertinent components have been formulated, we can now structure the subsequent risk flow throughout the network within the system. Conforming to Equation (6.6), all the initial risks are distributed to other information entities based on **CA** matrix, all the risk pathways between the four components are therefore characterized, and the results are shown both in the form of digraph and matrix (**R**) (Figure 6.5). Further, in light of Equation (6.7), the integral risk flows of all information entities within the system are addressed based on matrix RF to picture the ultimate risk scenario of the disturbed biosystem, wherein the network indirect risk flows are taken into consideration. All the overlapped risk flows are aggregated into single risk value for clarification, which also shown both in the form of digraph and matrix ($\underline{\mathbf{R}}$) (Figure 6.6).

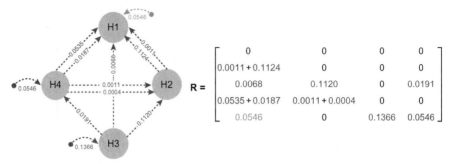

FIGURE 6.5

The direct risk flow digraph and matrix (**R**). To distinguish the risk flow from the material or energy flow, a dashed line is utilized in the risk flow digraph. The last row of the matrix indicates the instant input risk of each information entity. (For the color version of this figure, the reader is referred to the online version of this chapter.)

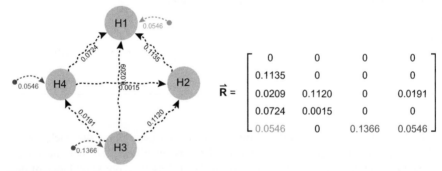

FIGURE 6.6

The integral risk flow digraph and matrix ($\vec{R}$). To make a distinction from the direct risk flow, crooked lines are employed to signify the pathways of integral risk flow. (For the color version of this figure, the reader is referred to the online version of this chapter.)

The network risk results show that the input risks infected the system through two independent ways originally, but then they are combined and operated together in the overlapped paths. On the basis of these risk networks, we can compare the input risk, network direct risk, and integral risk scenario of all information entities (Table 6.2). In order to match with r_{i0} (input risk of component i), R_i and $\vec{R}_i$ are used to signify the risks of component i received from its input environ in direct risk network and integral risk network, correspondingly. In the conventional assessment of instant ecological risk (i.e., r_{i0}), phytoplankton is most vulnerable of all, followed by piscivorous fish and zooplankton. However, the risk in the network

Table 6.2 A Comparison Among Input Risk, Direct Risk, and Integral Risk Condition of H_i

Risk condition	H_1	H_2	H_3	H_4	Pathways number
r_{i0}	0.0546	0	0.1366	0.0546	3
R_i	0.2471	0.1135	0.1366	0.0737	3+9
$\overline{R}_i$	0.2614	0.1135	0.1366	0.0737	3+9+5

Notes: 3 denotes the total number of input risks to the biosystem, 9 represents the number of direct risk pathways (pathway length = 1) in **R**, and 5 represents the number of indirect risk pathways (pathway length > 1) uncovered in **R**.

model turned out to be a quite different scenario from that, namely, piscivorous fish received the densest risks of all, followed by the phytoplankton and the detritus, while the zooplankton are the least affected one. Furthermore, there are significant differences between network direct risk and input risk in H_1 (piscivorous fish), H_2 (detritus), and H_4 (zooplankton), of which, the direct risk value of H_1 dramatically exceeded the initial value by 4.3 ×, and the direct risk value of H_4 is the 1.3 × bigger. The scenario of network integral risk significantly differs from the input risk. Besides, in the present observation scale of the biosystem, no observable differences between network direct risk and integral risk scenario are detected except that the integral risk value of H_1 is slightly higher than direct value, showing some evidence of indirect effect in the integral risk network. With respect to the number of risk flow pathways, a notable increase from the input situation, network direct situation to integral situation, is found, implicating that the deeper dynamics of the biosystem are presumably manifested through the network formulation. The proportions of different risk sources composing the received risk of four components are also identified in the integral risk network (Figure 6.7). The result showed that almost all components had multiple risk sources rather than only from the instant input value (except H_3, who only gave off risks but never received one from other components within this biosystem). That is to say, components that receive the largest risk instantly do not necessarily suffer the most in consideration of the entire system organization. As indicated by Dale et al. (2008), ERA techniques at high organization levels (i.e., community, habitat, and ecosystem scales) will be useful for revealing indirect ecological effects of disturbed biosystems. In terms of the network metric as developed, the system-wide risk dynamic and the component-scale microdynamics displayed a totally different picture from the conventional assessment both in the qualitative and quantitative aspects. In this sense, INA is important in unveiling the integral risk scenario associated with the ultimate fate of disturbed ecosystems, which otherwise will be hidden or neglected in the conventional ERA.

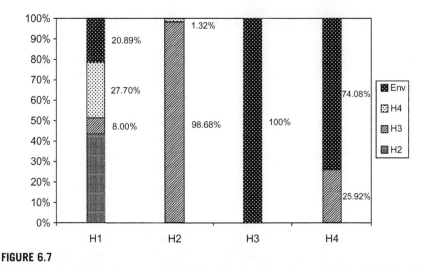

FIGURE 6.7

The proportions of different risk sources composing the received risk of H_i in the integral risk network ("Env" represents the external environment).

6.5 NETWORK INDICATORS FOR RISK MANAGEMENT

Most of the time, an ecosystem has to face a series of risk factors rather than a single hazard. Under these circumstances the existing ERAs fail to work effectively (Yeardley and Roger, 2000; Xu et al., 2004). Also, the impact on one component may induce chain effects on others in a complex fashion. In this context, the major challenges faced by risk modelers and assessors are to acquire the proper kind of multifactor environmental data for interpreting to efficient risk formulation and to quantify the multiprocess risks in an ecosystem-based and direct-viewing manner for future environmental protection activities (Bradbury et al., 2004; Hope, 2006).

Herein, in order to provide a candidate paradigm and methodology to address these challenges based on the developed information network, we made use of threshold theory to index the judgment of holistic risk condition for ecosystem evaluation. We parsed the present risk assessment into two scenarios according to different situations of risk factors and proposed some network indices for supporting system-wide ERA. Based on the results derived from the integral risk networks, the integral risk value of compartment i ($\vec{R}_i$) and the entire ecosystem ($\vec{R}_e$) are formulated as follows:

$$\begin{cases} \vec{R}_i = \sum_{x=1}^{l}\sum_{j=0}^{n} \vec{r}_{ij}^{(x)} \\ \vec{R}_e = \sum_{j=1}^{n} \vec{R}_i \end{cases} \qquad (6.8)$$

where $\overline{r}_{ij}^{(x)}$ is the integral risk flow from j to i induced by factor x. For now, based on risk propagation networks, the initial risks, the direct pathways (in **R**), and the indirect pathways (in $\underline{\mathbf{R}}$) can all be tracked, as well as the risk cycling flows. Three levels of risk that have been defined for each compartment are initial risk value (r_{i0}), direct risk value (R_i), and integral risk value ($\overline{R}_i$).

In addition, to provide early warning information to guide management actions, the maximum risk values that one functional guild and the ecosystem can tolerate to evade extinction or degradation are defined (threshold risk values). The threshold risk value for compartment i ($\overline{R}_{i(\text{thr})}$) and the entire ecosystem ($\overline{R}_{e(\text{thr})}$) are formulated as follows:

$$\begin{cases} \overrightarrow{R}_{i(\text{thr})} = \sum_{x=1}^{l} r_{i0\text{max}}^{(x)}, r_{i0\text{max}}^{(x)} = \Delta I_{\text{max}x} V_{ix} \\ \overrightarrow{R}_{e(\text{thr})} = \sum_{i=1}^{n} \overrightarrow{R}_{i(\text{thr})} \end{cases} \tag{6.9}$$

where $r_{i0\text{max}}^{(x)}$ is the maximum risk flow from the environment that will not cause sudden extinction. It is the result of the maximum change in intensity ($\Delta I_{\text{max}x}$), the relative vulnerability (V_{ix}), and the highest probability (1.00). The maximum change intensity of x factor is assumed to be the boundary of species extinction, which can be derived from experimental data and literature recommendations (as has been documented in other related ecosystems) for specific functional guilds.

An endangerment judgment with regard to a specific functional guild is proposed by comparing $\overline{R}_i$ with $\overline{R}_{i(\text{thr})}$, whereas the safety of the overall ecosystem is addressed through the comparison of $\overline{R}_e$ with $\overline{R}_{e(\text{thr})}$. If functional guild i crosses the line of threshold risk value, it is likely to become endangered in subsequent years and require ad hoc management strategies, and if the risk level of the entire ecosystem surpasses the threshold value, the ecosystem will be considered vulnerable after the disturbance, and the long-term impact of the construction should be prudently reconsidered.

The monitoring of indices could be more useful if environmental modelers use a risk-based approach to address important problems rather than simply tracking the changes of indicators (Suter, 2001). The evaluation of the disturbed ecosystem from a network point of view urges modelers to probe into the indirect effects of the risk flow within ecosystem dynamics, and the risk-based networks and their indices we defined here provide an overall yet succinct metric for doing so. Following the implication of the proposed network-oriented indices, the component-scale and ecosystem-wide risk information inherent in ecological networks can be extracted and formulated as potential metrics for risk assessment and management. Looking at the downstream ecosystem disturbed by dam construction example again, there is a pragmatic need to manifest the possible impacts on the overall river ecosystem before making management decisions, not only the impacts on single aquatic organisms but also their interactions crucial for risks dispersing. By turning to network analysis as proposed, managers are able to obtain the important knowledge on which species and whether the ecosystem are crossing the safety threshold and consequently endangered. In this sense, INA meets this very request for ecosystem management of aquatic ecosystems both before and after disturbance.

6.6 IDENTIFYING UNCERTAINTY IN NETWORK ANALYSIS

Uncertainties universally exist in all kinds of mathematical models. Frequently, ecological uncertainties resulting from spatial and temporal variation in one environmental factor will implicate other factors and thus complicate the evaluation of desired ecological processes or structures and their uncertainty. From the network point of view, an object's local environment consists of all the other objects in the system with which both direct and indirect interactions control the system's behaviors (von Uexükll, 1926; Odum, 1983). In this context, the indirect effect based on material or energy flows can be perceived and determined using a system technique such as network analysis. As a result, a new concept concerning the indirect effect of uncertainty propagation is developed as follows:

Indirect uncertainty (*IU*). The cumulative indeterminacy or variability when material/energy/information propagates through different environmental factors or species via more than one length (step) (e.g., from A to B, then B to C), different from the direct uncertainty (from A to B).

IU is closely relevant to the indirect effect identified in the system metrics. Actually, the propagation process of IU is deemed to be coupled with the indirect effect and treated as an ineluctable issue when we try to investigate system's operation. Perhaps a good example of IU is the game of chess. A player can almost spend all the time guessing what the competitor will do on the next step, that is, what options he or she will possibly take. But a cleverer player will penetrate into the scenario one step further by considering the combination of different options and related consequences on the next two or even more steps, whereby he or she can have in hand the holistic picture of the developing situation, and promisingly, win the game. This same reasoning can be perfectly applied to the apprehension of IU. That is, the system information about uncertain interactions among components can be explicitly clarified based on indirect uncertainty that otherwise will remain unseen in the simple indirect formulation.

In order to holistically elucidate the newly developed concept, the variable propagation fashion of IU built on system structure is decomposed into three scenarios in terms of different flow currencies, that is, material, energy, and information (see Figure 6.8, three-component systems are used for a succinct illustration). In a material-based system, substance will conceivably flow from component Á to component (material entity) Â with a variable rate f_a that is determined by their inherent relationships and affected by external intervention. The probability distribution of f_a is marked as pa, indicating the variability of the flow value. These two variables together represent the direct uncertainty about the propagation of material from Á to Â

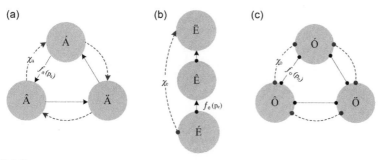

FIGURE 6.8

The propagation fashions of uncertainty with respect to material-based system (a), energy-based system (b), and information-based system (c) (Chen and Chen, 2012b). (For the color version of this figure, the reader is referred to the online version of this chapter.)

(signified as a black arrow pointing from the first term to the second). The aggregation of multiple direct uncertainties forms an indirect uncertainty. For example, the indirect uncertainty from Â to Á (marked as a red arrow with the value as $\chi_{a.}$) is formed by the direct flow from Â to Ä and then Ä to Á. On the circling dynamic as shown, components are interactional. Thus, on the narrow scale, Á affects Â through a material flow under direct certainty, whereas Â influences Á via two-step flows under indirect uncertainty on the large scale. The situations are similar with respect to the energy-based system and the information-based system. But in the energy-based system, energy is transferred but not circled in the system, thereby IU is one-direction oriented. And in the information-based system, IU of information is generally interconnected within the system (two-direction propagation between all components). The concept of IU should be incorporated into the network models for aquatic ecosystems in pursuit of a more complete and accurate understanding of ecosystem behaviors.

6.7 A SYSTEM-BASED ERA FRAMEWORK FOR AQUATIC ECOSYSTEMS

A decision-making friendly ERA for aquatic ecosystems requires good interpretation of food webs, ecosystem and socio-ecological systems. In light of this, an overall framework streamlining the modeling procedure of ERA from a systemic perspective is presented (Figure 6.9).

The framework consists of three main steps: problem formulation, risk characterization, and risk assessment. With respect to problem formulation, the purpose for the assessment is articulated, and stressors and risk endpoints are identified. The conceptual model and plan for analyzing and characterizing risks are also presented. As to the risk characterization, ERA can be modeled and evaluated at various levels. For example, food web analysis is applied to computing the changing condition of

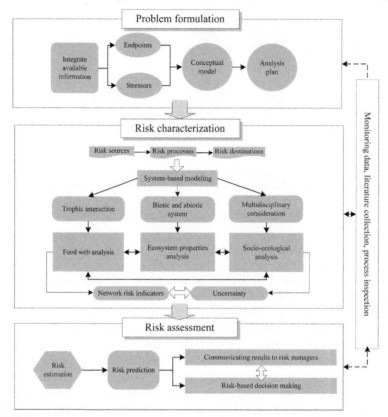

FIGURE 6.9

A system-based ERA framework for aquatic ecosystems.

After Chen et al. (2013).

trophic relations between different organisms subjected to a certain hazard. Ecosystem properties analysis is used to determine the interaction between the biotic system and abiotic systems, in which case environmental factors are incorporated in the assessment profile. And the socio-ecological analysis is applicable for holistic ERA with multidisciplinary considerations. On this basis, network risk indicators can be determined with the consideration of the simulation and inherent uncertainty. Finally, risk assessment first focuses on the estimation of risk based on model results, then makes predictions about the condition of ecosystem. These estimates and predictions are thereafter conveyed to risk managers in support of risk-based decision making. The key elements of ecological risk, that is, risk sources, risk process, and risk destination, are embedded into the whole ERA modeling procedure accordingly.

In this ERA framework, an integration of ERA modeling techniques (i.e., food web-based models, ecosystem-based models, and socio-ecological models) is introduced as the most important portion. Therefore, in order to achieve the goal of holistic assessment, the risk assessors should further consider the possibility and applicability of integrating the system-based risk assessment of different foci in the same analysis.

6.8 CONCLUSION

Assessments of large-scale human-induced perturbations entail the identification of multiple factors, destinations, and interaction processes associated with the perturbation (Bartell et al., 1992; Brismar, 2004; Hope, 2006). The network perspective has been recognized by general ecological interest and proved critical for deriving a deeper insight into ecosystem processes, especially the integral impact of disturbances involving vital environmental flows (DeAngelis et al., 1989; Neubert and Caswell, 1997; Manickchand-Heileman et al., 2004; Gattie et al., 2006; Tobor-Kapłon et al., 2007; Chen et al., 2012).

In this chapter, we introduced a system-based methodology for the holistic ERA by employing a new form of network analysis. In order to achieve this end, a conceptual conversion from the original energy-based network to the present information-based network is proposed. Also, a new distributed control metric termed CA is developed based on the existing control analysis. By taking a disturbed downstream ecosystem as a case study, we demonstrated the usefulness of risk-based networks in the unveiling of indirect effect embedded in the ecosystem dynamics. On this basis, the sorted network-based indicators and indirect uncertainty analysis for ecosystem management are proposed for assessing both system-wide integral condition and component-scale risk scenarios. By incorporating these aspects, a system-based ERA framework for aquatic ecosystems is also illustrated. It is concluded that as with energy and materials, the tracking of information flow also plays a same important role, if not a prior, in ecosystem assessment and management for combined human-natural systems. By turning to the INA, modelers can handily derive a straight picture accompanied by applicable indicators to trace the processing and operation of vital signals associated with system self-organization so long as the risk networks are established. The network models provide a novel perspective and methodology for assessing ecological risk at the system scale, and concurrently, serve as an elicitation of how we can effectively evaluate ecosystems on the same analytic basis of information networks.

References

Baird, D.J., Brink, P.J.V., 2007. Using biological traits to predict species sensitivity to toxic substances. Ecotoxicol. Environ. Saf. 67, 296–301.

Bartell, S.M., Gardner, R.H., O'Neill, R.V., 1992. Ecological Risk Estimation. Lewis, Boca Raton.

Bartell, S.M., Lefebvre, G., Kaminski, G., Carreau, M., Campbell, K.R., 1999. An ecosystem model for assessing ecological risks in Québec rivers, lakes, and reservoirs. Ecol. Modell. 124, 43–67.

Bedford, T., Cooke, R., 2001. Probabilistic Risk Analysis: Foundations and Methods. Cambridge University Press, Cambridge.

Bolger, P.M., 1997. Risk assessment: what is the question? In: Kamrin, M.A. (Ed.), Environmental Risk Harmonization. Wiley, Toronto.

Bradbury, S.P., Feijtel, T.C., Van Leeuwen, C.J., 2004. Meeting the scientific needs of ecological risk assessment in a regulatory context. Environ. Sci. Technol. 38, 463–470.

Brismar, A., 2004. Attention to impact pathways in EISs of large dam projects. Environ. Impact Assess. 24, 59–87.

Chen, G.Q., Zeng, L., Wu, Z., 2010. An ecological risk assessment model for a pulsed contaminant emission into a wetland channel flow. Ecol. Modell. 221, 2927–2937.

Chen, S.Q., Chen, B., 2012a. Network environ perspective for urban metabolism and carbon emissions: a case study of Vienna, Austria. Environ. Sci. Technol. 46 (8), 4498–4506.

Chen, S.Q., Chen, B., 2012b. Defining indirect uncertainty in system-based risk management. Ecol. Inform. 10, 10–16.

Chen, S.Q., Fath, B.D., Chen, B., 2011a. Information-based Network Environ Analysis: a system perspective for ecological risk assessment. Ecol. Indic. 11 (6), 1664–1672.

Chen, S.Q., Fath, B.D., Chen, B., Su, M.R., 2011b. Evaluation of the changed properties of aquatic animals after dam construction using ecological network analysis. Procedia Environ. Sci. 5, 114–119.

Chen, S.Q., Chen, B., Su, M.R., 2011c. An estimation of ecological risk after dam construction in LRGR, China: changes on heavy metal pollution and plant distribution. Procedia Environ. Sci. 5, 153–159.

Chen, S.Q., Chen, B., Dai, J., Su, M.R., 2012. Multi-dimension network consideration for eco-risk evaluation: from population to ecosystem. Procedia Environ. Sci. 13, 1882–1892.

Chen, S.Q., Chen, B., Fath, B.D., 2013. Ecological risk assessment on the system scale: a review of state-of-the-art models and future perspectives. Ecol. Model. 250, 25–33.

Christian, R.R., Brinson, M.M., Dame, J.K., Johnson, G., Peterson, C.H., Baird, D., 2009. Ecological network analyses and their use for establishing reference domain in functional assessment of an estuary. Ecol. Modell. 220, 3113–3122.

Dale, V.H., Biddinger, G.R., Newman, M.C., et al., 2008. Enhancing the ecological risk assessment process. Integr. Environ. Assess. Manag. 4 (3), 306–313.

DeAngelis, D.L., Bartell, S.M., Brenkert, A.L., 1989. Effects of nutrient recycling and food-chain length on resilience. Am. Nat. 134 (5), 778–805.

Fath, B.D., 1998. Network analysis: Foundations, Extensions, and Applications of a Systems Theory of the Environment. University of Georgia, Athens, Georgia, Ph.D. thesis.

Fath, B.D., 2004a. Network analysis in perspective: comments on WAND: an ecological network analysis user friendly tool. Environ. Model Softw. 19, 341–343.

Fath, B.D., 2004b. Distributed control in ecological networks. Ecol. Modell. 179, 235–246.

Fath, B.D., Borrett, S.R., 2006. A MATLAB function for Network Environ Analysis. Environ. Model Softw. 21 (3), 375–405.

Fath, B.D., Patten, B.C., 1998. Network synergism: emergence of positive relations in ecological systems. Ecol. Modell. 107, 127–143.

Fath, B.D., Patten, B.C., 1999. Review of the foundations of network environ analysis. Ecosystems 2, 167–179.

Fath, B.D., Scharler, U.M., Ulanowicz, R.E., Hannon, B., 2007. Ecological network analysis: network construction. Ecol. Modell. 208, 49–55.

Findlay, C.S., Zheng, L., 1999. Estimating ecosystem risks using cross-validated multiple regression and cross-validated holographic neural networks. Ecol. Modell. 119, 57–72.

Freedman, B., 1998. Environmental Science: A Canadian Perspective. Prentice Hall, Scarborough.

Gattie, D.K., Schramski, J.R., Bata, S.A., 2006. Analysis of microdynamic environ flows in an ecological network. Ecol. Modell. 28, 187–204.

Hallenbeck, W.H., 1986. Quantitative Risk Assessment for Environmental and Occupational Health. Lewis Publishers, Chelsea, MI.

Hannon, B., 1973. The structure of ecosystems. J. Theor. Biol. 41, 535–546.

He, D., Feng, Y., Gan, S., Magee, D., You, W., 2006. Transboundary hydrological effects of hydropower dam construction on the Lancang River. Chin. Sci. Bull. 51 (22), 16–24.

Helton, J.C., 1993. Risk, uncertainty in risk, and the EPA release limits for radioactive waste disposal. Nucl. Technol. 101 (1), 18–39.

Hope, B.K., 2006. An examination of ecological risk assessment and management practices. Environ. Int. 32, 983–995.

Kahneman, D., Tversky, A., 1979. Prospect theory: an analysis of decision under risk. Econometrica 47 (2), 263–292.

Kaplan, S., Garrick, B.J., 1981. On the quantitative definition of risk. Risk Anal. 1 (1), 11–27.

Kazanci, C., 2007. EcoNet, a new software for ecological model simulation and network analysis. Ecol. Modell. 208 (1), 3–8.

Knoben, R.A.E., Beek, M.A., Durand, A.M., 1998. Application of species sensitivity distributions as ecological risk assessment tool for water management. J. Hazard. Mater. 61, 203–207.

Kumamoto, H., Henley, E.J., 1996. Probabilistic Risk Assessment and Management for Engineers and Scientists. Wiley-IEEE Press, Piscataway, NJ.

Lee, C.J., Lee, K.J., 2006. Application of Bayesian network to the probabilistic risk assessment of nuclear waste disposal. Reliab. Eng. Syst. Saf. 91, 515–532.

Leontief, W.W., 1951. The Structure of American Economy, 1919–1939: An Empirical Application of Equilibrium analysis. Oxford University Press, New York, NY.

Leontief, W.W., 1966. Input–Output Economics. Oxford University Press, New York, NY.

Luczkovich, J.J., Ward, G.P., Johnson, J.C., Christian, R.R., Baird, D., Neckles, H., Rizzo, W.M., 2002. Determining the trophic guilds of fishes and macroinvertebrates in a seagrass food web. Estuaries 25 (6), 1143–1164.

Manickchand-Heileman, S., Mendoza-Hill, J., Kong, A.L., Aroch, F., 2004. A trophic model for exploring possible ecosystem impacts of fishing in the Gulf of Paria, between Venezuela and Trinidad. Ecol. Modell. 172, 307–322.

Nath, B., Hens, L., Compton, P., Devuyst, D., 1993. Environmental Management, vol. 3. VUB Press, Brussels, Belgium.

Neubert, M.G., Caswell, H., 1997. Alternatives to resilience for measuring the responses of ecological systems to perturbations. Ecology 78 (3), 653–665.

Novak, M., Wootton, J.T., Doak, D.F., et al., 2011. Predicting community responses to perturbations in the face of imperfect knowledge and network complexity. Ecology 92 (4), 836–846.

Odum, H.T., 1983. Systems Ecology: An Introduction. JohnWiley and Sons, New York.

Patten, B.C., 1978a. Systems approach to the concept of environment. Ohio J. Sci. 78, 206–222.

Patten, B.C., 1978b. Energy environments in ecosystems. In: In: Fazzolare, R.A., Smith, C.B. (Eds.), Energy Use Management, vol. 4. Pergamon Press, New York, NY.

Patten, B.C., 1982. Environs: relativistic elementary particles or ecology. Am. Nat. 119, 179–219.

Patten, B.C., 2006. Network perspectives on ecological indicators and actuators: enfolding, observability, and controllability. Ecol. Indic. 6 (1), 6–23.

Patten, B.C., Auble, G.T., 1981. System theory of the ecological niche. Am. Nat. 117, 893–922.

Patten, B.C., Witkamp, M., 1967. Systems analysis of 134Cesium kinetics in terrestrial microcosms. Ecology 48 (5), 813–824.

Pollino, C.A., Owen, W., Ann, N., Kevin, K., Barry, T.H., 2007. Parameterisation and evaluation of a Bayesian network for use in an ecological risk assessment. Environ. Model Softw. 22, 1140–1152.

Schramski, J.R., Gattie, D.K., Patten, B.C., Borrett, S.R., Fath, B.D., Thomas, C.R., Whipple, S.J., 2006. Indirect effects and distributed control in ecosystems; distributed control in the environ networks of a seven-compartment model of nitrogen flow in the Neuse River estuary, USA-Steady-state analysis. Ecol. Modell. 194, 189–201.

Schramski, J.R., Gattie, D.K., Patten, B.C., Borrett, S.R., Fath, B.D., Whipple, S.J., 2007. Indirect effects and distributed control in ecosystems; distributed control in the environ networks of a seven-compartment model of nitrogen flow in the Neuse River estuary USA-Time series analysis. Ecol. Modell. 206, 18–30.

Schramski, J.R., Kazanci, C., Tollner, E.W., 2011. Network environ theory, simulation, and EcoNet 2.0. Environ. Model Softw. 26, 419–428.

Slovic, P., 1999. Trust, emotion, sex, politics, and science: surveying the risk-assessment battlefield. Risk Anal. 19 (4), 689–701.

Straskraba, M., 2001. Natural control mechanisms in models of aquatic ecosytems. Ecol. Modell. 140 (3), 195–205.

Suter II, G.W., 2001. Applicability of indicator monitoring to ecological risk assessment. Ecol. Indic. 1, 101–112.

Suter, G.W.I.I., 2007. Ecological Risk Assessment. CRC Press, Boca Raton, p. 643.

Tiffan, K.F., Garland, R.D., Rondorf, D.W., 2002. Quantifying flow-dependent changes in subyearling fall Chinook salmon rearing habitat using two-dimensional spatially explicit modeling. North Am. J. Fish. Manag. 22, 713–726.

Tobor-Kapłon, M.A., Holtkamp, R., et al., 2007. Evaluation of information indices as indicators of environmental stress in terrestrial soils. Ecol. Model. 208, 80–90.

Tomsica, C.A., Granataa, T.C., Murphyb, R.P., Livchakc, C.J., 2007. Using a coupled eco-hydrodynamic model to predict habitat for target species following dam removal. Ecol. Eng. 30, 215–230.

Ulanowicz, R.E., 2004. Quantitative methods for ecological network analysis. Comput. Biol. Chem. 28, 321–339.

USEPA (US Environmental Protection Agency), 1992. Framework for ecological risk assessment. Risk Assessment Forum, Washington DC, USEPA. EPA/600/R-92-001.

USEPA (US Environmental Protection Agency), 1998. Guidelines for Ecological Risk Assessment. Risk Assessment Forum, Washington DC, USEPA. EPA/630/R095//002F.

von Uexükll, J., 1926. Theoretical Biology. Kegan, Paul, Trench, Tubner and Company, London.

Wu, J., Huang, J., Han, X., Xie, Z., Gao, X., 2003. Three-Gorges Dam—Experiment in Habitat Fragmentation? Science 300, 1239–1240.

Xu, X., Lin, H., Fu, Z., 2004. Probe into the method of regional ecological risk assessment—a case study of wetland in the Yellow River Delta in China. J. Environ. Manage. 70, 253–262.

Yeardley, J., Roger, B., 2000. Use of small forage fish for regional streams wildlife risk assessment: relative bioaccumulation of contaminants. Environ. Monit. Assess. 65 (3), 559–585.

Zhai, H.J., Cui, B.S., Hu, B., Zhang, K.J., 2010. Prediction of river ecological integrity after cascade hydropower dam construction on the mainstream of rivers in Longitudinal Range-Gorge Region (LRGR). China. Ecol. Eng. 36 (4), 361–372.

Zhang, Y.Z., Liu, J.L., Wang, L.Q., 2005. Changes in water quality in the downstream of Lancangjiang River after the construction of Manwan Hydropower Station (in Chinese). Resour. Environ. Yangtze Basin 14 (4), 501–506.

Modeling the Purification Effects of the Constructed *Sphagnum* Wetland on Phosphorus and Heavy Metals in Dajiuhu Wetland Reserve, China

Wei He[a], Fu-Liu Xu[a],*, Yanyan Zhang[a], Rong Tian[b], Hongxing Hu[c]

[a]*MOE Laboratory for Earth Surface Processes, College of Urban & Environmental Sciences, Peking University, Beijing, China*
[b]*Institute of Remote Sensing Applications, Chinese Academy of Sciences, Beijing, China*
[c]*School of Resource and Environmental Science, Wuhan University, Wuhan, China*
**Corresponding author: e-mail address: xufl@urban.pku.edu.cn*

7.1 INTRODUCTION

Wetland, known as the "earth's kidney," serves many functions such as water conservation, runoff regulation, peat accumulation, carbon sequestration, pollution purification, toxic substance transformation, and disaster prevention for drought and flood (Blanken and Rouse, 1996; Zhao, 1999; Chen and Lu, 2003; Peregon et al., 2007; Kayranli et al., 2010; Wang et al., 2010). The Dajiuhu *Sphagnum* wetland (SW), located in Shennongjia, is a rare type of subalpine wetland in China with a peat layer up to 2 m in depth (He et al., 2003). As the water source of the Duhe River, which is the largest tributary of the Hanjiang River that flows into the Danjiangkou Reservoir—the water source of the Middle Line Project of Water Transfer from the South to the North (MLPWTSN) in China—the Dajiuhu wetland plays a crucial role in safeguarding water quality in the Han River basin and Danjiangkou Reservoir (Yu et al., 2008). Hu et al. (2008)'s research indicated that the peat layer played an important part in purification ability of the SW because the peat has high specific surface area ($>200 \, m^2 \, g^{-1}$) (Asplund et al., 1972; Babel and Kurniawan, 2003; Hu et al., 2008) and contains a variety of polar functional groups such as aldehyde, carboxyl, keto, and phenolic hydroxyl (Adler and Lundquist, 1963; Bailey et al., 1999), which have great adsorption capacity in adsorbing heavy metal ions and some nonmetal contaminants such as cyanide, phosphate, and organic matter efficiently (Coupal and Lalancette, 1976; Ho and McKay, 2000; Ringqvist and Oborn,

2002). Since the 1950s, the Dajiuhu government has been more implementing reclamation policies such as digging drainage ditches on a large scale, planting forage for pasturage, and farming. These practices caused the gradual drying up of wetland and the degradation of the wetland area from 708 to 179 ha (Yin et al., 2007; Xiao et al., 2009). With a decrease in the water-holding capacity of wetland, the dominant species changed from *Carex argyi* into *Sphagnum palustre*, weakening the wetland's carbon fixation feature and pollutant-carrying capacity (Coupal and Lalancette, 1976; Yin et al., 2007). Hu et al. (2008) pointed out that once the wetland is destroyed, the peat resource will be lost, then the water purification ability (WPA) of the wetland would no longer exist.

Despite SW's function of water conservation, most researchers focused only on its application as a peat resource (Chen et al., 1990; Gardea et al., 1996; Ho et al., 1996; Sedeh et al., 1996; Ho and McKay, 2000; Ringqvist et al., 2002; Ho and McKay, 2004; Kalmykova et al., 2008). Some researchers were devoted to comparative studies of the constructed wetlands' WPA for phosphate when plants existed and when there were not any plants, concluding that quality of treated water with plants was better than those without plants (Brix, 1997; Lee and Scholz, 2007; Bindu et al., 2008). Some researchers used different substrata as purification material to obtain the best substratum (Mann, 1997; Sakadevan and Bavor, 1998; Brooks et al., 2000; Xu et al., 2006). Peng et al. (2007) studied the adsorption and release under aerobic (>-50 mV) and anaerobic conditions. Some researchers studied the accumulation and purification of the constructed wetlands to organic matter and heavy metals (von Felde and Kunst, 1997; Tanner et al., 1998; Cheng et al., 2002; Scholz, 2003; Peng et al., 2007).

This chapter studies the purification effects of the natural SW on nutritious elements of phosphorus (phosphate), cadmium (II), copper (II), lead (II), and zinc (II), using kinetic models to simulate its purification process, hoping to evaluate the purification value of the SW and to provide a scientific basis for the protection and proper utilization of the Dajiuhu SW (Hu et al., 2008, 2009; He et al., 2013).

7.2 MATERIAL AND METHODS

7.2.1 General situation of the study area

The Dajiuhu Wetland Park is a sub-alpine wetland park in China located west of the Shennongjia National Nature Reserve in western Hubei province of China (Figure 7.1). It has an elevation of 1600–1800 m, an annual mean temperature of 7.4 °C, and a monthly mean maximum temperature of 29.0 °C. The lowest temperature is −18 to −22 °C, and annual mean rainfall is 1585.4 mm with 85.9% occurring between April and October. The research site, Dajiuhu SW, belongs to the north subtropical mountain wetland. The peat deposit was formed in the early Holocene Age, which has high scientific research value and huge economic benefit (Pollino et al., 2007; Yu et al., 2008). The experimental site is at the edge of the basin, with

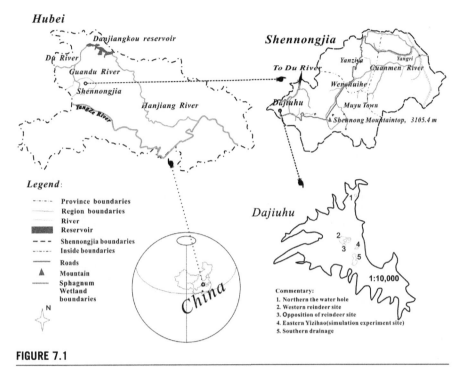

FIGURE 7.1

The geographic location of the experimental site (the lower right-hand part of the figure is the distribution map of the SW, on which simulation experiment is located).

flat terrain and hills nearby. Its geographic coordinates are 31°29′18.4″ N and 110°0′10.5″ E, with an elevation of 1754 m. The main vegetation type is Com. *C. argyi–S. palustre* (Zhao, 1999).

7.2.2 Simulation device

First, a hole with diameter 35 ($\pm$1) mm was drilled at the bottom of the lateral wall of a household plastic box (specifications: $300 \times 400 \times 260$ mm, wall thickness 2 mm, Jinzun Daily Necessities Co., Ltd., Shanghai, China). Second, the hole was blocked with a rubber plug containing a glass tube. Third, the tube outside of the box was covered with a latex tube with a spring water stopper, and the inside part was covered with gauze. Fourth, a 50-mm thickness of cobble layer was paved at the bottom of the box, which was then covered with nylon net bag (NB) containing small cobble. The design above was to prevent the outlet from blocking the cobble and to maintain a normal reflux. Finally, a peat layer of 10–20 mm thickness covered the NB evenly, and the *Sphagnum* plant layer with a volume of 300 ($\pm$20) $\times$ 400 ($\pm$20) $\times$ 150 ($\pm$10) mm^3 was covered above on the top (Figure 7.2).

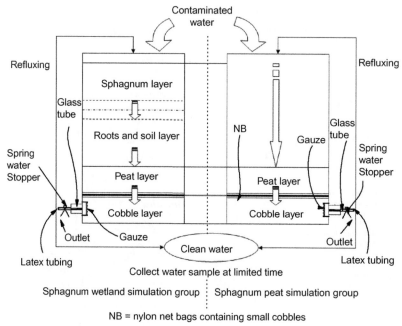

FIGURE 7.2

The process of the SW's purification effects on the contaminated water (the left part of this graph is the SW simulation group; the other is SW peat simulation group that was studied last year).

7.2.3 Simulation experimental method

7.2.3.1 Preliminary study of purification effects of the SW

A preliminary study of purification effects of the SW was performed in August 2007 (Hu et al., 2008). Purification of phosphorus and copper ions with a wide concentration level was studied using the above device. To be specific, 35 g l^{-1} stock solution of KH$_2$PO$_4$ and 40 g l^{-1} stock solution of Cu(NO$_3$)$_2$ were diluted into four concentrations of 0, 10, 100, 1000 mg l^{-1} with a volume of 10 l for each device, using the local well water. The polluted water containing total P (TP) or Cu^{2+} at four concentration levels was added to each of the four simulation devices respectively simultaneously. Meanwhile, purified water was refluxed using 600-ml plastic bottle from the outlet to the *Sphagnum* layer. The reflux speed was 2.5–5.0-l min^{-1}. Water was sprinkled evenly on the *Sphagnum* layer. Purified water samples were collected at time intervals of 20, 40, 60, 80, and 120 min. The sample at 0 min was collected after two refluxes. The sample volumes for TP and Cu^{2+} were 10 respectively for each sampling. The total sampling volume of each device was 60 ml, so changes in the volume of pollutants solution could be negligible, compared to the volume of 10,000 ml contaminated water before and after the experiment. A fixed volume of 6 mol l^{-1} sulfuric acid was added into metal sample to keep the water pH ≤ 1, while a certain volume of $1+9$ nitric acid was added into each TP sample to keep the water pH ≤ 2 for storage. After the experiment, the local well water was sampled and preserved by acid, which was used to determine the background value of the studied contaminated elements; clean *Sphagnum* paints and clean peat were collected to determine the metal content for the background value.

7.2.3.2 Further study of purification effects of the SW

A further study of purification effects of the SW was performed in July 2008 (Hu et al., 2009). Purification of phosphorus and various heavy ions with refined concentration levels was undertaken. To be specific, 20 g l^{-1} stock solution of KH_2PO_4 and the metal ions mixed solution (containing Cd^{2+}, Cu^{2+}, Pb^{2+}, Zn^{2+}) were diluted into four concentrations of 20, 40, 80, 160 mg l^{-1} with a volume of 10 l for each device, using the local well water. The polluted water containing TP or metals at four concentration levels was added to each of the four simulation devices respectively simultaneously. Meanwhile, purified water was refluxed using a 600 ml plastic bottle from the outlet to the *Sphagnum* layer. The reflux speed was 2.5–5.0 l min^{-1}. Water was sprinkled evenly on the *Sphagnum* layer. Purified water samples were collected at 0, 5, 10, 20, 40, 80, and 120 min. The sample at 0 min was collected after two refluxes. The sample volumes for TP and metals were 10 and 20 ml respectively for each sampling. The total sampling volume of each device was 80–140 ml. The preservation method of sample was the same as described in Section 7.2.3.1.

7.2.4 Storage and processing method for the *Sphagnum* and the peat

The peat samples obtained in Dajiuhu SW were dried by air on a plastic cloth in the laboratory, and big pieces of peat were crumbled into small ones. After that, animal and plant residues were picked out. The peat was ground, passed through a 100-mesh sieve, and stored in a 150-ml bottle. The *Sphagnum* plant samples were dried by air under the same conditions as the peat, then they were cut and ground into small pieces to pass through a 100-mesh sieve.

Approximately 1 g of the pretreated peat was placed in a 100 ml centrifuge tube containing 15 ml volume of 0.1 mol l^{-1} nitric acid, and the tube was put in the ultrasonic cleaner for 15 min. After that, the samples were centrifuged at 4500 rpm for 10 min, and the extracted liquid was placed into another tube. The same extraction was conducted two more times. The extract collected was filtered through double pieces of filter paper and diluted to a constant volume of 50 ml with 0.2% nitric acid. Approximately 0.2 g of the pretreated *Sphagnum* plants was placed into the conical flask containing 10 ml volume of 1:4 perchlorate-nitric acids, and the sample was shaken up. After standing 24 h, it was digested by heater until the volume reduced to about 2 ml. 0.2% nitric acid was used to dilute the sample, which had been filtered through three pieces of filter paper to a constant volume of 50 ml.

7.2.5 Chemical analysis of the sampling water

Every sample was centrifuged and diluted properly for measurement. All measurements were conducted in triplicate, and an average was applied. TP content of the samples was measured using molybdenum-antimony antispectrophotometric method by UV-2000 UV-Vis spectrophotometer (Shanghai Precision Instrument Co., Ltd., Shanghai, China). Metals were measured by TAS-990 atomic absorption spectrometry method (Beijing PuXi General Instrument Co., Ltd.) (Ding, 2006).

7.2.6 Purification capacity and purification rate

The purification amount (PA or q) was defined as the amount of contaminants (in micrograms) removed by SW by comprehensive effects including dilution of the wetland water, closure and absorption of the *Sphagnum*, microbial action, and adsorption to the peat and cobble. The unit of PA could be mg cm^{-2} when SW area was considered. The purification ratio (PR) was the ratio between the purification amount and the initial mount of contaminant. The dilution could only reduce the concentration of the contaminants, while removal from the water was more essential purification process. Therefore, the absorption effect mentioned below was the purification effects, which excluded the dilution effect.

The purification amount and PR were calculated by Equations (7.1) and (7.2):

$$q_t = (C_0 - C_t)V \text{ or } q_t = \frac{(C_0 - C_t)V}{A} \tag{7.1}$$

and

$$PR_t = \frac{q_t}{C_0 V} \times 100\% \tag{7.2}$$

where q_t is the PA of SW at t min (mg or mg cm^{-2}), PR$_t$ is the PR at t min (%), C_0 is the initial concentration of the contaminant (mg l^{-1}), C_t is concentration of the contaminant at t min (mg l^{-1}), V is the volume of the contaminant (l), and V is 10 l in this study.

7.2.7 Kinetic model

In the preliminary study, six models, namely, the pseudo first-order kinetic model (Model 1), the up-limitation-modified pseudo first-order kinetic model (Model 2), the pseudo second-order kinetic model (Model 3 or Model II), the Elovich model (Model 4), the two-constant rate model (Model 5), and the hyperbolic diffusion model (Model 6) were used to study the purification process of SW on TP and copper (Hu et al., 2008). In further study, the modified pseudo first-order kinetic model (Model I), which was different from the one mentioned above and the original because a new parameter was introduced to calculate the PA mainly caused by dilution of SW, and Model II were employed. The model equations are as follows (Ho and McKay, 2000; McDowell and Sharpley, 2003):

Model 1

$$q_t = q_e(1 - \exp(-k_1 t)) \tag{7.3}$$

Model 2

$$q_t = q_{max}\left(1 - \exp\left(-k_1' t\right)\right) \tag{7.4}$$

Model I

$$q_t = q_e - (q_e - q_0) \times \exp\left(-k_1'' t\right) \tag{7.5}$$

Model 3 or Model II

$$q_t = \frac{t}{\left(\frac{1}{k_2 q_e^2} + \frac{t}{q_e}\right)} \tag{7.6}$$

and linear form of Model II

$$\frac{t}{q_t} = \frac{1}{k_2 q_e^2} + \frac{t}{q_e} \tag{7.7}$$

Model 4

$$q_t = k_3^{-1} \ln(a_1 k_3) + k_3^{-1} \ln t \tag{7.8}$$

Model 5

$$\ln q_t = a_2 + k_4 \ln t \tag{7.9}$$

Model 6

$$q_t = a_3 k_5 + k_5 t^{0.5} \tag{7.10}$$

where q_t is the PA of SW at t min (mg or mg cm^{-2}), q_e is the PA of SW at time of the equilibrium point (mg or mg cm^{-2}), q_0 is the PA of SW at 0 min (mg or mg cm^{-2}), k_1, k_1', k_1'', k_2, k_3, k_4, and k_5 are the rate constants of Model 1, Model 2, Model I, Model 3 (Model II), Model 4, Model 5, and Model 6, respectively.

The contribution ratio of dilution effects (CRDEs) was defined as the ratio between q_0 and all the contaminants content added into the SW for each device. It gives:

$$\text{CRDE} = \frac{q_0}{C_0 V} \times 100\% \tag{7.11}$$

where C_0 is the initial concentration of the contaminant, (mg l^{-1}), V is 10 l, and $C_0 V$ is all the contaminant content.

To evaluate the model's fitting, the difference between fitted values and experimental data were expressed by the relative deviation between $PR_{120,c}$ and PR_{120} namely RD, whose equation is:

$$RD = \frac{|PR_{120,c} - PR_{120}|}{PR_{120}} \times 100\% \qquad (7.12)$$

where $PR_{120,c}$ is the PR of the contaminant at the time point of 120 min which can be calculated using the fitted values and PR_{120} is determined PR of the contaminant at the time point of 120 min.

Model II can be used to evaluate the beginning time of the purification equilibrium, and the equation is:

$$R_{\Delta t} = \frac{(q_{t_2} - q_{t_1})}{C_0 V} \quad (\Delta t = t_2 - t_1) \qquad (7.13)$$

where $R_{\Delta t}$ is the ratio between PA and $C_0 V$, and Δt is 5 min.

There is a certain relationship between the parameter q_e and C_0 (a linear relationship), and between k_2 and C_0 (a nonlinear relationship). q_e and k_2 are regressed against C_0 using Equations (7.14) and (7.15). The empirical formulas were derived using the fitting parameter and Equation (7.6) so that PA could be calculated when knowing the time t and the initial concentration C_0. Equations (7.14) and (7.15) were related to those used by Ho and McKay (1999, 2000):

$$q_e = A + BC_0 \qquad (7.14)$$

and

$$k_2 = \frac{C_0}{(aC_0 + b)} \qquad (7.15)$$

where A and B are parameters of Equation (7.10) (mg) and (l), a and b are parameters of Equation (7.15) (mg min^{-1}) and (mg^2 min l^{-1}).

7.3 RESULTS AND DISCUSSION

7.3.1 Background values

The background values of the dilution water (DW), *Sphagnum* plants, and peat were determined in Table 7.1. Copper (II) contents of the *Sphagnum* plants and peat were both higher than other metals, and the copper (II) in *Sphagnum* plants was higher than that in peat. The cadmium (II), lead (II), and zinc (II) contents of the *Sphagnum*

Table 7.1 The Background Values of Material Used in the Simulation Experiment

	TP	Cd	Cu	Pb	Zn
Dilution water (μg l^{-1})	13.2	1.5	0.0	48.8	35.5
Sphagnum plants (μg g^{-1})	–	5.1	1367.6	19.5	83.7
Peat (μg g^{-1})	–	5.2	207.9	41.5	70.1

plants and peat were all very low. The contaminants contents in the DW were too low to have any effects on the initial concentrations. The lead concentration was a little high, occupying 0.24% of the smallest concentration gradient 20 mg l^{-1}, and the effect of the DW was negligible.

7.3.2 Purification of the SW to TP

Average purification velocity (APV) as an important parameter to reflect SW's purification dynamic process was defined as the PA within a specified time. In Figure 7.3 the APV value between two adjacent sampling time points was the absolute slope value of the line between two points, and the APV value between nonadjacent sampling points was the absolute slope value of the regression line using the two points and that between them, neglecting the fitting degree. The APV values for TP during the first 5 min in ascending order were 23.89 mg min^{-1}, 30.03 mg min^{-1}, 83.69 mg min^{-1}, and 161.36 mg min^{-1}, indicating that with the increase of the initial concentration, the APV value at the beginning of the purification increased with the initial concentration. During every following adjacent time segment, the APV of the high concentration was larger than that of lower one. According to Figure 7.3, the obvious finding was that the absolute slope value between adjacent points decreased with time. This meant that the APV of the SW decreased with the increase of time. The 80 mg l^{-1} TP group was taken as an example: The APV values of 5–10 min, 10–20 min, 20–40 min, 40–80 min, and 80–120 min were 12.67 mg min^{-1}, 10.00 mg min^{-1}, 3.33 mg min^{-1}, 1.25 mg min^{-1}, and 0.75 mg min^{-1}, respectively, showing that the purification process approached equilibrium. The lower the initial concentration was, the more quickly the purification curve became flat, indicating that lowest concentration approached the equilibrium point fast.

7.3.3 Purification of SW to divalent metal ions

The same purification laws were observed in the divalent metal groups, similar to TP groups as shown in Figure 7.3. During the first 5 min, the APV values of every metal were apparently larger than those of the following time segments for the dilution effect played a vital role at beginning. During the time of 5–120 min, APV values reduced with the increase of time. The phenomena in the higher concentration groups (initial concentration of 80 mg l^{-1} and 160 mg l^{-1}) were especially obvious. For the lower concentration groups (initial concentration of 80 mg l^{-1} and 160 mg l^{-1}), and especially that of copper and lead groups, the curve became flat so that the changes could not be distinguished easily after 20 min. This means that the time taken to approach the equilibrium point was less for lower concentration groups, which was similar to TP groups. In addition, the time taken to approach the equilibrium for different metals varied at the same concentration gradient. Take APV values between 5 and 120 min as an example: The lower the APV value was, the quicker it was to approach the equilibrium. At the lower concentration groups, the time required to reach the equilibrium for the four metals could be ranked as follows:

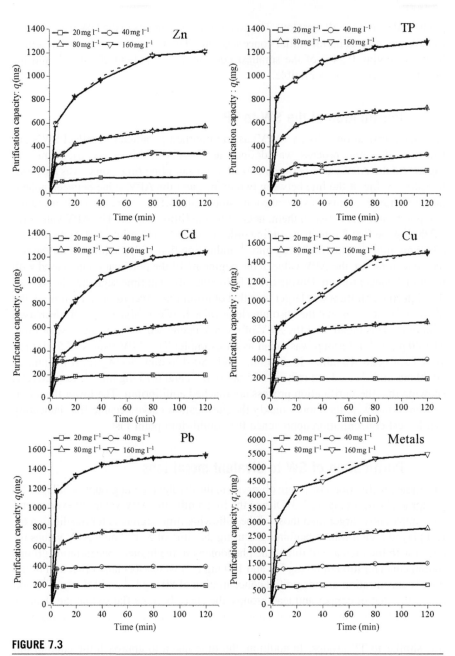

FIGURE 7.3

The purification amount of SW to different contaminants versus time (scatter) and the fitting curves using Model 1 (dashed line) at various initial concentrations.

$Pb^{2+} < Cu^{2+} < Cd^{2+} < Zn^{2+}$. And for the higher concentration groups it could be ranked as follows: $Pb^{2+} < Zn^{2+} < Cd^{2+} < Cu^{2+}$.

7.3.4 Purification ability of the SW to various contaminants

The PA was an essential parameter to reflect purification ability of SW, and the PA_{120} of SW to every contaminant was computed by Equation (7.2) when t was 120 as shown in Table 7.2. The PA_{120} of the simulated wetland tended to decrease with the increase of the initial setting concentration. It was found that the PR_{120} of metals was much larger than that of TP, suggesting that the SW had greater purification potential for metal ions than TP. By considering average value of various concentrations for every element, the purification ability of the SW to pollutants could be ranked as follows: $Zn^{2+} < Cd^{2+} < TP^{2+} < Cu^{2+} < Pb^{2+}$.

Owing to the limitation of the field simulation experiment, it was difficult to ensure that each SW in the simulation had the same content of peat, *Sphagnum* plants, and wetland water. As a result, the accurate quantitative relation between PR_{120} and initial concentration was not calculated, but only estimated. However, comparison among each ion in every gradient could be made. In all, the purification for lead and copper were both great, and PR_{120} values at all concentrations were more than 90%. Taking 40 mg l^{-1} copper liquor, for example, the PR_{120} was up to 99.83%. The effect for zinc was the poorest and PR_{120} at every concentration was smaller than that of the other three ions.

The results of an earlier study showed that the peat had a great contribution to the purification of SW for copper ion, accounting for more than 90% of the PA (Hu et al., 2008). Considering that peat had a strong adsorption of divalent metal ion (Rock et al., 1985; Ho and McKay, 2000; Ringqvist and Oborn, 2002), any comparison of purification effect of SW on different metal ions was actually a comparison to that of peat. Kalmykova et al. (2008) reported that the peat adsorption effect size order for four ions was in the order of $Pb^{2+} > Cu^{2+} > Cd^{2+} > Zn^{2+}$, and the same conclusion could be made in our study. The affinity between lead and peat and copper and peat

Table 7.2 The PR_{120} of Phosphorus and Metal Contaminants at 120 min

Concentration (mg l^{-1})	P	Zn	Pb	Cu	Cd	Metals
20	98.32	71.01	99.51	97.95	96.86	91.34
40	83.70	85.42	99.39	99.83	97.00	95.41
80	91.06	71.97	98.11	98.48	81.44	87.50
160	80.84	75.60	96.80	93.92	77.59	85.98
Average	88.48	76.00	98.45	97.55	88.22	90.06

Note: Metals (a code) represents total of every metal, and its value in this table is the average of all the metals at the same concentration.

was larger than that between cadmium and peat and zinc and peat. Ringqvist and Oborn (2002) suggested that the adsorption effect declined with the enhancement of the copper ion concentration. Since the adsorption sites probably existed for different ions on the peat surface and zinc did not have its own sites, resulting in adsorption competition with other three metal ions, it affected the adsorption of peat to zinc (Sprynskyy et al., 2006). In addition, when solution pH was less than 4.0, there were many hydrogen ions so that competition would occur between H^+ and metal ions for adsorption site (Chen and Wang, 2007). The poorest purification effect of SW to zinc was limited, resulting from the weakest affinity between peat and zinc and a large amount of H^+ (the pH of SW is 3–4) (Chen et al., 1990; Ringqvist and Oborn, 2002; Kalmykova et al., 2008).

7.3.5 The best kinetic model selection for purification effects of constructed *SW*

A pollution purification system includes dilution, chemical complexation, physical adsorption, and bioconcentration. Because adsorption is the major part of the purification, the preliminary study introduced six adsorption kinetic models to the model purification dynamic and the efficiency of SW (Table 7.3). The pseudo first-order kinetic model and the pseudo second-order kinetic model were employed to fit the experimental data. The coefficient of determination (R^2) for Model 1 ranged from 0.226 to 0.902, indicating poor fitness. After q_e was fixed as the maximum of PA, the R^2 for Model 2 ranged from 0.643 to 0.980, indicating better fitness. Although Model 2 reflected the dynamic process of purification to a certain extent, purification of wetland had multiple purification mechanism because of its heterogeneous system. The pseudo second-order kinetic model was often used in heterogeneous system. The results also proved that this model has the best fitness, indicated by R^2 ranging from 0.883 to 1.000.

In the SW purification system, the rates of both purification and release of pollutants should be considered. When approaching equilibrium, the two rates are close to each other. The Elovich model was employed to evaluate this process and to calculate the rates. We also obtained initial purification rates, and the results indicated that copper was purified much more easily by SW than phosphors. To study wetlands' mass transfer process, a hyperbolic diffusion model was employed. If the intercept $a_3 = 0$, the model fitted poorly. However, if $a_3 \neq 0$, the model fitted well, indicating that an initial purification effect might be obtained because of SW's dilution.

The average deviations between observed data and modeled data by six models at 120 min were 3.5%, 1.5%, 0.7%, 0.4%, 0.7%, and 1.2%. Considering both R^2 and the deviation, Models 3 ～ 6 all worked well for modeling the SW's pacification over the 2 h. However, only Model 3 could predict the purification equilibrium because other models will increase without upper limit, which is not reasonable for purification process (Figure 7.4). Therefore, Model 3 was used in our further study. Model 1 was also modified to study dilution in the further study.

Table 7.3 Parameters of Kinetic Equation of the Purification (the *Sphagnum* Wetland) to Different Pollutants

	$q_{120,e}$ (mg cm^{-2})	Model 1				Model 2					Model 3			
		q_e (mg cm^{-2})	k_1 (min^{-1})	R^2	$D_{120,c}$ (%)	q_e (mg cm^{-2})	B (mg cm^{-2})	k'_1 (min^{-1})	R^2	$D_{120,c}$ (%)	q_e (mg cm^{-2})	k_2 (cm^2 mg^{-1} min^{-1})	R^2	$D_{120,c}$ (%)
WP10	0.078	0.0751	0.0601	0.902	4.1	0.0833	0.8036	0.0168	0.964	0.1	0.0856	1592.8	0.998	0.7
WP100	0.579	0.5524	0.0667	0.897	4.7	0.8333	0.684	0.0044	0.910	1.1	0.6196	4.2038	0.998	1.4
WP1000	5.522	5.2938	0.0455	0.226	4.5	8.3333	0.4546	0.006	0.643	4.3	6.8517	0.0031	0.883	1.0
WCu10	0.083	0.0827	0.1754	0.719	0.8	0.0833	2.5236	0.0351	0.971	0.2	0.0841	1684.1	1.000	0.1
WCu100	0.813	0.7895	0.1129	0.564	2.9	0.8333	1.5639	0.0189	0.792	0.3	0.843	1.6694	0.998	0.2
WCu1000	7.335	7.0498	0.1023	0.711	3.9	8.3333	1.2663	0.0073	0.980	0.3	7.6185	0.0023	0.999	0.9

	$q_{120,e}$ (mg cm^{-2})	Model 4				Model 5				Model 6			
		a_1 (cm^2 mg^{-1} min^{-1})	k_3 (cm^2 mg^{-1})	R^2	$D_{120,c}$ (%)	a_2 (mg cm^{-2})	k_4 (mg cm^{-2} min^{-1})	R^2	$D_{120,c}$ (%)	a_3 (mg cm^{-2})	k_5 (mg cm^{-2} in^{-1})	R^2	$D_{120,c}$ (%)
WP10	0.078	0.0469	77.1605	0.966	0.6	-3.4778	0.1973	0.953	1.5	0.0415	0.0035	0.928	2.0
WP100	0.579	0.6739	11.8751	0.970	0.1	-1.3568	0.1704	0.961	0.5	0.3362	0.0227	0.931	1.1
WP1000	5.522	1.5518	0.9438	0.497	0.8	0.6222	0.2235	0.468	1.6	2.2011	0.3145	0.577	2.3
WCu10	0.083	1.00×10^{16}	574.713	0.980	0.2	-2.5846	0.0213	0.979	0.2	0.0785	0.0005	0.939	0.4
WCu100	0.813	547.1455	17.0969	0.882	0.2	-0.5702	0.0765	0.890	0.3	0.6439	0.0161	0.884	0.9
WCu1000	7.335	812.3113	1.6416	0.979	0.5	1.5576	0.0902	0.982	0.3	5.522	0.1678	0.978	0.3

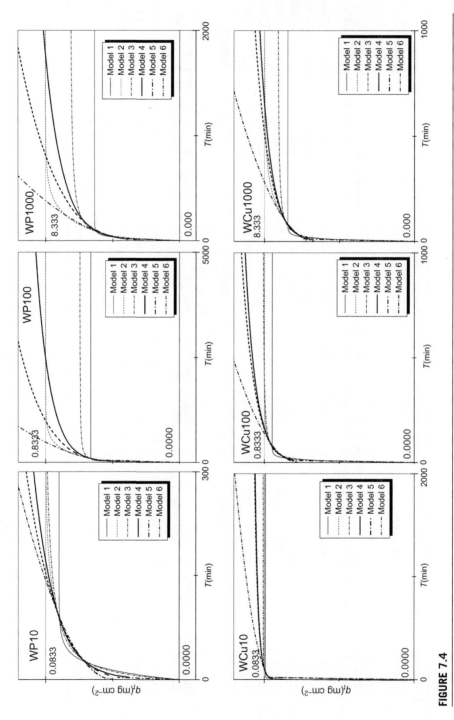

FIGURE 7.4

The prediction of the purification (the *Sphagnum* wetland) to different pollutants using six models.

7.3.6 **Modeling the purification effects of the constructed *SW***

Based on the research above, the dilution effect of the SW played an important role in the first 5 min of the purification. Under the assumption that the process of dilution was completed in 5 min, a nonlinear relationship was obtained using the experimental data after the time point of 5 min to construct a plot of q_t versus t according to Figure 7.3 (the dashed line). The constants in Model I (Equation 7.5) were presented in Table 7.4, in which the practical meaning of q_0 was the contribution of the dilution effect to the purification of the simulated wetland. The fitted values of the parameter q_e, the equilibrium purification amount, in Model I conformed to reality, that is to say, they were smaller than the initial set values. The values of rate constant k_1 were found to be greater at lower concentrations than that at higher ones. The data showed good compliance with the proposed modified pseudo first-order equation (Equation 7.5), because the regression coefficients were between 0.9069 and 0.9999, with most of them better than 0.99.

The CRDE to the purification effects of the SW could be calculated using Equation (7.11) in Table 7.5. For divalent metal ions, CRDE was the greatest at the initial concentration of 40 mg l^{-1}. Overall CRDE was greater at higher initial concentrations than at lower ones, which indicated that the adsorption of metal ions by the simulated wetland was greater with higher initial concentrations, that is, the value of the initial absorption rate of the SW to contaminants increased with the augmentation of the initial concentration. The average CRDE of the four metal ions at different concentrations decreased in the order: $Pb^{2+} > Cu^{2+} > Cd^{2+} > Zn^{2+}$.

The results once again suggested that peat had a different adsorption amount and initial adsorption rate for different metal ions, among which Pb^{2+} was the largest for both of them. Using the experimental data, a linear relationship was obtained from the plot of t/q_t versus t, which indicated the linear form of Model II (Equation 7.7), was more applicable to the system studied, because the regression coefficients were between 0.9855 and 1, shown in Table 7.4. The equilibrium purification amount q_e increased with the increase of initial concentrations of the contaminants, while the values of rate constant in Model II, k_2, were found to decrease when C_0 increased from 20 to 160 mg l^{-1}. The value of k_2 decreased by one order of magnitude when the initial concentration increased onefold for copper. Among different contaminants, the values of rate constant varied distinctly even at the same level of concentration, especially at lower concentrations.

The values of rate constant of copper and lead were about an order of magnitude higher than cadmium and zinc, when the concentrations were 20 and 40 mg l^{-1}. When it came to higher initial concentrations, that is, 80 and 160 mg l^{-1}, the value of rate constant of copper was at the same order of magnitude with cadmium and zinc with little difference between them, while the value of lead was obviously higher than that of the rest, suggesting the SW has the highest purification amount for lead.

Two parameters, the relative deviation (RD) and the average relative deviation (ARD), had been introduced when evaluating the two kinetic models. For Model I RD was between -1.5% and 13.0% and ARD was 2.1%, while for Model II,

Table 7.4 Parameters of Model I and Model II

Parameters		Model I						Model II				
C_0 (mg l⁻¹)	q_e (mg)	q_0 (mg)	$k_1 \times 10^{-2}$ (min⁻¹)	R^2	$PR_{120,c}$ (%)	RD (%)		q_e (mg)	$k_2 \times 10^{-4}$ (mg⁻¹min⁻¹)	R^2	$PR_{120,c}$ (%)	RD (%)
Cd												
20	193.23	134.99	8.80	0.995	96.6	−0.3		196.46	36.65	1.000	97.1	0.3
40	391.07	299.82	1.97	0.962	97.8	0.8		389.11	8.61	0.998	94.9	−2.1
80	655.59	292.40	2.82	0.991	81.9	0.6		684.93	1.72	0.997	80.0	−1.8
160	1269.63	499.34	2.87	0.999	79.4	2.3		1347.23	0.69	0.997	77.3	−0.4
Cu												
20	198.63	168.88	18.57	0.997	99.3	1.4		196.85	507.93	1.000	98.3	0.4
40	397.94	362.67	6.26	1.000	99.5	−0.3		400.00	26.15	1.000	99.2	−0.6
80	778.48	527.61	6.11	0.997	97.3	−1.2		819.67	2.22	1.000	98.0	−0.5
160	1551.19	1095.36	3.72	1.000	96.9	3.2		1603.69	0.64	0.986	92.7	−1.3
Pb												
20	197.22	163.95	17.19	0.984	98.6	−0.9		199.60	193.08	1.000	99.6	0.1
40	395.27	343.39	6.19	0.964	98.8	−0.6		400.00	53.42	1.000	99.6	0.2
80	773.66	345.23	5.27	0.992	96.7	−1.4		800.00	5.44	1.000	98.1	0.0
160	1726.52	630.54	1.46	0.984	107.9	11.5		1581.86	2.15	1.000	96.5	−0.3
Zn												
20	142.76	83.39	4.18	0.997	71.4	0.5		149.03	10.49	0.992	70.7	−0.4
40	386.14	234.51	1.22	0.907	96.5	13.0		355.87	5.75	0.993	85.5	0.1
80	583.74	285.20	2.50	0.981	73.0	1.4		602.41	2.09	0.997	70.6	−1.9
160	1249.96	501.24	2.64	0.996	78.1	3.3		1315.45	0.70	0.996	75.4	−0.3
TP												
20	197.55	93.12	5.14	0.992	98.8	0.5		204.50	11.03	0.999	98.6	0.3
40	335.38	95.08	5.15	1.000	83.8	0.2		355.87	2.70	0.986	81.9	−2.2
80	717.83	344.86	4.61	0.993	89.7	−1.5		757.58	2.36	0.999	90.5	−0.6
160	1315.61	746.83	2.64	0.997	82.2	1.7		1342.92	1.28	0.998	80.1	−1.0
Metals												
20	733.32	596.93	4.54	0.956	91.7	0.4		740.74	9.80	1.000	91.5	0.2
40	1556.87	1256.55	1.98	0.998	97.3	2.0		1543.45	3.25	1.000	94.9	−0.5
80	2771.03	1477.72	3.85	0.992	86.6	−1.0		2895.26	0.63	1.000	86.5	−1.1
160	5560.47	2783.17	3.00	0.972	86.9	1.0		5699.40	0.32	0.998	85.2	−1.0
ARD[a]						2.1						0.7

[a]ARD is average of the values above in RD column of each model.

Table 7.5 The Contribution Ratio of Dilution Effects to the Purification of the Simulated Wetland

Concentration mg l⁻¹	CRDE ($q_0/C_0 \cdot V$) (%)					
	P	Zn	Pb	Cu	Cd	Metals
20	45.47	41.69	84.44	81.98	67.49	68.90
40	36.61	58.63	90.67	85.85	74.95	77.53
80	43.11	35.65	65.95	43.15	36.55	45.33
160	46.68	31.33	68.46	39.41	31.21	42.60
Average	42.97	41.83	77.38	62.60	52.55	58.59

Table 7.6 The Expected Equilibration Time of Different Contaminants

Concentration (mg l⁻¹)	Equilibration Time (min)				
	Cd	Cu	Pb	Zn	P
20	40	15	20	65	70
40	55	35	25	65	90
80	80	75	50	75	75
160	90	95	55	90	70

RD was between −2.2% and 0.4% and ARD was 0.7%. When the regression coefficient was considered as well as RD and ARD, it could be concluded that Model II fit the purification system studied best, which was consistent with the results of an earlier year's study (Hu et al., 2008).

Model II fitted the experimental data very well, thus it could be used to estimate the equilibration time of the purification system. It was considered to have reached equilibrium if no changes in the purification amount greater than 0.5% of the initial set value were observed in a certain time segment (Kalmykova et al., 2008). The fitting effect of Model II was satisfactory, so the experimental data at every time point could be calculated by using parameters of Equation (7.7). Then, using Equation (7.13) to calculate the ratio of purification amount ($\Delta t = 5$ min) and initial set value, if $R_{\Delta t} < 0.5\%$, the equilibration time was t_2, according to Table 7.6. The equilibration time of every contaminant varied from 15 to 95 min, and it took a longer time to reach steady state for contaminants with higher initial concentrations. When the initial concentrations were 20 and 40 mg l⁻¹, lead and copper had the equilibration time of 15–35 min, while it took about two to three times longer for cadmium, zinc, and TP to reach equilibrium. When the initial concentrations were 80 and 160 mg l⁻¹, the equilibration time for all the contaminants were between 70 and 95 min, except for lead, the values of which were 50 and 55 min respectively. The time required to reach equilibrium for heavy metal ions had significant positive correlation with their initial concentrations ($p = 0.004 - 0.048 < 0.05$), while the correlation was poor for phosphorus ($p = 0.308 > 0.05$).

The relationship of the equilibrium PA, q_e, the rate constant, k_2, and the initial concentration, C_0, in Table 7.4 could be described by Equations (7.14) and (7.15) respectively, and the fitting parameters were calculated. The fitness was so good that correlation coefficient was 0.9953–1.000 for Equation (7.14), and it was 0.8629–0.9987 for Equation (7.15). Substituting the parameters into Equations (7.14) and (7.15) and then into Equation (7.6), the empirical formulas of q_t from C_0 and t were obtained, shown as Equations (7.16)–(7.21):

for Cd^{2+}:

$$q_t = \frac{(44.55 + 8.13C_0)^2 C_0 t}{3.43 \times 10^3 C_0 - 6.32 \times 10^4 + (44.55 + 8.13C_0)^2} \tag{7.16}$$

for Cu^{2+}:

$$q_t = \frac{(1.24 + 10.05C_0)^2 C_0 t}{1.26 \times 10^3 C_0 - 2.49 \times 10^4 + (1.24 + 10.05C_0)^2} \tag{7.17}$$

for Pb^{2+}:

$$q_t = \frac{(5.28 + 9.87C_0)^2 C_0 t}{0.60 \times 10^3 C_0 - 1.10 \times 10^4 + (5.28 + 9.87C_0)^2} \tag{7.18}$$

for Zn^{2+}:

$$q_t = \frac{(-8.48 + 8.19C_0)^2 C_0 t}{4.44 \times 10^3 C_0 - 6.99 \times 10^4 + (-8.48 + 8.19C_0)^2} \tag{7.19}$$

for TP:

$$q_t = \frac{(50.65 + 8.19C_0)^2 C_0 t}{6.50 \times 10^3 C_0 - 11.18 \times 10^4 + (50.65 + 8.19C_0)^2} \tag{7.20}$$

for metals:

$$q_t = \frac{(13.83 + 9.54C_0)^2 C_0 t}{9.00 \times 10^3 C_0 - 15.95 \times 10^4 + (13.83 + 9.54C_0)^2} \tag{7.21}$$

Equations (7.16)–(7.21) showed the contaminants purified by the simulated wetland as a function of contact time and initial concentration, thus the equations could be used to predict the purification amount of the five contaminants at any reaction time and at any given initial concentration between 20 and 160 mg l^{-1} using the system studied in this chapter.

7.3.7 **Importance of the *SW*, its purification mechanism, and further study**

Dajiuhu SW has recently confronted a great ecological environmental change such as the increase of discharge amount and planting of vegetables on the wetland, which leads to gradual degradation of wetland (Yu et al., 2008). All these activities caused great damage to the ecological function of the SW. Moreover, the impact of agricultural non-point source pollution to the water source area of MLPWTSN might happen in the near future unless we protect the remaining SW and recover the lost wetland. The conclusion of this paper emphasizes the importance of the Dajiuhu wetland such as SW to the preservation of the water source, suggesting that measures be taken to forbid further reclamation and to protect and restore the wetland as far as possible.

The SW's purification results from combination of chemical, physical, and biological and many other factors (Figure 7.5). The purification portion of the wetland includes a matrix part and a biological part. The matrix part consists of the soil layer (inorganic layer and organic humus layer), the peat layer, and gravel layer, while the biological part consists of leaves and roots (survival roots and rotted roots) of *Sphagnum* moss and microorganisms. Purification mechanisms of phosphorus are mainly *Sphagnum* moss leaves' uptake, microbial uptake, and adsorption of soil, peat, and gravel. Purification mechanisms of metals are uptake by leaves and roots of *Sphagnum* moss, complexation of organic humus soil, and adsorption of peat.

Considering the protection and sensitivity of the study area, a large-scale *in situ* study was not designed. Therefore, there were some disadvantages in a box simulation research near the area. They included the following: (1) The simulation time lasted only 120 min, so absorption of the *Sphagnum* plants and microbial action were not studied; (2) a small-scale study could not reflect the purification potential of the whole SW; (3) non-point resource of the farmland near the wetland was not considered in evaluation of the importance of the SW purification function; (4) the depth of

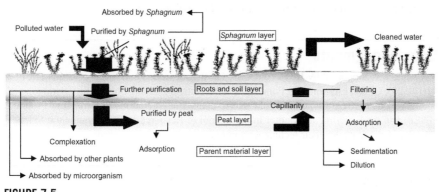

FIGURE 7.5

The purification mechanism of *Sphagnum* wetland.

the wetland, especial peat layer depth, was larger than the simulated depth in the box, so the purification of the wetland might be underestimated; (5) the concentration of the contaminated water in this study might be a little higher than that in the environment, hence, a lower concentration should be added. In consideration of the actual situation of Dajiuhu and disadvantage of the box simulation, a further study, comprising two directions, will be performed. One is determining the purification potential effect of SW on total phosphate from the farmland in large scale with a long duration, and the other is evaluation of the purification potential effect of the whole SW using the box simulation data, remote sensing data, geological structure data, and parameters of contamination transport model in the SW.

7.4 CONCLUSIONS

Six models, pseudofirst-order kinetic model, modified pseudo first-order kinetic model, pseudo second-order kinetic model, Elovich model, two-constant rate model, and hyperbolic diffusion model, were used to simulate the purification effect. Considering both the fitting determination coefficients and the prediction, the pseudo second-order kinetic model was the best for modeling the purification effects of the constructed SW on phosphorus and heavy metals. The purification ability of the SW was remarkable for all the contaminants: 80.8–98.3% of phosphorus, 71.0–85.4% of Zn^{2+}, 96.8–99.3% of Pb^{2+}, 93.9–99.8% of Cu^{2+}, and 81.44–96.9% of Cd^{2+} were removed from contaminated water after 2 h. The time heavy metal ions required to reach equilibrium had a significant positive correlation with their initial concentrations, while the correlation was poor for phosphorus. The dilution effect of the SW played an important role during the first 5 min; average CRDE value of every contamination is from 43.0% to 77.4%. Empirical formulas for predicting the purification ability of the SW were derived using parameters of Model II, which could be used in evaluating the purification value of the SW and providing the scientific basis for the protection and proper utilization of the Dajiuhu SW.

Acknowledgments

This work was supported by Key Project of National Science Foundation of China (NSFC) (41030529), Global Environment Fund (GEF), the National Foundation for the Distinguished Young Scholars (40725004). We would like to thank United Nations Environment Program (UNEP) and the Experimental Center in the School of Resource and Environmental Science of Wuhan University for their technical supports.

References

Adler, E., Lundquist, K., 1963. Spectrochemical estimation of phenylcoumaran elements in lignin. Acta Chem. Scand. 17 (1), 13–26.

Asplund, D., Ekman, E., Thun, R., 1972. Counter-current peat filtration of waste water. In: Proceedings of 4th International Peat Congress. vol. 5, pp. 358–371.

Babel, S., Kurniawan, T.A., 2003. Low-cost adsorbents for heavy metals uptake from contaminated water: a review. J. Hazard. Mater. 97 (1–3), 219–243.

Bailey, S.E., Olin, T.J., Bricka, R.M., Adrian, D.D., 1999. A review of potentially low-cost sorbents for heavy metals. Water Res. 33 (11), 2469–2479.

Bindu, T., Sylas, V.P., Mahesh, M., Rakesh, P.S., Ramasamy, E.V., 2008. Pollutant removal from domestic wastewater with Taro (*Colocasia esculenta*) planted in a subsurface flow system. Ecol. Eng. 33 (1), 68–82.

Blanken, P.D., Rouse, W.R., 1996. Evidence of water conservation mechanisms in several subarctic wetland species. J. Appl. Ecol. 33 (4), 842–850.

Brix, H., 1997. Do macrophytes play a role in constructed treatment wetlands? Water Sci. Technol. 35 (5), 11–17.

Brooks, A.S., Rozenwald, M.N., Geohring, L.D., Lion, L.W., Steenhuis, T.S., 2000. Phosphorus removal by wollastonite: a constructed wetland substrate. Ecol. Eng. 15 (1–2), 121–132.

Chen, Y.Y., Lu, X.G., 2003. The wetland function and research tendency of wetland science. Wetland Sci. 1 (1), 7–11 (in Chinese).

Chen, H., Wang, A.Q., 2007. Kinetic and isothermal studies of lead ion adsorption onto palygorskite clay. J. Colloid Interface Sci. 307 (2), 309–316.

Chen, X., Gosset, T., Thevenot, D.R., 1990. Batch copper-ion binding and exchange properties of peat. Water Res. 24 (12), 1463–1471.

Cheng, S.P., Grosse, W., Karrenbrock, F., Thoennessen, M., 2002. Efficiency of constructed wetlands in decontamination of water polluted by heavy metals. Ecol. Eng. 18 (3), 317–325.

Coupal, B., Lalancette, J.M., 1976. Treatment of waste-waters with peat moss. Water Res. 10 (12), 1071–1076.

Ding, G.B., 2006. Water quality-determination of copper, zinc, lead and cadmium-atomlc absorption spectrometry. In: Wei, J.B., Wu, F. (Eds.), Environment Monitoring Handbook. Chemical Industry Press, Beijing, pp. 25–26 (in Chinese).

Gardea, T.J.L., Tang, L., Salvador, J.M., 1996. Copper adsorption by esterified and unesterified fractions of *Sphagnum* peat moss and its different humic substances. J. Hazard. Mater. 48 (1–3), 191–206.

He, B.Y., Zhang, S., Cai, S.M., 2003. Climatic changes recorded in peat from the Dajiuhu lake basin in Shennongjia since the last 2600 years. Marine Geol. Quat. Geol. 23 (2), 109–115 (in Chinese).

He, W., Zhang, Y.Y., Tian, R., Hu, H.X., Chen, B., Chen, L.K., Xu, F.L., 2013. Modeling the purification effects of the constructed Sphagnum wetland on phosphorus and heavy metals in Dajiuhu Wetland Reserve, China. Ecol. Model. 252, 23–31.

Ho, Y.S., McKay, G., 1999. A kinetic study of dye sorption by biosorbent waste product pith. Resour. Conserv. Recycl. 25 (3–4), 171–193.

Ho, Y.S., McKay, G., 2000. The kinetics of sorption of divalent metal ions onto *Sphagnum* moss flat. Water Res. 34 (3), 735–742.

Ho, Y.S., McKay, G., 2004. Sorption of copper (II) from aqueous solution by peat. Water Air Soil Pollut. 158 (1), 77–97.

Ho, Y.S., Wase, D.A.J., Forster, C.F., 1996. Kinetic studies of competitive heavy metal adsorption by *Sphagnum* moss peat. Environ. Technol. 17 (1), 71–77.

Hu, H.X., He, W., Liu, Q.L., 2008. Study on simulating the purification effects of the *Sphagnum* wetland using the wastewater that contain phosphorus and cuprum in DaJiuhu. Resour. Environ. Yangtze Basin 17 (6), 917–923 (in Chinese).

Hu, H.X., Zhang, Y.Y., He, W., Tian, R., Zhong, X., Han, S.S., Li, S.S., Wang, J.J., Chen, W.F., Yang, Y., Chen, C., Deng, H., Wen, Y., Cui, Y.T., Li, X., Wang, X., Peng, J.J., Gao, X., Tang, Y., 2009. Study on the purification effects of the Sphagnum wetland on Cd2+, Cu2+, Pb2+, Zn2+ in Dajiuhu Basin of Shennongjia mountains. Resour. Environ. Yangtze Basin 18 (11), 1050–1057 (in Chinese).

Kalmykova, Y., Stromvall, A.M., Steenari, B.M., 2008. Adsorption of cd, cu, ni, pb and zn on *Sphagnum* peat from solutions with low metal concentrations. J. Hazard. Mater. 152 (2), 885–891.

Kayranli, B., Scholz, M., Mustafa, A., Hedmark, A., 2010. Carbon storage and fluxes within freshwater wetlands: a critical review. Wetlands 30 (1), 111–124.

Lee, B.H., Scholz, M., 2007. What is the role of *Phragmites australis* in experimental constructed wetland filters treating urban runoff? Ecol. Eng. 29 (1), 87–95.

Mann, R.A., 1997. Phosphorus adsorption and desorption characteristics of constructed wetland gravels and steelworks by-products. Austr. J. Soil Res. 35 (2), 375–384.

McDowell, R.W., Sharpley, A.N., 2003. Phosphorus solubility and release kinetics as a function of soil test P concentration. Geoderma 112 (1–2), 143–154.

Peng, J.F., Wang, B.Z., Song, Y.H., Yuan, P., Liu, Z.H., 2007. Adsorption and release of phosphorus in the surface sediment of a wastewater stabilization pond. Ecol. Eng. 31 (2), 92–97.

Peregon, A., Uchida, M., Shibata, Y., 2007. *Sphagnum* peatland development at their southern climatic range in West Siberia: trends and peat accumulation patterns. Environ. Res. Lett. 2 (4), 045014.

Pollino, C.A., Woodberry, O., Nicholson, A., Korb, K., Hart, B.T., 2007. Parameterisation and evaluation of a Bayesian network for use in an ecological risk assessment. Environ. Model Softw. 22 (8), 1140–1152.

Ringqvist, L., Oborn, I., 2002. Copper and zinc adsorption onto poorly humified *Sphagnum* and *Carex* peat. Water Res. 36 (9), 2233–2242.

Ringqvist, L., Holmgren, A., Oborn, I., 2002. Poorly humified peat as an adsorbent for metals in wastewater. Water Res. 36 (9), 2394–2404.

Rock, C.A., Fiola, J.W., Greer, T.F., Woodward, F.E., 1985. Potential of sphagnum peat to remove metals from landfill leachate. J. N. Engl. Water Pollut. Control Assoc. 19 (11), 32–47.

Sakadevan, K., Bavor, H.J., 1998. Phosphate adsorption characteristics of soils, slags and zeolite to be used as substrates in constructed wetland systems. Water Res. 32 (2), 393–399.

Scholz, M., 2003. Performance predictions of mature experimental constructed wetlands which treat urban water receiving high loads of lead and copper. Water Res. 37 (6), 1270–1277.

Sedeh, F., Igsell, P., Ringqvist, L., Lindström, E.B., 1996. Comparison of metal adsorption properties and determination of metal adsorption capacities of different peat samples. Resour. Environ. Biotechnol. 1, 111–128.

Sprynskyy, M., Buszewski, B., Terzyk, A.P., Namiesnik, J., 2006. Study of the selection mechanism of heavy metal (Pb2+, Cu2+, Ni2+, and Cd2+) adsorption on clinoptilolite. J. Colloid Interface Sci. 304 (1), 21–28.

Tanner, C.C., Sukias, J.P.S., Upsdell, M.P., 1998. Organic matter accumulation during maturation of gravel-bed constructed wetlands treating farm dairy wastewaters. Water Res. 32 (10), 3046–3054.

von Felde, K., Kunst, S., 1997. N- and COD-removal in vertical-flowsystems. Water Sci. Technol. 35 (5), 79–85.

Wang, M.N., Qin, D.Y., Li, Y.P., Wei, H.B., Shen, Y.Y., 2010. A study of the impact of wetlands on regional water cycle: the Qingdianwa Wetland example. Fresenius Environ. Bull. 19 (1), 9–19.

Xiao, F., Du, Y., Ling, F., Wang, X.L., Soc, I.C., 2009. Remote-sensing and modeling of the potential suitable sites for restoration in Dajiuhu sub-alpine wetland. In: 2009 International Conference on Environmental Science and Information Application Technology, vol. II, pp. 447–450, Proceedings.

Xu, D.F., Xu, J.M., Wu, J.J., Muhammad, A., 2006. Studies on the phosphorus sorption capacity of substrates used in constructed wetland systems. Chemosphere 63 (2), 344–352.

Yin, F.N., Wang, X.L., Yu, J., 2007. A study on landuse change of the Dajiu Lake and the impact on wetland ecological environment. J. Huazhong Normal University 41 (1), 148–151 (in Chinese).

Yu, J., Wang, X.L., Wu, Y.J., Yin, F.N., 2008. Changes of Shennongjia Dajiuhu landscape pattern and the strategies of wetland ecological restoration. J. Huazhong Agric. Univ. 27 (1), 122–126 (in Chinese).

Zhao, K.Y., 1999. Function of wetland ecosystem. In: Lang, H.Q., Zhao, K.Y., Chen, K.L. (Eds.), Wetland Vegetation in China. Science Press, Beijing, pp. 389–390 (in Chinese).

Ecological Accounting for a Constructed Wetland

Ling Shao[a], Z.M. Chen[b], G.Q. Chen[a,c],*

[a]*State Key Laboratory for Turbulence and Complex Systems, College of Engineering, Peking University, Beijing, China*
[b]*School of Economics, Renmin University of China, Beijing, China*
[c]*NAAM Group, King Abdulaziz University, Jeddah, Saudi Arabia*
Corresponding author: e-mail address: gqchen@pku.edu.cn

8.1 INTRODUCTION

Constructed wetlands were originally devised to treat wastewater with dependence on natural microbial, biological, physical, and chemical processes. The technical and economic performances of constructed wetlands have been extensively investigated from the engineering point of view (Brix, 1994; White, 1995; Cardoch et al., 2000). Some of the results have been compared with those of conventional chemical wastewater treatment systems for the purpose of making optimal engineering choices between alternatives. However, a constructed wetland has other ecological implications besides the improvement of water quality. As one of the most typical ecological engineering systems, it combines properties of both natural wetland and artificial engineering. The conventional economic accounting simply regarding the natural process as a free service fails to assess the overall performances of the constructed wetland (Krutilla, 1967). Owing to the rising concern for excessive consumption of natural resources, rapid growth of environmental emissions, and the increasing need of ecosystems restoration (Costanza et al., 1997), the ecological accounting method, which covers all biophysical flows of both natural resources and manmade products, has been proposed to reveal the real ecological implications of constructed wetlands.

The ecological accounting for constructed wetlands has been a hot research topic. The ecosystem service value of constructed wetlands has been measured by money in context of environmental economics, which has contributed a lot by providing an intelligible way to reveal the important role of a constructed wetland (Costanza et al.,

Developments in Environmental Modelling, Volume 26, ISSN 0167-8892, http://dx.doi.org/10.1016/B978-0-444-63249-4.00009-9

1997; Chen et al., 2009b).[1] However, using money, a derivative of human society, as an ecological equivalent may be unfair to other beings and not relevant to the sustainability assessment of the ecosystems (Odum, 1996).

A constructed wetland has direct and indirect ecological effects. The on-site direct employment of ecological endowments in terms of natural resources such as solar energy and environmental emissions such as greenhouse gases (GHGs) is obvious, while the indirect employment associated with various economic inputs to construct and sustain the engineering is somewhat implicit: Every economic input produced causes resource consumption and environmental emissions. The direct plus indirect ecological endowment required to produce a specific goods or service is defined as the embodied ecological endowment (Bullard and Herendeen, 1975; Costanza, 1980; Brown and Herendeen, 1996).

In the past decade, life cycle analysis (LCA) has been transplanted to estimate embodied ecological endowments of constructed wetlands (Dixon et al., 2003; Zuo et al., 2004; Machado et al., 2007; Fuchs et al., 2011; Pan et al., 2011), which has enormously extended our knowledge of constructed wetlands. However, the process-based LCA has inherent shortages. It tries to analyze each input of the constructed wetland individually and trace each stage of the production processes back to obtain the total environmental impacts. Generally, these traces have to be truncated after a few steps because it is time-consuming, infinite, and sometimes the tracing might even fall into a loop (Bullard et al., 1978; Treloar, 1997). The ignored upstream inputs can bring inevitable truncation errors. The current LCA studies do not need to trace each supply chain of the constructed wetland by itself, because the ecological endowment intensities of various products or services contributed by others are available. But due to the diverse technical efficiencies, the same products from various economies have different intensities. As a matter of fact, even the same

---

[1]A simple framework is shown below with reference to Costanza et al. (1997) and Chen et al. (2009b) to illustrate the ecosystem service value method as one of the most popular ecological accounting methods. The ecosystem services of a constructed wetland can be classified into six categories:

(1) Waste treatment value $V_w = V_p - V_a$, where V_p is the value gained from waste purification and V_a is the value lost from waste accumulation.

(2) Food and material production value $V_f = \sum P_i \times M_i$, where P_i is the price of the ith production and M_i is the net increase of the ith biomass.

(3) Water supply value $V_s = V_e - V_i$, where V_e and V_i are the values of the effluent and influent water, respectively.

(4) Gas regulation services due to the oxygen production and greenhouse gas emission $V_g = P_{ox} \times M_{ox} - P_e \times M_e$, where P_{ox} and P_e are the trade prices for oxygen and greenhouse gas, respectively; M_{ox} and M_e are the amounts of the oxygen production and greenhouse gas emission, respectively.

(5) Disturbance and water regulation value $V_d = C_d \times S$, where C_d is the construction cost for local disturbance and water regulation facility per unit storage volume and S is the water storage capacity of the ecosystem.

(6) Habitat and refugia provision value of biodiversity support $V_h = P_h \times A$, where P_h is the potential value for the biodiversity support service and A is the area that could currently fulfill the requirements as a conserved area for biodiversity and habitat.

kind of product produced by the same economy at a different time does not have the same intensity due to the varying economic structures. Moreover, the intensities are not consistent since the truncation of the trace is subjective and the accounting backgrounds and frameworks are quite different.

To cope with these problems, the systems ecological endowment accounting characterized as the combination of process analysis and ecological input–output analysis (IOA) has been performed by Chen and his collaborators (Chen and Chen, 2010; Chen, 2011) to assess various systems. The process analysis itemizes all the inputs and outputs of an economic or engineering system and then forms a complete inventory. Originating from the concept of embodiment in systems ecology by Odum (1983) as was explicitly illustrated in the embodied energy analysis by Costanza (1980), the ecological IOA, which integrates the consistent statistics for the macroscopic economy with various ecological impacts, provides a unified accounting base and avoids the truncation error.

This systems ecological accounting method is applicable to all kinds of engineering systems. Concerned ecological endowments include both the natural resources and environmental emissions such as energy, carbon, pollutants, water, land, minerals, or more fundamentally in systems ecological thermodynamics, solar energy (Odum, 1988, 1996) and cosmic exergy (Chen, 2005, 2006). This method has been concretely employed to assess the nonrenewable energy cost and GHG emissions by a solar power tower plant (Chen et al., 2011d), a wind power plant (Chen et al., 2011c), a "Pig-Biogas-Fish" system (Yang et al., 2012), as well as a building series (Chen et al., 2011a). This chapter presents a review for its application in the ecological accounting for a constructed wetland based on our abundant experiences in assessing the performances of ecological engineering (Chen et al., 2008, 2009a,b, 2010c, 2011b,e; Zhou et al., 2009; Shao et al., 2012).

8.2 METHODOLOGY

8.2.1 Accounting framework

Process analysis (LCA in particular) and IOA are two independent quantitative approaches for the embodied ecological endowment calculation of a target product or service in ecological accounting studies. Although process analysis can provide us with detailed information, it is labor intensive and suffers from the truncation error (Bullard et al., 1978; Treloar, 1997). The economic input–output tables-based IOA (Leontief, 1936, 1970) has been extended to calculate the embodied ecological endowment intensities of the produced goods and services within an economic system, especially their embodied energy and GHG emission intensities. It utilizes statistics and organizes matrices of all the intermediate inputs into goods and services of the whole economic network, making it capable of overcoming the truncation errors brought about by process analysis (Miller and Blair, 2009). However, the IOA gives an average quantity of ecological endowment embodied in the wastewater treatment industry, which is not detailed enough to elaborate the ecological implication of a specific constructed wetland system.

Given that, the hybrid approach combining process analysis and the IOA has been proposed and considered as an accurate, complete, quick, and widely applicable tool (Bullard et al., 1978; Joshi, 1999; Suh and Huppes, 2005). It enables us to account for the ecological endowment of a constructed wetland as a micro-system embedded in the macro-economy by tracing the indirect fluxes originated outside the process boundary with average sectoral intensities provided by proper IOA. As a consequence, the hybrid approach is utilized to account the ecological endowments of a constructed wetland in this work.

8.2.2 Inventory

Making a concrete inventory of inputs and outputs is the first and fundamental step of ecological accounting for a constructed wetland. A representative inventory with reference to daily practice of a constructed wetland is presented in Table 8.1.

8.2.3 Embodied ecological endowment intensity database

The embodied ecological endowment intensity is defined as the natural resources and environmental emission virtually contained in a unit of product or service. The embodied ecological endowment intensities of different natural resources are calculated according to the biophysical costs to generate those resources by natural processes. For example, the embodied solar energy intensity of raw coal is calculated as the total (direct plus indirect) solar energy cost by the earth process to transform organic matter into coal (Odum, 1988, 1996).

A systematic and consistent embodied ecological endowment intensity database of various economic products is vital for the ecological accounting of a constructed wetland. A number of intensity databases have been established by different researchers. Most of the intensities are obtained by process analysis and thus are principally associated with the issue of truncation error leading to systematic inconsistency. To avoid this limitation of process analysis, some other researchers apply IOA to calculate the average embodied ecological endowment intensities of economic products based on the consistent statistics for the macroscopic economy. However, in conventional IOA, also termed environmental IOA, the intensity is simply obtained by assigning direct environmental impacts delivered by the production to the final demand in terms of the Leontief inverse. In this way, logically the obtained intensity is only defined for the final demand and thus cannot be used in the assessment of any production process.

Recognizing this problem, Chen and his collaborators brought forward an alternative input–output based method, that is, the systems ecological IOA, to distinguish it from the aforementioned environmental IOA (Chen and Chen, 2010; Chen, 2011). Compared to the environmental input–output theory, which allocates environmental emissions and resources consumption to final demand, the systems input–output theory tries to simulate the balance and dynamics of internal embodied ecological endowment flows based on a systems perspective. As a result, the theory can be applied to analyze ecological endowment occupied by not only final demand but also

Table 8.1 Inventory of a Constructed Wetland

Note	Item		Input(+)/Output(−)	Unit	Raw Data
Environmental input					
	Sunlight		+	J/year	
	Wind, kinetic		+	J/year	
	Rain, chemical		+	J/year	
	Rain, geopotential		+	J/year	
	Earth cycle		+	J/year	
	Others		+/−	J/year	
Economic input					
	Soil excavating		−	t	
	Soil filling		+	t	
	Wastewater catchment		+	¥	
	Griller		+	kg	
	Valve		+	kg (¥)	
	Pump		+	kg (¥)	
	Electronic system		+	kg (¥)	
	Impervious membrane		+	kg	
	Bricks		+	t	
	Steel		+	kg	
	Concrete		+	t	
	Pipe		+	m	
	Substrate		+	t	
	Plant		+/−	kg	
	Waste				
		Wastewater	+	t/year	
		Solid waste	+/−	t/year	
		Others	+/−	t/year	
	Labor and others				
		Design and consult	+	¥	
		Maintenance	+	¥/year	
		Rent	+	¥/year	
		Others	+/−	¥	
Output					
	Food and material production		−	t/year	

Continued

Table 8.1 Inventory of a Constructed Wetland—cont'd

Note	Item		Input(+)/ Output(−)	Unit	Raw Data
	Waste treatment				
		Wastewater treatment	−	t/year	
		Solid waste treatment	+/−	t/year	
		Others	+/−	t/year	
	Water supply		−	t/year	
	Gas regulation				
		Greenhouse gas regulation	+/−	t/year	
		Oxygen regulation	+/−	t/year	
		Others	+/−	t/year	
	Disturbance and water regulation		−	m^3	
	Habitat and refugia provision		−	¥	
	Recycle material supply		−	t	
	Others		+/−	N/A	

intermediate productive activity, thus providing more comprehensive support for ecological endowment management policy. The systems ecological IOA-based databases are considered consistent, systematic, and accurate and therefore are proper to be applied in the ecological accounting of a constructed wetland.

This approach has been extensively extended by the multi-scale systems IOA of various ecological endowments for the world, Chinese and urban economies in different periods (Zhou, 2008; Chen and Chen, 2010, 2011a,b,c, 2013; Chen and Zhang, 2010; Chen et al., 2010b, 2011a, 2012; Zhou et al., 2010; Chen, 2011). Some databases aiming at different responsibility entities with distinguished ecological implications have also been obtained (Chen et al., 2013; Guo and Chen, 2013). Careful checks should be carried out at first in order to choose a proper database for the ecological accounting of a constructed wetland. It is recommended that the database adapted to the place and year where and when the constructed wetland was built should be the priority.

8.3 CASE STUDY

A case study has been carried out to illustrate the systems ecological accounting of a constructed wetland. The embodied GHG emissions, solar energy, and cosmic exergy have been accounted for a typical constructed wetland in Beijing (Zhou, 2004; Chen et al., 2008).

8.3.1 **Case description, inventory, and database**

Accompanied with the accelerating urbanization and industrialization of Beijing, the shortage of available water resources and heavy water pollution become critical problems hindering the development of the city. Under the theme of "Green Olympics," which was put forward for the Beijing Olympics 2008, a vertical subsurface flow constructed wetland for wastewater treatment was constructed in 2004 in a suburb near the estuary of the Longdao River, which has been heavily polluted by the municipal sewage. The designed lifetime of the case wetland is 20 years. With a vegetated bed of 602 m^2, the case system has a treatment capacity of 200 m^3 wastewater per day. A steel barrier pretreatment facility was installed to remove large-size suspended solids. A water pump was equipped to uplift water from the river to the wetland.

The construction and operation stages are the focus of this study. All the natural resources and economic products required in the life cycle of the case system have been itemized to form an inventory (see Table 8.2).

In the available Chinese databases that were published close to 2004, the supporting data for the Chinese economy in 2007 (135 sectors) is much more detailed than that for 2005 (42 sectors). Therefore the embodied endowment database for the Chinese economy in 2007 is adopted as the supporting data for estimating the ecological implications of various economic products (Chen and Chen, 2010). Emissions and resources considered have been classified into six categories as: (1) GHG in terms of CO_2, CH_4, and N_2O; (2) energy in terms of coal, crude oil, natural gas, hydropower, nuclear power, and firewood; (3) water in terms of freshwater; (4) resource exergy in terms of coal, crude oil, natural gas, grain, bean, tuber, cotton, peanut, rapeseed, sesame, jute, sugarcane, sugar beet, tobacco, silkworm feed, tea, fruits, vegetables, wood, bamboo, pulp, meat, egg, milk, wool, aquatic products, iron ore, copper ore, bauxite, lead ore, zinc ore, pyrite, phosphorite, gypsum, cement, nuclear fuel, and hydropower; (5) and (6) embodied solar energy and cosmic exergy in terms of sunlight, wind power, deep earth heat, chemical power of rain, geopotential power of rain, chemical power of stream, geopotential power of steam, wave power, geothermal power, tide power, topsoil loss, coal, crude oil, natural gas, ferrous metal ore, nonferrous metal ore, nonmetal ore, cement, and nuclear fuel. The database provides embodied ecological endowment intensities of 135 industrial sectors, including 5 sectors for agriculture; 5 for mining; 81 for manufacturing; 3 for electricity, gas, and water production and supply; 9 for transportation, storage, and post; 2 for construction and real estate; and 30 for other services.

8.3.2 **Embodied GHG emissions**

Three major GHGs, that is, CO_2, CH_4, and N_2O, are studied, and the Global Warming Potential factors of 1:21:310 for CO_2:CH_4:N_2O are used to account the equivalent CO_2 (CO_2-Eq.) emission according to IPCC (2006). As experimental data for the direct GHG emissions of the case system are not available, the data observed by

Table 8.2 Inputs and Outputs Inventory of the Case System

Note	Item	Quantity	Unit[a]
Environmental input			
	Sunlight	2.49E+12	J/year
	Wind, kinetic	3.65E+08	J/year
	Rain, chemical	1.15E+09	J/year
	Rain, geopotential	2.48E+07	J/year
	Earth cycle	1.14E+09	J/year
Economic input			
	Local organic substrate	690	m^3
	Mineral substrate	120	m^3
	Other substrate	220	m^3
	Geotextile	900	m^2
	Vegetation	28,500	Yuan
	Pump	2	
	Electric control	1	
	PP pipe	90	m
	PP valve	3	
	PE pipe	48	m
	PE valve	10	
	Steel griller	1	kg
	Bricks and cement	22	m^3
	Waste water	200	m^3/d
	Electricity	0.75	kW
	Maintenance	5.20	Yuan/d
Output			
	Treatment of wastewater	7.31E+04	m^3/year
	Plant biomass	3.09E+04	g/year
	Purified water	6.24E+04	m^3/year
	H_2O vapor	1.10E+04	t/year
	GHG	−5.82	t CO_2-Eq./year
	O_2	3.29E−02	t/year
	Regulation capacity	1.85E+02	m^3/year
	Vegetation area	6.02E+02	m^2/year

[a]All results without time unit refer to the whole lifetime of the case system.

Teiter and Mander (2005) as 0.72 mg N_2O/m^2 h, 4 mg CH_4/m^2 h, and 788.33 mg CO_2/m^2 h for vertical subsurface flow wetland beds have been adopted. Occupying a vegetated bed of 600 m^2, the direct GHG emissions for the whole lifetime of the case system are estimated as 0.08 t N_2O, 0.42 t CH_4, and 82.87 t CO_2, respectively.

After identifying the corresponding industrial sector for each economic input of the case system with reference to the economic input–output table, the GHG emission intensity can be promptly given. For example, electricity is provided by Sector 92 entitled "Production and supply of electric power and heat power," and the intensity of Sector 92 listed in the GHG emission intensity database is determined as the GHG emission intensity of electricity. The embodied GHG emissions of the case wetland are summarized in Table 8.3, in which some items of the same sector are combined for brevity.

The indirect GHG emissions, whose components are shown in Figure 8.1, account for 60.12% of the total GHG emissions. If only the direct emissions are considered, the GHG emissions of a constructed wetland will be greatly underestimated. About 60% of the indirect GHG emissions are caused by electricity. The GHG emission of every other input shares less than 10% of the total indirect emissions except for that of the local organic substrate. As ecological engineering, the constructed wetland relies a lot on natural resource inputs to treat the wastewater. The natural and waste resources as the vegetation and substrates account for about one-quarter

Table 8.3 The Embodied GHG Emissions of the Case System

Item	CO_2 (t)	CH_4 (t)	N_2O (t)	GHG (t CO_2-Eq.)	Fraction
Indirect GHG emission					
Geotextile	7.58E+00	4.97E−02	4.00E−04	8.75E+00	2.97%
Local organic substrate	5.72E+00	6.73E−02	4.08E−03	8.40E+00	2.85%
Mineral substrate	4.28E+00	2.73E−02	2.10E−04	4.93E+00	1.67%
Other substrate	1.57E+01	9.72E−02	4.53E−04	1.79E+01	6.07%
Vegetation	4.45E+00	1.11E−01	1.67E−02	1.20E+01	4.07%
Pump	1.07E+00	7.05E−03	4.26E−05	1.23E+00	0.42%
Electric control	3.40E−01	2.26E−03	1.67E−05	3.93E−01	0.13%
Pipe and valve	1.56E+00	1.02E−02	8.24E−05	1.80E+00	0.61%
Steel griller	2.11E−03	1.38E−05	8.30E−08	2.42E−03	0.00%
Bricks and cement	6.38E+00	2.13E−02	9.81E−05	6.86E+00	2.33%
Electricity	9.09E+01	6.15E−01	2.16E−03	1.05E+02	35.70%
Maintenance	8.24E+00	5.73E−02	8.47E−04	9.68E+00	3.29%
Subtotal	1.46E+02	1.07E+00	2.51E−02	1.77E+02	60.12%
Direct GHG emission	8.29E+01	4.21E−01	8.28E−02	1.17E+02	39.88%
Total GHG emission	2.29E+02	1.49E+00	1.08E−01	2.94E+02	100%

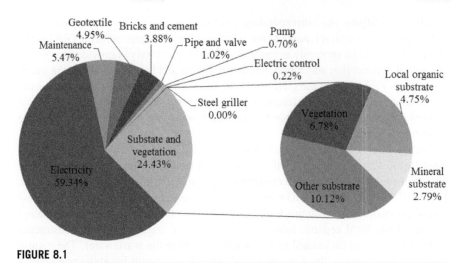

FIGURE 8.1

The components of the indirect GHG emission by the case system.

of the total indirect GHG emissions. Taking the direct emissions into account, the GHG emissions of natural processes and resources together would be a bit larger than a half of the total GHG emissions, which fully illustrates the importance of the natural resources for the constructed wetland.

8.3.3 Embodied solar energy

Based on the combination of ecological thermodynamics and systems ecology, the concept of emergy is put forward to measure the ecological value as the required solar energy to generate a product or service (Odum, 1983, 1988, 1996) or to neutralize an environmental pollutant (Chen, 2006). While a positive value of emergy means an ecological service, a negative one indicates the environmental impact associated with pollutant emission. This approach generally translates each environmental and economic flow, including energy, materials, labor, currency, and pollutant, into solar energy equivalents that reflects the quality value of the product or service. The natural resources, like sunlight, wind, rain, and tides, which used to be regarded as free and externalities of the production process, are included in emergy accounting as significant environmental supports, to which the classical emergy intensities contributed by Odum et al. have been applied (Odum, 1996; Odum et al., 2000).

Due to the fact that the treated wastewater is a useful resource, the ecological value of influent wastewater is valuated as the difference between the emergy embodied in the treated wastewater and the emergy cost in the treatment process (Chen et al., 2010c). It is a negative value of −8.60E+10 sej/t in this case. The emergy evaluation sheet of the case wetland is provided in Table 8.4.

Table 8.4 The Emergy Inventory of the Case Wetland

Item	Sector Code	Sector Contents	Emergy Intensity	Unit (sej/)	Emergy (sej)
Environmental (local) input					
Sunlight			1.00E+00	J	4.98E+13
Wind, kinetic			2.45E+03	J	1.79E+13
Rain, chemical			4.70E+04	J	1.08E+15
Rain, geopotential			3.05E+04	J	1.51E+13
Earth cycle			5.80E+04	J	1.33E+15
Subtotal					1.33E+15
External economic input					
Local organic substrate	5	Services in support of agriculture	3.03E+15	10^4 Yuan	1.45E+16
Geotextile	49	Manufacture of plastic	3.91E+15	10^4 Yuan	7.74E+15
Mineral substrate	10	Mining and processing of nonmetal ores and other ores	1.50E+16	10^4 Yuan	2.12E+16
Other substrate	52	Manufacture of brick, stone, and other building materials	6.29E+15	10^4 Yuan	1.25E+16
Vegetation	1	Farming	5.15E+15	10^4 Yuan	1.47E+16
Pump	67	Manufacture of pump, valve, and similar machinery	5.07E+15	10^4 Yuan	1.52E+15
Electric control	78	Manufacture of equipments for power transmission and distribution and control	5.08E+15	10^4 Yuan	5.59E+14
Pipe and valve	49	Manufacture of plastic	3.91E+15	10^4 Yuan	1.60E+15
Steel griller	63	Manufacture of metal products	6.05E+15	10^4 Yuan	3.03E+12
Bricks and cement	50	Manufacture of cement, lime, and plaster	5.67E+15	10^4 Yuan	2.49E+15

Continued

Table 8.4 The Emergy Inventory of the Case Wetland—cont'd

Item	Sector Code	Sector Contents	Emergy Intensity	Unit (sej/)	Emergy (sej)
Electricity	92	Production and supply of electric power and heat power	7.41E+15	10^4 Yuan	6.18E+16
Maintenance	122	Environment management	2.55E+15	10^4 Yuan	9.68E+15
Wastewater			−8.60E+10	t	−1.26E+17
Subtotal					2.26E+16
Total input					2.39E+16
Output					
Plant biomass			2.43E+09	g	1.50E+15
Purified water			1.46E+12	m^3	1.83E+18
H_2O vapor			4.84E+11	t	1.07E+17
GHG			2.04E+14	t CO_2- Eq.	−2.37E+16
O_2			2.77E+14	t	1.83E+14
Regulation capacity			8.07E+11	m^3	1.49E+14
Vegetation area			2.02E+10	m^2	1.22E+13
Total output					6.65E+17

The emergy associated with the wastewater influent amounts to −1.26E+17 sej, in magnitude about 84% of that of other inputs. The emergy of the purchased non-renewable resources is 1.19E+17 sej, almost equal in magnitude to purchased renewable resources' −1.11E+17 sej. The components of the emergy inputs with wastewater excluded are shown in Figure 8.2. Electricity, as one of the external economic inputs, contributes the largest share, while the environment inputs in terms of the direct emergy flows only contribute less than 1% to the total emergy input excluding wastewater.

Ecological engineering contains both natural ecosystem and artificial engineering works associated with renewable and nonrenewable energy and resources inputs. Some delicate emergy-based indices have been devised to illustrate the efficacy, renewability, environment cost and benefit, as well as the interaction between nature and human society for production systems. Their definitions, algorithms, and values for the case wetland are summarized in Table 8.5.

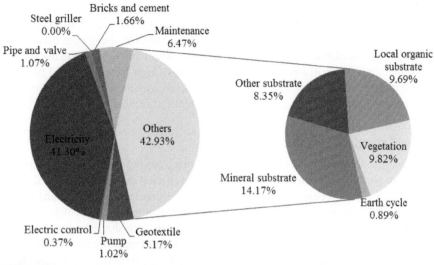

FIGURE 8.2

The components of emergy inputs excluding wastewater for the case wetland.

Table 8.5 The Emergy-Based Indices for the Case Wetland

Item	Symbol and Algorithm	Unit	Value
Item			
Wastewater emergy	W	sej	−1.26E+17
Local renewable emergy	R	sej	1.33E+15
Local nonrenewable emergy	N	sej	0
External renewable emergy	F_r	sej	−9.65E+16
External nonrenewable emergy	F_n	sej	1.19E+17
Purchased emergy	$F=F_r+F_n$	sej	2.26E+16
Total resource use	$U=R+N+F$	sej	2.39E+16
Emergy yield	Y	sej	6.65E+17
Index			
Emergy yield ratio (the efficacy of the purchased resources)	$EYR=U/F$		1.059
Environmental loading ratio (the stress of the given system on the environment)	$ELR=(F+N)/R$		1.70E+01
Environmental sustainability index (the ecological sustainability of the concerned system)	$ESI=EYR/ELR$		6.24E−02
Percent renewable index (the renewability of the concerned system)	$P_r=(R+F_r)/U$		20.40%

The EYR value of the case wetland is relatively higher than those of other wastewater treatment systems,[2] indicating a high efficacy of the purchased resources. Because a relatively small amount of external economic products have been used in the case wetland, the ELR and the ESI are respectively lower and higher than the former results indicating a small environmental loading and a high sustainability of the case system (Chen et al., 2009a). The case wetland obtains a higher Pr than the other systems as well, which confirms that the case system is sustainable.

8.3.4 Embodied cosmic exergy

Solar energy was previously believed as the primary driving force for the ecosphere in emergy synthesis (Odum, 1988). However, energy itself cannot be embodied since energy is always in conservative cycling and never consumed, which leads to the double accounting in emergy analysis at different ecological stages (Chen et al., 2010a). Exergy, a thermodynamic concept associated with the maximum amount of work that can be performed by a system in its process of reaching equilibrium with the reference environment, has been proposed to replace the energy as a unified measurement for a holistic description of the complex ecosystems (Brown and Ulgiati, 2002; Bastianoni et al., 2007). Chen (2005, 2006) proposed the concept of cosmic exergy, with cosmic background applied as the uniform thermodynamic equilibrium environment to evaluate the status and position of each ecological process in the universal exergy hierarchy. Cosmic exergy availability is regarded as the fundamental scarce natural resource of the earth system, with essential implication to the problem of global sustainability. The unit of cosmic exergy is cosmic Joule (Jc), in contrast to the solar Joule (sej) in emergy analysis. With reference to the Szargut's cumulative exergy and Odum's emergy (Odum, 1983; Szargut et al., 1988), the embodied cosmic exergy was further advanced as the total cosmic exergy consumption in the overall processes to obtain a product or service. Concrete diagramming symbols and calculation schemes for embodied cosmic exergy analysis have been contributed (Chen, 2005, 2006; Jiang, 2007; Ji, 2008, 2011; Chen et al., 2010a; Jiang et al., 2010).

The detailed embodied cosmic exergy flows of the case wetland are listed in Table 8.6. The ecological services, positive or negative, of the case system, for example, the treatment of wastewater and emission of GHG, are all treated as outputs and taken into account in this work. The embodied cosmic exergy intensities of the environmental inputs are available in Chen (2006), Jiang (2007), and Ji (2008).

The structure of the resources investment through the entire life cycle of the case system is provided in Figure 8.3.

The environment inputs share less than 1% of the total embodied cosmic exergy input of the case system, indicating a high dependency on the economic products of the artificial ecological engineering. The substrates are one of the most important components of a constructed wetland, and they contribute to about 30% of the total

---

[2]The detailed values of the indices of the other wastewater treatment systems can be found in Chen et al. (2009a).

Table 8.6 The Embodied Cosmic Exergy of the Case Wetland

Item	Embodied Cosmic Exergy Intensity	Unit (Jc/)	Embodied Cosmic Exergy (Jc)
Environmental (local) input			
Sunlight	1.02E−05	J	5.08E+08
Wind, kinetic	3.12E−02	J	2.28E+08
Rain, chemical	6.08E−01	J	1.40E+10
Rain, geopotential	3.51E−01	J	1.74E+08
Earth cycle	7.82E−03	J	1.79E+08
Subtotal			1.51E+10
External economic input			
Local organic substrate	5.45E+11	10^4 Yuan	2.60E+12
Geotextile	1.04E+12	10^4 Yuan	2.06E+12
Mineral substrate	3.29E+12	10^4 Yuan	4.66E+12
Other substrate	1.72E+12	10^4 Yuan	3.41E+12
Vegetation	9.23E+11	10^4 Yuan	2.63E+12
Pump	1.23E+12	10^4 Yuan	3.69E+11
Electric control	1.20E+12	10^4 Yuan	1.32E+11
Pipe and valve	1.04E+12	10^4 Yuan	4.24E+11
Steel griller	1.45E+12	10^4 Yuan	7.25E+08
Bricks and cement	1.56E+12	10^4 Yuan	6.86E+11
Electricity	1.97E+12	10^4 Yuan	1.64E+13
Maintenance	6.26E+11	10^4 Yuan	2.38E+12
Subtotal			3.58E+13
Total input			3.58E+13
Ecological output			
Treatment of wastewater	3.72E+08	m^3	5.44E+14
Plant biomass	6.45E+05	g	3.98E+11
Purified water	9.09E+07	m^3	1.13E+14
H_2O vapor	9.15E+07	t	2.02E+13
GHG	5.42E+10	t	−6.31E+12
O_2	7.36E+10	t	4.84E+10
Regulation capacity	1.74E+08	m^3	3.22E+10
Vegetation area	5.36E+06	m^2	3.23E+09
Total output			6.71E+14

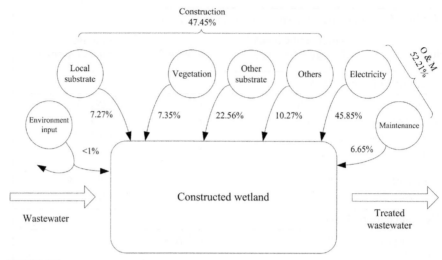

FIGURE 8.3

Overall resources investment of the case system.

embodied cosmic exergy input. The embodied cosmic exergy input during the operation stage is a little larger than that of the construction stage, showing the importance of the life cycle study. Electricity is the largest source of cosmic exergy input among all the resources. The ecological value of wastewater treatment contributes most to the overall ecological yield (80%), while that of water supply ranks the second (17%).

Some related indices, largely parallel to those employed in emergy analysis, have been constructed to illustrate different aspects of the production system with reference to the emergy synthesis; they are accordingly calculated and summarized in Table 8.7.

The net present ecological value of the case wetland is positive, reflecting that the case system is ecological profitable in the given scenario. Compared to the same indices of conventional wastewater treatment systems in Chen et al. (2011e), the case wetland has a higher local dependency in terms of the relatively large local resource utilization. About 15% of the embodied cosmic exergy input is provided by renewable resources. The EYR is much larger than those of the conventional systems, indicating a high ecological benefit of the ecological engineering. The ELR shows that the case system has a much smaller pressure on local ecological environment than the conventional ones. Finally, the high ESI shows that the case wetland achieves better performances to sustain the ecological environment.

8.4 CONCLUSION

The use of ecological accounting for constructed wetlands is urgently needed to reflect the multiple significant ecological roles of a constructed wetland. By reviewing the latest developments in related fields, the systems ecological endowment accounting framework has been presented to reveal the overall ecological implications of a

Table 8.7 The Embodied Cosmic Exergy-Based Indices of the Case Wetland

Item	Symbol	Unit	Value
Item			
Ecological input	I	sej	3.58E+13
Ecological yield	Y	sej	6.71E+14
Renewable input	R	sej	5.25E+12
Local input	L	sej	1.51E+10
Local nonrenewable input	I_{ln}	sej	0.00E+00
Local renewable input	I_{lr}	sej	1.51E+10
Purchased input	P	sej	3.58E+13
Index			
Net present ecological value (whether the system is ecologically profitable)	$NPEV = Y - I$	sej	6.36E+14
Local dependency (the fraction of the local resources)	$LD = L/I$		0.04%
Renewable dependency (the reliable degree on renewable resources)	$RD = R/I$		14.66%
Ecological yield ratio (the efficacy of the external inputs utilization)	$EYR = Y/P$		18.78
Ecological load ratio (the competitiveness of resources qualities)	$ELR = (P + I_{ln})/I_{lr}$		2.37E+03
Ecological sustainability index (the possible contribution of the system under certain environmental stress)	$ESI = EYR/ELR$		7.93E-03

constructed wetland on the basis of the hybrid method integrating process analysis with ecological IOA. A complete ecological inputs and outputs inventory and related systematic embodied ecological endowment intensity databases have been introduced as the basis of the ecological accounting of the constructed wetland.

The systems accounting for a typical constructed wetland in Beijing is performed as a case study. After compiling the specific inventory and adopting the embodied ecological endowment database for the Chinese economy in 2007 (Chen and Chen, 2010), the embodied GHG emissions, solar energy, and cosmic exergy of the construction and operation stages of the case system are tallied.

According to the results, electricity emerges as the largest ecological endowment among all the inputs excluding wastewater, making the negative ecological impacts of the operation stage larger than those of the construction stage. The embodied ecological endowment of substrates of three different categories ranks second (excluding wastewater), which reveal the important role of the substrates for the constructed wetland. Some related indices contributed by embodied solar energy (emergy) and cosmic exergy analyses have been calculated, which illustrates that the case system has relative high renewability and sustainability.

References

Bastianoni, S., Facchini, A., Susani, L., Tiezzi, E., 2007. Emergy as a function of exergy. Energy 32, 1158–1162.

Brix, H., 1994. Use of constructed wetlands in water-pollution control—historical development, present status, and future perspectives. Water Sci. Technol. 30, 209–223.

Brown, M.T., Herendeen, R.A., 1996. Embodied energy analysis and EMERGY analysis: a comparative view. Ecol. Econ. 19, 219–235.

Brown, M.T., Ulgiati, S., 2002. Emergy evaluations and environmental loading of electricity production systems. J. Clean. Prod. 10, 321–334.

Bullard, C.W., Herendeen, R.A., 1975. The energy cost of goods and services. Energ. Policy 3, 268–278.

Bullard, C.W., Penner, P.S., Pilati, D.A., 1978. Net energy analysis: handbook for combining process and input–output analysis. Resour. Energy 1, 267–313.

Cardoch, L., Day, J.W., Rybczyk, J.M., Kemp, G.P., 2000. An economic analysis of using wetlands for treatment of shrimp processing wastewater—a case study in Dulac, LA. Ecol. Econ. 33, 93–101.

Chen, G.Q., 2005. Exergy consumption of the earth. Ecol. Model. 184, 363–380.

Chen, G.Q., 2006. Scarcity of exergy and ecological evaluation based on embodied exergy. Commun. Nonlinear Sci. Numer. Simul. 11, 531–552.

Chen, Z.M., 2011. Analysis of Embodied Ecological Endowment Flow for the World Economy. Peking University, Beijing.

Chen, G.Q., Chen, Z.M., 2010. Carbon emissions and resources use by Chinese economy 2007: a 135-sector inventory and input–output embodiment. Commun. Nonlinear Sci. Numer. Simul. 15, 3647–3732.

Chen, G.Q., Chen, Z.M., 2011a. Greenhouse gas emissions and natural resources use by the world economy: ecological input–output modeling. Ecol. Model. 222, 2362–2376.

Chen, Z.M., Chen, G.Q., 2011b. Embodied carbon dioxide emission at supra-national scale: a coalition analysis for G7, BRIC, and the rest of the world. Energ. Policy 39, 2899–2909.

Chen, Z.M., Chen, G.Q., 2011c. An overview of energy consumption of the globalized world economy. Energ. Policy 39, 5920–5928.

Chen, Z.M., Chen, G.Q., 2013. Demand-driven energy requirement of world economy 2007: a multi-region input–output network simulation. Commun. Nonlinear Sci. Numer. Simul. 18, 1757–1774.

Chen, G.Q., Zhang, B., 2010. Greenhouse gas emissions in China 2007: inventory and input–output analysis. Energ. Policy 38, 6180–6193.

Chen, Z.M., Chen, B., Zhou, J.B., Li, Z., Zhou, Y., Xi, X.R., Lin, C., Chen, G.Q., 2008. A vertical subsurface-flow constructed wetland in Beijing. Commun. Nonlinear Sci. Numer. Simul. 13, 1986–1997.

Chen, B., Chen, Z.M., Zhou, Y., Zhou, J.B., Chen, G.Q., 2009a. Emergy as embodied energy based assessment for local sustainability of a constructed wetland in Beijing. Commun. Nonlinear Sci. Numer. Simul. 14, 622–635.

Chen, Z.M., Chen, G.Q., Chen, B., Zhou, J.B., Yang, Z.F., Zhou, Y., 2009b. Net ecosystem services value of wetland: environmental economic account. Commun. Nonlinear Sci. Numer. Simul. 14, 2837–2843.

Chen, H., Chen, G.Q., Ji, X., 2010a. Cosmic emergy based ecological systems modelling. Commun. Nonlinear Sci. Numer. Simul. 15, 2672–2700.

Chen, Z.M., Chen, G.Q., Zhou, J.B., Jiang, M.M., Chen, B., 2010b. Ecological input–output modeling for embodied resources and emissions in Chinese economy 2005. Commun. Nonlinear Sci. Numer. Simul. 15, 1942–1965.

Chen, Z.M., Xia, X.H., Tang, H.S., Li, S.C., Deng, Y., 2010c. Emergy based ecological assessment of constructed wetland for municipal wastewater treatment: methodology and application to the Beijing wetland. J. Environ. Inform. 15, 62–73.

Chen, G.Q., Chen, H., Chen, Z.M., Zhang, B., Shao, L., Guo, S., Zhou, S.Y., Jiang, M.M., 2011a. Low-carbon building assessment and multi-scale input–output analysis. Commun. Nonlinear Sci. Numer. Simul. 16, 583–595.

Chen, G.Q., Shao, L., Chen, Z.M., Li, Z., Zhang, B., Chen, H., Wu, Z., 2011b. Low-carbon assessment for ecological wastewater treatment by a constructed wetland in Beijing. Ecol. Eng. 37, 622–628.

Chen, G.Q., Yang, Q., Zhao, Y.H., 2011c. Renewability of wind power in China: a case study of nonrenewable energy cost and greenhouse gas emission by a plant in Guangxi. Renew. Sustain. Energy Rev. 15, 2322–2329.

Chen, G.Q., Yang, Q., Zhao, Y.H., Wang, Z.F., 2011d. Nonrenewable energy cost and greenhouse gas emissions of a 1.5 MW solar power tower plant in China. Renew. Sustain. Energy Rev. 15, 1961–1967.

Chen, Z.M., Chen, B., Chen, G.Q., 2011e. Cosmic exergy based ecological assessment for a wetland in Beijing. Ecol. Model. 222, 322–329.

Chen, Z.M., Chen, G.Q., Xia, X.H., Xu, S.Y., 2012. Global network of embodied water flow by systems input–output simulation. Front. Earth Sci. 6, 331–344.

Chen, G.Q., Guo, S., Shao, L., Li, J.S., Chen, Z.M., 2013. Three-scale input–output modeling for urban economy: carbon emission by Beijing 2007. Commun. Nonlinear Sci. Numer. Simul. 18, 2493–2506.

Costanza, R., 1980. Embodied energy and economic valuation. Science 210, 1219–1224.

Costanza, R., dArge, R., deGroot, R., Farber, S., Grasso, M., Hannon, B., Limburg, K., Naeem, S., Oneill, R.V., Paruelo, J., Raskin, R.G., Sutton, P., vandenBelt, M., 1997. The value of the world's ecosystem services and natural capital. Nature 387, 253–260.

Dixon, A., Simon, M., Burkitt, T., 2003. Assessing the environmental impact of two options for small-scale wastewater treatment: comparing a reedbed and an aerated biological filter using a life cycle approach. Ecol. Eng. 20, 297–308.

Fuchs, V.J., Mihelcic, J.R., Gierke, J.S., 2011. Life cycle assessment of vertical and horizontal flow constructed wetlands for wastewater treatment considering nitrogen and carbon greenhouse gas emissions. Water Res. 45, 2073–2081.

Guo, S., Chen, G.Q., 2013. Multi-scale input–output analysis for multiple responsibility entities: carbon emission by urban economy in Beijing 2007. J. Environ. Acc. Manag. 1, 43–54.

IPCC, 2006. IPCC Guidelines for National Greenhouse Gas Inventories. Institute for Global Environmental Strategies, Japan.

Ji, X., 2008. Theory of embodied cosmic exergy and its application in urban Ecosystem modelling and regulation, Ph.D. Dissertation. Peking University, Beijing, China (in Chinese).

Ji, X., 2011. Ecological accounting and evaluation of urban economy: taking Beijing city as the case. Commun. Nonlinear Sci. Numer. Simul. 16, 1650–1669.

Jiang, M.M., 2007. Embodied cosmic exergy analysis for urban ecosystem, Ph.D. Dissertation. Peking University, Beijing, China (in Chinese).

Jiang, M.M., Chen, Z.M., Zhang, B., Li, S.C., Xia, X.H., Zhou, S.Y., Zhou, J.B., 2010. Ecological economic evaluation based on emergy as embodied cosmic exergy: a historical study for the Beijing urban ecosystem 1978–2004. Entropy 12, 1696–1720.

Joshi, S., 1999. Product environmental life-cycle assessment using input–output techniques. J. Ind. Ecol. 3, 95–120.

Krutilla, J.V., 1967. Conservation reconsidered. Am. Econ. Rev. 57, 777–786.

Leontief, W., 1936. Quantitative input–output relations in the economic system. Econ. Stat. 18, 105–125.

Leontief, W., 1970. Environmental repercussions and the economic structure: an input–output approach. Rev. Econ. Stat. 52, 262–271.

Machado, A.P., Urbano, L., Brito, A.G., Janknecht, P., Salas, J.J., Nogueira, R., 2007. Life cycle assessment of wastewater treatment options for small and decentralized communities. Water Sci. Technol. 56, 15–22.

Miller, R.E., Blair, P.D., 2009. Input–Output Analysis: Foundations and Extensions, second ed. Cambridge University Press, New York.

Odum, H.T., 1983. Systems Ecology: An Introduction. John Wiley and Sons, New York.

Odum, H.T., 1988. Self-organization, transformity, and information. Science 242, 1132–1139.

Odum, H.T., 1996. Environmental Accounting: Emergy and Environmental Decision Making. Wiley, New York.

Odum, H., Brown, M., Brandt-Williams, S., 2000. Handbook of Emergy Evaluation: A Compendium of Data for Emergy Computation Issued in a Series of Folios; Folio #1 Introduction and Global Budget. Center for Environmental Policy, University of Florida, Gainesville, FL.

Pan, T., Zhu, X.D., Ye, Y.P., 2011. Estimate of life-cycle greenhouse gas emissions from a vertical subsurface flow constructed wetland and conventional wastewater treatment plants: a case study in China. Ecol. Eng. 37, 248–254.

Shao, L., Wu, Z., Zeng, L., Chen, Z.M., Zhou, Y., Chen, G.Q., 2012. Embodied energy assessment for ecological wastewater treatment by a constructed wetland. Ecol. Model. 252, 63–71.

Suh, S., Huppes, G., 2005. Methods for life cycle inventory of a product. J. Clean. Prod. 13, 687–697.

Szargut, J., Morris, D.R., Steward, F.R., 1988. Energy Analysis of Thermal, Chemical, and Metallurgical Processes. Hemisphere, New York.

Teiter, S., Mander, Ü., 2005. Emission of N_2O, N_2, CH_4, and CO_2 from constructed wetlands for wastewater treatment and from riparian buffer zones. Ecol. Eng. 25, 528–541.

Treloar, G.J., 1997. Extracting embodied energy paths from input–output tables: towards an input–output-based hybrid energy analysis method. Econ. Syst. Res. 9, 375–391.

White, K.D., 1995. Enhancement of nitrogen removal in subsurface flow constructed wetlands employing a 2-stage configuration, an unsaturated zone, and recirculation. Water Sci. Technol. 32, 59–67.

Yang, Q., Wu, X., Yang, H., Zhang, S., Chen, H., 2012. Nonrenewable energy cost and greenhouse gas emissions of a "Pig-Biogas-Fish" system in China. ScientificWorldJournal 2012, 7.

Zhou, Y., 2004. The research and application of wastewater treatment in the Longdao River: a demonstration project of the application of Danish Rootzone Technology. Project data.

Zhou, J.B., 2008. Embodied ecological elements accounting of national economy. Peking University, Beijing (in Chinese).

Zhou, J.B., Jiang, M.M., Chen, B., Chen, G.Q., 2009. Emergy evaluations for constructed wetland and conventional wastewater treatments. Commun. Nonlinear Sci. Numer. Simul. 14, 1781–1789.

Zhou, S.Y., Chen, H., Li, S.C., 2010. Resources use and greenhouse gas emissions in urban economy: ecological input–output modeling for Beijing 2002. Commun. Nonlinear Sci. Numer. Simul. 15, 3201–3231.

Zuo, P., Wan, S.W., Qin, P., Du, J., Wang, H., 2004. A comparison of the sustainability of original and constructed wetlands in Yancheng Biosphere Reserve, China: implications from emergy evaluation. Environ. Sci. Pol. 7, 329–343.

Modeling the Response of the Planktonic Microbial Community to Warming Effects in Maritime Antarctic Lakes: Ecological Implications

9

Antonio Camacho[a,*], Juan Antonio Villaescusa[a], Carlos Rochera[a], Sven Erik Jørgensen[b]

[a]*Cavanilles Institute of Biodiversity and Evolutionary Biology and Department of Microbiology and Ecology, Edificio de Investigacion, Campus de Burjassot, Universitat de Valencia, E-46100 Burjassot, Spain*
[b]*Professor Emeritus, Environmental Chemistry, University of Copenhagen, Copenhagen, Denmark*
[*]*Corresponding author: e-mail address: antonio.camacho@uv.es*

9.1 INTRODUCTION

Antarctica is one of the places on Earth where weather conditions are extremely harsh and unfavorable for life. Most of the Antarctic continent is permanently ice-covered due to its extreme meteorological conditions, remaining most of the year at temperatures below zero. In these systems the presence of liquid water is the main factor that enables the development of biological communities (Kennedy, 1993, 1995).

Maritime Antarctica, comprising the western side of the Antarctic Peninsula and nearby islands, is characterized by a less extreme climatic regime compared to continental Antarctica, because it displays relatively mild summers with higher mean temperatures and precipitation than the continent (Bañón et al., 2013). These climatic conditions result in the presence of a large amount of ice-free aquatic ecosystems during the austral summer, where conspicuous microbial communities develop (Izaguirre et al., 2003; Toro et al., 2007; Villaescusa et al., 2010). Besides small running water systems, during summer the ice cover of lakes and ponds melts, allowing higher light availability for photosynthesis and a temperature favoring biological processes. In deep lakes these temperatures also enhance vertical mixing of the water column. Additionally, snow melting on lake's catchment areas drives important inputs of inorganic nutrients and organic matter into the lake, resulting both from winter accumulation and from catchment's biotic activities (inorganic carbon and nitrogen fixation, birds' activity). These external inputs relatively enrich, within this short time period, these ultraoligotrophic lakes. Taken as a whole, all these processes allow for the development of conspicuous planktonic and benthic lacustrine communities during the austral summer (Ellis-Evans et al., 1998).

The area of maritime Antarctica, with special attention to South Shetland Islands and the Antarctic Peninsula, has been the main focus of in-depth ecological studies during the last decades. These studies have been performed with the aim of increasing the knowledge of different components of the biological communities focusing on different groups of organisms. The periphytic flora and benthic diatoms (e.g., Hansson and Håkansson, 1992; Jones, 1996; Kopalova and van de Vijver, 2013; Pla-Rabes et al., 2013), microbial communities (e.g., Tell et al., 1995; Vinocur and Pizarro, 2000; Pearce, 2005; Schiaffino et al., 2011; Villaescusa et al., 2013a,b), crustaceans (e.g., Paggi, 1996), and phytoplankton (e.g., Vinocur and Izaguirre, 1994; Izaguirre et al., 1998; Vinocur and Unrein, 2000; Allende and Mataloni, 2013), as well as the environmental features (e.g., Caulkett and Ellis-Evans, 1997), have been studied. Also, in other Antarctic freshwater ecosystems not located within the maritime Antarctic region, some extensive studies have been made, like in the McMurdo Dry Valleys area (Vincent, 1988; Hawes and Brazier, 1991; Priscu, 1998) or Vestfold Hills (Bell and Laybourn-Parry, 1999; Gibson, 1999).

During the last decades the aim to understand and describe with precision the ecological systems of Antarctica and predict possible scenarios for a warming climate has led the scientific community to the start using ecological models (Reid and Crout, 2008; Gutt et al., 2012). The development of models and predictions of future scenarios of Antarctic ecosystems has yet to grow; however, the importance of these models for monitoring and making prognoses on the effects of climate change make them highly demanded tools.

From the point of view of developing ecological models, if we focus on maritime Antarctica, the area known as Byers Peninsula, located in Livingston Island (South Shetland Islands), shows a huge number of freshwater ecosystems, including lakes and streams that have been intensively studied during the last decades. This area has been studied since the 1990s by some groups focusing on different topics such as benthic diatom communities (Jones et al., 1993), stream algal mats (Davey, 1993), or general limnological descriptions of freshwater ecosystems (Ellis-Evans, 1996a). However, during the last decade, a lot of studies have been conducted in the area of Byers Peninsula by the multidisciplinary research group LIMNOPOLAR (Quesada et al., 2013). The results of some of the different studies of the group are briefly reported here (Camacho, 2006; Fernández-Valiente et al., 2007; Toro et al., 2007; Quesada et al., 2009; Rochera et al., 2010, 2013; Villaescusa et al., 2010, 2013a,b; Velázquez et al., 2011, 2013). This research group has focused on the study of ecological processes and community structure of the different epicontinental freshwater ecosystems, with especial emphasis on Lake Limnopolar (being referred as a model lake), as well as on the description of the different microbial communities living in these ecosystems, such as terrestrial and aquatic microbial mats, lichens, and lake's planktonic and benthic communities (Benayas et al., 2013; Quesada et al., 2013).

Although many models of lakes have been built during the last decades, so far no model was developed to cover key aspects of the ecological functioning of Antarctic lakes. The high demand of ecological models for the prediction of ecosystem evolution under possible scenarios for climate change and warming in the area of maritime Antarctica has encouraged us to the development of this model. Our model (Villaescusa et al., submitted for publication) aims to describe the possible effects of warming during the austral summer on the plankton communities of Lake Limnopolar, especially on the bacterioplankton, as well as serving as a tool to describe and understand how this warming could affect this type of ecosystem. We focused our study on the dissolved organic carbon and inorganic nutrient dynamics after the ice melting in the lake catchment and on how these contributions affect the structure of the planktonic community in the lake. In brief, the model brings out the importance of the rise in temperature and how this forcing function influencing the inputs of nutrients increases biological activity in the lake, which is also shown as an increase in the system eco-exergy. The model has been designed to explore the effects on lake functioning caused by the high interannual meteorological variability in the area (Bañón et al., 2013) and, especially, to address how the current temperature increases affect several components of the microbial community as well as ecological processes within the lake and its catchment, thus revealing the ecological implications of warming on these highly sensitive Antarctic ecosystems (Camacho et al., 2012).

9.2 STUDY AREA: BYERS PENINSULA, AN ANTARCTIC SPECIAL PROTECTED AREA

Byers Peninsula is located in the western end of Livingston Island in the South Shetlands Islands (latitude 62°34′35″S, longitude 61°13′07″W), Antarctica (Figure 9.1). Livingston Island is close to the southeast side of Drake Passage. South Shetland holds the greatest number of scientific stations of any region in Antarctica. This group of islands is part of the area known as maritime Antarctica, comprising the coastal area of the western Antarctic Peninsula and surrounding islands.

Byers Peninsula is a periglacial region that covers an area of 62.6 km^2, extending 9 km from west to east and about 18.2 km from northeast to southwest. Due to the glacier retreat started around 4500 years BP, nowadays Byers became free of permanent ice, being one of the largest ice-free areas in the maritime Antarctica (SCAR, 2003). The center of the Peninsula is dominated by extensive gently undulating plateaus up to 105 m above sea level, interrupted by isolated Precambrian volcanic formations like Chester Cone—188 m—or Black Hill—143 m (Thomson and López-Martínez, 1996). The highest mountain formation is Start Hill (265 m) located

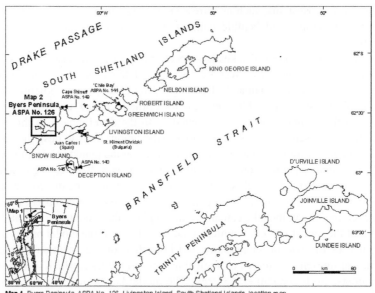

Map 1. Byers Peninsula, ASPA No. 126, Livingston Island, South Shetland Islands, location map.
Inset: location of Byers Peninsula on the Antarctic Peninsula

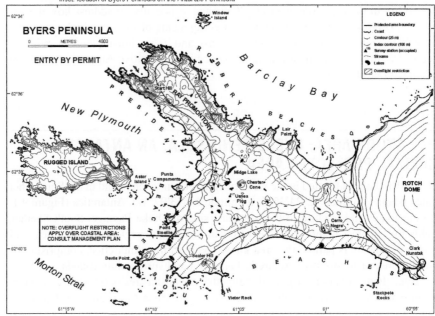

Map 2. Byers Peninsula, ASPA No. 126, topographic map.

FIGURE 9.1

Geographical location of Byers Peninsula in the maritime Antarctica. Lower map shows a detail of Byers Peninsula taken from the ASPA Management Plan.

From SCAR (2003).

northwest of the Peninsula. Geological formations of the Upper Jurassic and Lower Cretaceous marine type sedimentary and volcanic rocks are abundant in the area, being the site of the maritime Antarctica where most Mesozoic–Cenozoic rock formations are exposed. These formations have experienced strong glacial erosion and periglacial activity (Thomson and López-Martínez, 1996). Soils in this area display a special lithosol-fragmented stone layer that extends to about 100 cm above the permafrost. This permits an important underground water flow over this frost layer (Serrano et al., 1996).

The area of Byers displays a complex drainage network (Lyons et al., 2013) with many lakes, ponds, streams, and wetlands along the surface. These characteristics allow the appearance of extensive wetland areas covered by microbial mats and moss carpets. Also, the presence of a permafrost active layer situated around 50–70 cm depth in nearly all soils of Byers (De Pablo et al., 2013) is worth noting. Melting and warming processes during summer exert an intense effect in the permafrost layer, affecting most of the hydrologic and geochemical processes occurring in the lacustrine catchments (Fassnacht et al., 2013).

The climate in Byers Peninsula is strongly influenced by the proximity to the coast, being much less severe than that of continental Antarctica (Bañón et al., 2013). Weather is characterized by extreme windy conditions and relatively high precipitation. During summer, the average temperature is set around 1–3 °C with daily highs not exceeding 10 °C. In winter the minimum temperatures are rarely lower than −35 °C with peaks of about 0 °C. Rainfall and snowfall in the Peninsula is higher than in mainland of Antarctica with average annual values of around 700–1000 mm (Bañón, 2001), which mainly occur during summer. This area shows a characteristically changeable weather, which produces rapid shifts in winds and barometric pressure. The Peninsula is covered by snow for 7–8 months a year and is mostly snow-free, although with small remnants, during the austral summer. In spite of these general common features, a high meteorological variability within years occurs (Bañón et al., 2013), which influences biological communities within the aquatic ecosystems (Rochera et al., 2010). The abundant presence of epicontinental water bodies formed as a result of the retreat of the glacier that currently covers the major part of the island gives a marked limnological interest on the area.

Byers Peninsula is categorized as one of the 67 Antarctic specially protected areas (ASPA) in Antarctica. This area was first designated as ASPA No. 10 due to an Antarctic Treaty Recommendation (IV-10) in 1966. This recommendation was made due to a great scientific interest in geology, plants, and animal diversity, especially invertebrates, in such a small area. Part of the interest was also focused on the importance of vertebrate populations like elephant seals (*Mirounga leonina*) (Gil-Delgado et al., 2013) and small colonies of fur seals (*Arctocephalus gazella*). Later, Byers Peninsula underwent some changes in their protection form, being declared as a site of special scientific interest due to Recommendation VIII-2 and VIII-4 of the Antarctic Treaty in 1975. Due to the large geological, archaeological, and biological interest, Chile and the United Kingdom suggested the need to re-establish the statutory protection of ASPA (Recommendation XVI-5 Antarctic Treaty, 1991), but in this case only incorporating

Byers Peninsula and nearby coastal areas (Ellis-Evans, 1996a). The great and recognized biological value of this area was finally included in the management plan for the Special Protected Area No126 (SCAR, 2003). Byers Peninsula is a pristine area only opened to visits by small groups of scientists at a time, with stringent environmental protection measures. Also, the development of any kind of scientific activities within the area requires a permit issued by SCAR (Scientific Committee for Antarctic Research) and National Polar Committees.

9.3 ECOLOGICAL FEATURES: LAKE LIMNOPOLAR AND ITS CATCHMENT

Lakes situated in the Antarctic region are often seen as very unproductive systems because of physical limitations (low temperature and low light availability). However, during the austral summer, maritime Antarctic lakes, like those situated in Byers Peninsula, experience less harsh climatic conditions that permit the development of conspicuous microbial communities. Under these circumstances, the physical limitations of low temperature or light availability decrease, allowing other factors such as nutrient availability or biological interactions to exert a major role in controlling the productivity of these lakes. However, previous studies in Byers Peninsula reflected a high variability of the meteorological conditions for different years in this area of maritime Antarctica. This variation can determine the length of the ice-free period (Rochera et al., 2010; Bañón et al., 2013), which implies severe changes in the inputs of inorganic nutrients and organic matter from the catchment, affecting the microbial structure in the lakes (Pearce, 2005).

The biological interest of maritime Antarctic lakes deals mainly with the existence of simple planktonic and benthic microbial communities that mainly develop during the austral summer. Ice cover melting in the lakes helps break the winter reverse stratification and enables the transition to a mixed lake. In the area of Byers Peninsula, some of the lakes studied during the last decade, such as Lakes Limnopolar, Somero, Turbio, Chica, and Chester (Figure 9.2), are located in the central plateau of the Peninsula (Toro et al., 2007; Villaescusa et al., 2010). These lakes have many physical and chemical features in common. Unlike coastal lakes, those situated in the central plateau display deeper depths around 3–8 m and a marked oligotrophic status. However, some, like Lake Somero, are shallow, and this results in some ecological differences. Some others, like Lake Refugio (Figure 9.2), are located close to the coast, especially in the southern and western beaches. The latter are quite shallow, commonly less than 0.5 m max depth, and show a high trophic status due to strong inputs of inorganic nutrients and organic matter from birds and marine mammal activity.

Lakes in Byers Peninsula display a contrasting trophic status that is very much influenced by the external inputs of nutrients and organic matter. However, lake morphology and recycling processes linked to water–sediment interactions are also modulating the trophic status of these systems. Previous studies have demonstrated that the

FIGURE 9.2

Images of some of the lakes of Byers Peninsula. Top, from left to right: Lake Limnopolar, Lake Somero, and Lake Turbio. Bottom, from left to right: Lake Chica, Lake Chester, and Lake Refugio. (For color version of this figure, the reader is referred to the online version of this chapter.)

highly heterogeneous trophic statuses among Byers' lakes are highly reflected in the bacterioplankton community composition, the latter being a reflection of the trophic status of each lake (Villaescusa et al., 2010). Lake Limnopolar is situated in the central plateau of the Peninsula. Like other lakes, it is completely ice covered during most of the year. The relatively warm temperatures during the summer favor melting of the ice cover, keeping the lake without ice for a variable period. Studies over the summer period of 3 consecutive years (Rochera et al., 2010) revealed that the length of the ice-free period is highly variable and depends on the interannual meteorological heterogeneity in this area; this information has been confirmed for several additional years (Villaescusa et al., 2013b). A summary from continuous video camera coverage of the thaw and icing processes of the lake during the annual cycle 2011/2012 is given in Figure 9.3. As in other maritime Antarctic lakes, icing-melting processes highly influence in-lake light availability, with a high light extinction when the lake is covered by ice (Hawes and Schwarz, 2001), but increased light penetration and enhanced primary production activity when the ice layer melts.

　　Regarding inorganic nutrient availability and chlorophyll-*a* concentrations, Lake Limnopolar can be considered as ultraoligotrophic with very low concentrations of inorganic nutrients (nitrogen and phosphorus compounds) and dissolved organic matter. Inputs of these nutrients are mainly determined by the influence of the microbial mats and bryophytes that grow in the lake catchment, being higher during the first days of summer when snow melting accelerates. Even though the main general ecological characteristics remain unchanged year by year, the chemical and biological dynamics in Lake Limnopolar show different yearly patterns (Figure 9.4). These changing values are affected by the high interannual meteorological heterogeneity in the area or maritime Antarctica (Rochera et al., 2010; Bañón et al., 2013).

(a)

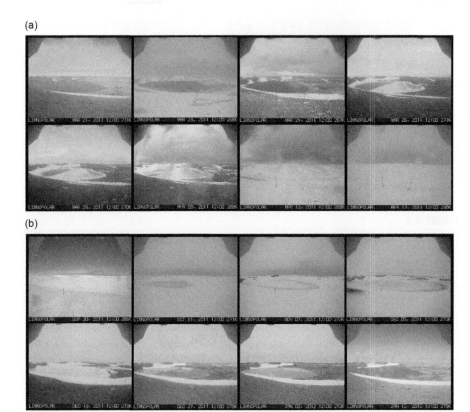

(b)

FIGURE 9.3

(a) Lake Limnopolar 2011 chronosequence from 21 March to 17 April showing the freezing process. (b) Lake Limnopolar 2011–2012 chronosequence from 30 September to 12 January showing the thawing process. All of these photographs were obtained through a continuous record with a video camera from February 2011 to February 2012. (For color version of this figure, the reader is referred to the online version of this chapter.)

Images provided by A.Quesada, M.A. de Pablo and M. Ramos, UAM & Universidad de Alcalá.

Unlike lakes situated in temperate systems, Antarctic lakes are characterized to be pristine systems with low-complexity food webs (Figure 9.5). These systems are dominated by different groups of microorganisms like bacterioplankton, nanozoo-plankton, phytoplankton, and crustacean zooplankton (Ellis-Evans, 1996b; Hansson et al., 1996; Wynn-Williams, 1996; Laybourn-Parry et al., 2001; Camacho, 2006; Sommer and Sommer, 2006). Due to these particular characteristics, the energy transfer in Lake Limnopolar between the planktonic organisms mostly circulates through the microbial loop (Azam et al., 1983; Laybourn-Parry, 1997). The phytoplankton community is mainly composed of chrysophytes, chlorophytes (Prasionophytes), diatoms (Camacho, 2006; Toro et al., 2007; Quesada

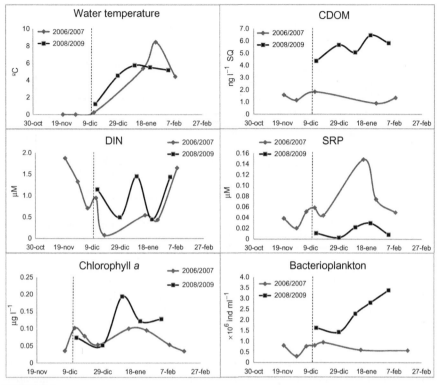

FIGURE 9.4

Some of the physical, chemical, and biological variables measured during austral summers 2006/2007 and 2008/2009 showing the high interannual variability in the lake. The vertical dashed line represents the date of total thawing for 2008/2009 and the vertical continuous line that of 2006/2007. CDOM (chromophoric dissolved organic matter, in ng l^{-1} of quinine sulfate equivalents), DIN (dissolved inorganic nitrogen), SRP (soluble reactive phosphorus).

et al., 2009), and a very small fraction of unicellular picocyanobacteria (Toro et al., 2007; Rochera et al., 2010). Heterotrophic protozoa are another component of the planktonic microbial community in the lake. Only nine species of ciliates have been registered in the lake, and only one is really euplanktonic, *Balanion planctonicum* (Petz et al., 2005), which can reach relatively high densities during some periods. Heterotrophic nanoflagellates show a high abundance in the planktonic community, although, as for ciliates, their diversity is higher in the benthic environment (Toro et al., 2007). Nanozooplankton (flagellated and ciliated protozoa) may play an important role as grazers on heterotrophic bacterioplankton (Camacho, 2006). The abundances of the heterotrophic bacterioplankton are higher than expected for such oligotrophic conditions, between 1×10^6 and 4×10^6 cell ml^{-1} (Toro et al., 2007;

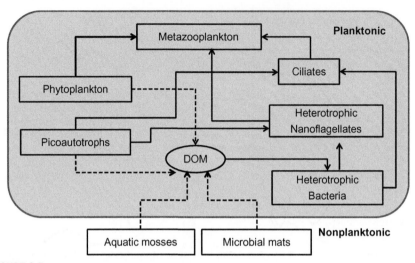

FIGURE 9.5

Lake Limnopolar planktonic microbial food web and external DOM (dissolved organic matter) inputs. Continuous arrows represent matter transfer from the donor to the consumer. Discontinuous arrows represent the main DOM inputs into the lake water.

Villaescusa et al., 2013b). The high abundance of bacterioplankton are supported by inputs of organic carbon from the microbial mats that cover part of the lake area (Camacho, 2006; Fernández-Valiente et al., 2007; Velázquez et al., 2013) and also from the aquatic mosses that grow in the bottom of the lake (Imura et al., 2003; Camacho, 2006; Toro et al., 2007). Crustacean zooplankton dominate the microbial food web as main consumers. As commonly found in Antarctic lakes, metazooplankton is composed mainly of the calanoid copepod *Boeckella poppei* and, to a lesser extent, by the fairy shrimp *Branchinecta gainii*, the latter mostly showing a benthic behavior (Paggi, 1987; Toro et al., 2007).

9.4 MODELING OF LAKE LIMNOPOLAR

The response of the lake's planktonic microbial community to warming effects was addressed through ecological modeling (Villaescusa et al., submitted for publication). As previously mentioned, Lake Limnopolar has been the subject of study and intensive monitoring during the last decade, and especially within the 2006–2009 austral summers. This has permitted the development of a carbon flux model that aims to describe the effects of temperature, light, inorganic nutrients, and organic carbon availability on the planktonic microbial community. The main objective of the model was to describe the carbon dynamics that mainly runs through the bacterioplankton community, being determined by the inputs of organic matter from the microbial mats and benthic mosses. Based upon a few years' data, the model, covering the microbial component of the biological community of this Antarctic lake, gave satisfactory results to represent the dynamics of the studied components, although revealing some problems related to its high sensitivity to within-years differences in environmental conditions. This model can be used to better understand the functioning of the maritime Antarctic lakes and their response to environmental changes, particularly to temperature, providing a useful tool to predict the possible response of these systems to changes in their environmental conditions linked to climate change, such as temperature, light availability, and organic matter inputs.

The model was developed using data for a period of 53 days, from 13 December to 3 February, obtained during 2008/2009 austral summer. The model included seven state variables, including *Phytoplankton abundance* (grouping all the planktonic primary producers in the lake, which include both planktonic algae and picocyanobacteria), *bacterioplankton abundance*, *dissolved organic carbon concentration* (DOC), and *particulate organic carbon concentration* (POC). Additionally, *soluble reactive phosphorus concentration* (SRP) was included to represent the inorganic nutrient limitation in the lake. Finally, two other state variables were modeled to estimate the influence of *benthic mosses*, as well as the carbon inputs from the catchment's *microbial mats*. The model flow diagram for the different state variables is shown in Figure 9.6. The model accounts for the dynamics of the different forms of organic carbon attending to their origin: (i) planktonic (autochthonous) organic carbon from processes produced within the pelagic zone of the lake and (ii) organic matter produced by benthic mosses distributed over the bottom of the lake and, especially, by the catchment's microbial mats.

To predict the system response against temperature and variables likely influenced by temperature changes, five forcing functions were included in the model. The different forcing functions were *temperature*, *photosynthetic active radiation* (PAR), *lake water dynamics* (Inflow and Outflow), and *lake volume*. These variables may exert an important effect on the lake microbial community, particularly in systems with a strong thawing process during the first steps of the austral summer, as the studied lake (Figure 9.3).

Using these variables and forcing functions, the model was constructed starting with different equations from the software Ecotox: Ecological modeling and Ecotoxicology (Jørgensen and Bendoricchio, 2001), which includes some literature references about ecological modeling. This software was also used to find the initial parameters for the model calibration; after that, all data were included in the Stella software to calibrate the model. The different equations describing the model can be found in Table 9.1. These equations describe the different processes occurring in a system mainly controlled by the above mentioned forcing factors.

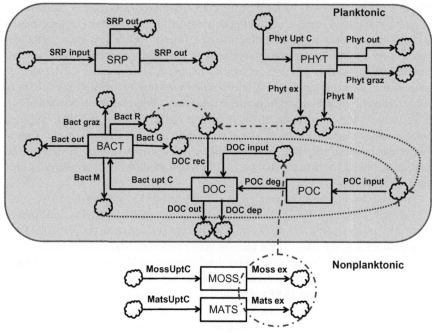

FIGURE 9.6

Model flow Stella diagram. For the interpretation of the different state variables and processes, check Table 9.1. The main POC and DOC flows are described with discontinuous arrows: (striped) allochthonous DOC; (striped and dotted) autochthonous DOC; (dotted) POC.

Modified from Villaescusa et al. (submitted for publication).

As remarked, maritime Antarctic lakes display a high interannual environmental heterogeneity (Figure 9.4). For the two studied periods (2006/2007 and 2008/2009) important differences appeared in the main state variables, such as bacterioplankton and dissolved organic matter. These are mainly linked to different inputs of organic matter in the system by runoff processes and by differences in bacterioplankton abundances. Temperature and the intensity of the ice thaw process together with the water inputs into the lake are the main functions that control these state variables. The existence of differences in the environmental parameters has been also described for previous years 2001/2002, 2002/2003, and 2003/2004 by Rochera et al. (2010). Meteorology in Byers Peninsula is very unstable and heterogeneous over different years; therefore, the duration of the ice-free period is conditioned mainly by temperature and snowfall during the annual cycle (Bañón et al., 2013; Fassnacht et al., 2013). The model acceptably described the dynamic of the different lake state variables for the studied summer 2008/2009. Therefore, modeled changes in temperature, light availability, and the inputs of nutrients describe the dynamics of part of the microbial community with an acceptable average error. Nevertheless, these

Table 9.1 Equations Used in the Model to Describe the Main Processes in the Lake

Process	Description	Equation
DOC input	External input of dissolved organic carbon	(Mats DOC*(Inflow/1000)/Vol)+(Moss DOC/Vol)
DOC rec	Internal input of dissolved organic carbon	Bact R+Phyt ex
DOC out	Lake output of dissolved organic carbon	DOC*R Outflow
DOC dep	Dissolved organic carbon deposition	0.15*DOC
POC input	External input of particulated organic carbon	0.15*DOC input
POC deg	Degradation of particulated organic carbon	0.2*POC
POC rec	Internal input of particulated organic carbon	Bact G+Bact M+Phyt M
Bact upt C	Bacterial uptake of dissolved organic carbon	0.28*BACT*((DOC−0.1)/(DOC+0.1))* 1.065^(Temperature−12)
Bact G	Bacterial grazing efficiency	0.15*Bact upt C
Bact R	Bacterial respiration	0.01*BACT
Bact M	Bacterial mortality	0.04*BACT
Bact out	Lake output of bacteria	BACT*R Outflow
Bact graz	Zooplankton grazing over bacterioplankton	0.18*((BACT−0.03)/(BACT+1.2))* 1.08^(Temperature−20)
Phyt upt C	Phytoplankton carbon fixation	PHYT*43.7*((SRP)/(SRP+0.01)* ((PAR*0.615)/((PAR*0.615)+0.17))+2* ((PAR*0.368)/((PAR*0.368)+0.17))+ ((PAR*0.265)/((PAR*0.265)+0.17)))* 1.08^(Temperature−20)
Phyt M	Phytoplankton mortality	0.6*PHYT
Phyt ex	Phytoplankton excretion	0.35*Phyt upt C
Phyt out	Lake output of phytoplankton	PHYT*R Outflow
Phyt graz	Zooplankton grazing over phytoplankton	0.18*((PHYT−0.0004)/(PHYT+0.12))* 1.08^(Temperature−20)
R inflow	Relative inflow	Inflow/Vol
R outflow	Relative outflow	Outflow/Vol
Moss upt C	Mosses carbon uptake	MOSS*0.05*((PAR*0.265)/ ((PAR*0.265)+0.1))* 1.06^(Temperature−12)
Moss ex	Mosses excretion	(0.0014*MOSS)*1.08^(Temperature−12)
Mats upt C	Microbial mats carbon uptake	MATS*0.008*(PAR/(PAR+0.1))* 1.065^(Temperature−16)

Continued

Table 9.1 Equations Used in the Model to Describe the Main Processes in the Lake—cont'd

Process	Description	Equation
Mats ex	Microbial mats excretion	$0.0025*MATS*1.09^{\wedge}(\text{Temperature}-12)$
Mats DOC	Input of DOC from microbial mats	Mats ex * Mats Surface
Moss DOC	Input of DOC from aquatical mosses	Moss ex * 15225
Mats Surface	Surface covered by microbial mats	MATS * 5742/276.09
SRP input	Input of soluble reactive phosphorus	(R Inflow * Mats ex/166) + (Moss ex/166) + (Bact R/43.7)
SRP out	Lake output of soluble reactive phosphorus	SRP * R_Outflow
SRP upt	Phytoplankton SRP uptake	Phyt upt C/43.7

Initial values for the parameter's development were obtained using Ecotox (ecological modeling and ecotoxicology) software (Jørgensen and Bendoricchio, 2001).

results should be interpreted with care, because our model can describe with precision the dynamics of years with similar meteorological characteristics as the studied year 2008/2009, but has bigger errors when compared to years with different meteorological characteristics (Villaescusa et al., submitted for publication).

According to the model objective, the sensitivity analysis revealed the strong effect of temperature as the main forcing function in the system. This analysis showed that the bacterioplankton community can be strongly favored by the increase in temperature, as it drives to a higher availability of organic carbon in the system (Table 9.2). The simulation predicts increases of up to 22.5% in bacterioplankton biomass with an average summer temperature increase of only 1 °C, and up to 144.7% if this increase is up to 4 °C. Although this bacterial stimulation would mainly be driven by DOC increases, the predicted pool of dissolved organic carbon does not increase very much because it is strongly consumed by heterotrophic bacteria, with only weak DOC accumulation within the system. Similarly, phytoplankton can also be stimulated by temperature increases up to 56.4% when temperature rises by 4 °C, and it is also enhanced by increases in solar radiation that are explained by an earlier melting of the lake's ice cap when temperature becomes milder.

The increase in temperature during the austral summer would make the system more complex, as described by eco-exergy (Jørgensen et al., 2010). This parameter describes the system complexity, addressing the relationships away from the thermodynamic equilibrium (Jørgensen and Svirezhev, 2004). As a result, eco-exergy can be used as a rapid tool to study the changes in the microbial community and the effects of changes in the different forcing functions like temperature. Additionally, sensitivity analysis showed a somewhat slight positive effect of PAR on the

Table 9.2 Increases (% Compared with the Value at the Standard Modeled Temperature) in the Different State Variables Modeled When Increasing Temperature by Intervals of 1 °C

Variable	Simulated temperature increase (°C)			
	1	2	3	4
BACT	22.5	52.6	92.9	144.7
PHYT	10.7	23.3	38.4	56.4
DOC	3.9	6.9	9.1	10.2
POC	15	34.4	58.5	88.5
MATS	0.1	0.2	0.3	0.4
MOSS	1.6	3.3	5.2	7.2

BACT, biomass of bacterioplankton; PHYT, biomass of phytoplankton; DOC, dissolved organic carbon; POC, particulate organic carbon; MATS, biomass of microbial mats within the catchment; MOSS, biomass of benthic mosses in the lake bottom.

photosynthetic communities, like phytoplankton and benthic mosses. Therefore, according to these results, the relevance of physical variables on the microorganism in Antarctic systems is clear (Fountain et al., 1999). However, in spite of these evidences, zooplankton can also play a main role of in these systems as a regulator of microbial communities during the milder meteorological conditions of the austral summer (Camacho, 2006). In this case, zooplankton might exert a strong top-down control on the food web by predation over protozoa, thus releasing part of the grazing pressure on bacterioplankton. In such cases, predation would modulate the main effects of the physical variables, like that of temperature.

9.5 CONCLUSIONS

Lake Limnopolar can be considered as a typical inland lake from the maritime Antarctica, displaying ultraoligotrophic conditions and a planktonic community that is dominated by bacteria, protists, and crustacean zooplankton. Microbial mats draping the catchment and mosses covering the lake bottom are the main providers of organic carbon, fueling the development of the planktonic community, whose functioning we have partly modeled. This region is nowadays experiencing a strong warming, which would affect the functioning of their microbially dominated communities through changes in the duration of the productive period and ecological processes, mediated by temperature.

As the model predicts, the microbial community of Lake Limnopolar displays a very sensitive response to temperature and, more slightly, to solar radiation, with inputs of organic carbon being modulated by temperature and strongly affecting the bacterioplankton community abundance. This reflects the importance of the microbial planktonic community in polar lakes compared to a relatively lower role of the microbial loop in temperate systems. The intense response to temperature increases make these aquatic systems effective sentinels to predict and interpret the climate changes occurring in the area of west Antarctica and Antarctic Peninsula (Camacho et al., 2012; Quesada et al., 2013). This area is suffering an important warming (Figure 9.7), as reflected by different satellite-based temperature records during the last decades (Shuman and Comiso, 2002; Turner et al., 2005). Therefore, the study of Antarctic epicontinental lake's dynamics together with modeling and monitoring systems, such as satellite-based studies, can provide a complete view of Antarctic climate changes. These tools can help the scientific community to predict and describe possible environmental scenarios in the future.

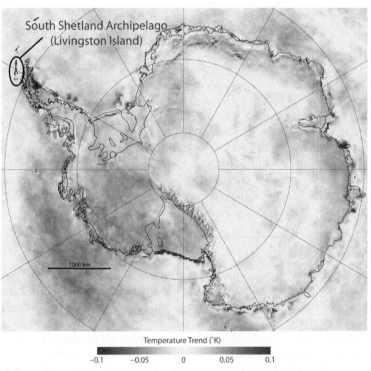

FIGURE 9.7

Long-term changes in yearly surface temperature in and around Antarctica between 1981 and 2007. (For color version of this figure, the reader is referred to the online version of this chapter.)

(Source: Modified from http://earthobservatory.nasa.gov/).

References

Allende, L., Mataloni, G., 2013. Short-term analysis of the phytoplankton structure and dynamics in two ponds with distinct trophic states from Cierva Point (maritime Antarctica). Polar Biol. 36, 629–644.

Azam, F., Fenchel, T., Field, J.G., Grey, J.S., Meyer-Reil, K.A., Thingstad, F., 1983. The ecological role of water-column microbes in the sea. Mar. Ecol. Prog. Ser. 10, 257–263.

Bañón, M., 2001. Observaciones meteorológicas en la Base Antártica Española Juan Carlos I. Monografía A-151, Instituto Nacional de Meteorología, Ministerio de Medio Ambiente. Madrid.

Bañón, M., Justel, A., Velázquez, D., Quesada, A., 2013. Regional weather survey on Byers Peninsula, Livingston Island, South Shetland Islands, Antarctica. Antarct. Sci. 25, 146–156.

Bell, E.M., Laybourn-Parry, J., 1999. Annual plankton dynamics in an Antarctic saline lake. Freshwater Biol. 41, 507–519.

Benayas, J., Pertierra, L., Tejedo, P., Lara, F., Bermudez, O., Hughes, K.A., Quesada, A., 2013. A review of scientific research trends within ASPA No. 126 Byers Peninsula, South Shetland Islands, Antarctica. Antarct. Sci. 25, 128–145.

Camacho, A., 2006. Planktonic microbial assemblages and the potential effects of metazooplankton predation on the food web of lakes from the maritime Antarctica and sub Antarctic islands. Rev. Environ. Sci. Biotechnol. 5, 167–185.

Camacho, A., Rochera, C., Villaescusa, J.A., Velázquez, D., Toro, M., Rico, E., Fernández-Valiente, E., Justel, A., Bañón, M., Quesada, A., 2012. Maritime Antarctic lakes as sentinels of climate change. Int. J. Des. Nat. Ecodyn. 7, 239–250.

Caulkett, A.P., Ellis-Evans, J.C., 1997. Chemistry of streams of Signy Island, maritime Antarctic: sources of major ions. Antarct. Sci. 9, 3–11.

Davey, M.C., 1993. Carbon and nitrogen dynamics in a maritime Antarctic stream. Freshwater Biol. 30, 319–330.

De Pablo, M.A., Blanco, J.J., Molina, A., Ramos, M., Quesada, A., Vieira, G., 2013. Interannual active layer variability at the Limnopolar Lake CALM site on Byers Peninsula, Livingston Island, Antarctica. Antarct. Sci. 25, 167–180.

Ellis-Evans, J.C., 1996a. Biological and chemical features of lakes and streams. In: López-Martínez, J., Thomson, M.R.A., Thomson, J.W. (Eds.), Geomorphological Map of Byers Peninsula, Livingston Island. In: BAS GEOMAP Series, Sheet 5-A, British Antarctic Survey, Cambridge, pp. 20–22.

Ellis-Evans, J.C., 1996b. Microbial diversity and function in Antarctic freshwater ecosystems. Biodiv. Conserv. 5, 1395–1431.

Ellis-Evans, J.C., Laybourn Parry, J., Bayliss, P.R., Perriss, S.J., 1998. Physical, chemical and microbial community characteristics of lakes of the Lasermann Hills, Continental Antarctica. Arch. Hydrobiol. 141, 209–230.

Fassnacht, S.R., López-Moreno, J.I., Toro, M., Hultstrand, D.M., 2013. Mapping snow cover and snow depth across the Lake Limnopolar watershed on Byers Peninsula, Livingston Island, Maritime Antarctica. Antarct. Sci. 25, 157–166.

Fernández-Valiente, E., Camacho, A., Rochera, C., Rico, E., Vincent, W.F., Quesada, A., 2007. Community structure and physiological characterization of microbial mats in Byers

Peninsula. Livingston Island (South Shetland Islands, Antarctica). FEMS Microbiol. Ecol. 59, 377–385.

Fountain, A.G., Berry Lyons, W., Burkins, M.D., Dana, G.L., Doran, P.T., Lewis, K.J., McKnight, D.M., Moorhead, D.L.L., Parsons, A.N., Priscu, J.C., Wall, D.H., Wharton, R.A., Ross, V.A., 1999. Physical control on the Taylor Valley Ecosystem, Antarctica. Bioscience 40, 961–971.

Gibson, J.A.E., 1999. The meromictic lakes and stratified marine basins of the Vestfold Hills, East Antarctica. Antarct. Sci. 11, 175–192.

Gil-Delgado, J.A., González-Solís, J., Barbosa, A., 2013. Populations of breeding birds in Byers Peninsula, Livingston Island, South Shetland Islands. Antarct. Sci. 25, 303–306.

Gutt, J., et al., 2012. Correlative and dynamic species distribution modeling for ecological predictions in the Antarctic: a cross-disciplinary concept. Polar Res. 31, 11091.

Hansson, L.A., Håkansson, H., 1992. Diatom community response along a productivity gradient of shallow Antarctic lakes. Polar Biol. 12, 463–468.

Hansson, L.A., Dartnall, H.J.G., Ellis-Evans, J.C., MacAlister, H., Tranvik, L.J., 1996. Variation in physical, chemical and biological components in the sub-Antarctic lakes of South Georgia. Ecography 19, 393–403.

Hawes, I., Brazier, P., 1991. Freshwater stream ecosystems of James Ross Island, Antarctica. Antarct. Sci. 3, 265–271.

Hawes, I., Schwarz, A.M.J., 2001. Absorption and utilization of irradiance by cyanobacterial mats in two ice-covered Antarctic lakes with contrasting light climate. J. Phycol. 37, 5–15.

Imura, S., Bando, T., Saito, S., Seto, K., Kanda, H., 2003. Benthic moss pillars in Antarctic lakes. Polar Biol. 22, 137–140.

Izaguirre, I., Vinocur, A., Mataloni, G., Pose, M., 1998. Phytoplankton communities in relation to trophic status in lakes from Hope Bay (Antarctic Peninsula). Hydrobiologia 369 (370), 73–87.

Izaguirre, I., Allende, L., Marinone, M.C., 2003. Comparative study of the planktonic communities of three lakes of contrasting trophic status at Hope Bay (Antarctic Peninsula). J. Plankton Res. 25, 1079–1097.

Jones, V.J., 1996. The diversity, distribution and ecology of diatoms from Antarctic inland waters. Biodivers. Conserv. 5, 1433–1449.

Jones, V.J., Juggins, S., Ellis-Evans, J.C., 1993. The relationship between water chemistry and surface sediment diatom assemblages in maritime Antarctic lakes. Antarct. Sci. 5, 339–348.

Jørgensen, S.E., Bendoricchio, G., 2001. Fundamentals of Ecological Modeling. Elsevier, Amsterdam, p. 530.

Jørgensen, S.E., Svirezhev, Y., 2004. Toward a Thermodynamic Theory for Ecological Systems. Elsevier, Amsterdam, p. 230.

Jørgensen, S.E., Ludovisi, A., Nielsen, S.N., 2010. The free energy and information embodied in the amino acid chains of organisms. Ecol. Model. 221, 2388–2392.

Kennedy, A.D., 1993. Water as a limiting factor in the Antarctic terrestrial environment: a biogeographical synthesis. Art Alp Res 25, 308–315.

Kennedy, A.D., 1995. Antarctic terrestrial ecosystem response to global environmental change. Annu. Rev. Ecol. Syst. 26, 683–704.

Kopalova, K., van de Vijver, B., 2013. Structure and ecology of freshwater benthic diatom communities from Byers Peninsula, Livingston Island, South Shetland Islands. Antarct. Sci. 25, 239–253.

Laybourn-Parry, J., 1997. The microbial loop in Antarctic lakes. In: Howards-Williams, C., Lyons, W.B., Hawes, I. (Eds.), Ecosystem Processes in Antarctic Ice-Free Landscapes. Balkema, Rotterdam, pp. 231–240.

Laybourn-Parry, J., Quayle, W.C., Henshaw, T., Ruddell, A., Marchant, H.J., 2001. Life on the edge: the plankton and chemistry of Beaver Lake, an ultra-oligotrophic epishelf lake, Antarctica. Freshwater Biol. 46, 1205–1217.

Lyons, W.B., Welch, K.A., Welch, S.A., Camacho, A., Rochera, C., Michaud, L., de Wit, R., Carey, A.E., 2013. Geochemistry of streams from Byers Peninsula, Livingston Island. Antarct. Sci. 25, 181–190.

Paggi, J.C., 1987. Limnological studies in the Potter Peninsula, 25 de Mayo Island, South Shetland Islands: biomass and spatial distribution of zooplankton. Biomass Sci. Ser. 7, 175–191.

Paggi, J.C., 1996. Feeding ecology of *Branchinecta gainii* (Crustacea: Anostraca) in ponds of South Shetland Islands, Antarctica. Polar Biol. 16, 13–18.

Pearce, D.A., 2005. The structure and stability of the bacterioplankton community in Antarctic freshwater lakes, subject to extremely rapid environmental change. FEMS Microbiol. Ecol. 53, 61–72.

Petz, W., Valbonesi, A., Quesada, A., 2005. Ciliate biodiversity in freshwater habitats of the maritime and continental Antarctica. Terra Antarct. 11, 43–50.

Pla-Rabes, S., Toro, M., van de Vijver, B., Rochera, C., Villaescusa, J.A., Camacho, A., Quesada, A., 2013. Stability and endemicity of benthic diatom assemblages from different substrates in a maritime stream on Byers Peninsula, Livingston Island, Antarctica: the role of climate variability. Antarct. Sci. 25, 254–269.

Priscu, J.C. (Ed.), 1998. Ecosystem Dynamics in a Polar Desert: the McMurdo Dry Valleys, Antarctica. In: Antarct Res Series 72, 72, American Geophysical Union, Washington, DC, p. 369.

Quesada, A., Camacho, A., Rochera, C., Velazquez, D., 2009. Byers Peninsula: a reference site for coastal, terrestrial and limnetic ecosystem studies in maritime Antarctica. Polar Sci. 3, 181–187.

Quesada, A., Camacho, A., Lyons, W.B., 2013. Multidisciplinary research on Byers Peninsula, Livingston Island: a future benchmark for change in Maritime Antarctica. Antarct. Sci. 25, 123–127.

Reid, T., Crout, N., 2008. A thermodynamic of freshwater Antarctic lake ice. Ecol. Model. 210, 221–241.

Rochera, C., Justel, A., Fernandez-Valiente, E., Bañón, M., Rico, E., Toro, M., Camacho, A., Quesada, A., 2010. Interannual meteorological variability and its effects on a lake from Maritime Antarctica. Polar Biol. 33, 1515–1628.

Rochera, C., Villaescusa, J.A., Velazquez, D., Fernandez-Valiente, E., Quesada, A., Camacho, A., 2013. Vertical structure of bi-layered microbial mats from Byers Peninsula, Maritime Antarctica. Antarct. Sci. 25, 270–276.

SCAR, 2003. Management Plan for Antarctic Specially Protected Area No.126 Byers Peninsula, Livingston Island, South Shetland Islands. SCAR Bulletin 150, July.

Schiaffino, M.R., Unrein, F., Gasol, J.M., Massana, R., Balague, V., Izaguirre, I., 2011. Bacterial community structure in a latitudinal gradient of lakes: the roles of spatial versus environmental factors. Freshwater Biol. 56, 1973–1991.

Serrano, E., Martínez de Pisón, E., López-Martínez, J., 1996. Periglacial and nival landforms and deposits. In: López-Martínez, J., Thomson, M.R.A., Thomson, J.W. (Eds.),

Geomorphological map of Byers Peninsula, Livingston Island. In: BAS Geompa Series, 5-A. British Antarctic Survey, Cambridge, pp. 28–34.

Shuman, C., Comiso, J., 2002. In situ and satellite surface temperature records in Antarctica. Ann. Glaciol. 34, 113–120.

Sommer, U., Sommer, F., 2006. Cladocerans versus copepods: the cause of contrasting top-down controls on freshwater and marine phytoplankton. Oecologia 147, 183–194.

Tell, G., Vinocur, A., Izaguirre, I., 1995. Cyanophyta of lakes and ponds of Hope Bay, Antarctic Peninsula. Polar Biol. 15, 503–509.

Thomson, M.R.A., López-Martínez, J., 1996. Introduction to the geomorphology quaternary geology of Byers Península. In: López-Martínez, J., Thomson, M.R.A., Thomson, M.R.A., Thomson, J.W. (Eds.), Supplementary text of the Geomorphological Map of Byers Peninsula. BAS Geomap Series, 5-A. British Antarctic Survey, Cambridge, pp. 1–4.

Toro, M., Camacho, A., Rochera, C., Rico, E., Bañón, M., Fernandéz-Valiente, E., Marco, E., Justel, A., Avedaño, M.C., Ariosa, Y., Vincent, W.F., Quesada, A., 2007. Limnological characteristics of the freshwater ecosystems of Byers Peninsula, Livingston Island, in maritime Antarctica. Polar Biol. 30, 635–649.

Turner, J., Colwell, S., Marshall, G., Lachlan-Cope, T., Carleton, A., Jones, P., Lagun, V., Reid, P., Lagovkina, S., 2005. Antarctic climate change during the last 50 years. Int. J. Climatol. 25, 279–294.

Velázquez, D., Rochera, C., Camacho, A., Quesada, A., 2011. Temperature effects on carbon and nitrogen metabolism in some Maritime Antarctic freshwater phototrophic communities. Polar Biol. 34, 1045–1055.

Velázquez, D., Lezcano, M.A., Frias, A., Quesada, A., 2013. Ecological relationships and stoichiometry within a Maritime Antarctic watershed. Antarct. Sci. 25, 191–197.

Villaescusa, J.A., Casamayor, E.O., Rochera, C., Velázquez, D., Chicote, A., Quesada, A., Camacho, A., 2010. A close link between bacterial community composition and environmental heterogeneity in maritime Antarctic lakes. Int. Microbiol. 13, 67–77.

Villaescusa, J.A., Casamayor, E.O., Rochera, C., Quesada, A., Michaud, L., Camacho, A., 2013a. Heterogeneous vertical structure of the bacterioplankton community in a non-stratified Antarctic lake. Antarct. Sci. 25, 229–238.

Villaescusa, J.A., Rochera, C., Velázquez, D., Rico, E., Quesada, A., Camacho, A., 2013b. Bacterioplankton summer dynamics in a Maritime Antarctic lake. Limnetica 32, 253–268.

Villaescusa, J.A., Jørgensen, S.E., Rochera, C., Velázquez, D., Quesada, A., Camacho, A., submitted for publication. Modelization of the microbial community sensitivity to temperature in an oligotrophic freshwater, Antarctic lake.

Vincent, W.F., 1988. Microbial Ecosystems of Antarctica. Cambridge University Press, Cambridge, p. 304.

Vinocur, A., Izaguirre, I., 1994. Freshwater algae (excluding Cyanophyceae) from nine lakes and pools of Hope Bay, Antarctic Peninsula. Antarct. Sci. 6, 483–489.

Vinocur, A., Pizarro, H., 2000. Microbial mats of twenty-six lakes from Potter Peninsula, King George Island, Antarctica. Hydrobiologia 437, 171–185.

Vinocur, A., Unrein, F., 2000. Typology of lentic water bodies at Potter Peninsula (King George Island, Antarctica) based on physical-chemical characteristics and phytoplankton communities. Polar Biol. 23, 858–870.

Wynn-Williams, D.D., 1996. Antarctic microbial diversity: the basis of polar ecosystem processes. Biodiv. Conserv. 5, 1271–1923.

Analytical Modeling for Environmental Dispersion in Wetland

Zi Wu[a], L. Zeng[b], G.Q. Chen[a,c],*

[a]*State Key Laboratory for Turbulence and Complex Systems, College of Engineering, Peking University, Beijing, China*
[b]*State Key Laboratory of Simulation and Regulation of Water Cycle in River Basin, China Institute of Water Resources and Hydropower Research, Beijing, China*
[c]*NAAM Group, King Abdulaziz University, Jeddah, Saudi Arabia*
Corresponding author: e-mail address: gqchen@pku.edu.cn

10.1 INTRODUCTION

Wetlands provide significant ecosystem services (Costanza et al., 1989, 1997; Chen et al., 2008, 2009a,b, 2011a,b; Zhou et al., 2009; Shao et al., 2013) in terms of contaminant degradation, water supply, climate regulation, flood storage, drought resistance, biodiversity conservation, and other functions (Mitsch and Gosselink, 1993). In wetlands, concentration transport is important for many aquatic organisms in dispersing nutrients and chemical signals (Lightbody and Nepf, 2006b). For application and engineering purposes, understanding the transport process is essential to environmental risk assessment, ecological restoration, and wastewater treatment engineering associated with wetlands, either natural or constructed (Carvalho et al., 2009). Specifically, when a hazardous contaminant is discharged into wetlands, an influenced region appears where pollutant concentration is beyond some given water quality standard level. An essential concern about this event is to predict the length and duration of the influenced region, which could be determined with profound and detailed knowledge on environmental dispersion (Zeng, 2010; Wu et al., 2011a; Zeng et al., 2011; Chen et al., 2012).

10.1.1 Methodology

In recent years, contaminant transport in wetland flows has been studied extensively by field and laboratory tests (Nepf et al., 1997; Lightbody and Nepf, 2006a), numerical simulations (Zhang et al., 2013), and analytical approaches (Zeng and Chen, 2009; Chen et al., 2010; Wu et al., 2011c, 2012; Zeng et al., 2011, 2012a,b). Focused on the detailed flow and transport processes at a length scale characterized by the stem diameter of the vegetation, the dispersion of the contaminant in wetland flows dominated by the bank-wall effect as well as the free-water-surface effect were, respectively, studied based on a laboratory experiment with plantlike cylinder arrays and a field test in a specific salt marsh: Nepf et al. (1997) measured velocity distribution with laser Doppler velocimetry and environmental diffusivity with a laser-induced fluorescence technique, and the contribution of stem wakes to the turbulent diffusivity were examined by a modified random-walk model; Lightbody and Nepf (2006b) measured stem frontal area, velocity, vertical diffusion, and longitudinal dispersion in a Spartina alterniflora salt marsh, and the dependents of the vertical diffusion coefficient and the longitudinal dispersion on the canopy morphology were discussed. Progress in the exploration of mean and turbulent flow and mass transport in the presence of aquatic vegetation was reviewed for a series of work as the aforementioned stem-scale experimental studies (Nepf, 2012).

Different from the researches on the stem-scale processes for specific wetland flows, Chen and his collaborators performed extensive analytical studies for contaminant transport in wetlands on the basis of the phase-average and Taylor's dispersion model (Chen et al., 2010, 2012; Wu et al., 2011a,b, 2012; Zeng et al., 2011, 2012a). While discontinuity of velocity and concentration caused by the existence of solid-phase material within the wetland flows is removed by the phase-average operation, the effects of the nonuniformity of the transverse flow velocity on the concentration transport at the intermediate scale is accounted for by the Taylor dispersion model at the environmentally concerned macroscopic scale with the transverse-average operation.

10.1.1.1 Phase average

Because of the existence of vegetation in the flow region, the transport process is very complicated and involves multiple spatial scales: the microscopic scale as characterized by the space among the stems or canopies of the vegetation, the intermediate scale as the phase-average scale, and the macroscopic scale as characterized by the geometry of the configurations under concern, such as the depth or the width of the wetland channel. Applications such as the environmental risk assessment and ecological restoration associated with wetlands are mostly concerned with the effective concentration distribution and its temporal evolution at the intermediate scale by the phase-average operation, which cancels out the microscopic concentration fluctuations and the discontinuity in space caused by the vegetation or the granular material in wetlands (Zeng, 2010). And the effective concentration dispersivity is an acquired quantity to account for the effects of microscopic fluctuations of both the velocity and the concentration. Actually, the theory and applications associated with phase average have been extensively performed in the field of fluid mechanics for porous media (Rajagopal and Tao, 1995; Liu and Masliyah, 2005; Zeng, 2010). With related physical parameters determined by the characteristics of scalar transport in wetland flows, the governing equations and constitutive relationships for the superficial flow resulting from the phase average can be well posed for the description of transport in wetland flows.

10.1.1.2 Taylor's dispersion model

Taylor's dispersion model is widely applied in describing the asymptotic concentration evolution in flows: After a transient initial stage of the transport, the concentration distribution across the cross-section tends to be the transverse mean concentration; if observed in a coordinate moving with the mean velocity of the cross-section, the transport process can be described by a one-dimensional diffusion equation, with a virtual coefficient known as Taylor dispersivity. This concept was first introduced by Taylor (1953) in his well-known research on soluble matter in solvent flowing slowly through a long, thin tube, and it refers to the process that solutes disperse toward longitudinal direction under the combined action of lateral solute diffusion and nonuniformity of transverse velocity distribution. More generally, Taylor dispersion can be defined as scalar substance dispersion in any direction, influenced by multiple factors, including molecular diffusion, turbulent diffusion, flow velocity, and biochemical reactions (Chatwin and Allen, 1985; Zeng, 2010).

Specifically in the field of wetland science, based on the theory of phase average, by the term environmental dispersion we refer to Taylor dispersion in porous media at an environmentally concerned scale where the contaminant cloud disperses toward the flow direction under the combined action of diffusion, cross-sectional concentration dispersion, and nonuniformity of transverse effective velocity distribution (Zeng, 2010; Wu et al., 2011a; Chen et al., 2012).

10.1.2 Analytical approaches

Environmental dispersion of contaminant in wetland flows has been studied analytically, with various techniques adopted to determine environmental dispersivities.

10.1.2.1 Concentration moment method

The concentration moment method by Aris (1956) is frequently applied for its ability in capturing the temporal variation of the dispersivity during the initial stage of the contaminant transport, even when the flow is steady. Since the moment equations can be solved exactly, concentration moments at different orders contain accurate statistical information on the evolution of the contaminant cloud. For example, the zeroth order moment reveals the conservation of the released contaminant; the first-order moment reveals the effective displacement of the contaminant cloud; the second-order moment reveals the property associated with the dispersion of the contaminant cloud. By obtaining the first three aforementioned moments, one can describe the contaminant transport process with the aid of the environmental dispersion model. Environmental dispersivity can be obtained either by directly considering its relations with the moments (Zeng, 2010; Wu et al., 2012) or by first adopting a closure model for the convection–diffusion equation on the basis of decoupling of the velocity and concentration, and then comparing the moment equations of the resulted dispersion model and the convection–diffusion equation (Wu et al., 2011a; Zeng et al., 2011).

To predict the critical length and duration of the contaminant cloud in wetland flows, the asymptotic evolution of the concentration at large time-scale is required. Thus determining only the steady part of the environmental dispersivity can substantially reduce the complexity in both the analytical solution procedure and the resulted analytical solutions.

10.1.2.2 Multiscale perturbation analysis

Expanding the contaminant concentration and time into multiple scales, the governing equations are different at each scale. By the first three perturbation problems the dominant concentration is found to be only the function of time and the longitudinal coordinate and could be governed by a pure diffusion equation. The dominant concentration is the transverse mean concentration and the environmental dispersion model is then recovered. This is the multiscale perturbation method for wetland flows (Wu et al., 2011b), which is an extension of Mei's homogenization technique for pure fluid flows (Mei et al., 1996). Recently, this method has been extended to find complete solutions for concentration distribution (Wu and Chen, 2014).

10.1.2.3 Mean concentration expansion method

Also applied is Gill's mean concentration expansion method (Gill, 1967), by which the concentration is expanded into transverse mean concentration and its longitudinal derivatives. Substituting this expansion into the governing convection–diffusion equation results in a generalized environmental dispersion model, containing second as well as higher-order derivation terms. According to previous discussions (Wu and Chen, 2012), the contribution of the third and higher-order terms could be neglected for long time evolution, and the resulting equation is exactly the classical Taylor dispersion model.

10.1.3 Progresses in the analytical modeling of environmental dispersion

10.1.3.1 Steady flow wetlands

Contaminant transport in a homogeneous wetland channel of a steady flow condition is the most fundamental and idealized case for analysis. The free-water-surface effect dominated wetland, the bank-wall effect dominated wetland, and the three-dimensional wetland channel dominated by both the free-water-surface and bank-wall effects have been considered for analytical solutions (Chen et al., 2010; Zeng, 2010; Wu et al., 2011b; Zeng and Chen, 2011). With the general equations for flow and contaminant transport in wetlands, a set of dimensionless characteristic parameters is determined, among which α is the most important one (Zeng et al., 2011). Describing the combined action of depth or width of the wetland, viscous friction of the vegetation or granular material, viscosity of the fluid, microscopic curvature of flow passage, and transverse momentum dispersion, α is the unique parameter for wetland flows and thus greatly affects the process of environmental dispersion. For all cases α plays the role as a damping factor to weaken transverse flow velocity nonuniformity as well as environmental dispersion (Wu et al., 2012). Another important parameter, Pe, is an analogy of that for transport in pure fluid flows, revealing the relative importance of flow convection and effective longitudinal diffusion in wetland flows (Wu et al., 2011a, 2012). We know from previous studies that for contaminant dispersion in wetland, the enhancement of environmental dispersivity is proportional to the square of Pe. For most cases Pe is large and the contribution of the effective longitudinal diffusion can be neglected.

10.1.3.2 Tidal flow wetlands

Tidal flow wetlands ranging from unvegetated tidal flats to salt marshes or mangroves exert a major influence on marine chemical cycles and contribute essentially to the balance of marine ecosystems (Wu et al., 2012). Different from that in steady flow wetlands, contaminant transport in tidal flow wetlands is much more complicated because of the additional complexity in the flow motion, which may vary periodically in magnitude or even change its direction repeatedly. For related discussions of dispersion in pure fluid flows with periodic conditions, two issues associated with the possible appearance of the negative dispersion coefficients and the effects of the period of the flow appear as new features in contrast to the steady cases (Wu et al., 2012). The negative dispersion coefficient has been referred to as physically unreasonable, because it implies the spontaneous development of infinite concentration (Yasuda, 1984). According to Yasuda (1984), this may be caused by the average operation, and by a different approach the appearance of the negative coefficient is avoided. Following Yasuda's average operation, environmental dispersion in tidal flow wetlands was analyzed (Wu et al., 2012). Environmental dispersivities oscillate around some constants after the initial stage, and their periodic averaged values perform as those for the steady wetland flow cases, increasing from zero to some constant. The only difference is in the enhancement of the dispersivities affected by the period of the flows. Similar to results for pure fluid flow dispersion, the effects of flow period are most remarkable at a low T_r (Wu et al., 2012), which is a dimensionless characteristic parameter inversely proportional to the flow period numerically (Zeng, 2010; Wu et al., 2012). The effects decrease as T_r increases and finally become negligible after some threshold (Wu et al., 2012).

10.1.3.3 Multizone flow wetlands

Another extension of the analysis on environmental dispersion in the single zone flow wetlands is to consider the media distribution heterogeneity. Media in wetlands refer to vegetation in flows or granular material upon the bed (Wu et al., 2011a; Chen et al., 2012; Wang et al., 2013). Typically, flow region in the wetland channel presents a multizone structure perpendicular to the flow direction: For example, vegetation density is different at each side of the wetland banks, or different kinds of vegetation are, respectively, grown near the bed and the water surface of the wetland. This media distribution heterogeneity can significantly alter the flow velocity distribution in wetlands. Some interesting patterns were observed by the obtained velocity profile (Chen et al., 2012), for example, the velocity nonuniformity is enhanced by the great velocity difference at each zone, and the maximum velocity is not necessarily at the free-water surface of the depth-dominated wetland. In the analysis of the two-zone and two-layer flow wetland (Wu et al., 2011a; Chen et al., 2012), new dimensionless parameters account for the difference between the zones are obtained, and all the parameters can thus be divided into two categories: the global ones and the relative ones, respectively, for revealing the integral property of the wetland as well as the relative magnitude of properties in different zones. Since the dispersion of contaminant in wetland flows is strongly related to the flow details, these parameters play important roles in affecting the variation of environmental dispersivities (Wu et al., 2011a; Chen et al., 2012). Most recently, environmental dispersion in a three-layer structured wetland was considered for analytical solutions (Wang et al., 2013).

10.1.3.4 Ecological degradation

For contaminant transport in wetland flows, ecological effects play a significant role in degrading the pollutant by various physical, chemical, or biological processes such as absorption, adsorption, hydrolysis, assimilation of plants, and bacteria metabolism (Zeng and Chen, 2011). The first-order reaction model is most extensively employed in environmental hydraulic design for constructed wetlands to predict the removal of contaminant, and it could be incorporated in the general formulation for contaminant transport to account for the effect of ecological degradation. Such attempts were made to study the depth-dominated wetland flows and later the two-zone wetland flow (Zeng and Chen, 2011; Chen, 2013). Technically, via an exponential transformation for the general formulation, the ecological degradation effects are removed to consider only the hydraulic part of the dispersion, and its contribution can be recovered when the analytical solution for the mean concentration under the hydraulic dispersion is determined (Zeng and Chen, 2011).

In the remaining text of this chapter, in Section 10.2, we introduce the basic equations for momentum and concentration transport for wetland flows on the basis of phase average, as well as three analytical approaches for environmental dispersion. Some results from studying the most typical wetlands dominated by the effects of free-water surface are presented in Section 10.3, which covers the steady flow wetlands, the tidal flow wetlands, and the two-layer flow wetlands; consideration of the ecological degradation effects and the prediction of the duration and the influenced region of the contaminant cloud are illustrated with an application example as well.

10.2 FORMULATION

After an instantaneous release of contaminant over the depth or the width of the wetlands, the temporal evolution of the contaminant cloud can be described by two stages. At the very beginning of the injection, the transport process is dominated by convection of the flow, and the concentration cloud is stretched out greatly by the transverse flow velocity difference. Before the diffusion could cancel out the transverse concentration difference, the concentration cloud is dispersed in large longitudinal distances, and the transverse mean concentration shows a skewed longitudinal distribution. The effects of diffusion, convection, and the present of the boundaries together contribute to the transport of contaminant. As time goes by, the longitudinal concentration gradient decreases, and the transverse diffusion gradually decreases the cross-sectional concentration difference. The mean concentration tends to show a Gaussian distribution. At the latter stage of the contaminant transport, the centroid of the cloud moves at the mean velocity of the flow, and the mean concentration disperses in the longitudinal direction by a virtual diffusion coefficient called Taylor dispersivity (Fischer et al., 1979; Zeng, 2010).

10.2.1 Momentum and concentration transport

For typical wetland flows, basic equations for momentum and concentration transport can be adopted generally at the phase average scale as (Liu and Masliyah, 2005; Zeng, 2010; Chen and Wu, 2012).

$$\rho\left(\frac{\partial \mathbf{U}}{\partial t} + \nabla \cdot \frac{\mathbf{U}\mathbf{U}}{\phi}\right) = -\nabla P - \mu F \mathbf{U} + \kappa \mu \nabla^2 \mathbf{U} + \kappa \nabla \cdot (\mathbf{L} \cdot \nabla \mathbf{U}), \tag{10.1}$$

$$\phi \frac{\partial C}{\partial t} + \nabla \cdot (\mathbf{U}C) = \nabla \cdot (\kappa \lambda \phi \nabla C) + \kappa \nabla \cdot (\mathbf{K} \cdot \nabla C), \tag{10.2}$$

where ρ is the density (kg m^{-3}), $\mathbf{U}$ velocity (m s^{-1}), t time (s), ϕ porosity (dimensionless), P pressure (kgm^{-1} s^{-2}), μ dynamic viscosity (kgm^{-1} s^{-1}), F shear factor (m^{-2}), κ tortuosity (dimensionless) to account for the spatial structure of aquatic plants, $\mathbf{L}$ momentum dispersivity tensor (kgm^{-1} s^{-1}), C concentration (kg m^{-3}), λ concentration diffusivity (m^2 s^{-1}), and $\mathbf{K}$ concentration dispersivity tensor (m^2 s^{-1}). Viscosity for momentum transfer and diffusivity for concentration transfer are valid for the description of the single phase flow at the microscopic passage scale; momentum and concentration dispersivities are properties valid for the description of the effective flow at the phase average scale out of the operation of phase average to cancel out the discontinuity between the two phases of the ambient water and the solid vegetation or granular material. The expression of the momentum equation for the effective flow through wetlands comes from a combination of the Navier–Stokes equation for single-phase fluid flows and Darcy's law for sweeping flows in porous media plus a term of second-order derivative to account for momentum dispersion. Similarly, the equation for concentration transfer is a combination of an advective–diffusive equation and a concentration dispersion law.

10.2.2 Analytical approaches for environmental dispersion in wetland flows

For the depth-dominated wetland flow, Equation (10.2) becomes

$$\frac{\partial C}{\partial t} + \frac{u}{\phi}\frac{\partial C}{\partial x} = k\left(\lambda + \frac{K}{\phi}\right)\frac{\partial^2 C}{\partial x^2} + k\left(\lambda + \frac{K}{\phi}\right)\frac{\partial^2 C}{\partial z^2}. \tag{10.3}$$

The nonpenetrate conditions at the bed wall of $z=0$ and free-water surface of $z=H$ read as

$$\frac{\partial C}{\partial z}\bigg|_{z=0} = \frac{\partial C}{\partial z}\bigg|_{z=H} = 0. \tag{10.4}$$

Consider a uniform and instantaneous release of contaminant with mass Q at the cross-section of $x=0$ at time $t=0$, the initial condition can be set as

$$C(x, z, t)|_{t=0} = \frac{Q\delta(x)}{\phi H}, \tag{10.5}$$

where $\delta(x)$ is the Dirac delta function.

Since the amount of released contaminant is finite, we have upstream and downstream boundary conditions as

$$C(x, z, t)|_{x=\pm\infty} = 0. \tag{10.6}$$

10.2.2.1 Concentration moment method

In a moving coordinate with $x_1 = x - U_0 t/\phi$, where U_0 characteristic velocity, Equations (10.3)–(10.6) can be expressed as

$$\phi\frac{\partial C}{\partial t} + u'\frac{\partial C}{\partial x_1} = \kappa(\lambda\phi + K)\frac{\partial^2 C}{\partial x_1^2} + \kappa(\lambda\phi + K)\frac{\partial^2 C}{\partial z^2}, \tag{10.7}$$

$$\frac{\partial C(x_1, z, t)}{\partial z}\bigg|_{z=0} = \frac{\partial C(x_1, z, t)}{\partial z}\bigg|_{z=H} = 0, \tag{10.8}$$

$$C(x_1, z, t)\bigg|_{t=0} = \frac{Q\delta(x_1)}{\phi H}, \tag{10.9}$$

$$C(x_1, z, t)|_{x=\pm\infty} = 0, \tag{10.10}$$

and $u' = u - U_0$.

Characteristic of an exponential decay in space, distribution of the concentration is subjected to the auxiliary relations (Aris, 1956; Fischer et al., 1979; Barton, 1983) as

$$x_1^p C(x_1, z, t)\bigg|_{x_1=\pm\infty} = \frac{\partial C}{\partial x_1}\bigg|_{x_1=\pm\infty} = x_1^p \frac{\partial^p C}{\partial x_1^p}\bigg|_{x_1=\pm\infty} = 0, \quad (p = 1, 2, \ldots) \tag{10.11}$$

The pth order concentration moment for contaminant transport in wetland flows is defined as

$$m_p^*(z, t) \equiv \int_{-\infty}^{\infty} C(x_1, z, t)x_1^p \mathrm{d}x_1. \tag{10.12}$$

Multiplying Equations (10.7)–(10.9) by x_1^p and integrating them with respect to x_1 in the interval of $(-\infty, \infty)$ with the aid of Equations (10.10) and (10.11), we have

$$\phi\frac{\partial m_p^*}{\partial t} = \kappa(\lambda\phi + K)\frac{\partial^2 m_p^*}{\partial z^2} + pu'm_{p-1}^* + p(p-1)m_{p-2}^*, \tag{10.13}$$

$$\left.\frac{\partial m_p^*}{\partial z}\right|_{z=0} = \left.\frac{\partial m_p^*}{\partial z}\right|_{z=H} = 0, \tag{10.14}$$

$$m_p^*(z,t)\big|_{t=0} = m_{p0}^*(z). \tag{10.15}$$

The governing equations with their initial and boundary conditions for the zeroth, the first-, and the second-order concentration moments can be obtained by setting p as 0, 1, and 2:

$$\phi\frac{\partial m_0^*}{\partial t} = \kappa(\lambda\phi + K)\frac{\partial^2 m_0^*}{\partial z^2}, \tag{10.16}$$

$$\left.\frac{\partial m_0^*}{\partial z}\right|_{z=0} = \left.\frac{\partial m_0^*}{\partial z}\right|_{z=H} = 0, \tag{10.17}$$

$$m_0^*(z,t)\big|_{t=0} = \frac{Q}{\phi H}, \tag{10.18}$$

$$\phi\frac{\partial m_1^*}{\partial t} = \kappa(\lambda\phi + K)\frac{\partial^2 m_1^*}{\partial z^2} + u'm_0^*, \tag{10.19}$$

$$\left.\frac{\partial m_1^*}{\partial z}\right|_{z=0} = \left.\frac{\partial m_1^*}{\partial z}\right|_{z=H} = 0, \tag{10.20}$$

$$m_1^*(z,t)\big|_{t=0} = 0, \tag{10.21}$$

$$\phi\frac{\partial m_2^*}{\partial t} = \kappa(\lambda\phi + K)\frac{\partial^2 m_2^*}{\partial z^2} + 2u'm_1^* + 2m_0^*\kappa(\lambda\phi + K). \tag{10.22}$$

$$\left.\frac{\partial m_2^*}{\partial z}\right|_{z=0} = \left.\frac{\partial m_2^*}{\partial z}\right|_{z=H} = 0, \tag{10.23}$$

$$m_2^*(z,t)\big|_{t=0} = 0. \tag{10.24}$$

These three sets of concentration moment equations can be solved exactly for m_0^*, m_1^*, m_2^*. The characteristics for the expansion of the contaminant cloud is revealed by the environmental dispersivity

$$D^* = \frac{1}{2}\frac{\overline{dv_2^*}}{dt} - \overline{v_1^*\frac{dv_1^*}{dt}} - \kappa(\lambda\phi + K), \tag{10.25}$$

where

$$v_p^*(z,t) = \frac{m_p^*(z,t)}{m_0^*(z,t)}, \quad (p = 1, 2, \ldots) \tag{10.26}$$

and the overline stands for the defined operation of vertical average

$$\bar{f} \equiv \frac{1}{H} \int_0^H f \, dz. \tag{10.27}$$

10.2.2.2 Multiscale perturbation analysis

Similar to Mei's multiscale analysis for the case with a single phase flow (Mei et al., 1996), three time scales can be identified. The relations between $T_1 = H^2/[\kappa(\lambda + (K/\phi))]$ as the diffusion time across the depth of the wetland H, $T_2 = L/u_m$ as the convection time across a longitudinal distance L which is far exceeding the depth of H, and $T_3 = L^2/[\kappa(\lambda + (K/\phi))]$ as the diffusion time across the distance of L can be generally expressed as

$$T_1 : T_2 : T_3 = 1 : \frac{1}{\epsilon} : \frac{1}{\epsilon^2} \tag{10.28}$$

by considering that the Péclet number

$$Pe = \frac{Hu_m}{\phi\kappa\left(\lambda + \dfrac{K}{\phi}\right)} \tag{10.29}$$

is of order unity, where u_m is the vertical mean velocity, $\epsilon = H/L \ll 1$.

With dimensionless parameters of

$$\psi = \frac{u}{u_m}, \quad \zeta = \frac{z}{H}, \quad \xi = \frac{x}{L}, \quad \tau = \frac{\kappa\left(\lambda + \dfrac{K}{\phi}\right)}{H^2} t, \tag{10.30}$$

the governing equation and boundary conditions for concentration can be rewritten as

$$\frac{\partial C}{\partial \tau} + \epsilon Pe\psi \frac{\partial C}{\partial \xi} = \epsilon^2 \frac{\partial^2 C}{\partial \xi^2} + \frac{\partial^2 C}{\partial \zeta^2}, \tag{10.31}$$

$$\left. \frac{\partial C}{\partial \zeta} \right|_{\zeta=0} = \left. \frac{\partial C}{\partial \zeta} \right|_{\zeta=1} = 0. \tag{10.32}$$

Here we introduce the multiple time coordinates

$$t_0 = \tau, \quad t_1 = \epsilon\tau, \quad t_2 = \epsilon^2\tau, \tag{10.33}$$

thus the variation of concentration C with dimensionless time τ can be expressed as

$$\frac{\partial C}{\partial \tau} \rightarrow \frac{\partial C}{\partial t_0} + \epsilon \frac{\partial C}{\partial t_1} + \epsilon^2 \frac{\partial C}{\partial t_2}, \tag{10.34}$$

associated with a multiple scale expansions

$$C(\xi, \zeta, t_0, t_1, t_2) = C_0(\xi, \zeta, t_0, t_1, t_2) + \epsilon C_1(\xi, \zeta, t_0, t_1, t_2) + \epsilon^2 C_2(\xi, \zeta, t_0, t_1, t_2) + O(\epsilon^3). \tag{10.35}$$

Substituting Equations (10.34) and (10.35) into Equations (10.31) and (10.32), different perturbation problems can be deduced.

The perturbation problem for $O(\epsilon^0)$ is obtained as

$$\frac{\partial C_0}{\partial t_0} = \frac{\partial^2 C_0}{\partial \zeta^2}, \tag{10.36}$$

$$\frac{\partial C_0}{\partial \zeta}\Big|_{\zeta=0} = \frac{\partial C_0}{\partial \zeta}\Big|_{\zeta=1} = 0. \tag{10.37}$$

For long time evolution, C_0 is independent of ζ and t_0

$$C_0(\xi, \zeta, t_0, t_1, t_2)|_{t_0 \to \infty} = C_0(\xi, t_1, t_2). \tag{10.38}$$

Applying depth-average to Equation (10.36), under long time evolution the result is

$$\frac{\partial C_0}{\partial t_0} = 0. \tag{10.39}$$

The perturbation problem for $O(\epsilon^1)$ is the following, neglecting a term smaller in order of magnitude (Wu et al., 2011b):

$$\frac{\partial C_0}{\partial t_1} + Pe\psi \frac{\partial C_0}{\partial \xi} = \frac{\partial^2 C_1}{\partial \zeta^2}, \tag{10.40}$$

$$\frac{\partial C_1}{\partial \zeta}\Big|_{\zeta=0} = \frac{\partial C_1}{\partial \zeta}\Big|_{\zeta=1} = 0. \tag{10.41}$$

We apply depth-average to Equation (10.40) and get

$$\frac{\partial C_0}{\partial t_1} + Pe\bar{\psi} \frac{\partial C_0}{\partial \xi} = 0. \tag{10.42}$$

Multiplying ϵ to each term of Equation (10.42), we have

$$\epsilon \frac{\partial C_0}{\partial t_1} + \epsilon Pe\bar{\psi} \frac{\partial C_0}{\partial \xi} = 0. \tag{10.43}$$

Subtracting Equation (10.40) from Equation (10.42) results in

$$Pe\psi' \frac{\partial C_0}{\partial \xi} = \frac{\partial^2 C_1}{\partial \zeta^2}, \tag{10.44}$$

where $\psi' = \psi - \bar{\psi}$ is the nonuniformity of the velocity.

Since C_0 is independent of ζ, we have

$$C_1 = Pe \frac{\partial C_0}{\partial \xi} g(\zeta). \tag{10.45}$$

Substituting Equation (10.45) into Equations (10.44) and (10.41) results in

$$\frac{d^2 g(\zeta)}{d\zeta^2} = \psi', \tag{10.46}$$

$$\frac{dg(\zeta)}{d\zeta}\bigg|_{\zeta=0} = \frac{dg(\zeta)}{d\zeta}\bigg|_{\zeta=1} = 0, \tag{10.47}$$

Then g and C_1 can be deduced if ψ' was found.

The perturbation problem for $O(\epsilon^2)$ is the following, neglecting a term smaller in order of magnitude (Wu et al., 2011b):

$$\frac{\partial C_0}{\partial t_2} + \frac{\partial C_1}{\partial t_1} + Pe\psi \frac{\partial C_1}{\partial \xi} = \frac{\partial^2 C_0}{\partial \xi^2} + \frac{\partial^2 C_2}{\partial \zeta^2}, \tag{10.48}$$

$$\frac{\partial C_2}{\partial \zeta}\bigg|_{\zeta=0} = \frac{\partial C_2}{\partial \zeta}\bigg|_{\zeta=1} = 0. \tag{10.49}$$

According to Equations (10.42) and (10.45),

$$\frac{\partial C_1}{\partial t_1} = Peg(\zeta)\frac{\partial}{\partial \xi}\frac{\partial C_0}{\partial t_1} = -Pe^2\bar{\psi}g(\zeta)\frac{\partial^2 C_0}{\partial \xi^2}. \tag{10.50}$$

Equation (10.48) can be rewritten as

$$\frac{\partial C_0}{\partial t_2} = Pe^2\psi' g(\zeta)\frac{\partial^2 C_0}{\partial \xi^2} = \frac{\partial^2 C_0}{\partial \xi^2} + \frac{\partial^2 C_2}{\partial \zeta^2}. \tag{10.51}$$

Applying the depth-average operation to Equation (10.51) and multiplying a quantity ϵ^2 at each term of this equation, we have

$$\epsilon^2 \frac{\partial C_0}{\partial t_2} = \epsilon^2 \left[1 - Pe^2\overline{\psi' g(\zeta)}\right]\frac{\partial^2 C_0}{\partial \xi^2}. \tag{10.52}$$

Adding Equations (10.52) and (10.43) to Equation (10.39) results in

$$\frac{\partial C_0}{\partial t_0} + \epsilon\frac{\partial C_0}{\partial t_1} + \epsilon^2\frac{\partial C_0}{\partial t_2} + \epsilon Pe\langle\psi\rangle\frac{\partial C_0}{\partial \xi} = \epsilon^2\left[1 - Pe^2\overline{\psi' g(\zeta)}\right]\frac{\partial^2 C_0}{\partial \xi^2}. \tag{10.53}$$

By the inverse relation presented in Equation (10.34), with $\zeta = x/H - Pe\langle\psi\rangle\tau$, Equation (10.53) can be written as

$$\frac{\partial C_0}{\partial \tau} = D\frac{\partial^2 C_0}{\partial \zeta^2}, \tag{10.54}$$

where

$$D = 1 - Pe^2\overline{\psi' G(\zeta)} \tag{10.55}$$

is the environmental dispersivity.

10.2.2.3 Mean concentration expansion method

With dimensionless parameters of

$$\xi = \frac{x - \overline{u_\phi} t}{l_c}, \quad \tau = t \Bigg/ \frac{H^2}{\kappa\left(\lambda + \dfrac{K}{\phi}\right)}, \quad \zeta = \frac{z}{H}, \tag{10.56}$$

where $\overline{u_\phi}$ is the averaged phase velocity based on

$$u_\phi = \frac{u}{\phi}, \tag{10.57}$$

and l_c is a characteristic length of the concentration cloud, defined as

$$l_c = \frac{H^2}{\kappa\left(\lambda + \dfrac{K}{\phi}\right)} u_c, \tag{10.58}$$

where u_c is a characteristic velocity of the flow by which the velocity can be dimensionalized as

$$\psi_\phi = \frac{u_\phi}{u_c}. \tag{10.59}$$

Ignoring the effective longitudinal diffusion effects, the governing equations and their boundary conditions for the concentration transport can be rewritten as

$$\frac{\partial C}{\partial \tau} + \psi'_\phi \frac{\partial C}{\partial \xi} = \frac{\partial^2 C}{\partial \zeta^2}, \tag{10.60}$$

$$\frac{\partial C}{\partial \zeta}\bigg|_{\zeta=0} = \frac{\partial C}{\partial \zeta}\bigg|_{\zeta=1} = 0, \tag{10.61}$$

where

$$\psi'_\phi = \psi_\phi - \overline{\psi_\phi}. \tag{10.62}$$

We apply the average operation to Equation (10.60) and get

$$\frac{\partial \bar{C}}{\partial \tau} + \overline{\psi'_\phi \frac{\partial C'}{\partial \xi}} = 0, \tag{10.63}$$

with the decompositions of $C = \bar{C} + C'$. Following Gill (1967), we have series expansions for the concentration deviations C' as (Wu and Chen, 2012)

$$C' = g^{(1)}(\zeta) \frac{\partial \bar{C}}{\partial \xi} + g^{(2)}(\zeta) \frac{\partial^2 \bar{C}}{\partial \xi^2} + \cdots \tag{10.64}$$

Substituting Equation (10.64) into Equation (10.63) results in

$$\frac{\partial \bar{C}}{\partial \tau} + \overline{\psi'_\phi \frac{\partial}{\partial \xi} \left(g^{(1)}(\zeta) \frac{\partial \bar{C}}{\partial \xi} + g^{(2)}(\zeta) \frac{\partial^2 \bar{C}}{\partial \xi^2} + \dots \right)} = 0, \tag{10.65}$$

which can be rewritten as

$$\frac{\partial \bar{C}}{\partial \tau} = \left(-\overline{\psi'_\phi g^{(1)}(\zeta)} \right) \frac{\partial^2 \bar{C}}{\partial \xi^2} + \left(-\overline{\psi'_\phi g^{(2)}(\zeta)} \right) \frac{\partial^3 \bar{C}}{\partial \xi^3} + \dots \tag{10.66}$$

After the initial stage of the contaminant transport in wetlands, as we discussed previously, the vertical mean concentration can be described by the Gaussian distribution (Wu and Chen, 2012). The terms with the third- and higher-order derivatives of the mean concentration can thus be neglected to simplify the formulation. Consequently, we get the following dispersion model:

$$\frac{\partial \bar{C}}{\partial \tau} = -\overline{\psi'_\phi g^{(1)}(\zeta)} \frac{\partial^2 \bar{C}}{\partial \xi^2}. \tag{10.67}$$

To determine the unknown functions $g^{(1)}(\zeta)$, Equations (10.64) and (10.67) are, respectively, substituted into Equation (10.63) to get

$$\begin{aligned} -\overline{\psi'_\phi g^{(1)}(\zeta)} &\left(\frac{\partial^2 \bar{C}}{\partial \xi^2} + g^{(1)}(\zeta) \frac{\partial^3 \bar{C}}{\partial \xi^3} + \dots \right) \\ +\psi'_\phi &\left(\frac{\partial \bar{C}}{\partial \xi} + g^{(1)}(\zeta) \frac{\partial^2 \bar{C}}{\partial \xi^2} + g^{(2)}(\zeta) \frac{\partial^3 \bar{C}}{\partial \xi^3} + \dots \right) \\ = \frac{\partial^2 g^{(1)}(\zeta)}{\partial \zeta^2} \frac{\partial \bar{C}}{\partial \xi} &+ \frac{\partial^2 g^{(2)}(\zeta)}{\partial \zeta^2} \frac{\partial^2 \bar{C}}{\partial \xi^2} + \frac{\partial^2 g^{(3)}(\zeta)}{\partial \zeta^2} \frac{\partial^3 \bar{C}}{\partial \xi^3} + \dots \end{aligned} \tag{10.68}$$

Comparing the terms associated with the first derivative gives

$$\psi'_\phi = \frac{\partial^2 g^{(1)}(\zeta)}{\partial \zeta^2}. \tag{10.69}$$

By integrating twice with the aid of Equation (10.61), Equation (10.69) becomes

$$g(\zeta) = \int_1^\zeta \int_0^\zeta \psi'_\phi d\zeta d\zeta. \tag{10.70}$$

In Equation (10.67), the quantity

$$D = -\overline{\psi'_\phi g^{(1)}(\zeta)} \tag{10.71}$$

can be called the enhancement of environmental dispersivity. This result is identical to Equation (10.55), including the additional constant 1, which stands for the overall effective longitudinal diffusion effect.

10.2.2.4 Ecological degradation, duration, and influenced region

The first-order reaction model remains as the most extensively employed model in environmental hydraulic design for constructed wetlands to predict the removal of contaminant. It could be incorporated in the general formulation for contaminant transport to account for the effect of ecological degradation. For a typical wetland flow, the basic equation for concentration transport can be adopted at the phase-average scale as (Zeng and Chen, 2011)

$$\phi\frac{\partial C}{\partial t}+\boldsymbol{\nabla}\cdot(\mathbf{U}C)=\boldsymbol{\nabla}\cdot(\kappa\lambda\phi\boldsymbol{\nabla}C)+\kappa\boldsymbol{\nabla}\cdot(\mathbf{K}\cdot\boldsymbol{\nabla}C)-\phi k_a C, \qquad (10.72)$$

where k_a is the apparent reaction rate (s^{-1}).

To separate the hydraulic effect, which could be considered independently with the concept of environmental dispersion, we introduce an exponential transformation as

$$C=C_h\exp(-k_a t). \qquad (10.73)$$

Substituting Equation (10.73) into Equation (10.72) yields an advection–diffusion equation for environmental dispersion

$$\phi\frac{\partial C_h}{\partial t}+\boldsymbol{\nabla}\cdot(\mathbf{U}C_h)=\boldsymbol{\nabla}\cdot(\kappa\lambda\phi\boldsymbol{\nabla}C_h)+\kappa\boldsymbol{\nabla}\cdot(\mathbf{K}\cdot\boldsymbol{\nabla}C_h). \qquad (10.74)$$

By dealing with the environmental dispersion problems based on Equation (10.74), the analytical solution for the hydraulic part of the mean concentration can be deduced when the environmental dispersivity is determined. Using the information in Section 10.2.2.1, for example, gives

$$\overline{C_h}(x_1,t)=\frac{Q}{\phi H\sqrt{4\pi G(t)}}\exp\left[-\frac{x_1^2}{4G(t)}\right], \qquad (10.75)$$

where

$$G(t)=\int_0^t D^*\mathrm{d}t. \qquad (10.76)$$

Referring to Equation (10.73), the analytical solution for the evolution of the mean concentration involving the ecological degradation effect is determined as

$$\bar{C}(x_1,t)=\frac{Q}{\phi H\sqrt{4\pi G(t)}}\exp\left[-\frac{x_1^2}{4G(t)}-k_a t\right]. \qquad (10.77)$$

Consequently, an influenced region of the contaminant cloud in the wetland flow can be defined as

$$S = 4\sqrt{-G(t)\left[k_{a}t + \ln\frac{C_r\phi H\sqrt{4\pi G(t)}}{Q}\right]}, \qquad (10.78)$$

where C_r is some given safe upper limit of the contaminant concentration.

10.3 ENVIRONMENTAL DISPERSION FOR DEPTH-DOMINATED WETLAND FLOWS

Since the environmental dispersion model is applied in the prediction of contaminant transport process, the key step is to obtain the analytical expressions for environmental dispersivities, as was done for the steady flow wetlands (Zeng and Chen, 2011), the tidal flow wetlands (Wu et al., 2012), and the two-layer flow wetlands (Chen et al., 2012), respectively. In this part, first we review some results previously obtained by the method of concentration moment on the environmental dispersivities for depth-dominated wetland flows; then an application example is given to illustrate the prediction of the duration and critical length of an influenced region after the release of the pollutant.

10.3.1 Steady flow wetlands

For this fundamental and idealized case, some basic properties for the variation of the environmental dispersivity in wetlands are revealed. As obtained by the method of concentration moment, the temporal evolution of the dispersivity can be divided into two stages: In the initial stage, the dispersivity increases from zero to approach some asymptotic constant; in the latter stage, it is actually steady at the limiting asymptotic constant. We mentioned that α acts as a damping factor, and by Figure 10.1a, it is obvious that dispersivity decreases as α increases. Dispersivity depends on Pe quadratically, and an increased value with the increase of Pe is observed in Figure 10.1b.

10.3.2 Tidal flow wetlands

The additional complexity for contaminant dispersion in tidal flow wetlands is due to the more complicated flow velocity distribution, which may vary periodically in magnitude or even change its direction repeatedly. As a result of the flow pattern, the environmental dispersivity appears to be in a periodic evolution with time. This is shown in Figure 10.2a and b. In Figure 10.2c, dispersivity increases as the apparent pressure gradient a_1 increases, and no negative values appeared. The dependence of dispersivity on the flow period is revealed in Figure 10.2d, in which some limiting cases can be seen (Wu et al., 2012).

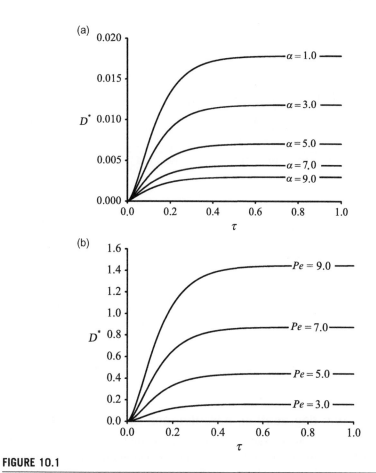

FIGURE 10.1

Variation of D^* with τ for steady flow wetlands.

10.3.3 Two-layer flow wetlands

Media distribution heterogeneity perpendicular to the flows causes the multizone structure of wetlands. To reveal the media distribution details, global as well as relative parameters are deduced. In Figure 10.3a–c, α, M, and N together gives a description for specific parameters, as α_1 and α_2 for layer 1 and 2, respectively. For two-layer flow wetlands, the dependence of dispersivity on Pe (Figure 10.3d) is much more complicated compared with the single-zone case, which is also brought by the different media distribution at different zones. Figure 10.3e shows the effects of another relative parameter, which is the square root of the ratio between vertical effective concentration dispersivity in layer 2 to that in layer 1. The contribution of the position of the interface of two layers is shown in Figure 10.3f.

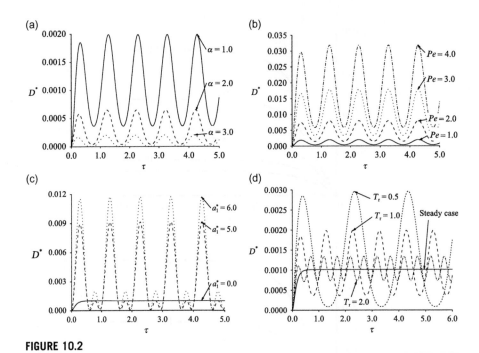

FIGURE 10.2

Variation of D^* with τ for tidal flow wetlands.

10.3.4 Ecological degradation, duration, and influenced region

For steady flow wetlands as an example, to assess the evolution of the contaminant cloud under hydraulic dispersion and ecological degradation effects after a pulsed emission, according to Equation (10.76) we have

$$G(t) = 2Pe^2 \sum_{n=1}^{\infty} \frac{1}{\beta_j^2} \left[\widetilde{\psi}(\beta_j)\right]^2 \frac{4e^{-t\beta_j^2} - e^{-2t\beta_j^2} + 2t\beta_j^2 - 3}{2\beta_j^2}, \qquad (10.79)$$

where $\beta_j = j\pi$, $(j = 0, 1, 2 \ldots)$, and

$$\widetilde{\psi}(\beta_j) = \frac{\alpha^2}{[1 - \alpha \coth(\alpha)]\left(\alpha^2 + \beta_j^2\right)}. \qquad (10.80)$$

Then the duration and the influenced region of the contaminant cloud can be evaluated by Equation (10.78).

Consider typical parameters of $\phi = 0.9$ (EPA, 1999), mean diameter of stem $d = 1.0 \times 10^{-2}$ m (Nepf and Ghisalberti, 2008), $H = 1.0$ m (Mitsch and Gosselink, 1993), and $\bar{u} = 0.15 \mathrm{ms}^{-1}$ (Stone and Shen, 2002).

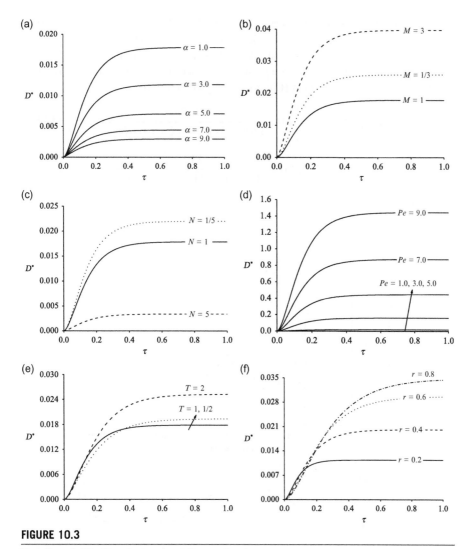

FIGURE 10.3

Variation of D^* with τ for two-layer flow wetlands.

The tortuosity can be calculated with the Bruggemann equation (Liu and Masliyah, 2005) as

$$\kappa = \sqrt{\phi}. \tag{10.81}$$

The shear factor F is calculated with permeability (Liu and Masliyah, 2005) derived from Ergun equation (Ergun, 1952) as

$$F = \frac{150 - (1 - \phi)^2}{d^2 \phi^3}. \tag{10.82}$$

With ambient water properties as $\rho = 1.0 \times 10^3$ kg m^{-3}, $\mu = 1.0 \times 10^{-3}$ kg m^{-1}s^{-1}, and $\lambda = 1.0 \times 10^{-5}$ m^2 s^{-1} (Lightbody and Nepf, 2006a), and the Reynolds and Péclet numbers at the stem scale defined as $Re_d \equiv \bar{u}d\rho/\mu$ and $Pe_d \equiv \bar{u}d/\lambda$, the vertical momentum microscopic dispersivity L_{zz} and the concentration microscopic dispersivity K can be calculated (Liu and Masliyah, 2005) as

$$L_{zz} = 160 Re_{d\mu\phi}^{11/3} (1-\phi)^{2/3} \left[4.0 + \frac{25Re_d}{(100+Re_d)(1+2\times 10^{-6}Re_d)} \right]^{-1}, \quad (10.83)$$

$$K = 160 Pe_{d\lambda\phi}^{11/3} (1-\phi)^{2/3} \left[4.0 + \frac{25Pe_d}{(100+Pe_d)(1+2\times 10^{-6}Pe_d)} \right]^{-1}, \quad (10.84)$$

respectively.

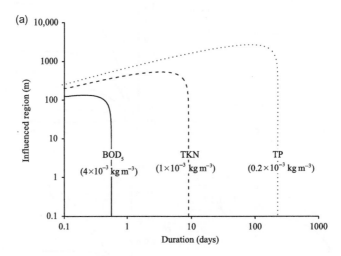

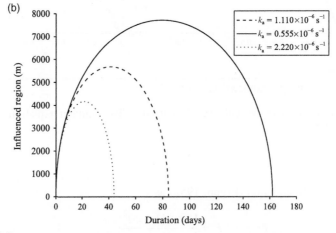

FIGURE 10.4

Variation of the influenced region with the duration of the contaminant cloud.

Then   we   have $\kappa = 0.95$, $F = 2.06 \times 10^4 \, \mathrm{m}^{-2}$, $L_{zz} = 1.28 \, \mathrm{kg \, m^{-1} \, s^{-1}}$,
$K = 1.85 \times 10^{-3} \, \mathrm{m^2 \, s^{-1}}$.

In this application, we consider four typical contaminant constituents of BOD_5 (5-day biochemical oxygen demand), TKN (total Kjeldahl nitrogen), TP (total phosphorus), and Hg (mercury) associated with industrial wastewater emissions, with the basic limit value of C_r given in National Standard of the People's Republic of China (GB 3838-2002) (CN EPA and CNGAQSIQ, 2002) as $4.0 \times 10^{-3} \, \mathrm{kg \, m^{-3}}$, $1.0 \times 10^{-3} \, \mathrm{kg \, m^{-3}}$, $0.2 \times 10^{-3} \, \mathrm{kg \, m^{-3}}$, and $1.0 \times 10^{-7} \, \mathrm{kg \, m^{-3}}$, respectively.

Set the instantaneous emission Q as $1 \, \mathrm{kg \, m^{-1}}$. The striking variation of the influenced region with duration of the contaminant cloud for the four pollutants is illustrated in Figure 10.4.

To understand the sole contribution of the hydraulic dispersion (with no ecological degradation effects), the evolution of the influenced region is shown in Figure 10.4a. For the first three pollutants with relatively higher permitted basic limit values, it is obvious that as limit value decreases, both the duration and the influenced region grow sharply as revealed in the log–log plot. Thus it can be concluded that for the pollutant Hg with several orders lower in the limit value, it is hard for the influenced region to disappear within a short period of time by the hydraulic dispersion mechanism alone in the given physical parameters. In Figure 10.4b, it is shown that with a typical ecological reaction rate, the influenced region of the Hg concentration vanishes quickly. For the same injected amount, the detrimental effect of constituent BOD_5 is the weakest. In contrast to the temporary, localized damage of BOD_5 on water quality, the heavy metal Hg causes greater damage, because the removal of Hg mainly depends on chemical decomposition, such as oxidation and reduction, and unlike some metals producing detrimental effect only in higher concentration, Hg is toxic for aquatic life and human health even at quite low concentration (EPA, 1999).

10.4 CONCLUSIONS

In this chapter, we reviewed some recent progresses in the analytical exploration of environmental dispersion in wetlands. To characterize the concentration transport in wetland flows, the methodology is based on the phase average and Taylor's dispersion model. Among the analytical approaches applied to determine the key quantity of environmental dispersivity, the concentration moment method is most frequently adopted for its ability to reveal the transient process before the steady status is approached.

As shown by previous results, the environmental dispersivities in some typical wetlands, such as the steady flow wetlands, the tidal flow wetlands, and the two-layer flow wetlands, are analyzed. For the unique dimensionless parameter in wetland flows, α plays as a damping factor to reduce the nonuniformity of flow velocity distribution and consequently decreases the environmental dispersivity D^*. The dependence of D^* on the Péclet number is quadratic, thus for typical physical parameters in wetland flows with a large Pe, the effective longitudinal diffusion effect can be neglected compared with the environmental dispersion. For tidal flow wetlands, D^* depends on the flow period heavily. In the multizone flow wetlands, as the two-layer flow wetlands are considered, the relative parameters accounting for the difference of properties at different zones contribute to the environmental dispersivity additionally.

For typical pollutant constituents such as BOD_5, TKN, TP, and Hg in wastewater emission, as an application example the evolution of contaminant cloud in the steady flow wetlands dominated by the free-water-surface effect is illustrated by the duration and influenced region of the contaminant cloud with concentration beyond some given environmental standard level, with essential implications for ecological impact assessment and environmental management.

References

Aris, R., 1956. On the dispersion of a solute in a fluid flowing through a tube. Proc. Roy. Soc. Lond. Ser. A—Math. Phys. Sci. 235, 67–77.

Barton, N.G., 1983. On the method of moments for solute dispersion. J. Fluid Mech. 126, 205–218.

Carvalho, F.P., Villeneuve, J.P., Cattini, C., Rendón, J., De Oliveira, J.M., 2009. Ecological risk assessment of PCBs and other organic contaminant residues in laguna de terminos, mexico. Ecotoxicology 18, 403–416.

Chatwin, P.C., Allen, C.M., 1985. Mathematical-models of dispersion in rivers and estuaries. Annu. Rev. Fluid Mech. 17, 119–149.

Chen, B., 2013. Contaminant transport in a two-zone wetland: dispersion and ecological degradation. J. Hydrol. 488, 118–125.

Chen, G.Q., Wu, Z., 2012. Taylor dispersion in a two-zone packed tube. Int. J. Heat Mass Transf. 55 (1–3), 43–52.

Chen, Z.M., Chen, B., Zhou, J.B., Li, Z., Zhou, Y., Xi, X.R., Lin, C., Chen, G.Q., 2008. A vertical subsurface-flow constructed wetland in Beijing. Commun. Nonlinear Sci. Numer. Simul. 13, 1986–1997.

Chen, B., Chen, Z.M., Zhou, Y., Zhou, J.B., Chen, G.Q., 2009a. Emergy as embodied energy based assessment for local sustainability of a constructed wetland in Beijing. Commun. Nonlinear Sci. Numer. Simul. 14, 622–635.

Chen, Z.M., Chen, G.Q., Chen, B., Zhou, J.B., Yang, Z.F., Zhou, Y., 2009b. Net ecosystem services value of wetland: Environmental economic account. Commun. Nonlinear Sci. Numer. Simul. 14, 2837–2843.

Chen, G.Q., Zeng, L., Wu, Z., 2010. An ecological risk assessment model for a pulsed contaminant emission into a wetland channel flow. Ecol. Model. 221 (24), 2927–2937.

Chen, G.Q., Shao, L., Chen, Z.M., Li, Z., Zhang, B., Chen, H., Wu, Z., 2011a. Low-carbon assessment for ecological wastewater treatment by a constructed wetland in Beijing. Ecol. Eng. 37 (4), 622–628.

Chen, Z.M., Chen, B., Chen, G.Q., 2011b. Cosmic exergy based ecological assessment for a wetland in Beijing. Ecol. Model. 222 (2), 322–329.

Chen, G.Q., Wu, Z., Zeng, L., 2012. Environmental dispersion in a two-layer wetland: Analytical solution by method of concentration moments. Int. J. Eng. Sci. 51, 272–291.

CN Environmental Protection Administration (EPA), 2002. CN General Administration of Quality Supervision, Inspection and Quarantine (GAQSIQ). Environmental Quality Standards for Surface Water. GB 3838-2002, China.

Costanza, R., Farber, S.C., Maxwell, J., 1989. Valuation and management of wetland ecosystems. Ecol. Econ. 1, 335–361.

Costanza, R., d'Arge, R., de Groot, R., Farber, S., Grasso, M., Hannon, B., Limburg, K., Naeem, S., O'Neill, R.V., Paruelo, J., Raskin, R.G., Sutton, P., van den Belt, M., 1997. The value of the world's ecosystem services and natural capital. Nature 387, 253–260.

Ergun, S., 1952. Fluid flow through packed columns. Chem. Eng. Prog. 48 (2), 89–94.

Fischer, H.B., List, E.J., Koh, R.C.Y., Imberger, J., Brooks, N.H., 1979. Mixing in Inland and Coastal Waters. Academic, New York.

Gill, W.N., 1967. A note on solution of transient dispersion problems. Proc. Roy. Soc. Lond. Ser. A—Math. Phys. Sci. 298 (1454), 335–339.

Lightbody, A.F., Nepf, H.M., 2006a. Prediction of near-field shear dispersion in an emergent canopy with heterogeneous morphology. Environ. Fluid Mech. 6 (5), 477–488.

Lightbody, A.F., Nepf, H.M., 2006b. Prediction of velocity profiles and longitudinal dispersion in emergent salt marsh vegetation. Limnol. Oceanogr. 51, 218–228.

Liu, S., Masliyah, J.H., 2005. Dispersion in porous media. In: Vafai, K. (Ed.), Handbook of Porous Media. CRC Press, USA, pp. 81–140.

Mei, C.C., Auriault, J.L., Ng, C.O., 1996. Some applications of the homogenization theory. Adv. Appl. Mech. 32, 277–348.

Mitsch, W.J., Gosselink, J.G., 1993. Wetlands. Van Nostrand Reinhold, New York, USA.

Nepf, H.M., 2012. Flow and transport in regions with aquatic vegetation. Annu. Rev. Fluid Mech. 44 (1), 123–142.

Nepf, H.M., Ghisalberti, M., 2008. Flow and transport in channels with submerged vegetation. Acta Geophysics 56, 753–777.

Nepf, H.M., Sullivan, J.A., Zavistoski, R.A., 1997. A model for diffusion within emergent vegetation. Limnol. Oceanogr. 42, 1735–1745.

Rajagopal, K.R., Tao, L., 1995. Mechanics of Mixtures. World Scientific Publishing Co Pte Ltd, Singapore.

Shao, L., Wu, Z., Zeng, L., Chen, Z.M., Zhou, Y., Chen, G.Q., 2013. Embodied energy assessment for ecological wastewater treatment by a constructed wetland. Ecol. Model. 252, 63–71.

Stone, B.M., Shen, H.T., 2002. Hydraulic resistance of flow in channels with cylindrical roughness. J. Hydraulic Eng. Asce 128 (5), 500–506.

Taylor, G., 1953. Dispersion of soluble matter in solvent flowing slowly through a tube. Proc. Roy. Soc. Lond. Ser. A-Math. Phys. Sci. 219 (1137), 186–203.

US Environmental Protection Agency (EPA), 1999. Manual: Constructed Wetlands Treatment of Municipal Wastewaters. EPA/625/R-99/010, EPA Office of Research and Development, Cincinnati, OH, USA.

Wang, P., Wu, Z., Chen, G.Q., Cui, B., 2013. Environmental dispersion in a three-layer wetland flow with free-surface. Communications in Nonlinear Science and Numerical Simulation: http://dx.doi.org/10.1016/j.cnsns.2013.04.027.

Wu, Z., Chen, G.Q., 2012. Dispersion in a two-zone packed tube: an extended Taylor's analysis. Int. J. Eng. Sci. 50 (1), 113–123.

Wu, Z., Chen, G.Q., 2014. Approach to transverse uniformity of concentration distribution of a solute in a solvent flowing along a straight pipe. J. Fluid Mech. 740, 196–213.

Wu, Z., Chen, G.Q., Zeng, L., 2011a. Environmental dispersion in a two-zone wetland. Ecol. Model. 222 (3), 456–474.

Wu, Z., Li, Z., Chen, G.Q., 2011b. Multi-scale analysis for environmental dispersion in wetland flow. Commun. Nonlinear Sci. Numer. Simul. 16 (8), 3168–3178.

Wu, Z., Li, Z., Zeng, L., Shao, L., Tang, H.S., Yang, Q., Chen, G.Q., 2011c. Environmental dispersivity in free-water-surface-effect dominated wetland: multi-scale analysis. Front. Environ. Sci. Eng. China 5 (4), 597–603.

Wu, Z., Zeng, L., Chen, G.Q., Li, Z., Shao, L., Wang, P., Jiang, Z., 2012. Environmental dispersion in a tidal flow through a depth-dominated wetland. Commun. Nonlinear Sci. Numer. Simul. 17 (12), 5007–5025.

Yasuda, H., 1984. Longitudinal dispersion of matter due to the shear effect of steady and oscillatory currents. J. Fluid Mech. 148, 383–403.

Zeng, L., 2010. Analytical Study on Environmental Dispersion in Wetland Flow. Ph.D. Thesis, Peking University, Beijing.

Zeng, L., Chen, G.Q., 2009. Notes on modelling of environmental transport in wetland. Commun. Nonlinear Sci. Numer. Simul. 14, 1334–1345.

Zeng, L., Chen, G.Q., 2011. Ecological degradation and hydraulic dispersion of contaminant in wetland. Ecol. Model. 222 (2), 293–300.

Zeng, L., Chen, G.Q., Tang, H.S., Wu, Z., 2011. Environmental dispersion in wetland flow. Commun. Nonlinear Sci. Numer. Simul. 16 (1), 206–215.

Zeng, L., Chen, G.Q., Wu, Z., Li, Z., Wu, Y.H., Ji, P., 2012a. Flow distribution and environmental dispersivity in a tidal wetland channel of rectangular cross-section. Commun. Nonlinear Sci. Numer. Simul. 17 (11), 4192–4209.

Zeng, L., Wu, Y.H., Ji, P., Chen, B., Zhao, Y.J., Chen, G.Q., Wu, Z., 2012b. Effect of wind on contaminant dispersion in a wetland flow dominated by free-surface effect. Ecol. Model. 237–238, 101–108.

Zhang, M., Li, C.W., Shen, Y., 2013. Depth-averaged modeling of free surface flows in open channels with emerged and submerged vegetation. Appl. Math. Model. 37 (1–2), 540–553.

Zhou, J.B., Jiang, M.M., Chen, B., Chen, G.Q., 2009. Emergy evaluations for constructed wetland and conventional wastewater treatments. Commun. Nonlinear Sci. Numer. Simul. 14, 1781–1789.

Trade-Offs Between Biodiversity Conservation and Nutrients Removal in Wetlands of Arid Intensive Agricultural Basins: The Mar Menor Case, Spain

11

Julia Martínez-Fernández[a],[*], Miguel-Angel Esteve-Selma[a], Jose-Miguel Martínez-Paz[b], María-Francisca Carreño[a], Javier Martínez-López[a], Francisco Robledano[a], Pablo Farinós[a]

[a]*Department of Ecology and Hydrology, University of Murcia, Murcia, Spain*
[b]*Department of Applied Economics, University of Murcia, Murcia, Spain*
[*]*Corresponding author: e-mail address: juliamf@um.es*

11.1 INTRODUCTION

Although wetlands provide a wide variety of ecosystem services such as biomass production, carbon storage, biodiversity, fish production, and nutrient removal (Okruszko et al., 2011), these ecosystems remain insufficiently valued in areas where they do not supply direct market goods, as is the case of many wetlands (Barbier, 2011; Maltby and Acreman, 2011), particularly those of arid and semi-arid environments. In this context, the concept of ecosystem services and their assessment may help to communicate the need for their conservation and sustainable management to policymakers and end-users (Maes et al., 2012).

In the last years, considerable efforts have been devoted to the assessment of ecosystem services at very different spatial, temporal, and conceptual resolutions using a variety of approaches, including monetary valuations of real or virtual markets for ecosystem goods and services, mapping of ecosystem services using remote-sensing monitoring data and modeling tools, among others. Model development usually constitutes a costly process in terms of required time and effort. Therefore, it is pertinent to ask whether modeling tools are useful for the assessment of ecosystem services, how they contribute to such an assessment, in what cases models are particularly needed, and what kind of models are more appropriate for this task.

In relation to the ecosystem services of wetlands, we consider that their assessment should adopt a perspective capable of accounting for the following issues:

(i) the relationships between wetlands and their watershed; (ii) the drivers of change across time and how they affect the wetlands and their services; (iii) the identification of less common services and goods that might be important for specific wetlands; and (iv) the potential interactions among wetland ecosystem services. We try to illustrate the above-mentioned questions regarding the role of modeling tools on the assessment of ecosystem services, in light of these issues, with the case of the Mar Menor lagoon and associated wetlands, located in southeastern Spain, one of the most arid areas in Europe.

The Mar Menor lagoon is a hypersaline Mediterranean coastal lagoon located in southeast Spain. Ramsar Site since 1994, it is the largest water surface of the western Mediterranean coast (135 km^2 surface area) and is almost closed by a sand bar 22 km long. Inside the lagoon there are five volcanic islands. The lagoon is characterized by its hypersaline, clear, and relatively oligotrophic waters, with a low phytoplanktonic biomass, since primary production is dominated by macrophytes (Perez Ruzafa et al., 2002). Close to its internal shore is a series of coastal wetlands, Marina del Carmolí, Playa de la Hita, and Saladar de lo Poyo (Figure 11.1), described as coastal crypto-wetlands (Vidal-Abarca et al., 2003).

The Mar Menor lagoon and associated wetlands are important sites for wintering and breeding waterbirds (Martínez-Fernández et al., 2005; Esteve et al., 2008; Robledano et al., 2011). The lagoon and wetlands maintain 18 habitats of European interest, according to the Habitat Directive (92/43/ECC). The ecological value of the Mar Menor lagoon and associated wetlands has been recognized in a series of rules and resolutions, at regional, national, and international levels (Ramsar site, Special Protection Area for Birds, Site of Community Importance (SCI) and Special Protection Area for the Mediterranean).

The Mar Menor watershed is a 1270 km^2 plain slightly inclined toward the lagoon and drained by several ephemeral watercourses (ramblas), most of which only flow into the lagoon after big rainfall events. The area has a Mediterranean arid climate, with warm winters, an annual mean temperature about 17 °C, annual mean rainfall of 330 mm, and a high interannual rainfall variation. As in other Mediterranean watersheds, heavy rainfall events and flash-floods play a major role (David et al., 1997; Xue et al., 1998), leading to the mobilization of important stocks of nutrients stored in the watershed. More than 80% of the total area of Campo de Cartagena is used for agriculture, especially for open-air horticultural crops, citrus fruits, and greenhouses (Figure 11.2).

The Mar Menor watershed has experienced profound land use changes, that is, urban and tourist development and increased areas of irrigated lands due to the Tagus-Segura water transfer. These changes have led to a noticeable increase in water and nutrient flows reaching the lagoon and associated wetlands. In this chapter we analyze (i) to what extent and how the ecosystem services of wetlands of arid environments, particularly nutrients removal and biodiversity conservation, can be affected by the land use and the hydrological changes at watershed scale; and (ii) whether there are interactions (synergistic effects or trade-offs) among such ecosystem services.

To this aim, we have developed and applied a dynamic model of the Mar Menor watershed to estimate the inflow of nutrients from diffuse and point sources reaching

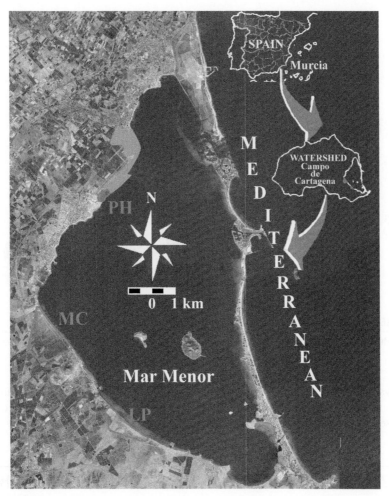

FIGURE 11.1

Location of Mar Menor wetlands. PH, Playa de la Hita; MC, Marina del Carmolí; LP, Saladar de Lo Poyo. (For the color version of this figure, the reader is referred to the online version of this chapter.)

the lagoon and associated wetlands. Then, some effects on the lagoon dynamics (jellyfish outbreaks) are described. Next, we assess the effects on some important ecosystem services, first regarding nutrient removal and second regarding biodiversity conservation, by means of the aquatic bird assemblages and the natural habitats of wetlands. Then we apply the dynamic model and environmental economic techniques to assess and value the wetland ecosystem service of nutrients removal and to compare among different measures to reduce nutrient inflows into the lagoon. We discuss interactions and potential trade-offs between the two studied wetland

FIGURE 11.2

Mar Menor watershed showing agriculture as dominant land use. (For the color version of this figure, the reader is referred to the online version of this chapter.)

ecosystem services (biodiversity conservation and nutrients removal) and finally make some concluding remarks on the role of models for the assessment of ecosystem services, particularly regarding wetlands.

11.2 DYNAMIC MODELING OF THE MAR MENOR WATERSHED
11.2.1 Model description

A watershed model has been developed with key environmental and socio-economic factors driving the dynamics of nutrient inputs into the lagoon (Chapelle et al., 2005; Martínez-Fernández and Esteve-Selma, 2007). It has a long-term time horizon, allowing the simulation of a 30-year time span on a daily basis. Several sectors have been considered: (i) nitrogen flows and compartments, accounting for N content in the soil solution, litterfall, live material, and humus; (ii) phosphorus flows and compartments, accounting for P content; (iii) land-use changes between natural areas, irrigated-tree crops, open-air horticultural crops, greenhouses, and urban areas; (iv) salty wastewater from desalination plants; (v) the role of wetlands on nutrient removal; (vi) nutrient inputs from urban sources; and (vii) the economic costs of main management measures. Information for model calibration, parameter estimation, and data inputs is provided by a range of sources, including empirical field studies, statistical databases, and literature. Figure 11.3 presents a simplified diagram of main model sectors.

The wetlands sector takes into account the active wetland area, the retention capacity for nitrogen and phosphorous, and the effect of water volume on nutrient retention. Empirical data from de Marina del Carmoli wetland were used to determine the retention ratio as a function of water volume (Figure 11.4) and the length of watercourse inside the wetlands.

The socioeconomic issues involved in the export of nutrients at watershed scale are not considered in a separate model or sector. Instead, they become part of the variables defining all model sectors, in close interaction with the environmental factors. The land use sector (Figure 11.5) takes into account the area and main land use changes between natural vegetation, drylands, urban areas, and each type of irrigated land (irrigated-tree crops, open-air horticultural crops, and greenhouses). The urban sector takes into account the resident population, the seasonal dynamics of tourist population, the efficiency of wastewater treatment plants, and the amount of wastewater reused for agriculture.

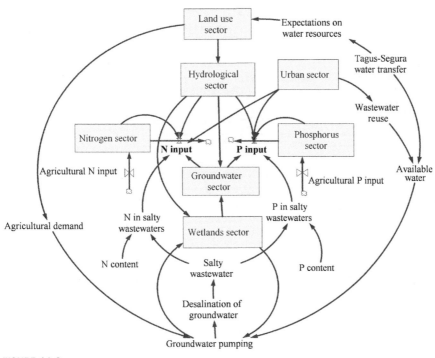

FIGURE 11.3

Simplified diagram of Mar Menor watershed model. (For the color version of this figure, the reader is referred to the online version of this chapter.)

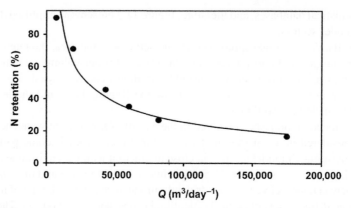

FIGURE 11.4

Nitrogen retention ratio as a function of total water inflow. Empirical data from Marina del Carmoli wetland.

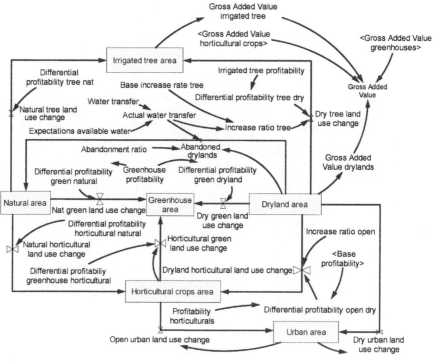

FIGURE 11.5

Simplified diagram of the land use model sector. (For the color version of this figure, the reader is referred to the online version of this chapter.)

11.2.2 Simulation results

The model simulation shows a good degree of agreement with observed values for available data series. The model tracks the noticeable increase in irrigated lands driven by two factors: the water transfer for irrigation, which started in 1979, and the higher profitability of such agriculture compared with drylands, especially in the case of greenhouses (Figure 11.6).

This has generated an increased input of nutrients from diffuse sources into the Mar Menor (Figure 11.7). The estimated load of nutrients shows strong fluctuations due to the high variability in rainfall and the occurrence of flash-flood events, when large amounts of nutrients and materials from the watershed are flushed out and enter the lagoon. This is expected, since the hydrological regime in this arid Mediterranean area shows extreme differences in water discharges. Flash-flood events play a key role for the water, nutrients, sediment, and pollution flows entering the lagoon. For example, it has been demonstrated that dissolved organic pollutant concentrations in the Albujon, the main watercourse, during a flash flood are several orders of magnitude higher than in regular periods and that heavy rainfalls account for more than 70% of total input of many pesticides through the Albujon watercourse (Moreno-González et al., 2013). In watersheds with intensive agriculture in arid environments, big rainfall events and flash floods can mobilize the nutrients accumulated during several months or years. The Mar Menor wetlands, located along the lagoon shore, provide important ecosystem services in case of flash-flood events, with the storing and slowing down of floodwaters, allowing the nutrient removal of overland flow waters, and acting as sediment traps. Nutrient loads, particularly during flash floods, are linked to one of the most important problems for the bathing

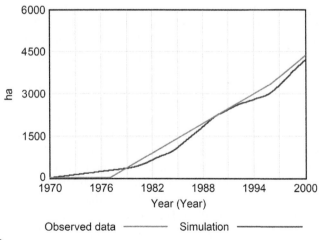

FIGURE 11.6

Area occupied by greenhouses in Mar Menor watershed. Observed and simulated values. (For the color version of this figure, the reader is referred to the online version of this chapter.)

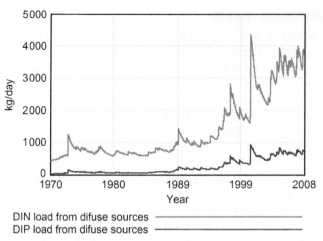

FIGURE 11.7

Simulated pattern of daily DIN (dissolved inorganic nitrogen) and DIP (dissolved inorganic phosphorus) input (kg/d) from diffuse sources using a 365-day moving average period. (For the color version of this figure, the reader is referred to the online version of this chapter.)

quality and the socio-tourist uses: the jellyfish outbreaks (described later). Although not explicitly considered in the model, the sediment trapping is also linked to the ecosystem service of bathing quality. Mar Menor wetlands contribute to reduced water turbidity, mudding, and loss of lagoon depth (a problem already identified in the Mar Menor lagoon).

According to the simulation results, an average annual load of 900 ton/year of DIN (dissolved inorganic nitrogen) and around 200 ton/year of DIP (dissolved inorganic phosphorus) from diffuse sources can be estimated for recent years. These values are coherent with the scarce empirical data available on nitrogen content and water flows in the watershed (Lloret et al., 2005; Velasco et al., 2006; García Pintado et al., 2007; Álvarez-Rogel et al., 2009; Serrano and Sironi, 2009).

Population in the Mar Menor area has also shown a quick growth in recent decades due to tourism (Figure 11.8). The existence of treatment plants has led to a large increase in wastewater and the input of nutrients from urban sources, especially during summer (Figure 11.9), where the population of some locations increases more than $10\times$ (Moreno-González et al., 2013). After big rainfall events, overflow water from wastewater treatment plants enters the lagoon (Alvarez-Rogel et al., 2006), sometimes causing the temporary closing of the affected lagoon bathing areas (Conesa and Jiménez-Cárceles, 2007). The estimated average urban input is around 130 ton/year of DIN and 17 ton/year of DIP, which represents 12% and 8% respectively of all sources (point and diffuse). A large number of studies have shown that in watersheds dominated by intensive agriculture the main contribution to water pollution of rivers, lagoons, and coastal waters comes from diffuse sources and, more

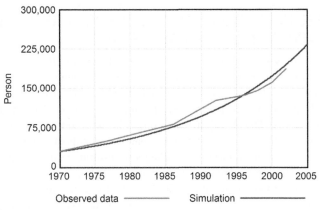

FIGURE 11.8

Peak summer population around Mar Menor. Observed and simulated values. (For the color version of this figure, the reader is referred to the online version of this chapter.)

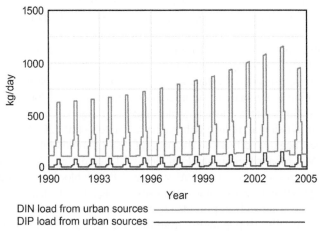

FIGURE 11.9

Simulated DIN (dissolved inorganic nitrogen) and DIP (dissolved inorganic phosphorus) input from urban sources. (For the color version of this figure, the reader is referred to the online version of this chapter.)

specifically, from agriculture (Jordan et al., 1997; Kronvang et al., 1999; Meissner et al., 2002; Lacroix et al., 2005). As shown, this is also the case for the Mar Menor watershed, one of the most important intensive agricultural areas in Europe (Moreno-González et al., 2013). Has this increase in nutrients inflow affected the lagoon ecosystem? The next section considers this question.

11.3 EFFECTS OF HYDROLOGICAL CHANGES ON THE MAR MENOR LAGOON: JELLYFISH OUTBREAKS

The increased inflows into the Mar Menor lagoon have modified its structure and dynamics. An increase in nitrogen content of the water column has been observed throughout the last decades (Perez Ruzafa et al., 2002; Lloret et al., 2005; Velasco et al., 2006), which in one decade shifted to values 10× higher (Perez Ruzafa et al., 2002). The Albujon watercourse is the main source for nitrate in the lagoon waters. There are clearly defined spatial gradients of water column transparency and nutrient concentrations in the lagoon, resulting as a consequence of the inputs from the Albujon watercourse (Lloret et al., 2005).

The increased nitrogen content has favored other changes in the lagoon and its biological assemblages. One of the main changes is the existence of jellyfish outbreaks during summer. In 1974 the opening of the Estacio channel caused strong hydrographical changes, with a higher water exchange with the Mediterranean, a salinity decrease, and strong alterations in the biota. In the mid 1980s, two allochtonous species of jellyfish, *Rhyzostoma pulmo* and *Cotylorhyza tuberculata*, entered into the Mar Menor. The resulting moderated lagoon temperature and salinity ranges allowed these species to complete their biological cycle inside the lagoon (Pérez Ruzafa and Aragón, 2003). Scyphomedusae species were recorded for the first time around 1990. The summer proliferation of the two allochtonous species of jellyfish started during mid 1990s, in response to the increased nutrient inflow (Perez Ruzafa et al., 2002; Lloret et al., 2005, 2008; Conesa and Jiménez-Cárceles, 2007). Both species attained large populations, especially in the case of *C. tuberculata*, which in 1997 reached around 46 million individuals in summer (Perez Ruzafa et al., 2002). These large jellyfish populations affect nutrient cycles and other lagoon compartments, since outbreaks of jellyfish may reduce fish biomass (Purcell et al., 2007).

Due to their negative effects on the bathing quality and therefore on the tourist value of the Mar Menor lagoon, the Fish General Directorate of Region de Murcia took some measures to try to control the summer jellyfish blooms, including the annual catch of jellyfish during summer by means of special boats. Data on annual jellyfish catches and fishing effort provided by the regional authorities, in combination with data of total jellyfish population from direct census available for some years, allowed the estimation of the maximum summer population of jellyfish between 1988 (the starting date for the allochtonous species) and 2012 (Figure 11.10). There are noticeable interannual changes of the summer population of jellyfish, which peaked in 2000 and then began to decline, with no outbreaks detected in year 2005 (Prieto et al., 2010). In the following years the jellyfish population peaks remained at very low values with the exception of year 2006, when a small recovery was observed (Dolores et al., 2009). However, in the last several years the summer jellyfish populations have increased again. The reasons for such shifts are still under research, although water temperature seems to play a key role (Prieto et al., 2010; Astorga et al., 2012). For example, the low values of jellyfish population in 2005 and 2006 have been related to very low temperatures in 2004–2005 winter, which

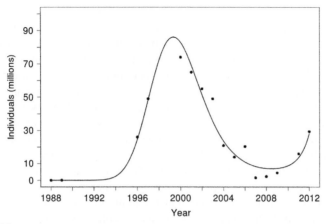

FIGURE 11.10

Estimated summer population of jellyfish in Mar Menor (dots) and Poisson adjustment (92.28% explained deviance, $p < 0.001$). Maximum summer populations were estimated from data on jellyfish catches, fishing effort, and total jellyfish population for available years provided by the Fish General Directorate of Murcia.

decreased polyp numbers in 2005, resulting in low jellyfish biomass also affecting jellyfish population in 2006 (Ruiz et al., 2012).

The increased nutrient inflows have caused other changes in the lagoon dynamics. The dominance shifts from the phanerogam *Cymodocea nodosa* to the macroalgae *Caulerpa prolifera* (Perez Ruzafa et al., 2002; Lloret et al., 2005; Conesa and Jiménez-Cárceles, 2007) are mainly explained by the inputs from the Albujon watercourse and associated changes in the water column parameters (Lloret et al., 2005). The observed changes in the macrophyte assemblages have other consequences for lagoon dynamics, such as the accumulation of organic matter under the meadows of *Caulerpa prolifera*; the subsequent appearance of anoxic conditions in some areas; and the decrease in populations of some commercial fish, mainly sparidae and mugilidae, which are negatively affected by the spread of the macroalga (Lloret et al., 2005).

11.4 ASSESSMENT OF ECOSYSTEM SERVICES: NUTRIENT REMOVAL

11.4.1 Role of wetlands and effects of measures to reduce nutrient inputs into the lagoon

At present, the main watercourse, the Albujon, is disconnected from the Marina del Carmoli wetland due to channeling works and, therefore, its full nutrient load enters the lagoon without benefiting from any removal function from the wetland.

However, overall, Mar Menor wetlands do intercept other ephemeral watercourses and, above all, they play a key role during big storms, reducing the nutrient load of flash-flood events. The watershed dynamic model has been used to estimate the proportion of nutrients being removed by the wetlands. According to simulation results, at present wetlands remove around 14% of total nutrients from diffuse sources.

As shown in previous sections, there is a clear need to achieve a more substantial reduction in the total load of nutrients reaching the lagoon. Moreover, according to model simulation results, the trend to an increase in the nutrient inflows would be maintained in the near future under a business-as-usual scenario characterized by urban and tourist development. Assuming that the overall performance of wastewater treatment plants is maintained, simulation results point to an increase in the input of nutrients from urban sources, especially during summer months. The need for a reduction in nutrient inputs is also a compulsory request by present European legislation: The Mar Menor watershed is designated as a Vulnerable Area according to the Nitrates Directive (91/676 ECC), the lagoon is designated as a Sensible Area according to the Urban Wastewater Directive (91/271 ECC), and the Mar Menor water should achieve and maintain good ecological status according to the Water Framework Directive (2000/60 EC).

The water body carried out a project for managing part of the agricultural drainage, consisting of several hydraulic facilities already built up, but not yet initiated, to collect part of the agricultural drainage coming from irrigated land. This drainage water will be collected and pumped to a desalination plant, after which it would be reused for irrigation. We have applied the dynamic model to simulate and assess the effectiveness of this management option to remove part of the nutrients reaching the lagoon and compare it with other management measures, particularly wetlands restoration. After implementing the measure of agricultural drainage management in the watershed dynamic model, simulation results suggest that this option might reduce the total nutrient inputs into the lagoon by around 10%.

The wetland restoration measure is based on (i) the restoration of part of the Marina del Carmoli wetland area lost due to land-use changes and (ii) the reconnection of the water flow of the Albujon watercourse to this restored area of the Marina del Carmoli wetland. According to model simulations, this measure would achieve around a 40% reduction in the amount of nutrients contributed by the Albujon watercourse. When all flows are considered, this management option still represents a noticeable additional reduction of around 20% in total nutrients input into the lagoon when compared with the base trend of urban and tourist development, doubling the reduction achieved by the management of agricultural drainage option (Figure 11.11).

11.4.2 Cost-effectiveness analysis

A cost-effectiveness analysis (CEA) has been applied to assess and compare the relative effectiveness of these two management measures (Martinez Paz et al., 2007). CEA is very useful to assess and select the management measures achieving the desired environmental goals at the lowest cost, which constitutes an important input for

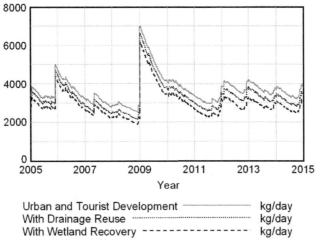

Urban and Tourist Development ——————— kg/day
With Drainage Reuse ··· kg/day
With Wetland Recovery – – – – – – – – – – – – – – kg/day

FIGURE 11.11

Simulated pattern of daily load of DIN (dissolved inorganic nitrogen) into the Mar Menor lagoon from surface water using a 365-day moving average period under the measures of management of agricultural drainage and restoration of wetlands, as compared with the base trend of urban and tourist development. (For the color version of this figure, the reader is referred to the online version of this chapter.)

the decision process. CEA has been applied in other studies on the performance and efficiency of using wetlands versus conventional measures to treat pollution processes (Schou et al., 2000; Kampas et al., 2002; Zanou et al., 2003; Lacroix et al., 2005). Figure 11.12 shows the methodological steps of the CEA analysis in the Mar Menor area.

Several factors were quantified to determine the capacity and overall performance of the two options, including the average flows in the drainage channels, the effect of overland flow on the nutrient removal efficiency of the wetland during flash-flood events and the maximum capacity of the hydraulic facilities of the drainage management option, including the drainage channels, pumping station, and desalination plant. The maximum capacity of the drainage management system fits the requirements under normal conditions but cannot manage floods. A 15-year time frame has been considered to calculate the CEA. All items have been valued according to market prices. The land to be purchased has been valued as agricultural and the attributed land profit, the opportunity costs of its use as wetland, corresponds to the gross added value of intensive horticultural crops. The water price imputed to the drainage reuse option constitutes an income, since the desalinated water will be sold for irrigation. To compute the added financial flows, the net present cost (NPC) has been calculated using a low discount rate of 2%, according to the proposal by Almansa and Martínez-Paz (2011) for this type of project. The relative efficiency of the two management options has been compared by means

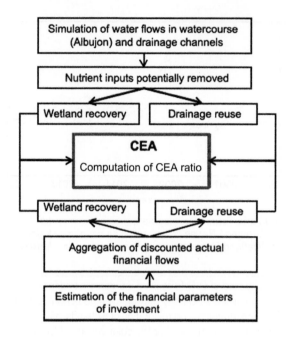

FIGURE 11.12

Basic methodological steps of the CEA applied in the Mar Menor site. (For the color version of this figure, the reader is referred to the online version of this chapter.)

of the cost-effectiveness ratio (CER, Zanou et al., 2003). All costs refer to euros in real terms (2007). The cost-effectiveness ratio of each option is determined as the quotient between the amount of nutrient removed by each option during the study period and the aggregated costs during such period.

Results indicate that wetlands restoration is more cost-effective than the agricultural drainage management, the solution adopted by the water body, since its unitary costs in terms of euros per kg nutrient being removed are around half of that corresponding to the drainage management measure. Each 100 € invested in the drainage reuse system would remove 7.6 kg of DIN and 2.7 kg of DIP, whereas the removed nutrients would be double under the wetland restoration option. In the case of nitrogen, CER values are around 13 and 6.5 €/kg for agricultural drainage management and wetlands restoration measures respectively. This constitutes an important insight for the water body, which had not previously considered the option of wetlands restoration and never assessed the efficiency of the management of agricultural drainage, despite the fact that hydraulic facilities were already built up.

These results agree with other studies (Gren et al., 1997; Turner et al., 1999; Gustafson et al., 2000; Zanou et al., 2003; Lacroix et al., 2005), showing that the construction and especially the restoration of wetlands is a highly cost-effective option to reduce diffuse pollution in agricultural watersheds. Moreover, as explained above, the restoration of wetlands also achieves the highest nutrient removal in absolute terms.

Lacroix et al. (2005) also point out that wetlands restoration, an option usually considered as being very expensive, is frequently more cost-effective than other strategies such as subsidies to reduce the amount of fertilizers in agriculture.

11.4.3 Economic valuation of the ecosystem service of nutrients removal of Mar Menor wetlands

Despite the considerable uncertainties in the assessment of ecosystem services (Johnson et al., 2012), and the difficulties inherent to a proper estimation of the total value of such services, the information provided, even by incomplete or partial valuations, may become a useful tool to build social support for conservation policies and to assess relevant services for different users and management decisions (Martin-López et al., 2014; Nemec and Raudsepp-Hearne, 2013), in particular for local and regional planning (Maes et al., 2012).

The combined use of the watershed dynamic model and CEA analysis has allowed a first quantitative economic assessment of the ecosystem service of nutrient removal in the case of Mar Menor wetlands. To this aim, we have taken into account the estimation of nitrogen being removed by wetlands under present conditions according to model simulation results, which amounts to an annual average of around 193 ton/year of DIN for the recent period. If present wetlands are lost and this service should be supplied by other processes, particularly with the management of agricultural drainage, the CEA analysis results shown earlier indicate that the value would be around 2,509,000 €/year, which provides a value of around 7169 €/year per ha of active wetland (salt marsh + reed bed) for this specific ecosystem service.

In the same way, it can be estimated the avoided costs achieved by the restoration of wetlands. If the goal of additional 20% nutrients reduction achieved by this measure should be obtained by the management of agricultural drainage, the estimated additional required budget should be around 2,000,000 €/year, taking into account the difference in the unitary costs (CER) of both considered options.

We have also carried out a preliminary assessment of some other avoided costs associated with the ecosystem service of nutrients removal of Mar Menor wetlands. As described earlier, one important effect of the increased inputs into the lagoon is the jellyfish outbreaks. The high population densities take place during summer months, causing important disturbances for bathing and other tourist activities, the main socioeconomic activity of the Mar Menor area. As in many other coastal areas, high densities of jellyfish are detrimental to tourist appeal (Purcell et al., 2007). In response to the high jellyfish density values of 1996 and 1997, the regional authorities started the following programs during summer months as measures to reduce the jellyfish population in the lagoon: (i) the installation of temporary devices (meshes) to isolate bathing areas from jellyfish and (ii) catching jellyfish using special boats. According to data provided by the Directorate of Fish and Aquaculture, an annual budget varying between 200,000 and 400,000 €/year has been appropriated for the measure of jellyfish catch. For years with an average summer jellyfish density of around 3–4 individuals per 100 m^3 in the lagoon water, with total summer jellyfish catch varying between 1000 and 3000 ton (fresh weight), an average estimated cost of around 93 €/Kg N being removed from the lagoon by the jellyfish catches. On the basis of such value, the avoided cost of the nutrient removal service of Mar Menor wetlands in relation to jellyfish outbreaks can be estimated. As indicated earlier, the hypothetical loss of present wetlands would increase the annual input of DIN into the lagoon by around 193 ton/year. Taking into account the relationship between such inputs and the jellyfish population, it can be estimated that to counteract the expected additional jellyfish increase would require increasing the annual economic budget allocated to jellyfish catch by around 50–60%.

It should be noted that all these figures are not intended to represent the actual value of Mar Menor wetlands since only some aspects of their ecosystem services are considered. Moreover, the existence of dependencies between results from the valuation of ecosystem services and the methodological approaches that are applied (Martin-López et al., 2014) should be taken into account. However, this exercise constitutes a minimum estimate of the economic value of some of the avoided costs generated by such ecosystem service.

In synthesis, the use of Mar Menor wetlands seems to be more effective and economically efficient than other measures as the management of agricultural drainage to achieve the required reduction in nutrient flows from the watershed. Therefore, the enhancement of such ecosystem service should become an important goal under an integrated management and policy in the Mar Menor area. In the next section we present the effects of hydrological and nutrient changes in another important ecosystem service of the Mar Menor lagoon and associated wetlands: biodiversity conservation.

11.5 ASSESSMENT OF ECOSYSTEM SERVICES: BIODIVERSITY CONSERVATION

Quantitative studies carried out at European scale (Maes et al., 2012) have corroborated the positive relationships between biodiversity and a variety of ecosystem services. Such studies provide evidence on that a good conservation status usually means richer biodiversity and higher levels of ecosystem services. Therefore, it is important to assess the trends in the conservation status of biodiversity, particularly in ecosystems under pressure as the Mar Menor lagoon and associated wetlands. In this section we assess the recent trends in some components of their biodiversity: the aquatic bird assemblages and the habitats of Mar Menor wetlands.

11.5.1 Changes in aquatic bird populations in relation to nutrient inputs and related trophic variables: Indicator species and guilds

Aquatic birds are a key component among the ecological values on which the various protection designations of the Mar Menor are based. Since the inclusion of the lagoon and its associated wetlands in the Ramsar List in 1994 (Robledano, 1998), waterbird numbers have been the most popular and affordable criterion for assessing the conservation value of this ecological complex. In Murcia, the Mar Menor lagoon is the site with the longest dataset of wintering waterbirds and the one with fewest gaps (Robledano et al., 2011). One decade after its listing under the Ramsar agreement, the use of these monitoring data started to shift from a mere accounting of numerical trends to the search of indicator responses of bird metrics to recorded or modeled series of environmental data (Martínez-Fernández et al., 2005; Robledano and Farinós, 2010; Robledano et al., 2011). This has allowed the creation of a database of waterbird censuses coupled with environmental variables related to the trophic status of the lagoon, which now has been updated.

11.5.1.1 Waterbird data and environmental variables

Waterbird data have been obtained from the January censuses carried out between 1972 and 2013 in the framework of the International Waterbird Census (IWC), coordinated in Murcia Region by ANSE (Hernández and Fernández-Caro, 2008, unpublished data). The analysis is restricted to five species belonging to the families *Podicipedidae* (2), *Phalacrocoracidae* (1), *Anatidae* (1), and *Rallidae* (1), the criteria being to select only diving or dabbling waterbirds able to exploit the water column and benthos at depths of 1 m or more (i.e., strictly littoral dabbling species, wading birds, and shorebirds were excluded). The five target species (Great Cormorant *Phalacrocorax carbo*, Black-necked Grebe *Podiceps nigricollis*, Great Crested Grebe *Podiceps cristatus*, Red-breasted Merganser *Mergus serrator,* and Common Coot *Fulica atra*) are also the most abundant wintering waterbirds and those more closely tied to using the lagoon as a feeding and roosting area, thus representing the bulk of the biomass of top-level avian consumers. All species are zoophagous (fish or invertebrate feeders) with the exception of Coot (mainly phytophagous). Among the animal consumers, all are mainly piscivorous, although including varying proportions of other animal prey in their diet. With the exception of the Great Cormorant, which can make regular flights to feed at varying distances around the lagoon, all the target species spend all or most of the winter time in open water within the lagoon.

Regarding environmental variables, the estimated annual inflow of nitrogen and summer jellyfish population presented earlier is used, as well as official data on fish landings declared by the two main fishing harbors of the lagoon as a proxy of fish production, focusing on the main potential prey species for piscivorous waterbirds (*Engraulis* sp., *Atherina* sp.). Further details on data sources and their processing can be found in the previous sections and in the cited publications. An overall graphical summary of the waterbird and environmental data is shown in Figure 11.13.

11.5.1.2 Statistical analysis

Generalized linear models (GLMs) with a Gaussian link were used to model the relationship between the biomass of waterbirds (computed as the product of census results by mean species' weight values from bibliography) and the environmental variables. Models were built for individual species and higher taxonomic or functional aggregations (families, guilds). As functional aggregations, we refer to groups of species with similar feeding ecology and distribution, of which we considered

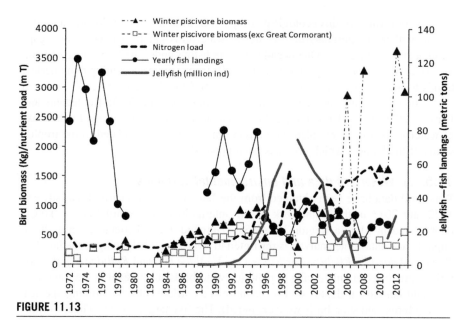

FIGURE 11.13

Changes in the environmental variables tested in regression models in relation to the variation of the main components (piscivores) of the waterbird assemblage of the Mar Menor lagoon.

two: (i) all piscivores and, nested within it, (ii) lagoon-restricted piscivores (excluding Great Cormorant, due to its mobile feeding strategy). Regression models were fitted in the freely distributed statistical software R. Model performance was evaluated using Akaike's information criterion (AIC), which seeks to optimize the trade-off between the explanatory ability of the model and its complexity, measured by the number of fitted parameters. The amount of variability explained by each model was assessed by means of its reduction in deviance with respect to the null (intercept-only) model (Saunders et al., 2013). Models were fitted sequentially by first selecting the best performing model with one of the two variables related to the trophic status of the lagoon (nutrients or fish), and later including jellyfish as potential modifiers of such status, as well as the interactions among the first and second variables. Quadratic terms were also tested for the trophic-related variables, in search of a best fit of the dependent variable. Nutrients and fish were included separately in the models due to their redundancy (Spearman's Rho $=-0.704$, $p<0.0001$). Nutrient load was tested with a lag of 2 years, considering the best response models of previous studies (Robledano et al., 2011), and jellyfish with 1- or 2-year lags.

11.5.1.3 Results and discussion

All species except the Red-breasted Merganser *Mergus serrator* showed a positive response to nutrient inputs and a negative one to fish catches (the latter was not tested for the Coot for its obvious lack of biological meaning). Models including the 2-year lagged estimate of annual nitrogen load usually explained more variation (or percent deviance) than those including the fish catch (Table 11.1). Although the Red-breasted Merganser can be considered a specialist piscivore, its relationship with the availability of its main prey (indicated by fish landings) was not statistically significant, while it showed a significant negative response to nutrient loading. The only model including an interaction is that of the Coot with nutrients and jellyfish (both with a 2-year lag). The net positive effect is not surprising since the control exerted by jellyfish on eutrophication is expected to affect mainly the growth of phytoplankton, but not necessarily other food sources preferred by phytophagous birds (e.g., filamentous algae).

The general picture, for most species and higher taxonomic or functional groupings, is an overall increase in biomass positively correlated with the increase in nutrient inputs. On the contrary, both waterbird biomass and nutrient inputs show a negative relationship with fish resources (Table 11.1). This global picture does not disagree with previous interpretations of waterbird responses, since the species contributing to the global trend vary along the period of study (Figure 11.13), allowing recognition of six temporal phases characterized by specific responses to the status of the lagoon (Figure 11.14). All except one of these phases have been described previously (Robledano et al., 2011) and can be summarized as:

> 1972–1979: Characterized by low values of nutrient inputs and high fishing yields, starting to decrease at the end of the period, probably due to overfishing; the waterbird assemblage is dominated by the specialist Red-breasted Merganser in all years except 1979.
>
> 1983–1987: Nutrient inputs are still low but steadily increasing, and a first positive response of generalist piscivores (mainly Great Cormorant) is detected.
>
> 1988–1995: Characterized by a gradual increase in nutrient inputs and by a growing numerical contribution of *Podicipedidae* with respect to Red-breasted Merganser and Great Cormorant.
>
> 1996–1998: A period characterized by low fishing yields, higher nutrient loads, and an incipient development of jellyfish, coincident with a diversification of the waterbird assemblage (the dominance of *Podicipedidae* decreases and Coot starts to be recorded in low numbers).
>
> 1999–2006: Shows additional increases in nutrient inputs and a higher abundance of jellyfish, which can exert some control on eutrophication, keeping a more fluctuant but relatively diverse waterbird assemblage (with lack of a clear dominance among species). The greatest abundance of Coot occurs during this phase, rising to a dominant position in 2006 and decreasing later.
>
> 2007–2013: The more recent phase, representing a return to a less diverse waterbird assemblage, dominated by two generalist piscivores: Black-necked Grebe and Cormorant. Red-breasted Merganser and Coot become scarce or almost disappear from the lagoon. If Cormorant is excluded, a stabilization of waterbird numbers around somewhat lower values (compared with precedent maxima) can also be stated.

Table 11.1 Selected GLMs Explaining Numbers and Biomass of Individual Waterbird Species and Higher Taxonomic or Functional Aggregations

Dependent Variable	Estimate	P	AIC	Explained Deviance (%)
Biomass of *Phalacrocorax carbo* (Null)			527.13	
Biomass of *Phalacrocorax carbo* ∼ Nload.2yr	1.40	***	500.74	58.82
Biomass of *Phalacrocorax carbo* ∼ Fish_land	−13.19	*	382.44	23.84
Biomass of *Podiceps nigricollis* (Null)			397.75	
Biomass of *Podiceps nigricollis* ∼ Nload.2yr	0.16	***	366.63	63.34
Biomass of *Podiceps nigricollis* ∼ Fish_land	−1.45	*	294.08	19.34
Biomass of *Podiceps cristatus* (Null)			391.98	
Biomass of *Podiceps cristatus* ∼ Nload.2yr	0.07	*	388.88	14.32
Biomass of *Podiceps cristatus* ∼ Fish_land	−0.20	NS		
Biomass of *Mergus serrator* (Null)			397.37	
Biomass of *Mergus serrator* ∼ Nload.2yr	−0.11	***	386.99	31.27
Biomass of *Mergus serrator* ∼ Fish_land	1.10	NS		
Biomass of *Fulica atra* (Null)			374.34	
Biomass of *Fulica atra* ∼ Nload.2yr	0.17	***	361.67	38.67
Biomass of *Fulica atra* ∼ Jfish.2yr	3.54	***	187.76	43.86
Biomass of *Fulica atra* ∼ Nload.2yr*Jfish_2			174.52	82.22
Nload_2	−0.15	0.08		
Jfish_2	−6.73	*		
Nload.2yr*Jfish	0.01	**		
Biomass of *Podicipedidae* (Null)			434.12	
Biomass of *Podicipedidae* ∼ Nload.2yr	0.23	***	417.36	43.36
Biomass of *Podicipedidae* ∼ Fish_land	−1.65	NS		
Biomass of Piscivores exc *Phalacrocorax carbo* (Null)			434.51	
Biomass of Piscivores exc *Phalacrocorax carbo* ∼ Nload.2yr	0.11	0.05	432.56	11.29
Biomass of Piscivores exc *Phalacrocorax carbo* ∼ Fish_land	−0.55	NS	330.15	
Biomass of Total Piscivores (Null)			547.82	
Biomass of Total Piscivores ∼ Nload.2yr	1.49	***	521.89	57.10
Biomass of Total Piscivores ∼ Fish_land	−12.77	*	403.18	19.07

Significance. ***: $p < 0.001$; **: $p < 0.01$; *: $p < 0.05$; NS: No Significant

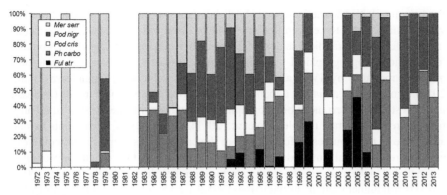

FIGURE 11.14

Percent contribution of the five species studied to the total abundance of waterbirds wintering in the Mar Menor Lagoon (January censuses).

The last phase is characterized by high nutrient inputs and similarly high numbers of the two dominant species: Great Cormorant and Black-necked Grebe. Also evident is the dramatic decrease that confirms the long-term decline of Red-breasted Merganser, once the most characteristic species of the lagoon. The two dominant species do not differ markedly in numbers (mean ± SE for the period 2007–2013, Cormorant: 861.3 ± 217.8, Black-necked Grebe: 814.5 ± 92.4), but do so in terms of biomass (Cormorant: 1879.4 ± 475.2, Black-necked Grebe: 814.5 ± 92.4).

Regarding the whole period of analysis, the apparent positive correlation with nutrient inputs is not exclusive of piscivores, since the only phytophagous species (Coot) shows the same response (Table 11.1). Interestingly, Coot biomass is also positively associated with the abundance of jellyfish in the lagoon (both thrive in the lagoon during the same period), which can be due to an independent response to a common underlying factor not captured by our environmental variables or to the interaction among them reflected by the models. Whatever the relationship between this short-term increase of herbivores and the blooms of jellyfish, the greater abundance of Coot could be related to a phase of high input of nutrients of urban origin (untreated sewage). This organic pollution has reached the lagoon through its main incoming watercourse (Albujón channel; García Pintado et al., 2007), and combined with the diffuse sources of nutrients of agricultural origin, may have favored the increase in wintering Coot, which are mainly restricted to the area of influence of the watercourse. After this phase, the increases in the two presently dominant species (Cormorant and Black-necked Grebe) would be explained by the persistent high loads of agricultural nitrogen, although the precise mechanisms by which such increased fertilization is transferred to waterbirds through the lagoon's trophic web are more difficult to establish. Clearly it is not a matter of increased fish resources, at least of the species accounted for by our variable.

In any case, correlative studies like that presented here do not necessarily demonstrate causal relationships (i.e., a direct effect of nutrient load on some waterbird food resource). Nor do the increase in bird abundances necessarily relate to local environmental factors. Recent studies have tried to discriminate which factors, operating at different geographical scales—from the site where the waterbirds are counted to the whole biogeographical population range—explain the change in waterbird numbers recorded by IWC and other monitoring schemes (Tománková et al., 2013). In our case, a statistical dependence on external demographic trends (increase in the western Mediterranean biogeographical population) had been established only for the Great Cormorant (Robledano et al., 2011). The ability of Cormorants to take advantage of new sources of food is one of the causes of its increase and could explain its trend without a more direct local trophic response. If the Great Cormorant is not included in the analysis, the total waterbird biomass during the last phase (2007–2013) is still higher than the initial values (405.9 ± 39.5 vs. 204.9 ± 50.2 kg in 1972–1975), but has decreased in relation to the peak values of the early 1990s (603.6 ± 56.4 kg, 1990–1995) and 2000s (617.7 ± 111.1 kg, 2000–2005).

On the other hand, the relationship of piscivorous waterbirds with fish resources might be obscured by the fact that we use a restricted set of fish catch statistics (two species). In the recent years there has been a recovery of commercial fish species not included in our dataset (e.g., *Sparus aurata*; García et al., 2001; Centro Regional de Estadística de Murcia, 2013), which could be exploited by large piscivores like the Great Cormorant. In fact, the contribution of the Mar Menor lagoon to the total number of wintering Cormorants in Murcia Region rose from ca. 30% to ca. 60% between 2003 and 2013. Since the regional population experienced also an overall increase during that period (Hernández and Fernández-Caro, 2008; ANSE, unpublished data), it is feasible that the trophic resources offered by the lagoon (or by the foraging range centered on it) have also increased. Certainly such an increase may be partially related to the increased fertilization of the lagoon, although a species as mobile as the Cormorant could also exploit other resources located within its daily flight range (e.g., fish stocked in irrigation infrastructures, open-sea aquaculture facilities, rivers, or reservoirs).

Different factors need to be invoked when explaining the recent numerical trend of the Black-necked Grebe, a species much more closely tied to the Mar Menor lagoon as feeding area and with a diet based on smaller prey. In this case the pathways through which the ongoing nutrient enrichment of the lagoon could be turned into increased food resources do not seem to flow through the fish compartment (at least the commercially exploited stock). The most plausible explanation for the recent numerical increase of Black-necked Grebe would be an enhanced supply of benthic macroinvertebrates and non-commercial fish thriving in *Caulerpa prolifera* algal beds, which are also thought to be responsible for the sequestration of large amounts of nutrients from the water column (Lloret and Marín, 2009). This is consistent with the fact that nutrient enrichment has not resulted yet in a hypereutrophic, phytoplankton-dominated status (even with jellyfish absent). It is also evident that much fewer food resources would be needed to sustain a population of Grebes than a equivalent one of Cormorants. Regarding the potential effect of external factors affecting waterbird populations at higher scales, the population of Black-necked

Grebes wintering in Murcia region is also increasing, which would suggest a common demographic response to higher-scale environmental drivers (e.g., hydrographic basin-level eutrophication or widespread artificial wetland creation). The Mar Menor population of Black-necked Grebes represents, on average, 60% of the whole regional population (2003–2013; Hernández and Fernández-Caro, 2008; ANSE, unpublished data), and this percentage has not changed markedly during the last decade. Even under a scenario of regional increase of its wintering population, it cannot be discarded that the capacity of the lagoon to host wintering grebes has been enhanced by internal nutrient loading.

11.5.1.4 Concluding remarks

As in previous studies, waterbirds provide some useful indication of changes in the trophic status of the Mar Menor lagoon. The replacement of dominant species and the resulting changes in the composition of the waterbird assemblage are interpretable on the basis of temporal phases characterized by specific syndromes of environmental pressures and impacts. To summarize the direction of change overriding such succession of waterbird assemblages, the biomass of avian consumers has increased over a 40-year period of variable but continuously increasing nutrient inputs, also characterized by a dramatic decline in fishing yields. *Mergus serrator*—the initially dominant and almost exclusive piscivore—has declined, while generalists like *Phalacrocorax carbo* and *Podiceps nigricollis* have continued to increase until becoming dominant in a simplified piscivore assemblage. While these two species could also be responding to more global environmental trends—e.g., recovery of previously endangered populations, background eutrophication at a regional or hydrographic basin scale—they doubtlessly portray a locally forced ecosystem status. The return to a more diverse assemblage of piscivorous waterbirds, characteristic of a less stressed situation, would indicate some recovery of local environmental quality. This intermediate assemblage, however, would still be far from the original community typical of oligotrophic and hypersaline waters. The species contributing to the diversification of the waterbird community during these intermediate stages (Great-crested Grebe, Coot) should be rather watched as early warnings of the disrupting effects of agricultural or urban eutrophication on the lagoon's ecology.

11.5.2 Effects of hydrological changes on habitats and vegetation dynamics

11.5.2.1 Habitat changes and their relationships with land use change in the watershed

Land use changes in the watershed and associated hydrological changes have caused a rise in water tables of aquifers and have increased the levels of groundwater, flooding periods, and soil water content in the wetlands (Alvarez-Rogel et al., 2007). We have studied the vegetation changes in the Marina del Carmoli wetland and to what extent such changes are caused by the increased inflow of water and nutrients from the watershed.

As all Mar Menor wetlands, the Marina del Carmolí wetland comprises salt steppe areas, salt marsh, reed beds, and a narrow sandy strip on the waterfront. Vegetation units are distributed according to the availability of water and its salinity. Salt steppes are located in areas with low water availability; reed beds are in areas with high water content and low salinity, whereas salt marsh occupies areas with intermediate water content and higher salinity. We studied the changes between 1984 and 2009 regarding the area occupied by the wetland and the major vegetation units and land cover classes: salt steppes, salt marsh, reed beds, crops, water bodies, bare soil, and infrastructures. Remote sensing techniques were applied based on Landsat TM and ETM+satellite images using supervised classification (Carreño et al., 2008).

Results show that in 1984 Marina del Carmolí was mainly covered by salt steppe, which comprised an area of 243 ha, whereas in 2009 this habitat had lost more than a half of its original area. On the contrary, reed beds, practically absent in 1984, occupied an important area in 2009 (165 ha), after an important expansion process since 1995 (Figure 11.15). The relative changes between salt steppe, salt marsh, and reed beds can be explained by the interaction between soil moisture and conductivity gradients. The initial increase of water inflows from the basin resulted in increased soil moisture and higher salinity, which favored the expansion of salt marsh at the expense of salt steppe. At a later stage, around 1995, the increased water inputs reduced water salinity and allowed the expansion of reed beds. Figure 11.16 shows this pattern of change across time.

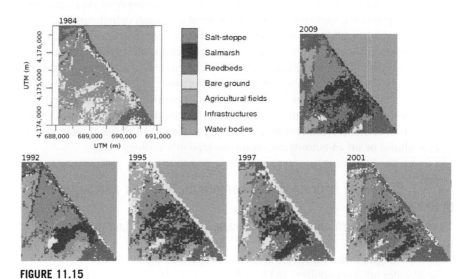

FIGURE 11.15

Maps of vegetation units and land cover of Marina del Carmolí wetland in 1984, 1992, 1995, 1997, 2001, and 2009, obtained by supervised classification of Landsat TM and ETM+images. (For the color version of this figure, the reader is referred to the online version of this chapter.)

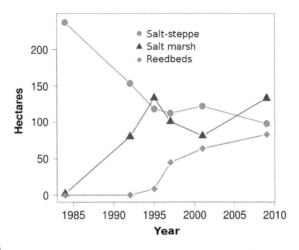

FIGURE 11.16

Area occupied by salt steppe, salt marsh, and reed beds in the Marina del Carmolí between years 1984 and 2009. Data obtained by supervised classification of Landsat TM and ETM + images. (For the color version of this figure, the reader is referred to the online version of this chapter.)

Changes in land use are the primary factor explaining the described changes in the habitats of Mar Menor wetlands, as found in many other studies (Gustafson and Wang, 2002; Liu et al., 2004; Parra et al., 2005; Olhan et al., 2010). Figure 11.17 shows the close relationship between the expansion of irrigation in the Mar Menor basin and the active wetland area (salt marsh and reed beds) in the set of wetlands associated with the inner shore of the Mar Menor lagoon (Marina del Carmolí, Lo Poyo and Playa de la Hita). The regression model supports this close relationship, especially when considering a 5-year time lag (Figure 11.18, $R^2_{adj}=0.945$, $p<0.001$), a period that can be considered as the time required for the habitat to respond to the increased water inflows (Carreño et al., 2008; Esteve et al., 2008).

An important question arises regarding whether the changes in the structure of vegetation of the Mar Menor wetlands (Marina del Carmolí, Lo Poyo, and Playa de la Hita) have also modified the biodiversity values supporting the designation of these wetlands as protected sites. We answer this question in the next section.

11.5.2.2 Assessment of changes in biodiversity and conservation value of Mar Menor wetlands

Despite the growing research on biodiversity and conservation issues, there are still important knowledge gaps and lack of empirical evidence on the role of species loss or changes in species composition in maintaining ecosystem services (Mertz et al., 2007; Bastian, 2013). However, the protection status of species or communities represents an important consensus on the importance of biodiversity and therefore it may constitute the basis for assessing the changes in biodiversity and conservation

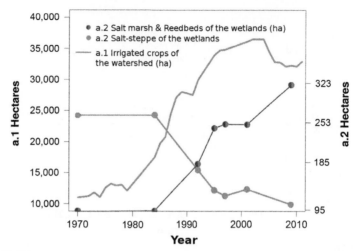

FIGURE 11.17

Axis a.1: Area of irrigated lands in the Mar Menor watershed; axis a.2: area occupied by salt steppe and area occupied by active wetland (salt marsh plus reed beds) in the wetlands associated with the inner shore of the Mar Menor lagoon (Marina del Carmolí, Lo Poyo and Playa de la Hita). (For the color version of this figure, the reader is referred to the online version of this chapter.)

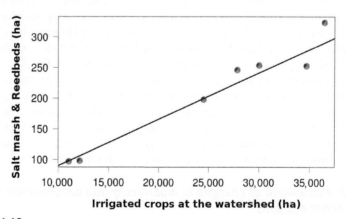

FIGURE 11.18

Regression model between the irrigated area of the Campo de Cartagena and the area occupied by salt marsh plus reed beds in the wetlands at the inner shore of the Mar Menor lagoon (Marina del Carmolí, Lo Poyo and Playa de la Hita) with a 5-year delay interval ($R^2_{adj}=0.945$, $p<0.001$). (For the color version of this figure, the reader is referred to the online version of this chapter.)

value. Protected species lists, biological value indices, species population modeling, and scenarios development can help economic valuation and decision making (Mertz et al., 2007; Bastian, 2013). Here we assess the changes in biodiversity and conservation value of Mar Menor wetlands from the point of view of the protection status of habitat and species regarding the EU directives, particularly the Habitats directive.

Along with the Birds Directive, the Habitats Directive constitutes the basis for the Nature 2000 network of protected sites, designed to conserve European biodiversity, particularly important habitat and endangered species. In the Mar Menor wetlands, following the typology of the Habitats Directive, the salt steppe unit is 95% composed by habitat 1510, "Mediterranean salt steppes." The salt marsh unit mostly consists of habitat 1420, "Mediterranean and thermo-Atlantic halophilous scrubs." Finally, the reed beds unit is dominated by *Phragmites australis*, which is not included in the Habitats Directive. Salt steppe is considered of priority interest by the Habitat Directive; the salt marsh is of community interest; and the reed beds are not included in the Directive.

According to the data obtained with remote sensing described above (Figure 11.16), between 1984 and 2009 the area of salt steppe, of priority interest, has been reduced to less than the half; the area of salt marsh, of community interest, has doubled; and reed beds, without interest from the point of view of the Directive, has multiplied fivefold. The net loss of salt steppe is very important since it is the habitat of Mar Menor wetlands with the highest interest from the point of view of the Directive. Moreover, salt steppe is a rare habitat, comprising only 12,976 ha in Spain. Therefore, any reduction in this habitat constitutes a noticeable loss, particularly taking into account that the conservation status of this priority habitat in Murcia is much higher than the average in Spain, with 83% and 37% in good condition status respectively (Esteve and Calvo, 2000).

To quantify the relative change in biodiversity and conservation value of wetland vegetation from the point of view of the EU Habitats Directive, it has been proposed and applied an index as the weighted average of the area occupied by each vegetation type, assigning the values 0 (no interest), 1 (community interest), and 2 (priority interest) to the reed beds, salt marsh, and salt steppe communities, respectively. This index ranges between 0 (minimum value) and 2 (maximum value). As shown (Figure 11.19), the changes have resulted in an overall reduction of 48% in the biodiversity value and conservation interest of the vegetation from the perspective of the EU Habitats Directive. This is a worrying issue since the Marina del Carmoli wetland as been designated as SCI on the basis of this directive.

Moreover, changes in the steppe bird community of Marina del Carmoli has been described in response to the increased flows affecting this wetland (Robledano et al., 2010). A Birds Directive ((79/409 ECC) based index was proposed and applied. This index of conservation status takes into account the abundance index (IKA) of species and their inclusion in Annex I of the EU Bird's Directive. Results showed a marked decline in this index along the 1984–2008 period (Robledano et al., 2010). Since the Carmoli wetland has also been designated as a Bird SPA (Special Protection Area) under the Birds Directive, this loss in biodiversity and conservation value from the

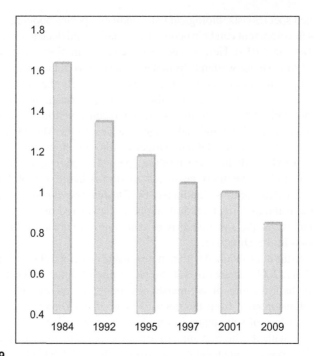

FIGURE 11.19

Evolution between 1984 and 2009 of the index that expresses the interest of vegetation at the wetlands of the inner shore of the Mar Menor lagoon (Marina del Carmolí, Playa de la Hita, and Lo Poyo) from the point of view of the EU Habitats Directive. (For the color version of this figure, the reader is referred to the online version of this chapter.)

perspective of the EU Birds Directive is also of special concern. The effects of hydrological changes on other biological assemblages of wetlands not included in the EU directives, such as ground beetles (Pardo et al., 2008), are also coherent with the trends of change shown by habitats and the steppe-bird community.

Besides, these results also show that conventional protection and conservation strategies usually do not take into account the close dependency of wetlands on the dynamics and management outside the protected area and that this may interfere on the protection and conservation goals of protected wetlands.

11.6 TRADE-OFFS BETWEEN ECOSYSTEM SERVICES OF WETLANDS

The described changes in the Mar Menor lagoon and its biological assemblages (jellyfish outbreaks, aquatic birds) show the need for measures aiming at reducing the nutrient flows. According to results using the dynamic model and the cost-effective analysis, the enhanced use of wetlands to remove part of such flows is more effective that the management of agricultural drainages, which is the solution initially considered by the water body.

However, the increased water and nutrient flows have reduced the biodiversity and conservation value of wetlands, as derived from the species and biological groups supporting the designation of such wetlands as protected sites on the basis of the European Bird and Habitat directives, as the indexes applied to the habitats and to the steppe-birds community have shown. Therefore, additional increases of water flows into the wetlands, as required to increase the amount of nutrients being removed and reduce the ongoing eutrophication process in the lagoon, might cause further losses in the biodiversity and conservation value of wetlands. This points to a potential trade-off among ecosystem services of wetlands, particularly in arid environments, since they frequently present particular characteristics such as hypersaline conditions and low water flows (crypto-wetlands), characteristics that are both highly vulnerable to hydrological changes and also the basis for a specific biodiversity, which is considered rare in the European context.

The importance and magnitude of the indicated trade-off might be higher in the short future since (i) water and nutrients flows might increase, as shown by the simulation of the base trend scenario; (ii) there is a need to substantially reduce the nutrient flows into the lagoon, according to the Wastewater, Nitrates, and Water Framework directives; (iii) the management of wetlands for nutrients removal seems to be the most effective and economically efficient measure to achieve such goal; and (iv) increased water flows into the wetlands, as required to enhance the ecosystem service of nutrients removal and reduce the flows into the lagoon, would further threaten the biodiversity of Mar Menor wetlands and their conservation value according to the Birds and Habitat directives.

Some strategies and management options may help to solve such potential trade-offs. One such strategy is to spatially differentiate—to some extent—both ecosystem services of wetlands by allocating further enhancements of the nutrient removal function outside the boundaries of the protected site. This can be carried out by means of the restoration of lost areas of wetlands or the creation of new wetland areas at the expense of marginal crops. This is the case of the wetlands restoration measure proposed in Marina del Carmoli (Figure 11.20) to restore part of the original wetland area, presently occupied by marginal crops and to reconnect the Albujon watercourse, at present disconnected from the wetland, in order to manage and remove the nutrients transported by this important watercourse before entering the lagoon.

There is considerable evidence about important trade-offs between some ecosystem services, particularly between provisioning services (e.g., food production from agriculture) and regulating services such as soil control and water quality, trade-offs that should be taken into account in ecosystem management (Maes et al., 2012). This has also been pointed out in the case of wetlands (Maltby and Acreman, 2011), including potential trade-offs between nutrient removal and species richness of wetlands (Zedler and Kercher, 2005; Verhoeven et al., 2006; Maltby et al., 2013). However, as shown in the Mar Menor case, these trade-offs are particularly

FIGURE 11.20

Area for proposed restoration of wetland area in Marina del Carmoli. Yellow: present protected site; red: crops where the original wetland area might be restored; blue: Albujon watercourse, at present disconnected from the wetland. (For interpretation of the references to color in this figure legend, the reader is referred to the online version of this chapter.)

important in the case of wetlands in arid environments, where the enhancement of the nutrients removal service may alter their low water and high saline conditions and associated biodiversity values.

This highlights the need for integrated approaches under which synergistic relationships and potential trade-offs between ecosystem services can be properly considered for better informed decisions. Modeling how land use changes and management decisions impact multiple ecosystem services remains as an important challenge (Nemec and Raudsepp-Hearne, 2013). As a contribution to such a challenge, ongoing research on Mar Menor wetlands will focus on a full integration of biodiversity and conservation status indicators and the assessment and valuation of ecosystem services within the integrated dynamic modeling framework for the Mar Menor area.

11.7 **CONCLUDING REMARKS**

In the light of results shown in the Mar Menor case, we can go back to the questions regarding the role of models for the assessment of ecosystem services. Results shown in this work illustrate how the Mar Menor modeling framework has been useful to assess the ecosystem services of wetlands, taking into account the issues outlined in the introduction: the wetland-watershed relationships, the drivers of change (irrigation), the identification of less common but site-important services (effect on jellyfish outbreaks through the control of nutrient loadings into the Mar Menor lagoon), and the potential interactions among wetland ecosystem services (trade-offs between nutrients removal and biodiversity conservation).

Models help to estimate important factors for ecosystem services assessment that are not empirically available. In the Mar Menor case, the modeling framework has allowed a first valuation of the nutrient removal service of their coastal wetlands. Models are not only suitable for describing the structure and function of ecosystems but also to show how external drivers like land use affect them and their ecosystem services (Galic et al., 2012), to reveal trade-offs and to assess the change of ecosystem services along time under different scenarios to inform adaptive management (Maltby et al., 2013). For example, models may help to address some of the identified knowledge gaps, such as how much can we increase the nutrients inflow in a certain wetland without compromising its biodiversity values (Zedler and Kercher, 2005). Models can also help to provide evidence regarding the benefits of more integrated wetland ecosystem management when compared to other policy options and measures (Maltby et al., 2013), as also shown in the Mar Menor case.

Assessment of ecosystem services using very simplified methods should be undertaken with caution. For example, some assessments are being carried out by means of remote sensing–based land cover maps, as a proxy of ecosystem types and constant monetary valuations per service of each land cover/ecosystem, leading to a final fixed economic value for a reduced set of general land covers/ecosystems (see, e.g., Cai et al., 2013). This type of approaches has the potential risk of misleading estimations due to the lack of consideration of interactions, dynamic changes, and the importance of context-specific factors. This may hide key processes and services that are not tracked with land cover maps and change along time or present complex interactions with other processes, leading to biased valuations. Moreover, the assessment of land-use changes in terms of ecosystem services by means of a final unique value per cover cannot identify complex trade-offs and the distribution of costs and benefits associated to such trade-offs, which might be relevant for stakeholders. This risk increases in the case of ecosystems less well represented by gross average values, as in the case of wetlands of arid environments. On the contrary, models, particularly context-based or context-adapted models, allow a more detailed consideration of ecosystem processes and their linkages with specific services, leading to reduced risks of bias in the valuation process.

Models are very useful tools to assess the ecosystem services, although it is important to identify the most appropriate approach for each case. For example, Vigerstol and Aukema (2011) review in relation to the assessment of services of freshwater ecosystems the advantages and limitations of well-established hydrological models (SWAT, VIC) and new model tools specifically devoted to assess ecosystem services (InVEST, ARIES), all of them being of interest depending of the specific purpose, available on-site information, and level of expertise, among other criteria.

A context-specific approach may be needed for assessing the services of less common ecosystems as wetlands of arid environments. For example, regarding the drivers of ecosystem change and their services, it is generally recognized that agricultural intensification usually reduces the hydrological flows supporting the wetlands (see, e.g., reviews by Zedler and Kercher, 2005; Maltby and Acreman, 2011; Maltby et al., 2013). However, as shown in the Mar Menor case, in arid environments the agricultural intensification may act in the opposite direction, increasing the hydrological flows reaching the wetlands and altering the oligotrophic and saline conditions of such systems. Therefore, the effects on the wetland ecosystem services of the same driver (agricultural intensification) may be rather different. This points to the need for context-specific approaches. Model outputs cannot be easily transferred between contexts without a reconsideration of model assumptions, structure, parameterization, and intended purpose (Galic et al., 2012). Context-specific or context-adapted models are useful not only for on-site better informed management and decisions regarding the concerned wetland and its services but also as illustrating examples for other cases.

Finally, since the assessment of ecosystem services is particularly useful for management and decision making, there is a need for transparency in the applied modeling tools (Galic et al., 2012) to develop confidence and to contribute to consensus building. The availability of confident modeling tools will help to show the linkages among apparently very different systems, processes, and services, as well as the benefits of more integrated approaches to wetlands, the watersheds they depend upon and—in the case of coastal wetlands—the coastal systems they are linked to. The modeling framework being developed in the Mar Menor tries to contribute to this final aim.

References

Almansa, C., Martínez-Paz, J.M., 2011. Intergenerational equity and dual discounting. Environ. Develop. Eco. 16, 685–707.

Alvarez-Rogel, J., Jiménez-Cárceles, F.J., Egea-Nicolás, C., 2006. Phosphorous and nitrogen content in the water of a coastal wetland in the Mar Menor lagoon (SE Spain): relationships with effluents from urban and agricultural areas. Water Air Soil Pollut. 173, 21–38.

Alvarez-Rogel, J., Jimenez-Carceles, F.J., Roca, M.J., Ortiz, R., 2007. Changes in soils and vegetation in a Mediterranean coastal salt marsh impacted by human activities. Estuar. Coast. Shelf Sci. 73, 510–526.

Álvarez-Rogel, J., Jiménez-Cárceles, F.J., Egea Nicolás, C., María-Cervantes, A., González-Alcaraz, M.N., Párraga Aguado, I., Conesa Alcaraz, H.M., 2009. Papel de los humedales costeros del Mar Menor en la depuración de aguas eutrofizadas: el caso de la Marina del Carmolí. In: Cabezas, F., Senent, M. (Eds.), Estado actual del conocimiento científico. Fundación Cluster-Instituto Euromediterráneo del agua, pp. 321–358.

Astorga, D., Ruiz, J., Prieto, L., 2012. Ecological aspects of early life stages of Cotylorhiza tuberculata (Scyphozoa: Rhizostomae) affecting its pelagic population success. Hydrobiologia 690, 141–155.

Barbier, E.B., 2011. Wetlands as natural assets. Hydrol. Sci. J. 56, 1360–1373.

Bastian, O., 2013. The role of biodiversity in supporting ecosystem services in Natura 2000 sites. Ecol. Indic. 24, 12–22.

Cai, Y.B., Zhang, H., Pan, W.B., Chen, Y.H., Wang, X.R., 2013. Land use pattern, socio-economic development, and assessment of their impacts on ecosystem service value: study on natural wetlands distribution area (NWDA) in Fuzhou city, southeastern China. Environ. Monit. Assess. 185, 5111–5123.

Carreño, M.F., Esteve, M.A., Martinez, J., Palazón, J.A., Pardo, M.T., 2008. Habitat changes in coastal wetlands associated to hydrological changes in the watershed. Estuar. Coast. Shelf Sci. 77, 475–483.

Centro Regional de Estadística de Murcia, 2013. Anuario Estadístico de la Región de Murcia. http://www.carm.es/econet/publica/catalogo_est_sintesis1.html.

Chapelle, A., Duarte, P., Fiandrino, A., Esteve, M.A., Galbiati, L., Marinov, D., Martínez, J., Norro, A., Plus, M., Somma, F., Tsirtsis, G., Zaldívar, J.M., 2005. Comparison between different modelling approaches for coastal lagoonsReport EUR 21817 EN. Institute for Environment and Sustainability. Ispra. Joint Research Centre. European Commission.

Conesa, H., Jiménez-Cárceles, F.J., 2007. The Mar Menor lagoon (SE Spain): a singular natural ecosystem threatened by human activities. Mar. Pollut. Bull. 54, 839–849.

David, M.B., Gentry, L.E., Kovacic, D.A., Smith, K.M., 1997. Nitrogen balance in and ex-port from an agricultural watershed. J. Environ. Qual. 26, 1038–1048.

Dolores, E., Bermúdez, L., Bas, I., Gómez, O., Viuda, E., López, J.I., Pla, M.L., Peñalver, J., 2009. Gestión de recursos pesqueros del Mar Menor. In: Cabezas, F., Senent, M. (Eds.), Mar Menor. Estado actual del conocimiento científico. Fundación Cluster-Instituto Euromediterráneo del agua, pp. 517–540.

Esteve, M.A., Calvo, J.F., 2000. Conservación de la naturaleza y biodiversidad en la Región de Murcia. In: Calvo, J.F., Esteve, M.A., López Bermúdez, F. (Eds.), Biodiversidad. Contribución a su conocimiento y conservación en la Región de Murcia. Instituto del Agua y Medio Ambiente. Servicio de Publicaciones Universidad de Murcia.

Esteve, M.A., Carreño, M.F., Robledano, F., Martínez-Fernández, J., Miñano, J., 2008. Dynamics of coastal wetlands and land use changes in the watershed: implications for the biodiversity. In: Russo, R.E. (Ed.), Wetlands: Ecology, Conservation and Restoration. Nova Science Publishers, New York, NY, pp. 133–175.

Galic, N., Schmolke, A., Forbes, V., Baveco, H., van den Brink, P.J., 2012. The role of ecological models in linking ecological risk assessment to ecosystem services in agroecosystems. Sci. Total Environ. 415, 93–100.

García Pintado, J., Martínez Mena, M., Barberá, G.G., Albaladejo, J., Castillo, V., 2007. Anthropogenic nutrient sources and loads from a Mediterranean catchment into a coastal lagoon: Mar Menor, Spain. Sci. Total Environ. 373, 220–239.

García, J., Rouco, A., García, B., 2001. Evolución del peso económico de la acuicultura marina. Importancia económica de la acuicultura. An. Veter. (Murcia) 17, 41–50.

Gren, I.M., Elofsson, K., Jannke, P., 1997. Cost-effective nutrient reductions to the Baltic Sea. Environ. Resour. Econ. 10, 341–362.

Gustafson, S., Wang, D., 2002. Effects of agricultural runoff on vegetation composition of a priority conservation wetland, Vermont, USA. J. Environ. Manag. 31, 350–357.

Gustafson, A., Fleischer, S., Joelsson, A., 2000. A catchment-oriented and cost-effective policy for water protection. Ecol. Eng. 14, 419–427.

Hernández, A.J., Fernández-Caro, A., 2008. Censo invernal de aves acuáticas de la Región de Murcia, 2003-2008. Memoria. In: Actas IV Congreso de la Naturaleza de la Región de Murcia & I del Sureste Ibérico. Asociación de Naturalistas del Sureste (ANSE), Murcia, pp. 131–158.

Johnson, K., Polasky, S., Nelson, E., Pennington, D., 2012. Uncertainty in ecosystem services valuation and implications for assessing land use tradeoffs: an agricultural case study in the Minnesota River Basin. Ecol. Econ. 79, 71–79.

Jordan, E., Correll, D., Weller, D., 1997. Effects of agriculture on discharges of nutrients from coastal plain watersheds of Chesapeake Bay. J. Environ. Qual. 26, 836–848.

Kampas, A., Edwards, A.C., Ferrier, R.C., 2002. Joint pollution control at a catchment scale: compliance costs and policy implications. J. Environ. Manag. 66, 281–291.

Kronvang, B., Svendsen, L.M., Jensen, J.P., Dørge, J., 1999. Scenario analysis of nutrient management at the river basin scale. Hydrobiologia 410, 207–212.

Lacroix, A., Beaudoin, B., Makowsk, D., 2005. Agricultural water nonpoint pollution control under uncertainty and climate variability. Ecol. Econ. 53, 115–127.

Liu, H.Y., Zhang, S.K., Li, Z.F., Lu, X.G., Yang, Q., 2004. Impacts on wetlands of large-scale land-use changes by agricultural development: the small Sanjiang Plain, China. Ambio 33, 306–310.

Lloret, J., Marín, A., 2009. The role of benthic macrophytes and their associated macroinvertebrate community in coastal lagoon resistance to eutrophication. Mar. Pollut. Bull. 58, 1827–1834.

Lloret, J., Marin, A., Marin-Guirao, L., Velasco, J., 2005. Changes in macrophytes distribution in a hypersaline coastal lagoon associated with the development of intensively irrigated agriculture. Ocean Coast. Manag. 48, 828–842.

Lloret, J., Marin, A., Marin-Guirao, L., 2008. Is coastal lagoon eutrophication likely to be aggravated by global climate change? Estuar. Coast. Shelf Sci. 78, 403–412.

Maes, J., Paracchini, M.L., Zulian, G., Dunbar, M.B., Alkemade, R., 2012. Synergies and trade-offs between ecosystem service supply, biodiversity and habitat conservation status in Europe. Biol. Conserv. 155, 1–12. http://dx.doi.org/10.1016/j.biocon.2012.06.016.

Maltby, E., Acreman, M.C., 2011. Ecosystem services of wetlands: pathfinder for a new paradigm. Hydrol. Sci. J. 56, 1341–1359.

Maltby, E., Acremanb, M., Blackwellc, M.S.A., Everardd, M., Morrise, J., 2013. The challenges and implications of linking wetland science to policy in agricultural landscapes—experience from the UK National Ecosystem Assessment. Ecol. Eng. 56, 121–133.

Martinez Paz, J.M., Martinez Fernández, J., Esteve Selma, M.A., 2007. Evaluación económica del tratamiento de drenajes agrícolas en el Mar Menor (SE España). Revista Española de Estudios Agrosociales y Pesqueros 215 (216), 211–231.

Martínez-Fernández, J., Esteve-Selma, M.A., 2007. Gestión integrada de cuencas costeras: dinámica de los nutrientes en la cuenca del Mar Menor (Sudeste de España). Revista de Dinámica de Sistemas 3, 2–23.

Martínez-Fernández, J., Esteve-Selma, M.A., Robledano-Aymerich, F., Pardo-Sáez, M.T., Carreño-Fructuoso, M.F., 2005. Aquatic birds as bioindicators of trophic changes and ecosystem deterioration in the Mar Menor lagoon (SE Spain). Hydrobiologia 550, 221–235.

Martin-López, B., Gómez-Baggethun, E., García-Llorente, M., 2014. Trade-offs across valuedomains in ecosystem services assessment. Ecol. Indicat. 37 (Part A), 220–228.

Meissner, R., Seeger, J., Rupp, H., 2002. Effects of agricultural land use changes on diffuse pollution of water resources. Irrig. Drain. 51, 119–127.

Mertz, O., Ravnborg, H.M., Lövei, G.I., Nielsen, I., Konijnendijk, C.C., 2007. Ecosystem services and biodiversity in developing countries. Biodivers. Conserv. 16, 2729–2737.

Moreno-González, R., Campillo, J.A., García, V., León, V.M., 2013. Seasonal input of regulated and emerging organic pollutants through surface watercourses to a Mediterranean coastal lagoon. Chemosphere 92, 247–257. http://dx.doi.org/10.1016/j.chemosphere.2012.12.022.

Nemec, K., Raudsepp-Hearne, C., 2013. The use of geographic information systems to map and assess ecosystem services. Biodivers. Conserv. 22, 1–15.

Okruszko, T., Duel, H., Acreman, M., Grygoruk, M., Florke, M., Schneider, C., 2011. Broad-scale ecosystem services of wetlands—overview of the current situation and future perspectives under different climate and water management scenarios. Hydrolog. Sci. J. 56 (8), 1501–1517.

Olhan, E., Gun, S., Ataseven, Y., Arisoy, H., 2010. Effects of agricultural activities in Seyfe Wetland. Sci. Res. Essays 5, 9–14.

Pardo, M.T., Esteve, M.A., Giménez, A., Martínez-Fernández, J., Carreño, M.F., Serrano, J., Miñano, J., 2008. Assessment of the hydrological alterations on wandering beetle assemblages (coleoptera: Carabidae and Tenebrionidae) in coastal wetlands of arid mediterranean systems). J. Arid Environ. 72, 1803–1810.

Parra, G., Jimnenez-Melero, R., Guerrero, F., 2005. Agricultural impacts on Mediterranean wetlands: the effect of pesticides on survival and hatching rates in copepods. Int. J. Limnol. 41, 161–167.

Pérez Ruzafa, A., Aragón, R., 2003. Implicaciones de la gestión y el uso de las aguas subterráneas en el funcionamiento de la red trófica de una laguna costera. In: Fornés, J.M., Llamas, R. (Eds.), Conflictos entre el desarrollo de las aguas subterráneas y la conservación de los humedales: litoral mediterráneo. Ediciones Mundi-Prensa, Madrid, pp. 215–245.

Perez Ruzafa, A., Gilabert, J., Gutiérrez, J.M., Fernández, A.I., Marcos, C., Sabah, S., 2002. Evidence of a planktonic food web response to changes in nutrient input dynamics in the Mar Menor coastal lagoon, Spain. Hydrobiologia 475 (476), 359–369.

Prieto, L., Astorga, D., Navarro, G., Ruiz, J., 2010. Environmental control of phase transition and polyp survival of a massive-outbreaker jellyfish. PLoS One 5 (11). http://dx.doi.org/10.1371/journal.pone.0013793.

Purcell, J.E., Uye, S., Lo, W., 2007. Anthropogenic causes of jellyfish blooms and their direct consequences for humans: a review. Mar. Ecol. Prog. Ser. 350, 153–174.

Robledano, F., 1998. Mar Menor. In: Bernúes, M. (Ed.), Zonas húmedas españolas incluídas en el Convenio de Ramsar. Dirección General de Conservación de la Naturaleza, Ministerio de Agricultura, Pesca y Alimentación, Madrid, pp. 323–334.

Robledano, F., Farinós, P., 2010. Waterbirds as bioindicators in coastal lagoons: background, potential value and recent research in mediterranean areas. In: Friedman, A.G. (Ed.), Lagoons: Biology, Management and Environmental Impact. Nova Science Publishers, Hauppauge, NY, pp. 153–183.

Robledano, F., Esteve, M.A., Farinós, P., Carreño, M.F., Martínez, J., 2010. Terrestrial birds as indicators of agricultural-induced changes and associated loss in conservation value of Mediterranean wetlands. Ecol. Indic. 10, 274–286.

Robledano, F., Esteve, M.A., Martínez-Fernández, J., Farinos, P., 2011. Determinants of wintering waterbird changes in a Mediterranean coastal lagoon affected by eutrophication. Ecol. Indic. 11, 395–406.

Ruiz, J., Prieto, L., Astorga, D., 2012. A model for temperature control of jellyfish (*Cotylorhiza tuberculata*) outbreaks: a causal analysis in a Mediterranean coastal lagoon. Ecol. Model. 233, 59–69.

Saunders, D.A., Wintle, B.A., Mawson, P.R., Dawson, R., 2013. Egg-laying and rainfall synchrony in an endangered bird species: implications for conservation in a changing climate. Biol. Conserv. 161, 1–9.

Schou, J.S., Skop, E., Jensen, J.D., 2000. Integrated agri-environmental modelling: a costs analysis of two nitrogen tax instruments in the Vejle Fjord watershed, Denmark. J. Environ. Manag. 58, 199–212.

Serrano, J.F., Sironi, J.S., 2009. Cuantificación y evolución de la carga contaminante de nutrientes y plaguicidas en aguas del Mar Menor y su relación con los aportes hídricos de la Rambla del Albujón y otros aportes subterráneos. In: Cabezas, F., Senent, M. (Eds.), Mar Menor. Estado actual del conocimiento científico. Fundación Cluster-Instituto Euromediterráneo del agua, pp. 245–284.

Tománková, I., Boland, H., Reid, N., Fox, A.D., 2013. Assessing the extent to which temporal changes in waterbird community composition are driven by either local, regional or global factors. Aquat. Conserv. Mar. Freshwat. Ecosyst. 23, 343–355.

Turner, K., Georgiou, S., Green, I.M., Wulff, F., Barret, S., Soderqviest, T., Bateman, I., Folke, C., Langaas, S., Zylicz, T., Maler, K.G., Markowska, A., 1999. Managing nutrient fluxes and pollution in the Baltic: an interdisciplinary simulation study. Ecol. Econ. 30, 333–352.

Velasco, J., LLoret, J., Millan, A., Marin, A., Barahona, J., Abellán, P., Sánchez-Fernández, D., 2006. Nutrient and particulate inputs into the Mar Menor lagoon (SE Spain) from an intensive agricultural watershed. Water Air Soil Pollut. 176, 37–56.

Verhoeven, J.T.A., Arheimer, B., Yin, C., Hefting, M.M., 2006. Regional and global concerns over wetlands and water quality. Trends Ecol. Evol. 21, 96–103.

Vidal-Abarca, M.R., Esteve Selma, M.A., Suárez Alonso, M.L., 2003. Los Humedales de la Región de Murcia: Humedales y Ramblas de la Región de Murcia. Consejería de Agricultura. Agua y Medio Ambiente, Murcia.

Vigerstol, K., Aukema, J.E., 2011. A comparison of tools for modeling freshwater ecosystem services. J. Environ. Manag. 92, 2403–2409.

Xue, Y., David, M.B., Gentry, L.E., Kovacic, D.A., 1998. Kinetics and modelling of dissolved phosphorous export from a tile-drained agricultural watershed. J. Environ. Qual. 27, 917–922.

Zanou, B., Kontogianni, A., Skourtos, M., 2003. A classification approach of cost effective managent measures for the improvement of watershed quality. Ocean Coast. Manag. 46, 957–983.

Zedler, J.B., Kercher, S., 2005. Wetland resources: status, trends, ecosystem services, and restorability. Annu. Rev. Environ. Resour. 30, 39–74.

Structurally Dynamic Model and Ecological Indicators to detect the crayfish invasion in a lake ecosystem

12

Michela Marchi[a],*, Sven Erik Jørgensen[b], Federico Maria Pulselli[a], Simone Bastianoni[a]

[a]*Department of Earth, Environmental and Physical Sciences, Ecodynamics Group, University of Siena, Siena, Italy*
[b]*Professor Emeritus, Environmental Chemistry, University of Copenhagen, Copenhagen, Denmark*
*Corresponding author: e-mail address: marchi27@unisi.it

12.1 INTRODUCTION

Man's technological capabilities have created an artificial system with an enormous potential to modify nature. Such modifications often destroy biological species and genetic heritage. This in turn means decreased biological complexity, diversification, and adaptation to change, as well as vulnerability and uncontrolled proliferation of certain populations (Tiezzi, 2003).

Introduction of species has been occurring for millennia, but biological invasions are now recognized as the greatest threat to biodiversity at global level, second only to loss and fragmentation of habitats. Transformation of an allochthonous species into an invader often has serious, insidious, and irreversible impacts on ecosystems, changing the ecological relationships between communities, altering evolutionary processes, causing dramatic changes in indigenous populations, and leading to many extinctions (Mack et al., 2000).

Increased exchange of goods and movement of people has had great benefits in some fields, but has increased the problem of diffusion of alien species, favoring their transport around the world. On the other hand, many invasive species are intentionally introduced, for example, pines and eucalypts for timber, decorative plants, and exotic animals.

Another path has now been added to these two "traditional" paths of invasion (coincidental and intentional): invasions induced by fast climate change. The phenomenon is not completely new (species have changed their geographical distribution throughout the history of the biosphere). What is new is the rate at which such changes are happening, too fast for species remaining in the same area to adapt. Moreover, a species that comes to a "new" area is already suited to its environmental and climatic conditions, since they are the same as in the original exotic environment. It is more likely that resident species have difficulty in adapting to a changing environment, while newcomers are already suited.

Adaptable to extreme environmental changes, alien species often have the advantage of better use of resources available in the habitat. They are therefore more efficient in lowering their internal entropy than native species, which on the contrary may reach thermodynamic equilibrium sooner because they are unable to regulate their homeostatic equilibrium rapidly. Many experimental studies have shown that invasive species may have better enzyme recovery than indigenous species (see, e.g., Corsi et al., 2007; Pastore et al., 2008) and greater ability to self-organize.

Invasions do not always have a negative result: In some cases, diversity increases, empty niches are (re-)occupied, and a more connected and efficient ecosystem structure is obtained.

The introduction and expansion of exotic species can be studied by evolutionary thermodynamics considering the entropy variations that are possible. Entropy is a concept introduced by Clausius to express the Second Law of Thermodynamics and can be seen as the degree of energy dissipation or disorder of a system. Biological and ecological systems only seem to violate the Second Law. The entropy balance must be total and must include biological and ecological systems as well as the outside environment with which they exchange energy and matter.

For all systems with evolutionary capacity (and thus also ecosystems), Morowitz's Equation (12.1) is pertinent:

$$-dS_{int} \leq dS_s \tag{12.1}$$

where dS_{int} is the change in entropy of an intermediate system (e.g., an ecosystem) capable of creating order out of chaos. The entropy of an intermediate system may increase or decrease, depending on its capacity to self-organize: The only limit to a system's entropy reduction is that the total entropy (intermediate system + environment) must always increase (Morowitz, 1978; Marchettini et al., 2008).

While entropy is easily calculated and/or measured in ideal systems, it is very difficult to encompass all the possible states in which an ecosystem can express its complexity. However, an ecosystem's entropy can be described in a qualitative way in various scenarios after invasion has occurred. Marchi et al. (2010) said that after the introduction of an invasive species, an ecosystem's entropy may vary according to the type of ecosystem and community structure (Figure 12.1).

A new species is introduced into a hitherto stable ecosystem at time t_0. This causes an immediate increase in system complexity due to an increase in biomass and biodiversity. The ecosystem therefore undergoes an overall decrease in entropy because its internal information content increases. At time t_1 the effects of the introduced species on the ecosystem begin to manifest, giving rise to three possible cases:

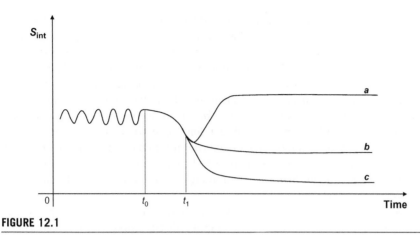

FIGURE 12.1

Concept diagram relating entropy and invasions.

(a) Increased entropy and ecosystem perturbation. An example is the effect of introduction of the Louisiana red crayfish *Procambarus clarkii* (Girard, 1852) into different Mediterranean freshwater ecosystems (Rodríguez et al., 2005).

(b) Integration of the species into the ecosystem and its ecological network by adaptation to new conditions. Introduction of the Mediterranean mussel *Mytilus galloprovincialis* to the coasts of South Africa is an example of scenario *b* (Branch and Steffani, 2004).

(c) The species adds new links, relations, and diversity to the ecosystem, decreasing its entropy. The dense colonies of the intertidal ascidian *Pyura praeputialis*, a recent invader of the shores of Chile, have become a resource, increasing the biomass and biodiversity of local ecosystems (see, e.g., Cerda and Castilla, 1999).

Evolutionary thermodynamics can be an interesting tool for investigation of this type, and since entropy is difficult to quantify in ecosystems, other functions can be used in an attempt to describe the evolution of invaded ecosystems in time: Thermodynamics-based *goal function* or *(ecological) orientors* (e.g., eco-exergy) can approximate what we indicated above as entropy, since they endeavor to describe how self-organized ecological systems develop by keeping their state as far as possible from thermodynamic equilibrium (Jørgensen et al., 2007).

12.2 THE CASE STUDY OF LAKE CHOZAS (SPAIN) TO DESCRIBE THE RESPONSE TO BIOLOGICAL INVASION

Lake Chozas is a small shallow water body (volume 17,551 m^3; maximum depth 1.8 m) in León (northwest Spain). Plant cover did not vary significantly in Lake Chozas from 1984 to 1996, with 95% of the bottom covered by a varied community of macrophytes (Fernández-Aláez et al., 1984, 1999). Loss of vegetation was recorded in 1997, when there was a sharp reduction in mean submerged vegetation density, although different species of macrophytes were affected differently (Fernández-Aláez et al., 2002). Destruction of the habitat of aquatic plants altered the trophic web of the lake, with serious loss of biodiversity and breakdown of submerged vegetation control mechanisms. In Lake Chozas, benthivory by the Louisiana red swamp crayfish *P. clarkii* led to decomposition of submerged vegetation and sharply increased nutrient load, due to intense bioturbation of sediment. The high concentrations of organic detritus in the water column increased turbidity. Resuspension of sediment by wind in the absence of vegetation also favored a cascade of mechanisms reinforcing water turbidity (see Figure 12.2; Rodríguez et al., 2003).

P. clarkii is recognized for its high invasive capacity in Mediterranean freshwater ecosystems; in fact, its tolerance to changes in abiotic conditions, lifestyle, rapid population development and ability to acquire food are all evidence of the ecological flexibility of the species, making it an invasive species described as "killer" crayfish.

To test the assumption that after 1997 Lake Chozas exemplified case (a) in Marchi et al. (2010), we developed a structurally dynamic model to verify whether an ecosystem perturbed by an alien species, in this case *P. clarkii*, showed a decrease in eco-exergy, which corresponds to an increase in entropy.

FIGURE 12.2

Photos of Lake Chozas before and after the biological invasion of the Louisiana red swamp crayfish, *Procambarus clarkii* (Girard, 1852), in 1997. (For the color version of this figure, the reader is referred to the online version of this chapter.)

Eco-exergy is defined as the amount of work that an (eco)system can perform by coming to thermodynamic equilibrium with its environment (Jørgensen, 1982, 1992a,b, 1997, 1999). It reflects how self-organized ecological systems develop by keeping their state as far as possible from thermodynamic equilibrium. In the structurally dynamic model presented in this chapter, we used eco-exergy to estimate complexity and order in Lake Chozas before and after the biological invasion. Eco-exergy is a measure of energy quality and provides an easier interpretation of ecological processes than entropy. It represents ecosystem changes well, because unlike entropy, it is not bound to heat variations, which are difficult to quantify in complex systems such as ecosystems. Entropy is an enigma of thermodynamics because it embodies time irreversibility, quality, and information, all properties that give it a central position in biology and ecology (Tiezzi, 2006). Like entropy, the eco-exergy of an ecosystem cannot be calculated exactly, since it is impossible to measure concentrations of all components and quantify all contributing factors precisely. However, the biomass of major ecosystem components can be estimated and eco-exergy can therefore be computed by Equation (12.2):

$$\text{Eco-exergy} = RT\sum_i \left[C_i \ln\left(\frac{C_i}{C_i^{\text{eq}}}\right) + (C_i - C_i^{\text{eq}}) \right] \tag{12.2}$$

where R is the gas constant, T ambient temperature, C_i and C_i^{eq} concentrations of ecosystem component i in the current state and at thermodynamic equilibrium, respectively (Jørgensen, 2008). Jørgensen (1997b) also proposed a relative eco-exergy index, Eco-ex (Equation 12.3):

$$\text{Eco-ex} = \sum_{i=1}^{n} \beta_i \cdot C_i \tag{12.3}$$

where C_i is the concentration of ecosystem component i ($1, \ldots, n$) (e.g., the biomass of a taxonomic or functional group) in the ecosystem, and β_i is a weighting factor, related to the information stored in the biomass. β-Values are closely correlated with the free energy of the information embodied in organisms (Jørgensen et al., 2010a). If $i = 1$ represents detritus, $\beta = 1$ corresponds to detritus, in other words β-values are normalized with respect to detritus. Since 1 mg of detritus contains 18.7 J, multiplying exergy by 18.7 according to Equation (12.2), we obtain eco-exergy in J/m^3 when C_i is in mg/m^3.

$\beta_i \equiv \ln C_i / C_i^{\text{eq}}$ has been calculated for various organisms, based on the number of non-nonsense genes coding the amino acid sequences of enzymes. Enzymes determine the life processes of organisms (Jørgensen et al., 2005). In any case, eco-exergy can provide an interpretation of the natural trend of ecosystems and is a good indicator of ecosystem services and sustainability (Jørgensen, 2010a). Thermodynamics-based *goal functions*, like eco-exergy and entropy, express the degree of energy dissipation or disorder of a system, and are therefore interesting tools to study the impacts of invasive species on ecosystems, highlighting time variations in stored information. The aim of the developed structurally dynamic model was to quantify the change in complexity of Lake Chozas resulting from introduction and proliferation of the exotic species, *P. clarkii*, using eco-exergy.

12.2.1 **The structurally dynamic model of Lake Chozas**

A dynamic model of the phosphorus cycle, based on 1996–1997 monitoring data, was developed for Lake Chozas using the program STELLA (version 8.1.4). The flow diagram of the model is shown in Figure 12.3.

The model had two layers, water and sediment, and state variables were reported in mg P/m^3 for both layers. The water data was used directly, and the sediment data was converted from mg P/m^2 to mg P/m^3 by assuming an active layer of sediment 0.1 m deep. The model had eight state variables related to phosphorus (P) concentrations, because P was assumed a limiting factor for the lake ecosystem. They were phosphorus in phytoplankton (PA), zooplankton (PZ), sediment (PESed), sediment pore water (P_I), submerged plants (Psp), detritus (P_Det), and exotic crayfish (P_crayfish), as well as soluble reactive phosphate in the water column (PS).

The model was applied to two scenarios:

1. Pre-1997, before introduction of the crayfish, when water quality was oligotrophic.
2. Post-1997, when the effects of the crayfish on the indigenous community of the lake had begun to manifest.

In the first scenario, we assumed zero phosphorus concentration in crayfish biomass and the lake had an abundance of macrophytes; in the second, monitoring data indicated an initial value of 100 mg P/m^3 or one specimen of *P. clarkii* per square meter and submerged plants were harvested in proportion to the density of crayfish (Rodríguez et al., 2003, 2005). Experimental measures of the concentrations of phosphorus in water, phytoplankton, zooplankton, submerged vegetation, detritus, and exotic crayfish were used to calibrate the model. The parameters and the equations used in the model, as well as the concentration of phosphorus present in the major elements of the sediment and the water column are shown in detail in Marchi et al. (2011a). Calibration was done using those parameter values that fit the monitoring data fairly well, in fact, a rigid structure and fixed set of parameters cannot reflect real changes in an ecosystem.

The seasonal fluctuations in phosphorus concentrations in the major state variables of the water column and sediment are shown in Figure 12.4 for scenario 1, before introduction of the alien crayfish.

Soluble reactive phosphorus concentration in water was low, also reflecting the low abundance of phytoplankton, zooplankton, and detritus in the water column. Phytoplankton showed a maximum in February and two other peaks in spring and summer, as expected. The curve of zooplankton concentration reflected phytoplankton dynamics. The low activity in the water phase was in equilibrium with activity in the lower layer, where phosphorus concentrations were high. Before introduction of *P. clarkii*, the principal phosphorus stock was benthic vegetation. Early in the year,

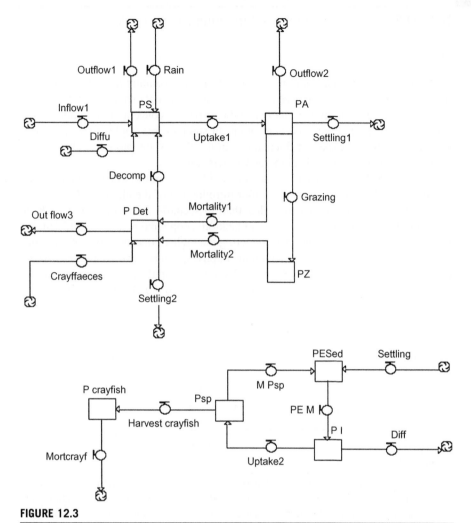

FIGURE 12.3

Flow diagram of model. The upper part of the diagram represents the water column: Inflow1 (phosphorus inflow into lake); diffu (phosphorus diffusion from pore water in the sediment to water phase); Outflow1 (outflow rate of soluble reactive phosphorus); PS (soluble reactive phosphorous concentration in lake water); Rain (phosphorus concentration in precipitation); Uptake1 (phosphorus uptake rate by phytoplankton); PA (phosphorus concentration in phytoplankton); Outflow2 (phosphorus uptake rate from lake water phase by submerged plants); Settling1 (settling of phytoplankton); Grazing (phytoplankton grazing by zooplankton); Mortality1 (mortality of phytoplankton); PZ (phosphorus concentration in zooplankton); Mortality2 (mortality of zooplankton); P Det (phosphorus concentration in detritus); Settling2 (settling detritus); Outflow3 (outflow rate of detritus); crayffaeces (crayfish faeces). The lower part of the diagram represents the layer of sediment: Settling (detritus and phytoplankton settling to sediment); PESed (phosphorus concentration in the sediment); PE M (phosphorus mineralization from exchangeable phosphorus in sediment); Uptake2 (phosphorus uptake rate from lake water phase by submerged plants); P I (Phosphorus concentration in pore water of sediment); Diff (Phosphorus diffusion from pore water in the sediment to water phase); Psp (phosphorus in submerged plants); M Psp (mortality of submerged plants); Harvest crayfish (submerged vegetation consumed by crayfish); P crayfish (phosphorus concentration in detritus); mortcrayf (mortality of crayfish). (For the color version of this figure, the reader is referred to the online version of this chapter.)

phosphorus was rather constant in sediment. When temperature increased in spring, the phosphorus concentration in sediment decreased and that in pore water increased. Later, increased temperature, radiation, and phosphorus concentration in pore water promoted growth of submerged vegetation. In summer, macrophyte concentration fell, followed by the expected fluctuations in the three state variables: phosphorus in sediment, pore water, and submerged vegetation.

Figure 12.5 shows seasonal fluctuations in elements in the water column and in the layer of sediment for scenario 2, after introduction of the alien crayfish.

Phosphate concentration was higher than in scenario 1, rising from a peak of 22–89 mg P/m^3. Thus, biomass of phytoplankton, zooplankton, and organic detritus were also higher than in scenario 1, reflecting the higher concentrations of nutrients in the water column. Seasonal fluctuations in phytoplankton and zooplankton were less evident than in scenario 1. Scenario 2 showed a rapid decrease in phosphorus concentration in submerged vegetation from 2500 to 30.38 mg P/m^3 in only 3

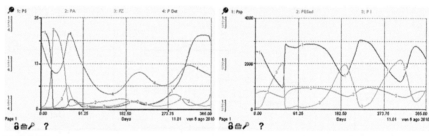

FIGURE 12.4

Scenario 1, before invasion of exotic crayfish: The graph on the left shows the seasonal fluctuations in phosphorus concentrations in the water column, while that on the right shows the fluctuations in phosphorus concentrations in the layer of sediment. (For the color version of this figure, the reader is referred to the online version of this chapter.)

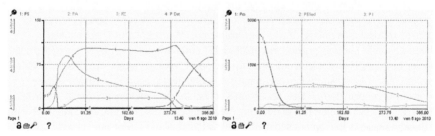

FIGURE 12.5

Scenario 2, after invasion of exotic crayfish: The graph on the left shows the seasonal fluctuation in phosphorus concentration in the water column, while that on the right shows the fluctuation in phosphorus concentrations in the layer of sediment. (For the color version of this figure, the reader is referred to the online version of this chapter.)

months. Phosphorus concentration in sediment was approximately constant in time, showing a slight decrease in autumn that reflected air temperature and the abundance of soluble reactive phosphate in the water column. Reactive phosphorus concentration in pore water was very low and without seasonal fluctuations. The dynamics of crayfish concentration reflected that of submerged vegetation, which decreased as the biomass of *P. clarkii* increased.

The eco-exergy of both the scenarios is shown in Figure 12.6. That of scenario 1 (on the left of the graph) had a mean value of 11,586 J per equivalent/(mg P/m^3 detritus) and its structure reflected the seasonal dynamics of submerged vegetation, which was the most abundant state variable in the lake before the invasion. After introduction of the crayfish, eco-exergy declined slowly (on the right of the graph). Mean eco-exergy in scenario 2 was 336 J per equivalent/(mg P/m^3 detritus), a significant decrease compared with scenario 1.

At the beginning of the invasion, the number of crayfish increased as eco-exergy decreased, because the complexity of the ecosystem was higher before the invasion due to richness of macrophytes, despite the fact that submerged vegetation is lower on the evolutionary scale than the crustacean (Bastianoni et al., 2010). The crayfish had a higher β-value, but its biomass was much less than that of submerged vegetation before the invasion. The decrease in biomass of submerged vegetation and eutrophic conditions made the lake vulnerable and this was quantified as a total decrease in eco-exergy, which also reflected the increase in entropy of the ecosystem caused by disappearance of the most abundant species. After the invasion, there was an increase in phosphorus concentrations of phytoplankton, zooplankton, and detritus, but the eco-exergy value confirmed the major role of submerged vegetation in this wetland. The amount of information embodied in phytoplankton and detritus was low, and although the energy quality of zooplankton is higher than that of submerged vegetation, the increase in total zooplankton biomass was not sufficient to re-establish the relatively stable state of the lake before the disappearance of macrophytes.

Development of the structurally dynamic model helped us to understand the seasonal dynamics of the major state variables of Chozas Lake and made it possible to weigh their effect on eco-exergy and entropy fluctuation over time.

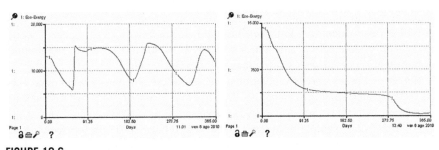

FIGURE 12.6

Eco-exergy in scenario 1 (on the left) and in scenario 2 (on the right). (For the color version of this figure, the reader is referred to the online version of this chapter.)

12.2.2 Re-organization of Chozas Lake in response to biological invasion

We used the previous structurally dynamic model (Marchi et al., 2011a) to examine the recovery of Chozas Lake when all the crayfish were removed and low phosphorus concentrations persisted in inflow water. To test reorganization of Lake Chozas, we first decreased the presence of the invader to limit its impact on the indigenous community. We did this by changing the mortality rate of crayfish (natural death and harvesting) from 0.1 to 0.2 mg P/m^3 per day to rapidly phase phosphorus concentration in crayfish out of the system. Although negative feedback could inhibit macrophyte recolonization, we expected reorganization of the chemicophysical parameters to gradually restore the ecosystem to almost the same community characteristics as before introduction of the alien species. We therefore simulated seasonal fluctuations of major state variables of the lake for 10 years after disappearance of the crayfish.

To model recovery of the wetland, we changed those parameters that were modified by the benthic pressure of the crayfish, such as the diffusion coefficient of phosphorus from sediment pore water to the water column and maximum growth rate of submerged plants.

Specifically, we postulated that the sediment reacquired the characteristics it had before the biological invasion almost 5 years after disappearance of the crayfish from the lake. After that, we allowed the indigenous community another 5 years of simulation to recover. The diffusion coefficient of phosphorus from sediment pore water to water column decreased from about 0.5 to 0.004 mg P/m^3 per day in the first 5 years after the end of the invasion and was then constant over the next 5 years of simulation. The decrease in diffusion could easily be explained by the growth of submerged vegetation, which increased resistance to the diffusion of phosphorus from pore water to water column. On the other hand, the maximum growth rate of the submerged plants dropped from 2 to 1.2 mg P/m^3 per day almost 5 years after disappearance of the crayfish and was then constant. The decrease in growth of submerged vegetation could be explained by its increasing difficulty in obtaining enough nutrients when vegetation increased (see Marchi et al., 2011b).

Figure 12.7 shows the disappearance of *P. clarkii* only 3 months after its introduction and reorganization of the indigenous community. The lake shifted slowly from a turbid to a clear state over the 10-year period. Mean eco-exergy was 12,955 J per equivalent/(mg P/m^3 detritus), showing an initial decrease, small oscillations in the middle period, and quite a fast increase in the last 5 years of simulation, a pattern that reflected seasonal fluctuations in benthic plants.

Moreover, the submerged vegetation increased when total phosphorus concentration in the water column was below 100 mg TP/m^3, in line with the findings of Scheffer et al. (1993), Scheffer (1997), Jeppesen et al. (1998), and Zhang et al. (2003). This is an example of hysteresis (Scheffer et al., 2001). In our scenario of ecosystem reorganization, we increased the mortality rate of *P. clarkii* to regulate population density within fairly narrow limits. Our aim was to minimize the impact of grazing on vegetation. In fact, submerged plants are harvested by these

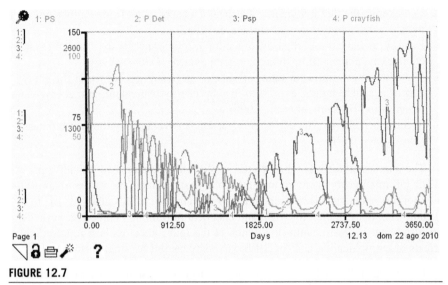

FIGURE 12.7

Ten-year simulation of fast disappearance in the invasice crayfish from Lake Chozas and reorganization of the indigenous community. (For the color version of this figure, the reader is referred to the online version of this chapter.)

crustaceans, especially adults, in proportion to their density. Besides natural death, we also postulated that some crayfish were harvested by humans, decreasing their concentration in the lake. Submerged plants recovered when the lake returned to oligotrophic conditions of water quality.

12.3 JOINT APPLICATION OF ECOLOGICAL INDICATORS TO ASSESS THE HEALTH STATUS OF TWO SPANISH LAKES

12.3.1 Ecological indicators

We analyzed all data monitored from 1984 to 2001 of two lakes (Sentiz and Chozas), both small water bodies in León (northwest Spain). Lakes Sentiz and Chozas were in the same district, having many similar characteristics, with the difference that Lake Chozas was invaded by the Louisiana red swamp crayfish in 1997. We examined three case studies: (1) Lake Sentiz, analyzed as control system by virtue of the absence of exotic crayfish, (2) Lake Chozas before, and (3) Lake Chozas after the biological invasion of *P. clarkii*.

To highlight the variations in the ecological complexity of Lake Chozas after the biological invasion, we applied four systemic ecological indicators able to estimate the health status of the two Spanish lakes by quantifying how far they are from thermodynamic equilibrium:

1. Eco-exergy, the thermodynamic distance of an ecosystem from equilibrium with the surrounding environment (Jørgensen, 1982).
2. Emergy, an indicator of the flow of natural resources necessary to sustain system complexity, expressed in solar energy equivalent (Odum, 1996).
3. The eco-exergy–empower ratio, or system efficiency in converting the energy cost of available inputs, expressed in solar energy equivalent, into ecosystem organization (Bastianoni et al., 2006).
4. Specific eco-exergy, the eco-exergy carried (in average) by the unit biomass (Jørgensen, 2008).

The joint use of different, but related, ecological indicators can provide quite realistic information about environmental health to supplement the values obtained by monitoring campaigns (Jørgensen et al., 2007). The aim of this analysis was to determine variations in the biological and chemical information embodied in the two Spanish water bodies, quantifying which of the four indicators best represents the effects of eutrophication and loss of biodiversity caused by alien species.

Eco-exergy reflects the survival of organisms in ecosystems, as demonstrated by dynamic models. In fact, it increases with increasing ecosystem complexity, biodiversity, and organization (Jørgensen and Nielsen, 2007).

Emergy is defined as the quantity of solar energy used directly or indirectly to obtain a product by a given process or to renew a resource that has been consumed (Odum, 1971a,b; Odum, 1996). It measures the cost in the ultimate energy sources, solar radiation. The units of emergy are *solar emergy joules* or *solar emjoules* (sej). Emergy enables different inputs into a system to be measured on a common basis; by expressing all inputs and outputs in solar energy equivalent, a system's economic and environmental aspects can be considered simultaneously. The total emergy (Em) of a product produced by a given process is given approximately by Equation (12.4):

$$Em = \sum_{i=1}^{n} Tr_i E_i \qquad (12.4)$$

To convert all flows of energy (E_i) involved in the process into solar energy equivalent, a conversion factor is used. *Unit emergy value* is defined as the emergy per unit flow or per unit product (sej/unit) and is called *solar transformity* (Tr_i, in sej/J) when the inputs are quantified in energy terms (in J). Emergy flows per unit time, known as empower, are expressed in sej/year. In the present study, we considered renewable flows of energy and matter entering the two Spanish ecosystems in order to estimate the empower, expressed in solar energy equivalent per year, supporting the lakes.

The eco-exergy–empower ratio index represents the ecosystem complexity (eco-exergy) obtained by an emergy flow through a system. It is the eco-exergy of the whole system, divided by the emergy flow that drives it. The units of the eco-exergy–empower ratio are (kJ year)/sej, indicating ecosystem efficiency in transforming direct and indirect solar energy inputs into organization (Bastianoni and Marchettini, 1997; Bastianoni et al., 2006). Indeed, the eco-exergy–empower ratio reflects the state (age, internal complexity, and health status) of the system per unit input.

Specific eco-exergy, SpEco-ex, is calculated by applying Equation (12.5):

$$\text{SpEco-ex} = \frac{\text{Eco-ex}_{tot}}{\text{Biom}_{tot}} \tag{12.5}$$

where Eco-ex_{tot} represents the total eco-exergy value (expressed in kJ/m^3), and Biom_{tot} the total biomass contained in the elements of the ecosystem (in g/m^3). Specific eco-exergy is independent of the biomass concentration present in the ecosystem.

The eco-exergy–empower ratio can be seen as an indicator of ecosystem services, able to express the ecosystem efficiency in generation services from direct and indirect solar input. Jørgensen (2010a) determines ecosystem services and thereby sustainability from eco-exergy or work capacity. The economic value of the work capacity of various ecosystems can easily be calculated using energy costs:

1. 1 MJ costs €0.01 or $0.014.
2. 1 GJ therefore costs €10 or $14.

These are the market prices of electric power in the year 2009 and electricity is assumed to be 100% exergy (work energy). The monetary value of work capacity covers all possible services of natural capital, because it provides information added to statistical analysis and direct measurements.

12.3.2 Application of ecological indicators to Lakes Sentiz and Chozas

The concentration of biomass in the two lakes for the three case studies and the β-values related to every taxonomic group, considered in the eco-exergy evaluation, are shown in Table 12.1. The functional groups considered were phytoplankton, zooplankton, submerged vegetation, benthic and epiphytic macroinvertebrates, fish, birds, and detritus, as well as *P. clarkii* in Lake Chozas after its introduction.

The eco-exergy (Table 12.2) obtained for Lake Sentiz was lower than that of Lake Chozas before the invasion (3.30E+09 and 4.78E+10 kJ, respectively), which differed by virtue of the massive presence of submerged vegetation and clear water. After the invasion, eco-exergy was lower than in Lake Sentiz (5.22E+08 against 3.30E+09 kJ), indicating the dramatic effects of proliferation of *P. clarkii*. Lake Chozas showed a sharp loss of biodiversity and an overall decrease in biomass that reflect a considerable loss of information.

Although some studies report a positive correlation between the increase in complexity of a lacustrine system and its eutrophic status (Aoki, 2006), the present results showed a different trend. Lake Sentiz had a lower level of information than Lake Chozas before the invasion; indeed, the latter was characterized by relatively little phytoplankton, zooplankton, and detritus in the water phase, as well as a massive

Table 12.1 Biomass of the Different Taxonomic Groups in Lake Sentiz and in Lake Chozas Before and After the Invasion

| Taxonomic Group | Biomass (mg/m^3) | | | β-value |
	Lake Sentiz	Lake Chozas Before the Invasion	Lake Chozas After the Invasion	
Phytoplankton	9.52E+02	3.97E+02	5.39E+02	20
Zooplankton	6.33E+02	8.80E+01	1.07E+02	163
Submerged vegetation	2.00E+02	1.56E+06	1.44E+02	92
Benthic macroinvertebrates				
Oligochaeta	1.08E+02	1.08E+02	5.32E+02	133
Chironomidae	9.84E+02	5.41E+03	2.49E+03	163
Others[a]	4.08E+01	4.08E+01	0.00E+00	120
Epiphytic macroinvertebrates				
Hydra	6.75E−03	2.81E−03	0.00E+00	91
Nematoda	3.33E−05	1.39E−05	0.00E+00	133
Oligochaeta	2.43E−02	1.01E−02	5.72E−04	133
Gastropoda	3.33E−01	1.39E−01	0.00E+00	310
Acari	1.76E−01	7.34E−02	0.00E+00	167
Ostracoda	6.07E−03	2.53E−03	2.11E−04	39
Ephemeroptera	1.28E+00	2.53E−03	7.24E−03	167
Odonata	4.55E−01	1.90E−01	4.37E−02	167
Heteroptera	4.55E−01	2.53E−03	5.00E−05	167
Lepidoptera	3.23E−01	1.35E−01	0.00E+00	221
Trichoptera	6.07E−03	2.53E−03	0.00E+00	167
Diptera	2.96E−01	1.23E−01	3.77E−03	322
Fish	1.11E+02	4.13E+02	4.13E+02	499
Birds				
Ducks	1.07E+02	2.95E+02	1.92E−02	980
Coots	1.85E+02	3.08E+02	2.31E−02	980
Detritus				
Detritus in the water column	5.08E+06	1.03E+03	2.65E+04	1
Detritus in the sediment	6.03E+04	8.65E+04	1.68E+05	1
Crayfish	0.00E+00	0.00E+00	1.11E+03	232
Total	5.14E+06	1.66E+06	2.01E+05	

The column to the right shows the corresponding β-value, as estimated by Jørgensen et al. (2005).
[a]*Others: Platyhelminthes (flatworms).*

Table 12.2 Eco-Exergy of the Different Taxonomic Groups Studied in the Three Case Study

Taxonomic Group	Lake Sentiz (kJ)	Lake Chozas Before the Invasion (kJ)	Lake Chozas After the Invasion (kJ)
Phytoplankton	1.08E+07	2.61E+06	3.54E+06
Zooplankton	5.87E+07	4.71E+06	5.72E+06
Submerged vegetation	1.05E+07	4.72E+10	4.34E+06
Benthic macroinvertebrates			
Oligochaeta	8.13E+06	4.69E+06	2.32E+07
Chironomidae	9.12E+07	2.89E+08	1.33E+08
Others[a]	2.78E+06	1.60E+06	0.00E+00
Epiphytic macroinvertebrates			
Hydra	3.49E+02	8.40E+01	0.00E+00
Nematoda	2.52E+00	6.06E−01	0.00E+00
Oligochaeta	1.84E+03	4.42E+02	2.50E+01
Gastropoda	5.87E+04	1.41E+04	0.00E+00
Acari	1.67E+04	4.02E+03	0.00E+00
Ostracoda	1.35E+02	3.24E+01	2.70E+00
Ephemeroptera	1.22E+05	1.39E+02	3.97E+02
Odonata	4.32E+04	1.04E+04	2.40E+03
Heteroptera	4.32E+04	1.39E+02	2.74E+00
Lepidoptera	4.06E+04	9.76E+03	0.00E+00
Trichoptera	5.76E+02	1.39E+02	0.00E+00
Diptera	5.42E+04	1.30E+04	3.99E+02
Fish	3.16E+07	6.77E+07	6.77E+07
Birds			
Ducks	5.94E+07	9.50E+07	6.18E+07
Coots	1.03E+08	9.90E+07	7.42E+07
Detritus			
Detritus in the water column	2.89E+09	3.37E+05	8.70E+06
Detritus in the sediment	3.43E+07	2.84E+07	5.52E+07
Crayfish	0.00E+00	0.00E+00	8.46E+07
Total	3.30E+09	4.78E+10	5.22E+08

[a]Others: Platyhelminthes (flatworms).

concentration of underwater plant biomass. After the invasion, Lake Chozas not only became eutrophic but underwent a sharp decrease in biomass. Almost complete disappearance of underwater vegetation affected plant-related taxonomic groups, especially epiphytic macroinvertebrates and water birds (see Table 12.1). We also noted a sharp increase in phytoplankton, zooplankton, and detritus in the water column; these were indices of eutrophic status.

Table 12.3 shows the inputs to the two ecosystems (three case studies) with their transformities. Lake Chozas had the same intensity of inputs before (year 1996) and after the invasion (year 1997). There were no substantial changes in only 1 year, although Lake Chozas after the invasion contained an extra input, namely the flow of exotic crayfish, estimated by the calibration of the previous structurally dynamic model at about $1.11E+03$ mg/m^3 (Marchi et al., 2011a,b). This concentration was converted into J/m^3 using energy content per gram of crayfish ($3.89E+03$ J/g wet weight; Villarreal, 1991). The initial input of *P. clarkii* into Lake Chozas was small, indicating the invasive vocation of this species. The notes necessary for the calculations of raw data were reported in Appendix A in Marchi et al. (2012).

The flows of emergy through Lake Sentiz and Lake Chozas before and after the invasion are shown in Table 12.4.

The solar emergy calculated for Lake Chozas was moderately greater than that of Lake Sentiz. The latter had a solar emergy of $6.20E+15$ sej/year, compared to $6.71E+15$ sej/year for Lake Chozas before and after the introduction of the exotic crayfish. Emergy analysis of Lake Chozas showed identical results, despite the contribution of exotic crayfish, amounting to $2.31E+12$ sej/year, after the biological invasion. The flow of solar emergy varied in Lake Sentiz and Lake Chozas before and after the invasion due to the different physical dimensions of the two systems. Lake Sentiz is smaller in area than Lake Chozas, but larger in volume. The smaller emergy flow of Lake Sentiz was due to a lower input of rainfall (chemical and geopotential energy), even if Sentiz was characterized by a higher water inflow and concentration of nutrients than Chozas. The empower of Lake Chozas was the same before and after the invasion. In fact, the extra input of crayfish did not have a very significant weight, and did not change the emergy flow sustaining internal processes of the system. In other words, the emergy input to the lake remained the same, irrespective of the presence of crayfish.[1]

The resulting solar emergy contrasted with the values obtained by eco-exergy analysis, which ranked highest for Lake Chozas before the invasion, when it was rich in submerged vegetation, followed by Lake Sentiz, which was mesotrophic, and last of all by Lake Chozas after the introduction of *P. clarkii*. Indeed, these crustaceans are distinguished by a high capacity for internal self-organization at the expense of the health of the ecosystems into which they are introduced (Gherardi and Acquistapace, 2007; Cardoso and Free, 2008; Marchettini et al., 2011).

---

[1]Note that here we did not consider the emergy of the initial input of genetic information embodied in each individual, and this could change the results significantly. The solar transformity had been estimated as up to $E+13$ sej/J (Odum, 1988, 1996); that of DNA in species was $1.20E+12$ sej/J, and that of generating a new species was $8.00E+15$ sej/J (Odum, 1996 as quoted by Ulgiati and Brown, 2009).

Table 12.3 Flows of Natural Inputs into Lake Sentiz and into Lake Chozas Before and After Introduction of the Invasive Crayfish, with Their Respective Transformities

No.[a]	Inputs	Units	Sentiz Lake (unit/year)	Chozas Lake Before the Invasion (unit/year)	Chozas Lake After the Invasion (unit/year)	Solar Transformity (sej/unit)	References Solar Transformity
1	Solar energy	J	8.57E+11	1.02E+12	1.02E+12	1.00E+00	By definition
2	Rain						
2a	Rain: chemical	J	1.33E+11	1.58E+11	1.58E+11	3.10E+04	Odum et al. (2000) (Folio#1)
2b	Rain: geopotential	J	1.91E+11	2.27E+11	2.27E+11	1.76E+04	Campbell (2003)
3	Geothermal heat	J	8.53E+10	1.02E+11	1.02E+11	1.20E+04	Odum et al. (2000) (Folio#1)
4	Inflow water	g	9.65E+07	5.57E+07	5.57E+07	1.45E+05	Odum et al. (2000) (Folio#1)
5	Phosphorus	g	3.34E+03	6.86E+02	6.86E+02	2.20E+10	Brandt-Williams (2002) (Folio#4)
6	Nitrogen	g	4.00E+04	2.31E+04	2.31E+04	2.41E+10	Brandt-Williams (2002) (Folio#4)
7	Crayfish	J	0.00E+00	0.00E+00	7.59E+07	3.04E+04	Brown et al. (2006)

[a]See Marchi et al. (2012).

Table 12.4 Solar Emergy in the Three Scenarios of the Two Spanish Lakes

No. of inputs	Lake Sentiz (sej/year)	Lake Chozas Before the Invasion (sej/year)	Lake Chozas After the Invasion (sej/year)
Solar energy	8.57E+11	1.02E+12	1.02E+12
Rain			
Rain: chemical	4.12E+15	4.91E+15	4.91E+15
Rain: geopotential	3.36E+15	4.00E+15	4.00E+15
Geothermal heat	1.02E+15	1.22E+15	1.22E+15
Inflow water	1.40E+13	8.08E+12	8.08E+12
Phosphorus	7.36E+13	1.51E+13	1.51E+13
Nitrogen	9.65E+14	5.57E+14	5.57E+14
Crayfish	0.00E+00	0.00E+00	2.31E+12
Total	6.20E+15	6.71E+15	6.71E+15

Figure 12.8 shows variations in emergy density, eco-exergy density and the eco-exergy–empower ratio for the three case studies. The estimated emergy density was 2.04E+12 sej/(dm^3 year) for Lake Sentiz, and 3.82E+12 sej/(dm^3 year) for Lake Chozas before and after the invasion. Eco-exergy density, on the other hand, was highest in Lake Chozas before the invasion (2.72E+07 kJ/dm^3), followed by Lake Sentiz (1.08E+06 kJ/dm^3). Lake Chozas after the invasion had an eco-exergy density of 2.98E+05 kJ/dm^3, the lowest of the three. As expected, the eco-exergy–empower ratio showed a similar trend to eco-exergy density. Indeed, we obtained a value of 7.12E−06 (kJ year)/sej for Lake Chozas in oligotrophic conditions, against 5.31E−07 and 7.78E−08 (kJ year)/sej for Sentiz Lake and Chozas after 1997, respectively.

The eco-exergy–empower ratio of Lake Chozas was obviously higher before the invasion, indicating much higher internal organization than the other two case studies, and therefore superior ecosystem health status. This showed that eutrophication leads to an increase in elements of lower rank, such as detritus and phytoplankton (Bastianoni et al., 2010), determining a decrease in efficiency to transform direct and indirect solar energy inputs into organization.

Specific eco-exergy was equal to 2.11E+01 kJ/g for Lake Sentiz, 1.64E+03 and 1.48E+02 kJ/g for Lake Chozas before and after the invasion, respectively (Figure 12.9).

Lake Sentiz showed the lower specific eco-exergy value than the other lake analyzed, despite that Sentiz was characterized by the biggest biomass concentration. In fact, Lake Chozas before the biological invasion was more complex than Lake Chozas after the invasion (depleted in biodiversity) and Lake Sentiz, which was characterized by a biomass mainly composed of elements at a low level of the evolutionary scale (such as phytoplankton and detritus).

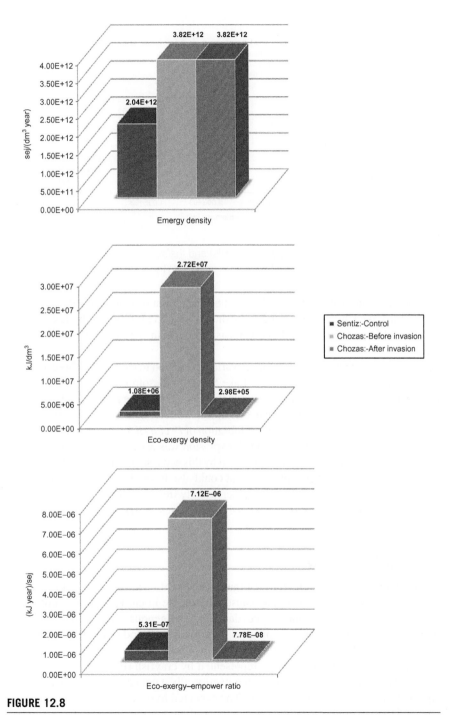

FIGURE 12.8

Variations in emergy density, eco-exergy density, and eco-exergy–empower ratio for the three scenarios.

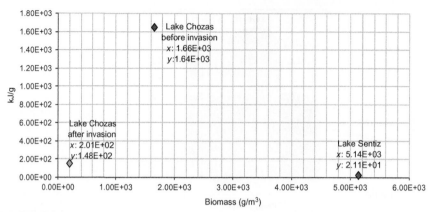

FIGURE 12.9

Comparison between the biomass values (*x*-axis) and the specific eco-exergy value (*y*-axis). (For the color version of this figure, the reader is referred to the online version of this chapter.)

The economic value quantified from the energy cost of work capacity was much higher for Lake Chozas before the invasion (6.68E+05 $), followed by Lake Sentiz (4.57E+04 $) and finally by Lake Chozas after the invasion (6.54E+03 $), as in Table 12.5.

The economic value of the two lakes varied in relation to case study. The economic value of work capacity was lower for Lake Sentiz in mesotrophic state than for Lake Chozas before the invasion. This suggests that in our study the phenomenon of eutrophication led to economic loss related to lower welfare. The monetary value of the work capacity of Lake Chozas after proliferation of *P. clarkii*, however, was the lowest of the three, suggesting that the crayfish explosion was associated with an enormous loss of potential goods and services provided by this ecosystem. This was reflected by a drop in gross local product, a problem that could also have global effects, because invasions of exotic species are becoming increasingly frequent all over the world due to globalization and the rise in mean annual temperatures (Mack et al., 2000).

12.4 CONCLUSION

Despite their greater capacity to self-organize than native non-invasive crustaceans, Louisiana red swamp crayfish may cause an increase in entropy of the whole ecosystem into which they are introduced, damaging indigenous communities. The case study of Lake Chozas after the biological invasion was a good example in support of scenario (a) in Marchi et al. (2010).

The removal of the phosphorus represented by crayfish (by harvesting) implied complete recovery of the lake and its eco-exergy, albeit not necessarily with the same organisms having the same properties. The expected hysteresis created by introduction and harvesting of crayfish was observed under the following conditions:

Table 12.5 Value of Annual Ecosystem Services from the Three Scenarios, Determined from Lake Ecosystem Capacity to Do Work Using Costs of Energy

Taxonomic Group	Lake Sentiz ($[a])	Lake Chozas Before the Invasion ($[a])	Lake Chozas After the Invasion ($[a])
Phytoplankton	1.52E+02	3.65E+01	4.95E+01
Zooplankton	8.21E+02	6.59E+01	8.01E+01
Submerged vegetation	1.46E+02	6.61E+05	6.08E+01
Benthic macroinvertebrates			
Oligochaeta	1.14E+02	6.57E+01	3.25E+02
Chironomidae	1.28E+03	4.05E+03	1.87E+03
Others[b]	3.89E+01	2.25E+01	0.00E+00
Epiphytic macroinvertebrates			
Hydra	4.89E−03	1.18E−03	0.00E+00
Nematoda	3.53E−05	8.48E−06	0.00E+00
Oligochaeta	2.57E−02	6.18E−03	3.50E−04
Gastropoda	8.22E−01	1.98E−01	0.00E+00
Acari	2.34E−01	5.63E−02	0.00E+00
Ostracoda	1.88E−03	4.53E−04	3.78E−05
Ephemeroptera	1.71E+00	1.94E−03	5.55E−03
Odonata	6.05E−01	1.46E−01	3.35E−02
Heteroptera	6.05E−01	1.46E−01	3.84E−05
Lepidoptera	5.68E−01	1.37E−01	0.00E+00
Trichoptera	8.07E−03	1.94E−03	0.00E+00
Diptera	7.59E−01	1.83E−01	5.58E−03
Fish	4.42E+02	9.47E+02	9.47E+02
Birds			
Ducks	8.31E+02	1.33E+03	8.65E+02
Coots	1.44E+03	1.39E+03	1.04E+03
Detritus			
Detritus in the water column	4.04E+04	4.72E+00	1.22E+02
Detritus in the sediment	1.55E−02	1.23E−01	1.10E−01
Crayfish	0.00E+00	0.00E+00	1.18E+03
Total	4.57E+04	6.68E+05	6.54E+03

[a]$ (2009 value).
[b]Others: Platyhelminthes (flatworms).

phytoplankton dominance at total phosphorus about $\geq 200\text{--}250$ mg TP/m^3 and submerged vegetation returned at total phosphorus <100 mg TP/m^3.

Eco-exergy quantifies the sustainability of ecosystems (see Jørgensen, 2006, 2010b; Jørgensen et al., 2010a,b) as it expresses the work capacity required to maintain all internal processes. Eco-exergy is a very good indicator of ecosystem health as it faithfully represents the health of Lake Sentiz, compared to Lake Chozas, including conditions before and after the biological invasion. The results obtained by the application of eco-exergy as ecosystem indicator in this case were consistent with general lake management viewpoints.

Emergy gives the environmental cost of the inputs supporting the ecosystem, expressed in solar energy equivalent. It shows the (pre)conditions that enable the system to survive, but it does not assess the health status or integrity of the ecosystem.

The eco-exergy–empower ratio is a particularly good indicator for expressing ecosystem changes because it quantifies the efficiency in gaining eco-exergy from solar energy equivalent. It reflected the effects of eutrophication and perturbation caused by alien species in the two freshwater systems better than the other two indicators.

Specific eco-exergy values confirmed the importance played by the macrophytes, which have stored a higher degree of internal information than phytoplankton and detritus. The presence of the exotic crayfish created an ecosystem characterized by greater internal vulnerability, despite that *P. clarkii* is more evolved in internal complexity than the submerged vegetation.

Accounting of ecosystem services on an eco-exergy basis provides good indications of monetary gains or losses possible in perturbed systems, including eutrophic or invaded ecosystems.

Thermodynamic holistic indicators have been used before, and it is common and useful to apply them for environmental management, pollution treatment, and, as we have seen, biological invasion evaluation.

References

Aoki, I., 2006. Min–max principle of entropy production with time in aquatic communities. Ecol. Complex. 3 (1), 56–63.

Bastianoni, S., Marchettini, N., 1997. Emergy/exergy ratio as a measure of the level of organization of systems. Ecol. Model. 99 (1), 33–40.

Bastianoni, S., Pulselli, F.M., Rustici, M., 2006. Exergy versus emergy flow in ecosystems: is there an order in maximizations? Ecol. Indic. 6, 58–62.

Bastianoni, S., Marchi, M., Tiezzi, E., 2010. Invaders decrease the structural complexity of the indigenous community of Chozas Lake. In: Brebbia, C., Carpi, A. (Eds.), Design & Nature V. WIT Press Publications, Southampton and Boston, pp. 517–526.

Branch, G.M., Steffani, C.N., 2004. Can we predict the effects of alien species? A case-history of the invasion of South Africa by *Mytilus galloprovincialis* (Lamarck). J. Exp. Mar. Biol. Ecol. 300, 189–215.

Brandt-Williams, S.L., 2002. Handbook of Emergy Evaluation: A Compendium of Data for Emergy Computation Issued in a Series of Folios. Folio#4.

Brown, M.T., Cohen, M.J., Bardi, E., Ingwersen, W.W., 2006. Species diversity in the Florida Evergaldes, USA: a system approach to calculating biodiversity. Aquat. Sci. 68, 254–277.

Campbell, D.E., 2003. A note on the uncertainty in estimates of transformities based on global water budgets. In: Brown, M.T., Odum, H.T., Tilley, D.R., Ulgiati, S. (Eds.), Emergy Synthesis 2. Proceedings of the Second Biennial Emergy Analysis Conference. Center for Environmental Policy, University of Florida, Gainesville, pp. 349–353.

Cardoso, A.C., Free, G., 2008. Incorporating invasive alien species into ecological assessment in the context of the Water Framework Directive. Biol. Invasions 3 (4), 361–366.

Cerda, M., Castilla, J.C., 1999. Diversidad biológica y biomasa de macro-invertebratos en matrices intermareales de *Pyura praeputialis*. Rev. Chil. Hist. Nat. 74, 841–853.

Corsi, I., Pastore, A.M., Lodde, A., Palmerini, E., Castagnolo, L., Focardi, S.E., 2007. Potential role of cholinesterases in the invasive capacity of the freshwater bivalve, *Anodonta woodiana* (Bivalvia: Unionacea): a comparative study with the indigenous species of the genus, *Anodonta* sp. Comp. Biochem. Physiol. C 145, 413–419.

Fernández-Aláez, M., Luis, E., Fernández-Aláez, C., 1984. Distribution y analisis de la vegetacion macrofitica en las lagunas de Chozas da Arriba, León. Limnetica 1, 101–110.

Fernández-Aláez, M., Fernández-Aláez, C., Rodríguez, C.F., Bécares, E., 1999. Evaluation of the state of conservation of shallow lakes in the Province of León (northwest Spain) using botanical criteria. Limnetica 17, 107–117.

Fernández-Aláez, M., Fernández-Aláez, C., Rodríguez, S., 2002. Seasonal changes in biomass of charophytes in shallow lakes in the northwest of Spian. Aquat. Bot. 72, 335–348.

Gherardi, F., Acquistapace, P., 2007. Invasive crayfish in Europe: the impact of *Procambarus clarkii* on the littoral community of a Mediterranean lake. Freshw. Biol. 52 (7), 1249–1259.

Jeppesen, E., Søndergaard, M., Jensen, J.P., Mortensen, E., Hansen, A.M., Jørgensen, T., 1998. Major perturbation in biological structure and dynamics of a shallow hypertrophic lake following a reduction in sewage loading: an 18-year study in Lake Søbygaard, Denmark. Ecosystems 1, 250–267.

Jørgensen, S.E., 1982. A holistic approach to ecological modeling by application of thermodynamics. In: Mitsch, W. et al., (Eds.), Energetics and Systems, 176, Ann Arbor Science, Ann Arbor, MI, pp. 72–82.

Jørgensen, S.E., 1992a. Development of models able to account for changes in species composition. Ecol. Model. 62, 163–170.

Jørgensen, S.E., 1992b. Parameters, ecological constraints and exergy. Ecol. Model. 62, 163–170.

Jørgensen, S.E., 1997. Integration of Ecosystem Theories. A Pattern, second ed. Kluwer Academic Publishers, Dordrecht, The Netherlands, p. 386.

Jørgensen, S.E., 1999. State-of-the-art of ecological modelling with emphasis on development of structural dynamic models. Ecol. Model. 120, 72–89.

Jørgensen, S.E., Ladegaard, N., Debeljak, M., Marques, J.C., 2005. Calculations of exergy for organisms. Ecol. Model. 185, 165–175.

Jørgensen, S.E., 2006. Eco-Exergy as Sustainability. WIT Press, Southampton, UK, p. 207.

Jørgensen, S.E., Fath, B.D., Bastianoni, S., Marques, J.C., Müller, F., Nielsen, S.N., Pattern, B.C., Tiezzi, E., Ulanowicz, R.E., 2007. A New Ecology: System Perspective. Elsevier, Amsterdam, p. 288.

Jørgensen, S.E., Nielsen, S.N., 2007. Application of exergy as thermodynamic indicator in ecology. Energy 32 (5), 673–685.

Jørgensen, S.E., 2008. Specific exergy as ecosystem health indicator. In: Jørgensen, S.E., Fath, B.D. (Eds.), Encyclopedia of Ecology. Elsevier, Oxford, pp. 3332–3333.

Jørgensen, S.E., Ludovisi, A., Nielsen, S.N., 2010a. The free energy and information embodied in the amino acid chains of organisms. Ecol. Model. 221, 2388–2392.

Jørgensen, S.E., 2010a. Ecosystem services, sustainability and thermodynamic indicators. Ecol. Complex. 7, 311–313.

Jørgensen, S.E., 2010b. Eco-exergy as ecological indicator. In: Jørgensen, S.E., Xu, F., Costanza, R. (Eds.), Handbook of Ecological Indicators for Assessment of Ecosystem Health, second ed. CRC, Baton Rouge, Florida, pp. 77–89.

Jørgensen, S.E., Xu, F., Costanza, R., 2010b. Handbook of Ecological Indicators for Assessment of Ecosystem Health, second ed. CRC, Baton Rouge, Florida.

Mack, R.N., Simberloff, D., Lonsdale, W.M., Evans, H., Clout, M., Bazzaz, F.A., 2000. Biotic invasions: causes, epidemiology, global consequences and control. Ecol. Appl. 10 (3), 689–710.

Marchettini, N., Pulselli, R.M., Rossi, F., Tiezzi, E., 2008. Entropy. In: Jørgensen, S.E., Fath, B.D. (Eds.), Encyclopedia of Ecology. Elsevier, Oxford, pp. 1297–1305.

Marchettini, N., Marchi, M., Corsi, I., Tiezzi, E., 2011. A thermodynamic approach to biological invasions. Int. J. Des. Nat. Ecodyn. 6 (1), 10–19.

Marchi, M., Corsi, I., Tiezzi, E., 2010. Biological invasions and their threat to ecosystems: two ways to thermodynamic euthanasia. Ecol. Model. 221, 882–883.

Marchi, M., Jørgensen, S.E., Bécares, E., Corsi, I., Marchettini, N., Bastianoni, S., 2011a. Dynamic model of Lake Chozas (León, NW Spain)—decrease in eco-exergy from clear to turbid phase due to introduction of exotic crayfish. Ecol. Model. 222 (16), 3002–3010.

Marchi, M., Jørgensen, S.E., Bécares, E., Corsi, I., Marchettini, N., Bastianoni, S., 2011b. Resistance and re-organization of an ecosystem in response to biological invasion: some hypotheses. Ecol. Model. 222 (16), 2992–3001.

Marchi, M., Jørgensen, S.E., Bécares, E., Fernández-Aláez, C., Rodríguez, C.F., Fernández-Aláez, M., Pulselli, F.M., Marchettini, N., Bastianoni, S., 2012. Effect of eutrophication and exotic crayfish on health status of two Spanish lakes: a joint application of ecological indicators. Ecol. Indic. 20, 92–100.

Morowitz, H., 1978. Foundation of Bioenergetics. Academic Press, New York, p. 344.

Odum, H.T., 1971a. Environment, Power and Society. John Wiley, New York, USA, p. 336.

Odum, H.T., 1971b. An energy circuit language for ecological and social systems: its physical basis. In: Patten, B. (Ed.), Systems Analysis and Simulation in Ecology, vol. 2. Academic Press, New York, USA, pp. 139–211.

Odum, H.T., 1988. Self-organization, transformity, and information. Science 242, 1132–1139.

Odum, H.T., 1996. Environmental Accounting: Emergy and Environmental Decision Making. John Wiley & Sons, Inc., New York, USA, p. 370.

Odum, H.T., Brown, M.T., Brandt-Williams, S., 2000. Handbook of Emergy Evaluation: A Compendium of Data for Emergy Computation Issued in a Series of Folios. Folio#1.

Pastore, A.M., Corsi, I., Marchi, M., Castagnolo, L., Della Torre, C., Focardi, S., 2008. Sensitivity of *Procambarus clarkii* (Girard, 1852) to different organophosphorus pesticides and two heavy metals in relation to correct use of the species as a bioindicator. Abstract from Fourteenth International Symposium on Pollutant responses in Marine Organisms (PRIMO 14) – Neurotoxicity. Mar. Environ. Res. 66, 69.

Rodríguez, C.F., Bécares, E., Fernández-Aláez, M., 2003. Shift from clear to turbid phase in Lake Chozas (NW Spain) due to the introduction of American red swamp crayfish (*Procambarus clarkii*). Hydrobiologia 506 (509), 421–426.

Rodríguez, C.F., Bécares, E., Fernández-Aláez, M., Fernández-Aláez, C., 2005. Loss of diversity and degradation of wetlands as a result of introducing exotic crayfish. Biol. Invasions 7, 75–85.

Scheffer, M., Hosper, S.H., Meijer, M.L., Moss, B., Jeppesen, E., 1993. Alternative equilibria in shallow lakes. Trends Ecol. Evol. 8, 275–279.

Scheffer, M., 1997. Ecology of Shallow Lakes. Chapman & Hall, London.

Scheffer, M., Carpenter, S., Foley, J.A., Folke, C., Walker, B., 2001. Catastrophic shifts in ecosystems. Nature 413, 591–596.

Villarreal, H., 1991. A partial energy budget of the Australian Crayfish *Cherax tenuimanus*. J. World Aquacult. Sci. 22 (4), 252–258.

Tiezzi, E., 2003. The End of Time. WIT Press Publications, Southampton and Boston, p. 200.

Tiezzi, E., 2006. Steps Towards an Evolutionary Physics. WIT Press Publications, Southampton and Boston, p. 157.

Ulgiati, S., Brown, M.T., 2009. Emergy and ecosystem complexity. Commun. Nonlinear Sci. Numer. Simul. 14, 310–321.

Zhang, J., Jørgensen, S.E., Beklioglu, M., Ince, O., 2003. Hysteresis in vegetation shift—Lake Mogan prognoses. Ecol. Model. 164, 227–238.

Development of Ecological Models for the Effects of Macrophyte Restoration on the Ecosystem Health of a Large Eutrophic Chinese Lake (Lake Chaohu)

13

Fu-Liu Xu[a],*, Sven Erik Jørgensen[b], Xiang-Zhen Kong[a], Wei He[a], Ning Qin[a]

[a]*MOE Laboratory for Earth Surface Processes, College of Urban & Environmental Sciences, Peking University, Beijing, China*

[b]*Professor Emeritus, Environmental Chemistry, University of Copenhagen, Copenhagen, Denmark*

*Corresponding author: e-mail address: xufl@urban.pku.edu.cn

13.1 INTRODUCTION

Since the 1980s, the term *ecosystem health* has appeared with increasing frequency in the literature. Costanza (1992) best summarized the key elements behind the conceptual definition of ecosystem health as follows: (1) homeostasis, (2) absence of disease, (3) diversity or complexity, (4) stability or resilience, (5) vigor or scope for growth, and (6) balance between system components. These definitional elements cover not only the biophysical aspects of ecosystems, but the socioeconomic aspects as well. Ecosystem health focuses on the maintenance of ecological integrity, given the impacts of human activity on natural systems, and how these systems withstand change. It also considers the transformation of systems and the implications such transformations have on economic opportunity, the maintenance of social systems, and on human health in general (Rapport, 1999). Ecosystem health is as much about implementing strategies in environmental management as it is about fostering a new integrative science (Costanza, 1992).

Eutrophication is a serious "disease" of lakes and ranks as one of the most pervasive water quality problems around the world. It has badly damaged lake ecosystem health and resulted in imbalance between biological components and a decrease in biodiversity and resilience. To deal with eutrophication one can either attack the cause (the load of nutrients) or the symptoms (Rast and Hdland, 1988). Many different techniques have been suggested since the 1970s (Rast and Hdland, 1988; Clasen et al., 1989; Ryding and Rast, 1989; Cooke et al., 1993). A recent development for the control of eutrophication is the application of ecological engineering (Mitsch and Jørgensen, 1989), involving designed wetlands for water treatment and growing macrophytes in eutrophic waters (Reed et al., 1988; Brix and Schierup, 1989). The use of natural or artificial wetlands to restore and protect lakes becomes more and more common, since there is a recognized purification phenomenon in natural wetlands (Tilton and Kadlec, 1979; Nichols, 1983) and in artificial wetlands (Wolverton, 1987; Reed et al., 1988; Brix and Schierup, 1989; Hammer, 1989, 1992; Cooper and Findlater, 1990; Reed, 1990; Conley et al., 1991; Gumbricht, 1992, 1993; Mitsch, 1992). Macrophytes play a very important role in constructed wetlands for NPS control by uptaking nutrients and toxic substances and by creating a favorable environment for a variety of complex chemical, biological, and physical processes that contribute to the removal and degradation of pollutants (Gumbricht, 1993). Macrophytes growing in eutrophic lakes are also crucial in regulating lake biological structure, because macrophytes may limit algal growth by competing for nutrients with algae, by blocking light, and because they may increase herbivorous fish biomass by providing fodder and places of rest and refuge.

Lake Chaohu (Table 13.1) is one of the five largest freshwater lakes in China. Before the 1950s, it was well known for its scenic beauty and for the richness of its aquatic products. It was covered by aquatic macrophytes, and the water was clear enough to see the bottom. The water column was mesotrophic and nearly saturated with oxygen. Low concentration of phytoplankton was present, and it was dominated by diatoms. More than 190 species of zooplankton were recovered. It had rich benthic invertebrates and piscivorous fish (Wang et al., 1995). However, from

Table 13.1 Basic Limnological Characteristics of the Lake Chaohu

Parameters	Units	Data
Geographical position		117°16′542″–117°51′46″
Catchment area	km^2	12,938
Mean surface area	km^2	760
Mean volume	m^3	1,900,000,000
Mean depth	m	3.06
Maximum depth	m	6.78
Mean annual inflow	m^3	4,120,000,000
Mean annual outflow	m^3	3,490,000,000
Mean retention time	day	136
Mean phosphorus load	t/year	777.8
Mean nitrogen load	t/year	22,556
Annual average T-P concentration	mg/l	0.204
Annual average T-N concentration	mg/l	2.3
Annual average Chl-a concentration	mg/m^3	14.979
Annual average primary production	gC/m^2year	123

the early 1950s to the early 1980s, seasonal riparian wetlands decreased from 150 to 16 km^2, due to farming. In the late 1960s, grass carp were introduced to the lake. The coverage of macrophytes greatly dropped from 30% to 2.5% of the area, only less than 10% of the lake's primary productivity is from macrophytes, owing to the loss of riparian wetlands and the grazing pressure from grass carp. The herbivorous fish declined significantly from 38.4% to 3.5%, and carnivorous fish increased from 32.6% to 83%. The total catch of fish reduced from 3395 to 3000 tone. In total catch, carnivorous fish like *Coilia ectenes* and *Neosalanx* spp. became absolutely dominant, while herbivorous fish like *Hypophthalmichthys molitrix* and omnivorous fish like *Cyprinus carpio*, *Carassius*, and *Xenocypris* spp. dropped to 10.3%. The proportion of large carnivorous fish like *Erythroculter ilishaeformis* fell to 6.7% (Wang et al., 1995; Xu, 1997). The serious eutrophication has caused negative ecological, health, social, and economic effects on the lake and its utilization.

The significant increase of algal biomass, the great decrease of macrophytes, and the change in fish population caused disharmony of the lake's biological structure. This further aggravated the lake eutrophication (Xu et al., 1999c). Since the later 1980s, in order to control the lake eutrophication, except for point pollution treatment, ecological engineering was applied to Lake Chaohu. The expectations are to decrease nutrient input, to increase nutrient output, and to regulate the lake's biological structure (more macrophytes, zooplankton, and herbivorous fish, less algae and plantivorous fish) by restoring riparian wetlands and macrophytes. The experiments were carried out in 1987 and 1988 for restoring seasonal riparian wetlands and replanting macrophytes (weed and water lettuce). The results demonstrated that the water quality inside the water lettuce community and the reed community were better than that outside. Compared with the water far away from the communities (open waters), N concentration within the reed community decreased from 5.39 to 5.19 mg/l, and P concentration from 0.44 to 0.38 mg/l. N concentration within the water lettuce community decreased from 5.4 to 4.8 mg/l and P concentration from 0.44 to 0.36 mg/l. Transparency in Sechi Disk depth (SD) inside the water lettuce community and the reed community were significantly increased to 70 and 61 cm from 38 cm in open waters. Chl-*a* concentration inside water the lettuce community was sharply dropped to 0.04 mg/l from 0.57 mg/l in open waters (Tu et al., 1990; Xu et al., 1999c).

However, there are some uncertainties in eutrophication control by ecological engineering, for example, the choice of macrophyte species and the effects when ecological engineering is applied to whole lakes. Experiment and monitoring of all changes in lake biota and ecosystem is a time-consuming and very laborious process. Ecological modeling may help to overcome this problem to a certain extent. Modeling of ecological processes can be considered as a low-cost solution compared to other experimental methods. This chapter presents the effects of macrophytes on the Lake Chaohu ecosystem health determined by modeling the dynamics of some ecological indicators in the lake ecosystems (Xu et al., 1999b).

13.2 METHODS

13.2.1 Procedures for ecosystem health assessment based on ecological model

Several procedures for ecosystem health assessment have been reported and discussed in the literature. Schaeffer et al. (1988) suggested seven guidelines. Haskell et al. (1992), on the other hand, proposed the following sequence: (1) identifying symptoms, (2) identifying and measuring vital signs, (3) making a provisional diagnosis, (3) conducting tests to verify the diagnosis, (5) making a prognosis, and (6) prescribing a treatment. Jørgensen (1995a) proposed a tentative procedure for a practical assessment of ecosystem health based on ecological indicators (exergy, structural exergy, and buffering capacity) as follows: (1) establishing questions relevant to the health of an ecosystem; (2) assessing the most important mass flows and balances related to these questions; (3) creating a conceptual diagram of the ecosystem, including those components important to the mass flows defined under (2) above; (4) developing a dynamic model (if the data are not sufficient, a steady-state model should be applied) using conventional procedures (Jørgensen, 1994); (5) using the model to calculate exergy, structural exergy, and the relevant buffering capacities (if the model is dynamic it will be possible to identify the seasonal changes in exergy, structural exergy, and buffer capacities); and (6) assessing the ecosystem's health whereby high exergy, structural exergy, and buffering capacity imply a healthy ecosystem.

In present study, the procedures for ecosystem health assessment based on ecological model are illustrated in Figure 13.1. The following are the necessary steps: (1) to analyze the ecosystem structure of a lake for determination of the structure and complexity of lake ecological model; (2) to develop a steady or a dynamic model with ecological health indicators, by designing the conceptual diagram, setting up model equations, estimating model parameters, and integrating with ecological indicators; (3) to compare the simulated and observed values of important state variables and process rates (i.e., model calibration) so as to evaluate the applicability of the model to ecosystem health assessment for the lake; (4) to calculate ecosystem health indicators using the developed model; and (5) to assess lake ecosystem health according to the values of ecosystem health indicators (Xu et al., 2001).

13.2.2 Development of ecological model for ecosystem health assessment

Two ecological models were developed in this study to predict the effects of macrophytes on the Lake Chaohu ecosystem health. Models 1 and 2 describe, respectively, the nutrient-food web dynamics of the lake ecosystems with and without macrophytes. The methods for the prediction of the effects of macrophytes on water quality and ecosystem health of the Lake Chaohu were shown in Figure 13.2.

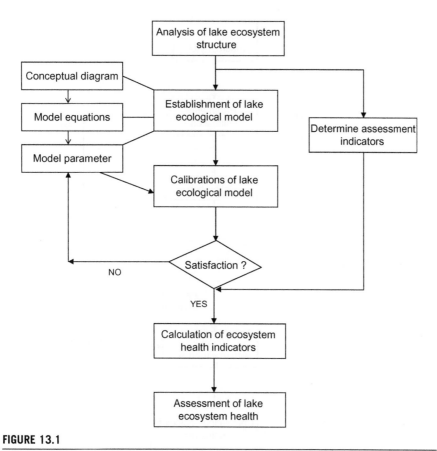

FIGURE 13.1

The procedures for ecosystem health assessment based on the ecological model.

13.2.2.1 *Model structure*

Model 1 includes the following submodels: nutrient, phytoplankton, zooplankton, fish, detritus, and sediments submodel. Model 2 has seven submodels including the submodels in model 1 plus the macrophytes submodel. The conceptual diagram for models 1 and 2 are illustrated in Figures 13.3 and 13.4, respectively. The state variables and forcing functions for model 2 are shown in Table 13.2.

13.2.2.2 *Model equations*

The equations for each submodel in model 2 are given in Tables 13.3–13.9. The details about the phytoplankton submodel are described in the next paragraph.

Equations of the phytoplankton submodel are shown in Table 13.3. State variables are phytoplankton biomass (BA) (g/m^3) and the internal phosphorus concentration (g/m^3). Phytoplankton dynamics includes the following processes: growth,

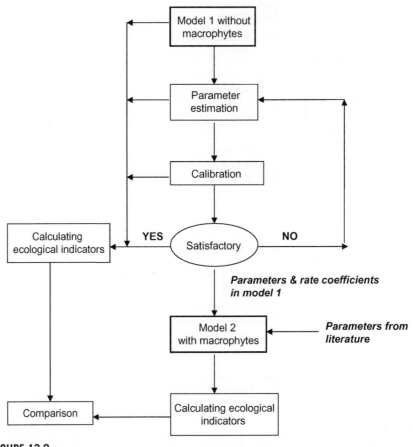

FIGURE 13.2

The methods for modeling the effects of macrophyte restoration on ecosystem health of the Lake Chaohu.

respiration, settling, grazing loss by zooplankton, nonpredatory mortality, and hydraulic washout. A differential equation that includes these processes can be expressed as Table 13.3, Equation (13.1).

Phytoplankton growth rate (GA) (Table 13.3, Equation 13.2) is obtained by combining in a multiplicative way the maximum growth rate and the function for temperature limitation, light limitation, and nutrient limitation (Bowie et al., 1985).

Numerous temperature adjustment functions have been used to model algal growth. Most of them fall into one of three major categories: (i) linear increases in growth rate with temperature, (ii) exponential increase in growth rate with temperature, and (iii) temperature optimum curves in which the growth rate increases with temperature up to the optimum temperature and then decreases with higher

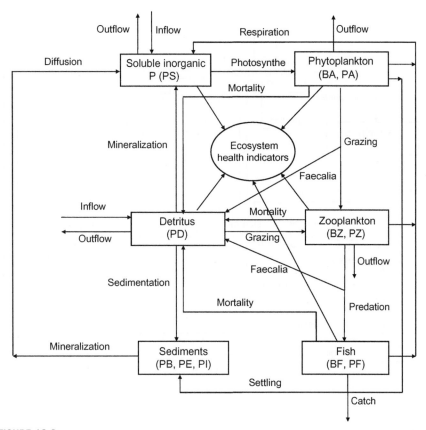

FIGURE 13.3

Conceptual diagram for model 1 without macrophyte submodel and relevant process.

temperatures (Bowie et al., 1985). There are different expressions for each category (Bowie et al., 1985; Jørgensen, 1992c). This chapter uses the temperature optimum curves equation suggested by Jørgensen (1976) to demonstrate the temperature effects on maximum growth rate (Table 13.3, Equation 13.3). T_{opt} is considered to be the actual temperature in the summer because the phytoplankton has maximum growth in the summer.

The light limitation (Table 13.3, Equation 13.4) is given as a saturation expression, analytically integrated over depth (Bowie, et al., 1985). The light extinction coefficient (λ) is defined as the linear sum of several extinction coefficients representing each component of light absorption, including the phytoplankton coefficient, the macrophyte coefficient, and the base light extinction coefficient for water without particulate or dissolved organic matter. This expression includes the shading effects of macrophytes on phytoplankton growth.

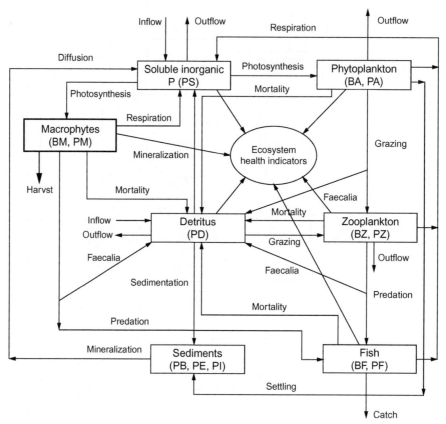

FIGURE 13.4

Conceptual diagram for model 2 with macrophyte submodel and relevant process.

Nutrient limitation for algal growth is treated as a two-step process, and only phosphorus is considered as a limiting nutrient. From Jørgensen (1976), the phosphorus limitation for phytoplankton growth is a function of the internal phosphorus levels (PA) and minimum phosphorus cell quota (PA_{min}) (Table 13.3, Equation 13.5). Internal phosphorus content is simulated as a state variable, and it depends on the phosphorus uptake rate and phytoplankton net growth rate (Table 13.3, Equation 13.6). The uptake rate is a function of both the internal and external phosphorus contents (Table 13.3, Equation 13.7).

The loss processes include respiration, settling, grazing by zooplankton, nonpredatory mortality, and hydraulic washout. Nonpredatory mortality accounts for all phytoplankton algal losses except for grazing, settling, and respiration. It includes processes such as senescence, bacterial decomposition of cells, and stress-induced mortality due to severe nutrient deficiencies, extreme environmental conditions,

Table 13.2 The State Variables and Forcing Functions for Model 2	
Variables	
BA	Phytoplankton biomass (g/m^3)
BM	Macrophyte biomass (g/m^3)
BZ	Zooplankton biomass (g/m^3)
BF	Fish biomass (g/m^3)
PA	Phosphorus in phytoplankton (g/m^3)
PM	Phosphorus in macrophytes (g/m^3)
FPZ	Proportion of phosphorus in zooplankton (kg P/kg ZA)
FPF	Proportion of phosphorus in fishes (kg P/kg BF)
PD	Phosphorus in detritus (g/m^3)
PB	Phosphorus in biologically active layer in sediment (g/m^3)
PE	Exchangeable phosphorus in sediment (g/m^3)
PI	Phosphorus in interstitial water (g/m^3)
PS	Soluble phosphorus in lake water (g/m^3)
Forcing functions (all given as time Table 14.13)	
QTRI	Inflow from tributaries (m^3/d)
PSTRI	Soluble inorganic P concentration in inflow (mg/l)
PDTRI	Detritus P concentration in inflow (mg/l)
QPREC	Precipitation to the lake (m^3/d)
PSPRES	Soluble inorganic P concentration in precipitation (mg/l)
Q	Outflow (m^3/d)
V	Lake volume (m^3)
D	Lake depth (m)
T	Temperature of lake water (°C)
I_0	Light irradiance on the surface of lake water ($kcal/m^2$ d)

or toxic substances (Bowie et al., 1985). It is specified as a maximum mortality rate multiplied with a temperature adjustment function (Table 13.3, Equation 13.8).

Respiration includes all metabolic losses and excretory process, and it is expressed as a function of maximum respiration rate and temperature (Table 13.3, Equation 13.9).

The phytoplankton settling rate depends on the density, size, shape, and physiological state of the phytoplankton cells, the viscosity and density of the water, and the turbulence and velocities of the flow field (Bowie et al., 1985). It is generally expressed as a function of settling velocity (SVS) (m/day), water depth (D) (m), and temperature adjustment function for the settling velocity (Table 13.3, Equation 13.10).

The grazing loss depends on the zooplankton grazing rate (GZ) and assimilation efficiency (Y_0). The grazing rate is defined as a function of the zooplankton density, which in turn varies dynamically with the food supply (algal concentration)

Table 13.3 Phytoplankton Submodel Equations

$$\frac{d}{dt}BA = \left(GA - MA - RA - SA - \frac{GZ}{Y_0} - \frac{Q}{V}\right)BA \tag{13.1}$$

$$GA = GAmax \cdot F(TA) \cdot F(PA) \cdot F(L) \tag{13.2}$$

$$F(TA) = \exp\left(-2.3\left|\frac{T - TAopt}{TAopt - TAmin}\right|\right) \tag{13.3}$$

$$F(L) = \frac{1}{\lambda D} \ln\left[\frac{I_0 + KI}{KI + I_0 \cdot e^{(-\lambda D)}}\right] \tag{13.4}$$

where $\lambda = \alpha + \beta BA + \gamma BM$

$$F(PA) = 1 - \frac{PAmin}{PA} \tag{13.5}$$

$$\frac{d}{dt}PA = AUP \cdot BA - \left(MA + RA + SA + \frac{GZ}{Y_0} + \frac{Q}{V}\right)PA \tag{13.6}$$

$$AUP = AUPmax \cdot F(Pint) \cdot F(Pext) \tag{13.7}$$

where $F(Pint) = \dfrac{FPAmax - FPA}{FPAmax - FPAmin}$

$$F(Pext) = \frac{PSA}{PSA + KPA}$$

$$FPA = \frac{PA}{BA}$$

$$MA = MAmax \cdot F(TA) \tag{13.8}$$

$$RA = RAmax \cdot F(TA) \tag{13.9}$$

$$SA = \frac{SVS}{D}\sqrt{\theta^{(T-20)}} \tag{13.10}$$

$$GZ = MYZ\frac{BZ}{BA} \tag{13.11}$$

where BZ, MYZ see Table 13.5, Equations (13.23) and (13.24)

Table 13.4 Macrophyte Submodel Equations

$$\frac{\mathrm{d}}{\mathrm{d}t}\mathrm{BM} = (\mathrm{GM} - \mathrm{MM} - \mathrm{RM} - \mathrm{HRVST})\mathrm{BM} - \frac{\mathrm{PRED2}}{Y_1}\mathrm{BF} \qquad (13.12)$$

$$\mathrm{GM} = \mathrm{GMmax} \cdot F(\mathrm{TM}) \cdot F(\mathrm{PM}) \cdot F(\mathrm{LM}) \qquad (13.13)$$

$$F(\mathrm{TM}) = \exp\left(-2.3\left|\frac{T - \mathrm{TMopt}}{\mathrm{TMopt} - \mathrm{TMmin}}\right|\right) \qquad (13.14)$$

$$F(\mathrm{LM}) = \frac{1}{\lambda D}\ln\left[\frac{I_0 + \mathrm{KI}}{\mathrm{KI} + I_0 \cdot \mathrm{e}^{(-\lambda D)}}\right] \qquad (13.15)$$
where $\lambda = \alpha + \beta\mathrm{BA} + \gamma\mathrm{BM}$

$$F(\mathrm{PM}) = 1 - \frac{\mathrm{PMmin}}{\mathrm{PM}} \qquad (13.16)$$

$$\left(\frac{\mathrm{d}}{\mathrm{d}t}\mathrm{PM} = \mathrm{UMP} \cdot \mathrm{BM} - \left[(\mathrm{MM} + \mathrm{RM} + \mathrm{HRVSY}) \cdot \mathrm{PM} + \frac{\mathrm{PRED2}}{Y_1} \cdot \mathrm{BF} \cdot \mathrm{FPM}\right]\right) \qquad (13.17)$$
where $\mathrm{FPM} = \dfrac{\mathrm{PM}}{\mathrm{BM}}$

$$\mathrm{MUP} = \mathrm{MUPmax} \cdot F(\mathrm{Pint}) \cdot F(\mathrm{Pext}) \qquad (13.18)$$
where $F(\mathrm{Pint}) = \dfrac{\mathrm{FPMmax} - \mathrm{FPM}}{\mathrm{FPMmax} - \mathrm{FPMmin}}$

$$F(\mathrm{Pext}) = \frac{\mathrm{PSM}}{\mathrm{PSM} + \mathrm{KPM}}$$

$$\mathrm{MM} = \mathrm{MMmax} \cdot F(\mathrm{TM}) \qquad (13.19)$$

$$\mathrm{RM} = \mathrm{RMmax} \cdot F(\mathrm{TM}) \qquad (13.20)$$

$$\mathrm{HRVST} = \mathrm{MAX}(\mathrm{BM} - H0, 0) \qquad (13.21)$$

$$\mathrm{PRED2} = \mathrm{PRED2max} \cdot F(\mathrm{TM}) \cdot F(\mathrm{BM}) \qquad (13.22)$$
where $F(\mathrm{BM}) = \mathrm{MAX}\left(0, \dfrac{\mathrm{BM} - \mathrm{KSM}}{\mathrm{BM} + \mathrm{KM}}\right)$

Table 13.5 Zooplankton Submodel Equations

$$\frac{d}{dt}BZ = \left(MYZ - RZ - MZ - \frac{Q}{V} \right)BZ - \frac{PRED1}{Y_1}BF \tag{13.23}$$

$$MYZ = MYZmax \cdot F(TZ) \cdot F(BA) \cdot FZCC \tag{13.24}$$

$$F(TZ) = \exp\left(-2.3 \left| \frac{T - TZopt}{TZopt - TZmin} \right| \right) \tag{13.25}$$

$$F(BA) = MAX\left(0, \frac{BA - KSA}{BA + KA} \frac{BA}{BA + DETR \cdot GZperf} + \frac{DETR - KSDT}{DETR + KDT} \frac{DETR \cdot GZperf}{BA + DETR \cdot GZperf} \right) \tag{13.26}$$

where GZperf = 1 if BA ≥ 1; GZperf = 10 if BA < 1;

$$FZCC = 1 - \frac{BZ}{KZ00} \tag{13.27}$$

$$RZ = RZmax \cdot F(TZ) \tag{13.28}$$

$$MZ = MZbas + MZtox$$

$$MZbas = MZmax \cdot F(TZ)$$

$$MZtox = TOXZ \cdot \exp(Ktoxz \cdot (T - 25))$$
$$\text{if } T \geq 25 \text{ and } BA \geq 30 \tag{13.29}$$

$$MZtox = 0 \text{ if } T < 25 \text{ and } BA < 30$$

$$PRED1 = PRED1max \cdot F(TZ) \cdot F(BZ) \tag{13.30}$$

$$\frac{d}{dt}FPZ = MYZ \cdot (FPA - FPZ) \tag{13.31}$$

where $F(BZ) = MAX\left(0, \frac{BZ - KSZ}{BZ + KZ} \right)$

where $FPA = \frac{PA}{BA}$

Table 13.6 Fish Submodel Equations

$$\frac{d}{dt}BF = (GF - RF - MF - CATCH) \cdot BF \tag{13.32}$$

$$GF = GFmax \cdot F(TF) \cdot F(BZBM) \cdot KFCC \tag{13.33}$$

$$F(TZ) = \exp\left(-2.3 \cdot \left|\frac{T - TFopt}{TFopt - TFmin}\right|\right) \tag{13.34}$$

$$F(BZBM) = MAX\left(0, \frac{BZ - KSZ}{BZ + KZ} \frac{BZ}{BZ + BM \cdot Cperf} + \frac{BM - KSM}{BM + KM} \frac{BM \cdot Cperf}{BZ + BM \cdot Cperf}\right) \tag{13.35}$$

where Cperf = 1 if BZ ≥ 1; Cperf = 10 if BZ < 1;

$$KFCC = 1 - \frac{BF}{KFC0} \tag{13.36}$$

$$RF = RFmax \cdot F(TF) \tag{13.37}$$

$$MF = MFbas + MFtox$$

$$MFbas = MFmax \cdot F(TF)$$

$$MFtox = TOX \cdot \exp(Ktox \cdot (T - 25))$$

if T ≥ 25 and BA ≥ 30

$$MZtox = 0 \text{ if } T < 25 \text{ and } BA < 30 \tag{13.38}$$

$$\frac{d}{dt}FPF = \left(\frac{PRED1}{Y1} \cdot FPZ + \frac{PRED2}{Y2} \cdot FPM\right) - \left(\frac{PRED1}{Y1} + \frac{PRED2}{Y2}\right) \cdot FPF \tag{13.39}$$

where $FPM = \frac{PM}{BM}$

Table 13.7 Detritus Submodel Equations

$$\frac{d}{dt}PD = \left(\frac{1}{Y_0}-1\right) \cdot GZ \cdot PA - \left(\frac{1}{Y_1}-1\right) \cdot PRED1 \cdot PZ$$

$$+ \left(\frac{1}{Y_2}-1\right) \cdot PRED2 \cdot PM + MA \cdot PA + MM \cdot PM$$

$$+ MZ \cdot PZ + MF \cdot PF + QPDIN \tag{13.40}$$

$$- \left(KDF + SD\frac{Q}{-V}\right) \cdot PD$$

$$QPDIN = \frac{QTRI \cdot PDTRI}{V} \tag{13.41}$$

$$KDF = KDP10 \cdot FT3 \text{ where } FT3 = \phi^{(T-10)} \tag{13.42}$$

$$SD = \frac{SVD}{D} \cdot \sqrt{FTS} \text{ where } FTS = \theta^{(T-20)} \tag{13.43}$$

$$PZ = FPZ \cdot BZ \tag{13.44}$$

$$PF = FPF \cdot BF \tag{13.45}$$

(see Table 13.5, Equation 13.2). It is expressed as zooplankton growth rate (MYZ) multiplied with the ratio of zooplankton to phytoplankton (FZP) (Table 13.3, Equation 13.11), since the zooplankton is grazing proportionally to their own compartment size, but unrelated directly to the phytoplankton compartment size.

13.2.2.3 Parameters
The data from April 1987 to March 1998 for the model forcing function are presented in Table 13.10. The parameters used in the models are determined in three different ways: (1) Some parameter values are taken from the literature or other model studies; (2) some values are found from laboratory experiments; and (3) some values are determined during calibration. The symbol, description, literature range, value in the models, and sources are listed in Table 13.11.

13.2.2.4 Calibration
The calibrations are made for model 1. The comparisons between simulated values and observed values of important variables and process rates include growth rate, respiration rate, mortality rate, settling rate, biomass of phytoplankton, internal phosphorus concentration in phytoplankton cells, growth rate and biomass of zooplankton and fish.

Table 13.8 Sediment Submodel Equations

$$\frac{d}{dt}PB = \frac{QSED}{AB} - QBIO - QDSORP \qquad (13.46)$$

where $AB = \left(\frac{DB}{D}\right)DMU$

$$QSED = MIN(SA \cdot PA, 0.00506) \qquad (13.47)$$

$$QBIO = FT6 \cdot 0.00563 \cdot \frac{PB/1800}{1000 \cdot DB} \qquad (13.48)$$

where $FT6 = \exp(0.203 \cdot T)$

$$QDSORP = \frac{0.6 \cdot \log(PS) - 2.27}{1000 \cdot DB} \qquad (13.49)$$

$$\frac{d}{dt}PE = \frac{KEX \cdot (SA \cdot PA - QSED + SD \cdot PS)}{AE} - KE \cdot PE \qquad (13.50)$$

where $AE = \frac{LUL \cdot DMU}{D}$

$$KE = KE20 \cdot FTS \qquad (13.51)$$

where $FTS = \theta^{(T-20)}$

$$\frac{d}{dt}PI = \frac{AE}{AI} \cdot KE \cdot PE - \frac{QDIFF}{AI} \qquad (13.52)$$

where $AI = \frac{LUL \cdot (1 - DMU)}{D}$

$$QDIFF = \frac{FT4 \cdot (KDIF \cdot (PI - PS) - 1.7)}{1000 \cdot DB} \qquad (13.53)$$

Table 13.9 Soluble Inorganic Phosphorus Submodel Equations

$$\frac{d}{dt}PS = RA \cdot PA + RM \cdot PM + RZ \cdot PZ + RF \cdot PF + QPSIN$$
$$+ KDP \cdot PD + QDIFF + AB \cdot (QBIO + QDSORP) \qquad (13.54)$$
$$- AUP \cdot BA - MUP \cdot BM - \frac{Q}{V} \cdot PS$$

$$QPSIN = \frac{(QTRI \cdot PSTRI + QPREC \cdot PSPREC)}{V} \qquad (13.55)$$

KDP, PZ, PF see Table 13.6, Equations (13.42), (13.44), and (13.45);
QBIO, QDSORP, QDIFF see Table 13.8, Equations (13.48), (13.49), and (13.53)

Table 13.10 The Data from April 1987 to March 1998 for the Model Forcing Function[a]

	T (°C)	I_0 (Kcal/m²)	D (m)	V (10⁸ m³)	QTRI (10⁶m³/d)	PDTRI (mg/l)	PSTRI (mg/l)	Q (10⁶m³/d)	QPREC (10⁶m³/d)
April 1987	23.80	4063.7	2.27	17.20	29.17	0.028	0.013	23.15	3.50
May	24.03	3794.2	2.28	19.20	13.15	0.022	0.022	24.48	1.82
June	27.40	4200.0	2.80	21.40	56.85	0.040	0.022	10.26	8.55
July	32.25	4500.0	4.30	33.70	39.36	0.067	0.022	17.73	4.43
August	28.90	3491.6	4.30	33.40	2.40	0.026	0.022	39.08	0.34
September	24.00	3506.7	3.37	25.90	13.29	0.024	0.022	31.94	2.99
October	18.08	2074.6	3.07	23.50	8.73	0.021	0.022	26.91	1.52
November	17.90	1788.9	2.28	17.20	0.87	0.018	0.022	24.32	0.00
December	6.21	2051.6	1.99	14.80	0.82	0.019	0.022	1.66	0.47
January 1988	5.90	1480.5	1.99	14.80	7.59	0.024	0.024	0.90	2.32
February	5.20	1541.3	2.29	17.20	9.66	0.021	0.024	16.66	1.89
March	8.40	2244.6	2.11	15.70	3.96	0.021	0.024	8.13	0.82

[a]The model forcing functions include inflow from tributaries (10⁶m³/d) (QTRI), soluble inorganic P concentration in inflow (mg/l) (PSTRI), detritus P concentration in inflow (mg/l) (PDTRI), precipitation to the lake (10⁶m³/d) (QPREC), outflow (10⁶m³/d) (Q), lake volume (10⁸m³) (V), lake depth (m) (D), temperature of lake water (°C) (T), light radiation on the surface of lake water (kcal/m² d) (I_0).

Table 13.11 Model Parameters

Symbol	Description	Unit	Literature Range	Value Used	Sources
Phytoplankton submodel					
GAmax	Maximum growth rate of phytoplankton	1/day	1–5	4.042	Measurement
MAmax	Maximum mortality rate of phytoplankton	1/day		0.96	Measurement
RAmax	Maximum respiration rate of phytoplankton	1/day	0.005–0.8	0.6	Measurement
AUPmax	Maximum P uptake rate of phytoplankton	1/day	0.0014–0.01	0.003	Calculation
TAopt	Optimal temperature for phytoplankton growth	°C		28	Measurement
TAmin	Minimum temperature for phytoplankton growth	°C		5	Measurement
FPAmax	Maximum kg P per kg phytoplankton biomass	–	0.013–0.03	0.013	Jørgensen (1976)
FPAmin	Minimum kg P per kg phytoplankton biomass	–	0.001–0.005	0.001	Jørgensen (1976)
KI	Michaelis constant for light	kcal/m^2 d	173– 518	400	Jørgensen (1976)
KPA	Michaelis constant of P uptake for phytoplankton	mg/l	0.0005–0.08	0.06	Measurement
SVS	Settling velocity of phytoplankton	m/day	0.1–0.8	0.19	Jørgensen (1976)
α	Extinction coefficient of water	1/m		0.27	
β	Extinction coefficient of phytoplankton	1/m		0.18	
γ	Extinction coefficient of macrophyte	1/m		0.15	Calibration
θ	Temperature coefficient for phytoplankton settling	–		1.03	

Continued

Table 13.11 Model Parameters—cont'd

Symbol	Description	Unit	Literature Range	Value Used	Sources
Macrophyte submodel					
GMmax	Maximum growth rate of macrophyte	1/day	0.14–0.289	0.15	Calibration
MMmax	Maximum mortality rate of macrophyte	1/day		0.01	Calibration
RMmax	Maximum respiration rate of macrophyte	1/day	0.01–0.5	0.01	Calibration
PRED2max	Maximum feeding rate of fish on macrophyte	1/day		0.012	Calibration
MUPmax	Maximum P uptake rate of macrophyte	1/day	0.01–0.03	0.005	
TMopt	Optimal temperature for macrophyte growth	C		28	Measurement
TMmin	Minimum temperature for macrophyte growth	C		5	Measurement
FPMmax	Maximum internal P ratio in macrophyte	kg/kg	0.00423–0.00543	0.0045	
FPMmin	Minimum internal P ratio in macrophyte	kg/kg	0.0009–0.00107	0.001	
KM	Michaelis constant for fish predation	mg/l		0.85	
KSM	Thershold macrophyte biomass for fish predation	mg/l		0.075	Calibration
KPM	Michaelis constant of P uptake for macrophyte	mg/l		0.015	
Zooplankton submodel					
MYZmax	Maximum growth rate of zooplankton	1/day	0.1–0.8	0.35	Jørgensen (1976)
MZmax	Maximum basal mortality rate of zooplankton	1/day	0.001–0.125	0.125	Jørgensen (1976)
TOXZ	Toxic mortality rate	1/day		0.075	Calibration
Ktoxz	Toxic mortality adjustment coefficient	–		0.5	Calibration
RZmax	Maximum respiration rate of zooplankton	1/day	0.001–0.36	0.02	Jørgensen (1976)
PRED1max	Maximum feeding rate of fish on zooplankton	1/day	0.012–0.06	0.04	Calibration

		Unit	Range	Value	Source
TZopt	Optimal temperature for zooplankton growth	°C		28	Measurement
TZmin	Minimum temperature for zooplankton growth	°C		5	Measurement
KZ	Michaelis constant for fish predation	mg/l		0.75	Jørgensen et al. (1991)
KSZ	Threshold zooplankton biomass for fish predation	mg/l		0.75	Jørgensen (1976)
KA	Michaelis constant for zooplankton grazing	mg/l	0.01–2	0.5	
KSA	Threshold phytoplankton biomass for zooplankton	mg/l	0.01–0.2	0.2	
KZCC	Zooplankton carrying capacity	mg/l		30	Calculation
KSDT	Threshold detritus biomass for zooplankton	mg/l		0.005	Calibration
KDT	Michaelis constant for grazing on detritus	mg/l		0.05	Calculation
Y_0	Assimilation efficiency for zooplankton grazing	–	0.5–0.8	0.63	Jørgensen (1976)

Fish submodel

		Unit	Range	Value	Source
GFmax	Maximum growth rate of fish	1/day		0.015	Measurement
MFmax	Maximum basal mortality rate of fish	1/day		0.003	Jørgensen et al. (1991)
Ktoxf	Toxic mortality rate	1/day		0.05	Calibration
TOXF	Toxic mortality adjustment coefficient	–		0.015	Calibration
RFmax	Maximum respiration rate of fish	1/day	0.00055–0.0055	0.002	Calculation
TFopt	Optimal temperature for fish growth	°C		22	Measurement
TFmin	Minimum temperature for fish growth	°C		5	Measurement
CATCH	Catch rate of fish	1/day		0.001	Calibration
KFCC	Fish carrying capacity	mg/l		40	Calculation
Y1	Assimilation efficiency for fish predation	–		0.5	Calibration

Continued

Table 13.11 Model Parameters—cont'd

Symbol	Description	Unit	Literature Range	Value Used	Sources
Detritus, sediment, and soluble inorganic phosphorus submodel					
DB	Depth of biologically active layer in sediment	m		0.005	Measurement
LUL	Depth of unstable layer in sediments	m		0.16	Measurement
DMU	Dry matter weight of upper layer in sediment	kg/kg		0.3	Measurement
KE20	Mineralization rate of PE at 20 °C	1/day		0.0673	Measurement
KDIFF	Diffusion coefficient of P in interstitial water	–		1.21	Jørgensen (1976)
KEX	Ratio of exchangeable P to total P in sediments	–		0.18	Measurement
SVD	Settling velocity of detritus	m/day		0.002	Jørgensen (1976)
KDP10	Decomposition rate of detritus P at 10°C	1/day		0.1	Calculation
ϕ	Temperature coefficient for detritus degradation			1.072	Jørgensen (1976)
θ	Temperature coefficient for PE decomposition			1.03	

13.2.3 Ecosystem health indicators used in the ecological models

In order to assess ecosystem health quantitatively, various indicators covering different aspects of ecosystem health have been suggested. These indicators range from single species indicators (e.g., Kerr and Dickey, 1984) to composites of species (e.g., Karr et al., 1986; Karr, 1991) to measures of biodiversity, to system level measures of ecosystem structure, function, and organization (e.g., Hannon, 1985; Ulanowicz, 1986; Schindler, 1990; Costanza, 1992; Jørgensen, 1995a,b; Xu et al., 1999a) to very broad measures that go beyond the biophysical realm to include a number of human socioeconomic factors (e.g., Rapport et al., 1985; Rapport, 1992). Ecosystem health assessment requires an analysis of the linkages between the human pressures on ecosystems and landscapes, alterations in ecosystem structure and function, alterations in ecosystem services and service levels, and societal responses to changes in any one or more of these linkages (Daily, 1997; Rapport et al., 1998a,b; Rapport, 1999). Indeed, effective diagnosis requires an exploration for and identification of the most critical of these links. Evaluating ecosystem health in relation to ecological, economic, and human realms requires an integration of human values with biophysical processes, an integration that, to date, has been explicitly avoided by conventional science (Cairns and Pratt, 1995; Cairns, 1997; Costanza et al., 1997; Daily, 1997; Rapport et al., 1998b).

In the present study, the ecosystem health indicators used in two models include exergy (Ex), structural exergy (Ex_{st}), the ratio of zooplankton to phytoplankton (R_{BZBA}), and transparency in Secchi Disc depth (SD). The formulas for the calculation of ecological indicators are shown in Table 13.12.

13.2.3.1 Exergy and structural exergy

Exergy, a thermodynamic concept, was first applied to ecology in the late 1970s, and it is defined as the amount of work a system can perform when it is brought to thermodynamic equilibrium with its environment. The environment or reference state could be defined as the inorganic soup on earth 4 billion years ago before the

---

Table 13.12 Formula for the Calculation of Ecosystem Health Indicators[a]

(1)	Exergy (MJ/m^3)	$Ex = \sum_{i=1}^{n} (B_i * W_i)$
(2)	Structural exergy (MJ/g)	$Ex_{st} = Ex/B_{total}$
(3)	Ratio of zooplankton to phytoplankton biomass	$RBZBA = BZ/BA\ 1.8278$
(4)	Transparency in Secchi depth (cm)	$SD = \exp(3.4382 - 0.0838 * (BA))$

[a]*(1) From Jørgensen (1995a,b); B_i is biomass of ith organic component in the considered ecosystem (g dw/m^3); W_i is weighting factor of ith organic component. It can be computed from the information genes stored in various organisms (see Jørgensen, 1995a,b; Jørgensen et al., 1995a,b for details).*
(2) From Jørgensen (1995a,b); B_{total} is the total biomass in the considered ecosystem, which is the sum of all B_i.
(3) From Tu et al. (1990). BA is phytoplankton biomass (g dw/m^3).

biological evolution started (Jørgensen and Mejer, 1977, 1979). Exergy is expected to increase as ecosystems mature and develop away from the thermodynamic equilibrium. Exergy also expresses the energy expended in the organization and construction of living organisms by accounting for the genetic information accumulated within organisms. Organization of an organism is seen as being expressed through the information contained in its genes. The higher the organization of an organism, the higher its exergy, because it has cost more exergy to construct a more complex organization.

The exergy of an ecosystem cannot be measured but may be computed for each system component by multiplying its concentration, measured in terms of its average standing biomass, with its genetic information content using a conversion factors. When added up, the sum of individual exergies expresses the exergy of the whole system (see Table 13.11). For instance, in the lake ecosystem, if the concentration biomass of phytoplankton, zooplankton, and fish are, respectively, 5, 3, and 2 mg dw/l, the weighting factors for phytoplankton, zooplankton, and fish are respectively 34, 144, 344 (Jørgensen, 1995a,b), then exergy $= 5*34 + 3*144 + 2*344 = 1290$ MJ/l.

Structural exergy (Ex_{st}) is exergy calculated relative to the total biomass (see Table 13.9). The structural exergy measures the ability of the ecosystem to utilize the available resources. The more complex a network and the better the niche utilization, the higher structural exergy we expect (Jørgensen, 1992a, 1994).

Exergy has been used as a goal function in ecological models to account for the changes in composition of various organisms (Jørgensen, 1982, 1986, 1988, 1992a,b,c, 1994; Nielsen, 1990, 1992, 1994, 1995, 1997; Jørgensen and Nielsen, 1994; Coffaro et al., 1997). As ecological indicators, exergy and structural exergy were used to assess the ecological condition and ecosystem health of lake ecosystems (Jørgensen, 1995a,b; Marques et al., 1997; Xu, 1996, 1997) and to compare quantitatively the ecological state of agroecosystems (Dalsgaard, 1995, 1996; Dalsgaard et al., 1995).

13.2.3.2 *The ratio of zooplankton/phytoplankton (R_{BZBA})*

Algal biomass is a crucial indicator of lake eutrophication. With increasing eutrophication, algal biomass is of course increased. However, in very eutrophic lake ecosystems, zooplankton biomass is very low. Consequently, the ratio of zooplankton/phytoplankton is very low. McCauley and Kalff (1981) found that the ratio of zooplankton to phytoplankton decreases with increasing trophy in seventeen Canadian lakes with Chl-*a* value ranging from 10 to 100 mg/m^3. In the Lake Chaohu ecosystem in September 1987, phytoplankton biomass was very high (53.9 mg/l); zooplankton biomass was only 1.72 mg/l; and the lake was very eutrophic (Xu, 1996, 1997). The low zooplankton biomass in very eutrophic lake ecosystems may be caused by the toxicity of high phytoplankton biomass. Algal blooms are also commonly observed indirect effects of other chemical pollution such as acidification and some pesticides (e.g., Duban, 24D-DMA, bifenthrin, and Carbaryl; e.g., Hurlbert et al., 1972; Boyal, 1980; Drenner et al., 1993; Havens, 1994). Probably, this is mainly attributable to the release from grazing pressure, since zooplankton communities are more sensitive to chemical stress than phytoplankton communities, and zooplankton biomass is absolutely decreased due to the direct toxic effects of chemical pollutants. In any event, the ratio of zooplankton/phytoplankton would always decline. It can therefore serve as an ecological indicator for the effects of chemical pollution, including eutrophication, on lake ecosystem health.

13.2.3.3 Transparency in Secchi Disc depth (SD)

Transparency is an important physical characteristic in terms of water quality. Secchi depth measurements are normally used to predict the transparency. The Secchi depth is very low due to the high algal biomass and high inorganic resuspended matters. From Tu et al. (1990), there is a close relationship between transparency and phytoplankton biomass in Lake Chaohu. The equation is shown in Table 13.12.

13.3 RESULTS

The fourth-order Runge–Kutta method was used for the simulation. The entire computational period was from May 1987 to April 1988. A time step of one day was adopted for the simulations. The calibration was performed for model 1 in order to determine some parameters. The results from model 1 are compared with the measured values (Figure 13.5a–h). Model 2 used the same parameters and rate coefficients as model 1 for phytoplankton, zooplankton, fish, detritus, sediment, and soluble inorganic phosphorus submodels. The macrophyte submodel in model 2 was given realistic parameters according to the literature and the model studies for the similar lakes (e.g., Lake Tai, Lake Daishan in China). The lake ecosystem health indicators are calculated by means of models 1 and 2 and are compared in Figures 13.6–13.8. The changes in ecological health indicators following the increase of macrophyte initial value found by model 2 are illustrated in Figures 13.9–13.12.

13.3.1 Model 1 calibration and present health state of Lake Chaohu ecosystem

Figure 13.5a–f shows the comparisons of simulated and measured growth rate, respiration rate, nonpredatory mortality rate, settling rate, biomass, and internal phosphorus concentration of phytoplankton from May 1987 to April 1988. There are good agreements between observations and simulations. The results reveal that, in summer season 1987, phytoplankton growth rate, respiration rate, mortality rate, settling rate, and its biomass and internal phosphorus are higher than at any other time, which may be due to temperature effects. Figure 13.5g demonstrates the comparisons of simulated and measured growth rate of zooplankton and fish from May 1987 to April 1988. There is a good accordance between observations and simulations of zooplankton growth rate. Figure 13.5h gives the comparisons of simulated and measured fish growth rate from May 1987 to April 1988. The results show that the fish growth rates are higher in spring and autumn 1987 and are lower in summer 1987 and winter 1988. The low growth rate in summer is probably due to the aerobic environment, which is

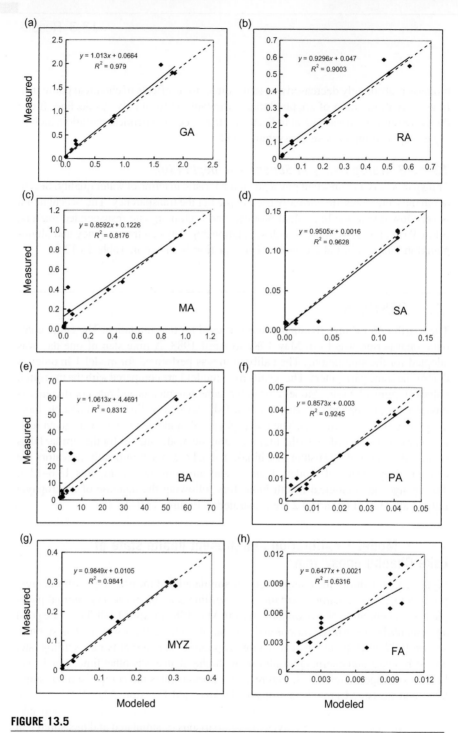

FIGURE 13.5

Comparisons of modeled values by model 1 and measured data. (a) GA: phytoplankton growth rate; (b) RA: phytoplankton respiration rate; (c) MA: phytoplankton mortality rate; (d) SA: phytoplankton settling rate; (e) BA: phytoplankton biomass; (f) PA: phosphorus in phytoplankton cells; (g) MYZ: zooplankton growth rate; (h) GF: fish growth rate. (For color version of this figure, the reader is referred to the online version of this chapter.)

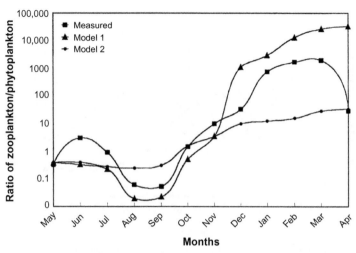

FIGURE 13.6

The ratio of zooplankton:phytoplankton found by means of model 1 and model 2 with initial value of macrophyte biomass 2 g/m^3 compared with measured data (from May 1987 to April 1988).

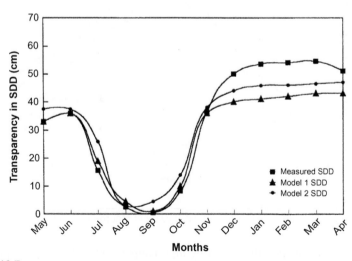

FIGURE 13.7

Transparency in Secchi disk depth found by means of model 1 and model 2 with initial value of macrophyte biomass 2 g/m^3 compared with measured data (from May 1987 to April 1988).

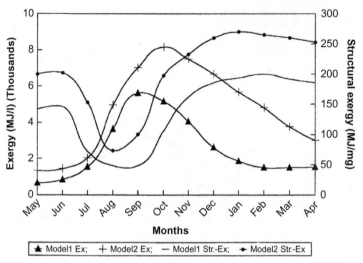

FIGURE 13.8

Comparison of exergy and structural exergy found by means of model 1 and model 2 with initial value of macrophyte biomass (g/m^3).

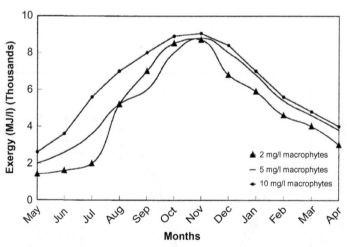

FIGURE 13.9

Exergy changes in model 2 with different initial value of macrophyte biomass.

not suitable for fish growth, caused by high temperature and high phytoplankton biomass (Table 13.10). The low growth rate in the winter season may be due to limitations of low temperatures.

Figures 13.6 and 13.7 demonstrate the comparisons of simulated and calculated values (from measured data) of zooplankton/phytoplankton ratio (R_{BZBA}) and transparency in Secchi Disc depth (SD) from May 1987 to April 1988. There are good

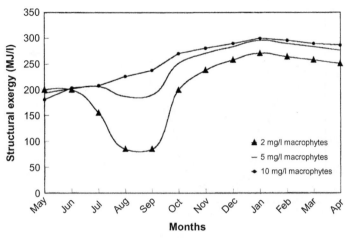

FIGURE 13.10

Structural exergy changes in model 2 with different initial value of macrophyte biomass.

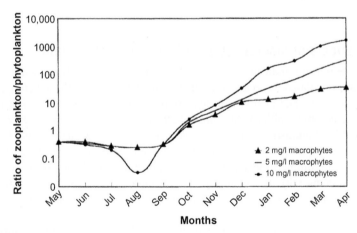

FIGURE 13.11

The changes in the ratio of zooplankton:phytoplankton in model 2 with different initial value of macrophyte biomass.

agreements between observations and simulations. The results show that the ratio of zooplankton to phytoplankton and transparency in summer 1987 have the lowest values, with the highest values in winter 1988.

The above sets of comparisons between simulated values by model 1 and observations suggest that model 1 well represents the pelagic ecosystem structure and function in Lake Chaohu. Model 1 can reproduce observed state-variable concentrations and process rates using model equations and coefficients.

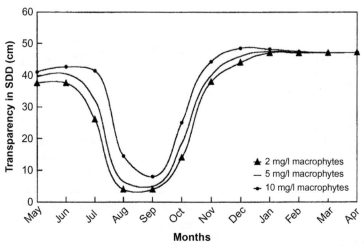

FIGURE 13.12

The changes in the transparency in Secchi disk depth in model 2 with different initial value of macrophyte biomass.

13.3.2 Effects of macrophytes on lake ecosystem health

The ecosystem health indicators in Lake Chaohu ecosystems with and without macrophytes are calculated by means of models 1 and 2, and the results are compared in Figures 13.6–13.8. The changes of ecosystem health indicators following the increase of macrophyte initial values are demonstrated in Figures 13.9–13.12.

Figure 13.6 demonstrates that the ecosystems with initial macrophyte biomass 2 mg/m^3 have similar zooplankton/phytoplankton ratio (R_{BZBA}) in springtime, higher R_{BZBA} in summertime, and lower R_{BZBA} in autumn and wintertime compared to the ecosystems without macrophytes. The increase of R_{BZBA} in summertime may be caused by decreased phytoplankton biomass and increased zooplankton biomass.

Figure 13.7 shows that the transparency in Secchi Disc depth (SD) is similar in two situations with and without macrophytes. The lowest SD takes place in the summer season, which may be due to the effect of high phytoplankton biomass. The highest SD occurs in the winter season, which means that phytoplankton biomass is lowest. Compared with the lake ecosystem without macrophytes, SD in the lake ecosystems with initial macrophyte biomass 2 mg/m^3 is increased in most times of the year, which means macrophyte restoration can increase transparency in the lake ecosystems.

Figure 13.8 shows that the highest exergy value in the lake ecosystem without macrophytes takes place in the summer season (e.g., September 1987) due to algal bloom; but the highest exergy value in the lake ecosystem with initial macrophyte biomass 2 mg/m^3 happens in the autumn season (e.g., October–November 1987) owing to high fish biomass (see later explanation). Exergy in the lake ecosystem with macrophytes is higher than that in the lake ecosystem without macrophytes, which means that the total biomass is increased following macrophyte restoration.

Figure 13.8 also reveals that structural exergy dynamics are similar in the two situations with and without macrophytes. Lowest structural exergy occurs in the summer season, which may be interpreted as worst health of lake ecosystem. Highest structural exergy takes place in the winter season, which means best health of lake ecosystem. Compared with the lake ecosystem without macrophytes, structural exergy in the lake ecosystem with initial macrophyte biomass $2 \, \text{mg/m}^3$ is increased in each season, which means macrophytes can increase structural exergy in the lake ecosystems.

Figure 13.9 shows that exergy is generally increased following the increase of initial macrophyte biomass from 2 to 5, $10 \, \text{mg/m}^3$ (with some exception in August–November). The highest exergy happens in autumn, which is probably due to the effects of the increased fish biomass.

Figure 13.10 reveals that structural exergy is generally increased following the increase of initial macrophyte biomass from 2 to 5, $10 \, \text{mg/m}^3$ (with some exception in May–June); of special note is that structural exergy in summertime is significantly increased, which may be the effect of the decreased phytoplankton biomass and increased fish biomass.

Figure 13.11 demonstrates that there are different effects of the increased macrophyte initial biomass at different times on the ratio of zooplankton/phytoplankton (R_{BZBA}). In May–June, macrophytes have fewer effects on R_{BZBA}. Following the increase of initial macrophyte biomass from 2 to 5, $10 \, \text{mg/m}^3$, R_{BZBA} is slightly decreased in July–September, R_{BZBA} is slightly increased in October–December, and R_{BZBA} is obviously increased in January–April.

Figure 13.12 shows that transparency in Secchi Disc depth (SD) is increased following the increase of initial macrophyte biomass from 2 to 5, $10 \, \text{mg/m}^3$, which means that the phytoplankton biomass decreases when the macrophyte biomass increases.

In general, compared with ecosystems without macrophytes, the ecosystems with initial macrophyte biomass $2 \, \text{mg/m}^3$ have higher exergy, higher structural exergy, higher transparency, and higher zooplankton/phytoplankton ratio, which means better ecosystem health. The increase of initial macrophyte biomass from 2 to 5, $10 \, \text{mg/m}^3$ means the increase of exergy, structural exergy, transparency, and zooplankton/phytoplankton ratio. This implies that macrophyte restoration can improve the lake ecosystem health measured by the here used ecological indicators.

13.4 DISCUSSION

The effects of macrophytes on lake ecosystems have been stressed by many researchers, for example, macrophytes can decrease the abundance of phytoplankton biomass and increase Secchi Disc depth (Van Donk et al., 1989; Scheffer, 1990; Li, 1996). They can stabilize a lake ecosystem, especially by the uptake of nutrients, allelopathy, and offering shelter to zooplankton and piscivorous fish (Moss et al., 1988; Grimm, 1989; Gulati, 1989; Van Donk et al., 1989). The models developed in this chapter predict well the effects of macrophytes on the lake ecosystem health by use of the ecological indicators exergy, structural exergy, transparency, and the ratio of zooplankton to phytoplankton biomass. The models can also predict the effects of macrophytes on the biological structure of the lake. It can be seen from Figures 13.13 and 13.14 that, following the increase of initial macrophyte biomass, phytoplankton biomass decreased and fish biomass increased. The decrease of phytoplankton biomass may be due to a reduction of nutrient and light availability in the water, caused by the competition of macrophytes. The increase of fish may be responsible for the increase of herbivorous fish and piscivorous fish (e.g., pike), since herbivorous fish need vegetation as a food resource, and piscivorous (e.g., pike population) need vegetation for successful recruitment and use of herbivorous fish as food source. There is a well-documented positive relation between the standing stock of herbivorous and piscivorous fish and the presence of aquatic vegetation (see Moss et al., 1988; Grimm, 1989; Gulati, 1989; Van Donk et al., 1989).

The results show that the model calibration is good, but both for measurements and simulated values, there are large differences from starting values in May 1987 to the ending values April 1988 for most variables. This may be mainly caused by the changes of water temperature and light radiation, since they are forcing functions for the lake models and main limiting factors for most biochemical and ecological progresses. Table 13.10 shows that there are great changes in temperature (T) and radiation (I_0) between May 1987 ($T = 23.8\ °C$, $I_0 = 4063.7\ kcal/m^2\ d$) and April 1988 ($T = 8.4$, $T_0 = 2244.6\ kcal/m^2\ d$). This results in the similar changes of simulated

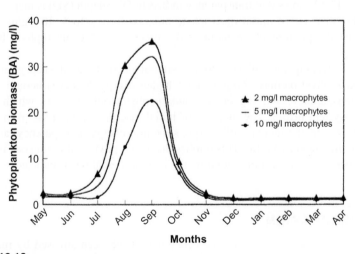

FIGURE 13.13

The effects of different initial value of macrophyte biomass on phytoplankton biomass the Lake Chaohu.

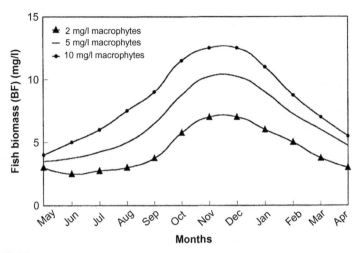

FIGURE 13.14

The effects of different initial value of macrophyte biomass on fish biomass the Lake Chaohu.

values to measurements for most parameters, progress rates, and state variables, especially, the phytoplankton growth rate, respiration rate, mortality rate, settling rate, biomass, and internal phosphorus concentration.

The equation for the calculation of transparency in Secchi Disc depth (SD) in Table 13.12 should be improved. Except for phytoplankton, macrophytes should be included in the equation for the calculation of transparency. As we know, there are close relationships between macrophyte biomass concentration and the light penetration in lake water. Normally, macrophytes disappear because the light penetration is reduced due to high phytoplankton concentrations. However, from Figure 13.7, which is drawn according to calculations from Equation (13.4) in Table 13.12, it can be seen that the Secchi depth does not change much with and without macrophytes. This is because Equation (13.4) in Table 13.12 does not include macrophytes. The equation only presents the effects of phytoplankton on Secchi depth, and does not present the direct effects of macrophytes. We have already shown that the phytoplankton concentrations are reduced following the restoration of macrophytes (Figure 13.13). The increase of macrophyte biomass can also result in the decrease of inorganic resuspended matters in the lake water (see Wang et al., 1995; Xu, 1997). So the restoration of macrophytes should make lake water transparency greatly increase.

The increase in model complexity often causes an increase in the number of parameters, and model parameter estimation is often difficult and time consuming if the number of parameters to be calibrated is high. The conventional method for parameter estimation is calibration, that is, adjusting most sensitive parameters within some range in the literature until a good agreement between observations and model predictions of the most important state variables and process rates are obtained. During model calibration, it is necessary to compare the modeled and observed rates of processes, except for the comparisons of modeled and measured state-variable values. Models can reproduce concentrations of ecosystem variables; however, models may reproduce some under- and overestimated ratios of material flow. For instance, phytoplankton dynamics are controlled by growth and loss processes. It is easy to imagine situations when both algal growth and loss rate are low or high, and the resulting rates of algal biomass changes are the same. Thus, we could get the right results (state-variable concentration) for the probable wrong reasons (compensation rate processes).

The results from this study show that exergy (Ex) and structural exergy (Ex_{st}) as ecological indicators can be applied to characterize the ecosystem, and the changes of exergy and structural exergy are highly related to the eutrophication gradient. Jørgensen (1992a,b,c, 1995a,b) showed that exergy increases with increased eutrophication, because eutrophication implies more biomass mainly in the form of phytoplankton. The structural exergy will also increase with eutrophication at a low to medium level because at all the trophic levels, phytoplankton, zooplankton, plantivorous fish, and top-carnivorous fish will increase in concentration. When, however, the lake becomes eutrophic, very eutrophic, or maybe hypereutrophic, the phytoplankton concentration will still increase, but the eutrophication of the top-carnivorous fish will decrease, because they hunt by sight, which is made difficult by the increased turbidity of the lake water. Because the weighting factor is significantly higher for fish than for phytoplankton, the structural exergy will decrease at a high and increasing level of eutrophication. The results from the research on the changes of exergy and structural exergy in Lake Chaohu reveal that structural exergy has a clearly negative correlation to eutrophication state and lake ecosystem health (Xu, 1996, 1997).

It should be pointed out that the restoration of macrophytes in eutrophic waters is not an easy task, and it requires comprehensive studies. First, it is necessary to find the limiting factors of macrophyte growth and the quantitative relationship between structure and environmental function of macrophytes. Second, artificial vegetation reproduction and plantation skills are needed in order to restore a macrophyte community on a large scale in a relatively short time to make macrophytes win in the competition with algae. Third, the studies on stability of macrophytes and on the competition between macrophytes and algae are necessary. Fourth, optimization initial density and harvest density for the restoration and harvest of macrophytes have to be determined in order to get the greatest ecological, economic, and social benefits.

13.5 CONCLUSIONS

Two ecological models that describe phosphorus-food web dynamics with and without macrophyte in Lake Chaohu ecosystems have been developed in this chapter. There are good agreements between observed and simulated values of important state variables and process rates. The models predict well the effects of macrophyte restoration on biological structure and ecosystem health of the Lake Chaohu, that is, following the increase of macrophyte biomass, phytoplankton biomass declines, and fish biomass, exergy, structural exergy, transparency, and the ratio of zooplankton to phytoplankton biomass are increased. This means that macrophyte restoration can regulate the lake's biological structure and improve the lake ecosystem health. The models may be used as a tool for design and management of macrophyte restoration.

Acknowledgments

Funding for this study was provided by the National Science Foundation of China (NSFC) (41030529, 41271462), the National Foundation for Distinguished Young Scholars (40725004), and the National Project for Water Pollution Control (2012ZX07103-002).

References

Bowie, G.L., Mills, W.B., Porcella, D.B., et al., 1985. Rates, Constant, and Kinetics Formulations in Surface Water Quality Modeling. US EPA, Athens, GA.

Boyle, T.R., 1980. Effects of the aquatic herbicide 2,4-DMA on the ecology of experimental ponds. Environ. Pollut. 21, 35–49.

Brix, H., Schierup, H.-H., 1989. The use of aquatic macrophytes in water-pollution control. Ambio 18, 100–107.

Cairns Jr., J., 1997. Protecting the delivery of ecosystem services. Ecosyst. Health 3, 185–194.

Cairns Jr., J., Pratt, J.R., 1995. The relationship between ecosystem health and delivery of ecosystem services. In: Rapport, D.J., Gaudet, C., Calow, P. (Eds.), Evaluating and Monitoring the Health of Large-Scale Ecosystems. Springer-Verlag, Berlin, pp. 63–76.

Clasen, J., Rast, W., Ryding, S.O., 1989. Available techniques for treating eutrophication. In: Ruding, S.O., Rast, W. (Eds.), The Control of Eutrophication of Lakes and Reservoirs. Man and the Biosphere Series, vol. 1. UNESCO, Paris.

Coffaro, G., Bocci, M., Bendoricchio, G., 1997. Application of structural dynamic approach to estimate space variability of primary producers in shallow marine water. Ecol. Model. 102, 97–114.

Conley, L.M., Dick, R.I., Lion, L.M., 1991. An assessment of the root zone method of wastewater treatment. Res. J. Water Pollut. Control Fed. 63, 239–247.

Cooke, G.D., Welch, E.B., Peterson, S.A., Newroth, P.R., 1993. Restoration and Management of Lakes and Reservoirs, second ed. Lewis Publishing, Boca Raton, p. 548.

Cooper, P.F., Findlater, B.C. (Eds.), 1990. Constructed Wetlands in Water Pollution Control. Proceedings of the International Conference on the Use of Constructed Wetlands in Water Pollution Control. Pergamon Press, Oxford, pp. 77–84.

Costanza, R., 1992. Toward an operational definition of ecosystem health. In: Costanza, R., Norton, B.G., Haskell, B.D. (Eds.), Ecosystem Health: New Goals for Environmental Management. Island Press, Washington, DC, Covelo, pp. 239–256.

Costanza, R., d'Arge, R., de Groot, R., Farber, S., Grasso, M., Hannon, B., Limburg, K., Naeem, S., O'Neill, R.V., Paruelo, J., Raskin, R.G., Sutton, P., Van den Belt, M., 1997. The value of the world's ecosystem services and natural capital. Nature 387, 253–260.

Daily, G., 1997. Nature's Services: Societal Dependence on Natural Ecosystems. Island Press, Washington, DC.

Dalsgaard, J.P.T., 1995. Applying systems ecology to the analysis of integrated agricultur-aquaculture farms. NAGA, The ICLARM Quarterly 18 (2), 15–19.

Dalsgaard, J.P.T. (1996) An ecological modeling approach towards the determination of sustainability in farming system. Ph.D.thesis, Royal Agriculture & Veterinary. University.

Dalsgaard, J.P.T., Lightfoot, C., Christensen, V., 1995. Towards quantification of ecological sustainability in farming systems analysis. Ecol. Eng. 4, 181–189.

Drenner, R.W., Hoagland, K.D., Smith, D., Barcellona, W.J., Johnson, P.C., Palmieri, M.A., Hobson, J.F., 1993. Effects of sediment-bound bifenthrin on gizzard shad and plankton in experimental tank mesocosms. Environ. Toxicol. Chem. 12, 1297–1306.

Grimm, M.P., 1989. Northern pike (*Esox lucius L.*) and aquatic vegetation, tool in the management of fisharies and water quality in shallow waters. Hydrobiol. Bull. 23, 59–65.

Gulati, R.D., 1989. Structure and feeding activities of zooplankton community in Lake Zwemlust, in the two years after biomanipulation. Hydrobiol. Bull. 23, 35–48.

Gumbricht, T., 1992. Tertiary wastewater treatment using root-zone method in temperature climate. Hydrobiologia 170, 245–266.

Gumbricht, T., 1993. Nutrient removal capacity in submersed macrophyte pond systems in a temperate climate. Ecol. Eng. 1, 49–61.

Hammer, D.A., 1989. Constructed Wetlands for Wastewater Treatment: Municipal, Industrial and Agricultural. Lewis Publishers, Chelsea, Michigan, p. 831.

Hammer, D.A., 1992. Designing constructed wetlands to treat agricultural nonpoint source pollution. Ecol. Eng. 1, 49–82.

Hannon, B., 1985. Ecosystem flow analysis. Can. J. Fish. Aquat. Sci. 213, 97–118.

Haskell, B.D., Norton, B.G., Costanza, R., 1992. What is ecosystem health and why should we worry about it? In: Costanza, R., Norton, B.G., Haskell, B.D. (Eds.), Ecosystem Health: New Goals for Environmental Management. Island Press, Washington, DC, Covelo, pp. 3–20.

Havens, K.E., 1994. An experimental comparison of the effects of two chemical stress on a freshwater zooplankton assemblage. Environ. Pollut. 84, 245–251.

Hurlbert, S.H., Mulla, M.S., Wilson, H.R., 1972. Effects of organophosphorous insecticide on the phytoplankton, zooplankton, and insect populations of freshwater ponds. Ecol. Monogr. 42, 269–299.

Jørgensen, S.E., 1994. Fundamentals of Ecological Modeling, second ed. Elsevier, Amsterdam, p. 630.

Jørgensen, S.E., 1976. A eutrophication model for a lake. Ecol. Model. 2, 147–165.

Jørgensen, S.E., 1982. Exergy and buffering capacity in ecological system. In: Mitsch, W., Ragade, R., Bosserman, R., Dillon, J. (Eds.), Energetics and Systems. Ann Arbor Science Publishers, Ann Arbor, MI, pp. 61–72.

Jørgensen, S.E., 1986. Structural dynamic model. Ecol. Model. 31, 1–9.

Jørgensen, S.E., 1988. Use of models as experimental tool to show that structural changes are accompanied by increased exergy. Ecol. Model. 41, 117–126.

Jørgensen, S.E., 1992a. Integration of Ecosystem Theories: A Pattern. Kluwer Academic Publishers, Dordrecht/Boston/London.

Jørgensen, S.E., 1992b. Parameters, ecological constraints and exergy. Ecol. Model. 62, 163–170.

Jørgensen, S.E., 1992c. Development of models able to account for changes in species composition. Ecol. Modell. 62, 195–208.

Jørgensen, S.E., 1994. Review and comparison of goal functions in system ecology. VIE MILIEU 44 (1), 11–20.

Jørgensen, S.E., 1995a. Exergy and ecological buffer capacities as measures of ecosystem health. Ecosyst. Health 1 (3), 150–160.

Jørgensen, S.E., 1995b. The application of ecological indicators to assess the ecological condition of a lake. Lakes Reserv. Res. Manag. 1, 177–182.

Jørgensen, S.E., Mejer, H., 1977. Ecological buffer capacity. Ecol. Model. 3, 39–61.

Jørgensen, S.E., Mejer, H.F., 1979. A holistic approach to ecological modeling. Ecol. Model. 7, 169–189.

Jørgensen, S.E., Nielsen, S.N., 1994. Models of the structural dynamics in lakes and reservoirs. Ecol. Model. 74, 39–46.

Jørgensen, S.E., Nielsen, S.N., Jørgensen, L.A., 1991. Handbook of Ecological Parameters and Ecotoxicology. Elsevier Press, Amsterdam.

Jørgensen, S.E., Nielson, S.N., Mejer, H.F., 1995a. Emergy, environ, exergy and ecological modeling. Ecol. Model. 77, 99–109.

Jørgensen, S.E., Halling-Sorensen, B., Nielsen, S.N., 1995b. Handbook of Environmental and Ecological Modeling. CRC Press, Inc., New York, p. 660.

Karr, J.R., 1991. Biological integrity: a long neglected aspect of water resource management. Ecol. Appl. 1, 66–84.

Karr, J.R., K.D. Fausch, P.L. Angermeier, P.R. Yant, & I.G. Schlosser (1986). Assessing Biological Integrity in Running Waters: A Method and its Rationale. Champaign: Illinois Natural History Survey, Special Publication 5.

Kerr, S.R., Dickey, L.M., 1984. Measuring the health of aquatic ecosystems. In: Carins, V.W., Hodson, P.V., Nriagu, J.O. (Eds.), Contaminant Effects on Fisheries. Wiley, New York, NY.

Li, W.C., 1996. Ecological restoration of shallow eutrophic lakes-experimental studies on the recovery of aquatic vegetation in Wuli Lake. J. Lake Sci. 8 (Suppl.), 1–10 (in Chinese).

Marques, J.C., Pardal, M.A., Nielsen, S.N., Jørgensen, S.E., 1997. Analysis of the properties of exergy and biodiversity along an estuarine gradient of eutrophication. Ecol. Model. 102, 155–167.

McCauley, E., Kalff, J., 1981. Empirical relationships between phytoplankton and zooplankton biomass in lakes. Can. J. Fish. Aquat. Sci. 38, 458–463.

Mitsch, W.J., 1992. Landscape design and the role of created, restored, and natural riparian wetlands in controlling nonpoint source pollution. Ecol. Eng. 1, 27–47.

Mitsch, W.J., Jørgensen, S.E., 1989. Ecological Engineering: An Introduction to Technology. John Wiley & Sons, New York, p. 472.

Moss, B., Irvine, K., Stanfield, J., 1988. Approaches to the restoration of shallow eutrophicated lakes in England. Verh. Int. Ver. Limnol. 23, 414–418.

Nichols, D.S., 1983. Capacity of natural wetlands to remove nutrients from wastewater. J. Water Pollut. Control Fed. 55, 495–505.

Nielsen, S.N., 1990. Application of exergy in structural-dynamical modeling. Verh. Int. Ver. Limnol. 24, 641–645.

Nielsen, S.N., 1992. Application of maximum exergy in structural dynamical models. Ph.D. thesis. Danish National Environmental Research Institute. 52 pp.

Nielsen, S.N., 1994. Modeling structural dynamical changes in a Danish shallow lake. Ecol. Model. 73, 13–30.

Nielsen, S.N., 1995. Optimization of exergy in a structural dynamical model. Ecol. Model. 77, 111–122.

Nielsen, S.N., 1997. Examination and optimization of different exergy forms in macrophyte societies. Ecol. Model. 102, 115–127.

Rapport, D.J., 1992. Evaluating ecosystem health. J. Aquat. Ecosyst. Health 1, 15–24.

Rapport, D.J., 1999. On the transformation from healthy to degraded aquatic ecosystems. Aquat. Ecosyst. Health Manag. 2 (2), 97–103.

Rapport, D.J., Regier, H.A., Hutchinson, T.C., 1985. Ecosystem Behavior Under Stress. Am. Nat. 125, 617–640.

Rapport, D.J., Costanza, R., Epstein, P.R., Gaudet, C., Levins, R. (Eds.), 1998a. Ecosystem Health. Blackwell, Boston.

Rapport, D.J., Costanza, R., McMichael, A.J., 1998b. Assessing ecosystem health. Trends Ecol. Evol. 13 (10), 397–402.

Rast, W., Hdland, M., 1988. Eutrophication of lakes and reservoirs: a framework for making management decisions. Ambio 17 (1), 2–12.

Reed, S.C. (Ed.), 1990. Natural Systems for Wastewater Treatment. Manual of Practice FD-16. Water Pollution Control Federation, Alexandria, USA, p. 260.

Reed, S.C., Middlebrooks, E.J., Crites, R.W., 1988. Natural Systems for West Management & Treatment. McGran Hill, New York.

Ryding, S.O., Rast, W., 1989. In: The Control of Eutrophication of Lakes and Reservoirs. Man and the Biosphere Series, vol. 1. UNESCO, Paris.

Schaeffer, D.J., Herricks, E.E., Kerster, H.W., 1988. Ecosystem Health: 1. Measuring Ecosystem Health. Environ. Manag. 12, 445–455.

Scheffer, M., 1990. Multiplicity of state states in freshwater ecosystems. Hydrobiologia 200 (201), 367–377.

Schindler, D.W., 1990. Experimental perturbations of whole lakes as tests of hypotheses concerning ecosystem structure and function. Oikos 57, 25–41.

Tilton, D.L., Kadlec, R.H., 1979. The utilization of a freshwater wetland for nutrient removal from secondary treated waste water effluents. J. Environ. Qual. 8, 328–334.

Tu, Q.Y., Gu, D.X., Yi, C.Q., Xu, Z.R., Han, G.Z., 1990. The Researches on the Lake Chaohu Eutrophication. The publisher of University of Science and Technology of China, Hefei, p. 225 (in Chinese).

Ulanowicz, R.E., 1986. Growth and Development: Ecosystems Phenomenology. Springer-Verlag, New York, NY.

Van Donk, E., Gulati, R.D., Grimm, M.P., 1989. Food web manipulation in Lake Zwemlus: positive and negative effects during the first two years. Hydrobiol. Bull. 23, 19–34.

Wang, S.Y., Jin, C.S., Meng, R.X., Xu, F.L., 1995. Environmental research for the Lake Chao in Anhui Province. In: Jin, X.C. (Ed.), Lakes in China, vol. 1. China Ocean Press, Beijing, p. 580.

Wolverton, B.C., 1987. Aquatic plants for wastewater treatment: an overview. In: Reddy, K.R., Smith, W.H. (Eds.), Aquatic Plants for Water Treatment and Resources Recovery. Magnolia Pub., Inc., Orlando, FL, pp. 3–16.

Xu, F.L., 1996. Ecosystem health assessment of Lake Chao, a shallow eutrophic Chinese lake. Lakes Reserv. Res. Manag. 2, 101–109.

Xu, F.L., 1997. Exergy and structural exergy as ecological indicators for the development state of the Lake Chao ecosystem. Ecol. Modell. 99, 41–49.

Xu, F.-L., Jørgensen, S.E., Tao, S., 1999a. Ecological indicators for assessing freshwater ecosystem health. Ecol. Model. 116, 77–106.

Xu, F.-L., Tao, S., Xu, Z.-R., 1999b. The restoration of riparian wetlands and macrophytes in the Lake Chao, an eutrophic Chinese lake: possibility and effects. Hydrobiologia 405, 169–178.

Xu, F.-L., Jørgensen, S.E., Tao, S., Li, B.-G., 1999c. Modeling the effects of ecological engineering on ecosystem health of a shallow eutrophic Chinese lake (Lake Chao). Ecol. Model. 117, 239–260.

Xu, F.-L., Dawson, R.W., Tao, S., Cao, J., Li, B.-G., 2001. A method for lake ecosystem health assessment: an ecological modeling method (EMM) and its application. Hydrobiologia 443, 159–175.

Development of Structural Dynamic Model for the Ecosystem Evolution of a Large Shallow Chinese Lake (Lake Chaohu)

Xiang-Zhen Kong[a], Sven Erik Jørgensen[b], Fu-Liu Xu[a,*], Wei He[a], Ning Qin[a], Qing-Mei Wang[a], Wen-Xiu Liu[a]

[a]*MOE Laboratory for Earth Surface Processes, College of Urban & Environmental Sciences, Peking University, Beijing, China*
[b]*Professor Emeritus, Environmental Chemistry, University of Copenhagen, Copenhagen, Denmark*
Corresponding author: e-mail address: xufl@urban.pku.edu.cn

14.1 INTRODUCTION

14.1.1 Alternative states of shallow lakes

Numerous restoration projects in the last decades of the twentieth century have inspired lake ecologists that hysteresis does exist in shallow lakes, and simply reducing the nutrient loading cannot always help the eutrophicated lakes recover to their original clear state (Scheffer and van Nes, 2007). It was proposed that shallow lakes have two alternative equilibriums over a range of nutrient concentrations: a clear state dominated by submerged vegetation and a turbid state characterized by high algal biomass (Scheffer et al., 1993). For a submerged-vegetation-dominated lake with clear water, stochastic events or fluctuations generated by internal processes (sediment recycling) and external forcing (nutrient loading) make the ecosystem hit a "sudden change" threshold nutrient concentration and switch to an alternative turbid state (Scheffer et al., 2001; Zhang et al., 2003a). Once lakes turned to a turbid state, they subsequently resisted restoration efforts (Phillips et al., 1978; Meijer et al., 1989) because they are stabilized by several mechanisms (Moss, 1988; Scheffer, 1998) that act as positive feedback that can cause the turbid state to be alternative attractors (Scheffer et al., 1993). For instance, fish communities turned from herbivorous to planktivorous (Xie, 2009), which promotes phytoplankton growth by recycling nutrients and controlling the development of zooplankton. Also fish and waves may stir up sediments in shallow lakes without macrophytes, which make it difficult for submerged plants to settle due to light limitation and disturbance of the sediment (Scheffer and van Nes, 2007).

14.1.2 Lake Chaohu as a eutrophicated shallow lake

Lake Chaohu is the fifth largest freshwater lake in China. Some basic limnological characteristics of Lake Chaohu are listed in Table 14.1. It has suffered from serious eutrophication since the 1980s. In 1987, the total phosphorus input amounted to 1050 t, 34% of which had remained in the lake. The high phosphorus loading resulted in a high annual average total phosphorus concentration of 0.23 mg/L (Tu et al., 1990). In the early 1960s, a dam was erected on the only outflow river of Lake Chaohu, which largely changed the hydrology of the lake and had an adverse effect on the macrophytes in the lake (Xu et al., 1999b). The coverage area fraction of macrophytes decreased significantly from 30% in 1950s to less than 1% today (Xie, 2009). In addition, the dam also blocked the input of juvenile fish from Yangtze River to Lake Chaohu, which led to a structural change in the fish community of the lake. The percentage composition of big fish such as silver carp and large culters in the annual fish yield declined from 38.2% to 2.6% while the fraction of small fish such as tapertail anchovies increased from 11.2% to 74% (Xie, 2009). The algae have

Table 14.1 Basic Limnological Characteristics of Lake Chaohu

Items	Units	Values
Latitude position	–	177°16′54″–177°51′46″
Average length (E–W)	km	55
Average width (N–S)	km	15
Catchment area	km^2	13,000
Mean surface area	km^2	760
Mean volume	$10^9 m^3$	1.9
Mean depth	m	3.06
Maximum depth	m	6.78
Mean annual inflow	$10^9 m^3$	4.1
Mean annual outflow	$10^9 m^3$	3.5
Mean retention time	day	136
Mean solar radiation	kJ/(cm^2 a)	490.69
Mean temperature	°C	16
Mean phosphorus load	t/a	780
Mean nitrogen load	t/a	23,000
Average T-P concentration	mg/l	0.20
Average T-N concentration	mg/l	2.3
Average Chl-*a* concentration	g/m^3	15.0
Average primary production	mgC/(m^2 d)	336
Population density	Ind./$10^6 m^3$	2774

bloomed every summer in recent years due to sufficient nutrients and lack of preda-tion pressure. The concentration of chlorophyll-a can be as high as 100 µg/L in June (Deng et al., 2007). Lake Chaohu had already switched to the turbid state (Scheffer et al., 1993), which is generally unacceptable due to lake eutrophication, the disap-pearance of macrophytes, changes of fish fauna, and the associated decrease in biodiversity (Janse et al., 2010).

14.1.3 Modeling approaches for restoration of Lake Chaohu

Restoration is urgent for the Lake Chaohu ecosystem. Methods such as water level management (Xu et al., 1999b), sediment dredging, or silver carp release have been proposed or even conducted recently in the lake. To assess the long-term effective-ness of different types of lake restoration methods, ecological models for lakes are of great importance. Ecological models for shallow lakes have been developed from the static TP and chlorophyll models (Vollenweider, 1975, 1979) to much more complex dynamic models. The latter have been applied to lakes all over the world (Jørgensen, 1976; Jørgensen et al., 1978; Kuusisto et al., 1998; Sagehashi et al., 2000; Arhonditsis and Brett, 2005; Hu et al., 2006). For Lake Chaohu, Xu et al. (1999a) developed an ecological model by combining the phosphorus cycles, plankton, mac-rophytes, and fish biomass dynamics. Increasing the macrophyte area was the resto-ration method that was discussed, but it did not include the structural dynamics (Xu et al., 1999a). However, it is obvious that the lake restoration should be associated with structural variations in lake ecosystems, which will result in parameter changes in the model. Structural dynamic models (SDMs) are models that consider the var-iation of the parameters to account for adaptations and shifts in the species compo-sition (Jørgensen, 1986). The idea is based on the maximum exergy storage, which indicates that an ecosystem tends to move away from thermodynamic equilibrium as far as possible (Jørgensen, 1999; Zhang et al., 2003a,b), namely, toward higher exergy in the system, which is defined as the amount of work that a system can per-form when it is brought into thermodynamic equilibrium with its environment (Jørgensen, 1992a,b, 1997, 1999). The parameters change according to expert and empirical knowledge or the optimization of a goal function of exergy, which de-scribes the fitness based on the changing condition (Jørgensen et al., 2002a). There-fore, SDMs can capture the drastic changes in species and property variations in response to changes in the forcing function. SDMs have been successful in at least 23 case studies (Jørgensen, 1986, 1988, 1992a, 1999, 2002; Jørgensen and Nielsen, 1994; Jørgensen and Padisak, 1996; Jørgensen and de Bernardi, 1997, 1998; Zhang et al., 2003b; Zhang et al., 2004; Gurkan et al., 2006; Zhang et al., 2010; Marchi et al., 2011). Lake Chaohu has experienced a shift from the clear state dominated by sub-merged vegetation to the turbid state characterized by a high algal biomass within the last several decades. Therefore, SDMs are important for predicting the restoration effects qualitatively, including structural changes (Gurkan et al., 2006).

Among the SDM studies concerning the drastic changes in lakes under significant changes in forcing function, only a few of them investigated and discussed parameter variations according to exergy optimization (Jørgensen and de Bernardi, 1998). On the other hand, the development of exergy in the model during the lake restoration is usually examined to test whether it is in accordance with maximum exergy storage hypothesis. Those studies focused on the lake shift from the clear state to the turbid state (Zhang et al., 2003a,b); parameter dynamics were investigated in seasonal simulation, while the parameter variations during the shift were not discussed. In addition, most SDM approaches selected a set of parameters for variation, but there is possibility that when including different sets of parameters in the model, the results of the SDMs will be slightly or largely different. Probably the system will provide a higher exergy with more parameters to be varied, which will give a wider possibility for adaptation.

In this study, because phosphorus is the limiting nutrient (Xu et al., 1999a) for Lake Chaohu, an ecological model focusing on the phosphorus cycle was developed, which contains essential compartments and processes to simulate the lake state shift from the turbid state to the clear state. The model was calibrated based on seasonal data over a year when the lake was eutrophicated. After calibration, the model was applied to predict the long-term effects of six different lake restoration strategies. Subsequently, structural dynamic approaches were conducted based on the most successful restoration strategies, which showed rapid and significant lake state shift. The development of the key state variables, parameters, and exergy were studied and compared with those of non-SDM approaches. Different sets of parameters were also included in the SDMs and the results are compared and discussed.

14.2 MODEL DEVELOPMENT

14.2.1 Model framework

A dynamic phosphorus cycling model was developed for Lake Chaohu. The model was developed based on the assumption that the lake is well mixed, that is, we treated the lake as a whole. Therefore, spatial heterogeneity was not considered in this study. The conceptual diagram of the model is shown in Figure 14.1. There are ten state variables in this model. The symbols, units, and initial values of these state variables are provided in Table 14.2. The acceptable calibration of the model is largely due to several improvements during the development of the model. The equations for each sub-model and these improvements are discussed in the following section.

14.2.1.1 Soluble phosphorus sub-model

$$\frac{d}{dt}PS = Inflow\_PS + Pdif + Decomp - Out\_PS - Upt\_PA - Upt\_PSP2 \quad (14.1)$$

where $Inflow\_PS$ (mg/L/d) is the inflow of soluble phosphorus of the lake and $Inflow\_PS = (PSPREC \cdot QPREC + PSRIVER \cdot QRIVER)/V$; $Pdif$ (mg/L/d) is the

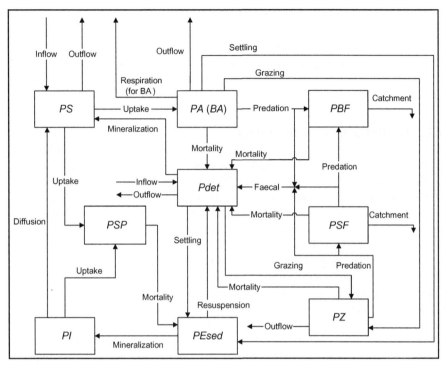

FIGURE 14.1

Conceptual diagram of the model.

Table 14.2 Details of the State Variables in the Model

Symbol	Description	Unit	Initial Value
TP^a	Total phosphorus	mg P/L	0.255
PS	Soluble phosphorus in lake water	mg P/L	0.22
BA	Phytoplankton biomass	mg P/L	1
PA	Phosphorus in phytoplankton	mg P/L	0.01
PZ	Phosphorus in zooplankton	mg P/L	0.005
PSP	Phosphorus in macrophytes	g P/m²	2
Pdet	Phosphorus in detritus	mg P/L	0.02
PEsed	Exchangeable phosphorus in sediment	g P/m²	0.0288
PI	Phosphorus in interstitial water	g P/m²	0.064
PSF	Phosphorus in small fish	mg P/L	0.2
PBF	Phosphorus in big fish	mg P/L	1.5

^{a}TP = PS + PA + PZ + Pdet.

phosphorus diffusion from the pore water and $Pdif = Pdiffu/D \cdot LUL/D$. Out_PS (mg/L/d) is the outflow of soluble phosphorus of the lake, and $Out\_P = PS \cdot Qout/V$. $Decomp$ (mg/L/d^{-1}) is the decomposition of detritus and can be found in detritus sub-model. Upt_PA (mg/L/d^{-1}) is the phosphorus uptake by phytoplankton and can be found in phytoplankton sub-model. Upt_PSP2 (mg/L/d^{-1}) is the phosphorus uptake by macrophyte and can be found in submerged plant sub-model.

14.2.1.2 Phytoplankton sub-model

A two-step process (Jørgensen, 1976) was included for the growth of phytoplankton: (1) the uptake of nutrients following Monod's kinetics and (2) the growth determined by the internal substrate concentration. This process was also used by Xu et al. (1999a). It was found that the two-step process gave a better fit of the model to the observed data for PA than the traditional one-step modeling (Jørgensen et al., 1978), and the results turned out to be unacceptable without this process in this study. To apply this process, a state variable named BA was included in the model, representing the biomass of phytoplankton. In addition, respiration was considered as influential only on the biomass of phytoplankton while little effect was found on the internal phosphorus concentration. Thus, the respiration process was added only to the BA sub-model, which was different from the model of Xu et al. (1999a). The parameter of FPA, which is the ratio of the PA and BA in phytoplankton, had a maximum boundary (FPA_{max}) in the range of 0.025–0.030 and a minimum boundary (FPA_{min}) in the range of 0.001–0.005. The limitation factor in the uptake of PA was $(FPA_{max} - FPA)/(FPA_{max} - FPA_{min})$ while that of biomass growth in BA was $(FPA - FPA_{min})/(FPA_{max} - FPA_{min})$ (Jørgensen and Fath, 2011) or $(FPA - FPA_{min})/FPA$ (Jørgensen, 1976). Both the boundary values of FPA and the equation of the limitation factor were open to calibration and were found to have a significant impact on the outcomes of PA and BA in simulation.

Temperature limitation factors were found to be influential not only on PA and BA but also on the calibration of other state variables. It was concluded that the exponential expression of $\exp(-2.3*abs(T - T_{opt}/T_{opt} - T_{min}))$, which has an optimized temperature with a peak value, should be applied in uptake, growth, and respiration processes, etc., while the monotone increasing function of Arrhenius' equation, $\theta^{(T-20)}$, was supposed to be used in mortality and mineralization processes, etc. The optimized temperature in the exponential expression was also open to calibration.

$$\frac{d}{dt}PA = Upt\_PA - Out\_PA - Mor\_PA - Settling\_PA - Grazing\_Z1 - PR\_BF1$$

(14.2)

where Upt_PA (mg/L/d) is the phosphorus uptake by phytoplankton and $Upt\_PA = Up\_rate \cdot PA$.

$$Up\_rate = Upmax\_PA \cdot PSlimit \cdot TlimitUPA \cdot FPint$$

$$PSlimit = \frac{PS}{PS + KPA}, \quad TlimitUPA = \exp\left(-2.3 \cdot \left|\frac{T - TUAopt}{TUAopt - TUAmin}\right|\right),$$

$$FPint = \frac{FPAmax - FPA}{FPAmax - FPAmin}, \quad FPA = PA/BA$$

Out_PA (mg/L/d) is the outflow of phytoplankton and *Out_PA = PA*Qout/V*
Mor_PA (mg/L/d) is the phytoplankton mortality and *Mor_PA = MorA_rate · PA*
    MorA_rate = MAmax · TlimitMA; TlimitMA = 1.08$^{(T-20)}$
Settling_PA (mg/L/d) is the phytoplankton settlement and
Settling_PA = SetPA_rate · PA
    SetPA_rate = SetPAmax/D · TlimitSA; TlimitSA = θ$^{(T-20)}$
Grazing_Z1 (mg/L/d) is the grazing on phytoplankton by zooplankton and can be found in zooplankton sub-model.
PR_BF1 (mg/L/d) is the grazing on phytoplankton by fish and can be found in fish sub-model.

$$\frac{d}{dt}BA = Growth\_BA - Out\_BA - Mor\_BA - Res\_BA - Settling\_BA \\ -(Grazing\_Z1 + PR\_BF1) \cdot 100 \tag{14.3}$$

where *Growth_BA* (mg/L/d) is the phytoplankton growth and
Growth_BA = GA_rate · BA
    GA_rate = GAmax · TlimitGA · Llimit · PAlimit

$$TlimitGA = \exp\left(-2.3 \cdot \left|\frac{T - TGAopt}{TGAopt - TGAmin}\right|\right),$$

$$Tlimit = \frac{1}{\lambda \cdot D} \ln\left(\frac{KI + I_0}{KI + I_0 \cdot \exp(-\lambda \cdot D)}\right), \quad \lambda = \alpha + \beta \cdot PA$$

$$PAlimit = 1 - \frac{PAmin}{PA} = 1 - \frac{BA \cdot FPAmin}{PA}$$

Out_BA (mg/L/d) is the outflow of phytoplankton and *Out_BA = BA · Qout/V*.
Mor_BA (mg/L/d) is the phytoplankton mortality and
Mor_BA = MorBA_rate · BA
    MorBA_rate = MAmax · TlimitMA
Res_BA (mg/L/d) is the phytoplankton respiration and
Res_BA = ResBA_rate · BA

$ResBA\_rate = RAmax \cdot TlimitRA, \; TlimitRA = TlimitMA$

Settling_BA (mg/L/d) is the phytoplankton settlement and

$Settling\_BA = SetBA\_rate \cdot BA$

$\quad SetBA\_rate = SetBAmax/D \cdot TlimitSA, \; SetBAmax = SetPAmax$

Grazing_Z1 (mg/L/d) is the grazing on phytoplankton by zooplankton and can be found in zooplankton sub-model.

PR_BF1 (mg/L/d) is the grazing on phytoplankton by fish and can be found in fish sub-model.

14.2.1.3 Submerged vegetation sub-model

The submerged vegetation was also modeled in this study (PSP). The submerged vegetation uptakes phosphorus from both water and sediment. Sediment is supposed to act as the major source. The growth of the PSP is limited by the light, phosphorus concentration, and temperature. As the submergents are dead, they enter the PEsed compartment in the sediment. Subsequently, they can be resuspended into water phase or mineralized into pore water (PI) and diffused into water phase. Therefore, the model in this study is suggesting that when the submerged plants are dead, they actually pump P from sediment to water phase. In addition, *PSP* was difficult to determine. The value used in the literature (2 g/m^2) (Zhang et al., 2003b) was a typical value of *PSP* under the phytoplankton-dominant state of Lake Chaohu. It was reported that the coverage of macrophytes in Lake Chaohu decreased from 30% in the 1950s to less than 1% today. Therefore, it was expected that the PSP in the submerged vegetation-dominant state should be approximately 60 g/m^2.

$$\frac{d}{dt}PSP = Upt\_PSP1 + Upt\_PSP2 - Mor\_PSP \quad (14.4)$$

where *Upt_PSP1* (g/m^2/d) is the phosphorus uptake of submerged vegetation from sediment and $Upt\_PSP1 = Upt\_1rate \cdot PSP$

$Upt\_1rate = Gmax1PSP \cdot Llimit \cdot PIlimit \cdot TlimitUp1$

$$PIlimit = \frac{PI}{PI+KPI}; \quad TlimitUp1 = \exp\left(-2.3 \cdot \frac{T - TSPopt1}{TSPopt1 - TSPmin1}\right)$$

Upt_PSP2 (g/m^2/d) is the phosphorus uptake of submerged vegetation from water and $Upt\_PSP2 = Upt\_2rate \cdot PSP$

$\quad Upt\_2rate = Gmax2PSP \cdot Llimit \cdot PSlimit2 \cdot TlimitUp2$

$$PSlimit2 = \frac{PS}{PS+KPSP}; \quad TlimitUp2 = \exp\left(-2.3 \cdot \frac{T - TSPopt2}{TSPopt2 - TSPmin2}\right)$$

Mor_PSP (g/m^2/d) the submerged vegetation mortality and

$Mor\_PSP = MorSP\_rate \cdot PSP$

$\quad MorSP\_rate = Mmax\_SP \cdot TlimitMSP; \; TlimitMSP = 1.08^{(T-20)}$

14.2.1.4 Zooplankton sub-model

Given that the zooplankton could graze both phytoplankton and detritus, two grazing processes were applied in the sub-model for zooplankton, and grazing preference functions $PA/(PA + Pdet)$ and $Pdet/(PA + Pdet)$ were added to the two processes, respectively. When phytoplankton is insufficient, zooplankton has a tendency to switch to feeding mainly on detritus. In addition, it was found that cyanobacteria dominated in the lake, which was not preferred by most zooplankton (Suzuki et al., 2000), and PA was much smaller than $Pdet$ in this study. As a result, the zooplankton mainly grazed detritus rather than phytoplankton in this model, which was in accordance with the situation in the lake.

$$\frac{d}{dt}PZ = Grazing\_Z1 + Grazing\_Z2 - Mor\_PZ - PR\_SF \qquad (14.5)$$

where $Grazing\_Z1$ (mg/L/d) is the grazing on phytoplankton by zooplankton and
$Grazing\_Z1 = Gr1\_rate \cdot PZ$
  $Gr1\_rate = Gr1max/Gr\_coef \cdot TlimitGrA \cdot BAlimit \cdot Pref\_PA$

$$TlimitGrA = \exp\left(-2.3 \cdot \frac{T - TGrAopt}{TGrAopt - TGrAmin}\right); \quad BAlimit = \frac{BA - KSA}{BA + KA};$$

$$Pref\_PA = \frac{PA}{PA + Pdet}$$

$Grazing\_Z2$ (mg/L/d) is the grazing on detritus by zooplankton and
$Grazing\_Z2 = Gr2\_rate \cdot PZ$
  $Gr2\_rate = Gr2max/Gr\_coef \cdot TlimitGrD \cdot PDlimit \cdot Pref\_PD$

$$TlimitGrD = \exp\left(-2.3 \cdot \frac{T - TGrDopt}{TGrDopt - TGrDmin}\right); \quad PDlimit = \frac{Pdet \cdot 100 - KSD}{Pdet \cdot 100 + KD};$$

$$Pref\_PD = \frac{Pdet}{PA + Pdet}$$

Mor_PZ (mg/L/d) is the zooplankton mortality and
$Mor\_PZ = MorZ\_rate \cdot PZ + (1/Gr\_coef\text{-}1) \cdot (Grazing\_Z1 + Grazing\_Z2)$
  $MorZ\_rate = MZmax \cdot TlimitMZ \cdot FiZ; \quad TlimitMZ = TlimitMA;$
$$FiZ = \begin{cases} 1.6, & \text{if } PA > 0.025 \\ 1, & \text{if } PA \leq 0.025 \end{cases}$$

14.2.1.5 Detritus sub-model

Detritus in this model is the degradable organic matter in the water column. Various sources for the detritus exist in the model structure, including water inflows; faecaj of the fish predation; mortality of phytoplankton, zooplankton, and fish; and also the resuspension from the sediment. It also subjects to the decomposition and settling processes.

$$\frac{d}{dt}Pdet = Inflow\_Pdet + PREDin + Mor\_PA + Mor\_PZ + Mor\_BF + Mor\_SF$$
$$+ REsusp - Decomp - Settling\_Pdet$$

(14.6)

where $Inflow\_Pdet$ (mg/L/d) is the inflow of detritus in the lake and
$Inflow\_Pdet = QRIVER \cdot PdetRIVER/V$
$PREDin$ (mg/L/d) is the residues of the fish predation that becomes detritus and
$PREDin = (1/Y1-1) \cdot Pr\_SF + (1/Y2-1) \cdot Pr\_BF1 + (1/Y3-1) \cdot Pr\_BF2$
$Decomp$ (mg/L/d) is the decomposition of detritus and
$Decomp = Decomp\_rate \cdot Pdet$
$\quad Decomp\_rate = Decomp10 \cdot TlimitDD; TlimitDD = \phi^{(T-10)}$
$Settling\_Pdet$ (mg/L/d) is the detritus settlement and
$Settling\_Pdet = SetPdet\_rate \cdot Pdet$
$\quad SetPdet\_rate = SetPdet20/D \cdot TlimitSD; TlimitSD = \eta^{(T-20)}$
$REsusp$ (mg/L/d) is the resuspension from sediment and $REsusp = REsuspension/ D \cdot LUL/D$

Mor_PA, Mor_PZ, Mor_SF and Mor_BF (mg/L/d) are the mortality of phytoplankton, zooplankton, small fish, and big fish and can be found in sub-models of phytoplankton, zooplankton, and fish.

14.2.1.6 Fish sub-model

Two state variables named PBF and PSF, which represent the "big fish" (piscivorous and herbivorous fish) and the "small fish" (planktivorous fish), respectively, were added to the model according to the ecological food web of Lake Chaohu. The "big fish" consume both the small fish and phytoplankton while the "small fish" feed on the zooplankton. The fish sub-model was built according to the food web study in Lake Chaohu (Zhang and Lu, 1986). The fish harvesting process is included in the model. The catch rate of fish is calibrated and set as 0.001 (1/d), which is the same as that in Xu et al. (1999a). According to the research conducted by Guo (2005) on the fishery industry in Lake Chaohu in 2002, the fishery potentiality was amounted to 2.6×10^4 t, and the catchment was about 1.1×10^4 t. Therefore, the catchment rate was approximately 0.001 (1/d). The catchment rate value in 1987 could be a little different from this, but it is believed that this value is reliable for the model simulation. Although fish were not a focus in this study, they have a significant influence on the lake ecosystem. It is possible that the fish sub-model was not sufficiently complex, and a more detailed development of the interaction between fish and other components should be involved for improvement in this model in further study.

$$\frac{d}{dt}P\_SF = Pr\_SF - Pr\_BF1 - Mor\_PSF - Catchment1 \tag{14.7}$$

where Pr_SF (mg/L/d) is the predation on zooplankton by small fish and
$Pr\_SF = Pr\_rate \cdot P\_SF$
$\quad Pr\_rate = Prmax \cdot TlimitPrF \cdot FPZF/Y$

$$TlimitPrF = \exp\left(-2.3 \cdot \frac{T - TPrFopt}{TPrFopt - TPrFmin}\right); \quad FPZF = MAX\left(0, \frac{PZ - KS}{PZ + KZF}\right)$$

Mor_PSF (mg/L/d) is the mortality of small fish and
$Mor\_PSF = MorSF\_rate \cdot P\_SF$
$\quad MorSF\_rate = MorSFmax \cdot TlimitMSF; \ TlimitMSF = 1.08^{(T-20)}$
$Catchment1$ (mg/L/d) is the catchment of small fish in the lake and
$Catchment1 = Catch\_eff \cdot P\_SF$

$$\frac{d}{dt}P\_BF = Pr\_BF1 + Pr\_BF2 - Mor\_PBF - Catchment2 \tag{14.8}$$

where Pr_BF1 (mg/L/d) is the predation on small fish by big fish and
$Pr\_BF1 = Pr1\_rate \cdot P\_BF$
$\quad Pr1\_rate = Pr1max \cdot TlimitPrF \cdot FPSF/Y1; \ FPSF = MAX\left(0, \frac{P\_SF - KS1}{P\_SF + KSF}\right)$

Pr_BF2 (mg/L/d) is the predation on phytoplankton by big fish and
$Pr\_BF2 = Pr2\_rate \cdot P\_BF$
$\quad Pr2\_rate = Pr2max \cdot TlimitPrF \cdot FPAF/Y2; \ FPAF = MAX\left(0, \frac{PA - KS2}{PA + KPAF}\right)$
Mor_PBF (mg/L/d) is the mortality of big fish and
$Mor\_PBF = MorBF\_rate \cdot P\_BF$
$\quad MorBF\_rate = MorBFmax \cdot TlimitMBF; \ TlimitMBF = 1.08^{(T-20)}$
$Catchment1$ (mg/L/d) is the catchment of big fish in the lake and
$Catchment2 = Catch\_eff \cdot P\_BF$

14.2.1.7 Sediment sub-model

Lake Chaohu is a typical shallow lake, with an average water depth of approximately 3.06 m. Therefore, there is strong resuspension flux from the sediment due to large winds, which had already been included in the model here according to the equation in Tu et al. (1990). It was found that this process has a profound impact on the results of both *PEsed* and *PI*. In addition, the resuspension flux (REsusp) is elevated when the water depth (*D*) gets lower, which indicates that more phosphorus has been released to water phase.

$$\frac{d}{dt}PEsed = Settling + Mor\_PSP - PE\_M - REsuspension \qquad (14.9)$$

Settling (g/m^2/d) is the input of settlement from water to sediment and
Settling = (Settling_PA + Settling_Pdet) · D/KTE · D/LUL
PE_M (g/m^2/d) is the mineralization of sediment and PE_M = PEM_rate · PEsed
   PEM_rate = PEM20 · TlimitPEM; TlimitPEM = $\sigma^{(T-20)}$
REsuspension (g/m^2/d) is the resuspension from sediment and
REsuspension = REsus_rate · PEsed
   REsus_rate = REsus0 · $w^{0.4}$

$$\frac{d}{dt}PI = PE\_M - Pdiffu - Upt\_PSPI \qquad (14.10)$$

Pdiffu (g/m^2/d) is the diffusion to water from pore water and
Pdiffu = Diff_coef · (PI · (1-DMU)/LUL)-PS · D · TlimitPID

$$TlimitPID = \exp\left(-2.3 \cdot \frac{T - TPIDopt}{TPIDopt - TPIDmin}\right)$$

14.2.2 Parameter determination

The symbols, descriptions, and units of the forcing functions in the model are given in Table 14.3. The data on the wind speed (w) were from the China Meteorological Data Sharing Service System (CMDSSS, 1987–1988), while the data for all other forcing functions were collected from Tu et al. (1990). The parameters in the model were restricted to their ranges according to experiments or experiences reported in various literatures (Chen and Orlob, 1975; Lehman et al., 1975; Jørgensen, 1976; Tu et al., 1990; Asaeda and Van Bon, 1997; Hu et al., 1998; Xu et al., 1999a,b; Sagehashi et al., 2001; Zhang et al., 2003a,b; Bruce et al., 2006; Amemiya et al., 2007). Subsequently, these parameters were calibrated based on the observations with a trial and error process and also within their theoretical ranges. The description for the specific parameters can be found in other references (Xu et al., 1999a; Kong et al., 2013).

14.2.3 Seasonal simulation and validation

The simulation duration was from May 1, 1987, to April 30, 1988, with a time step of 1 day. The monthly observed data were also collected from Tu et al. (1990). We applied a trial-and-error method for model calibration and parameter determination. Three quantitative fit criteria were applied in the model performance assessment: (1) h index, which was used in Zhang et al. (2003b), and the results are considered as acceptable if $0 \leq h \leq 1$; (2) S.D., that is, the standard deviation between the simulated and observed data, which was used in Gurkan et al. (2006). It is considered as

Table 14.3 Details of the Forcing Functions in the Model

Symbol	Description	Unit
QRIVER	Inflow from tributaries	m^3/d
PSRIVER	Soluble inorganic P concentration in inflow	$g\ P/m^3$
PdetRIVER	Detritus P concentration in inflow	$g\ P/m^3$
QPREC	Precipitation to the lake	m^3/d
PSPREC	Soluble inorganic P concentration in precipitation	$g\ P/m^3$
Qout	Outflow from tributaries	m^3/d
V	Lake Volume	m^3
D	Lake Depth	m
T	Temperature of lake water	°C
I_0	Light irradiance on the surface of lake water	$kcal/m^2/d$
w	Wind speed	m/s

acceptable when *S.D.* is less than 45% (Jørgensen et al., 2002b; Zhang et al., 2004; Gurkan et al., 2006). (3) The correlation coefficients between the simulated and observed data (*R*).

The results of the calibrated seasonal simulation in the state variables are presented in Figures 14.2 and 14.3. The simulations agreed acceptably with the observations, where all of the *h* values were less than 1 (Table 14.4). The *S.D.* values were all less than 45% except for *PZ*, indicating a relatively poor calibration for the zooplankton (Table 14.4). The most important state variable—the phosphorus in phytoplankton (*PA*)—is well calibrated, and the corresponding *R* value is the highest, suggesting a similar seasonal variation in the simulation as in the observations.

The factors leading to the seasonal characteristics in each variable can be revealed by the model. The elevated soluble phosphorus in water (*PS*) in spring and summer was largely a result of the decomposition of the detritus (Decompin, Figure 14.4), which was much higher than other input pathways to the *PS*. Subsequently, the decrease of the *PS* in winter was also related to the lower input from the decomposition of the detritus. It is suggested that the phosphorus in the detritus (*Pdet*) was an important source of the soluble phosphorus in the water column. The detritus originated from the water inflow, the death and predation of the zooplankton and fish, and the resuspension of sediment, which are shown in Figure 14.4 simultaneously. The high value of phosphorus in detritus may due to the high inflow rate of detritus as well as a significant resuspension from sediment. The temperature in spring and summer was higher, resulting in a higher decomposition rate. These two factors explain the elevated decomposition of the detritus in spring and summer.

The concentration of phosphorus in phytoplankton (*PA*) peaked in June and October, remained low in winter, and started to increase in spring (Figure 14.2). This pattern was determined by the uptake process (Figure 14.4), which was influenced by many factors, including the limitation of light, the temperature of soluble phosphorus

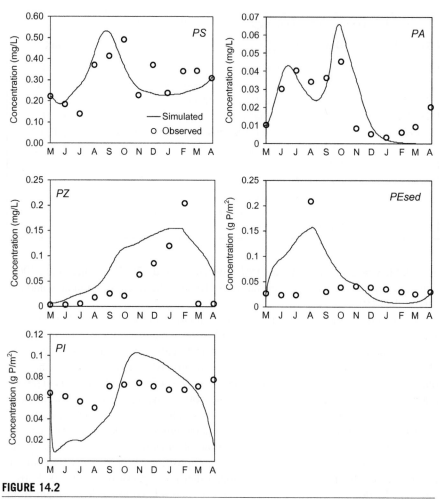

FIGURE 14.2

Comparison of the observed and simulated data.

in water, and the phosphorus concentration of the phytoplankton. It can be found that the temperature limitation was the dominant factor, which largely contributed to the seasonal patterns of the uptake flux and subsequently determined the seasonal pattern of *PA*. The seasonal variations of the concentration of phosphorus in zooplankton (*PZ*) were higher in the winter and remained low in other seasons, which were different than those of the phytoplankton. The zooplankton grazed on the detritus (*Pdet*) and phytoplankton (*PA*), the fluxes of which were Grazing1 and Grazing2, respectively (Figure 14.4). Grazing1 was higher than Grazing2 throughout the simulation duration, indicating that the zooplankton mainly grazed on detritus rather than phytoplankton. This result was not only due to the higher grazing rate of zooplankton on

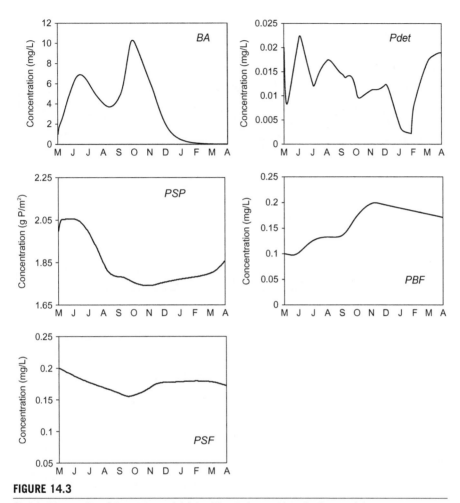

FIGURE 14.3

Simulation results of the state variables without observations in the model.

Table 14.4 The Quantitative Fit Criteria of the State Variables in the Calibrated Model

Compartments[a]	h	S.D. (%)	R
PS	0.14	29.95	0.59
PA	0.22	38.66	0.80
PZ	0.33	50.72	0.73
PEsed	0.34	39.51	0.55
PI	0.25	43.70	0.46

[a]Units are shown in Table 14.2.

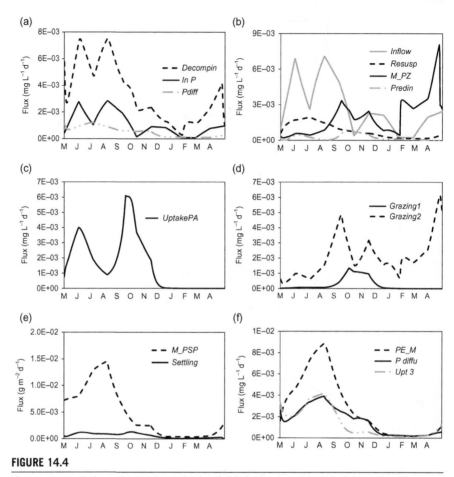

FIGURE 14.4

Key processes in important state variables. (a) Soluble phosphorus in lake water (*PS*);
(b) phosphorus in detritus (*Pdet*); (c) phosphorus in phytoplankton (*PA*); (d) phosphorus in
zooplankton (*PZ*); (e) phosphorus in macrophytes (*PSP*); (f) phosphorus in interstitial
water (*PI*).

detritus but also because the concentration of phosphorus in phytoplankton was significantly lower than that in detritus. In addition, the dominating algae in Lake
Chaohu during the summer bloom was cyanobacteria, which is not usually grazed
by zooplankton (Gurkan et al., 2006). This was also one of the reasons for the domination of cyanobacteria in the lake during the summer bloom (Xie, 2009).

The concentration of exchangeable phosphorus in sediment (*PEsed*) peaked in
August and remained low and stable in other seasons. According to the model results,
the elevated input from the mortality of submerged vegetation (*M_PSP*) in August
largely contributed to this peak (Figure 14.4). Meanwhile, the flux of the setting was

low and of little seasonal variation. The concentration of phosphorus in pore water (*PI*) was relatively more stable than that in other compartments. It decreased in spring and summer but remained at a higher value in autumn and winter. This pattern was captured by the model simulation, but the variation was relatively exaggerated. The input of *PI* was the mineralization of the *PEsed* while the outputs were the diffusion to the water column and the uptake by submerged vegetation. All of these processes peaked in summer (Figure 14.4).

Figure 14.3 shows the other results of the simulation in state variables, which cannot be calibrated due to lack of observation data. The biomass of phytoplankton (*BA*) was similar in variation to *PA*. The phosphorus in detritus (*Pdet*) fluctuated in the simulation but with a similar value at the beginning and the end. The phosphorus in submerged vegetation and fish was relatively stable in the simulation, which was consistent with expectation.

The model developed in this study was sufficient to simulate the phosphorus cycling in Lake Chaohu because the key processes in the lake were incorporated. The calibration results in the model were relatively acceptable (Table 14.4), which increased confidence in the prediction of the restoration effect by this model and also the following structural dynamic approaches.

14.3 RESTORATION METHODS AND THE POSSIBLE EFFECTS

14.3.1 Potential restoration methods

The duration of the model simulation was extended to 25 years after calibration, starting May 1, 1987, and ending April 30, 2012, with the same seasonal environmental conditions as in 1987.5–1988.4. This simulation, which is referred to as "Origin," provided a stable model performance throughout the 25 years in all of the state variables. The stability outcomes in this long-term simulation depended on the former annual cycle simulation, in which all of the state variables were numerically close at the beginning and in the end. Six different lake restoration methods conducted on the model are described as follows and their results are compared with those in "Origin."

14.3.1.1 Ecological economy water level

The dam that was erected in 1961 on Lake Chaohu had significantly changed the seasonal water levels, which are shown in Figure 14.5. After 1961, the water levels in spring were higher than that before 1961, which left little opportunities for the germination of submerged vegetation, while in summer, the water levels were lower than those before 1961. Water management by decreasing water level to meet the light level for submerged-vegetation development might be a solution for lake restoration (Scheffer et al., 1993). The ecological economy water level (EEWL) was based on the study of Xu et al. (1999b), in which the water level in late winter and early spring (January to March) should be 7.5 m, which is high enough to guarantee water-taking and ensures that 1000-ton ships can pass, while also low enough

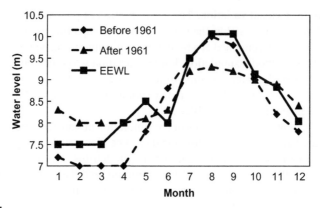

FIGURE 14.5

Monthly average water level in Lake Chaohu before and after 1961 when the sluice gate was built and also the EEWL.

for the safety of low-lying paddy fields and keeping the wetland area up to about 10% of total lake surface area. In addition, the water level in April and June should not exceed 8.0 m to ensure fish fry migration from the Yangtze River to the lake and in May should be as high as 8.5 m for irrigation purposes in August. Moreover, the highest water level, as in August or September, should not exceed 10.5 m in order to prevent floods and keep the dikes safe. Details of EEWL have been illustrated in Figure 14.5. The EEWL was applied in the model simulation, and the corresponding water outflow was also redesigned to make this management possible.

14.3.1.2 Phosphorus loading reduction by 40% (P_fold%40)

It was reported that approximately 40% of the phosphorus loading of Lake Chaohu was from point source pollution while the remaining 60% was from non-point source pollution (Tu et al., 1990). Therefore, a reduction of 40% phosphorus loading was implemented in this study. However, due to hysteresis, lake restoration by reducing the nutrient level may have little effect (Scheffer et al., 1993).

14.3.1.3 Phosphorus loading reduction by 40% and Ecological Economy Water Level (EEWL + P_fold%40)

Both restoration methods of EEWL and P_fold%40 were implemented simultaneously. Because merely reduction of nutrient loading may be of little effect, it is interesting to test if the combination of water level management and nutrient loading will have better effect.

14.3.1.4 Dredging of the surface sediment (Dredge)

Dredging the sediment from the lakebed has already been conducted in Lake Chaohu. Here, the thickness of the upper sediment layer (LUL) was changed from 0.16 to 0.05 to see if this method could help to restore the lake.

14.3.1.5 *Release of "big fish" (BFup)*

Scheffer et al. (2001) proposed that a temporary significant reduction of fish biomass as "shock therapy" can bring such lakes back into a permanent clear state if the nutrient level is not too high. Similarly, the initial value of *PBF* was increased from 0.2 to 2 mg/L to simulate the fish release in Lake Chaohu, and the effects were evaluated.

14.3.1.6 *Water level decreased by 1 m (De1m)*

This method is actually an idealistic method wherein the water level was decreased by 1 m throughout the year. This can lead to a much stronger change in water level than EEWL. The reasons to conduct this restoration in the model were (1) to show the importance of the water depth in the lake and (2) to provide a rapid case in which the status of the lake was changed significantly. Therefore, it was possible to build a SDM with less calculation required.

14.3.2 Possible effects of various restoration methods

The calibrated model was applied to predict the results of the six lake restoration strategies and assess the corresponding effectiveness. The development scenarios of the two most important state variables—*PA* and *PSP*—are presented in Figures 14.6 and 14.7 under different restoration strategies. The peak values of all of the state variables in the model in the first and last years of the 25-year simulation are also listed in Table 14.5, where it can be seen that the peak values of all the state variables were very close in the first and last years when no restoration was implemented, referred to as "Origin." In addition, the results of *PA* and *PSP* both provided stable outcomes (Figures 14.6 and 14.7), indicating relatively stable results throughout the long-term simulation. This "Origin" result provided a reference for the results of the six restoration methods.

EEWL has a significant impact on the phytoplankton, which decreased immediately in the second year and gradually decreased afterwards (Figure 14.6). The *PA* value decreased to 0.0044 mg/L in the end of the simulation, which was approximately 1/10 of initial value (0.0467 mg/L), while the *BA* also declined by approximately 90% (Table 14.5). At the same time, the *PSP* gradually increased (Figure 14.6) and ended with the value of 4.6155 g/m^2, which was more than two times greater than the initial value. The total phosphorus (*TP*) decreased significantly (from 0.8142 to 0.4529 mg/L). *TP* did not enter the theoretical range for lake shift, probably because the volume of the lake water was concentrated in EEWL (Zhang et al., 2003a). The *PEsed* rose due to the elevated setting rate of water particles, while the *PI* declined due to higher uptake by the submerged vegetation and the higher diffusion to the water column resulting from the lower *PS*. Lake restoration by reducing the nutrient loading was not necessarily effective due to internal cycling. Here, P_fold%40 resulted in a less significant improvement in the lake status than EEWL. The *PA* decreased from 0.0423 to 0.0168 mg/L, suggesting a 60% reduction in the phytoplankton. Meanwhile, the *TP* value also experienced a decline both at the beginning and in the end of the simulation in comparison to the results of the EEWL.

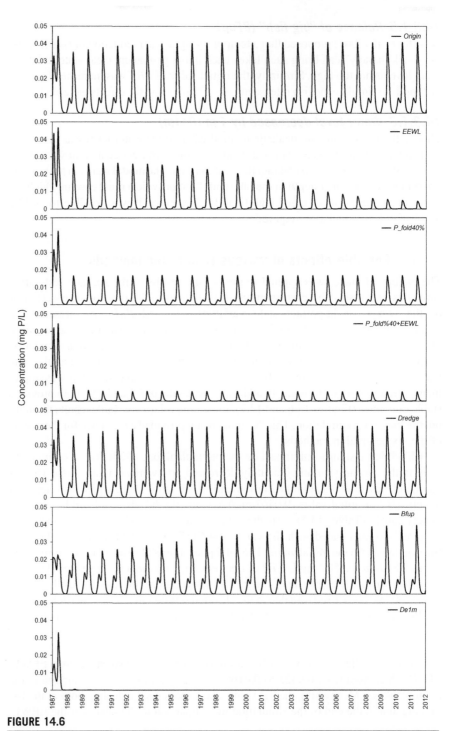

FIGURE 14.6

Response of the phosphorus in phytoplankton (*PA*) to different lake restoration methods.

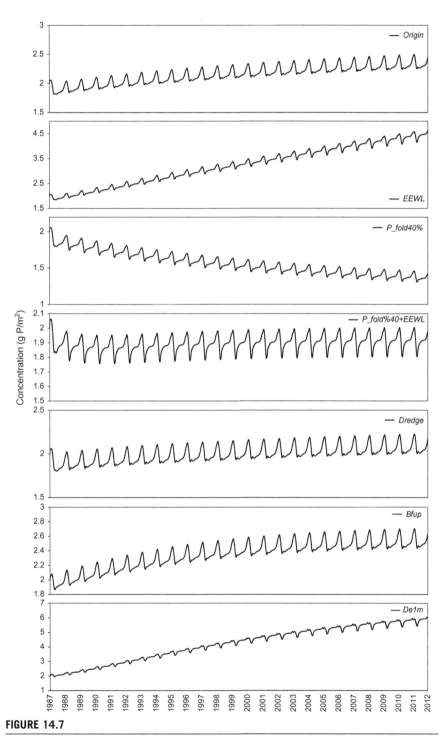

FIGURE 14.7

Response of the phosphorus in submerged vegetation (*PSP*) to different lake restoration methods.

Table 14.5 Comparison of the Results of the State Variables at the Start and the End of the Long-Term Simulation Under Different Restoration Strategies

Variables[a]	Origin		EEWL		P_fold%40		P_fold%40 + EEWL		Dredge		Bfup		De1m	
	Start	End	Start	End	Start	End	Start	End	Start	End	Start	End	Start	End
TP^b	0.7793	0.7169	0.8142	0.4529	0.5867	0.3451	0.5977	0.3541	0.7821	0.6248	0.8975	0.7143	0.819	0.4633
PS	0.5895	0.5453	0.6105	0.3167	0.4094	0.2085	0.4081	0.2275	0.5923	0.4530	0.7173	0.5438	0.6314	0.3148
PA	0.0442	0.0406	0.0467	0.0044	0.0423	0.0168	0.0444	0.0053	0.0442	0.0409	0.0226	0.0394	0.0330	0.0000
PZ	0.1198	0.1139	0.1273	0.1103	0.1150	0.1109	0.1224	0.1113	0.1198	0.1142	0.1265	0.1137	0.1221	0.1086
Pdet	0.0258	0.0171	0.0297	0.0215	0.0200	0.0089	0.0228	0.0100	0.0258	0.0167	0.0311	0.0174	0.0325	0.0399
PEsed	0.1584	0.1902	0.1590	0.3456	0.1567	0.1103	0.1573	0.1507	0.1579	0.1697	0.1609	0.2054	0.1653	0.4505
PI	0.0720	0.0635	0.0670	0.0253	0.0684	0.0331	0.0640	0.0239	0.0702	0.0616	0.0640	0.0627	0.0640	0.0852
PSP	2.0614	2.5004	2.0659	4.6155	2.0585	1.4605	2.0636	2.0049	2.0601	2.2285	2.0816	2.7056	2.1865	6.0554
PBF	0.3028	0.2102	0.3318	0.0110	0.2915	0.0023	0.3191	0.0052	0.3030	0.2127	0.2264	0.2258	0.2047	0.0078
PSF	1.5000	1.8377	1.5000	3.2471	1.5000	1.5505	1.5000	2.3001	1.5000	1.7783	1.5000	1.8804	1.5623	3.8728
Exergy	1,667,032	1,954,509	1,676,418	3,180,318	1,667,031	1,490,128	1,676,417	2,209,084	1,667,032	1,894,795	3,343,365	2,014,526	1,795,953	4,021,070
BA	7.2625	6.7879	7.6938	0.8175	7.0126	3.0232	7.3996	0.9855	7.2645	6.8340	3.8065	6.6109	5.7709	0.0000

Note: All data are the peak values of the first or the last year in the simulation.

[a]Units are shown in Table 14.2 (exergy as J/L).

[b]$TP = PS + PA + PZ + Pdet$.

However, the lower level of phosphorus in the system led to a lower level of *PSP* and provided a value decreased from 2.0585 to 1.4605 g/m^2. Therefore, based on this model, even though the reduction of the phosphorus loading may have a positive effect on the *TP* and *PA*, the restoration of the submerged vegetation was not necessarily successful. The P_fold%40 + EEWL provided a reduction in the level of *TP* to a similar extent as P_fold%40, as well as a reduction in the phytoplankton (*PA* and *BA*) close to the extent of EEWL (Table 14.5). It seems that the combination of the restoration strategies also combined their positive effects. However, the submerged vegetation was not restored under this scenario. The *PSP* remained constant throughout the simulation period and ended at 2.0049 g/m^2. The reduction of the phosphorus loading limited the growth of the submerged vegetation, which was represented as the phosphorus fraction to the biomass (1%). However, submerged vegetation gets the majority of P from sediment and the sediment P level is high, therefore it was not largely suppressed when the external P loading was reduced.

The dredging of the lake did not show a significant effect on the model outcomes in comparison to the restoration strategies of P_fold%40 or EEWL. The *TP* and *PA* slightly declined by approximately 20% and 7.5%, respectively. The *PSP* increased by 8.2% under this restoration. Bfup had a nearly 25% reduction in *TP* from 0.8975 to 0.7143 mg/L. However, the initial level of *TP* was significantly higher than that in other scenarios due to the addition of fish while the *PS* value in the end was almost the same as the "Origin" results. The initial level of *PA* was strongly suppressed due to elevated predation from the "big fish," but, as the *PBF* decreased during the long-term simulation, the *PA* value gradually increased to the same level as before the biomanipulation. The submerged vegetation was restored by 30%, with the peak value of 2.7056 g/m^2 during the last several years of the long-term simulation.

The final restoration was an idealistic method called De1m, which decreased the water level by 1 m in each month. Under this strategy, the phytoplankton almost disappeared while the submerged vegetation gradually increased to almost three times of the initial level in the last year (6 g/m^2), which turned out to be the best outcome among all of the restoration methods. This strategy is probably impracticable in lake restoration but provided a significant and immediate shift in lake status, which will be applied in the following structural dynamic approach together with EEWL.

The parameters of the model are calibrated based on the data from 1987 to 1988 and are applied in the long-term simulation. However, the eco-environment of the lake is changing over time, thus it requires updated monitoring data for parameter calibration in the 25 years. But to the best of our knowledge, monitoring data in Lake Chaohu are scarce. There are no reported data sets as abundant and reliable as those documented in Tu et al. (1990). Therefore, we only used this data for model calibration. Further field observation studies will be conducted in Lake Chaohu so that we can calibrate and validate our model to the current situation.

It was proposed that the changes in the hydrological conditions due to the construction of the sluice caused the loss of riparian wetlands, the macrophytes, and the blooms of algae in Lake Chaohu (Xu et al., 1999b). Here, the model ascertained the positive effect of water level management on the lake restoration, especially EEWL, which is believed to be more practical than De1m. The water depth in lake is one of the most important factors that affect the probability of a lake to be dominated by submerged vegetation (Scheffer and van Nes, 2007). The decreased water level reduced the light limitation of the submerged vegetation growth (Middelboe and Markager, 1997), allowed the plants to grow to the surface of the water (Scheffer and van Nes, 2007), and provided conditions for recovery. In addition, the shading effect was weakened because the phytoplankton was strongly suppressed under EEWL. A pronounced change induced by nature water level manipulation has also been observed in other lakes, such as the Swedish lakes Krankesjön and Tåkern (Blindow, 1992). Therefore, water level manipulation is a potential tool for managing the ecosystem state of shallow eutrophic lakes (Scheffer et al., 1993). A government project in China, which diverts the relatively cleaner water from Yangtze River into Lake Chaohu, has been conducted. The purpose of this project was to increase the water quality of the lake. It was recommended that better outcomes may be obtained if this project can also be associated with the water level management, such as EEWL. On the other hand, the reduction of the nutrient loading by 40% (strategy of P_fold%40) showed a positive effect on the reduction of *PA*. However, the *PSP* was not restored and even slightly decreased. The combination of the loading reduction and the water level management of EEWL showed an enhanced and rapid suppression of *PA*, but *PSP* was not restored under this method either. Moreover, the final values of *PA* in the combination restoration were close to that in EEWL. Therefore, EEWL should be the most practical and effective restoration method.

Submerged vegetation recovery was not very satisfactory in this study. The amount recovered by the model prediction is not sufficient. This is probably because submerged vegetation was expressed as phosphorus concentration in this model (1% of the biomass), which could be limited by the reduction in the system input. In addition, we did not observe a clear shift in submerged vegetation (*PSP*; Figure 14.7). Lake Chaohu has a surface area of 760 km^2, resulting in a much slower recovery speed of the submerged vegetation. Zhang et al. (2003a) developed a similar model and obtained a clear shift in submerged vegetation under nutrient loading variations. However, the lake that was modeled was much smaller than Lake Chaohu (less than 1% of surface area). It has also been shown by several ecological models that spatial heterogeneity of the environment tends to reduce the chance for large-scale shifts between alternative stable states (Van Nes and Scheffer, 2005). Unfortunately, there is also possibility that the submerged vegetation sub-model deviated to some extent from the real conditions in Lake Chaohu, resulting in unsatisfactory prediction results. Therefore, caution is needed when assessing the accuracy of the predicted restored concentration of submerged vegetation.

As a large lake, it was impossible to implement the fish stock reduction (Scheffer et al., 1993) in Lake Chaohu. The biomanipulation method by increasing the value of *PBF*, which was called "Bfup," did not show a very successful lake restoration. In addition, the "Dredge" method in the surface of the sediment did not provide positive results. There were already similar projects in Lake Chaohu, such as the release of fish fry by the fishery authorities and sediment dredge by ships in the lake. These projects may not aim at lake restoration but do benefit the lake environmental health protection. However, the results in this study suggested that these strategies may not be sufficient for lake restoration toward the submerged vegetation-domination state.

14.4 STRUCTURAL DYNAMIC APPROACHES
14.4.1 Structural dynamic approaches on the lake shift

The switching of the lake from algae domination to submerged vegetation domination was obvious in the lake restoration simulations by EEWL and De1m (see Section 14.3). The shift in the lake state should definitely be associated with the change of certain parameters in the model. However, these parameters remained constant in the non-SDM approaches. The structural dynamic approach was introduced into the lake model simulation of the phytoplankton to submerged vegetation switch. In this section, two structural dynamic approaches were conducted under the restoration method of both EEWL and De1m because rapid and significant transitions in the state variables within the first several years were found under both scenarios. The model was performed for 5 years, which can save the calculation work substantially and was also sufficient for change of the lake state because the retention time of the lake was 136 days (Table 14.1). It was checked once in every 25 days whether the present parameter combination or those with $\pm 10\%$ change yielded the highest exergy. The exergy (kJ/L) was calculated as in Equation (14.11) (Jørgensen, 1997; Xu, 1997):

$$\text{Exergy} = 18.7 \cdot \sum_{i=1}^{n} \beta_i \cdot C_i \qquad (14.11)$$

where Ex (J/L) is the exergy (Jørgensen, 1997). β_i is the weighting factor of the i-th component in the system and C_i (mg P/L) is the corresponding concentration of this component (Zhang et al., 2003b). $\beta_1 = 1$ represents detritus, and β_i equals to 1 for sediment, 20 for phytoplankton, 40 for zooplankton, 50 for macrophytes, and 499 for fish. These values are made on basis of the number of informative genes in the organisms (Jørgensen and de Bernardi, 1998; Jørgensen and Fath, 2011). In addition, the exergy in detritus has been estimated to be 18.7 kJ/g (Jørgensen and de Bernardi, 1998), so that the equation above for Ex was multiplied by a factor of 18.7 to obtain the exergy in J/L.

In each restoration method, two structural dynamic approaches, referred to as SDM1 and SDM2, were conducted and their results were compared.

- SDM1: Two dynamic parameters (*GAmax* and *UPmaxrate*) were involved.
- SDM2: Four dynamic parameters (*GAmax*, *UPmaxrate*, *Mmax_PA* and *Mmax_PSP*) were involved.

The definitions for these parameters are as follows, *GAmax*: maximum growth rate of phytoplankton; *UPmaxrate*: maximum phosphorus uptake rate of phytoplankton; *Mmax_PA*: maximum mortality rate of phytoplankton; *Mmax_PSP*: maximum mortality rate of submerged vegetation.

The parameter variation during the restoration process of Lake Chaohu was interpreted by the optimization of the exergy as the goal function. The development of exergy was calculated for non-SDM and two SDMs under EEWL and De1m to verify if changes in the parameters could provide higher exergy, which indicated a better solution and a higher possibility for survival in lake ecosystems. The results in both SDM1 and SDM2 under two restoration methods were also compared to test whether including more parameters in SDMs will result in different parameter variations, higher flexibility, and exergy or not. In addition, because EEWL and De1m were both controlling the water level with different intensities, we compared the results to evaluate the model response under different scale of changes in forcing function.

14.4.2 **SDMs results and discussion**

Before incorporating the structural dynamic approach into the lake restoration scenarios, it was necessary to see the development of the exergy in non-SDMs under different restoration, which is presented in Figure 14.8. De1m provided the highest exergy, indicating the best restoration effect in the lake ecosystem. EEWL had the second highest exergy value. It can therefore be concluded that the control of the water level in the lake has a significant impact on the lake system in terms of exergy, which was influenced by the value of state variables such as *PA* or *PSP*. The other restoration methods suggested a relatively insufficient recovery of the lake from the

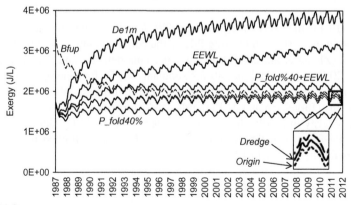

FIGURE 14.8

Development of exergy in response to different lake restoration methods. (For the color version of this figure, the reader is referred to the online version of this chapter.)

point of view of the system. P_fold%40+EEWL showed exergy even lower than that of the "Origin." The initial level of exergy in Bfup was higher than that in the other methods due to the elevated *PBF* values. However, the exergy in this scenario gradually decreased and ended at almost the same level as that in the "Origin," suggesting little improvement in the lake status.

The developments of *PA, PSP* in non-SDM, SDM1, and SDM2 were presented in Figure 14.9 for the scenario of EEWL and Delm while the variations of the parameters in the SDMs and the corresponding development of exergy are shown in Figures 14.10 and 14.11, respectively. The details of the results can be found in Table 14.6. In SDM1 under EEWL and De1m, *PA* decreased much faster than the results of non-SDM, but *PSP* did not show significant changes (Figure 14.9) because the related parameters were not included in the structural dynamic approach and remained constant. *UPmaxrate* and *GAmax* gradually decreased to low values under

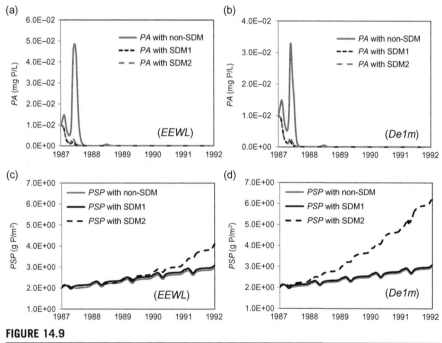

FIGURE 14.9

Development of phosphorus in phytoplankton (*PA*) and submerged vegetation (*PSP*) in non-SDM, SDM1, and SDM2 approaches under the restoration method of EEWL (a and c) and De1m (b and d) during 1987 to 1992. SDM1: structural dynamic model with two dynamic parameters (*GAmax* and *UPmaxrate*); SDM2: structural dynamic model with four dynamic parameters (*GAmax, UPmaxrate, Mmax_PA*, and *Mmax_PSP*); *GAmax*: maximum growth rate of phytoplankton; *UPmaxrate*: maximum phosphorus uptake rate of phytoplankton; *Mmax_PA*: maximum mortality rate of phytoplankton; *Mmax_PSP*: maximum mortality rate of submerged vegetation.

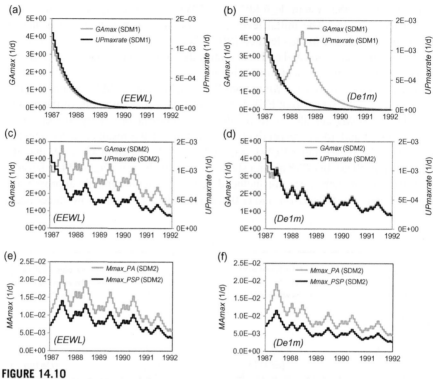

FIGURE 14.10

Parameter changes in the SDM1 (a and b) and SDM2 (c, d, e, and f) under the restoration methods of EEWL (a, c, and e) and De1m (b, d, and f) during 1987–1992. SDM1: structural dynamic model with two dynamic parameters (*GAmax* and UPmaxrate); SDM2: structural dynamic model with four dynamic parameters (*GAmax*, *UPmaxrate*, *Mmax_PA*, and *Mmax_PSP*); *GAmax*: maximum growth rate of phytoplankton; *UPmaxrate*: maximum phosphorus uptake rate of phytoplankton; *Mmax_PA*: maximum mortality rate of phytoplankton; *Mmax_PSP*: maximum mortality rate of submerged vegetation.

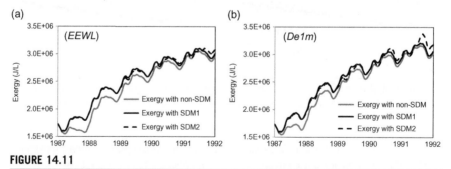

FIGURE 14.11

Development of exergy in non-SDM, SDM1, and SDM2 approaches under the restoration method of EEWL (a) and De1m (b) from 1987 to 1992.

Table 14.6 Comparison of the Results of SDMs at the Start and the End of the 5-Year Simulation Under Restoration Strategy of EEWL and De1m

Method	Models	PA (mg P/L)		PSP (g P/m²)		GAmax (1/d)		UPmaxrate (1/d)		Mmax_PA (1/d)		Mmax_PSP (1/d)		Exergy (J/L)	
		Start	End	Start	End	Start	End	Start	End	Start	End	Start	End	Start	End
EEWL	Non-SDM	0.01	0.00	0.01	2.98	3.60	3.60	0.00126	0.00126	0.01080	0.01080	0.00720	0.00720	1,734,404.4	2,961,197.0
	SDM1	0.01	0.00	2.00	3.08	3.60	0.02	0.00126	0.00000	0.01080	0.01080	0.00720	0.00720	1,734,404.4	3,011,323.9
	SDM2	0.01	0.00	2.00	4.10	3.60	1.25	0.00126	0.00024	0.01080	0.00554	0.00720	0.00369	1,734,404.4	3,079,095.1
De1m	Non-SDM	0.01	0.00	2.00	3.01	3.60	3.60	0.00126	0.00126	0.01080	0.01080	0.00720	0.00720	1,734,404.4	3,019,496.5
	SDM1	0.01	0.00	2.00	3.09	3.60	0.01	0.00126	0.00000	0.01080	0.01080	0.00720	0.00720	1,734,404.4	3,068,046.5
	SDM2	0.01	0.00	2.00	6.18	3.60	0.83	0.00126	0.00022	0.01080	0.00453	0.00720	0.00275	1,734,404.4	3,183,701.6

Note: SDM1, structural dynamic model with two dynamic parameters (GAmax and UPmaxrate); SDM2, structural dynamic model with four dynamic parameters (GAmax, UPmaxrate, Mmax_PA, and Mmax_PSP); GAmax, maximum growth rate of phytoplankton; UPmaxrate, maximum phosphorus uptake rate of phytoplankton; Mmax_PA, maximum mortality rate of phytoplankton; Mmax_PSP, maximum mortality rate of submerged vegetation.

EEWL while fluctuation was found in *GAmax* under De1m, but it also approached low values at the end of the 5-year simulation (Figure 14.10). The exergy in SDM1 under both restorations was slightly higher than that in non-SDM (Figure 14.11 and Table 14.6), indicating a higher probability of survival in the lake and a better "solution" for model parameters. In SDM2, the inclusion of the maximum mortality rate of phytoplankton and submerged vegetation (*Mmax_PA* and *Mmax_PSP*) under both restorations had little impact on the outcomes of *PA*. In contrast, the inclusion significantly increased the value of *PSP* (Figure 14.9), which ended at 4.10 g P/m^2 under EEWL and 6.18 g P/m^2 under De1m. These values are significantly higher than the final values in the corresponding SDM1 (Table 14.6). The four parameters all decreased under both restorations, among which *GAmax* and *UPmaxrate* had a slower descending trend and much more fluctuations when compared to SDM1 (Figure 14.10). In addition, under De1m, all four parameters showed a faster decreasing speed (Figure 14.10) and ended at lower values (Table 14.6) than those in EEWL, suggesting a stronger adaptation under a stronger change in the forcing function. Furthermore, under both restorations, the development of exergy in SDM2 was similar to that in SDM1 to a great extent but became slightly higher in the last part of the simulation (Figure 14.11), leading to a higher final exergy (Table 14.6). The final exergy was also higher under De1m than EEWL, suggesting that the lake could recover to a better condition under De1m than EEWL.

From an ecological perspective, because the simulation was associated with the disappearance of phytoplankton and the recovery of submerged vegetation, it was reasonable to observe the decreasing trends of the four parameters (maximum growth rate and the phosphorus uptake rate of phytoplankton, the maximum mortality rates of phytoplankton and submerged vegetation) under both restorations. Jørgensen (1999) provided a case study based on the results from Søbygaard Lake (Jeppesen et al., 1990). The study found that parameters variations were associated with radical changes in the lake 3 years after the nutrient loading was reduced. Particularly, the maximum growth rate of phytoplankton also decreased significantly. Ecosystems belong to the complex adaptive systems (Brown, 1995). With large numbers of feedbacks and regulations, the living organisms and populations can survive and reproduce under changes in external conditions (Jørgensen, 1999). Adaptation happens when the actual properties of the species are changed (Jørgensen, 1999). Indeed, the changes in the above four parameters are the adaptation of phytoplankton and submerged vegetation in response to the drastic changes in water levels.

At the same time, these variations in parameters contributed to the elevation in exergy (Figure 14.11) when compared to the non-SDM approaches, indicating that the new combination for the parameters were a better "solution" of the ecosystem, which will give more possibilities to survive in nature (Jørgensen, 1999). In addition, the maximum exergy storage is based on the hypothesis that an ecosystem tends to move away from thermodynamic equilibrium as far as possible (Zhang et al., 2003b). Result in this study can be considered as the support for maximum exergy storage hypothesis (Zhang et al., 2003a).

No doubt those structural changes could occur in lake state transitions from phytoplankton-domination to submerged-vegetation domination. This will result in a significant variation in the parameters in model language (Jørgensen, 1999). These changes should be quantitatively evaluated by a SDM as in this study, which is more convincing than the artificial assignment of the key parameters variation in the model. In addition, Scheffer et al. (1993) have illustrated that many ecological mechanisms are probably involved in maintaining the lake system in one of the two alternative stable states, or more precisely, two alternative regimes (Scheffer and Carpenter, 2003; Scheffer and van Nes, 2007). The lake system may have resistance in state transition. Therefore, the parameter variations were not necessarily linear during the shift. Instead, the fluctuations should occur (Figure 14.9) due to system feedbacks or regulations. These characteristics of parameter variations can only be captured by structural dynamic approaches. Moreover, it is worth noting that in SDMs that simulated for five years, the final values of *PSP* were almost equal to the final values in non-SDM approaches with much longer duration (25 years; Table 14.5). Therefore, there is a possibility that the results of non-SDMs in Section 14.4 had largely underestimated the recovery pace of the submerged vegetation, because parameter variations were not included. Instead, SDMs could evaluate the effect of restoration method by estimating the speed of the submerged vegetation recovery with higher accuracy.

The results also showed that water level management had a significant impact on the lake while De1m provided a stronger change in forcing function than EEWL, resulting in higher final PSP and exergy values, a faster decreasing speed and lower final values of the parameters in both SDM1 and SDM2 (Figures 14.9–14.11 and Table 14.6). Therefore, the SDMs can provide a stronger change in the model parameters because the intensity of the forcing function variations is enhanced.

In SDM2 under both EEWL and De1m, when two more parameters (*Mmax_PA*, *Mmax_PSP*) were incorporated in the simulation compared to SDM1, the dynamics of the two former parameters (*GAmax*, *UPmaxrate*) presented different variation processes (Figure 14.10a–d) with a slower decreasing speed and more fluctuations, suggesting that the results can be very likely different than each other when a different parameter set is included in the SDM. Moreover, the SDM2 provided reasonable values for the four parameters in the end of the simulation, all of which were within the theoretical ranges or slightly deviated from the lower boundaries. Meanwhile, the final values of *GAmax* and *UPmaxrate* were extremely low in SDM1, which were far from the parameter ranges in literature (Kong et al., 2013). In addition, SDM2 was better than SDM1 because a higher exergy was obtained (Figure 14.11), which should be attributed to the significant elevation of the *PSP* under both restoration methods. SDM2 gives an opportunity for more parameters to be selected, which provided a wider possibility in adaptation. As a result, unrealistic outcomes were provided in SDM1 because insufficient parameters were included while a better and more realistic "solution" was obtained by SDM2.

Theoretical speaking, all parameters in the model should be included in the SDM due to the top-down and bottom-up effect in the lake ecosystem. However, it is impossible to do so because of the calculation capacity limitation. Sensitivity analysis could be a useful tool in parameter selection, but it is more important to study the parameters associated with the state variables of most concern in the model, such as *PA* and *PSP* in this study. Moreover, the maximum growth rate of submerged vegetation was considered in SDM. This idea was finally discarded for two reasons: First, a decline of the mortality rate would give the same effect as an increase of the growth rate, and adding one more parameters would increase the amount of calculations significantly. Second, the maximum growth rate of the submerged vegetation was relatively stable in reality. Little variation could cause a disaster in the model simulation. The parameters in model were limited by the realistic ranges that were possible in nature. Therefore, the parameter variation in SDM should be extremely careful, realistic, and restricted to certain ranges. It is more convincing if there are in situ measurements of the parameter changes for validation. The data in a reference lake or data from mesocosm (Suzuki et al., 2000) can be alternatives when it is too difficult to obtain the values in Lake Chaohu.

14.4.3 Conclusion

An ecological model focusing on the dynamics of phosphorus cycling was developed for Lake Chaohu, China. The model was satisfactory calibrated based on monthly observations from 1987.5 to 188.4, when the lake was already suffering from eutrophication. The model provided stable results in a long-term simulation of 25 years after calibration, which is consistent with the real situation. Subsequently, the long-term effects of six restoration strategies proposed for Lake Chaohu were predicted, and the outcomes were compared with the stable results when no restoration was implemented. The competition between the submerged vegetation and the phytoplankton was successfully modeled. The lake water level manipulations by applying EEWL had a significant impact on the phytoplankton (reduced by 90%), while the submerged vegetation was gradually restored by a factor of two. Decreasing the water level by 1 m (De1m) resulted in the disappearance of phytoplankton, while the submerged vegetation ended at almost three times the initial level but with little practicability. Phosphorus loading reduction by 40% (P_fold%40) resulted in a reduction in the level of soluble phosphorus in lake water (*PS*) but a less significant improvement in the reduction of phytoplankton in comparison to EEWL. The combination of the restoration strategies (P_fold%40 + EEWL) combined the positive effects, but the submerged vegetation was not restored under this scenario. Dredging the lake sediment and biomanipulation by adding "big fish" (Dredge and BFup) did not show a significant restoration in the lake. EEWL was recommended as the most practical and effective restoration method. In addition, the indicator of exergy generally increased during the restoration processes, suggesting a better condition of the lake ecosystem. Among the six restoration methods, De1m and EEWL had the first and second highest exergy values.

A structural dynamic approach was introduced into the lake model simulation of the phytoplankton–submerged vegetation switch. The variation of the parameters of phosphorus uptake, growth and mortality rate associated with phytoplankton, and submerged vegetation were captured while the exergy in the SDMs was generally higher than that in non-SDM, indicating a better solution for the parameters under changing forcing functions, which will lead to more chances of survival in nature. The results also showed that in the SDMs, De1m provided higher final PSP and exergy values, a faster decreasing speed and lower final values of parameters than EEWL. As De1m gave a stronger change than EEWL, the SDMs can provided stronger changes in the model structural because the intensity of the forcing function variations were enhanced.

In SDM1 with two parameters and SDM2 with two more parameters, the first two parameter variations had generally similar trends. More fluctuations and more realistic final values appear when more parameters were included. Exergy was even higher with more parameters in SDMs because it gives an opportunity for more parameters to be selected, which provided a wider possibility of adaptation and a more realistic "solution" for the ecosystem. Therefore, the parameter selection in SDMs should be very careful and realistic.

The results of the prediction on the effects of the lake restoration strategies require further validation in the future. In addition, an improvement in the model is expected in further study, especially in the fish and submerged vegetation sub-model that were insufficient in their complexity. Nevertheless, this study has provided valuable information for policy-making in Lake Chaohu ecosystem restoration management.

References

Amemiya, T., Enomoto, T., Rossberg, A.G., Yamamoto, T., Inamori, Y., Itoh, K., 2007. Stability and dynamical behavior in a lake-model and implications for regime shifts in real lakes. Ecol. Model. 206 (1–2), 54–62.

Arhonditsis, G.B., Brett, M.T., 2005. Eutrophication model for Lake Washington (USA) Part I. Model description and sensitivity analysis. Ecol. Model. 187, 140–178.

Asaeda, T., Van Bon, T., 1997. Modelling the effects of macrophytes on algal blooming in eutrophic shallow lakes. Ecol. Model. 104 (2–3), 261–287.

Blindow, I., 1992. Long-term and short-term dynamics of submerged macrophytes in 2 shallow eutrophic lakes. Freshwater Biol. 28, 15–27.

Brown, J.H., 1995. Macroecology. The University of Chicago Press, Chicago, IL, p. 269.

Bruce, L.C., Hamilton, D., Imberger, J., Gal, G., Gophen, M., Zohary, T., Hambright, K.D., 2006. A numerical simulation of the role of zooplankton in C, N and P cycling in Lake Kinneret, Israel. Ecol. Model. 193 (3–4), 412–436.

Chen, C.W., Orlob, G.T., 1975. Ecologic Simulation for Aquatic Environments. Academic Press, New York, NY.

CMDSSS (China Meteorological Data Sharing Service System), 1987–1988. Daily Meteorological Data. http://www.cma.gov.cn/2011qxfw/2011qsjgx/index.htm.

Deng, D.G., Xie, P., Zhou, Q., Yang, H., Guo, L.G., 2007. Studies on temporal and spatial variations of phytoplankton in Lake Chaohu. J. Intergr. Plant Biol. 49, 409–418.

Guo, L.G., 2005. Studies on Fisheries Ecology in a Large Eutrophic Shallow Lake, Lake Chaohu: Doctoral dissertation. Institude of Hydrobiology, Chinese Acadamy of Sciences, Wuhan.

Gurkan, Z., Zhang, J.J., Jørgensen, S.E., 2006. Development of a structurally dynamic model for forecasting the effects of restoration of Lake Fure, Denmark. Ecol. Model. 197, 89–102.

Hu, W.P., Salomonsen, J., Xu, F.L., Pu, P.M., 1998. A model for the effects of water hyacinths on water quality in an experiment of physico-biological engineering in Lake Taihu, China. Ecol. Model. 107 (2–3), 171–188.

Hu, W.P., Jørgensen, S.E., Zhang, F.B., 2006. A vertical-compressed three-dimensional ecological model in Lake Taihu, China. Ecol. Model. 190, 367–398.

Janse, J.H., Scheffer, M., Lijklema, L., Van Liere, L., Sloot, J.S., Mooij, W.M., 2010. Estimating the critical phosphorus loading of shallow lakes with the ecosystem model PCLake: sensitivity, calibration and uncertainty. Ecol. Model. 221, 654–665.

Jeppesen, E.J., Mortensen, E., Sortkjaer, O., Kristensen, P., Bidstrup, J., Timmermann, M., Jensen, J.P., Hansen, A.M., Søndergaard, M., Müller, J.P., Jensen, J., Riemann, B., Lindegaard-Petersen, C., Bosselmann, S., Christoffersen, K., Dall, E., Andersen, J.M., 1990. Fish manipulation as a lake restoration tool in shallow, eutrophic temperate lakes. Cross-analysis of three Danish case studies. Hydrobiologia 200:201, 205–218.

Jørgensen, S.E., 1976. An eutrophication model for a lake. Ecol. Model. 2, 147–165.

Jørgensen, S.E., 1986. Structural dynamic-model. Ecol. Model. 31, 1–9.

Jørgensen, S.E., 1988. Use of models as experimental tool to show that structural-changes are accompanied by increased exergy. Ecol. Model. 41, 117–126.

Jørgensen, S.E., 1992a. Development of models able to account for changes in species composition. Ecol. Model. 62, 195–208.

Jørgensen, S.E., 1992b. Parameters, ecological constraints and exergy. Eco. Model. 62, 163–170.

Jørgensen, S.E., 1997. Integration of Ecosystem Theories: A Pattern, second ed. Kluwer Academic Publishers, Dordrecht, The Netherlands, p. 386.

Jørgensen, S.E., 1999. State-of-the-art of ecological modelling with emphasis on development of structural dynamic models. Ecol. Model. 120, 75–96.

Jørgensen, S.E., 2002. Integration of Ecosystem Theories: A Pattern, third ed. Kluwer Academic Publishers, Dordrecht, The Netherlands.

Jørgensen, S.E., de Bernardi, R., 1997. The application of a model with dynamic structure to simulate the effect of mass fish mortality on zooplankton structure in Lago di Annone. Hydrobiologia 356, 87–96.

Jørgensen, S.E., de Bernardi, R., 1998. The use of structural dynamic models to explain successes and failures of biomanipulation. Hydrobiologia 379, 147–158.

Jørgensen, S.E., Fath, B.D., 2011. Fundamentals of Ecological Modelling, fourth ed. Elsevier, Amsterdam.

Jørgensen, S.E., Nielsen, S.N., 1994. Models of the structural dynamics in lakes and reservoirs. Ecol. Model. 74, 39–46.

Jørgensen, S.E., Padisak, J., 1996. Does the intermediate disturbance hypothesis comply with thermodynamics? Hydrobiologia 323, 9–21.

Jørgensen, S.E., Mejer, H., Friis, M., 1978. Examination of a lake model. Ecol. Model. 4, 253–278.

Jørgensen, S.E., Marques, J., Nielsen, S.N., 2002a. Structural changes in an estuary, described by models and using exergy as orientor. Ecol. Model. 158, 233–240.

Jørgensen, S.E., Ray, S., Berec, L., Straskraba, M., 2002b. Improved calibration of a eutrophication model by use of the size variation due to succession. Ecol. Model. 153, 269–277.

Kong, X.Z., Jørgensen, S.E., He, W., Qin, N., Xu, F.L., 2013. Predicting the restoration effects by a structural dynamic approach in Lake Chaohu, China. Ecol. Model. 266, 73–85.

Kuusisto, M., Koponen, J., Sarkkula, J., 1998. Modelled phytoplankton dynamics in the Gulf of Finland. Environ. Model. Softw. 13, 461–470.

Lehman, J.T., Botkin, D.B., Likens, G.E., 1975. Assumptions and rationales of a computer model of phytoplankton population-dynamics. Limnol. Oceanogr. 20 (3), 343–364.

Marchi, M., Jørgensen, S.E., Bécares, E., Corsi, I., Marchettini, N., Bastianoni, S., 2011. Dynamic model of Lake Chozas (León, NW Spain)-Decrease in eco-exergy from clear to turbid phase due to introduction of exotic crayfish. Ecol. Model. 222, 3002–3010.

Meijer, M.L., Raat, A.J., Doef, R.W., 1989. Restoration by biomanipulation of Lake Bleiswijkse Zoom the Netherlands first results. Hydrobiol. Bull. 23, 49–58.

Middelboe, A.L., Markager, S., 1997. Depth limits and minimum light requirements of freshwater macrophytes. Freshw. Biol. 37, 553–568.

Moss, B., 1988. Ecology of Fresh Waters: Man & Medium, second ed. Blackwell Scientific, Oxford, pp. 1–400.

Phillips, G.L., Eminson, D., Moss, B., 1978. A mechanism to account for macrophyte decline in progressively eutrophicated fresh waters. Aquat. Bot. 4, 103–126.

Sagehashi, M., Sakoda, A., Suzuki, M., 2000. A predictive model of long-term stability after biomanipulation of shallow lakes. Water Res. 34, 4014–4028.

Sagehashi, M., Sakoda, A., Suzuki, M., 2001. A mathematical model of a shallow and eutrophic lake (the Keszthely Basin, Lake Balaton) and simulation of restorative manipulations. Water Res. 35 (7), 1675–1686.

Scheffer, M., 1998. Ecology of Shallow Lakes. Chapman and Hall, London, p. 357.

Scheffer, M., Carpenter, S.R., 2003. Catastrophic regime shifts in ecosystems: linking theory to observation. Trends Ecol. Evol. 18, 648–656.

Scheffer, M., van Nes, E.H., 2007. Shallow lakes theory revisited: various alternative regimes driven by climate, nutrients, depth and lake size. Hydrobiologia 584, 455–466.

Scheffer, M., Hosper, S.H., Meijer, M.L., Moss, B., Jeppesen, E., 1993. Alternative equilibria in shallow lakes. Trends Ecol. Evol. 8, 275–279.

Scheffer, M., Carpenter, S., Foley, J.A., Folke, C., Walker, B., 2001. Catastrophic shifts in ecosystems. Nature 413, 591–596.

Suzuki, M., Sagehashi, M., Sakoda, A., 2000. Modelling the structural dynamics of a shallow and eutrophic water ecosystem based on mesocosm observations. Ecol. Model. 128, 221–243.

Tu, Q.Y., Gu, D.X., Yi, C.Q., Xu, Z.R., Han, G.Z., 1990. The Researches on the Lake Chaohu Eutrophication. Publisher of University of Science and Technology of China, Hefei (in Chinese).

Van Nes, E.H., Scheffer, M., 2005. Implications of spatial heterogeneity for regime shifts in ecosystems. Ecology 86, 1797–1807.

Vollenweider, R.A., 1975. Input-output models with special reference to the phosphorus loading concept. Schweiz. Z. Hydrol. 37, 53–84.

Vollenweider, R.A., 1979. Concept of nutrient load as a basis for the external control of the eutrophication process in lakes and reservoirs. J. Water Wastewater Res. 12, 46–56.

Xie, P., 2009. Reading About the Histories of Cyanobacteria, Eutrophication and Geological Evolution in Lake Chaohu. Science Press, Beijing (in Chinese).

Xu, F.L., 1997. Exergy and structural exergy as ecological indicators for the development state of the Lake Chaohu ecosystem. Ecol. Model. 99, 41–49.

Xu, F.L., Jørgensen, S.E., Tao, S., Li, B.G., 1999a. Modeling the effects of ecological engineering on ecosystem health of a shallow eutrophic Chinese lake (Lake Chao). Ecol. Model. 117, 239–260.

Xu, F.L., Tao, S., Xu, Z.R., 1999b. The restoration of riparian wetlands and macrophytes in Lake Chao, an eutrophic Chinese lake: possibilities and effects. Hydrobiologia 405, 169–178.

Zhang, T.F., Lu, X.P., 1986. Investigation and evaluation of water environmental quality of Lake Chaohu. In: Lake Chaohu Water Environment, Ecological Evaluation and Countermeasures-Special Report (No.11), Hefei (in Chinese).

Zhang, J.J., Jørgensen, S.E., Beklioglu, M., Ince, O., 2003a. Hysteresis in vegetation shift - Lake Mogan prognoses. Ecol. Model. 164, 227–238.

Zhang, J.J., Jørgensen, S.E., Tan, C.O., Beklioglu, M., 2003b. A structurally dynamic modelling—Lake Mogan, Turkey as a case study. Ecol. Model. 164, 103–120.

Zhang, J.J., Jørgensen, S.E., Mahler, H., 2004. Examination of structurally dynamic eutrophication model. Ecol. Model. 173, 313–333.

Zhang, J.J., Gurkan, Z., Jørgensen, S.E., 2010. Application of eco-exergy for assessment of ecosystem health and development of structurally dynamic models. Ecol. Model. 221, 693–702.

Exploring the Mechanism of Catastrophic Regime Shift in a Shallow Plateau Lake: A Three-Dimensional Water Quality Modeling Approach

15

Rui Zou[a], Yuzhao Li[b], Lei Zhao[c], Yong Liu[b],*

[a]*Tetra Tech, Inc., Fairfax, Virginia, USA*
[b]*Key Laboratory of Water and Sediment Sciences (MOE), College of Environmental Science and Engineering, Peking University, Beijing, China*
[c]*Yunnan Key Laboratory of Pollution Process and Management of Plateau Lake-Watershed, Kunming, China*
*Corresponding author: e-mail address: yongliu@pku.edu.cn

15.1 INTRODUCTION

Eutrophication is the disturbance of aquatic ecosystems with excessive nutrients or an impaired ecosystem, leading to algae blooms and anoxic events. It has been widely acknowledged that eutrophication is a stubborn condition for surface waters worldwide (Harper, 1992; Carpenter, 2005). Massive lake ecological changes can affect water supplies, fisheries, rangeland productivities, as well as other ecosystem services. Lake eutrophication is also believed to be a regime shift from the clean status to the turbid one. Previous studies proved that such shifts can be attributed to alternative stable states (Scheffer et al., 2001; Scheffer and Carpenter, 2003; Turner et al., 2008; Carpenter et al., 2011). Specifically, an oligotrophic lake is in a clean regime, while an eutrophic lake is a stable regime symptomized by turbid water and an algae-dominant aquatic ecosystem (Scheffer et al., 2001; Scheffer and Carpenter, 2003). Regime shifts always occur unexpectedly and follow nonlinear trends. The theories of ecological resilience and recovery revealed that once lake eutrophication occurred, it usually took more time or higher costs to restore it to its predisturbance oligotrophic state or clean water (Gunderson, 2000; Walker and Salt, 2006). Apparently, it is essential to explore the mechanisms behind the regime shift from the clean state to the turbid state in lakes in order to provide guidance for practical management decision making for eutrophication prevention and control.

Understanding the mechanism of regime shift and providing management guidance would rely highly on the development and application of mathematical models. In general, two kinds of models, namely statistic models and mechanistic models, have been widely applied in exploring lake ecological systems. Statistical analysis has been popularly applied to explore mechanism of eutrophication among different lakes or reveal their similarities as well as differences (Solow and Beet, 2005; Gal and Anderson, 2010). However, such models suffer the major limitation of lacking predictive capability due to their data-oriented nature. An alternative to these data-oriented models is the process-oriented water quality models (WQM), which have a more explicit mechanistic basis and allow for comprehensive prediction of system behavior under a wide variety of conditions (Chau and Jin, 1998; Arhonditsis and Brett, 2005). As for a specific lake, it is more likely to achieve insight and foresight of a catastrophic regime shift in the ecosystem using a mechanistic model rather than using statistical models. During the last two decades, there have been significant achievements in the development and application of mechanistic lake WQM (Carpenter et al., 1999; Guo et al., 2001; Chan et al., 2013). The applications of these models successfully helped understand key processes of aquatic ecosystems, hence providing guidelines for controlling the exogenous or ingenuous input for eutrophication control.

Since algal dynamics in eutrophic waters is highly related to vertical light structure, three-dimensional (3D) models are often developed and used in lake or coastal regimes studies (Drago et al., 2001; Moll and Radach, 2003). Cerco and Cole (1993) applied a 3D eutrophication model CE-QUAL-ICM to Chesapeake Bay over a 3-year period (1984–1986). The model successfully simulated water-column and sediment processes that affect water quality. Phenomena simulated include formation of the spring algal bloom subsequent to the annual peak in nutrient runoff, onset and breakup of summer anoxia, and coupling of organic particle deposition with sediment water nutrient and oxygen fluxes (Cerco and Cole, 1993). Mao et al. (2008) developed a 3D eutrophication model to investigate the eutrophication dynamics in Lake Taihu, China. The model fully coupled the biological processes and hydrodynamics and also took into account the effects of sediment release and the external loads from the tributaries. The comparisons between model results and field data in year 2000 indicated that the model is able to simulate the eutrophication dynamics in Lake Taihu with reasonable accuracy (Mao et al., 2008). The previous studies demonstrated that 3D WQM coupled with hydrodynamics could provide information about temporal variations of the chemical and biological processes. Integration of ecological and hydrodynamic processes could certainly provide insightful understanding of detailed eutrophication features.

The environmental fluids dynamics code (EFDC) was selected as the computational platform for developing the model in this study. EFDC is a widely applied 3D hydrodynamic and water quality simulation framework capable of simulating water circulation, temperature dynamics, and advanced eutrophication processes involving nutrients, phytoplankton, macrophytes, and predation/grazing processes. Compared with other analog systems, obvious advantages of the EFDC model include its strong ability to adapt to the needs of the problem. Specifically, the EFDC model can be used for zero-dimensional, one-dimensional, two-dimensional, and 3D water environment simulation. Currently, the EFDC has been successfully applied in river, lake, estuary, bay, and wetland water environment systems (Seo et al., 2010; Li et al., 2011; Peng et al., 2011; Shi et al., 2011; Wu and Xu, 2011; Alarcon and McAnally, 2012). Jeong et al. (2010) used EFDC to analyze the salinity intrusion characteristics downstream in the Geum River. The numerical simulation was performed to investigate the influence range for salinity intrusion when the gates were fully opened. The results showed that the EFDC model used for numerical simulation has high accuracy (Jeong et al., 2010). Wu and Xu (2011) applied a 3D EFDC model to predict the algae bloom in Lake Daoxiang, Beijing. The results showed that the simulated chlorophyll *a* (Chl *a*) concentration basically agrees with the observed concentration (Wu and Xu, 2011). Zhao et al. (2012) developed a 3D hydrodynamic and water quality model based on EFDC for Lake Fuxian. The model accurately reproduced the hydrodynamic as well as water quality process and suggested a reasonable numerical representation of the prototype system for further TMDL analyses. Previously, we developed a 3D water quality model in Lake Yilong for understanding the water quality responses to various load reduction intensities and ecological restoration measures (Zhao et al., 2013). It verified that the EFDC model may perform well for Chl *a* simulation and algae bloom prediction. An EFDC-based model was developed in this study to simulate the response of water quality and aquatic ecosystem to external perturbation in Lake Yilong.

15.2 MATERIALS AND METHODOLOGY

15.2.1 Study area: Lake Yilong

Lake Yilong is one of the nine largest plateau lakes in Yunnan Province, southwestern China (Figure 15.1; Zhao et al., 2013). It has a normal elevation of 1414 m, average lake depth of 3.9 m, and maximum depth of 5.7 m. The lake area is 28.37 km^2 and the volume is 114.9 million m^3. Historical water quality data from 1998 to 2009 and previous studies (i.e., the *Lake Yilong Total Maximum Daily Load*, YLTMDL) showed that lake water quality deteriorated sharply in 2009 (Figure 15.2). The lake function was also severely impaired, as evidenced by a large area of degraded submerged aquatic vegetation, algae blooms, and surges in Chl *a* concentration in 2009. To facilitate effective decision making on eutrophication control and ecological restoration in Lake Yilong, it is necessary to quantitatively reveal the external factors driving the catastrophic shift in the lake. Considering the necessity of resolving spatially variable hydrodynamics and complex water quality and phytoplankton/macrophyte interactions, the EFDC was customized to simulate the main water quality and ecological processes in Lake Yilong. The Shiping County Environmental Monitoring Center is responsible for monthly water quality monitoring. Data were collected from three observation stations (*Lake West Station, Lake East Station, Lake Middle Station*).

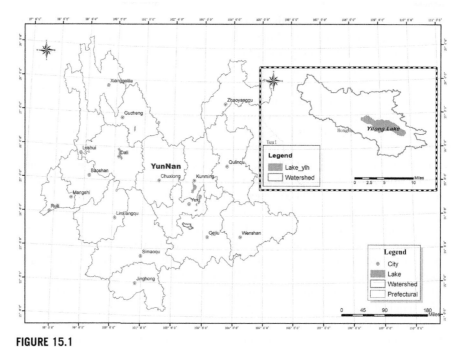

FIGURE 15.1

Lake Yilong Watershed (Zhao et al., 2013). (For the color version of this figure, the reader is referred to the online version of this chapter.)

15.2.2 Modeling framework

As mentioned before, we have previously developed a 3D water quality model to analyze the load reduction for different water quality accomplishments (Zhao et al., 2013). A 3D hydrodynamic and water quality model was thereby developed in this study for Lake Yilong to reveal a catastrophic regime shift in aquatic ecosystem, based on what have done in Zhao et al. (2013). It is a multitask, highly integrated modular computational environmental fluid dynamics package (Hamrick, 1996). In 2009, the previously flourishing macrophyte communities in the lake had nearly been eliminated by stocking a large number of grass carp. Since then, the water had deteriorated and a catastrophic regime shift occurred and persisted. Thus, a submerged macrophyte module as well as fish predation module have been developed in Lake Yilong model in addition to the basic hydrodynamic process, water quality module (Figure 15.3).

The general governing equations of EFDC are a set of partial differential equations, including a variety of hydrodynamic processes, water quality and eutrophication models of the 21 state variables, and sediment geochemistry models of 27 variables. The water quality model simulates the spatial and temporal distribution of water quality parameters including dissolved oxygen (DO); suspended algae; various components of carbon, nitrogen, phosphorus, and silica cycles; and fecal

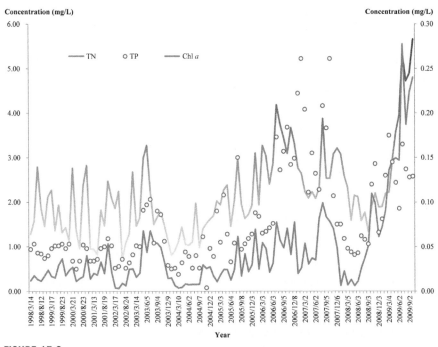

FIGURE 15.2

Trend of Chl *a* and TN, TP, in Lake West station. The data were from Lake West station from 1998 to 2009; Chl *a* and TP concentrations were enlarged by ten times for better comparisons. (For the color version of this figure, the reader is referred to the online version of this chapter.)

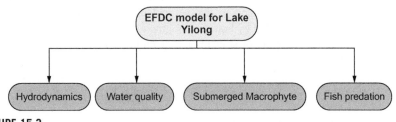

FIGURE 15.3

Basic composition of EFDC model for Lake Yilong. (For the color version of this figure, the reader is referred to the online version of this chapter.)

coliform bacteria. The water quality model includes a sediment model. The sediment process model, upon receiving the particulate organic matter deposited from the overlying water column, simulates their diagenesis and resulting fluxes of inorganic substances and sediment oxygen demand back to the water column (Park and Lee,

2002; Park et al., 2005). The hydrodynamic module of the model solves 3D, vertically hydrostatic, free surface, and turbulent averaged equations of motion for a variable-density fluid. Dynamically coupled transport equations for turbulent kinetic energy, turbulent length scale, salinity, and temperature are also solved (Hamrick and Wu, 1997).

The general governing equations for EFDC are (Park et al., 2005):

$$\frac{\partial(m_x m_y HC)}{\partial t} + \frac{\partial}{\partial x}(m_y HuC) + \frac{\partial}{\partial y}(m_x HvC) + \frac{\partial}{\partial z}(m_x m_y wC)$$

$$= \frac{\partial}{\partial x}\left(\frac{m_y HA_x}{m_x}\frac{\partial C}{\partial x}\right) + \frac{\partial}{\partial y}\left(\frac{m_x HA_y}{m_y}\frac{\partial C}{\partial y}\right) + \frac{\partial}{\partial z}\left(m_x m_y \frac{A_z}{H}\frac{\partial C}{\partial z}\right) + m_x m_y HS_C \quad (15.1)$$

where C is the concentration of a water quality state variable; u, v, w are velocity components in the curvilinear, sigma x-, y-, and z-directions, respectively; A_x, A_y, A_z are the turbulent diffusivities in the x-, y-, and z-directions, respectively; S_C is the internal and external sources and sinks per unit volume; H is the water column depth; m_x, m_y are the horizontal curvilinear coordinate scale factors. Water temperatures are needed for computation of the water quality state variables, and they are provided by the internally coupled hydrodynamic model (Park et al., 2005).

15.2.2.1 Formulation of hydrodynamic processes

The formulation of the governing equations for ambient environmental flows characterized by horizontal length scales that are orders of magnitude greater than their vertical length scales begins with the vertically hydrostatic, boundary layer form of the turbulent equations of motion for an incompressible, variable density fluid (Ji et al., 2001). The equations in hydrodynamic process are formulated in horizontal coordinates, x and y, curvilinear and orthogonal so that they could accommodate realistic horizontal boundaries well. A time variable mapping or stretching transformation is desirable so that uniform resolution could be provided in the vertical direction. To be aligned with gravitational vector and bounded by bottom topography and a free surface permitting long wave motion, the mapping or stretching is given by in EFDC (Hamrick, 1996; Wool et al., 2003):

$$z = \frac{(z^* + h)}{(\zeta + h)} \quad (15.2)$$

where * denotes the original physical vertical coordinates and h and ζ are the physical vertical coordinates of the bottom topography and the free surface, respectively (Vinokur, 1974; Hamrick, 1986; Blumberg and Mellor, 1987). Transforming the vertically hydrostatic boundary layer form of the turbulent equations of motion and utilizing the Boussinesq approximation for variable density result in the momentum and continuity equations and the transport equations for salinity and temperature in the following form (Hamrick, 1996):

$$\partial_t(mHu) + \partial_x(m_y Huu) + \partial_y(m_x Hvu) + \partial_z(mwu) - (mf + v\partial_x m_y - u\partial_y m_x)Hv$$

$$= -m_y H\partial_x(g\zeta + p) - m_y(\partial_x h - z\partial_x H)\partial_z p + \partial_z(mH^{-1}A_v\partial_z v) + Q_u \quad (15.3)$$

$$\partial_t(mHv) + \partial_x\left(m_yHuv\right) + \partial_y(m_xHvv) + \partial_z(mwv) + \left(mf + v\partial_xm_y - u\partial_ym_x\right)Hu$$
$$= -m_xH\partial_y(g\xi + p) - m_x\left(\partial_yh - z\partial_yH\right)\partial_zp + \partial_z(mH^{-1}A_v\partial_zv) + Q_v \tag{15.4}$$

$$\partial_zp = -gH(\rho - \rho_0)\rho_0^{-1} = -gHb \tag{15.5}$$

$$\partial_t(m\zeta) + \partial_x\left(m_yHu\right) + \partial_y(m_xHv) + \partial_z(mw) = 0 \tag{15.6}$$

$$\partial_t(m\zeta) + \partial_x\left(m_yH\int_0^1 u\,dz\right) + \partial_y\left(m_xH\int_0^1 v\,dz\right) = 0 \tag{15.7}$$

$$\rho = \rho(p, S, T) \tag{15.8}$$

$$\partial_t(mHS) + \partial_x\left(m_yHuS\right) + \partial_y(m_xHvS) + \partial_z(mwS) = \partial_z\left(mH^{-1}A_b\partial_zS\right) + Q_S \tag{15.9}$$

$$\partial_t(mHT) + \partial_x\left(m_yHuT\right) + \partial_y(m_xHvT) + \partial_z(mwT) = \partial_z\left(mH^{-1}A_b\partial_zT\right) + Q_T \tag{15.10}$$

Where, u and v are the horizontal velocity components in the curvilinear, orthogonal coordinates x and y, m_x and m_y are the square roots of the diagonal components of the metric tensor, $m = m_x \cdot m_y$ is the Jacobian or square root of the metric tensor determinant (Hamrick, 1996). The vertical velocity is related to the physical vertical velocity w^* by:

$$w = w^* - z\left(\partial_t\zeta + um_x^{-1}\partial_x\zeta + vm_y^{-1}\partial_y\zeta\right) + (1-z)\left(um_x^{-1}\partial_xh + vm_y^{-1}\partial_yh\right) \tag{15.11}$$

In Equations (15.3) and (15.4), f is the Coriolis parameter, A_v is the vertical turbulent or eddy viscosity, and Q_u and Q_v are momentum source-sink terms. The density, ρ, is a function of temperature, T, and salinity or water vapor, S, in hydrospheric and atmospheric flows, respectively. The buoyancy, b, is defined from the reference value. The continuity equation (15.6) has been integrated with respect to z using the vertical boundary conditions, $w = 0$, at $z = (0,1)$, which follows from Equation (15.11). Q_S and Q_T include subgrid scale horizontal diffusion and thermal sources and sinks, while A_b is the vertical turbulent diffusivity. It yields the equivalent of the rigid lid ocean circulation equations employed by Semtner (1974) and equations similar to the terrain following equations used by Clark (1977) to model mesoscale atmospheric flow (Semtner, 1974; Clark, 1977; Hamrick, 1992; Hamrick, 1996).

15.2.2.2 Water quality process

The water quality module in EFDC is carbon based. Cyanobacteria biomass, diatom algae, green algae as well as stationary algae are the four algae species which are represented in carbon units (Jin et al., 2000; Zou et al., 2006). The biochemical oxygen demand (BOD) is contributed by three organic carbon variables equivalently. Organic carbon, nitrogen, and phosphorous can be represented by up to three reactive subclasses, refractory particulate, labile particulate, and labile dissolved. The use of the subclasses allows a more realistic distribution of organic material by reactive classes when data is intended to estimate distribution factors. The 21 simulating water column state variables in water quality module are listed in Table 15.1 and their interactions are illustrated in Figure 15.4 (Hamrick, 1994; Tech, 1999).

15.2.2.3 Submerged macrophyte module enhancement

The submerged macrophyte module was firstly conducted and compiled in this study so that the effects of submerged macrophyte on lake nutrients cycle as well as phytoplankton could be understanding more clearly. The biomass of submerged macrophyte is carbon based, which is consistent with other modules in EFDC. The model lumped all submerged macrophyte species into one "superspecies to be in consistent with what implemented in WQM". The kinetic balance equation for submerged macrophyte is as below:

$$\frac{\partial M}{\partial t} = (P_M - R_M - L_M - L_G)M \qquad (15.12)$$

where M is macrophyte biomass per unit area (g carbon/m^2); t is time; P_M is macrophyte growth rate (day^{-1}); R_M is macrophyte respiration rate (day^{-1}); L_M is nonrespiratory macrophyte loss rate (day^{-1}); and L_G is macrophyte loss rate from fish grazing (day^{-1}).

Table 15.1 EFDC Model Water Quality State Variables

Number	State Variables	Representation in EFDC
1	Cyanobacteria biomass	BC
2	Diatom algae	BD
3	Green algae	BG
4	Stationary algae	SA
5	Refractory particulate organic carbon	RPOC
6	Labile particulate organic carbon	LPOC
7	Dissolved organic carbon	DOC
8	Refractory particulate organic phosphorus	RPOP
9	Labile particulate organic phosphorus	LPOP
10	Dissolved organic phosphorus	DOP
11	Total phosphate	PO4
12	Refractory particulate organic nitrogen	RPON
13	Labile particulate organic nitrogen	LPON
14	Dissolved organic nitrogen	DON
15	Ammonia nitrogen	NH_4
16	Nitrate nitrogen	NO_3
17	Particulate biogenic silica	SU
18	Dissolved available silica	SA
19	Chemical oxygen demand	COD
20	Dissolved oxygen	DO
21	Total active metal	TAM

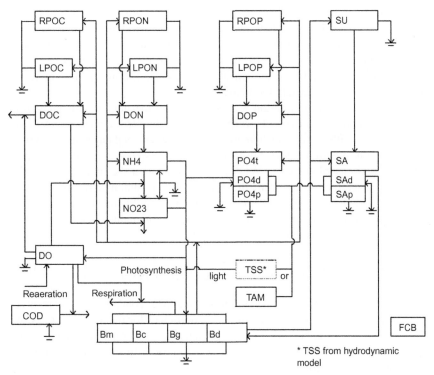

FIGURE 15.4

Schematic diagram of EFDC water quality model structure.

Modified from Park et al. (1995).

The equation take into consideration four mainly factors that regulate the submerged macrophyte growth including nutrients, light, temperature and self-shading. The growth rate P_M is calculated as:

$$P_M = PP_M \cdot f_1(N) \cdot f_2(I) \cdot f_3(T) \cdot f_4(C) \tag{15.13}$$

where PP_M is the maximum growth rate under optimal conditions for macrophyte (day^{-1}); $f_1(N)$ is the nutrient limiting function (unitless); $f_2(I)$ is the light-limiting function (unitless); $f_3(T)$ is temperature correction function (unitless); and $f_4(C)$ is the density-limiting factor representing self-shading effect (unitless).

Although the actual nutrient-limiting factor should be both the water column and the bed nutrient sources, the standard EFDC as well as other complex WQM generally calculate limiting factor for periphytons/macrophytes based on water column nutrient concentration alone. This formulation may not be the case in reality for the reason that macrophytes could continue to fuel growth by nutrient sources from the bed sediment when water column nutrients are too low to support healthy growth. A remedy for the limitation in this formulation is to simulate the macrophyte by parts, that is, roots and shoots/leaves, and quantify the limiting factor based on both the water column and bed sediment nutrient concentrations (Ji et al., 2002; Ji, 2008). This formulation, although it provides higher mechanistic resolution in the macrophyte formulation, can make the parameterization of the water quality model too complex to be calibrated because in most real cases data for quantifying the related roots–shoots/leaves parameters are unavailable.

A slightly modified formulation was proposed considering the deficiency of the standard formulation and issue of overparameterization for the roots-shoots/leaves formulation to handle the cases where macrophytes uptake bed nutrients when water column nutrients are diminished. This formulation is based on a general assumption used in the majority of advanced WQM, which imply that the first-choice nutrient sources for macrophytes are from the water column, although macrophytes are allowed to continue growing at a depressed but nonnegligible rate by using bed sediment as the primary nutrient sources. The corresponding mathematical form for the nutrient limiting factor is thus:

$$f_1(N) = \text{Max}\left[\text{Min}\left(\frac{NH_4 + NO_3}{KHN + NH_4 + NO_3}, \frac{PO_4}{KHP + PO_4}\right), f_{1min}\right] \tag{15.14}$$

where NH_4 and NO_3 are the water column NH_4 and NO_3 concentration, respectively (mg/L); KHN is the nitrogen half-saturation coefficient for macrophytes (g N/m^3); KHP is the phosphorus half-saturation coefficient for macrophytes (g P/m^3); and f_{1min} is the specified minimum threshold of the nutrient function. In this formula, when water column nutrient concentration is sufficiently high, the nutrient-limiting factor would be controlled by the water column concentration; otherwise, the - nutrient-limiting factor will be fixed at the prespecified threshold value, while the corresponding primary nutrient sources would be shifted from the water column to the bed. With this formulation, only one additional calibration parameter, f_{1min}, is added to the model. Therefore, the data requirement for reliable calibration is not significantly higher than the standard formulation.

The fraction of nutrient uptake from bed sediment when the water column nutrient is diminished is calculated as:

$$RB = 1.0 - \text{Min}\left[1.0, \frac{\text{Min}\left(\frac{NH_4 + NO_3}{KHN + NH_4 + NO_3}, \frac{PO_4}{KHP + PO_4}\right)}{f_{1min}}\right] \tag{15.15}$$

Apparently, when water column nutrient is depleted, macrophytes would uptake nutrients predominantly from the bed.

Since macrophytes in the standard EFDC model are simulated at the bottom layer alone, the nutrient and light conditions are calculated also at the bottom layer (Bai and Lung, 2005; Ji, 2008). To account for the conditions that macrophytes can extend from the bottom of the water up to multiple layers, the nutrients and light-limiting factors should be calculated for the corresponding water column extending beyond the bottom layer. Therefore, the EFDC code is modified through incorporating the following equations:

$$NH_4 + NO_3 = \frac{\int_0^{h_1} (NH_{4_z} + NO_{3_z}) \, dz}{h_1} \tag{15.16}$$

where NH_{4_z} and NO_{3_z} are the concentration of NH_4 and NO_3 at distance z measured from the bottom of the water (mg/L); h_1 is the height of macrophytes (m).

The light limitation factor in EFDC is correspondingly modified to account for the macrophyte height as:

$$f_2(I) = \frac{2.718 \cdot FD}{Kess \cdot h_1} \left(e^{-\alpha} - e^{-\beta} \right) \tag{15.17}$$

$$\alpha = \frac{I_0}{FD \cdot I_{opt}} e^{-Kess \cdot H} \tag{15.18}$$

$$\beta = \frac{I_0}{FD \cdot I_{opt}} e^{-Kess \cdot (H - h_1)} \tag{15.19}$$

where $FD = 1$ for instantaneous solar radiation or $0 \leq FD < 1$ for daylight fraction of daily averaged solar radiation; Kess is total light extinction coefficient (m^{-1}); I_0 is water surface instantaneous solar radiation when $FD = 1$ or average solar radiation when $0 \leq FD < 1$ (Langley/day); I_{opt} is the optimal light intensity for macrophyte growth (Langley/day); H is the water depth (m); and h_1 is the average height of macrophytes (m).

The equation for temperature limitation is given by

$$f_3(T) = \begin{cases} \exp\left(-KTP1[T - TP1]^2\right) & \text{if } T \leq TP1 \\ 1 & \text{if } TP1 < T < TP2 \\ \exp\left(-KTP2[T - TP2]^2\right) & \text{if } T \geq TP2 \end{cases} \tag{15.20}$$

where T is water temperature calculated by hydrodynamics module (°C); $TP1 < T < TP2$ is the optimal range for macrophytes growth (°C); $KTP1$ is the effect coefficient of temperature below $TP1$ for macrophytes growth (°C); $KTP2$ is the effect coefficient of temperature above $TP2$ for macrophytes growth (°C). The equation for self-shading limitation is given by

$$f_4(C) = e^{-K_S \cdot M} \tag{15.21}$$

where K_S is the light extinction coefficient for macrophytes' self-shading (m^2/g).

The respiration rate of macrophytes is assumed to be temperature dependent and given by

$$R_M = BMR_M \cdot \exp(KTB_M[T - TR_M]) \tag{15.22}$$

where BMR_M is the basal respiration rate at TR_M (day^{-1}); KTB_M is the effect of temperature on the respiration rate $(°C^{-1})$; TR_M is the reference temperature for basal respiration (°C).

15.2.2.4 Fish predation parameterization

In the early spring of 2009, local authorities had been stocking a large number of herbivorous fish (grass carp). To evaluate the effects of stocking, a key process was developed to simulate the impact of fish predation on aquatic vegetation. Limited by the observation data and the complexity of the process, it was not feasible to simulate grass carp stocking, growth, predation, and excretion kinetics. Therefore, the process was simplified in the EFDC model as: (a) an additional predator was introduced to the existing EFDC mathematical equation, which included the parameters of the mathematical methods of grass carp into the system, and (b) assuming that with the growth of fish, predation increased, thus, additional predation further parameterized into a linear growth variable dating from the fish stocking day. The equation is listed as follows:

$$L_M^t = \begin{cases} 0 & \text{if } t \le t1 \\ PR_L + (t - t1)(PR_U - PR_L) & \text{if } t1 < t < t2 \\ PR_U & \text{if } t \ge t2 \end{cases} \qquad (15.23)$$

where L_M^t is real-time predation coefficient on time t; PR_L is minimum predation coefficient (1/day); PR_U is the maximum predation coefficient (1/day); $t1$ means start time for grass carp stocking (day), and $t2$ means end time for grass carp stocking (day).

15.2.2.5 Coupling with EFDC

15.2.2.5.1 Coupling with water column nutrient dynamics

$$\frac{\partial C_i}{\partial t} = \frac{S_i + \delta \cdot (FR_i \cdot R_M + FLM_i \cdot L_M + FLG_i \cdot L_G - (1 - RB) \cdot FPM_i \cdot P_M) \cdot M}{h_1}$$

$$(15.24)$$

where C_i is concentration of nutrient species i; S_i is sources and sinks for nutrient species i due to all processes other than macrophyte dynamics; δ is the delta function, which takes value 1 at cells with macrophytes, or 0 at cells without macrophyte impact; FR_i is the conversion factor between respired macrophyte biomass and the generated mass of nutrient species i; FLM_i is the conversion factor between the nonrespiratory macrophyte loss and the generated mass of nutrient species i; FLG_i is the conversion factor between the fish-related macrophyte loss and the generated mass of nutrient species i; FPM_i is the conversion factor between the macrophyte growth and the uptaken mass of nutrient species i. It should be noted that all the conversion factors have accounted for the stoichiometric ratio between macrophyte biomass and the pertinent nutrient species.

In the standard EFDC framework, the nutrient species include three organic carbon constituents (LPOC, RPOC, and DOC), four phosphorus constituents (LPOP, RPOP, DOP, and PO$_4$), and five nitrogen species (LPON, RPON, DON, NH$_4$, and NO$_2$/NO$_3$), and two silicon constituents (SU and SA), so the subscript i can take numbers up to 14. Similarly, for interactions between macrophyte and DO, the equation is:

$$\frac{\partial C_{DO}}{\partial t} = \frac{S_{DO} + \delta \cdot (-FR_{DO} \cdot R_M + FPM_{DO} \cdot P_M) \cdot M}{h_1} \tag{15.25}$$

where the notations of each variable follows the convention described above.

Another mechanism incorporated into the model is to relate the local algal predation rate to the macrophyte biomass. The basic assumption is that when the macrophyte community is damaged, the habitat for zooplankton is impaired, hence depressing the predation pressure on phytoplankton. The corresponding Monod-type equation is as:

$$PR = PR_b + \delta \cdot PR_z \cdot \frac{M}{M + AHSM} \tag{15.26}$$

where PR is the total nonrespiratory loss term of phytoplankton (day^{-1}); PR$_b$ is the background nonrespiratory loss term of phytoplankton (day^{-1}); PR$_z$ is the maximum predation rate from zooplankton that relies on macrophyte for habitat (day^{-1}); and AHSM is the half-saturation coefficient for the predation rate.

15.2.2.5.2 Coupling with bed nutrient dynamics

Macrophytes impact the sediment bed nutrient budget mainly through two pathways: (a) macrophyte nonrespiratory and fish grazing loss result in particulate organic matters that would settle to the bed as nutrient sources; and (b) macrophyte uptake would consume inorganic nitrogen and phosphorus from the bed, hence serving as a sink of nutrient in the bed (Park et al., 1995; Bai and Lung, 2005).

Since the first pathway is handled in the coupling of macrophytes with water column water quality, the coupling with bed sediment only addresses the second pathway. The EFDC sediment diagenesis model is formulated using a two-layer scheme to differentiate a very thin (approximately 0.1 cm) aerobic layer and a thicker (>10 cm) anaerobic layer, where the sediment diagenesis process occurs. Since the surface, aerobic layer is so thin that it only contains insignificant amount of nutrient mass in comparison to the lower anaerobic layer, it is reasonable to assume that the macrophyte roots would uptake nutrient only from the anaerobic layer. Considering that in the anaerobic layer inorganic nitrogen is dominated by NH$_4$, it is further assumed that only NH$_4$ serves as the nitrogen sources for macrophyte root, leading to the following equations:

$$\frac{\partial NH_{4_B}}{\partial t} = \frac{(S_{B-NH_4} - \delta \cdot FPM_{B-NH_4} \cdot P_M) \cdot M \cdot RB}{h_B} \tag{15.27}$$

where NH$_{4_B}$ is the concentration of NH$_4$ in the anaerobic layer of the bed (mg/L); S_{B-NH_4} is the sink and sources of NH$_4$ in the bed from all other processes (g/m^2/

day); FPM_{B-NH_4} is the conversion factor between NH_4 uptake and macrophyte growth; and h_B is the thickness of the bed (m).

Similarly, the equation for PO_4 in the bed can be written as:

$$\frac{\partial PO_{4_B}}{\partial t} = \frac{(S_{B-PO_4} - \delta \cdot FPM_{B-PO_4} \cdot P_M) \cdot M \cdot RB}{h_B} \qquad (15.28)$$

where PO_{4_B} is the concentration of PO_4 in the anaerobic layer of the bed (mg/L); S_{B-PO_4} is the sink and sources of PO_4 in the bed from all other processes (g/m^2/day); FPM_{B-PO_4} is the conversion factor between dissolved PO_4 uptake and macrophyte growth.

15.2.3 Model development for Lake Yilong

Similar to what was done in Zhao et al. (2013), it is a multistep process to develop the model for Lake Yilong, including (a) determination of EFDC complexity; (b) grid generation, configuration of initial and boundary conditions; (c) model calibration; and (d) scenario analysis.

15.2.3.1 Determination of EFDC model complexity

The EFDC can be applied with different levels of complexity depending on the purpose of the study as well as the validity of the data. The purpose of the current study is to simulate the dynamic factors causing the catastrophic shift in the aquatic ecosystems, thus it is necessary to adopt a model that could reflect complex interactions between nutrients and phytoplankton, aquatic plants, DO, and sediment, as well as other components. It thereby requires a construct with maximum complexity in EFDC. Specifically, the EFDC model for Lake Yilong should include all relevant water quality indicators, such as carbon, nitrogen, phosphorus, algae, and DO, to fully characterize the process of eutrophication. Furthermore, the competitive relationship between the aquatic vegetation and phytoplankton nutrients was also analyzed.

The sediment model will not only describe chemical and biological interactions but will also describe the association between water quality and sediment composition. Specifically, the sediment model can be used to predict changes in sediment nutrient flux historically or in certain scenarios, such as watershed management and restoration. Meanwhile, to reflect the submerged vegetation succession, a large vascular aquatic plants module has also been developed.

15.2.3.2 Grid generation

A curvilinear grid method was used to describe the shoreline of Lake Yilong. The horizontal cross-sectional curve was generated for a discrete water body using lake terrain data to specify lattice depth. The lake was represented by using a total of 241 horizontal grids where the smallest grid was approximately 0.05 million m^2 and the largest was approximately 0.18 million m^2. The average depth at the shallowest grid was about 1.5 m; while the average depth of the deepest grid was about 5.3 m when the water surface elevation was at 1414 m above sea level (Zhao et al., 2013). Although Lake Yilong is shallow, with no significant thermal stratification, phytoplankton and aquatic vegetation are still influenced by light and nutrient dynamics; therefore, it is desirable to resolve variability in vertical light intensity and nutrient use using 3D spatial resolution. In this model, the grid level was further cut into five layers, and a total of 1205 computational grids were generated from top to bottom to represent Lake Yilong in its entirety (Figure 15.5).

15.2.3.3 Initial conditions

Initial conditions were chosen as the starting point for the model simulation. In this study, the period simulated was from the summer of 2008 to the summer of 2009. A water height of 1723.28 m, recorded July 21, 2008, was set as the initial elevation. The initial temperature was set as 20.0 °C based on a value observed in early July. The three velocity vectors were initialized at 0.0 m/s by conventional hydrodynamics. In order to gain reasonably representative initial hydrodynamic conditions, such as temperature and flow field, to drive the water quality model, the hydrodynamic model was running for a month in advance as a warm-up period.

Comparing to hydrodynamic model, the initial water quality conditions were more complex because of a long time needed to eliminate the impact of initial conditions during the run process of the model. To get a more accurate water quality model initialization, the observational data from August 22, 2008, was inserted to each calculation unit, thus forming the initial field of water quality modeling; the model parameters were showed in Table 15.2.

15.2.3.4 Boundary conditions

The horizontal and surface boundary conditions were included in the boundary conditions reflecting external driving forces for the model. The lateral boundary conditions consisted of tributary flow rate into the lake and associated temperature and water

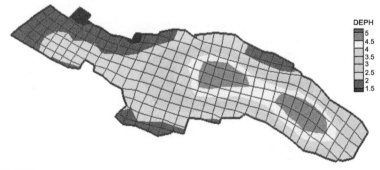

FIGURE 15.5

Computational grids of hydrodynamic water quality simulation in Lake Yilong (Zhao et al., 2013). (For the color version of this figure, the reader is referred to the online version of this chapter.)

Table 15.2 Parameters Calibration of the EFDC Model for Lake Yilong

	Parameter	Description	Value of Yilong Model	References
1	P_a	Max growth rate of algae group 1	2.3	0.2–9.0
2	P_b	Max growth rate of algae group 2	2	0.2–9.0
3	P_c	Max growth rate of algae group 3	2.2	0.2–9.0
4	R_a	Basal respiration rate of algae group 1	0.1	0.01–0.92
5	R_b	Basal respiration rate of algae group 2	0.11	0.01–0.92
6	R_c	Basal respiration rate of algae group 3	0.11	0.01–0.92
7	D_a	Predation death rate of algae group 1	0.01	0.03–0.3
8	D_b	Predation death rate of algae group 2	0.05	0.03–0.3
9	D_c	Predation death rate of algae group 3	0.02	0.03–0.3
10	KE_b	Background light extinction coefficient	0.3	0.25–0.45
11	KE_c	Chlorophyll a induced light extinction coefficient	0.012	0.002–0.02
12	KHN_a	Nitrogen half saturation coefficient for algae group 1	0.03	0.006–4.32
13	KHN_b	Nitrogen half saturation coefficient for algae group 2	0.03	0.006–4.32
14	KHN_c	Nitrogen half saturation coefficient for algae group 3	0.03	0.006–4.32
15	KHP_a	Phosphorus half saturation coefficient for algae group 1	0.005	0.001–1.52
16	KHP_b	Phosphorus half saturation coefficient for algae group 2	0.005	0.001–1.52
17	KHP_c	Phosphorus half saturation coefficient for algae group 3	0.005	0.001–1.52
18	TMR_a	Optimal temperature range for algae group 1	25–30	N/A
19	TMR_b	Optimal temperature range for algae group 2	10–15	N/A
20	TMR_c	Optimal temperature range for algae group 3	22–23	N/A
21	S_a	Settling rate of algae group 1	0.1	0.001–13.20
22	S_b	Settling rate of algae group 2	0.2	0.001–13.20
23	S_c	Settling rate of algae group 3	0.15	0.001–13.20
24	SRP	Settling rate of refractory organic matters	0.15	0.02–9.0
25	SLP	Settling rate of labile organic matters	0.15	0.02–9.0
26	KRN	RPON hydrolysis rate	0.001	0.001
27	KLN	LPON hydrolysis rate	0.04	0.01–0.63

Continued

Table 15.2 Parameters Calibration of the EFDC Model for Lake Yilong—cont'd

	Parameter	Description	Value of Yilong Model	References
28	KDN	DON decay rate	0.05	0.01–0.63
29	KN	Base nitrification rate	0.18	0.001–1.3
29	KRP	RPOP hydrolysis rate	0.001	0.001
30	KLP	LPOP hydrolysis rate	0.04	0.01–0.63
31	KDP	DOP decay rate	0.05	0.01–0.63
32	KRC	RPOC hydrolysis rate	0.001	0.001
33	KLC	LPOC hydrolysis rate	0.04	0.01–0.63
34	KDC	DOC decay rate	0.06	0.01–0.63

quality components of the tributaries. The surface boundary conditions were described by temporal meteorological conditions. In the Lake Yilong model, the horizontal boundary conditions and the nutrient level were configured based on the monitoring data from main tributaries during 2008–2009. The geographical coordinates in the grid for the estuary determined the dimensions of the horizontal boundary conditions. For those tributaries with available inflow rate and water quality concentrations, appropriate boundary conditions were set based on observed data, while those rivers without detailed data relied on hydrodynamic model calibration processes for the estimation of boundary conditions. In addition to tributaries, atmospheric deposition was another major source of nutrient loading. The concentrations of total phosphorus (TP) and total nitrogen (TN) in the study area from dry deposition were 0.00019 $g/m^2/day$ and 0.0069 $g/m^2/day$, respectively, whereas concentrations of TP and TN from wet deposition were 0.039 and 0.95 mg/L, respectively (Zhao et al., 2013).

Atmospheric boundary conditions required data from the following parameters to drive the hydrodynamic model simulation, including atmospheric pressure, air temperature, relative humidity, precipitation, evaporation, solar radiation, cloud cover, wind speed, and wind direction. In the modeling process, hourly data were obtained from the Shiping County Weather Station (SPWS) and were further modified to a compatible format of EFDC, which could be used to configure the atmospheric boundary conditions.

15.2.3.5 Simulation and calibration

Calibration of the hydrodynamic and water quality model in Lake Yilong was implemented in a phased manner as literatures and what Zhao et al has done in 2013 (Lung, 2001; Zhao et al., 2013): (a) Developing and calibrating the hydrodynamic processes; (b) Running the water quality module; and (c) activating the water column nutrient-phytoplankton-macrophyte simulation modules and sediment diagnosis module to calibrate the WQM.

15.3 RESULTS AND DISCUSSIONS

15.3.1 Hydrodynamic simulation and calibration

Hydrodynamic simulation and calibration was carried out before the water quality model was activated. The premise of Lake Yilong hydrodynamic and water quality model was the simulation of hydrodynamics and flow balance. The simulation lasted from July 21, 2008, to October 18, 2009. The simulation variables were flow field, water level, and temperature. The flow field in lake Yilong was mainly affected by temperature, light as well as wind. The model calibration parameters were lake water level and temperature (Figures 15.6 and 15.7).

Figure 15.6 showed the difference between simulated and observed water levels of Lake Middle Station. The simulation results were in good agreement with the observed daily water level data, indicative of an overall balance in water. Temperature was also an important calibration variable because it affects physical process in fluid dynamics and thermal balance process simulation in lake. Therefore, the temperature data from three different stations across the lake were used for hydrodynamic calibration. The model simulation time step is 270 s. Model simulated temperature was output for every 6 h and compared against observed data at the three stations (Figure 15.7). The model results showed that the 3D hydrodynamic model configured in this study accurately simulated the hydrodynamic processes of Lake Yilong to successfully lay a solid foundation for exploring the mechanism of catastrophic regime shift of Lake Yilong.

15.3.2 Water quality simulation and calibration

The water quality model calibration was conducted through fine-tuning the model parameters to reproduce observed water quality at the three monitoring stations. The basic constituents for calibration were Chl a, DO, NH_4, TN, and TP.

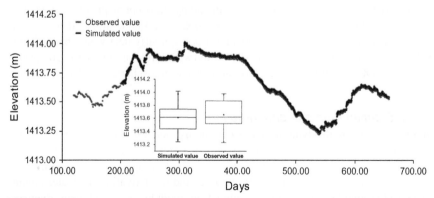

FIGURE 15.6

Comparison of simulated and observed elevations of Lake Yilong. (For the color version of this figure, the reader is referred to the online version of this chapter.)

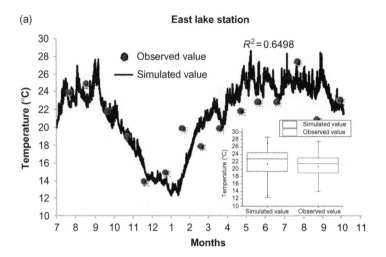

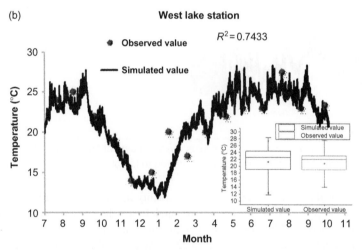

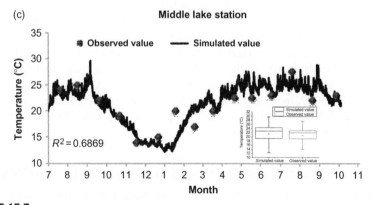

FIGURE 15.7

Comparison of simulated and observed temperatures. (For the color version of this figure, the reader is referred to the online version of this chapter.)

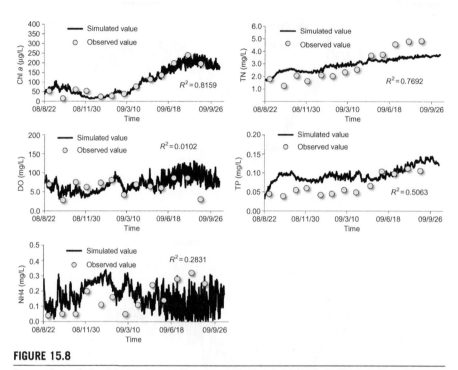

FIGURE 15.8

Comparison of simulated water quality versus observed data at Lake East station. (For the color version of this figure, the reader is referred to the online version of this chapter.)

The observed data showed that the water quality in 2009 was significantly worse than that in the past few years. The EFDC model reproduced the trends accurately. Water quality model calibration is based on the previous calibration of hydrodynamic model. The simulation lasted from August 22, 2008, to October 15, 2009. This process was repeated about 100 times until the simulated values could reproduce the observed temporal and spatial patterns. In the eutrophication model, the key parameters related to phytoplankton, nitrogen, phosphorus, and carbon needed to be calibrated. Observed data from the Lake East station and the simulated water quality results were plotted in Figure 15.8. Trends as well as the magnitude of changes in water quality from the model results were consistent with the observational data. According to the model's predictions for 2009, more severe algal blooms were expected, and a growth trend in the TN and TP concentrations was also predicted by Lake Yilong model that was also consistent with observed data.

15.3.3 Simulation of the catastrophic regime shift in Lake Yilong

Simulated results revealed that algae bloom or the sharp increase in the concentration of Chl *a* in the lake was caused by the reduction or disappearance of aquatic vegetation rather than by TN and TP concentration. In other words, it is disappearance of aquatic plants that caused algae bloom rather than the opposite. The depletion of aquatic plants would deprive zooplankton communities of living habitats, hence depressing the biomass and activities of zooplankton communities. This in turn would remove the controlling stress on phytoplankton, causing algae bloom to be more intensive. Meanwhile, the loss of macrophyte biomass would be converted into organic matter through the metabolism of fish as well as the mortality process itself, therefore increasing the water column as well as the benthic nutrient concentration. The increase in water nutrient concentration can then further stimulate algal bloom, causing worse entrophic problems and more turbid situations. The simulation results showed that fish stocking is a key factor causing drastically worsened eutrophication in 2009. Specifically, Chl *a* concentration with stocked fish rose to much higher values than those of the year without stocked fish. Therefore, a question naturally to be asked is: What if there were no fish stocking activities in early 2009? Will there still be higher algal bloom in 2009 than in 2008 due to difference in loading and climate condition? To answer this question, a scenario was analyzed and the result is shown in Figure 15.9. As shown, without stocking of grass carp in 2009, the Chl *a* concentration would remain at essentially the similar level as that in 2008. Such results indicate that human activities can have significant impact on water quality as well as aquatic ecosystem, therefore, it is important to wisely plan any human activities to avoid producing negative or even disastrous consequences.

Historical water quality data from 1998 to 2009 and previous studies (i.e., YLTMDL) showed that the aquatic system in Lake Yilong experienced significant interannual fluctuations, while a clear signal of regime shift is identifiable for 2009.

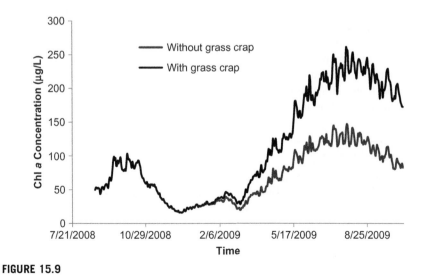

FIGURE 15.9

Model-based hindcasting of Chl *a* concentrations with versus without grass carp. (For the color version of this figure, the reader is referred to the online version of this chapter.)

In 2009, the previously flourishing macrophyte communities in the lake had nearly been eliminated, and a sharp increase in nutrient and phytoplankton concentration occurred. Therefore, taking into consideration EFDC simulation results and previous studies, it could be speculated that catastrophic shift in Lake Yilong from 2008 to 2009 was probably caused by stocking a large number of herbivorous fish.

15.4 CONCLUSIONS

(1) A 3D hydrodynamic and water quality model was developed for Lake Yilong, China, to reveal the underlying mechanism of the catastrophic regime shift in the lake ecosystem. The model accurately reproduced the observed water surface elevation, water temperature, and nutrient and algal conditions, indicating a reasonable numerical representation of the actual hydrodynamics and eutrophication dynamics in the lake. Besides that, the macrophyte module in EFDC was enhanced to better resolve the vertical light–nutrient–phytoplankton–macrophyte interactions, and a gross fish grazing parameterization was introduced to represent the impact of fish on macrophyte dynamics. The model-simulated water quality status was consistent with the observed data in both trend and magnitude, suggesting that it might be used to explore the mechanisms underlying the catastrophic regime shift and environmental management of Lake Yilong.

(2) The model results showed that the existence of aquatic vegetation had significant impacts on algae blooms in Lake Yilong. The observed algae bloom intensification in 2009 was caused by the reduction or disappearance of the aquatic vegetation. The model results suggest that the fish stocking is a key factor causing eutrophication and catastrophic regime shift in the lake. The modeling study indicated that the lake ecological system is fragile with regard to the impact from human activities, therefore, wise planning and intelligent analysis of human activities are necessary to assure the health of lake ecological systems. To serve such a purpose, the mechanistic water quality modeling approach provides an effective tool for quantifying natural and human activities and their impact on lake ecosystems.

Acknowledgments

This study was supported by National Natural Science Foundation of China (No. 41222002) and the Scientific Research Foundation for the Returned Overseas Chinese Scholars, MOE. We acknowledge Zhao et al. (2013) for citing Figures 15.1 and 15.5.

References

Alarcon, V.J., McAnally, W.H., 2012. Using hydrodynamic modeling for estimating flooding and water depths in Grand Bay, Alabama. In: Computational Science and Its Applications – ICCSA 2012, vol. 7334. Springer, Verlag Berlin Heidelberg, pp. 578–588.

Arhonditsis, G.B., Brett, M.T., 2005. Eutrophication model for Lake Washington (USA): part I: Model description and sensitivity analysis. Ecol. Model. 187, 140–178.

Bai, S., Lung, W.-S., 2005. Modeling sediment impact on the transport of fecal bacteria. Water Res. 39, 5232–5240.

Blumberg, A.F., Mellor, G.L., 1987. A description of a three-dimensional coastal ocean circulation model. Coast. Estuar. Sci. 4, 1–16.

Carpenter, S.R., 2005. Eutrophication of aquatic ecosystems: bistability and soil phosphorus. Proc. Natl. Acad. Sci. U.S.A. 102, 10002–10005.

Carpenter, S.R., Ludwig, D., Brock, W.A., 1999. Management of eutrophication for lakes subject to potentially irreversible change. Ecol. Appl. 9, 751–771.

Carpenter, S., Cole, J., Pace, M., Batt, R., Brock, W., Cline, T., Coloso, J., Hodgson, J., Kitchell, J., Seekell, D., 2011. Early warnings of regime shifts: a whole-ecosystem experiment. Science 332, 1079.

Cerco, C.F., Cole, T., 1993. 3-dimentional eutrophication model of cheaspeakebay. J. Environ. Eng. ASCE 119, 1006–1025.

Chan, S.N., Thoe, W., Lee, J.H.W., 2013. Real-time forecasting of Hong Kong beach water quality by 3D deterministic model. Water Res. 47, 1631–1647.

Chau, K.W., Jin, H.S., 1998. Eutrophication model for a coastal bay in Hong Kong. J. Environ. Eng. ASCE 124, 628–638.

Clark, T.L., 1977. A small-scale dynamic model using a terrain-following coordinate transformation. J. Comput. Phys. 24, 186–215.

Drago, M., Cescon, B., Iovenitti, L., 2001. A three-dimensional numerical model for eutrophication and pollutant transport. Ecol. Model. 145, 17–34.

Gal, G., Anderson, W., 2010. A novel approach to detecting a regime shift in a lake ecosystem. Methods Ecol. Evol. 1, 45–52.

Gunderson, L.H., 2000. Ecological resilience–in theory and application. Annu. Rev. Ecol. Syst. 425–439.

Guo, H.C., Liu, L., Huang, G.H., Fuller, G.A., Zou, R., Yin, Y.Y., 2001. A system dynamics approach for regional environmental planning and management: a study for the Lake Erhai Basin. J. Environ. Manage. 61, 93–111.

Hamrick, J.M., 1986. Long-term dispersion in unsteady skewed free surface flow. Estuar. Coast. Shelf Sci. 23, 807–845.

Hamrick, J.M., 1992. Estuarine environmental impact assessment using a three-dimensional circulation and transport model. In: Spaulding, M.L. et al., (Ed.), Estuarine and Coastal Modeling, Proceedings of the 2nd International Conference. American Society of Civil Engineers, New York, NY, pp. 292–303.

Hamrick, J.M., 1994. Application of the EFDC, environmental fluid dynamics computer code to SFWMD Water Conservation Area 2A. *Report. No. JMH-SFWMD-94-01*, Williamsburg, VA.

Hamrick, J.M., 1996. User's Manual for the Environmental Fluid Dynamics Computer Code. Department of Physical Sciences, School of Marine Science, Virginia Institute of Marine Science, College of William and Mary, Virginia 23062, USA.

Hamrick, J., Wu, T., 1997. Computational Design and Optimization of the EFDC/HEM3D Surface Water Hydrodynamic and Eutrophication Models. Society of Industrial and Applied Mathematics, Philadelphia, PA, pp. 143–161.

Harper, D., 1992. Eutrophication of Freshwaters: Principles, Problems and Restoration. Springer Netherland, published online ISDN:978-94-011-3082-0. http://link.springer.com/content/pdf/10.1007/978-94-011-3082-0.pdf.

Jeong, S., Yeon, K., Hur, Y., Oh, K., 2010. Salinity intrusion characteristics analysis using EFDC model in the downstream of Geum River. J. Environ. Sci. (China) 22, 934–939.

Ji, Z.G., 2008. Hydrodynamics and Water Quality: Modeling Rivers, Lakes, and Estuaries. Wiley-Interscience, Hoboken, NJ.

Ji, Z.G., Morton, M.R., Hamrick, J.M., 2001. Wetting and drying simulation of estuarine processes. Estuar. Coast. Shelf Sci. 53, 683–700.

Ji, Z.-G., Hamrick, J.H., Pagenkopf, J., 2002. Sediment and metals modeling in shallow river. J. Environ. Eng. ASCE 128, 105–119.

Jin, K.-R., Hamrick, J.H., Tisdale, T., 2000. Application of three-dimensional hydrodynamic model for Lake Okeechobee. J. Hydrol. Eng. 126, 758–771.

Li, Y.P., Acharya, K., Yu, Z.B., 2011. Modeling impacts of Yangtze River water transfer on water ages in Lake Taihu, China. Ecol. Eng. 37, 325–334.

Lung, W.-S., 2001. Water Quality Modeling for Wasteload Allocations and TMDLs. Wiley, New York, NY, USA.

Mao, J.Q., Chen, Q.W., Chen, Y.C., 2008. Three-dimensional eutrophication model and application to thihu lake, China. J. Environ. Sci. (China) 20, 278–284.

Moll, A., Radach, G., 2003. Review of three-dimensional ecological modelling related to the North Sea shelf system: part 1: models and their results. Prog. Oceanogr. 57, 175–217.

Park, S.S., Lee, Y.S., 2002. A water quality modeling study of the Nakdong River, Korea. Ecol. Model. 152, 65–75.

Park, K., Kuo, A.Y., Shen, J., Hamrick, J.M., 1995. A Three-Dimensional Hydrodynamic-Eutrophication Model (HEM-3D): Description of Water Quality and Sediment Process Submodels. Virginia Institute of Marine Science, Gloucester Point, VA.

Park, K., Jung, H.S., Kim, H.S., Ahn, S.M., 2005. Three-dimensional hydrodynamic-eutrophication model (HEM-3D): application to Kwang-Yang Bay, Korea. Mar. Environ. Res. 60, 171–193.

Peng, S., Fu, G.Y.Z., Zhao, X.H., Moore, B.C., 2011. Integration of Environmental Fluid Dynamics code (EFDC) model with geographical information system (GIS) platform and its applications. J. Environ. Inform. 17, 75–82.

Scheffer, M., Carpenter, S.R., 2003. Catastrophic regime shifts in ecosystems: linking theory to observation. Trends Ecol. Evol. 18, 648–656.

Scheffer, M., Carpenter, S., Foley, J.A., Folke, C., Walker, B., 2001. Catastrophic shifts in ecosystems. Nature 413, 591–596.

Semtner, A.J., 1974. An oceanic general circulation model with bottom topographyNumerical Simulation of Weather and Climate, Technology Report 9. Department of Meteorology, University of California, Los Angeles, CA, p. 41.

Seo, D., Sigdel, R., Kwon, K.H., Lee, Y.S., 2010. 3-D hydrodynamic modeling of Yongdam Lake, Korea using EFDC. Desalin. Water Treat. 19, 42–48.

Shi, J.H., Li, G.X., Wang, P., 2011. Anthropogenic influences on the tidal prism and water exchanges in Jiaozhou Bay, Qingdao, China. J Coastal Res. 27, 57–72.

Solow, A.R., Beet, A.R., 2005. A test for a regime shift. Fish. Oceanogr. 14, 236–240.

Tech, T., 1999. Three-dimensional hydrodynamic and water quality model of Peconic estuary. Technical Report. Prepared for Peconic Estuary Program, Suffolk County, NY, Tetra Tech, Inc., Fairfax, VA.

Turner, R.E., Rabalais, N.N., Justic, D., 2008. Gulf of Mexico hypoxia: alternate states and a legacy. Environ. Sci. Technol. 42, 2323–2327.

Vinokur, M., 1974. Conservation equations of gasdynamics in curvilinear coordinate systems. J. Comput. Phys. 14, 105–125.

Walker, B., Salt, D., 2006. Resilience Thinking: Sustaining Ecosystems and People in a Changing World. Island Press, 1718 Connecticut Avenue, NW, Suite 300, Washington, DC 20009, USA.

Wool, T.A., Davie, S.R., Rodriguez, H.N., 2003. Development of three-dimensional hydrodynamic and water quality models to support total maximum daily load decision process for the Neuse River estuary, North Carolina. J. Water Resour. Plan. Manag. 129, 295–306.

Wu, G.Z., Xu, Z.X., 2011. Prediction of algal blooming using EFDC model: case study in the Daoxiang Lake. Ecol. Model. 222, 1245–1252.

Zhao, L., Zhang, X., Liu, Y., He, B., Zhu, X., Zou, R., Zhu, Y., 2012. Three-dimensional hydrodynamic and water quality model for TMDL development of Lake Fuxian, China. J. Environ. Sci. (China) 24, 1355–1363.

Zhao, L., Li, Y., Zou, R., He, B., Zhu, X., Liu, Y., Wang, J., Zhu, Y., 2013. A three-dimensional water quality modeling approach for exploring the eutrophication responses to load reduction scenarios in Lake Yilong (China). Environ. Pollut. 177, 13–21.

Zou, R., Carter, S., Shoemaker, L., Parker, A., Henry, T., 2006. Integrated hydrodynamic and water quality modeling system to support nutrient total maximum daily load development for Wissahickon Creek, Pennsylvania. J. Environ. Eng. ASCE 132, 555–566.

Floating Treatment Wetlands for Nutrient Removal in a Subtropical Stormwater Wet Detention Pond with a Fountain

16

Ni-Bin Chang*, Martin P. Wanielista, Zhemin Xuan, Zachary A. Marimon

Department of Civil, Environmental, and Construction, University of Central Florida,
Orlando, FL, USA
*Corresponding author: e-mail address: nchang@ucf.edu

16.1 OVERVIEW OF STORMWATER FLOW AND QUALITY IMPACT

Rainfall happens periodically and results in runoff and flooding impacts especially in urban regions. Many watershed management plans are designed to reduce stormwater runoff and flooding. Stormwater runoff from roadway and adjacent communities in urban and suburban regions contains numerous contaminants in varying but significant concentrations (Göbel et al., 2007). The chemical composition of stormwater runoff may be influenced by watershed characteristics such as land use, traffic volume, and percent impervious cover (Muthukrishnan, 2006). Impervious surfaces have long been implicated in the decline of stormwater quality and increase of flooding in urban areas (Brattebo and Booth, 2003). These contaminants in stormwater runoff may include but are not limited to motor oil, gasoline, sediment, fertilizer, pesticides, bacteria, heavy metals, nutrients, hydrocarbons (e.g., polycyclic aromatic hydrocarbons, PAHs), and others. Due to the presence of various contaminants, stormwater runoff without proper treatment and control has been found to be harmful to the aquatic environment, with potential negative ecological impacts on receiving waters, such as groundwater aquifers, streams, lakes, and coastal waters (Fisher et al., 1995; Marsalek et al., 1999). To date, research efforts of independent stormwater ponds have tended to focus on a small number of ponds and/or ponds with suspected performance failures (DHEC, 2007). Additionally, there have been no large scale, comprehensive analyses of stormwater pond performance in the southeastern United States (DHEC, 2007). Given the fact that there are many stormwater treatment areas (i.e., retention and detention ponds) in Florida and some other states, a thorough, systematic study to overhaul some unique stormwater ponds, quantify the pollutant

Developments in Environmental Modelling, Volume 26, ISSN 0167-8892, http://dx.doi.org/10.1016/B978-0-444-63249-4.00018-X

loads and existing maintenance practices, and determine the essential changes of maintenance methods to reduce these contaminants is an acute need.

The leading water quality problem in many urban watersheds is pollutants in sediment and turbidity coming from stormwater. Erosion and sedimentation control associated with stormwater runoff can be controlled by improving the proper design, construction, and operation/maintenance of relevant best management practices (BMPs) during construction and long-term operation stages of stormwater systems. Some BMPs, such as silt fence barriers and floating turbidity barriers for use during and after construction, can not only minimize erosion but also properly manage runoff for controlling both stormwater quantity and quality in terms of nutrients, heavy metals, pesticide, and bacteria.

Lawn and garden activities can contaminate stormwater with fertilizer, pesticide, and sediment. Large quantities of the organophosphorous (OP) pesticides such as diazinon and chlorpyrifos are oftentimes applied to urban and agricultural watersheds. These pesticides that are transported in stormwater runoff either dissolved in water or attached to soil particles (sediment) may directly affect the health of aquatic organisms due to toxic effect (DHEC, 2007). Persistent concentrations of pesticides and herbicides can bioaccumulate and biomagnify and can significantly impact ecosystems within the stormwater detention ponds (DHEC, 2007). Their presence in drinking water sources also threatens human health (DHEC, 2007).

Some common nutrients associated with fertilizer application in the water body are ammonia, nitrite, nitrate, and phosphorus. The negative impacts can be eutrophication, oxygen depletion, and toxic effects due to the presence of algal toxin toward flora and fauna (Göbel et al., 2007). These nutrients can impact on both public health and ecosystem integrity. For example, without proper treatment, ammonia in the leaking sewage flushed out from stormwater can stimulate phytoplankton growth, exhibit toxicity to aquatic biota, and exert an oxygen demand in surface waters (Beutel, 2006). Undissociated ammonia is extremely volatile and in aqueous solution either ionizes or volatizes. Ionized ammonia is very toxic for fish species (Tarazona et al., 2008). Fish mortality, health, and reproduction can be affected by the presence of a minute amount of ammonia-N (Servizi and Gordon, 2005). Furthermore, nitrate can cause human health problems such as liver damage and even cancers (Gabel et al., 1982; Huang et al., 1998). Nitrate can also bind with hemoglobin and create a situation of oxygen deficiency in an infant's body called methemoglobinemia (Kim-Shapiro et al., 2005). Nitrite that is an intermediate product of nitrification can react with amines chemically or enzymatically to form nitrosamines that are very potent carcinogens (Sawyer et al., 2003).

Elevated bacteria in stormwater runoff and during wet weather flow conditions in urban streams have been well documented (Schueler and Holland, 2000; Pitts, 2004a,b; Bossong et al., 2005). Either *Escherichia coli* or fecal coliform has been selected as a pathogen indicator of bacterial contamination. Typical concentrations of bacteria (whether measured as *E. coli* or fecal coliform) in urban stormwater are often two orders of magnitude greater than instream primary contact recreational standards (Schueler and Holland, 2000). Leaking septic tanks can be major sources of human fecal contamination in aquatic ecosystems in addition to animal scat, droppings, and feces causing high level of *E. coli* or fecal coliform.

Heavy metals in stormwater are found to have both acute and chronic toxic effects toward aquatic flora and fauna (Marsalek et al., 1999; Wium-Andersen et al., 2011). Stormwater and pond water may show high levels of heavy metals, such as zinc and copper (Göbel et al., 2007). Toxicity of heavy metals depends on physical and chemical properties (e.g., pH, redox, complexing ligands, and ion strength) as the mobility and bioavailability of the metals are affected by these conditions (Ure and Davidson, 2002).

Very recently, studies by the U.S. Geological Survey (USGS) (Bommarito et al., 2010a,b; Mahler et al., 2010) have identified coal tar-based sealcoat as a major source of PAH contamination in urban areas for large parts of the nation. Coal tar-based sealcoat is the black, viscous liquid sprayed or painted on asphalt pavement such as parking lots, causing PAH pollution country wide. Friction from vehicle tires abrades pavement sealcoat into small particles. These particles with PAH content are washed off pavement by rain and carried down storm drains and into lakes, streams, and stormwater retention and detention basins. Several PAHs are suspected human carcinogens and are toxic to aquatic life (Pinkney et al., 2009). PAH analytes that the USEPA considers to be "priority pollutants" were significantly higher in the sediments of commercial ponds compared to that of reference, low-density residential and high-density residential ponds (Weinstein et al., 2008). However, PAHs have a relatively low solubility and high affinity toward organic carbon (Simon and Sobieraj, 2006). That is, PAHs are often found attached to settlable particles (Marsalek et al., 1999), and therefore, mainly found in the sediment and not the water phase of wet detention ponds (Wium-Andersen et al., 2011). Hence, the sediments are often found to be highly toxic as pollutants settle and accumulate in the sediments (Grapentine et al., 2008). Therefore, erosion and sediment control at the watershed scale has multiple implications in stormwater pollution prevention and control.

16.2 STORMWATER TREATMENT CAPACITY

To minimize the stormwater impact, a flow control (FC) facility (i.e., stormwater structural controls) can be either a pond (e.g., either detention or retention pond), an underground tank or vault, or an infiltration system specifically designed to capture, store, and then slowly release stormwater runoff downstream or into the ground. Conventional stormwater retention/detention ponds designed for flood control are often used along with other low impact development (LID) options (i.e., such as bioswales, green roofs, rain gardens) to provide stormwater treatment and thus protect natural water bodies from development-pollution runoff. Such stormwater treatment practices are the most effective LID for removing particulate forms of pollutants and restoring the natural hydrological cycle, especially the solids fraction to be settled, but removal of dissolved, or colloidal, pollutants is minimal. Therefore, pollution prevention or control at the sources offers a more effective way to control the dissolved pollutants and source tracking at the watershed level turns out to be critical. Fortunately, most toxic stormwater contaminants (heavy metals and organic compounds) are mostly associated with stormwater particulates. The removal of the solids will also remove much of the pollutants of interest.

Understanding the functionality of stormwater structural controls in relation to municipal separate storm sewer system (MS4) permits is also helpful in promoting the inspection and maintenance of these FC facilities. A family of structural controls can be highlighted, in which stormwater is retained onsite and allowed to infiltrate into the soil and evaporate rather than being discharged. These infiltration controls, a part of "Green Infrastructure," may reduce the stormwater volume and pollutant load too. These controls may include but are not limited to infiltration basins, exfiltration trenches, French drains, treatment swales, and detention/retention ponds.

The following description provides definitions of the major types of structural controls contained within MS4 permits by Florida Department of Environmental Protection (FDEP) (FDEP, 2011), which are relevant to this project.

(1) *Dry retention systems* (*basins*): The stormwater retention (dry) ponds are infiltration systems that are excavated into the ground for flood control mainly. Typically, they are vegetated to minimize erosion, and the roots help to maintain the permeability of the soils.

(2) *Exfiltration trenches/French drains*: They are shallow, excavated trenches in which stormwater is stored in perforated or slotted pipes and percolates out through the surrounding gravel envelope and filter fabric into the soil. In South Florida, these systems are called French drains.

(3) *Grass treatment swales*: Swales designed for stormwater treatment can be classified into the following two categories: (a) swales with swale blocks or raised driveway culverts and (b) swales without swale blocks or raised driveway culverts. Swales that are online retention systems are defined in Chapter 403.803 (14), Florida statutes, as follows: (a) has a top width to depth ratio of the cross section equal to or greater than 6:1, or side slopes equal to or flatter than 3 ft horizontal to 1-ft vertical; (b) contains contiguous areas of standing or flowing water only following a rainfall event; (c) is planted with or has stabilized vegetation suitable for soil stabilization, stormwater treatment, and nutrient uptake; and (d) is designed to take into account the soil erodibility, soil percolation, slope, slope length, and drainage area so as to prevent erosion and reduce pollutant concentration of any discharge.

(4) *Underdrain systems*: These systems consist of a dry basin underlain with perforated drainage pipe that collects and conveys stormwater following percolation from the basin through suitable soil. Underdrain systems are generally used where high water table conditions dictate that recovery of the stormwater treatment volume cannot be achieved by natural percolation (i.e., retention systems) and suitable outfall conditions exist to convey flows from the underdrain system to receiving waters.

(5) *Dry detention systems*: They are designed to store a defined quantity of runoff and slowly release the collected runoff through an outlet structure to adjacent surface waters. After drawdown of the stored runoff is completed, the storage

basin does not hold any water, thus the system is normally "dry." Dry detention basins are similar to retention systems in that the basins are normally dry. They are used in areas where the soil infiltration properties or seasonal high water table elevation will not allow the use of a retention basin.

(6) *Detention ponds with effluent filtration systems*: These systems usually are permanently wet ponds, but can include dry ones, that have a sand filter to provide effluent filtration. The filters usually are in the side banks of the pond, but they may be in the bottom of the basin. The filters are maintenance intensive and do a poor job of removing nutrients.

(7) *Wet detention systems*: These systems are permanently wet ponds that are designed to slowly release a portion of the collected stormwater runoff through an outlet structure. Wet detention systems are the recommended BMP for sites with moderate to high water table conditions. Wet detention systems provide removal of both dissolved and suspended pollutants by taking advantage of physical, chemical, and biological processes within the pond.

(8) *Weirs, channel control structures, and other control structures*: They are structural components of a stormwater system that detain stormwater and allow its controlled release.

Floating treatment wetlands (FTWs) are an innovative variant on these systems and a possible solution to the problems. Additionally, plants grow on floating mats rather than being rooted in the sediments (Figure 16.1), so water depth is not a concern, and the mats are unlikely to be affected by fluctuations in water levels.

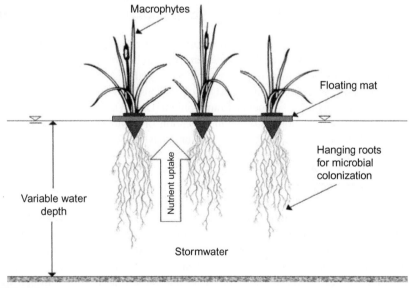

FIGURE 16.1

Cross-section of a typical floating treatment wetland (Chang et al., 2013).

FTW is one of the emerging BMPs for stormwater treatment where macrophytes provide a suitable root-zone environment for microorganisms that allow the plants to remove nutrients through direct uptake into their tissue (Billore and Prashant, 2008; Breen, 1990). In FTWs, plants grow on floating mats rather than rooted in sediments, thus avoiding the detrimental effects of severe water level fluctuation or inundation (Headley et al., 2006). In addition, a smaller land area requirement and additional aesthetic values make this a promising ecotechnology favored by water quality management practitioners.

In the past few years, Hart et al. (2003) measured uptake by macrophytes, microbes, and abiotic mechanisms conduced on septic tank effluent. Hubbard et al. (2004) measured nutrient uptake by macrophytes on swine lagoon wastewater using a modified batch process. Stewart et al. (2008) mainly focused on the microbial contribution in a tank-scale environment with floating mats, whereas White et al. (2009) transferred the tank-scale results to actual ponds with floating mats for 100% coverage on the pond surface. Although these studies investigated the effectiveness of FTWs, the variability between the case studies with varying time and spatial scales makes direct comparisons difficult.

Biologically, aquatic macrophyte-based wastewater treatment systems were far more diverse than present-day mechanical treatment systems (Hammer, 1989; Moshiri, 1993). Free floating macrophytes provide shade for the water column, resulting in a cooler habitat for fish and macroinvertebrates (Nahlik and Mitsch, 2006). The hanging roots provide a large surface area for denitrifying bacteria; the resulting anaerobic environment can remove nitrate through the denitrification process (Govindarajan, 2008). The macrophyte roots can also entrap fine suspended particulates that would otherwise remain in suspension in a conventional pond system (Headley et al., 2006).

The removal of nitrogen (N) in aquatic ecosystems is one of the greatest interests in stormwater treatment. Much of the nitrate removal can be attributed to two key processes under similar conditions of low oxygen concentrations. They include denitrification, which is the reduction of nitrate (NO_3^-) to gaseous N compounds (NO, N_2O, and N_2), and dissimilatory NO_3^- reduction to ammonium (DNRA). The latter is also termed fermentative NO_3^- reduction, NO_3^- ammonification, or fermentative ammonification. There is evidence that many soil bacteria and fungi have the ability to perform DNRA and redox status and C/NO_3^- ratio have been identified as the most important factors regulating DNRA in soil (Burgin and Hamilton, 2007; Rütting et al., 2011).

In floating wetlands, the plants are not rooted in sediments, and they are forced to acquire nutrition directly from the water column (Headley et al., 2006; Vymazal, 2007). Because the reduction of nitrate (NO_3^-) to gaseous N compounds (NO, N_2O, and N_2) may occur in the soil pores within the floating mat, microbes that live on the surface of plant roots in a wetland can remove 10 times more nitrate than the plants themselves due to DNRA process (Adams, 1992). Moreover, a biofilm can form on the surface of root particles allowing microbes to assimilate nitrogen species, although nitrogen may not able to be removed by sorption directly. The nutrient and other element uptake into biomass increases as physiological growth continues.

Total nitrogen (TN) and total phosphorus (TP) are removed when the plants are regularly harvested. Finally, algal toxins are not present in the pond because the lack of nutrients prevents their growth. In addition to the above attributes, they can potentially improve tissue culture responses, including somatic embryogenesis, organogenesis, adventitious shoot production and growth, and the rooting of micro-propagated tissues (Van Winkle and Pullman, 2005).

To date, little information has been published on FTWs. This chapter presents a field-scale testing of FTWs provided by Floating Islands International to show the effectiveness of nutrient removal at a stormwater wet detection pond with a fountain (Wanielista et al., 2012).

16.3 FIELD POND STUDY

16.3.1 Study site

A stormwater wet detention pond, "Pond 5," located in a community near the main campus of University of Central Florida (UCF) in Orlando, Florida, was used to inves-tigate the application potential of FTWs for water quality improvement (Figure 16.2). The pond near the natural forest has a surface area around 340 m^2 at discharge control elevation (23.0 m) in a local community-based watershed with a drainage area of about 1.64 ac. Inflow and outflow pipes were constructed at 22.1 m elevation. A con-crete structure at 21.9 m in the adjacent wetland receives the outflow discharge from the pond. It has a 3.2-cm diameter orifice at 23.0 m and a fiberglass skimmer top at 23.4 m so that when the water level in Pond 5 rises over 23.4 m, the flood water will spill away from the top of the concrete structure directly toward the nearby wetland.

The wet pond has been serving the community for 14 years with a smaller wa-tershed area and pond size in communities in Central Florida. Some emergent mac-rophytes have emerged along the bank of Pond 5 for years, making it ideal to examine how vegetation planted on floating mats assimilates nutrients from pond water in a mature stormwater wet detention pond ecosystem. A thick sediment layer has formed at the bottom of the pond, which also contributes to the removal of nu-trients from the pond system.

To support a better ecosystem service design to this community, fibrous matrix FTWs were applied in Pond 5, and a fountain in the center provided aeration through-out the entire monitoring period. Temporal observation elucidated ecological evolu-tion and interactions in the established ecosystem and provided the knowledge basis for application of FTWs in mature stormwater ponds.

16.3.2 Materials and methods

16.3.2.1 Water balance analysis

16.3.2.1.1 Water level

The storage for Pond 5 was represented by water level data and recorded by the same water level sensor model (Global Water WL400) installed at the mouth of the circu-lar outlet culvert (i.e., 0 m in raw water level data is equivalent to 21.7 m). The data

FIGURE 16.2

Location of the Pond 5 off UCF campus. (For color version of this figure, the reader is referred to the online version of this chapter.)

Adapted from Google Earth.

logger (Global Water GL500-2-1) was connected to the water level sensor and set to record the water level data at 10-min intervals.

16.3.2.1.2 Rainfall

During the experiment period from 2011 to 2012, rainfall (the direct amount falling into the pond) was measured and read from a 15-cm Tipping Bucket rain gauge (Figure 16.3: RG200, Global Water) on site. The radar rainfall data from The St. Johns River Water Management District was used as a backup rainfall data source when the rain gauge was not functioning due to unpredictable factors.

FIGURE 16.3

Rain gauge. (For color version of this figure, the reader is referred to the online version of this chapter.)

Table 16.1 Watershed Area and Empirical Runoff Coefficients Used for Pond 5 in Florida

	Runoff Coefficient (C)	C, Used Value	Watershed Area (acre)	Weighted Runoff Fraction
Lawns	0.05–0.35	0.20	0.1950	0.024
Roofs	0.75–0.95	0.85	0.5957	0.309
Concrete streets	0.7–0.95	0.83	0.7615	0.386
Pond	1.00	1.00	0.0849	0.052
Total			1.6371	0.771

16.3.2.1.3 Inflow

The amount of surface runoff, considered as the principal component of the inflow, depended on the land size of the watershed that produced runoff flowing into Pond 5. Due to budget limitations, no flowmeter was installed at the inlet; instead, the rational runoff was used to estimate the inflow amount. In rational equation, the watershed area and the empirical runoff coefficients used for the Pond 5 in Florida were summarized (Table 16.1):

FIGURE 16.4

Evaporation pan. (For color version of this figure, the reader is referred to the online version of this chapter.)

$$Q = CIA \qquad (16.1)$$

where Q, peak discharge, in cfs; C, runoff coefficient; I, rainfall intensity (in./h); and A, drainage area (ac).

16.3.2.1.4 Evaporation

For Pond 5, one evaporation pan (Figure 16.4) located in the UCF stormwater laboratory was used to measure evaporation rate, which was further converted to the pond evaporation rate by multiplying by a coefficient of 0.7 which is an empirical value being applied in Central Florida (Kohler et al., 1955).

16.3.2.1.5 Infiltration

Directly measuring infiltration to the groundwater table with time for the whole pond area was unfeasible; therefore, a period of time when the water level was lower than the level of orifice on the concrete structure was selected to estimate the infiltration amount. For simplification, the infiltration rate was considered a constant for the water balance calculation. Once infiltration was determined, the outflow (the unknown term in the Pond 5 study) in the water balance equation could be calculated.

16.3.2.1.6 Outflow

A concrete structure was constructed at 21.87 m to connect Pond 5 to the adjacent wetland. The structure had a 15-cm diameter orifice at 22.67 m and a fiberglass skimmer on the top at 23.04 m. When the water level in Pond 5 rose above 22.67 m, outflow would discharge, and when the water level rose higher than 23.04 m, the flood water would spill away from the top of the concrete structure directly toward the nearby wetland.

16.3.2.1.7 Water balance

A storm event-based water balance model for Pond 5 was developed that includes the following terms:

$$\Delta \text{Storage} = \text{Direct Rainfall} + \text{Inflow} - \text{Evaporation} - \text{Infiltration} - \text{Outflow}$$

$$(16.2)$$

The storage for Pond 5 was represented by water level data, recorded by a water level sensor (Global Water WL400) installed at the mouth of the circular outlet culvert (i.e., 0 m in raw water level data is equivalent to 22 m). The data logger (Global Water GL500-2-1) was connected with the water level sensor and set to record the water level data at intervals of 10 min. The data were exported via its USB port to a laptop computer as an Excel compatible *.CSV file.

During the experimental period, rainfall, measured as the direct amount falling into the pond, was recorded from a 15-cm Tipping Bucket rain gauge (RG200, Global Water) on site. The radar rainfall data from The St. Johns River Water Management District were used as a backup rainfall data source when the rain gauge was not functioning due to some unpredictable factors. Surface runoff, considered the principal component of the inflow, is the water flow that occurs when the soil reaches the saturated level; therefore, the runoff amount depends on the size of the watershed that produces runoff flowing into Pond 5. Because the drainage area is small, the rational method was used to estimate inflow amount based on the varying rainfall data with respect to fixed watershed area and the runoff coefficient. Evaporation, the direct amount evaporated from the pond water surface, is dependent on many factors, such as temperature, wind, and atmospheric pressure. In our study, an evaporation pan located nearby was used to measure evaporation rate, which is further converted to the pond evaporation rate by multiplying by a coefficient of 0.7.

Directly measuring the infiltration to the groundwater table with time for the whole pond area was unfeasible; therefore, a period of time when the water level was lower than the level of the orifice on the concrete structure was selected to estimate the infiltration amount. During that time, the terms of direct rainfall, inflow, and outflow can be considered to be zero, and then the water balance equation can be simplified as:

$$\Delta \text{Storage} = -\text{Evaporation} - \text{Infiltration}. \qquad (16.3)$$

That is, the infiltration term can be calculated as the water level loss after subtracting the evaporation amount. For simplification, the infiltration rate was considered a constant in the context of water balance calculation. Once the infiltration term was determined, the outflow term in the water balance equation could be calculated. As for the outflow measurements, a concrete structure was constructed at an elevation of 21.87 m connecting Pond 5 to the adjacent wetland that receives the outflow discharge from the pond. Outflow can also be estimated based on the other parameter values in Equation (16.2).

16.3.2.2 Nutrient removal evaluation of FTWs

As a preanalysis, water quality analysis was conducted for three storm events and three nonstorm events in the first half of July 2011. Nonstorm event analysis was used to produce an instantaneous snapshot of nutrient distribution throughout the pond and a nutrient reduction between inlet and outlet. Event-based sampling efforts were done in parallel with the nonstorm events sampling campaign. Both types of analysis with different time scales will be presented in Section 16.3.1.

To estimate removal efficiencies (REs) using fibrous matrix FTWs, a postanalysis at Pond 5 was conducted for 9 months after the floating wetland deployment. Water quality parameters were monitored to calculate the nutrient REs of the FTWs. The postanalysis was further divided into two parts, nonstorm based and event based. The data in postanalysis were used to calculate the additional water quality improvement due to the fibrous matrix FTWs.

16.3.2.3 Operating hydraulic residence time

Design HRT is the ratio of the pond volume and the inflow rate:

$$HRT = V/Q \tag{16.4}$$

where HRT, hydraulic residence time, d; V, pond volume, m^3; and Q, inflow rate, $m^3\,d^{-1}$.

RE is related to holding or reaction time and is thus primarily dependent on the pond's HRT at a particular moment in time. The operating HRT is not equivalent to a constant HRT value, however, because influent flow varies over time and the rate never becomes steady, so the operating HRT must be defined another way.

We selected 40 studies to include in a database to identify runoff event mean concentration (EMC) values for single land use categories in Florida (Harper, 2011). For Pond 5, the geometric means of 2.102 mg L^{-1} for TN and 0.497 mg L^{-1} for TP (particulate plus dissolved) were used for multifamily residential runoff, as well as the initial nutrient concentration in the runoff. Because the event-based sampling efforts were carried out in parallel with the monthly sampling campaign, the operating HRT can be defined as: (1) the time interval between the occurrence of the storm and the time of sampling (which was converted to a daily basis as a matter of convenience) and (2) the time interval on the daily basis between the end of last storm event and the time of the subsequent nonstorm sampling. The event-based data therefore revealed how much of the nutrient content was removed by the physical sedimentation process within a short HRT (event based), and the monthly based data implied how much of the nutrient content was removed by the biological treatment during a long HRT. RE varied with different operating HRT; thus, a plot of operating HRT versus REs was formed to provide another perspective of nutrient removal performance of FTWs, which is presented in Section 16.3.2.

16.3.2.4 REs and credit of floating wetlands

In addition to the self-purification capacity via a natural process, floating wetlands were introduced to further improve water quality, an essential component to quantify additional credit for floating wetlands in terms of (1) assumed value based (outlet value vs. assumed runoff value) and (2) inlet value-based (outlet value vs. inlet value) nutrient control. Particulates are known to settle out during a short HRT, and therefore, floating islands contribute little to particulate removal. Over a long period of time via biological processes, however, the mostly dissolved fraction of nitrogen and phosphorus can be removed. The procedure for assessing the performance credit of floating wetlands is described as:

(1) Runoff concentration based (RE):

(A) Short-term settling-dominated RE (RE_S):

$$RE_S = \frac{C_{assumed} - \overline{C_{I-S}}}{C_{assumed}} \times 100\%, \tag{16.5}$$

where assumed inputs of TN and TP are 1.068 and 0.179 mg L^{-1}, respectively, for Pond 4M and 2.102 and 0.497 mg L^{-1} for Pond 5; and $\overline{C_{I-S}}$ is the geometric mean of nutrients concentration at the inlet for the storm events.

(B) Overall removal efficiency (RE_O):

$$RE_O = \frac{C_{assumed} - \overline{C_{O-N}}}{C_{assumed}} \times 100\%, \tag{16.6}$$

where $\overline{C_{O-N}}$ is the geometric mean of nutrients concentration at the outlet in the nonstorm events.

(C) Long-term biologically dominated removal efficiency (RE_B):

$$RE_B = RE_O - RE_S = \left(\frac{C_{assumed} - \overline{C_{O-N}}}{C_{assumed}} - \frac{C_{assumed} - \overline{C_{I-S}}}{C_{assumed}} \right) \times 100\%$$
$$= \frac{\overline{C_{I-S}} - \overline{C_{O-N}}}{C_{assumed}} \times 100\% \tag{16.7}$$

RE_B in terms of TN and TP was calculated for both preanalysis (without FTWs) and postanalysis (with FTWs) for two types of FTWs. A marginal concentration-based improvement was used to estimate the credit of floating wetlands as RE_B (with FTWs)—RE_B (without FTWs).

(2) Pond concentration based:

Concentration reduction percentage (CRP) was used to estimate the difference in nutrient levels near the inlet and outlet (note that the fountain in Pond 5 operated throughout the entire monitoring period):

$$EMC = \frac{\sum C_i V_i}{\sum V_i}, \tag{16.8}$$

where V is volume of flow during period i, and C is the average concentration associated with study period, and

$$CRP = \frac{C_{inlet} - C_{outlet}}{C_{inlet}} \times 100\%, \tag{16.9}$$

where C_{inlet} and C_{outlet} stand for the EMC at the inlet and outlet when a storm event occurs.

16.3.2.5 FTWs deployment in Pond 5

Four floating mats (BioHaven® Floating Island) were deployed at Pond 5 on July 15, 2011 (Figure 16.1) and covered collectively 8.7% of the pond surface area to ensure the reliability of the engineering practices for nutrient removal. The floating mats are primarily composed of fine (0.018 cm diameter) intertwined polymer strands (made from recycled materials) of 5% fiber and 95% pore space (by volume) bonded to provide a three-dimensional nonwoven matrix with an area of 7.4 m^2 (80 ft^2). Each mat holds about 160 pots for plants. As a whole, a single floating mat can provide sufficient porosity and permeability, and the individual polymer strands provide an ideal substrate for colonization by microbial biofilm. The substrate also provides an excellent growth medium for the roots of aquatic, riparian, and terrestrial plants (Stewart et al., 2008). Nitrogen and carbon dioxide generated by microbial metabolism along with gas-filled plant roots provide buoyancy for the islands. Plant species were selected according to a previous mesocosm-scale study by Chang et al. (2013) in which the adaptability of roots to long-term submergence in wetland conditions was prioritized. Both soft rush (*Juncus effusus*) and pickerelweed (*Pontederia cordata*) are indigenous wetland plants of southeastern United States that have high survival potential in the Central Florida climate; consequently, these two plant species were used on floating islands in this full-scale field experiment. The four floating mats were tied together to form a ring surrounding the fountain, away from the inlet and outlet (Figure 16.5). Pots in the floating mat were filled with peat moss as the plant substrate.

16.3.2.6 Sampling and analysis

During storm and nonstorm events, water samples were collected in triplicate close to the inlet and the outlet at Pond 5 and then mixed as composite samples (Figure 16.6). All the composite samples were stored at 4 °C and delivered to a certified laboratory for chemical analysis of nutrients using various methods (Table 16.2). Note that the fountain in Pond 5 operated throughout the entire monitoring period.

16.4 RESULTS AND DISCUSSION

16.4.1 Temporal and spatial nutrients distributions

16.4.1.1 Preanalysis

The preanalysis period was defined as the study period before the deployment of the floating wetland. Within the preanalysis period, three storm and three nonstorm events were investigated in the first half of July to determine the background of this pond. The CRP results for nutrient levels at the inlet and outlet (Figure 16.7; Table 16.3) showed that for storm events, the nutrient levels for TP and OP in inflow and outflow were similar (Table 16.4). Three forms of nitrogen in the outflow were even higher than those in the inflow. Low concentrations of NH_3 and $NO_2 + NO_3$

FIGURE 16.5

Preparation and deployment of floating wetland (July 15, 2011) (Chang et al., 2013). (For color version of this figure, the reader is referred to the online version of this chapter.)

FIGURE 16.6

Sampling locations of water columns (marked by inlet and outlet) and sediment beds (marked by red circles) at the Pond 5. (a) Storm events. (b) Nonstorm events (Chang et al., 2013). (For interpretation of the references to color in this figure legend, the reader is referred to the online version of this chapter.)

Table 16.2 Outline of Analytical Methods

Parameter	Analytical Method
TN	SM21 4500-N C
$NO_2 + NO_3$	EPA 353.2/SM21 4500-NO_3 F
NH_3	EPA 350.1/SM21 4500-NH_3 G
TP	EPA 365.1/SM21 4500-P B
OP	EPA 365.1/SM21 4500-P F

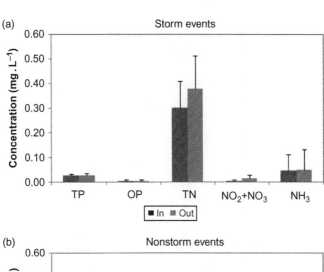

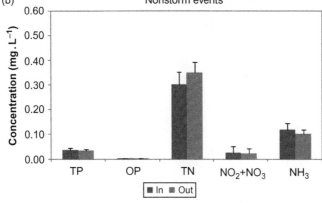

FIGURE 16.7

Nutrients concentration during preanalysis. (a) Storm event. (b) Nonstorm event (Chang et al., 2013). (For color version of this figure, the reader is referred to the online version of this chapter.)

Table 16.3 Nutrients Concentration for Storm Events During Preanalysis (mg L^{-1})

Date	TP		OP		TN		NO$_2$+NO$_3$		NH$_3$	
	In	Out	In	Out	In	Out	In	Out	In	Out
7/2/11	0.032	0.034	0.008	0.008	0.223	0.332	0.011	0.032	0.012	0.009
7/7/11	0.030	0.032	0.009	0.009	0.427	0.528	0.003	0.017	0.008	0.001
7/12/11	0.023	0.016	0.001	0.001	0.251	0.272	0.005	0.003	0.123	0.146
Average	0.028	0.027	0.006	0.006	0.300	0.377	0.006	0.017	0.048	0.052
CRP (%)	3.5		0.0		−25.6		−173.7		−9.1	

Table 16.4 Nutrients Concentration for Nonstorm Events During Preanalysis (mg L^{-1})

Date	TP		OP		TN		NO$_2$+NO$_3$		NH$_3$	
	In	Out	In	Out	In	Out	In	Out	In	Out
7/8/11	0.044	0.038	0.003	0.002	0.362	0.388	0.054	0.045	0.149	0.114
7/9/11	0.040	0.036	0.004	0.002	0.265	0.302	0.007	0.016	0.114	0.110
7/11/11	0.026	0.027	0.001	0.001	0.281	0.358	0.015	0.006	0.100	0.086
Average	0.037	0.034	0.003	0.002	0.303	0.349	0.025	0.022	0.121	0.103
CRP (%)	8.2		37.5		−15.4		11.8		14.6	

indicated that the dominant N form was organic nitrogen; however, the smaller difference in TN levels between the inlet and outlet, along with a positive CRP of TP, OP, NH_3, and nitrite-nitrogen + nitrate-nitrogen (NO_2-N + NO_3-N), indicated that a moderate self-purification occurred in Pond 5. In nonstorm events, organic nitrogen was partially converted to NH_3, which led to the increase of $NO_2 + NO_3$ due to the aeration by the fountain when compared to the counterparts in storm events.

16.4.1.2 Postanalysis

The postanalysis period is defined as the study period after the deployment of the floating wetland. During the postanalysis period, *in situ* data for water quality analysis at Pond 5 were monitored continuously to test if the deployment would function as we expected in the two scenarios, storm versus nonstorm events. Water samples in four storm and four nonstorm events were collected, and nutrient samples were delivered to the same certified laboratory off campus for chemical analysis. The overall performance of the fibrous matrix FTWs between storm and nonstorm events was investigated and compared to the preanalysis and postanalysis conditions.

Six storm events were monitored after the deployments on August 16 and 28, September 19, and October 8 and 29 in 2011 and on April 6 in 2012. The nutrient levels in runoff during postanalysis (Table 16.5) were much higher than those during preanalysis (Table 16.6), even though a high removal of TN and $NO_2 + NO_3$ was observed (Figure 16.8a), which confirmed the credit of the floating wetland performance. In addition to the analysis for storm events, sampling for seven nonstorm events was conducted on July 27, August 23, September 2, November 17, and December 14, 2011, and on February 2 and March 27, 2012. Positive removal was observed in terms of all forms of nutrients (Figure 16.8b). The overall CRP of phosphorus was substantial; 46.3% TP and 79.5% OP were removed, probably by the combination of adsorption through peat moss in the floating wetlands and sedimentary process in the pond. The overall reduction of TN, $NO_2 + NO_3$, and NH_3 reached 16.9%, 16.7%, and 53.0%, respectively. In short, significant improvements were found in postanalysis (Tables 16.5 and 16.6).

16.4.2 Operating HRT and REs

The operating HRT associated with nutrient REs during the postanalysis was summarized in Tables 16.7 and 16.8 for the Pond 5 study. Similarly, the logarithmic trends (Figures 16.9 and 16.10) indicated that longer operating HRTs lead to higher REs. During postanalysis, TP removal was stable over 68% when the operating HRT was longer than a few hours. In comparison, TN removal was a more complicated dynamic process due to the involvement of nitrogen and denitrification processes. Furthermore, the operation of the fountain introduced more dissolved oxygen, interrupting denitrification and sedimentation, both of which influence the removal of TN. This then led to the decreased REs with a longer operating HRT.

Table 16.5 Nutrients Concentration for Storm Events During Postanalysis at Pond 5 (mg L^{-1})

Date	TP		OP		TN		NO$_2$+NO$_3$		NH$_3$	
	In	Out	In	Out	In	Out	In	Out	In	Out
8/16/11	0.052	0.035	0	0.001	0.853	0.645	0.194	0.017	0.186	0.325
8/28/11	0.015	0.046	0	0.001	0.638	0.431	0	0.003	0.194	0.159
9/19/11	0.004	0.027	0	0.002	0.465	0.649	0.006	0.007	0.073	0.143
10/8/11	0.053	0.055	0.028	0.027	0.324	0.320	0.056	0.049	0.044	0.043
10/29/11	0.035	0.038	0	0.001	0.253	0.215	0.054	0.01	0.034	0.026
4/6/12	0.160	0.060	0.094	0.036	0.941	0.455	0.008	0.004	0.092	0.003
Average	0.053	0.042	0.020	0.010	0.579	0.505	0.053	0.020	0.104	0.100
CRP (%)	21.3		51.5		12.7		62.3		3.3	

Table 16.6 Nutrient Concentration for Nonstorm Events During Postanalysis at Pond 5 (mg L^{-1})

Date	TP		OP		TN		NO$_2$+NO$_3$		NH$_3$	
	In	Out	In	Out	In	Out	In	Out	In	Out
7/27/11	0.196	0.033	0.112	0.001	1.154	0.481	0.028	0.02	0.468	0.137
8/23/11	0.031	0.028	0.002	0.004	0.514	0.542	0	0	0.169	0.176
9/2/11	0.093	0.054	0.039	0	0.841	0.751	0.014	0	0.447	0.348
11/17/11	0.017	0.02	0.001	0.001	0.47	0.827	0.061	0.063	0.017	0.029
12/14/11	0.052	0.037	0.025	0.011	0.780	0.512	0.032	0.043	0.183	0.022
2/2/12	0.030	0.028	0.019	0.016	0.737	0.611	0.012	0.014	0.011	0.009
3/27/12	0.018	0.013	0.000	0.000	0.151	0.150	0.094	0.060	0.017	0.016
Average	0.057	0.030	0.023	0.005	0.666	0.553	0.034	0.029	0.224	0.105
CRP (%)	46.3		79.5		16.9		16.7		53.0	

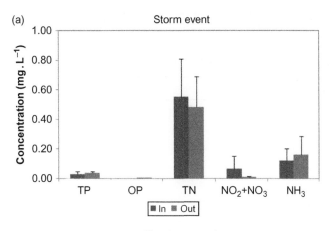

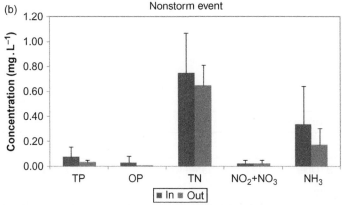

FIGURE 16.8

Nutrient concentration during postanalysis (Chang et al., 2013). (For color version of this figure, the reader is referred to the online version of this chapter.)

16.4.3 Hydrological processes

Hydrological analysis began with the first rainfall event on July 2, 2011. The changing water levels showing a zigzag temporal trend in response to the rainfall and runoff trigger subsequent biogeochemical processes over the time horizon (Figure 16.11). As the nutrients were flushed out to the pond, the natural settling process gradually reduced suspended solids in the water column, while dissolved nutrients in the water column and attached nutrients at the surface of the particles experienced nitrification. The settling particles added more nutrients to the sediment, while the sediment layer continuously contributed to the denitrification and stabilized more phosphorus. Even so, the resuspension of nitrogen from the sediment layer could outweigh the water column nitrification in the entirely study period; however, the deployment of the FTWs may offer a relatively stable hydrodynamic pattern in the pond, thereby reducing the resuspension effect.

Table 16.7 Operating HRT Associated with TN Removal with a FTW

Sampling Date (dd-mm-yy)	Operating HRT (d)	TN (mg L^{-1})	Removal (%)
Event based			
16-08-11	0.06	0.853	59.4
28-08-11	N/A[a]	0.638	69.6
19-09-11	N/A[a]	0.465	77.9
08-10-11	N/A[a]	0.324	84.6
29-10-11	0.43	0.253	88.0
06-04-12	0.02	0.941	55.2
Monthly based			
27-07-11	4	0.481	77.1
23-08-11	4	0.542	74.2
02-09-11	2	0.751	64.3
17-11-11	17	0.470	77.6
14-12-11	27	0.512	75.6
02-02-12	37	0.611	70.9
27-03-12	16	0.150	92.9

[a]These data were omitted for formula fitting due to the missing rainfall data.

Table 16.8 Operating HRT Associated with TP Removal with a FTW

Sampling Date (dd-mm-yy)	Operating HRT (d)	TN (mg L^{-1})	Removal (%)
Event based			
16-08-11	0.06	0.853	59.4
28-08-11	N/A[a]	0.638	69.6
19-09-11	N/A[a]	0.465	77.9
08-10-11	N/A[a]	0.324	84.6
29-10-11	0.43	0.253	88.0
06-04-12	0.02	0.941	55.2
Monthly based			
27-07-11	4	0.481	77.1
23-08-11	4	0.542	74.2
02-09-11	2	0.751	64.3
17-11-11	17	0.470	77.6
14-12-11	27	0.512	75.6
02-02-12	37	0.611	70.9
27-03-12	16	0.150	92.9

[a]These data were omitted for formula fitting due to the missing rainfall data.

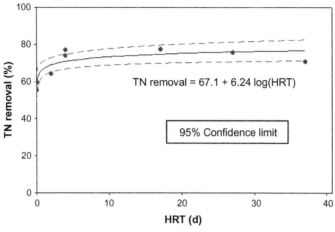

FIGURE 16.9

Operating HRT versus TN removal efficiencies at Pond 5. (For color version of this figure, the reader is referred to the online version of this chapter.)

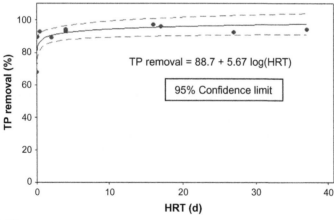

FIGURE 16.10

Operating HRT versus TP removal efficiencies at Pond 5. (For color version of this figure, the reader is referred to the online version of this chapter.)

16.5 **BMP CREDIT ASSESSMENT**

In addition to flood control and downstream erosion prevention, nutrient removal is also a major function of a wet detention pond. Besides its self-purification capacity via a natural process, FTW technology was introduced to further improve the water quality via additional nutrient removal. It was noted in the sampling of the influent

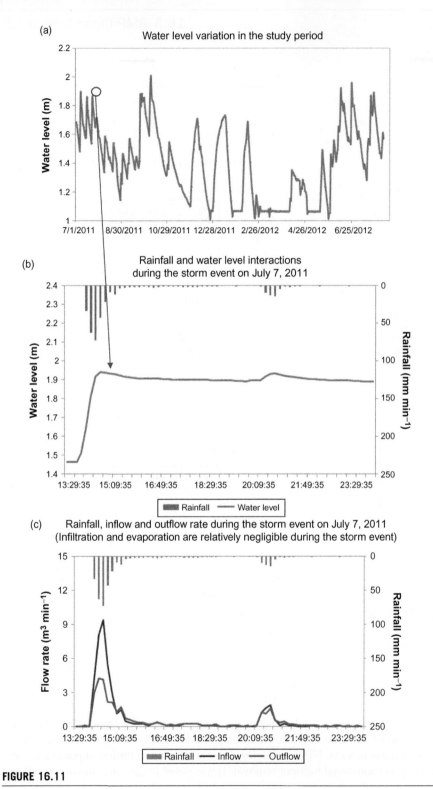

FIGURE 16.11

Hydrological control of the stormwater wet detention pond (Chang et al., 2013).

Table 16.9 Credit of Fibrous Matrix FTWs in Pond 5 with Aeration

		TN		TP	
		Without FTW	**With FTW**	**Without FTW**	**With FTW**
$\overline{C_{I-S}}$		0.288	0.519	0.028	0.031
$\overline{C_{O-N}}$		0.347	0.498	0.033	0.028
Runoff concentration based[a]	RE_B (%)[b]	−2.83 (−1.78 to −4.49)	0.99 (0.63 to 1.58)	−1.06 (−0.76 to −1.46)	0.65 (0.47 to 0.90)
	Credit (%)	3.82 (2.41to 6.06)		1.71 (1.23 to 2.37)	
Pond concentration based	RE (%)	−20.6	4.0	−18.7	10.4
	Credit (%)	24.6		29.1	

[a]Multifamily residential land use.
[b]Geometric average and (±1 standard deviation).

that the type of sampling minimizes the inclusion of particulate material. This is done so that it is recognized that particulates will most likely settle out and floating islands do not remove particulates, but only a dissolved fraction of nitrogen and phosphorus. Table 16.9 summarizes the credit estimation for fibrous matrix FTWs. Since aeration introduced resuspension of sediment, it is noticeable that less TN can be removed by the settling effect at Pond 5. A smaller pond size usually results in a greater variation of surface area, given similar input flow rates. The fibrous matrix FTWs were applied at Pond 5 with 5% coverage in July when the pond level was at its highest. Since Pond 5 was relatively small, the coverage percentage increased proportionally to the drop of water level over time, which might result in the higher removals.

16.6 CONCLUSIONS

The FTW in Pond 5 had fair removals due to greater concentrations of N and P in the water column, presumably because of the fountain. For nonstorm events, phosphorus removal was substantial because of the increase in the initial concentrations, presumably due to resuspension of nutrients into the water column from the fountain operation; about 47.7% TP and 79.0% orthophosphate were removed. The removal rates of TN, nitrite- and nitrate-nitrogen $(NO_x - N = NO_2^- - N + NO_3^- - N)$, and ammonia-nitrogen $(NH_3–N)$ were also calculated as 15.7%, 20.6%, and 51.1%, respectively. Considering plant species, nutrient uptake and assimilation by soft rush (*J. effusus*) was much higher than that by pickerelweed (*P. cordata*) through both leaves and roots in this case. For soft rush, uptake rate in spring is much higher than that in fall. About 77.0 g N and 8.8 g P were removed from pond water via uptake and assimilation during the second phase. Despite organic nitrogen accumulation due to the pickerelweed leaf debris sedimentation, the organic nitrogen concentration in pond water was still kept at a low level, which implies that the ecosystem is capable of efficiently managing the withered plants and circulating nutrients. The operating HRT was calculated to demonstrate the FTWs performance in both ponds and indicated that longer operating HRTs generally led to higher REs. According to the pond concentration measurements, the credit for using a FTW was 12–25% for nitrogen and phosphorus, based on operating data from the wet detention pond before and after the introduction of a FTW. Because fountain aeration introduced resuspension of nutrients, more removal by a FTW can be expected with higher concentrations, as demonstrated in Pond 5. For this pond location and water fountain, however, the effluent concentration was higher for both nitrogen and phosphorus.

Notwithstanding, Pond 5 had additional loading of nutrients from the mud at the bottom, presumably caused by the fountain, which resulted in a higher removal by the FTW. Nevertheless, the effluent concentration for the aerated pond was higher than the nonaerated one. FTWs removed more dissolved pollutants with higher starting concentrations. Therefore, in rare cases when stormwater concentrations increased the biological available concentrations, an FTW helped to reduce the concentrations even further than shown during average operation.

Acknowledgments

The authors are grateful for the financial support provided by the Florida Department of Transportation (FDOT) (Grant No. BDK78 985-01) and Floating Island International (Grant No. 16208092). Special thanks to Mr. Lee Mullon, Ms. Liu Zhang, Ms. Jamie Jones, and Mr. Benjamin Vannah for their assistance when deploying the FTW at Pond 5.

References

Adams, E.B., 1992. Wetlands: Nature's water purifiers. Cooperative Extension. Washington State University, Pullman, WA.

Beutel, M.W., 2006. Inhibition of ammonia release from anoxic profundal sediments in lakes using hypolimnetic oxygenation. Ecol. Eng. 28, 271–279.

Billore, S.K., Prashant, S.J.K., 2008. Restoration and conservation of stagnant water bodies by gravel-bed treatment wetlands and artificial floating reed beds in tropical India. In: The 12th World Lake Conference, pp. 981–987.

Bommarito, T., Sparling, D.W., Halbrook, R.S., 2010a. Toxicity of coal-tar pavement sealants and ultraviolet radiation to *Ambystoma Maculatum*. Ecotoxicology 19 (6), 1147–1156.

Bommarito, T., Sparling, D.W., Halbrook, R.S., 2010b. Toxicity of coal-tar and asphalt sealants to Eastern Newts, *Notophthalmus viridescens*. Chemosphere 81 (2), 187–193.

Bossong, C., Stevens, M., Doerfer, J., and Glass, B., 2005. Summary and evaluation of the quality of stormwater in Denver, Colorado, water years 1998–2001, U.S. geological survey scientific investigations report 2005–5150 (http://pubs.usgs.gov/sir/2005/5150/).

Brattebo, B.O., Booth, D.B., 2003. Long-term stormwater quantity and quality performance of permeable pavement systems. Water Res. 37 (18), 4369–4376.

Breen, P.F., 1990. A mass balance method for assessing the potential of artificial wetlands for wastewater treatment. Water Res. 24, 689–697.

Burgin, A.J., Hamilton, S.K., 2007. Have we overemphasized the role of denitrification in aquatic ecosystems? A review of nitrate removal pathways. Front. Ecol. Environ. 5 (2), 89–96.

Chang, N.B., Xuan, Z., Marimon, Z., Islam, K., Wanielista, M.P., 2013. Exploring hydrobio-geochemical processes of floating treatment wetlands in a subtropical stormwater wet pond. Ecol. Eng. 54, 66–76.

Department of Health and Environmental Control (DHEC), 2007. State of Knowledge Report: Stormwater Ponds in the Coastal Zone. South Carolina Office of Ocean and Coastal Resource Management in cooperation with the South Carolina Sea Grant Consortium.

Fisher, D.J., Knott, M.H., Turley, S.D., Turley, B.S., Yonkos, L.T., Ziegler, G.P., 1995. The acute whole effluent toxicity of storm water from an International Airport. Environ. Toxicol. Chem. 14 (6), 1103–1111.

Gabel, B., Kozicki, R., Lahl, U., Podbielski, A., Stachel, B., Struss, S., 1982. Pollution of drinking water with nitrate. Chemosphere 11, 1147–1154.

Göbel, P., Dierkes, C., Coldeway, W.G., 2007. Storm water runoff concentration matrix for urban areas. J. Contam. Hydrol. 91, 26–42.

Govindarajan, B., 2008. Nitrogen Dynamics in a Constructed Wetland Receiving Plant Nursery Runoff in Southeastern United States. Dissertation, University of Florida, Gainesville, FL.

Grapentine, L., Rochfort, Q., Marsalek, J., 2008. Assessing urban stormwater toxicity: methodology evolution from point observations to longitudinal profiling. Water Sci. Technol. 57 (9), 1375–1381.

Hammer, D.A., 1989. Constructed Wetlands for Wastewater Treatment: Municipal, Industrial, and Agricultural. CRC Press, Boca Raton, Florida.

Harper, H.H., 2011. New Updates to the Florida Runoff Concentration (emc) Database. Florida Stormwater Association, Tampa, Florida, Tallahassee, Florida, USA.

Hart, B., Cody, R., Truong, P., 2003. Hydroponic vetiver treatment of post septic tank effluent. In: Proceedings—The Third International Conference on Vetiver (ICV3), October 6–9, 2003, Guangzhou, P.R. China.

Headley, T.R., Tanner, C.C., Council, A.R., 2006. Application of Floating Wetlands for Enhanced Stormwater Treatment: A Review: NIWA Client Report. Auckland Regional Council, New Zealand.

Huang, C.P., Wang, H.W., Chiu, P.C., 1998. Nitrate reduction by metallic iron. Water Res. 32, 2257–2264.

Hubbard, R.K., Gascho, G.J., Newton, G.L., 2004. Use of floating vegetation to remove nutrients from swine lagoon wastewater. Trans. ASCE 47, 1963–1972.

Kim-Shapiro, D.B., Gladwin, M.T., Patel, R.P., Hogg, N., 2005. The reaction between nitrite and hemoglobin: the role of nitrite in hemoglobin-mediated hypoxic vasodilation. J. Inorg. Biochem. 99, 237–246.

Kohler, M.A., Nordenson, T.J., and Fox, W.E., 1955. Evaporation from Pans and Lakes. U.S. Dept. Com., Weather Bur. Res. Paper 38. 21 pp.

Mahler, B.J., Van Metre, P.C., Wilson, J.T., Musgrove, M., Burbank, T.L., Ennis, T.E., Bashara, T.J., 2010. Coal-tar-based parking lot sealcoat—an unrecognized source of PAH to settled house dust. Environ. Sci. Technol. 44, 894–900.

Marsalek, J., Rochfort, Q., Brownlee, B., Mayer, T., Servos, M., 1999. An exploratory study of urban runoff toxicity. Water Sci. Technol. 39, 33–39.

Moshiri, G., 1993. Constructed Wetlands for Water Quality Improvement. Lewis Publishers, Boca Raton, Florida.

Muthukrishnan, S., 2006. Treatment of heavy metals in stormwater runoff using wet pond and wetland mesocosm. In: Proceedings of the Annual International Conference on Soils, Sediments, Water and Energy, 11. 9, Available at: http://scholarworks.umass.edu/soilsproceedings/vol11/iss1/9.

Nahlik, A.M., Mitsch, W.J., 2006. Tropical treatment wetlands dominated by free-floating macrophytes for water quality improvement in Costa Rica. Ecol. Eng. 28, 246–257.

Pinkney, A.E., Harshbarger, J.C., Rutter, M.A., 2009. Tumors in brown bullheads in the Chesapeake Bay Watershed—analysis of survey data from 1992 through 2006. J. Aquat. Anim. Health 21, 71–81.

Pitts, R., 2004a. Control of microorganisms in Urban waters (http://unix.eng.ua.edu/rpitt/Class/ExperimentalDesignFieldSampling/MainEDFS.html).

Pitts, R., 2004b. Detention pond design and analysis (http://rpitt.eng.ua.edu/Class/Water%20Resources%20Engineering/M9c2%20WinTR55%20ponds%20docs.pdf).

Rütting, T., Boeckx, P., Müller, C., Klemedtsson, L., 2011. Assessment of the importance of dissimilatory nitrate reduction to ammonium for the terrestrial nitrogen cycle. Biogeosciences 8, 1779–1791.

Sawyer, C.N., McCarty, P.L., Parkin, G.F., 2003. Chemistry for Environmental Engineering Science. fifth ed. McGraw-Hill, New York.

Schueler, T., Holland, H., 2000. Microbes and Urban Watersheds: Concentrations, Sources and Pathways, The Practice of Watershed Protection. The Center for Watershed Protection, Ellicott City, MD.

Servizi, J.A., Gordon, R.W., 2005. Acute lethal toxicity of ammonia and suspended sediment mixtures to chinook salmon (Oncorhynchus tshawytscha). B. Environ. Contam. Toxicol. 44, 650–656.

Simon, J., Sobieraj, J., 2006. Contributions of common sources of polycyclic aromatic hydrocarbons to soil contamination. Remediation J. 16, 25–35.

Stewart, F.M., Mulholland, T., Cunningham, A.B., Kania, B.G., Osterlund, M.T., 2008. Floating islands as an alternative to constructed wetlands for treatment of excess nutrients from agricultural and municipal wastes results of laboratory-scale tests. Land Contam. Reclamation 16, 25–34.

Tarazona, J.V., Munoz, M.J., Ortiz, J.A., Nunez, M.O., Camargo, J.A., 2008. Fish mortality due to acute ammonia exposure. Aquac. Res. 18, 167–172.

Ure, A.M., Davidson, C.M. (Eds.), 2002. Chemical Fractionation in the Environment. Blackwell Science, Oxford.

Van Winkle, S.C., Pullman, G.S., 2005. Achieving desired plant growth regulator levels in liquid plant tissue culture media that include activated carbon. Plant Cell Rep. 24, 201–208.

Vymazal, J., 2007. Removal of nutrients in various types of constructed wetlands. Sci. Total Environ. 380, 48–65.

Wanielista, M.P., Chang, N.B., Chopra, M., 2012. Floating Wetland Systems for Nutrient Removal in Stormwater Ponds: Final Report FDOT Project BDK78 985–01. Florida Department of Environmental Protection, Florida, USA.

Weinstein, J.E., Crawford, K.D., and Garner, T.R., 2008. Chemical and biological contamination of stormwater detention pond sediments in coastal South Carolina, report submitted to South Carolina Sea Grant Consortium & South Carolina Department of Health and Environmental Control—office of Ocean and Coastal Resource Management Charleston, SC.

White, S.A., Seda, B., Cousins, M., Klaine, S.J., Whitwell, T., 2009. Nutrient remediation using vegetated floating mats. In: SNA Research Conference McMinnville, TN, 54, pp. 39–43.

Wium-Andersen, T., Nielsen, A.H., Hvitved-Jakobsen, T., Vollertsen, J., 2011. Heavy metals, PAHs and toxicity in stormwater wet detention ponds. Water Sci. Technol. 64, 503–511.

System Dynamics Modeling for Nitrogen Removal in a Subtropical Stormwater Wet Pond

17

Ni-Bin Chang*, Zachary A. Marimon, Zhemin Xuan, Benjamin Vannah, Jamie Jones

Department of Civil, Environmental, and Construction, University of Central Florida,
Orlando, FL, USA
*Corresponding author: e-mail address: nchang@ucf.edu

17.1 INTRODUCTION

Nutrients such as ammonia, nitrite, nitrate, and phosphorus in stormwater effluents are commonly known water body contaminants that threaten public health and aquatic ecosystem integrity. Excessive inputs of these nutrients may exceed the pond's treatment capacity and result in eutrophication. This progression can reduce oxygen for aquatic organisms and organic matter decomposition that restrain a healthy aquatic ecosystem. As a result of this harmful cycle, algal blooms gradually cover the entire pond surface, blocking any penetration of sunlight into the water column. Of the two most influential nutrients in eutrophication, phosphorus is removed significantly more (averaging 60%), while nitrogen removals are much poorer for nitrogen (30%) (Harper and Baker, 2007). Without proper treatment, ammonium-nitrogen in aquatic systems can stimulate phytoplankton growth, exhibit toxicity to aquatic biota, and exert an oxygen demand (Beutel, 2006). Nondisassociated ammonia can be extremely volatile and become either ionized or volatized in aqueous solutions, and ionized ammonia has been proved to be very toxic for fish species (Tarazona et al., 2008). The presence of even a minute amount of ammonia-N has been shown to affect fish mortality, health, and reproduction (Servizi and Gordon, 2005). In addition to ammonia, nitrate-nitrogen can cause many health problems as well, particularly liver damage and even some cancers in humans (Gabel et al., 1982; Huang et al., 1998). In infants, nitrate can bind with hemoglobin to create a life-threatening oxygen deficiency called methemoglobinemia (Kim-Shapiro et al., 2005). In addition, nitrite can react with amines, chemically or enzymatically, to form nitrosamines, which are very potent carcinogens (Sawyer et al., 2003).

This study evaluates the FTWs as a technology to improve nitrogen removal and create a hydrobiogeochemical model to evaluate the system dynamics. Then, the model will be used to evaluate current vegetation management practices to measure the effect on the biological processes affected, which in turn alter the effectiveness of the ecosystems for pollution abatement.

17.2 LIMITATIONS OF TRADITIONAL STORMWATER PONDS

Traditionally, stormwater wet detention ponds were essentially built to provide esthetic and recreational benefits as well as flood and downstream erosion control. Stormwater management mainly dealt with conveying the excess runoff through a drainage system to the nearest waterway. However, due to increase of human activity, many possible nutrient sources are infused into the ponds with the surface runoff including fertilizers, animal excrement, and organic debris. Today, stormwater management is evolving to integrate stormwater infrastructure planning with relevant water quality and ecosystem impacts.

Wet detention ponds, which are actually constructed wetlands, are currently a very popular stormwater management technique. These permanently wet ponds are designed to slowly release collected runoff through an outlet structure. Pollutant removal processes in wet detention ponds occur through a variety of mechanisms, including physical processes such as sedimentation, chemical processes such as precipitation and adsorption, and biological uptake from algae, bacteria, and rooted vegetation.

Wet ponds have been extensively monitored under a wide variety of conditions. If well-designed and properly maintained, suspended solids removals of 70–90% can be obtained. BOD_5 and COD removals of about 70%, nitrogen removals of about 20–30%, and heavy metal removals of about 60–95% can also be obtained (Pitts, 2004a; Chang et al., 2012a,b). Nevertheless, possible wet pond maintenance problems may include (Pitts, 2004b):

(1) Wet ponds can require about 3–6 years to achieve an ecological balance. During the initial unstable period, excessive algal growths, fish kills, and nuisance odors may occur.
(2) Wet ponds can have poor water quality, thus discouraging any water contact, recreation, and consumptive fishing.

The ecosystem dynamics in those wet ponds, which could disturb or enhance the nutrient uptake, remains neglected in stormwater system designs. Proposed solutions have included larger basins with longer residence times (Harper and Baker, 2007), but these basins would increase costs, and land requirements and are generally inapplicable postdevelopment. In addition to constructed wetlands that are artificially built to utilize various aquatic plants to remediate nutrient-rich surface flows of both stormwater and wastewater (Belmont and Metcalfe, 2003; Iamchaturapatra et al., 2007; Baldwin et al., 2009; White et al., 2009). Removal processes include storage in plants; microorganisms; and other chemical, biochemical, and physical mechanisms such as adsorption, precipitation, and sedimentation. The littoral zone area may act as either a sink for pollutants, removing them from incoming water, or as a source, adding them to the water creating a long-standing controversy in aquatic ecology for the use of plants to aid in nutrient removal for these systems (Mickle and Wetzel, 1978; Wetzel, 1983; Carpenter and Lodge, 1986; Gersberg et al., 1986). The inclusion of macrophytes should reduce algae and cyanobacteria by competing for nutrient resources and possibly from the release of allelopathic substances that inhibit algal growth (Harrison and Durance, 1985).

The overgrowth of plants (Figure 17.1) and the abrupt algal blooms (Figure 17.2) due to eutrophication can deteriorate water quality and create maintenance problems that reduce pollution abatement efficiencies in stormwater ponds. This leads to the reduction of flood mitigation capacity, the increase of public health risk, and the deterioration of water quality. Therefore, the maintenance of a detention pond should

(a) (b)

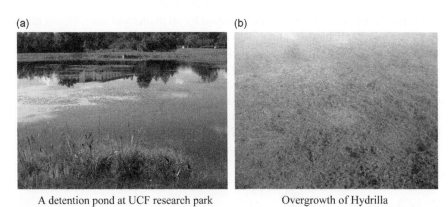

A detention pond at UCF research park   Overgrowth of Hydrilla

FIGURE 17.1

A case of the overgrowth of plant species at UCF Research Park: (a) A Detention pond at UCF Research Park, (b) Overgrowth of Hydrilla. (For color version of this figure, the reader is referred to the online version of this chapter.)

(a) (b)

A detention pond at UCF main campus   Overgrowth of algae

FIGURE 17.2

A case of the overgrowth of algae in the Pond 4M at UCF campus: (a) A detention pond at UCF main campus, (b) Overgrowth of algae. (For color version of this figure, the reader is referred to the online version of this chapter.)

include the removal of excess plant growth in the littoral zones, the cleanup of inlet and outlet facilities, and the removal of algal blooms when necessary (Figure 17.3).

When deciding on the postconstruction management practices of a wet detention pond, such as herbicide application, it is necessary to consider the biological interactions within the pond ecosystem. Excessive nutrient retention in stormwater ponds can produce outbreaks of algal blooms. To control this phenomenon, Florida Fish and Wildlife Conservation Commission (FWC) permits the use of copper-based algaecides and herbicides when alternative management options are not available. Introduced in the early 1900s, copper-based algaecides (e.g., copper sulfate, $CuSO_4$) are the lowest-cost and most widely herbicide to control algae and other nuisance vegetation in surface water systems (Gettys et al., 2009).

Although copper is a trace element required by all plants and animals, too much copper can kill plants by disrupting plant metabolism by replacing cofactors in key enzymes (e.g., magnesium, an essential element in chlorophyll molecules) and disrupting photosynthetic activity and other cellular processes (Brown and Rattigan, 1979; Rabe et al., 1982; Stoyanova and Tchakalova, 1993; Küpper et al., 1996). Improper use of copper sulfate is likely toxic to most invertebrates, including beneficial nitrogen-cycling bacteria. In any circumstances, copper-based algaecides and herbicides could be directly responsible for reducing the nitrogen removal efficiencies of

FIGURE 17.3

Severe algal bloom event and emergency response action in a pond, South Florida. (For color version of this figure, the reader is referred to the online version of this chapter.)

Courtesy of Beemats, Inc.

detention ponds in Florida (Harper and Baker, 2007). More serious negative effects exist from the use of copper algaecides, including fish and zooplankton toxicity, the depletion of dissolved oxygen, copper accumulation in the sediments, increased internal nutrient cycling, and increase tolerance to copper by some nuisance species of blue-green algae (Tiedemann et al., 2009).

Studies have found nitrification inhibition occurs at just 0.3 mg/L Cu^{2+} (Yanong, 2009). Literature values for residential stormwater ponds are limited, but Hu et al. (2004) found that a concentration of Cu^{2+} about 0.01 mM equated to about a 20% reduction in nitrification. The involvement of chemical inhibition would affect and complicate the inherent aquatic nitrogen cycle and ecosystem balance within stormwater wet ponds.

17.3 FLOATING TREATMENT WETLAND TECHNOLOGIES

Stormwater runoff is highly variable in both intensity and duration due to the erratic nature of storm events; thus, sediment-rooted plants in conventional treatment wetlands experience a range of water depths and periods of inundation (Greenway and Polson, 2007). The duration of inundation, the depth of water, the frequency of flooding, and occurrence of droughts are known to affect plant growth, establishment, and survival (Greenway and Polson, 2007). Long periods of flooding are stressful to some bottom-rooted wetland plants (Ewing, 1996; Headley et al., 2006). To manage this issue, wetland size might be increased to buffer against extremes during water level fluctuations, or high flows can be bypassed. With bypasses, however, there is a significant portion of incoming stormwater is excluded from treatment (Headley et al., 2006), and large land area requirements for installation can limit their applicability.

The floating treatment wetland (FTW), a complementary best management practice (BMP) technology, has emerged as a sustainable solution for nitrogen control. FTWs deployed for stormwater wet detention ponds are deemed an innovative BMP. The buoyancy of plastic intertwined fibers/mats of FTWs is sufficient to keep plants afloat at the surface of the water body rather than having rooted plants relying on underlying pond sediment. Terrestrial or emergent aquatic wetland plants, macrophytes, are planted on the floating mats in a wet pond. Additional nutrient removal can be achieved by the biofilm of fungi, bacteria, and algae that form at the surface along the roots submerged in the water column. Microbes that live on the surface of plant roots remove 10 times more nitrate than do the plants themselves (Adams, 1992).

The implementation of FTWs may be a good alternative to maintain the balance of ecosystem in a wet pond with sustainability implications to avoid using the copper-based algaecides to control algal bloom. This observation can be evidenced by two recent algal bloom events in Pond 4M at UCF campus recorded during the experimental periods: March 15, 2011 (before the deployment, Figure 17.4) and April 26, 2011 (after the deployment, Figure 17.5). The comparisons between these two events clearly reveal the FTW's effectiveness at controlling algal blooms after its deployment. Simply by visual inspection (Figures 17.4 and 17.5), the FTWs show

FIGURE 17.4

Algal bloom #1 (whole area) on March 15, 2011 (before the FTW deployment) (Wanielista and Chang, 2012). (For color version of this figure, the reader is referred to the online version of this chapter.)

FIGURE 17.5

Algal bloom #2 (two ends) on April 26, 2011 (after the FTW deployment) (Wanielista and Chang, 2012). (For color version of this figure, the reader is referred to the online version of this chapter.)

the potential of suppressing the presence of algal blooms to some extent at Pond 4M. The effectiveness of FTWs is further confirmed numerically by comparing the removal efficiencies of nutrients between the two algal bloom events while seasonal variations/climate changes during these periods had no impact on these two phases of the experiment (Figure 17.6).

17.4 FIELD CAMPAIGN FOR INVESTIGATING THE COPPER IMPACT

The standard methods used for chemical analyses of metals, nutrients, and others in this study are summarized in Table 17.1. The water column and sediment of five stormwater ponds having a history of $CuSO_4$ application were sampled on March 8, 2013, and analyzed for copper (Table 17.2 and Figure 17.7). Water sample

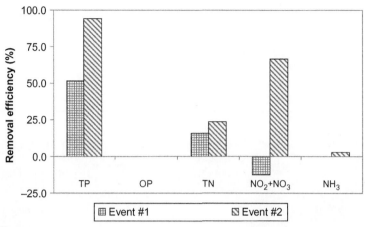

FIGURE 17.6

Comparisons of nutrients removal efficiencies of two algal bloom events: Event #1 stands for the case before deployment and event #2 stands for the case after deployment (Wanielista and Chang, 2012). (For color version of this figure, the reader is referred to the online version of this chapter.)

Table 17.1 Water Quality Sampling and Analysis Standards

Parameter		Sediment	Water Column
Metal	Ag, Ba, Cd, Cr, Cu, Ni, Pb Se, Zn	SW846 Method 6020 ICP/MS	SW846 Method 6020 ICP/MS
	As	SW846 Method 6020 ICP/MS	SW846 Method 6020 ICP/MS
E. coli		N/A	SM 9222 D
TSS		N/A	EPA 160.2
Alkalinity		N/A	EPA 310.1
Total nitrogen (TN)		N/A	SM21 4500-N C
Total phosphorus (TP)		N/A	EPA Method 365.4
Ortho phosphate as OP		NA	SW 365.1
Ammonia (NH_3)		N/A	EPA 350.1
Nitrite+nitrate (NO_x)		N/A	EPA 353.2
pH		Hach HQ40d portable meter	
Conductivity		Hach HQ40d portable meter	
Dissolved oxygen		Hach HQ40d portable meter	
Turbidity		Turbidimeter	
Chl-*a*		Aquafluor™ Handheld Fluorometer	

Table 17.2 Copper Concentrations at the Tested UCF Ponds

Pond Site	Pond Coordinates	Water Sample, Cu (μg/L)	Sediment Sample, Cu (μg/g)
Cu Pond 1	28.609366°, −81.190540°	8	36.9
Cu Pond 2	28.606522°, −81.189168°	4	8.2
Cu Pond 3	28.593055°, −81.196691°	4	1.4
Cu Pond 4	28.591832°, −81.204670°	6	2.1
Cu Pond 5	28.607946°, −81.205112°	11	25.8

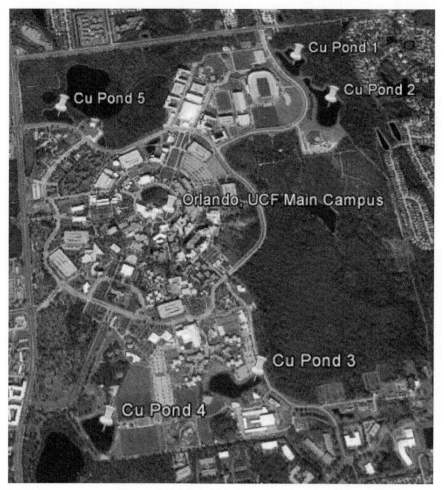

FIGURE 17.7

Locations of the five UCF ponds tested for monitoring metal concentrations "Pond 5." 2013.28°32′49.42″N and 81°12′03.64″W. GOOGLE EARTH. January 1, 2012. June 23, 2013. (For color version of this figure, the reader is referred to the online version of this chapter.)

Adapted from Google Earth.

Table 17.3 Heavy Metal Concentrations Observed in the First Flush at the College of Nursing (Pond 3)

College of Nursing Pond		
Pollutant	**Concentration**	**Limit**
Cu (µg/L)	14	20
Pb (µg/L)	2	142.5
Zn (µg/L)	21	680
Cr (µg/L)	53	23
Cd (µg/L)	2	1
Ni (µg/L)	2	56

analyses from the five study ponds indicated that the 20 µg/L threshold effect for copper (EPA, 1986; Pitts, 2004a,b) was not exceeded (Table 17.3). This result was expected because $CuSO_4$ does not persist in the water column, but quickly settles to the bottom over the course of 2 h where it is trapped in the sediment (Button et al., 1977). The minimal threshold effect is the concentration below which no effect on sediment-dwelling organisms is expected, and the sediment is considered to be clean or marginally polluted.

On the other hand, the toxic threshold effect is the concentration at which the organisms in the sediment are adversely affected, and the sediment is considered heavily polluted (28 and 86 µg/g, respectively, for copper) (EC & MENVIQ, 1992). The highest concentration detected in the ponds' sediment occurred at Pond 1 with a concentration of 36.9 µg/g, a value that exceeds the minimal threshold effect, yet it is still far below the toxic threshold effect. Although this concentration is significantly less than the toxic threshold effect, the copper concentration in both the water column and sediment still present health concerns to the organisms. This in turn provides a threat to the entire ecosystem balance, which may lead to further water quality issues in the future.

Analysis of the first flush sample for heavy metals at an additional pond located at the College of Nursing yielded the results reported in Table 17.3 and Figure 17.8. In Figure 17.8, the threshold effect limits are denoted by a horizontal red line. If a pollutant does not contain a visible red line (threshold effect), then the limit is well above the current concentration, and, for the sake of clarity and proper scaling, it has been left off the graph.

All of the heavy metals are below the threshold effect with the exception of chromium. At a concentration of 53 µg/L, this exceeds the 23 µg/L threshold effect for organisms of the order Cladocera (small crustaceans commonly referred to as water fleas), and it violates the 50 µg/L drinking water quality criterion (US EPA, 1986; Pitts, 2004a,b). Yet, this concentration was achieved from a first flush sample and will be diluted across the entire pond volume. The concentration in the pond is likely much lower and does not warrant cause for concern or additional sampling to verify the status of the pond.

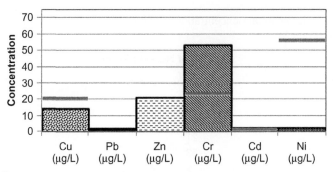

FIGURE 17.8

First flush heavy metal concentrations in the pond inlet at College of Nursing, UCF (i.e., red lines are the threshold of each heavy metal). (For interpretation of the references to color in this figure legend, the reader is referred to the online version of this chapter.)

17.5 COLLECTION OF NUTRIENT DATA

Field-scale applications of FTWs were implemented in a stormwater wet detention pond, located in a community near the main campus of UCF in Orlando, Florida, for water quality improvement (Figure 17.9). Four floating mats (BioHaven® Floating Island) were deployed, covering collectively 8.7% of the pond surface area and holding about total 640 pots for plants (Chang et al., 2013). The macrophyte species used in the project, Soft Rush (*Juncus effusus*) and Pickerelweed (*Pontederia cordata*), were selected according to a previous mesocosm-scale study in order to establish some standardization in the assessments made by The Stormwater Management Academy at UCF (Chang et al., 2012a,b).

Water sampling was conducted from August 2011 to May 2012 during the experimental period (Tables 17.4 and 17.5). These measurements not only indicate the performance of FTW technology in this subtropical stormwater wet pond but also serve as the input data for model calibration and the reference value of output data for model validation in the system dynamics model (see the following section).

Plants were sampled three times during the study period to determine the increase of nitrogen content over time (expressed as milligrams per plant sample). The nutrient content in different sampling times could be used to calculate the increased biomass and estimate the nitrogen uptake rate in the system dynamics model. Nitrogen uptake and assimilation by soft rush was much greater than that by pickerelweed, especially through leaves (Table 17.6). Uptake rate by soft rush in spring is much greater than that in fall (Table 17.6), and the few surviving pickerelweed lost biomass in winter and did not rebound during the study period. About 77.0 g N were removed from pond water via uptake and assimilation during the study period (Chang et al., 2013).

Sediment samples in the pond were collected in parallel with the plant tissue samples from five predetermined locations in the study period. The content of different nitrogen species was monitored to determine the spatial distribution of nitrogen in

(a)

(b)

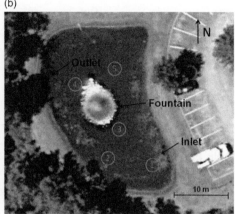

The community   The pond with sampling locations 1–5

FIGURE 17.9

Location of the stormwater wet detention pond ("Pond 5." 2013.28°32′49.42″N and 81°12′03.64″W. GOOGLE EARTH. January 1, 2012. June 23, 2013): (a) The community, (b) The pond with sampling locations 1–5. (For color version of this figure, the reader is referred to the online version of this chapter.)

Adapted from Google Earth.

Table 17.4 Nitrogen Concentrations (mg/L) of Storm Events During the Study Period (Chang et al., 2013)

Date	TN		NO$_x$—N		NH$_3$—N	
	In	Out	In	Out	In	Out
8/28/11	0.638	0.431	0	0.003	0.194	0.159
9/19/11	0.465	0.649	0.006	0.007	0.073	0.143
10/8/11	0.324	0.320	0.056	0.049	0.044	0.043
10/29/11	0.253	0.215	0.054	0.01	0.034	0.026
4/6/12	0.941	0.455	0.008	0.004	0.092	0.003
4/26/12	0.825	0.823	0.037	0.050	0.014	0.004
5/15/12	0.709	0.671	0.141	0.095	0.181	0.125

the sediment throughout the bottom of pond (Table 17.7) and to evaluate the rate values of the different forms of nitrogen transformation in the sediment layer for the system dynamics model. The numbers from 1 to 5 in this table represent sampling location in this pond (Figure 17.9b). Similar to the varying nutrient concentrations

Table 17.5 Nitrogen Concentrations (mg/L) of Nonstorm Events (Regular Sampling) During the Study Period (Chang et al., 2013)

Date	TN		NO_x—N		NH_3—N	
	In	Out	In	Out	In	Out
9/2/11	0.841	0.751	0.014	0	0.447	0.348
11/17/11	0.47	0.827	0.061	0.063	0.017	0.029
12/14/11	0.780	0.512	0.032	0.043	0.183	0.022
2/2/12	0.737	0.611	0.012	0.014	0.011	0.009
3/27/12	0.151	0.150	0.094	0.060	0.017	0.016
5/21/12	0.575	0.541	0.012	0	0.097	0.077

Table 17.6 Nutrient Content (mg per Plant Sample) in the Roots and Leaves of Plant Species During the Study Period (Chang et al., 2013)

		Soft Rush		Pickerelweed	
		Mean	St. Dev.	Mean	St. Dev.
Leaf	9/2/11	16.61	12.01	28.43	8.68
	12/14/11	48.18	18.90	24.40	13.68
	5/21/12	239.55	91.34	5.99	3.77
Root	9/2/11	9.95	5.70	17.77	8.85
	12/14/11	18.38	3.75	38.40	7.74
	5/21/12	35.78	18.91	23.79	8.68

Table 17.7 Nitrogenous Species (mg/kg) in Each of the Five Sediment Sample Locations During the Study Period (Chang et al., 2013)

Nitrogen Species	mDate	Sample Locations in Evaluation				
		1	2	3	4	5
NH_3—N	9/2/11	6.41	4.09	12.4	12.5	2.44
	12/14/11	5.20	6.71	5.68	7.63	2.36
	5/21/12	6.45	10.2	5.35	32.3	9.11
NO_x—N	9/2/11	0.01	0.335	0.272	0.352	0.302
	12/14/11	2.03	0.443	13.9	0.0282	0.812
	5/21/12	3.23	1.17	1.12	1.54	1.09
Organic N—N	9/2/11	346.59	101.91	976.60	759.50	579.56
	12/14/11	148.80	969.29	1264.32	1022.37	285.64
	5/21/12	251.55	160.8	2604.65	7957.7	4260.89
TN	9/2/11	353.01	106.34	989.27	772.35	582.30
	12/14/11	156.03	976.44	1283.90	1030.03	288.81
	5/21/12	261.23	172.17	2611.12	7991.54	4271.09

observed in water columns, low concentrations of NH_3—N and NO_x—N indicate that the dominant N form is organic nitrogen in sediment. A remarkable increase of nitrogen contents in sediment from December 2011 to May 2012 was observed at locations 3 through 5, which occurred exactly beneath the FTWs. This observation coincides with the dramatic decrease of pickerelweed leaf biomass during the same time window. Despite organic nitrogen accumulation due to the pickerelweed leaf debris sedimentation, the organic nitrogen concentration in pond water was maintained at a low level, which implies that the ecosystem is capable of self-regulating to manage the withered plants.

17.6 INVESTIGATION OF AQUATIC NITROGEN CYCLING

In this study, a field-scale study to collect nutrient data in a pond nearby UCF campus (Pond 5 hereafter) was carried out based on a Before–After assessment design model (Box and Tiao, 1975) to better understand the influence of chemical inhibition due to the presence of copper on nitrogen control in such a residential wet detention pond with FTW applications. The various processes embedded in the nitrogen cycle (Figure 17.10 and Table 17.8) can be linked through a system dynamics model that

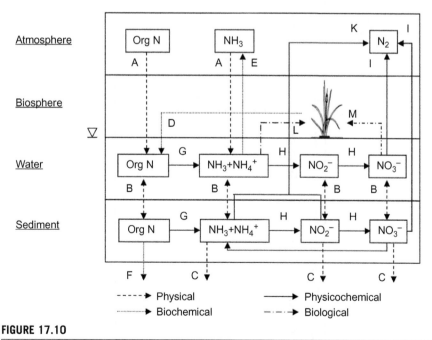

FIGURE 17.10

Possible pathways of nitrogen species (Kadlec and Wallace, 2009).

Table 17.8 Legend of Pathways in Figure 17.11 (Marimon et al., 2013)

Pathway	Symbol	Description	Pathway	Symbol	Description
Physical	A	Wet/dry deposition of organic nitrogen and ammonia	Biochemical	G	Ammonification in water column and sediment
	B	Settling and diffusion		H	Nitrification in water column and sediment
	C	Groundwater infiltration		I	Denitrification in water column and sediment
Physicochemical	D	Leaching of leaf litter		J	Dissimilatory reduction of *Nitrate* to ammonium (DRNA)
				K	Anaerobic ammonium oxidation (Anammox)
	E	Ammonia volatilization	Biological	L	Ammonium uptake
	F	Accretion of organic nitrogen		M	Nitrate uptake

includes the majority and/or the most significant biogeochemical processes. A few mass balance approaches apply conservation principles as modeling constraints to elemental translocation within lake ecosystems and have revealed close approximations to actual nutrient cycling in lakes and stormwater ponds (Vezjak et al., 1998; Mayo and Bigambo, 2005). The conceptual model constructed in this study (Figure 17.10) shows how these variables are connected by flows (Jørgensen, 1995). The next phase of the conceptual stage is to choose and/or develop the mathematical equations describing the functional relationships between state variables and decipher the best-fit values through parameterization. However, the interplay of atmosphere, biosphere, water column, and sediment may affect the balance of local ecosystems in the pond. Proper coupling of physical, chemical, and biological processes to enhance nitrogen removal by assimilation and through the nitrogen cycle (nitrification and denitrification) across different environmental media in the pond with FTW control periodically may be addressed by the system dynamics model to be addressed in section 17.7 for predicting the effluent water quality.

17.7　SYSTEM DYNAMICS MODELING

The method of system thinking has been used for over 30 years (Forrester, 1961). It provides us with effective tools for better understanding those large-scale complex management problems. System dynamics, designed based on system thinking, is a well-established methodology for studying and managing complex feedback systems. It requires constructing the unique "causal loop diagrams" or "stock and flow diagrams" to form a system dynamics model for applications. Relevant work on how to develop system dynamics models can be found in the literature (Forrester, 1961, 1968; Randers, 1980; Mohapatra, 1994).

To build a system dynamics model, one should identify a problem and develop a dynamic hypothesis explaining the cause of the problem. The mode formulation is normally designed to test a computer simulation model with regard to alternative policies in the problem. Simulation runs in a system dynamics model is governed entirely by the passage of time. Such a time-step simulation analysis takes a number of simulation steps along the timeframe to update the status of system variables of concern as a result of system activities. When the initial conditions are assigned for those variables, which denote the state of the system, the model may start to produce the related consequences for those system variables based on the initiation of action and the flow of information.

Most computer simulation applications using system dynamics models rely on the use of software, such as Vensim® or Stella®, in which the mechanisms of system dynamics can be handled by a user-friendly interface. These model development procedures are designed based on a visualization process that allows model builders to conceptualize, document, simulate, and analyze models of dynamic systems. They offer a flexible way for building a variety of simulation models from causal loops or stock and flow. The dynamic relationships between the elements, including variables, parameters, and their linkages, can be created onto the interface using user-friendly visual tools. The feedback loops associated with these employed variables can be visualized at every step throughout the modeling process. Simulation runs are carried out entirely along the prescribed timeline. At the end, some designated system variables of interest are brought up to date for demonstration and policy evaluation.

While dynamic systems models may of necessity be complex, their complexity is achieved through combinations of simpler submodels linked to simulate the system in question. These submodels are themselves dynamic systems models exhibiting specific systems behaviors such as linear, exponential, and logistic growth or decay, overshoot and collapse, and oscillation (Deaton and Winebrake, 2000). In this study, the dynamic models presented characterize solid waste generation as exhibiting the behavior of linear growth. In these models, the concept of feedback within the system is not explored due to the difficulty of linking waste generation *per se* directly back to consumption activities.

The system dynamics modeling software, STELLA® v.9.4.1., was chosen in this study for its known usefulness in biogeochemical processes modeling related to nitrogen across a multimedia environment of an existing stormwater wet pond in Orlando (Jørgensen et al., 2009). The complete conceptual model is illustrated in Figure 17.10, which exhibits the connectivity among the multimedia environment. Precipitation and water level were measured continuously with the Global Water WL400 and Global Water GL500-2-1 installed at the mouth of the circular outlet culvert of the stormwater wetland pond. In Figure 17.12, the volume stock (*Pond Vol*) is controlled by maximum and minimum levels through runoff, discharge, evaporation, and infiltration regulated by *if/then* functions. Stocks for discharge and infiltration volume (*Discharge Total and Infilt Vol*) allow mass balance computations. The model input is based on load mass of nitrogen species in storm-event data. Dry deposition data come from the EPA's Clean Air Status and Trends Network (CASTNET).change in. Conversely, ammonia volitization from the system to the atmosphere was calculated as a first-order removal of the ammonia+ammonium stock. The calibrated coefficient value was modified from literature findings that ammonia may represent between 1% and 10% of ammonia:ammonium proportion in the water. The value was extrapolated to fit the average 7.9 pH of Pond 5. Nitrogen fixation also was not included because no persistent algal blooms occurred to signal the need for further analysis. The nitrogen loading of individual species inflows in terms of organic nitrogen (*Org N Load*), ammonia+ammonium (*NH Load*), and nitrite+nitrate (*NOx Load*). The runoff quantity (*RUN*) was multiplied by the loads to determine the dynamic inputs proportional to the runoff rates reflecting the landscapes. Infiltration connects the product of the infiltration converter (*Infiltration Volume*) and nitrite+nitrate concentration (*NOx Conc*). Overland losses were assumed as quantities above the holding capacity and were connected to all species of nitrogen the discharging surface water as *TN Discharge*. The discharge load would be equivalent to the product of the discharge volume and nitrogen concentration.

Biochemical transformations were modeled through first-order reaction rates. Ghost stocks (*Pond Vol*, *Sed Vol*) were used in each reaction to derive concentrations for the underlying equations. Ammonification and nitrification only occur in aerobic layers (*AE* %), and denitrification only occurs in the anaerobic layers (*AN%*). Temperature adjustments with the Arrhenius constant were unnecessary because the southern North American climate minimally affects microbial activity (Knight, 1986; Gearheart, 1990). Similar to mineralization constants, nitrification constants in the aerobic sediment layer were greater than those of the water column (Martin and Reddy, 1997) and assumed to be greater in the FTW as well. Absorption of ammonium and nitrate flows (*NH Assimilation* and *NOx Assmiliation*) to the macrophytes was converted with the product of the assimilation rate (*Assim Rate*), macrophyte quantity (*Macrophyte Pop*), and proportion of $NH_4 : NO_3^-$ (*NH:NOx rate*). The multiphase sampling campaign allows biomass changes to represent total nutrient translocation through biosynthesis. Plants utilize both inorganic forms of ammonium and nitrate, but because the macrophyte biomass is dried and analyzed for total nitrogen, the data provide no indication of preference. This model assumes uptake of each species based on the average proportions in the water, modified from findings by Mayo and Bigambo (2005). Growth rates are related to the change in biomass over time to determine the growth rates, which averaged out to 0.95 mg/d in the calibration period, respectively, for *J. effusus*. (See Figure 17.11 of the Stella model and Table 17.9 for summary of STELLA model symbol descriptions.) Stock, flow, and converter symbols used in the Stella model are defined and delineated in Appendixes A, B, and C, respectively.

17.8 RESULTS AND DISCUSSION

17.8.1 Model calibration and validation

The calibration of each nitrogen species was performed by calibrating the *k* values (Figure 17.12) through parameterization as close as possible to the actual nonstorm sample data available (Jørgensen, 1995) to reach the R^2 value equal to 0.83 (Figure 17.13). The *k* values for the volume within the FTW were increased relative to the open water column, which would support less dense microorganism communities. The final step for modeling is validation, conducted by testing the model under a different set of forcing functions or initial conditions (Jørgensen, 1995) across independent datasets from different time periods that truly validates the effectiveness using a goodness-of-fit analysis (Rykiel, 1996). The FTWs were removed from the pond on May 21, 2012, but four water column samples were taken during the 3 months afterward and used to finish validating the model. The biogeochemical effects of the FTWs were derived from the model coinciding with the removal, and the regression of model validation produced an $R^2 = 0.97$ (Figure 17.14).

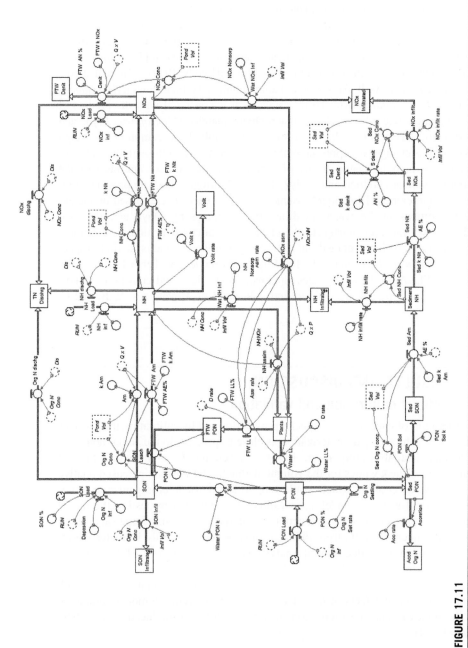

FIGURE 17.11

The STELLA model of nitrogen cycling in the studied pond (Marimon et al., 2010). (For color version of this figure, the reader is referred to the online version of this chapter.)

Table 17.9 Description of Variables Used in the STELLA Model (Marimon et al., 2013)

STELLA Symbols	Parameter	Equation	Source
Run Vol	Runoff volume	$Q = CiA$ Q = peak discharge, cfs C = rational method runoff coefficient i = rain intensity (in/hour) A = area	Calculated
Org N Load	Organic nitrogen loading	$L = C*V$ L = load C = concentration (mg/L) V = volume (L)	Measured
NH Load	Ammonia/ammonium loading	$L = C*V$	Measured
NOx Load	Nitrite/nitrate loading	$L = C*V$	Measured
Deposition	Dry deposition of soluble organic nitrogen	$r_{Deposition} = 6000$ mg/d	CASTNET (Indian River Lagoon, FL, USA)
Evap	Evaporation	$dV - E/dt$ V = volume E = evaporation t = time	Measured
NOx inf	Infiltrated nitrate + nitrite	$r_{NO_x,Inf} = C*V*r_{NO_x,Nonsorp}$ (mg) $r_{NO_x,Inf}$ = Infiltrated NO_x (mg) $r_{NO_x,Nonsorp}$ = portion that cannot adsorb in soil	Measured
NHx inf	Infiltrated ammonia/ammonium	$R_{NH\ Inf} = C*V*r_{NH\ Nonsorp}$ (mg) $r_{NO_x,Inf}$ = Infiltrated NH (mg) $r_{NO_x,Nonsorp}$ = portion that cannot be adsorbed	Measured
SON inf	Soluble organic nitrogen infiltrated	$R_{Org\ N\ Inf} = C*V*r_{NO_x,Nonsorp}$ $R_{Org\ N\ Inf}$ = Infiltrated NO_x (mg) $R_{Org\ N\ Nonsorp}$ = portion that cannot adsorb in soil	Measured

Continued

Table 17.9 Description of Variables Used in the STELLA Model (Marimon et al., 2013)—cont'd

STELLA Symbols	Parameter	Equation	Source
Org N Settling	Settling of particulate organic nitrogen	$dS/dt = k*S$ S = storage $k = 5 \times 10^{-4}$	Loiselle et al. (2006)
Org N dischg	Organic nitrogen surface discharge	$L = C*V$	Wilde (1994)
NH dischg	Ammonium surface discharge	$L = C*V$	Wilde (1994)
NOx dischg	Nitrate surface discharge	$L = C*V$	Wilde (1994)
Water LL	Leaf litter that falls into water column and degrades in the sediment	$r_{dc} = k_{Decay}*N_{Plants}$ R_{dc} = rate R_{Decay} = constant	Mayo and Bigambo (2005)
Accretion	Accretion	$r_{accretion} = C_{Org\,N}*r_{acc}$ $r_{accretion}$ = accretion rate (mg) $C_{Org\cdot N}$ = concentration of organic nitrogen (mg/L) r_{acc} = accretion rate	Kadlec et al. (2012)
Am, Nit	Water column processes: ammonification and nitrification	$r_{Am} = C*k$ k = reaction rate constant	Beran and Kargi (2005)
NH asm	Macrophyte assimilation of ammonium	$r_{NH} = N_{Plants}*(NH:NO_3)$ r_{NH} = ammonium assimilation rate N_{Plants} = plant nitrogen content $(NH:NO_3)$ = Proportion of ammonium:nitrate in water column	Agro Services International; Empirical
NOx asm	Macrophyte assimilation of nitrate	$r_{NO_3} = N_{Plants}*(NH:NO_3)$ r_{NO_3} = Nitrate assimilation rate	Agro Services International; Empirical
Volit	Volitization	$r_{Am} = C*k$	Kadlec and Wallace (2009)
FTW LL	Leaf litter degrading in the FTW aerobic region	$r_{dc} = k_{Decay}*N_{Plants}$	Empirical/optimized
FTW LL	Leaf Litter that falls onto the top of the FTW	$r_{dc} = k_{Decay}*N_{Plants}$	Mayo and Bigambo (2005)
SON Leach	Leaching of soluble organic nitrogen from FTW LL	$r_{dc} = k_{Leach}*N_{Plants}$ k_{Leach} = rate of leaching	Calibrated
FTW Am, FTW Nit, Denit	Ammonification, nitrification, and denitrification occurring within the aerobic zone of FTW	$r_{Nit} = Cnk$	Beran and Kargi (2005)

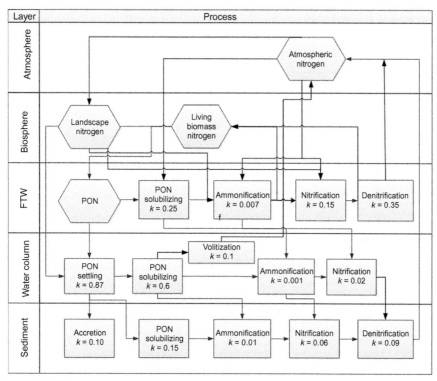

FIGURE 17.12

k values calibrated in the STELLA model. (For color version of this figure, the reader is referred to the online version of this chapter.)

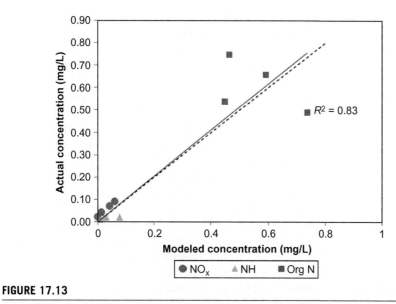

FIGURE 17.13

Correlation between the measured and simulated values in model calibration. (For color version of this figure, the reader is referred to the online version of this chapter.)

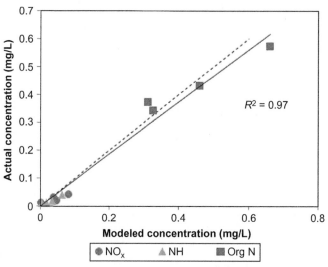

FIGURE 17.14

Correlation between the measured and simulated values in model validation. (For color version of this figure, the reader is referred to the online version of this chapter.)

17.8.2 Sensitivity analysis for addressing copper-related ecotoxicity

Heavy metals that can create toxic conditions in aquatic systems (i.e., copper) can enter stormwater ponds through stormwater runoff and algaecide applications that disrupt the pollution abatement via ecotoxicity to microorganisms and vegetation (Jørgensen, 1995). To quantify the copper impact on nitrogen dynamics in the Pond 5, the biochemical reaction rates were adapted based on findings by Hu et al. (2004) for inhibition testing to the nitrogen cycle. The simulation utilized reduced nitrogen cycling rates to inhibit processes by 10%, 20%, and 30%. The results of the sensitivity analysis showed that total nitrogen in the water column increases as organic nitrogen builds up (Tables 17.10 and 17.11). Further systematic bottlenecking at the organic nitrogen is evident by the reduced ammonia + ammonium and nitrite + nitrate loads, following a linear pattern of decrease in the water column related to copper inhibition increases. Long-term organic nitrogen accumulation will likely reduce volume and increase oxygen demand as a result of the lower nitrogen processing rates that alter the biogeochemical processing that regulated the system's efficiency. The general increasing trend of organic nitrogen levels signifies the long-term source of organic buildup that will reduce volume and increase discharge in the long run, although not as significantly in the short operational period of the current modeling analysis. Increased sedimentation can also create larger anaerobic zones where biochemical reaction rates are naturally slower, which exponentially increases the bottleneck effect at the ammonification stage.

Table 17.10 Nitrogen Mass (mg) in the Water Column, Discharged from the Water Column, Denitrified from the FTW with Copper Inhibition of Biochemical Reactions at 0%, 20%, 50%, and 75%

Copper Inhibition (%)	TN	Org N	NH	NO$_x$	FTW Denit	TN Discharged
0	494,212	468,805	13,704	11,703	1,796,034	21,157
10	495,461	472,204	13,109	10,148	1,790,984	17,974
20	496,752	475,647	12,441	8664	1,790,984	15,028
30	498,100	479,135	11,692	7274	1,779,284	12,338

Table 17.11 Change in Nitrogen from Copper Ecotoxicity in the Water Column, Discharged from the Water Column, Denitrified from the FTW with Copper Inhibition of Biochemical Reactions at 0%, 20%, 50%, and 75%

Copper Inhibition (%)	TN (%)	Org N (%)	NH (%)	NO$_x$ (%)	FTW Denit (%)	TN Discharged (%)
10	0	1	−4	−13	0	−15
20	1	1	−10	−30	0	−34
30	1	2	−16	−51	−1	−59

17.9 CONCLUSIONS

This ecological engineering practice has illuminated water quality issues at a few stormwater wet detention ponds and carried out a stormwater treatment BMP study utilizing the FTW. The use of FTW resulted in significant changes in TN mass within the water column of the pond through systematic mechanisms of nitrogen transformations. The unique characteristics of the FTW, including the cubic form and available pore space, allowed the retention and degradation of leaf litter and promoted a more thorough nitrification–denitrification process. By stabilizing the pond ecosystem, the FTW provides a biologically efficient removal BMP that may provide long-lasting remediation effects unlike the commonly used chemical algaecides that disrupt the nitrification/denitrification processes. The system dynamics model developed in this study can effectively address the global trend when some of the nitrification and denitrification processes were altered under the impact of heavy metals. For future studies, some standardization for evaluating biological treatment products such as algaecides will be necessary to establish comparisons without confounding effects to the bio-chemical processes via ecotoxicity or other ecological factors involved in natural systems for stormwater pollution control. Altering the FTW designs to facilitate conditions that promote organic matter degradation or use other technologies to reduce organic nitrogen from the influent could be more efficient. This should in turn allow the pond to maintain the volumetric operating capacity and promote efficient biochemical nitrogen removal.

APPENDIX A. STOCK SYMBOLS USED IN THE STELLA MODEL

STOCK □	Description	Units
Accd_Org_N	Accreted nitrogen	mg
Discharg_Total	Total volume of discharge	L
FTW Denit	Nitrogen removed from the FTW by denitrification	mg
FTW PON	Particulate organic nitrogen within the FTW	mg
Infilt Vol	Volume of water infiltrated	L
NH	Ammonia + ammonium in the water column	mg
NH Infiltrated	Ammonia + ammonium infiltrated to groundwater from the water column	mg
NOx	Nitrite + nitrate in the water column	mg
NOx Infiltrated	Nitrite + nitrate infiltrated to groundwater from the water column	mg
Plants	Nitrogen content of plant biomass	mg
PON	Particulate organic nitrogen in the water column	mg
Pond Vol	Volume of the pond	L
Sediment NH	Ammonia + ammonium in the sediment	mg
Sed NOx	Nitrite + nitrate in the sediment	mg
Sed PON	Particulate organic nitrogen in the sediment	mg
Sed SON	Soluble organic nitrogen in the sediment	mg
Sed Denit	Nitrogen removed from the sediment by denitrification	mg
Sed Vol	Volume of the sediment (Pore space + sediments)	L
SON	Soluble organic nitrogen in the water column	mg
SON Infiltrated	Soluble organic nitrogen infiltrated to groundwater from the water column	mg
TN Dischrg	Total nitrogen discharged from the water column	mg
Volit	Volatilization of ammonia nitrogen from water column	mg

APPENDIX B. FLOW SYMBOLS USED IN THE STELLA MODEL

FLOW ⊏⊃▷	Equation	Units	Description
Accretion	Sed_PON*Acc_rate		Accretion
Am	k_Am*(Org_N_Conc*(Pond_Vol-Q_x_V))	mg/d	Ammonification
Denit	(NOx_Conc*FTW_k_NOx)*(Q_x_V*FTW_AN_%)	mg/d	FTW denitrification
Discharge	Influent—(discharge + evaporation + infiltration)	mg/d	Surface discharge
Evap	Change in volume—(discharge + infiltration)	mg/d	Evapotranspiration
FTW LL	(Q_x_P*Plants*D_rate)*FTW_LL%	mg/d	Leaf litter falling onto the FTW
FTW Nit	(NH_Conc*FTW_k_Nit)*(Q_x_V*FTW_AE%)	mg/d	Nitrification within the FTW
FTW Am	(Org_N_Conc*FTW_k_Am)*(Q_x_V*FTW_AE%)	mg/d	Ammonification within the FTW
Infiltration	Change in volume—(evapotranspiration + infiltration)	L/d	Infiltration
NH Assim	if((Q_x_P*NH:NOx*Asm_rate) > NH) THEN(0)ELSE(Q_x_P*NH:NOx*Asm_rate)	mg/d	Vegetative assimilation of ammonium
NH dischg	NH_Conc*Dis	mg/d	
	Ammonia + ammonium discharge	NH infilt	
	(Sed_NH_ConcNH_Nonsorp)*Infil_Vol	mg/d	
	Ammonia + ammonium infiltrated from sediment		
NH Load	(NH_inf*RUN)	mg/d	
	Ammonia + ammonium load from runoff		
Nit	(NH_Conc*k_Nit)*(Pond_Vol-Q_x_V)	mg/d	Nitrification in the water column
NOx asm	if((Asm_rate*Q_x_P*NOx:NH) > NOx) THEN(0)ELSE(Asm_rate*Q_x_P*NOx:NH)	mg/d	Assimilation rate of nitrate
NOx infilt	(Sed_NOx_Conc*NOx_Nonsorp)*Infil_Vol	mg/d	Nitrite + nitrate infiltrated into groundwater from the sediment
NOx Load	(NOx_inf*RUN)	mg/d	Nitrite + nitrate load from runoff

Continued

FLOW ⟿	Equation	Units	Description
NOx dischg	NOx_Conc*Dis	mg/d	Nitrite + nitrate load discharged
Org N dischg	Org_N_Conc*Dis	mg/d	Organic nitrogen load discharged
Org N Settling	PON*Org_N_Set_rate	mg/d	PON settling rate
PON Load	(RUN*Org_N_inf*PON_%)	mg/d	Particulate organic nitrogen in influent
PON Sol	Sed_PON*PON_Sol_k	mg/d	Solubilization rate of PON in sediment
Runoff	RUN	L/d	Volume of runoff entering the pond
Sed Am	(Sed_Org_N_conc*k_Am)*(Sed_Vol*AE_%)	mg/d	Ammonification in the sediment
Sed Nit	Sed_NH_Conc*k_Nit*(Sed_Vol*AE_%)	mg/d	Nitrification in the sediment
Sol	PON*Water_PON_k	mg/d	PON solubilization in the water column
SON Infil	(Infil_Vol*Org_N_Conc)*SON_Nonsorp	mg/d	Soluble organic nitrogen infiltrated to groundwater from the water column
SON Leach	PON_k*FTW_PON	mg/d	SON leaching from the FTW
SON Load	(RUN*Org_N_inf*SON_%)	mg/d	Soluble organic nitrogen in influent
S denit	Sed_NOx_Conc*k_NOx*(AN_%*Sed_Vol)	mg/d	Nitrates removed by denitrification in sediment
Water LL	(Plants*D_rate)*Water_LL%	mg/d	Leaf litter falling into the water column
Wat NH Inf	(NH_Conc*Infil_Vol)*(NH_Nonsorp)	mg/d	
	Ammonia + ammonium infiltrated to groundwater from water column		
Wat NOx Inf	(NOx_Conc*Infil_Vol) *(NOx_Nonsorp)	mg/d	Nitrite + nitrate infiltrated to groundwater from water column
Water SON Infil	(Infil_Vol*Org_N_Conc)*SON_Nonsorp	mg/d	Soluble organic nitrogen infiltrated to groundwater from the water column
Volit rate	NH*Volit_k	mg/d	Volatilization rate form water column to atmosphere

APPENDIX C. **CONVERTER SYMBOLS USED IN THE STELLA MODEL**

Converter ◯	Description	Units	Value
Acc rate	Accretion reaction rate	mg/d	0.000379
AE %	Aerobic fraction of sediment	N/A	0.08
AN %	Anaerobic fraction of sediment	N/A	0.92
Asm rate	Total nitrogen assimilation rate by Soft Rush	mg/d	0.945
Deposition	Dry atmospheric nitrogen deposition	mg/d	4000
Dis	Discharge	L/d	Measured
D rate	Vegetative decay rate	d^{-1}	0.006
Evap Vol	Evaporation	L/d	Measured
FTW AE%	Aerobic fraction of FTW	N/A	0.45
FTW LL%	Fraction of leaf litter falling onto the FTW	N/A	0.9
FTW Quant	Quantity of 7.4 m^2 BioHaven© FTWs	N/A	4.0
FTW Vol	Volume of submerged portion of BioHaven© FTW	L	1460.0
FTW AN %	Anaerobic fraction of FTW	N/A	0.55
FTW k Am	Ammonification reaction rate within FTW	d^{-1}	0.06
FTW k Nit	Nitrification reaction rate within FTW	d^{-1}	2
FTW k NOx	Denitrification rate within FTW	d^{-1}	5
Infil Vol	Volume infiltrated to groundwater	L/d	573.03
K Am	Ammonification reaction rate in the water column	d^{-1}	0.01
K Nit	Nitrification reaction rate in the water column	d^{-1}	0.095
NH:NOx	Proportion of ammonium: nitrates for assimilation	N/A	Varies over time
NH Conc	Ammonia+ammonium concentration in water column	mg/L	Varies over time
NH Nonsorp	Portion of ammonia+ammonium that cannot adsorb to sediment	mg/L	0.8
NOx:NH	Proportion of Nitrates: ammonium in water column	N/A	Varies over time
		N/A	0.95

Continued

Converter ◯	Description	Units	Value
NOx Nonsorp	Portion of nitrite + nitrate that cannot adsorb to sediment		
Org N Conc	Organic nitrogen concentration in the water column	mg/L	Varies over time
Org N Set rate	Settling rate of PON in the water column	d^{-1}	0.87
Plant Pop	Quantity of plants per FTW	N/A	160.0
PON %	Fraction of PON within TON of influent	N/A	0.8
PON k	Solubilization rate of PON to SON from FTW	d^{-1}	0.09
PON Sol k	Solubilization rate of PON to SON in sediment	d^{-1}	0.15
Q x P	Total quantity of plants on FTWs	N/A	FTW_Quant*Plant__Pop
Q x V	Total submerged volume of FTWs in Pond	L	FTW_Vol*FTW_Quant
RUN	Runoff influent	L	Measured
Sed k Am	Ammonification rate in sediment	d^{-1}	0.02
Sed k Nit	Nitrification rate in sediment	d^{-1}	0.025
Sed k denit	Denitrification reaction rate in the sediment	d^{-1}	0.1
Sed NH Conc	Ammonia + ammonium concentration in sediment	mg/L	Varies over time
Sed Org N conc	Organic nitrogen concentration in sediment	mg/L	Varies over time
Sed NOx Conc	Nitrite + nitrate concentration in sediment	mg/L	Varies over time
SON %	Fraction of PON within TON of influent	N/A	0.2
Water LL%	Leaf litter rate falling into water column from FTW	d^{-1}	0.5
Water PON k	Solubilization rate of PON in water column	d^{-1}	0.2
Volit k	Reaction rate coefficient for volatilization	d^{-1}	0.03

References

Adams, E.B., 1992. Wetlands: Nature's Water Purifiers. Cooperative Extension, Washington State University, Seattle, WA.

Baldwin, A.H., Simpson, T.W., Weammert, S.E., 2009. Urban wet ponds and wetlands best management practice. Final report, December 2009.

Belmont, M.A., Metcalfe, C.D., 2003. Feasibility of using ornamental plants (Zantedeschia aethiopica) in subsurface flow treatment wetlands to remove nitrogen, chemical oxygen demand and nonylphenol ethoxylate surfactants a laboratory-scale study. Ecol. Model. 21, 233–247.

Beran, B., Kargi, F., 2005. A dynamic mathematical model for wastewater stabilization ponds. Ecol. Model. 181, 39–57.

Beutel, M.W., 2006. Inhibition of ammonia release from anoxic profundal sediments in lakes using hypolimnetic oxygenation. Ecol. Eng. 28, 271–279.

Box, G.E.P., Tiao, G.C., 1975. Intervention analysis with applications to economic and environmental problems. J. Am. Stat. Assoc. 70, 70–79.

Brown, B.T., Rattigan, B.M., 1979. Toxicity of soluble copper and other metal ions to Elodea canadensis. Environ. Pollut. 20, 303–314.

Button, K.S., Hostetter, H.P., Mair, D.M., 1977. Copper dispersal in a water supply reservoir. Water Res. 11, 539–544.

Carpenter, S.R., Lodge, D.M., 1986. Effects of submersed macrophytes on ecosystem processes. Aquat. Bot. 26, 341–370.

Chang, N.B., Islam, K., Wanielista, M., 2012a. Floating wetland mesocosm assessment of nutrient removal to reduce ecotoxicity in stormwater ponds. Int. J. Environ. Sci. Technol. 9, 453–462.

Chang, N.B., Islam, K., Marimon, Z., Wanielista, M.P., 2012b. Assessing biological and chemical signatures related to nutrient removal by floating islands in stormwater mesocosms. Chemosphere 88, 736–743.

Chang, N.B., Xuan, Z., Marimon, Z., Islam, K., Wanielista, M.P., 2013. Exploring hydrobiogeochemical processes of floating treatment wetlands in a subtropical stormwater wet detention pond. Ecol. Eng. 54, 66–76.

Deaton, M.L., Winebrake, J.J., 2000. Dynamic Modeling of Environmental Systems. Springer-Verlag, New York.

EC, MENVIQ (Enviornment Canada, Ministere de l'Envionment du Quebec), 1992. Interim Criteria for Quality Assessment of St. Lawrence River Sediment. Environment Canada, Ottawa, Canada.

Environmental Protection Agency (EPA), 1986. Quality criteria for water, EPA-440/5-86-001. Washington, DC.

U.S. Environmental Protection Agency (US EPA), 1986. Quality Criteria for Water, EPA-440/5-86-001, Washington, DC.

Ewing, K., 1996. Tolerance of four wetland plant species to flooding and sediment deposition. Environ. Exp. Bot. 36, 131–146.

Forrester, J.W., 1961. Industrial Dynamics. The MIT Press, Cambridge, MA, USA.

Forrester, J.W., 1968. Principles of System. Productivity Press, Cambridge, MA.

Gabel, B., Kozicki, R., Lahl, U., Podbielski, A., Stachel, B., Struss, S., 1982. Pollution of drinking water with nitrate. Chemosphere 11, 1147–1154.

Gearheart, R.A., 1990. Nitrogen removal at the Arcata constructed wetlands. In: Presentation at the Water Pollution Control Federation Meeting, Washington, DC.

Gersberg, R.M., Elkins, B.V., Lyon, S.R., Goldman, C.R., 1986. Role of aquatic plants in wastewater treatment by artificial wetlands. Water Res. 20, 363–368.

Gettys, L.A., Haller, W.T., Bellaud, M., 2009. Biology and Control of Aquatic Plants: A Best Management Practices Handbook. Aquatic Ecosystem Restoration Foundation, Marietta, GA.

Greenway, M., Polson, C., 2007. Macrophyte establishment in stormwater wetlands: coping with flash flooding and fluctuating water levels in the subtropics. In: American Society of Civil Engineers. http://ascelibrary.org/doi/abs/10.1061/40927(243)635.

Harper, H.H., Baker, D.M., 2007. Evaluation of Current Stormwater Design Criteria Within the State of Florida. Orlando, Florida. Environmental Research & Design, Inc., Orlando, FL.

Harrison, P.G., Durance, C.D., 1985. Reductions in photosynthetic carbon uptake in epiphytic diatoms by water-soluble extracts of leaves of Zostera Marina. Mar. Biol. 90, 117–119.

Headley, T.R., Tanner, C.C., Council, A.R., 2006. Application of Floating Wetlands for Enhanced Stormwater Treatment: A Review: NIWA Client Report. Auckland Regional Council, New Zealand.

Hu, Z., Chandran, K., Grasso, D., Smets, B.F., 2004. Comparison of nitrification inhibition by metals in batch and continuous flow reactors. Water Res. 38, 3949–3959.

Huang, C.P., Wang, H.W., Chiu, P.C., 1998. Nitrate reduction by metallic iron. Water Res. 32, 2257–2264.

Iamchaturapatra, J., Yi, S.W., Rhee, J.S., 2007. Nutrient removals by 21 aquatic plants for vertical free surface-flow (VFS) constructed wetland. Ecol. Eng. 29, 287–293.

Jørgensen, S.E., 1995. State of the art ecological modeling in limnology. Ecol. Model. 78, 101–115.

Jørgensen, S.E., Chon, T.S., Recknagel, F., 2009. Handbook of Ecological Modelling and Informatics. WIT Press, Billerica, MA.

Kadlec, R.H., Wallace, S.D., 2009. Treatment Wetlands, second ed. Taylor and Francis Group, LLC, Boca Raton, FL.

Kadlec, R.H., Pries, J., Lee, K., 2012. The Brighton treatment wetlands. Ecol. Eng. 47, 56–70.

Kim-Shapiro, D.B., Gladwin, M.T., Patel, R.P., Hogg, N., 2005. The reaction between nitrite and hemoglobin: the role of nitrite in hemoglobin-mediated hypoxic vasodilation. J. Inorg. Biochem. 99, 237–246.

Knight, R.L., 1986. Florida effluent wetlands, Total Nitrogen. CH2M Hill wetland technical reference document series no. 1, Gainesville, FL.

Küpper, H., Küpper, F., Spiller, M., 1996. Environmental relevance of heavy metal-substituted chlorophylls using the example of water plants. J. Exp. Bot. 47, 259–266.

Loiselle, S., Cózar, A., Van Dam, A., Kamsiime, F., Kelderman, P., Saunders, M., Simonit, S., 2006. Tools for wetland ecosystem resource management in east Africa: focus on the Lake Victoria papyrus wetlands. Wetland Nat. Res. Manage. 190, 97–121.

Marimon, Z.A., Xuan, Z., Chang, N.B., 2013. System dynamics modeling with sensitivity analysis for floating treatment wetlands in a stormwater wet pond. *Ecol. Model*, 267, 66–79.

Martin, J.F., Reddy, K.R., 1997. Interaction and spatial distribution of wetland nitrogen processes. Ecol. Model. 105, 1–21.

Mayo, A.W., Bigambo, T., 2005. Nitrogen transformation in horizontal subsurface flow constructed wetlands I: model development. Phys. Chem. Earth 30, 658–667.

Mickle, A.M., Wetzel, R.G., 1978. Effectiveness of submersed angiosperm epiphyte complexes on exchange of nutrients and organic carbon in littoral systems. I. Inorganic Nutrients. Aquat. Bot. 4, 303–316.

Mohapatra, P.K., 1994. Introduction to System Dynamics Modeling. Orient Longman Ltd., Hyderabad, India/New York.

Pitts, R., 2004a. Control of microorganisms in urban waters. http://unix.eng.ua.edu/rpitt/Class/ExperimentalDesignFieldSampling/MainEDFS.html.

Pitts, R., 2004b. Detention pond design and analysis. http://rpitt.eng.ua.edu/Class/Water%20Resources%20Engineering/M9c2%20WinTR55%20ponds%20docs.pdf.

Rabe, R., Schuster, H., Kohler, A., 1982. Effects of copper chelate on photosynthesis and some enzyme activities of Elodea canadensis. Aquat. Bot. 14, 167–175.

Randers, J. (Ed.), 1980. Elements of the System Dynamics Method. Productivity Press, Cambridge, MA.

Rykiel Jr., E.J., 1996. Testing ecological models: the meaning of validation. Ecol. Model. 90, 229–244.

Sawyer, C.N., McCarty, P.L., Parkin, G.F., 2003. Chemistry for Environmental Engineering Science, fifth ed. McGraw-Hill, New York.

Servizi, J.A., Gordon, R.W., 2005. Acute Lethal Toxicity of Ammonia and Suspended Sediment Mixtures to Chinook Salmon (Oncorhynchus tshawytscha). Bull. Environ. Contam. Tox. 44, 650–656.

Stoyanova, D.P., Tchakalova, E.S., 1993. The effect of lead and copper on the photosynthetic apparatus in Elodea canadensis Rich. Photosynthetica 28, 63–74.

Tarazona, J.V., Munoz, M.J., Ortiz, J.A., Nunez, M.O., Camargo, J.A., 2008. Fish mortality due to acute ammonia exposure. Aquat. Res. 18, 167–172.

Tiedemann, J.A., Witty, M., Souza, S., 2009. The future of coastal lakes in Monmouth County. Technical report submitted to National Oceanic and Atmospheric Administration (NOAA).

Vezjak, M., Savsek, T., Stuhler, E.A., 1998. System dynamics of eutrophication processes in lakes. Eur. J. Oper. Res. 109, 442–451.

Wanielista, M., Chang, N.B., 2012. Floating wetland systems for nutrient removal in storm-water ponds. Final report submitted to Florida Department of Transportation, FL.

Wetzel, R.G., 1983. Attached algal-substrata interactions: fact or myth, and when and how? In: Wetzel, R.G. (Ed.), Periphyton of Freshwater Ecosystems. The Hague, The Netherlands.

White, S.A., Seda, B., Cousins, M., Klaine, S.J., Whitwell, T., 2009. Nutrient remediation using vegetated floating mats. In: SNA Research Conference, vol. 54, pp. 39–43.

Wilde, F.D., 1994. Geochemistry and factors affecting ground-water quality at three storm-water-management sites in Maryland: Report of investigations No. 59. Department of Natural Resources Maryland Geological Survey, Baltimore, MD.

Yanong, R.P.E., 2009. Use of copper in marine aquaculture and aquarium systems. IFAS Publication FA165. Gainesville, FL.

Modeling Management Options for Controlling the Invasive Zebra Mussel in a Mediterranean Reservoir

18

Carles Alcaraz[a],*, Nuno Caiola[a], Carles Ibáñez[a], Enrique Reyes[b]

[a]*IRTA Aquatic Ecosystems, E-43540 Sant Carles de la Ràpita, Catalonia, Spain*
[b]*Department of Biology, East Carolina University, Greenville, North Carolina, USA*
*Corresponding author: e-mail address: carles.alcaraz@gmail.com; carles.alcaraz@irta.cat

18.1 INTRODUCTION

The zebra mussel, *Dreissena polymorpha* (Pallas, 1771), is a freshwater dreissenid mussel only native to the Eurasian Ponto-Caspian region, but during and after the Industrial Revolution in the 1800s, it was introduced to many Eurasian countries due to the construction of several canals connecting most important Eurasian watersheds (Karatayev et al., 2003; Ram and Palazzolo, 2008). The species is still expanding its invasive range worldwide, and recently (in the last 20 years), the species has been reported in Italy, the United Kingdom, and Spain (Minchin et al., 2005; Ram and Palazzolo, 2008; Rajagopal et al., 2009). In North America, the zebra mussel was first discovered in the mid-1980s in the Great Lakes region, and subsequently, during the 1990s, it had spread into other Atlantic watersheds such as the Hudson and the Mississippi basins; and at the end of the 2000s, it was found in the Pacific coast (Bunt et al., 1993; Strayer et al., 1996; Ram and Palazzolo, 2008). While former introductions are mainly related to commerce through the different canals built to connect the different European river basins, the recent introductions are associated with other mechanisms, such as ballast waters, angling, or the movement of recreational boats (Karatayev et al., 2003; Ram and Palazzolo, 2008).

Considered as one of the most aggressive freshwater pests worldwide, the large ecological and economical impacts of the invasive zebra mussel are well known. After introduced and established, the zebra mussel often becomes the most abundant benthic organism, and it has the ability to act as an ecosystem engineer, profoundly modifying both ecosystem structure and function of the invaded ecosystem (Arnott and Vanni, 1996; Karatayev et al., 1997; Chase and Bailey, 1999; Nicholls et al., 1999; Karatayev et al., 2005; Strayer and Malcom, 2006; Ram and Palazzolo, 2008). The presence of zebra mussel reduces the available benthic habitat for some organisms but provides a suitable habitat for others, thus changing ecosystem community structure and trophic interactions; furthermore, its intense feeding activity reduces

Developments in Environmental Modelling, Volume 26, ISSN 0167-8892, http://dx.doi.org/10.1016/B978-0-444-63249-4.00020-8

phytoplankton and zooplankton abundances, affecting water transparency, oxygen concentration, and modifying nutrient cycling through trophic cascades (Arnott and Vanni, 1996; Karatayev et al., 1997; Ram and Palazzolo, 2008). The direct and indirect feedbacks created between the zebra mussel and the rest of species in the water system affect the full water community and ecosystem processes, leading to shifts in plank-tonic, benthic, and fish communities (Arnott and Vanni, 1996; Strayer et al., 1996; Karatayev et al., 2003). Zebra mussel fouling in artificial structures has also a severe impact on human water supplies and industrial systems, thus imposing strong economic impact (i.e., environmental costs) to its invasions because of its negative effects on ecosystem services and tourism (Pimentel et al., 2005; Connelly et al., 2007). Consequently, the zebra mussel is considered to be one of the world's 100 worst invasive alien species by the ISSG (Invasive Species Specialist Group, http://www.issg.org/database/welcome/) and the DAISIE (Delivering Alien Invasive Species Inventories for Europe; http://www.europe-aliens.org/speciesTheWorst.do).

In the Iberian Peninsula, the zebra mussel was first introduced to the Ebro River, in the Riba-roja Reservoir (see Figure 18.1). A few individuals were found reaching an average density of ca. 328 individuals/m^2 close to the dam in summer 2001 (Altaba et al., 2001). Since its introduction, the zebra mussel has spread and expanded through the main Ebro River and is currently present in seven tributaries (Durán et al., 2010). In the last few years, it has also colonized other Mediterranean river basins such as the Llobregat, Millars, Xúquer, or Segura basins. Although the zebra mussel may present large point densities (>20,000 individuals/m^2) within the Ebro River, these densities are usually associated with specific artificial structures at the river facilities such as water intakes (Palau, 2009), probably due to the reason of increasing water flow, food source, and fixing substrate availability (Mellina and Rasmussen, 1994; Jones and Ricciardi, 2005). In natural substrates in the Riba-roja

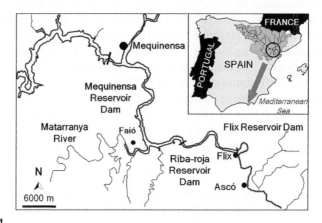

FIGURE 18.1

Location of the Riba-roja Reservoir in the Ebro River basin (northeastern Iberian Peninsula).

Reservoir, mussel population has fluctuated in a five- to six-year cycle with density peaks of ca. 3000–3500 individuals/m^2, and its density is also lower under the dam.

Both economic loss (up to ⪕11.6 millions in the Ebro Basin for the period 2005–2009) and ecological impacts of zebra mussel are mainly determined by population density and its variation across space and time (Karatayev et al., 2003; Ram and Palazzolo, 2008; Durán et al., 2012). Previous predictive efforts for abundance studies of the zebra mussel focused on water chemistry descriptors such as nutrients and substratum physical characteristics (e.g., Ramcharan et al., 1992; Mellina and Rasmussen, 1994; Wilson and Sarnelle, 2002; Jones and Ricciardi, 2005), but the factors causing and controlling the different zebra mussel population dynamics are poorly known, although this knowledge is crucial to understand fluctuations, make predictions, and develop adequate management plans. Based on a previously presented model (Alcaraz et al., 2014), the objectives of this chapter are (i) to qualitatively predict future mussel population dynamics under different scenarios, and (ii) to assess the potential use of the model for the exploration of different management measures on mussel dynamics, thus helping in the control of this invasive species.

18.2 METHODS

18.2.1 Study area

The population dynamics of zebra mussel were modeled in the Riba-roja Reservoir, located in the lower section of the Ebro River (NE Iberian Peninsula) (Figure 18.1). The Ebro River, with 930 km in length and a watershed of ca. 85,362 km^2, crosses the northeastern part of the Iberian Peninsula and discharges into the Mediterranean Sea originating a delta of 320 km^2. It is the river with the highest flow in the Iberian Peninsula (426 m^3/s of average flow), but there are significant differences between dry (118 m^3/s) and wet (569 m^3/s) years. The Ebro Basin is a highly regulated system with at least 190 reservoirs, most of them built between 1940 and 1970, managed for multiple uses such as drinking water supply, agriculture, industrial, hydropower, and flood control (see also Suárez-Serrano et al., 2010; Alcaraz et al., 2011; Rovira et al., 2012). The Riba-roja Reservoir (ca. 100 km upstream the mouth), with a capacity of 210 hm^3 and 11 days of residence time, is a middle-sized reservoir classified as mesotrophic. As in the rest of the basin, the Riba-roja Reservoir supports different and intense human activities, with one of these activities, angling, associated the mussel introduction through the movement of recreational boats or by the introduction of exotic fish species from central European Basins (Rajagopal et al., 2009).

18.2.2 Model description

The model used here is the same as described in Alcaraz et al. (2014), with minor modifications, and we just recall here its main features. The aim of the model is to predict the long-term demography of the zebra mussel population (individuals/m^2

by month) in the Riba-roja Reservoir and to assess the effects of different environmental factors on these dynamics. A monthly time step was used instead of previous models (Strayer and Malcom, 2006; Casagrandi et al., 2007), using a year time step to better fit mussel population dynamics and to simulate both inter- and intra-annual population trends, thus increasing its usefulness in species management.

18.2.2.1 Zebra mussel life cycle, number of age classes, life-span, and fecundity

The model included the three main stages of mussel life cycle: larvae, juvenile, and adult (Ackerman et al., 1994) (Figure 18.2). We considered the most common case in which lifespan does not exceed years (Chase and Bailey, 1999; Karatayev et al., 2006), thus besides veliger and juveniles, we included four adult age-classes (AA^{1+}, AA^{2+}, AA^{3+}, and AA^{4+}; from one to four years old, respectively) (Figure 18.2). Furthermore, according to previous works (e.g., Neumann et al., 1993), juvenile mussels can grow and mature within the same year of birth (AA^{0+}), making them thus capable of producing new veliger larvae, if allowed by environmental conditions. Consequently, a second reproduction peak can be observed, in agreement with previous works, where a second veliger peak is reported (e.g., Neumann et al., 1993; Strayer et al., 1996; Jantz and Neumann, 1998). This second peak is

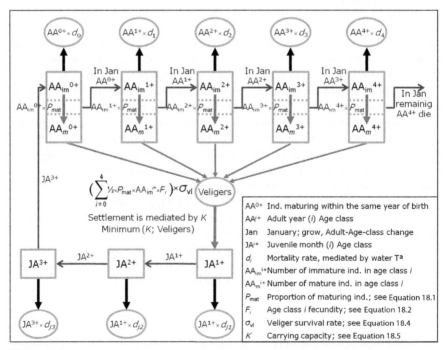

FIGURE 18.2

Simplified STELLA diagram of the model applied.

associated with larvae present in early to mid spring, which settle and mature in summer. According to this, we modeled the zebra mussel reproductive cycle by considering that eggs were fertilized immediately, veliger settled at the start of the next month; and after 4 months from settlement, if water temperature allowed it, juvenile mussels matured (AA^{0+}) and were able to reproduce causing a second veliger peak (Figure 18.2).

Temperature is one of the major environmental factors controlling and limiting zebra mussel reproduction. Spawning is commonly observed between 12 and 30 °C (Ram et al., 1996), and consequently, we modeled mussel reproduction by initiating when water temperature exceeded 12 °C (April or May). However, the maximum recorded temperature during the modeled period was 26.8 °C, thus reproduction was inhibited only at water temperatures below 12 °C (from December to March). A logistic function was used to estimate maturity (P_{mat}) as a function of water temperature (Equation 18.1). Veliger presence and water temperature from 2004 to 2008 were obtained from the Ebro Water Authority (CHE; http://www.chebro.es).

$$P_{mat} = \frac{1}{1 + e^{(-15.192 + 0.806T^a)}}; \text{ Nagelkerke's } R^2 = 0.84, n = 74, \chi_1^2 = 54.56, \tag{18.1}$$
$$P < 0.0001$$

Adult mussels were able to reproduce only once by spawning season and the number of eggs produced per spawning event and adult female was derived from Sprung (1991):

$$\text{Fecundity} = 10^6 \times 0.606 \text{SL}^{4.42}; \text{ where SL is shell length, in mm, for age-class } i \tag{18.2}$$

A demographic analysis was performed to establish average shell length (SL) (Conides et al., 1995; Strayer and Malcom, 2006). Adult SL were set to 7.20, 12.81, 18.73, 24.05, and 29.30 mm for 0–4 adult age classes, respectively; and juvenile SL were set to 3.0, 4.4, and 5.8 mm for the three monthly classes, respectively.

We considered a sexual proportion of 1:1 (Ram et al., 1996; Ram and Palazzolo, 2008) for all age classes, thus the total egg production, at time t, was computed as:

$$\text{TN}_{eggs}(t) = \sum_{i=0}^{4} \frac{1}{2} P_{mat} \text{Nr}_i F_i \tag{18.3}$$

where P_{mat} is the probability to find a mature individual (see Equation 18.1), whereas Nr_i and F_i are, respectively, the number of remaining individuals ready to reproduce and the number of eggs produced by females of age class i (see Equation 18.2).

18.2.3 Zebra mussel mortality rates and system carrying capacity

Generally, zebra mussel population dynamics suggests a density-dependent regulatory mechanism including cannibalism on larvae (MacIsaac et al., 1991; Strayer and Malcom, 2006; Casagrandi et al., 2007). Therefore, following Ricker (1954), when cannibalism is exerted by adults on their own larvae, veliger survival rate (σ_{vl}) is an exponentially decreasing function of the adult number, which can be expressed as:

$$\sigma_{vl}(N) = \sigma_0 e^{\left(\Sigma_{i=2}^{4} - \beta_i^* n_i(t)\right)}, \text{ where } \sigma_0 = \sigma_E \sigma_v \qquad (18.4)$$

σ_E is the number of eggs externally fertilized and reaching the veliger stage and σ_v is the survival of veligers from the time of hatching to the adult stage at low adult densities. So, σ_0 can be interpreted as the environmental survival rate, the average fraction of released eggs that reach veliger stage and are ready to settle in absence of adult cannibalism. n_i is the number of adults in age class I, and β_i is a unitless scaling parameter determining the order of magnitude of the filtration rate and accounting for the veliger removal by adults in age class i. Mussel size selection is mainly regulated by siphon diameter, so we considered that filtration population was only constituted by larger individuals (AA^{2+}, AA^{3+}, and AA^{4+}); and filtering activity as an exponentially increasing function of the mussel shell size; thus filtering parameter was set to β_2; $\beta_3 = 2\beta_2$; and $\beta_4 = 2\beta_3$ (Bunt et al., 1993; MacIsaac et al., 1995; Strayer and Malcom, 2006).

The monthly survival of both juvenile and adult age classes can be computed as: $n_i(t+1) = \sigma_i n_i(t)$ and the range of survival rates was obtained from literature. We selected an optimum temperature range between 16 and 24 °C (e.g., Karatayev et al., 1998; Griebeler and Seitz, 2007; van Nes et al., 2008), mortality rate was doubled outside the optimal range and set to 0.99 when threshold tolerance limits were reached, at 0 and 34 °C (Karatayev et al., 1998; Griebeler and Seitz, 2007).

We also introduced a monthly density-regulation function in the model, expressed as:

$$K = 712.20 + 3479.50 \times TP \qquad (18.5)$$

where K is the carrying capacity (in Ind/m^2) and TP is the total phosphorus concentration (in mg/L). The model fit (R^2) between observed mussel densities and predicted densities by the model was 0.65. Veligers successfully settled only if the total number of attached mussels was smaller than carrying capacity (K), dying otherwise. Since the aim of the simulations were to test mussel density variations following a total phosphorus concentration diminution, we selected a carrying capacity only based on total phosphorus, because it has been proved to be more effective in estimating mussel density under small phosphorus concentrations.

18.2.4 Model formulation and calibration

The software STELLA 9.1 was used to compute the model differential-equations to estimate monthly changes in zebra mussel population density (individuals/m^2) (Figure 18.2). Monthly averaged physicochemical data were obtained from the water quality monitoring program performed by the Ebro Water Authority (CHE; www.chebro.es). Zebra mussel density values in the study area were mainly obtained from the Grup Natura Freixe and CHE databases (for more details, see Ibáñez et al., 2012).

Because the zebra mussel was first detected in summer of 2001, mussel density was set to 0 before June 2000, when density simulations started. Model parameters such as mortality rates or filtration rate were calibrated to fit model simulations to observed mussel population densities. Because the available mussel density data is scarce, all data gathered were fully used in the model calibration process. However, although the model was not validated, according to the results, we obtained reliable simulations coherent with bibliographic data.

18.3 SCENARIO ANALYSES

Two simulation scenarios were developed to model the effect of nutrient loading and reservoir water-level fluctuation on mussel population dynamics. In the lower Ebro River, the TP–chlorophyll relationship is particularly narrow (Ibáñez et al., 2012), and in the last two decades, a marked phosphorus reduction has been followed by a chlorophyll reduction (and hence mussel food availability). Therefore, the effect of this nutrient reduction on mussel density was simulated by a regular reduction of TP over time and by a nutrient removal in a given time step. The results of these simulations may help us to better understand future mussel population dynamics if TP reduction in the river basin continues.

Mussel population density varies across space and time, resulting in irregular demographic patterns; however, its distribution in depth is more stable and the maximum density is usually found at the same water depth within a given water body (Mellina and Rasmussen, 1994; Mackie and Schloesser, 1996; Burla and Ribi, 1998; Karatayev et al., 1998; Navarro et al., 2006). Therefore, by only reducing water level in the reservoir, we can kill a significant proportion of the mussel population. We simulated the effect of increasing mussel mortality rate, at different moments (e.g., before spawning season, during density peaks or population minimums), on population dynamics. K and temperature were set to be constant to avoid misleading results.

18.4 RESULTS AND DISCUSSION

18.4.1 Population dynamics and model calibration

Among the main factors determining zebra mussel impacts are its large population growth and ability to expand immediately after introduction (Arnott and Vanni, 1996; Strayer and Malcom, 2006). Zebra mussel population density may vary over time, displaying irregular demographic patterns that can be grouped into four main types: boom-bust (population dynamics is characterized by larger densities for a few years after colonization followed by lower densities sustained over the long-term), cyclic (mussel population follows regular density peaks), in equilibrium (characterized by small density fluctuations from year to year), and chaotic, where the population shows irregular density fluctuations with no long-term trend (reviewed in

Strayer and Malcom, 2006). According to this, the zebra mussel model generated different population dynamics (see more details in Alcaraz et al., 2014), and by modifying two of the model parameters (σ_0 and/or β_i both affecting veliger survival rate), we obtained all those mussel dynamics (i.e., boom-bust, cyclic, equilibrium, and chaotic) previously described in the literature (e.g., Ramcharan et al., 1992; Strayer and Malcom, 2006; Casagrandi et al., 2007). Overall, when population growth was unconstrained (i.e., population growth not limited by K), for a fixed value of β_i, larger mussel density peaks and variations in the population peaking cycle interval can be observed by simulating an increase in σ_0, usually showing a transition from cyclic to chaotic dynamics. However, in a population-limited growth situation, by fixing K in the model, only a certain number of the surviving larvae were allowed to settle, thus reducing the effects of both σ_0 and β_i on population dynamics and showing that K is the most important factor determining population dynamics. Upon fixing K, larger values of σ_{vl} (by increasing σ_0 or reducing β_i) generated cyclic dynamics, while lower σ_{vl} values (i.e., lower σ_0 and larger β_i values) determined chaotic dynamics characterized by irregular fluctuations in mussel population densities. Furthermore, since σ_{vl} is related to both β_i and σ_0 through the equation: $\sigma_{vl} = \sigma_0 * e^{\left(-\beta_i{}^* N_i\right)}$, the model generated similar population dynamics by increasing σ_0 while reducing β_i; and vice versa, and a marked reduction in adult mortality rates (i.e., increasing N_i) showed similar effects to increasing β_i (see more details in Alcaraz et al., 2014).

Since modeled mussel fecundity is dependent on SL, a sensitivity analysis was carried out based on the estimates of mussel SL. The analysis was performed by using SL ± 2 mm for age 0^+ individuals; and SL ± 5 mm for the rest of adult age classes, but we did not observe significant effects on population dynamics. Although larger individuals produced more larvae, when population growth is restricted by K, larvae are in excess so they are not able to settle and die. Finally, the same population dynamics were observed when more age classes (five and six different age classes) were incorporated into the model (see more details in Alcaraz et al., 2014).

18.4.2 **Total phosphorus effect**

In the Iberian Peninsula, the zebra mussel was first detected in the Lower Ebro River, near the Riba-roja Reservoir dam, and currently, it is one of the most abundant organisms in the lower basin (Ram and Palazzolo, 2008; Rajagopal et al., 2009). However, when comparing observed density values to values reported in the literature for other rivers and lakes from North America and Europe (e.g., MacIsaac and Sprules, 1991; Chase and Bailey, 1999; Wilson and Sarnelle, 2002; Strayer and Malcom, 2006), and also to simulations with population growth unrestricted by K, mussel density peaks are lower than expected. Therefore, the growth of the mussel population in the Lower Ebro River is not only mediated by an adult-density-regulation mechanism but also by environmental restrictions (i.e., K). In our model, we calculated reservoir carrying capacity (K) as a function of TP water concentration. TP concentration is highly correlated with chlorophyll, since it is one of the main factors controlling phytoplankton production, that is, food availability (Dillon and Rigler, 1974; Wilson and Sarnelle, 2002; Ibáñez et al., 2012).

Carrying capacity and hence TP concentration mainly affected mussel population dynamics by both controlling population growth (i.e., dynamic pattern) and limiting the height of the density peaks. In the baseline scenario (i.e., using TP concentrations in the reservoir), mussel population showed two density peaks around 3000 individuals separated by five years (Figure 18.3); however, population density diminished when the percentage of available TP concentration (Figure 18.4) was reduced. Phosphorus concentration is strongly correlated with chlorophyll, but the TP-chlorophyll relationship is particularly tight in the lower Ebro River (Ibáñez et al., 2012), where a

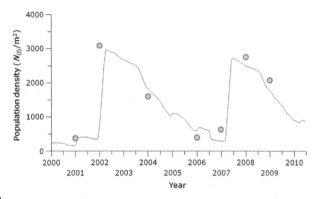

FIGURE 18.3

Simulated mussel density by the model; observed mussel densities (gray spots) are also shown.

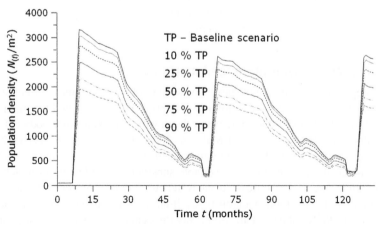

FIGURE 18.4

Simulated effects of phosphorus reduction (%) on mussel density population.

drastic phosphorus reduction in the last 20 years has been followed by a reduction in chlorophyll concentration (Ibáñez et al., 2012); between 1990 and 2005, both phosphorus and chlorophyll concentration has experienced a reduction of ca. 5 and 20 times, respectively. It is expected that this situation will not change in the future, since the TP reduction has also caused a regime shift from a turbid phytoplankton dominated system to a clear macrophyte-dominated system (Ibáñez et al., 2012), therefore further TP reduction is probably expected in the next years.

After fixing water temperature by averaging the last five years' monthly data, to avoid misleading results (e.g., density responding to water temperature instead to K), we simulated the effect of reduced TP concentration (between 0% and 90%) in the mussel density trend (Figure 18.4). Mussel density peaks decreased when reducing the TP concentration available, but this reduction did not modify the cyclic population pattern with density peaks separated by five years (Figure 18.4). When reducing the TP concentration, population K is lower and consequently fewer surviving larvae are able to settle, thus affecting maximum density peaks; however, the relationship between TP concentration and mussel density peaks was not lineal since population peaks were ca. 40% lower in the 90% TP reduction scenario than in the baseline scenario (Figure 18.4). According to previous studies, both the density and biomass of benthic organisms are influenced by several factors such as depth, bottom slope, substrate type, turbidity, water flow, nutrient concentration, and phytoplankton density (Rasmussen and Kalff, 1987; Rasmussen, 1988; Mellina and Rasmussen, 1994). Because of its effect in limiting water productivity, TP concentration is often considered a major factor structuring the abundance of benthic fauna (e.g., Rasmussen and Kalff, 1987; Cloern and Jassby, 2008; Souza et al., 2013) and the zebra mussel in particular (e.g., Ramcharan et al., 1992; Wilson and Sarnelle, 2002). Our simulations are in concordance with those relationships previously reported in the literature for both benthic macroinvertebrates and zebra mussels, and also with results presented by Ibáñez et al., (2012) in the lower Ebro River. Consequently, in the Ebro Basin, TP has been shown to be a good predictor of mussel population dynamics and maximum density.

18.4.3 Scenario with increased mortality

Mussel population density is not stable, showing marked variations and fluctuations across space and time, but its distribution in depth is more stable and the maximum density is usually found at the same water depth within a given water body (Mackie and Schloesser, 1996; Strayer and Malcom, 2006; Karatayev et al., 1998). Zebra mussels are restricted to the epilimnion (where oxygen is sufficient), and their maximum density is a function of a suitable substrate for settlement or attachment (e.g., avoiding fine silt substrate or strong water current), oxygen concentration, temperature, turbidity, and food availability (Mellina and Rasmussen, 1994; Mackie and Schloesser, 1996; Karatayev et al., 1998; Navarro et al., 2006). Consequently, one of the most important factors affecting mussel mortality is unpredictable events such as floods or water-level fluctuations.

In the Riba-roja Reservoir, the epilimnion is about 11 m deep and maximum mussel biomass is found between 0 and 5 m (Navarro et al., 2006; Ecohydros, 2008) since water level is usually constant; so a variation in the reservoir water level, by selectively releasing water from the dam, can have a significant effect on mussel density by exposing population to desiccation. We simulated what would happen if we increased mussel mortality rate by varying the reservoir level. After fixing both temperature and K to avoid misleading results, mussel population density was decreased by increasing mortality rate for a given time step (Figure 18.5). However, population response was related to several factors such as the increase of the mortality rate (i.e., the percentage of the population affected), the population density when applied to the mortality event, the population age-class composition, as well as to the mussel reproduction cycle. Overall, mussel population rapidly recovered density values when mortality rate was increased before the reproductive season, even if mortality was set to 99% (Figure 18.5a). One key aspect determining mussel invasiveness is its spawn capacity (Arnott and Vanni, 1996; Ram and Palazzolo, 2008), thus if the population is perturbed during the reproductive season, more surviving veliger larvae are able to settle and population may easily recover from it. In concordance with this, population took longer to recover from a mortality event simulated before the reproductive season, since mussels did not release more larvae until the next spawn (Figure 18.5b). The probability that the population became extinct increased when the mortality event occurred in a situation of minimum population density (i.e., increasing the effect of density reduction) (Figure 18.5c) and enhanced if the mortality event occurred out of the spawning season. Finally, the cyclic population dynamics was modified by increasing mortality rate in a given month, and marked dynamic changes were observed when mussel population was dominated by small population age-class individuals (i.e., AA^{1+} or AA^{2+}), just before the spawning season (Figure 18.5d). Our simulations showed that mussel management was improved by combining more than one mortality event (e.g., by combining a water-level diminution and a level increment or two water-level reductions), and the best results were obtained then the first event was simulated just before the spawning season and the second after the next spawning season (Figure 18.6), thus reducing the population's ability to respond to perturbations by limiting larvae production.

In the Riba-roja Reservoir, as in other water bodies, the depth distribution pattern of zebra mussel is determined by water column characteristics. The reservoir stratification and the water column characteristics are determined by the interaction between the chemical characteristics of water inputs from the Mequinensa Reservoir and two tributaries flowing into the reservoir (Figure 18.1), and also by management practices of the reservoir (i.e., electricity production) (Navarro et al., 2008). The reservoir is mainly used for hydroelectric production, and it is possible to release water at different dam heights. Therefore, by simply releasing water from the dam, we can alter the reservoir water level, thus modifying the reservoir stratification pattern and both the height and depth of water layers (i.e., epilimnion, mesolimnion, and hypolimnion). By managing reservoir water release, it is possible to regulate the volume and the location of the water layers providing the optimal conditions for the development

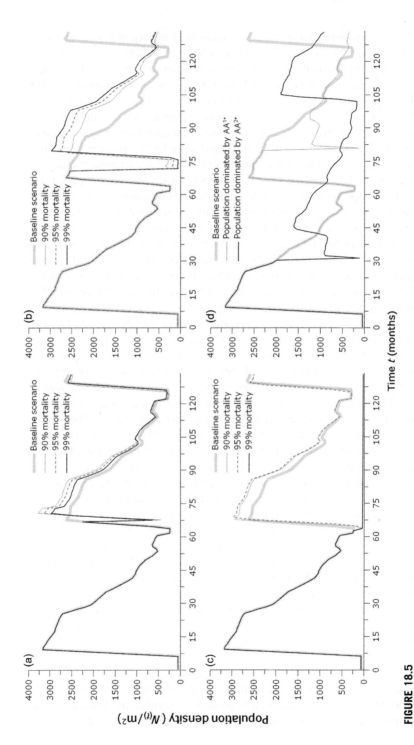

FIGURE 18.5

Simulated mussel density by the model under different population mortality scenarios. Mortality event occurring before (a) and after (b) the start of the spawning season, and (c) during a minimum population density peak. (d) Population mortality event fixed at 95%.

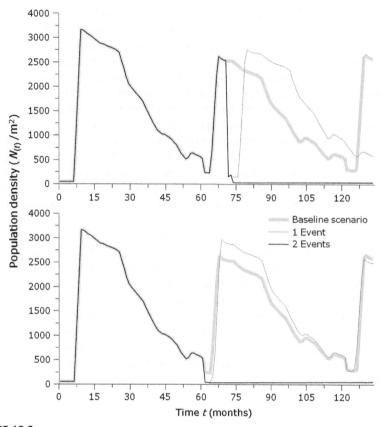

FIGURE 18.6

Simulated mussel density by the model when two consecutive events were simulated, one after the spawning season and the second before the start of the next reproductive season. Population mortality event fixed at 95%.

of zebra mussel population. Our simulations showed that mussel population can be strongly affected by increasing population mortality rate, although mussel population usually quickly recovered from crashes. Consequently, our simulations suggest that water release from the reservoir may be an interesting and simple way to control *D. polymorpha* density within the reservoir. Furthermore, mussel population management may be improved by combining a water-level diminution followed by an increment, reducing suitable habitat and hence reducing population density.

18.5 FINAL REMARKS

The zebra mussel, *D. polymorpha* (Pallas, 1771), is one of the most widely distributed freshwater organisms and considered to be one of the most aggressive invaders worldwide. With well-documented economic and ecological impacts on ecosystem functioning and native fauna, mussel density is proposed to be the key factor determining its invasiveness, but factors responsible for population dynamics are poorly known. By using a demographic model where mussel population growth was regulated by both an adult-density mechanisms and a system carrying capacity (calculated from water reservoir data), we were able to reproduce all mussel population dynamics previously described in the literature. We also obtained coherent simulation-results with literature data by using a K based on total phosphorus, thus suggesting its usefulness in species management. Our model suggests that if the phosphorous reduction observed in the basin continues or remains stable in the near future, an increase in mussel density is not expected. Although there are strong variations in mussel population density across space and time, its depth-distribution is highly limited by oxygen concentration; consequently, the maximum population density is usually observed at the same water depth within a given water body. Based on this research, we propose a simple management option consisting of regulating reservoir water level by releasing water from the dam, thus increasing population mortality rate by exposing mussels to desiccation. Simulations showed larger effects when mortality event occurred outside of the spawning season (i.e., reducing population response capability) and when two water releases were combined, one before the spawning season and a second after the next spawning season. Given the experimental nature of our simulations and their potential usefulness as a management tool, we concluded therefore that specific research is needed to improve the species control plan. Model predictions at a local scale may be helpful in developing adequate control plans by both predicting future species trends and forecasting management effects.

Acknowledgments

Financial support was provided by the Spanish Ministry of Environment (MMA), through the project 072/SGTB/2007/1.1, and the Government of Catalonia (Catalan Water Agency, ACA). C. A. held a postdoctoral contract (DOC-INIA program) granted by the Spanish National Institute for Agricultural and Food Research and Technology (INIA), Spanish Ministry of Economy and Competitiveness. We thank Prof. Sven Erik Jørgensen for his valuable comments on an earlier version of the model. Specials thanks are also due to CHE, Dirección General de Aragón and Grup Natura Freixe staff for data provided.

References

Ackerman, J.D., Sim, B., Nichols, S.J., Claudi, R., 1994. A review of the early life history of zebra mussels (*Dreissena polymorpha*): comparisons with marine bivalves. Can. J. Zool. 72, 1169–1179.

Alcaraz, C., Caiola, Ibáñez, C., Muñoz-Camarillo, G., Reyes, E., 2014. Modeling zebra mussel population dynamics in a reservoir and its application for the management of this invasive species. Submitted manuscript.

Alcaraz, C., Caiola, N., Ibáñez, C., 2011. Bioaccumulation of pollutants in the zebra mussel from hazardous industrial waste and evaluation of spatial distribution using GAMs. Sci. Total Environ. 409, 898–904.

Altaba, C.R., Jiménez, P.J., López, M.A., Rofes, J., Viñado, J., 2001. Localización y evaluación de una nueva invasión biológica: el mejillón cebra (*Dreissena polymorpha*) en El Ebro. Spanish Ministry of Environment.

Arnott, D.L., Vanni, M.J., 1996. Nitrogen and phosphorus recycling by the zebra mussel (*Dreissena polymorpha*) in the western basin of Lake Erie. Can. J. Fish. Aquat. Sci. 53, 646–659.

Bunt, C., MacIsaac, H., Sprules, W.G., 1993. Pumping rates and projected filtering impact of juvenile Zebra Mussels (*Dreissena polymorpha*) in Western Lake Erie. Can. J. Fish. Aquat. Sci. 50, 1017–1022.

Burla, H., Ribi, G., 1998. Density variation of the zebra mussel *Dreissena polymorpha* in Lake Zürich, from 1976 to 1988. Aquat. Sci. 60, 145–156.

Casagrandi, R., Mari, L., Gatto, M., 2007. Modelling the local dynamics of the zebra mussel (*Dreissena polymorpha*). Freshw. Biol. 52, 1223–1238.

Chase, M.E., Bailey, R.C., 1999. The ecology of the zebra mussel (*Dreissena polymorpha*) in the lower Great Lakes of North America: I. Population dynamics and growth. J. Great Lakes Res. 25, 107–121.

Cloern, J.E., Jassby, A.D., 2008. Complex seasonal patterns of primary producers at the land–sea interface. Ecol. Lett. 11, 1294–1303.

Conides, A., Koussouris, T., Gritzalis, K., Bertahas, I., 1995. Zebra mussel, *Dreissena polymorpha*: population dynamics and notes on control strategies in a reservoir in western Greece. Lake Reserv. Manag. 11, 329–336.

Connelly, N.A., O'Neill, C.R., Knuth, B.A., Brown, T.L., 2007. Economic impacts of zebra mussels on drinking water treatment and electric power generation facilities. Environ. Manag. 40, 105–112.

Dillon, P.J., Rigler, F.H., 1974. The phosphorus–chlorophyll relationship in lakes. Limnol. Oceanogr. 19, 767–773.

Durán, C., Lanao, M., Anadón, A., Touyá, V., 2010. Management strategies for the zebra mussel invasion in the Ebro River basin. Aquat. Invasions 5, 309–316.

Durán, C., Lanao, M., Pérez, L., Moreu, C.C., Anadón, A., Touya, V., 2012. Estimación de los costes de la invasión del mejillón cebra en la cuenca del Ebro (periodo 2005–2009). Limnetica 31, 213–230.

Ecohydros, 2008. Aplicación de tecnologías acústicas de y evaluación de poblaciones de peces y mejillón cebra en los embalses de la cuenca del Ebro. Tomo III. Cartografía del Mejillón Cebra. MARM & CHE.

Griebeler, E.M., Seitz, A., 2007. Effects of increasing temperatures on population dynamics of the zebra mussel *Dreissena polymorpha*: implications from an individual-based model. Oecologia 151, 530–543.

Ibáñez, C., Alcaraz, C., Caiola, N., Rovira, A., Trobajo, R., Alonso, M., Durán, C., Jiménez, P.J., Munné, A., Prat, N., 2012. Regime shift from phytoplankton to macrophyte dominance in a large river: top-down versus bottom-up effects. Sci. Total Environ. 416, 314–322.

Jantz, B., Neumann, D., 1998. Growth and reproductive cycle of the zebra mussel in the River Rhine as studied in a river bypass. Oecologia 114, 213–225.

Jones, L.A., Ricciardi, A., 2005. Influence of physicochemical factors on the distribution and biomass of invasive mussels in the St. Lawrence River. Can. J. Fish. Aquat. Sci. 62, 1953–1962.

Karatayev, A.Y., Burlakova, L.E., Padilla, D.K., 1997. The effects of *Dreissena polymorpha* (Pallas) invasion on aquatic communities in Eastern Europe. J. Shellfish Res. 16, 187–203.

Karatayev, A.Y., Burlakova, L.E., Padilla, D.K., 1998. Physical factors that limit the distribution and abundance of *Dreissena polymorpha* (Pall.). J. Shellfish Res. 17, 1219–1235.

Karatayev, A.Y., Burlakova, L.E., Padilla, D.K., Johnson, L.E., 2003. Patterns of spread of the zebra mussel (*Dreissena polymorpha* (Pallas)): the continuing invasion of Belarussian lakes. Biol. Invasions 5, 213–221.

Karatayev, A.Y., Burlakova, L.E., Padilla, D.K., 2005. Contrasting distribution and impacts of two freshwater exotic suspension feeders, *Dreissena polymorpha* and *Corbicula fluminea*. In: The Comparative Roles of Suspension-feeders in Ecosystems. Springer, The Netherlands, pp. 239–262.

Karatayev, A.Y., Burlakova, L.E., Padilla, D.K., 2006. Growth rate and longevity of *Dreissena polymorpha* (Pallas): a review and recommendations for future study. J. Shellfish Res. 25, 23–32.

MacIsaac, H.J., Sprules, W.G., Leach, J.H., 1991. Ingestion of small-bodied zooplankton by zebra mussels (*Dreissena polymorpha*): can cannibalism on larvae influence population dynamics? Can. J. Fish. Aquat. Sci. 48, 2051–2060. http://dx.doi.org/10.1139/f91-244.

MacIsaac, H.J., Lonnee, C.J., Leach, J.H., 1995. Suppression of microzooplankton by zebra mussels: importance of mussel size. Freshw. Biol. 34, 379–387.

Mackie, G.L., Schloesser, D.W., 1996. Comparative biology of zebra mussels in Europe and North America: an overview. Am. Zool. 36, 244–258.

Mellina, E., Rasmussen, J.B., 1994. Patterns in the distribution and abundance of zebra mussel (*Dreissena polymorpha*) in rivers and lakes in relation to substrate and other physicochemical factors. Can. J. Fish. Aquat. Sci. 51, 1024–1036.

Minchin, D., Lucy, F., Sullivan, M., 2005. Ireland: a new frontier for the zebra mussel *Dreissena polymorpha* (Pallas). Oceanol. Hydrobiol. Stud. 34, 19–30.

Navarro, E., Bacardit, M., Caputo, L., Palau, T., Armengol, J., 2006. Limnological characterization and flow patterns of a three-coupled reservoir system and their influence on *Dreissena polymorpha* populations and settlement during the stratification period. Lake Reserv. Manag. 22, 293–302.

Neumann, D., Borcherding, J., Jantz, B., 1993. Growth and seasonal reproduction of *Dreissena polymorpha* in the Rhine River and adjacent waters. In: Nalepa, T.F., Schloesser, D.W. (Eds.), Zebra Mussels: Biology, Impacts, and Control. Lewis Publishers, Ann Arbor, Michigan, pp. 95–109.

Nicholls, K.H., Hopkins, G.J., Standke, S.J., 1999. Reduced chlorophyll to phosphorus ratios in nearshore Great Lakes waters coincide with the establishment of dreissenid mussels. Can. J. Fish. Aquat. Sci. 56, 153–161.

Palau, A., 2009. Basic ecology of zebra mussel and water masses vulnerability. L'Atzavara 18, 61–66.

Pimentel, D., Zuniga, R., Morrison, D., 2005. Update on the environmental and economic costs associated with alien-invasive species in the United States. Ecol. Econ. 52, 273–288.

Rajagopal, S., Pollux, B.J.A., Peters, J.L., Cremers, G., Moon-van der Staay, S.Y., van Alen, T., Eygensteyn, J., van Hoek, A., Palau, A., Bij de Vaate, A., van der Velde, G., 2009. Origin of Spanish invasion by the zebra mussel, *Dreissena polymorpha* (Pallas, 1771) revealed by amplified fragment length polymorphism (AFLP) fingerprinting. Biol. Invasions 11, 2147–2159.

Ram, J.L., Palazzolo, S.M., 2008. Globalization of an aquatic pest: economic costs, ecological outcomes, and positive applications of zebra mussel invasions and expansions. Geogr. Compass 2, 1755–1776.

Ram, J.L., Fong, P.P., Garton, D.W., 1996. Physiological aspects of zebra mussel reproduction: maturation, spawning and fertilization. Am. Zool. 36, 326–338.

Ramcharan, C.W., Padilla, D.K., Dodson, S.I., 1992. Models to predict potential occurrence and density of the zebra mussel, *Dreissena polymorpha*. Can. J. Fish. Aquat. Sci. 49, 2611–2620.

Rasmussen, J.B., 1988. Littoral zoobenthic biomass in lakes, and its relationship to physical, chemical and trophic factors. Can. J. Fish. Aquat. Sci. 45, 1436–1447.

Rasmussen, J.B., Kalff, J., 1987. Empirical models for zoobenthic biomass in lakes. Can. J. Fish. Aquat. Sci. 44, 990–1001.

Ricker, W.E., 1954. Stock and recruitment. J. Fish. Res. Board Can. 11, 559–623.

Rovira, A., Alcaraz, C., Ibáñez, C., 2012. Spatial and temporal dynamics of suspended load at a-cross-section: the Lowermost Ebro River (Catalonia, Spain). Water Res. 46, 3671–3681.

Souza, F.M., Brauko, K.M., Lana, P.C., Muniz, P., Camargo, M.G., 2013. The effect of urban sewage on benthic macrofauna: a multiple spatial scale approach. Mar. Pollut. Bull. 67, 234–240.

Sprung, M., 1991. Costs of reproduction: a study on the metabolic requirements of the gonads and fecundity of the bivalve *Dreissena polymorpha*. Malacologia 33, 63–70.

Strayer, D.L., Malcom, H.M., 2006. Long-term demography of a zebra mussel (*Dreissena polymorpha*) population. Freshw. Biol. 51, 117–130.

Strayer, D.L., Powell, J., Ambrose, P., Smith, L.C., Pace, M.L., Fischer, D.T., 1996. Arrival, spread, and early dynamics of a zebra mussel (*Dreissena polymorpha*) population in the Hudson River estuary. Can. J. Fish. Aquat. Sci. 53, 1143–1149.

Suárez-Serrano, A., Alcaraz, C., Ibáñez, C., Trobajo, R., Barata, C., 2010. *Procambarus clarkii* as a bioindicator of heavy metal pollution sources in the lower Ebro River and Delta. Ecotoxicol. Environ. Saf. 73, 280–286.

van Nes, E.H., Noordhuis, R., Lammens, E.H., Portielje, R., Reeze, B., Peeters, E.T., 2008. Modelling the effects of diving ducks on zebra mussels *Dreissena polymorpha* in lakes. Ecol. Model. 211, 481–490.

Wilson, A.E., Sarnelle, O., 2002. Relationship between zebra mussel biomass and total phosphorus in European and North American lakes. Arch. Hydrobiol. 153, 339–351.

SubWet 2.0. Modeling the Performance of Treatment Wetlands

19

Annie Chouinard[a],*, **Gordon C. Balch[b]**, **Brent C. Wootton[b]**, **Sven Erik Jørgensen[c]**,
Bruce C. Anderson[a],*

[a]*Department of Civil Engineering, Queen's University, Kingston, Ontario, Canada*
[b]*Centre for Alternative Wastewater Treatment, Fleming College, Lindsay, Ontario, Canada*
[c]*Professor Emeritus, Environmental Chemistry, University of Copenhagen, Copenhagen, Denmark*
**Corresponding authors: e-mail address: a.chouinard@queensu.ca; anderson@civil.queensu.ca*

19.1 INTRODUCTION

Treatment wetlands are either natural or constructed systems managed in a specific manner for the treatment of wastewaters. Although traditionally applied for the treatment of domestic and municipal sewage from both separate and combined sewerage, treatment wetlands have been applied globally since the late 1980s to treat various types of wastewaters, including agricultural wastewaters (cattle, swine, poultry, dairy), mine drainage, food processing wastewaters (winery, abattoir, fish, potato, vegetable, meat, cheese, milk, sugar production), heavy industry wastewaters (polymers, fertilizers, chemicals, oil refineries, pulp and paper mills), landfill leachate and runoff waters (urban, highway, field, airport, nursery, greenhouse) (Hammer, 1989; Vymazal, 1998; Mander and Jenssen, 2003; Kadlec and Wallace, 2008; Babatunde et al., 2010). Among the treatment wetlands, horizontal subsurface flow constructed wetlands are a widely applied design. Treatment is achieved by a variety of physical, chemical, and biological processes, such as sedimentation, filtration, precipitation, sorption, plant uptake, microbial decomposition, and nitrogen transformations (Wetzel, 2000; Kadlec and Wallace, 2008).

The increasing application of treatment wetlands for wastewater treatment together with increasingly strict water quality standards is an ever-growing motive for the development of numerical models to be used as predictive process design tools. The main objective of the modeling effort is to increase the predictive insight into the functioning of complex treatment wetlands by using process- or mechanistic-based models that describe in details transformation and degradation processes (Langergraber et al., 2009). Once reliable numerical models are developed and validated against experimental data, they can be used for improving and evaluating existing design criteria. To date only a few numerical models are available and able to describe treatment processes in horizontal subsurface flow treatment wetlands. Most of the literature on models refers to simple first-order decay models (e.g., Rousseau et al., 2004; Stein et al., 2006) or describes the treatment wetland as a black box (e.g., Pastor et al., 2003; Tomenko et al., 2007), acknowledging only a limited understanding of the studied facility. The number of mechanistic or process-based models is very limited. This chapter reviews the SubWet 2.0 model, a horizontal subsurface flow modeling program designed to predict the level of treatment that can be expected based on the characteristics of several parameters known to influence treatment (e.g., wetland size, loading rates). The SubWet model is based on 16 rate constants specific to a variety of processes involved in the treatment of BOD, ammonia, nitrate, Org-N, and total phosphorus, and uses an integrated approach and performance-based data to calibrate the model to site conditions.

19.2 MODEL STRUCTURE
19.2.1 General considerations

SubWet, a software program used in the design of subsurface horizontal flow artificial wetlands for water quality improvement and treatment, was originally developed by UNEP–DTIE–IETC. After being successfully used as a design tool in 15 cases in Tanzania, it was felt that the model should be upgraded for cold climate application. The Centre for Alternative Wastewater Treatment of Fleming College further developed a new version in collaboration with UNEP–DTIE–IETC, creating SubWet 2.0 to accommodate temperate and cold climatic conditions including summer arctic and temperate winter conditions.

Cold climate wetlands are defined as those where the surface temperature range varies from well below freezing in winter months to temperatures above 20 °C during the summer (applicable to temperate and arctic climates). This modification was accomplished by calibrating the model with data collected from natural tundra wetlands currently in use for the treatment of municipal effluents within the Kivalliq region of Nunavut, Canada. The application of this software to natural tundra wetlands is beyond the original purpose it was designed for. However, the calibration of this model with arctic data has demonstrated its ability to model treatment performance within natural tundra wetlands and thus provide an additional predictive tool to aid northern stakeholders in the treatment of municipal effluents.

The SubWet model can consider the influence of several factors at one time, while empirical equations are generally not able to consider more than two factors at one time and usually in isolation of the other influential parameters. The model has causality, which means that we know the process behind the model and can therefore put these processes into an equation. SubWet can be used to allow managers to predict the impact to treatment efficiency based on an alteration to the aerial loading rates, hydraulic retention time (HRT), and the desired level of effluent treatment. The model can also be used as a predictive tool to help managers determine the size of wetland needed to meet treatment objectives. This will assist managers in determining if the current wetland size can accommodate projected growth in population and anticipated effluent volumes. The model can be used to predict treatment performance anticipated from alterations to the size of the treatment area that could be increased through the construction of infiltration/dispersion ditches and structures that divert flow to other parts of the wetland that are not currently involved in treatment of the effluent. Ultimately, SubWet can be used by resource managers to demonstrate the treatment benefit acquired from the use of designated treatment wetlands. It can also be used as a predictive tool to forecast the potential these areas could provide from the application of selected management operations. This will help resource managers in cost–benefit analysis when planning for future needs.

19.2.2 **Model parameter ranges and default values**

SubWet 2.0 has been developed to model both cold and warm climate wetlands. This has been accomplished by determining the most appropriate rate coefficient values for operation as either a cold climate or a warm climate model. As such, SubWet has been programmed with a specific set of cold climate default coefficient values and a specific set of warm climate default coefficient values. The choice to operate Sub-Wet in a cold climate mode or a warm climate mode will dictate which of the two default sets (cold or warm) will be used. Modification of specific coefficient values within either the cold or warm set can be undertaken to calibrate SubWet to an individual wetland by comparing the simulated treatment values (i.e., modeled water quality parameters) to the measured values for that particular wetland. Slight modifications to specific coefficient values will often improve the simulation by making the simulated output values closer to the measured values. Table 19.1 summarizes the cold climate default coefficient values while Table 19.2 lists the warm climate default coefficient values.

19.3 **MODEL CALIBRATION**

The default coefficient parameters developed for operation of SubWet in either a cold or warm climate mode are generally a good starting point for operation of the SubWet program. It should be understood that the simulated results will in most cases vary from observed (measured) results. The reasons for this are many, but most often related to the inability of mathematical formula to model complex environmental processes and, for the sake of simplicity, most of these models rely heavily on relatively few parameters; thus many influential parameters are either not measured or not known. It is generally accepted that the standard deviation around sampling and analytical procedures is typically between 10% and 12% and thus the standard deviation to be expected for comparisons between measured values and model simulated values can generally be expected to be in the range of 15–20% (e.g., $(12^2 + 12^2)^{0.5} = 17\%$). With this in mind, simulated results that are within 80% of the measured values are generally considered to be reasonable approximations. However, if greater agreement is desired or if the agreement is less than with 80%, then

Table 19.1 The Default Coefficient Values for the Operation of SubWet in a Cold Climate Mode

Abbreviations (See Glossary Where Full Name Is Provided)	Coefficient Range	Default Coefficient Values
AC	0.05–2.0	0.9 (1/24 h)
NC	0.1–2.5	0.9 (1/24 h)
OC	0.05–2.0	0.25 (1/24 h)
DC	0.00–5	3.5 (1/24 h)
TA	1.02–1.06	1.05 (no unit)
TN	1.02–1.09	1.07 (no unit)
TO	1.02–1.06	1.04 (no unit)
TD	1.05–1.12	1.07 (no unit)
KO	0.01–2	0.01 (mg/L)
OO	0.01–2	0.05 (mg/L)
MA	0.05–2	0.1 (mg/L)
MN	0.01–1	0.1 (mg/L)
PA	0.00–1	0.01 (1/24 h)
PN	0.00–1	0.001 (1/24 h)
PP	0.00–1	0.001 (1/24 h)
AF	0–100	0.36

SubWet can be calibrated to the conditions of a specific wetland by modification of the coefficient parameters embedded within SubWet. The cold climate default coefficient parameter values within SubWet are based on five natural wetland data sets from Nunavut within the Canadian Arctic, while the warm climate parameter values are based on the average of nine constructed wetlands from the United Republic of Tanzania, eastern Africa.

SubWet, as is the case with virtually all other wetland models, was developed for constructed wetlands that have engineered boundaries, known volumes, defined media, and regulated flow rates. Natural wetlands are completely different in many ways. For example, the boundaries of the natural wetlands are not only irregular in shape, but often unknown in terms of which part of the wetland is actually involved in the treatment (i.e., what portion is active). Second, primary features such as media depth, porosity, and hydraulic conductivity can be quite variable and generally unknown. There is also the aspect of changing preferential flow paths, the infiltration of unknown volumes of ground, and surface waters from the surrounding landscape that all affect HRTs, dilution, and flow rates. The use of measured performance data to calibrate the SubWet model integrates, in a limited manner, some of the unknown processes influencing treatment performance, thus lessening the need to know specific details concerning individual influencing processes. Therefore, the burden to know precise details regarding such factors as soil depth and the influence of melt waters becomes less demanding since the model is comparing a simulated integrated

Table 19.2 The Default Coefficient Values for the Operation of SubWet in a Warm Climate Mode (Jørgensen and Gromiec, 2011)

Abbreviations (See Glossary Where Full Name Is Provided)	Coefficient Range	Default Coefficient Values
AC	0.05–2.0	0.5 (1/24 h)
NC	0.1–2.5	0.8 (1/24 h)
OC	0.05–2.0	0.5 (1/24 h)
DC	0.00–5	2.2 (1/24 h)
TA	1.02–1.06	1.04 (no unit)
TN	1.02–1.09	1.047 (no unit)
TO	1.02–1.06	1.04 (no unit)
TD	1.05–1.12	1.09 (no unit)
KO	0.01–2	1.3 (mg/L)
OO	0.01–2	1.3 (mg/L)
MA	0.05–2	1 (mg/L)
MN	0.01–1	0.1 (mg/L)
PA	0.00–1	0.01 (1/24 h)
PN	0.00–1	0.01 (1/24 h)
PP	0.00–1	0.003 (1/24 h)
AF	0–100	1

treatment response to a measured integrated treatment response. Calibration of the model to measure data is accomplished by modifying the coefficient parameter values known to be operative in the water quality parameter being examined. Thus some knowledge regarding the treatment process for each water quality parameter is required in order to correctly calibrate SubWet. It is, however, recognized that factors such as inter-year variability in climate, loading rates, and the composition of the waste stream can still introduce variability in the predicted year-to-year values. However, it is anticipated that increased monitoring and the generation of additional measured data will better identify the level of year-to-year variability and overall improve the predictive capability of the model and identify the level of uncertainty to be expected.

The default parameter values for the cold climate mode and the warm climate mode represent average or typical parameter values for operation of SubWet under cold or warm climate conditions. These cold climate and warm climate default parameters are an initial good start choice when first attempting to simulate the treatment of municipal effluents. However, each wetland is unique, and the user may be able to find a better agreement between observed (measured) values and simulated values after model calibration. This implies that some initial testing of the effluent exiting the wetland needs to be undertaken so that observed values are available for the user to assess how closely the simulated values are matching the observed values.

19.3.1 Field trials from a cold climate environment (arctic Canada, natural tundra wetland examples)

The two data sets below are used to illustrate how the SubWet model can be calibrated to natural tundra wetlands. The first data set was generated from the natural tundra wetland located near the hamlet of Chesterfield Inlet situated in the Kivalliq region (western Hudson Bay) of Nunavut, Canada. The second data set originates from the natural tundra wetland near the hamlet of Baker Lake, also in the Kivalliq region of Nunavut, Canada.

19.3.1.1 Chesterfield Inlet data set

The following example with the Chesterfield Inlet data set illustrates how SubWet 2.0 can be calibrated to this tundra wetland. Table 19.3 compares the values simulated by SubWet to the values observed in the field. An examination of these values reveals that for most parameters there is a close approximation of the simulated to the observed results. In this example, the greatest discrepancy between simulated and observed is for the parameters BOD_5 and ammonium. The values in Table 19.3 indicate that SubWet overestimated the decomposition of organic matter (e.g., removal of BOD_5) and overestimated the rate of nitrification (e.g., conversion of ammonium to nitrate). This suggests that the coefficients for the decomposition of organic matter and nitrification are too high and should be lowered.

The magnitude by which these coefficients are altered will most likely be approached by those just beginning to use SubWet in a trial-and-error manner where one coefficient at a time is altered, the simulation is rerun, and the graphical expression of the simulated to observed values are reexamined. SubWet performs all simulations rapidly, thus allowing the operator the ability to quickly investigate several scenarios in a short period of time in order to identify the modifications to the coefficients that provide the greatest correlation between simulated and observed results. It has been determined that for this data set a change in the decomposition rate of organic matter (OC) from 0.25 to 0.235, a change in the nitrification rate (NC) from 0.9 to 0.42, and a change in the half-saturation constant for nitrification (KO) from 0.01 to 1.4 produces simulation values for BOD_5 and ammonium-N that are much closer to the observed values. Table 19.4 shows the simulation results after

Table 19.3 Comparison of Simulated and Observed Values for the Chesterfield Inlet Data Set

	Unit	Simulation Results	Observed Values
BOD_5	mg O_2/L	9.4	10.5
Ammonium-N	mg N/L	0.19	1.1
Nitrate-N	mg N/L	0.02	0.01
Total N	mg N/L	0.22	1.1
Phosphorous	mg P/L	0.42	0.4

Table 19.4 Simulation Results After Calibration for the Chesterfield Inlet Data Set

	Unit	Simulation Results	Observed Values
BOD_5	mg O_2/L	10.4	10.5
Ammonium-N	mg N/L	1.1	1.1
Nitrate-N	mg N/L	0.06	0.01
Total N	mg N/L	1.2	1.1
Phosphorous	mg P/L	0.42	0.4

Table 19.5 Initial and Modified Design Values for Chesterfield Inlet

	Initial Design Values	Modified Design Values
Width (m)	70	70
Length (m)	720	360
Depth (m)	0.3	0.3
Area (m^2)	50,000	25,000
Volume (m^3)	15,000	7,500
Selected flow (m^3/24 h)	36	36
RTB	22.9	11.45

calibration. The simulated ammonium-N value is now closer to the observed value. The nitrate concentration has increased, but it is still minor (< 0.1 mg N/L). Nevertheless, the default values gave acceptable results that could be used for wetland design and predictive management needs; however, as shown, calibration can improve model's ability to produce simulated values that are very close to the observed values.

Overall the effluent treatment by the Chesterfield Inlet wetland is good, primarily since the wetland is relatively large. The SubWet model can, however, be used to predict wetland performance should land use factors alter the original size of the wetland such as the construction of a roadway through the wetland that impedes natural flow paths. In this type of scenario, SubWet could be used to determine the treatment performance should the active treatment portion of the wetland shrink in size. The following example is designed to demonstrate how SubWet could be used to predict treatment performance when the size of the Chesterfield Inlet wetland is reduced by half. Table 19.5 summarizes the initial and modified design values for this test. Notice that the length of the wetland has been reduced from 720 to 360 m, which in effect reduces the wetland and effluent holding capacity by 50% to 7,500 m^3 and the effluent volume to 2,100 m^3. Notice that the hydraulic retention time (RTB) has been reduced from 23 to 11.5 days.

With these changes, SubWet predicts that BOD_5 will rise to >30 mg/L, while the ammonium-N, nitrate-N, phosphorus, and organic nitrogen treatment remain acceptable. The phosphorus concentration remains below 1 mg P/L and the nitrate-N is also very low due to the effective denitrification. Further simulations of SubWet predict that if the wetland size was reduced by 25% instead of the above example of 50%, the BOD_5 concentrations of the treated effluent exiting the wetland would be predicted to be approximately 15 mg/L while the corresponding treatment efficiency of ammonium, nitrate, phosphorus, and organic nitrogen would remain high.

19.3.1.2 Baker Lake data set

The data set from Baker Lake is interesting in that the BOD_5, ammonium, and organic nitrogen concentrations of the effluent entering the wetland are much higher than normally encountered in municipal wastewater effluents. In addition, the physical size of the Baker Lake wetland is typically smaller than many other wetlands in Nunavut, Canada. Because of this, the default coefficient parameters for the cold climate operation of the SubWet model do not provide an adequate prediction for most of the simulated wastewater parameters. The initial simulations were performed with the default coefficient parameters set for the operation of SubWet in the cold climate mode. As shown below, the simulated BOD_5 and total phosphorus values are relatively close to the observed values for these two parameters. However, the values for nitrate, ammonium, and organic nitrogen are not acceptable but can be improved when SubWet is calibrated for this specific site. Table 19.6 summarizes the differences between the observed and simulated results for BOD_5, ammonium, nitrate, organic nitrogen, and total phosphorus, along with the rates of ammonification (expressed as mg organic nitrogen converted to ammonium), nitrification (expressed as mg ammonium converted to nitrate), and denitrification (expressed as mg nitrate converted to dinitrogen gas). The values represented in Table 19.6 correspond to values prior to calibration of the program.

Although the simulated results for BOD_5 and total phosphorus are relatively close to the actual observed concentration measured in the effluent exiting the wetland, the nitrogen compounds show less agreement between simulated and observed results, suggesting that SubWet requires calibration for these compounds. For example, the

Table 19.6 Comparison of Simulated and Observed Concentrations and Rates for the Baker Lake Data Set

	Unit	Simulation Results	Observed Values	Deviation %
BOD_5	mg O_2/L	281	247	15
Ammonium-N	mg N/L	84.3	61.9	36
Nitrate-N	mg N/L	4.5	0.52	–
Organic-N	mg N/L	5.5	0.0	–
Phosphorus	mg P/L	8.3	9.4	13
Ammonification	mg N/L	51.9	57.4	10
Nitrification	mg N/L	49.8	76.4	36
Denitrification	mg N/L	44	75.5	42

simulated value for nitrate is approximately 4.5 mg/L, yet the observed value is 0.52 mg/L. This large difference is unacceptable and is likely caused by an underestimation of the denitrification rate or an overestimation of the nitrification rate. The simulated nitrate concentration is a product of both denitrification (conversion to nitrogen gas), which removes nitrate from the effluent stream, and nitrification (conversion of ammonium to nitrate), which produces nitrate. Likewise, the model results for ammonium once again show a large discrepancy between the simulated ammonium concentrations ($\sim$84 mg/L) in comparison to the observed value, which is closer to 62 mg/L. The initial concentration of organic nitrogen within the effluent entering the wetland was approximately 57 mg/L, and although the difference between the simulated (5.5 mg/L) and observed results (0 mg/L) of the effluent exiting the wetland is closer than found for nitrate and ammonium, calibration of SubWet should improve the predictability for organic nitrogen.

A comparison of the observed (measured) concentrations of the nitrogenous compounds within the effluent (pre- and posttreatment) to SubWet simulated results can provide insight into which processes (e.g., ammonification, nitrification, and denitrification) require calibration within the model. Once identified, the cold climate default coefficient parameters can be modified and the simulation rerun until the best calibration is achieved. Table 19.7 summarizes the concentrations of organic nitrogen, ammonium, and nitrate within the effluent both entering the wetland (pretreatment) and exiting the wetland (posttreatment) and identifies the net change in these compounds (e.g., mg/L ammonium loss during treatment). A review of this data indicates that the simulation is underestimating the rate of nitrification, since the predicted concentration of ammonium increases (e.g., 81–84 mg/L) when in fact the observed values indicate it actually declines (e.g., 81–62 mg/L). Likewise, the rate of denitrification is also too low. The observed loss of nitrate is low, changing only 0.03 mg/L in concentration; however, the SubWet simulation predicted a dramatic increase from 0.55 to 4.5 mg/L. The rate of ammonification predicted by SubWet appears to be more appropriate, and simulated results are approximately 10% of the observed results and with the range of acceptability. However, the observed results indicate a complete removal of organic nitrogen from the effluent stream, while the model predicts a lower removal rate that results in a final effluent concentration of approximately 5.5 mg/L. Thus the rate of ammonification could also be improved slightly through calibration, although it is not completely necessary. A review of the overall loss of all nitrogenous compounds (e.g., denitrification) determined through observation (measured) indicates that the concentration was reduced by approximately 75.5 mg/L, whereas SubWet predicts only a concentration decrease of 44 mg/L, a significant underestimation of the overall capacity of the wetland's efficiency. In summary, the data suggest that the greatest calibration gains can be made by altering the coefficient parameters associated with both nitrification and denitrification, and to a lesser extent ammonification, in a manner that increases the rate of conversion. However, one should understand the interplay between nitrification and the decomposition of organic matter (e.g., BOD_5) remembering that both these processes are competing for oxygen and thus increasing the rate of nitrification may

Table 19.7 Comparison of Observed (Measured) and Simulated Removal Rates for Nitrogen Compounds Within Baker Lake Effluent After Wetland Treatment Prior to Calibration

	Ammonification		Nitrification		Denitrification	
	Org-N	→	Ammonium	→	NO_2^-/NO_3^-	→	N_2 (gas)
Obs Δ (mg/L)	57.4 to 0		81 to 62		0.55 to 0.52		
Obs mass Δ (mg/L)	(+57.4)		(+19)		(+0.03)		
Total Obs loss from ammonification (mg/L)		57.4					
Total Obs loss from nitrification (mg/L)				(57.4 + 19) = 76.4			
Total Obs loss from denitrification (mg/L)						(57 + 19 + 0.03) = 75.5	
Sim Δ (mg/L)	57.4 to 5.5		81 to 84		0.55 to 4.5		
Obs Sim mass Δ (mg/L)	(+51.9)		(−3)		(−3.95)		
Total Sim loss from ammonification (mg/L)		51.9					
Total Sim loss from nitrification (mg/L)				(51.9 − 3) = 48.9			
Total Sim loss from denitrification (mg/L)						(51.9 − 3 − 3.95) = 44	

This comparison is being made to determine which processes within the SubWet 2.0 could be improved through calibration to this site.

limit the availability of oxygen for the decomposition of organic matter and result in a higher predicted BOD_5 concentration.

The best way to approach the required modifications for the Bake Lake example is to start by modifying the coefficient parameters associated with denitrification, and once completed, then to increase the coefficient rate associated with nitrification and lastly with ammonification. The rate of denitrification can be increased by (i) increasing the value for the "denitrification rate" (DC) parameter, (ii) reducing the "half-saturation constant for denitrification" (MN), and (iii) reducing the "temperature coefficient of denitrification" (DC) to better reflect the wetland temperature of 8.4 °C. [Note: The temperature coefficient influences only the rate of the reaction, and thus reducing this coefficient makes the model less sensitive to temperature changes.] The rate of nitrification can be increased by increasing the value for the "nitrification rate" (NC) parameter. The rate of ammonification can be increased by increasing the "decomposition rate of organic nitrogen" (AC) parameter. It should be noted that increasing nitrification will mean that more oxygen is consumed in this process, resulting in less oxygen available for the decomposition of organic matter, which will eventually be expressed as higher BOD_5 values. This would not be desired because the simulated BOD_5 values are already close to the observed values and any increasing of the BOD may mean that the simulated values become unacceptably high. This effect can be partially overcome by slightly increasing the value of the coefficient parameter governing the "decomposition rate of organic matter" (OC).

A series of simulations were run with modified coefficient parameter values using a trial-and-error approach to determine which modifications provided the best overall simulations. The trial-and-error approach determined that the parameter modifications shown in Table 19.8 provided the best results. Table 19.8 lists the parameters changed and the final values chosen.

Table 19.8 Coefficient Parameters for the Calibration of SubWet 2.0 to the Baker Lake Wetland Data Set

Rate Coefficient Parameters	Abbreviations	Initial Values	Final Chosen Values	Units
Max. nitrification rate	DC	3.5	5.0	1/24 h
Half-saturation constant for denitrification	MN	0.1	0.01	mg N/L
Temperature coefficient of denitrification	TD	1.07	1.05	Unitless
Max. nitrification rate	NC	0.9	1.7	Unitless
Max. decomposition rate of organic nitrogen	AC	0.9	1.2	Unitless
Max. decomposition rate of organic matter	OC	0.2	0.22	1/24 h

The calibration efforts significantly improved the nitrification of ammonium. The simulation of nitrate did improve (e.g., down from a simulated concentration of 5.5–3.5 mg/L) but did not reach the observed value of 0.52 mg/L. However, Table 19.9 shows that the overall denitrification rate improve significantly. The data shown in Table 19.9 indicate that the overall observed loss of nitrogen from the waste stream was 75.5 mg/L and the simulated loss calculated by SubWet after calibration was 70.9 mg/L, a significant improvement from the 44 mg/L (see Table 19.7) predicted by SubWet prior to calibration.

It is generally advisable to assess the success of the calibration effort by monitoring the rates of denitrification, nitrification, and ammonification rather than the concentrations of individual wastewater parameters. A comparison of the rate values between observed and simulated results is a more robust way to assess the success of the calibrations. For example, in the Baker Lake data set, the final calibration values chosen indicated that the overall removal of nitrogenous compounds from the waste stream was between 4% and 6% of the observed values (see Table 19.10). This is well within the acceptable limit for models, despite the still relatively high dissimilarity between the simulated and observed values for nitrate. Although the difference for nitrate does appear unacceptably great, the overall proportional contribution of nitrate is small in comparison to the improved removal of ammonium, which was a larger component of the overall loss of the total nitrogenous compounds. It should be noted that these efforts did lower the simulated value for BOD_5. However, even with the lowering of this value, the simulated BOD_5 result was still within 7% of the observed value and thus an acceptable estimate.

The Baker Lake data set provides an example of one of the more challenging calibration exercises. However, despite the unusually high strength of the waste stream, the calibration of SubWet demonstrated that this model can provide a reasonable approximation of treatment efficiencies.

Overall, the treatment efficiency of the Baker Lake wetland was poor, in part due to the relatively small size of the wetland, but also as a result of the unusually high-strength effluent that is being treated. SubWet can be used to predict the treatment efficiency if the Baker Lake wetland were increased in size by a factor of four. The simulated output indicates that the BOD_5 concentration would have been reduced from 250 mg/L to about 75 mg/L, a sharp increase in efficiency but still at a BOD_5 concentration that is too great for most regulatory regimes. In reality however, the Baker Lake wetland is only one small portion of the Baker Lake complex through which the municipal effluent flows. This complex contains other ponds and wetland systems that ensure the overall treatment of the effluent at the point of discharge is good.

19.3.2 Field trial from a warm climate environment (Iringa, Tanzania—a constructed wetland example)

Iringa is a large village situated 350 km southwest of Dar es Salaam, Tanzania. The warm climate wetland located in Iringa is a constructed wetland and not a natural tundra wetland as were the two previous examples from Chesterfield Inlet and Baker

Table 19.9 A Comparison of Observed (Measured) and Simulated Removal Rates for Nitrogen Compounds Within Baker Lake Effluent After Wetland Treatment

	Ammonification		**Nitrification**		**Denitrification**	
	Org-N	→	Ammonium	→	NO_2^-/NO_3^-	→	N_2 (gas)
Obs Δ (mg/L)	57.4 to 0		81 to 62		0.55 to 0.52		
Obs mass Δ (mg/L)	(+57.4)		(+19)		(+0.03)		
Total Obs loss from ammonification (mg/L)		57.4					
Total Obs loss from nitrification (mg/L)				$(57.4+19)=76.4$			
Total Obs loss from denitrification (mg/L)						$(57+19+0.03)=75.5$	
Sim Δ (mg/L)	57.4 to 3.1		81 to 61.5		0.55 to 3.45		
Obs Sim mass Δ (mg/L)	(+54.3)		(+19.5)		(−2.9)		
Total Sim loss from ammonification (mg/L)		54.3					
Total Sim loss from nitrification (mg/L)				$(54.3+19.5)=73.8$			
Total Sim loss from denitrification (mg/L)						$(54.3+19.5)-2.9=70.9$	

Table 19.10 Comparison of Simulated and Observed Concentrations and Rates

	Unit	Simulation Results	Observed Values	Deviation %
BOD_5	mg O_2/L	230	247	7
Ammonium-N	mg N/L	61.5	61.9	0.6
Nitrate-N	mg N/L	3.5	0.52	–
Organic-N	mg N/L	3.1	0.0	–
Phosphorous	mg P/L	8.3	9.4	13
Ammonification	mg N/L	54.3	57.4	6
Nitrification	mg N/L	73.8	76.4	4
Denitrification	mg N/L	70.9	75.5	6

Table 19.11 Simulated Results Are Compared with Observed Values for Treated Wastewater for the Tanzania Data Set

	Unit	Simulation Results	Observed Values
BOD_5	mg O_2/L	26.7	26.0
Nitrate-N	mg N/L	0.45	0.5
Ammonium-N	mg N/L	9.2	9.1
Phosphorous-P	mg P/L	5.2	5.0
Organic-N	mg N/L	0.45	0.5

Lake. Thus the characteristics of this wetland have been engineered and as such the volume, bed type, bed depth, flow rates, and many other parameters are well known and in many cases under greater control (e.g., flow rates). This example is being provided to demonstrate how SubWet performs within a constructed wetland and within a warm climate. It also provides an insight into the flexibility of the SubWet program by demonstrating how a temporary change in influent strength influences the overall performance and how quickly the treatment processes within the wetland can be expected to respond.

Table 19.11 compares the SubWet simulated values to the observed values measured at the wetland site. In this example, it should be noted that there is better agreement for this wetland between the SubWet simulated values and the observed values than for either of the previous natural wetland examples prior to model calibration. This should be expected since the Iringa wetland is a constructed wetland where parameters are better known and less variable than found in natural tundra wetlands. Within this table it can be seen that the simulated nitrate-N and organic-N are within 10% of the measured values, while these difference are even less for BOD_5, ammonium-N, and phosphorus at approximately 3%, 1%, and 4%, respectively. Therefore, good agreement was achieved with the model's warm climate default values with no need for calibration of SubWet to this specific wetland.

The BOD$_5$ for the treated water in this example was found to be 26 mg/L, but 15 mg/L is often the required standard. SubWet can be used to determine to what extent the wetland would need to be increased in order to achieve a BOD$_5$ of 15 mg/L or lower. With some trial and error it was found that increasing the volume of the wetland to 1000 m^3 (e.g., 40 m long * 25 m wide * 1 m deep) increases the HRT from 1.25 days to 2 days and reduces the BOD$_5$ of the treated wastewater from an original 26 to 13.7 mg/L. Likewise nitrate-N is reduced from 0.45 to 0.26 mg/L; ammonium-N from 9.2 to 5.8 mg/L; total phosphorus from 5.2 to 4.4 mg/L; and organic nitrogen from 0.45 to 0.13 mg/L. Larger treatment wetlands generally result in:

- Greater nitrification resulting in reduced ammonium-N.
- Greater denitrification can occur even when more nitrate is formed from the nitrification of ammonium. More phosphorus is removed due to longer retention times that facilitate greater adsorption and more uptake by plants.
- Organic nitrogen generally decreases in response to the solubilization of organic matter.

SubWet can also be used to predict changes in treatment performance should the organic loading to the wetland be temporarily altered. For example, if the loading to a wetland is temporarily increased, SubWet can be used to model how treatment is impacted in terms of efficiency and duration. In a simulation (not shown here) the volume and strength of the effluent entering the Iringa wetland was temporarily increased for a 5-day period and then monitored to determine how the treatment efficiency was impacted and for how long. During this 5-day period, the flow rate was increased from 50 m^3/24 h to 60 m^3/24 h, the BOD$_5$ concentration was increased from 123 to 200 mg/L, and ammonium-N was increased from 12 to 20 mg/L. Using these values, SubWet predicted that the HRT would be reduced from 1.25 to 1.04 days. SubWet also predicted that the removal of BOD$_5$ would decrease and the concentration exiting the wetland would increase from 26 to 75 mg/L. However, the normal treatment efficiency of the wetland would be restored back to the 26 mg/L level within 1–2 days after the overloading had ceased. In a similar fashion, the overloading caused ammonium-N concentrations within the treated effluent to rise from 9.2 to 14.5 mg/L; however, once again, these peak levels dropped back down to the normal treatment level within 1–2 days after overloading. SubWet can therefore be used by managers in a predictive manner to *a priori* determine the outcome of operational procedures such as increasing the rate at which the effluent from a sewage lagoon is decanted to a treatment wetland.

19.4 ADVANTAGES OF SUBWET IN COMPARISON TO OTHER PREDICTIVE TOOLS

The options available for modeling the performance of natural tundra wetlands are limited. This is primarily due to the fact that these wetlands are not engineered, and because of this, much less is known regarding media depth, flow rates, the influence of preferential flow paths and infiltration of surface or ground waters, and many other characteristics required for model input. The options available beside SubWet 2.0 include relatively simple design models like heuristics and regression equations along with first-order kinetic models or sophisticated two-dimensional (2D) or three-dimensional (3D) models.

Heuristics are considered the fastest but also the least precise design tool. Since they are based on observations from a wide range of systems, wastewater types, and climatic conditions, they inherently contain a large degree of uncertainty and are therefore better be used after more extensive calculations and designs have been made as a rapid check of anticipated results (Rousseau et al., 2004). Moreover, most of the heuristic approaches have been generated for warmer climates and based on constructed wetlands, not natural wetlands.

The first-order k-C^* models are based on areal rate constants (k), flow rates, and wastewater concentrations entering the wetland. Once they were considered being the state-of-the-art in the modeling of treatment wetlands. They consist of first-order equations that under the influence of ideal plug-flow behavior and constant conditions (e.g., influent, flow, and concentrations) predict an exponential profile between inlet and outlet (Kadlec and Wallace, 2008). The parameters k, C^*, and θ group a large number of other characteristics representing a complex matrix of interactions in a treatment wetland as well as external influences like weather conditions. Therefore, there can be high variability in reported values for k_A, k_V, C^*, and θ (Rousseau et al., 2004). Many, if not most, of the areal rate constants used for these models have been developed in more southern locations under warmer climatic conditions and with data generated from constructed wetlands, not natural wetlands similar to those found in the Arctic.

Sophisticated 2D and 3D models such as HYDRUS require model input parameters that are often unknown for natural wetlands, and thus their overall precision suffers greatly from a reliance on default parameters that most likely do not adequately reflect tundra wetland conditions.

The majority of the investigations on treatment wetlands have mainly been focused on input–output data rather than on data reflective of internal processes. Despite this fact, regression equations seem to be a useful tool in interpreting and applying these input–output data (Rousseau et al., 2004). However, these black-box models cluster complex system processes of the treatment wetland into only two or three parameters, which clearly is an oversimplification. Influential factors such as climate and bed design (length, width, depth) are neglected, which in turn leads to a wide variety of regression equations and a large degree of uncertainty in the design and predicted treatment performance. Furthermore, most of these regression equations rely on wastewater concentrations with only a limited number of these regression equations that rely on both hydraulic loading rates and influent concentration in the forecast of effluent concentration. As a result, in order to predict the maximum allowable hydraulic loading rates based on a given influent concentration and a desired effluent standard, only those regression equations that combine both hydraulic loading rates and influent concentrations can be used, which effectively limits the number of regression equations currently available for use (Rousseau et al., 2004).

SubWet 2.0 provides what is believed to be a good compromise between first-order kinetic models and the more sophisticated 2D and 3D models. SubWet utilizes sixteen rate constants in an integrated manner to predict the treatment of BOD, organic nitrogen, ammonium, nitrate, and total phosphorus. SubWet also provides the user the ability to calibrate these rate constants to site conditions in order to better reflect actual measured values. The calibration method to some extent accommodates for some of the influential processes that could be occurring within the wetland for which input data does not exist. For example, in most northern tundra wetlands, the area involved in the actual treatment process is likely smaller than the physical borders of the wetland. Modification (calibration) of specific rate constants within SubWet can therefore be used to ensure that simulated results closely match measured results. Obviously, variability between seasons and years may require more frequent calibrations. It is anticipated that the accuracy of the SubWet predictions will only increase as the data set for the wetland increases, thus providing greater insight into seasonal and yearly variability.

19.5 SUMMARY AND CONCLUSIONS

Calibration of this model with arctic data has clearly demonstrated its ability to model treatment performance within natural tundra wetlands and thus provide an additional predictive tool to aid northern stakeholders in the treatment of municipal effluents. The example with the Chesterfield Inlet data set has illustrated how SubWet 2.0 can be used to predict treatment performance anticipated from alterations to the size of the treatment area. The data set from Baker Lake showed that it is generally advisable to assess the success of the calibration effort by monitoring the rate of denitrification, nitrification, and ammonification rather than the concentrations of individual wastewater parameters. Despite the unusually high strength of the waste stream in Baker Lake, the calibration of SubWet has proved that this model can provide a reasonable approximation of treatment efficiencies.

The warm climate constructed wetland located in Iringa provided an insight into the flexibility of the SubWet program by demonstrating how a temporary change in influent strength influences the overall performance and how quickly the treatment processes within the wetland can be expected to respond. In this example, a better agreement was achieved between the SubWet simulated values and the observed values than for either of the previous natural wetland examples prior to model calibration; this should be expected since the Iringa wetland is a constructed wetland where parameters are better known and less variable than found in natural tundra wetlands.

To date only a few numerical models are available and able to describe treatment processes in horizontal subsurface flow treatment wetlands. It has been shown that the SubWet 2.0 model can consider the influence of several factors at one time, where empirical equations are generally not able to consider more than two factors at one time and usually in isolation of the other influential parameters. The model has causality, which means that we know the processes behind the model and can therefore put these processes into an equation.

Ultimately, SubWet will allow resource managers to demonstrate the current treatment benefit acquired from the use of individual wetlands and will be suitable to be used as a predictive tool to forecast the potential these areas could provide and determine the size of wetland needed to meet treatment objectives. This in turn will help resource managers in cost–benefit analysis when planning for future needs such as projected growth in population and anticipated effluent volumes.

References

Babatunde, A.O., Zhao, Y.Q., Zhao, X.H., 2010. Alum sludge-based constructed wetland system for enhanced removal of P and OM from wastewater: concept, design and performance analysis. Bioresour. Technol. 101 (16), 6576–6579.

Hammer, D.A., 1989. Constructed Wetlands for Wastewater Treatment—Municipal, Industrial and Agricultural. Lewis Publishers, Chelsea, MI.

Jørgensen, S.E., Gromiec, M.J., 2011. Mathematical models in biological waste water treatment – Chapter 7.6. In: Jørgensen, S.E., Fath, B.D. (Eds.), Fundamentals of Ecological Modelling, Volume 23, 4th Edition: Applications in Environmental Management and Research. Elsevier, Amsterdam, Netherlands, 414 pages.

Kadlec, R.H., Wallace, S.D., 2008. Treatment Wetlands, second ed. CRC Press, Boca Raton, FL.

Langergraber, G., Giraldi, D., Mena, J., Meyer, D., Peña, M., Toscano, A., Brovelli, A., Asuman, E.A., 2009. Recent developments in numerical modelling of subsurface flow constructed wetlands. Sci. Total Environ. 407 (13), 3931–3943.

Mander, U., Jenssen, P., 2003. Constructed Wetlands for Wastewater Treatment on Cold Climates. WIT Press, Southampton, England.

Pastor, R., Benqlilou, C., Paz, D., Cardenas, G., Espuña, A., Puigjaner, L., 2003. Design optimisation of constructed wetlands for wastewater treatment. Resource Conservion Recycling 37 (3), 193–204.

Rousseau, D.P.L., Vanrolleghem, P.A., De Pauw, N., 2004. Model-based design of horizontal subsurface flow constructed treatment wetlands: a review. Water Res. 38 (6), 1484–1493.

Stein, O.R., Biederman, J.A., Hook, P.B., Allen, W.C., 2006. Plant species and temperature effects on the k–C* first-order model for COD removal in batch-loaded SSF wetlands. Ecol. Eng. 26 (2), 100–112.

Tomenko, V., Ahmed, S., Popov, V., 2007. Modelling constructed wetland treatment system performance. Ecol. Model. 205 (3–4), 355–364.

Vymazal, J., 1998. Wetland treatment in czech republic. In: Vymazal, J., Brix, H., Cooper, P.F., Green, M.B., Haberl, R. (Eds.), Constructed Wetlands for Wastewater Treatment in Europe. Backhuys Publishers, Leiden, The Netherlands, pp. 67–76.

Wetzel, R.G., 2000. Fundamental processes within natural and constructed wetland ecosystems: short-term vs. long-term objectives. In: Proceedings of the Seventh International Conference on Wetland Systems for Water Pollution Control, 11–16 November 2000, Lake Buena Vista, FL, pp. 3–12.

Glossary

AC ammonification rate coefficient (1/24 h)

AF inverse phosphorus adsorption capacity (mg/L)

DC denitrification rate coefficient (1/24h)

KO Michaelis–Menten constant for the influence of oxygen on the nitrification rate (mg/L)

MA Michaelis–Menten constant for nitrification (mg/L)

MN Michaelis–Menten constant for denitrification (mg/L)

NC nitrification rate coefficient (1/24 h)

OC oxidation rate coefficient for organic matter, expressed as BOD_5 (1/24 h)

OO Michaelis–Menten constant for influence of oxygen on the oxidation rate of organic matter, expressed as BOD_5 (mg/L)

PA plant uptake rate coefficient for ammonium (1/24 h)

PN plant uptake rate coefficient for nitrate (1/24 h)

PP plant uptake rate coefficient for phosphorus (1/24 h)

TA temperature coefficient for ammonification (no units)

TD temperature coefficient for denitrification (no units)

TN temperature coefficient for nitrification (no units)

TO temperature coefficient for oxidation of organic matter expressed as BOD_5 (no units)

Framing the Need for Applications of Ecological Engineering in Arctic Environments

20

Colin N. Yates[a],*, Gordon C. Balch[b], Brent C. Wootton[b]

[a]*Faculty of Environment, Waterloo, Ontario, Canada*
[b]*Centre for Alternative Wastewater Treatment, Fleming College, Lindsay, Ontario, Canada*
**Corresponding author: e-mail address: cyates@uwaterloo.ca*

20.1 INTRODUCTION

The Arctic tundra biome is regarded as one of the most inhospitable places on earth, with extremely short growing seasons, low mean annual temperatures, low precipitation, and soils dominated by permafrost. It is also has one of the lowest net primary productivity values of any ecosystem globally (Houghton and Skole, 1990). As a result, the Arctic tundra makes up a small portion of global species diversity, approximately 1% (Chernov, 2002).

Because of the stressful abiotic factors, high and low Arctic tundra species, in particular vegetation, experience infrequent sexual reproduction, low rates of recruitment (Mcgraw and Shaver, 1982), and slow growth rates, together with long life spans (Hobbie, 2007). Tundra vegetation also has low species diversity (Forbes et al., 2001). The influence of abiotic factors of the Arctic tundra also are reflected in the low availability of nutrients (many of which is permanently locked in the soil) and low rates of organic decomposition. The cold temperature of the Arctic limits nutrient availability because the processes required for decomposition (weathering and recycling) are inhibited (Nadelhoffer et al., 1991). Cold temperatures have also been found to reduce inputs of nutrients from N_2 fixation (Chapin and Bledsoe, 1992). Therefore, the formation of soil in the Arctic is very slow. To further complicate Arctic terrestrial ecosystems, the tundra has a diverse array of mesoscale geomorphological features (Walker and Walker, 1991).

Similarly, aquatic environments in the Arctic have low productivity and respond and recover from anthropogenic stresses differently than more temperate localities (Douglas and Smol, 2000; Benstead et al., 2007; Michelutti et al., 2007). Because of these characteristics, Arctic ecosystems are thought to have trouble recovering from disturbances and therefore, natural regeneration from human disturbance in the Arctic is slow (Reynolds and Tapper, 1996; Forbes et al., 2001; Filler et al., 2006).

Developments in Environmental Modelling, Volume 26, ISSN 0167-8892, http://dx.doi.org/10.1016/B978-0-444-63249-4.00022-1

It has been approximately four decades since anthropogenic disturbances to the tundra were first investigated and attempts to rehabilitate patches of various ecosystems began (Forbes et al., 2001). But more important to note is that disturbance through resource extraction and development within circumpolar regions has become even more intense the past two decades (Forbes and Jefferies, 1999; Kumpula et al., 2011). Walker and Walker (1991) noted that although there has been heightened attention to minimize disturbance since the 1980s, the overall size of impacted areas is increasing. Therefore, it has been no surprise that subsequent studies have shown that it is expected that direct human activities, such as resource exploitation, development, land management, and reindeer husbandry will have more impact on the Arctic environment in the next several decades than climate change (Cramer, 1997; Forbes et al., 2001). These relatively small-scale patch disturbances caused by development are thought to have implications to wildlife much greater beyond their physical footprint (Forbes et al., 2001). Similar small-scale changes in aquatic environments have been observed, particularly with respect to increased nutrient loads entering surface water. Benstead et al. (2007) suggest that nutrient enrichment from anthropogenic influences on Arctic oligotrophic streams may have dramatic and persistent long-term effect on the primary producers. Earlier work (Benstead et al., 2005) demonstrated that continued anthropogenic impacts from activities such as mining and drilling will likely have long-term influence on these systems, even with just low-level nutrient enrichment. Similar results were observed in long-term fertilization studies in Arctic streams, where shifts in habitat structure, species abundance, and distribution were observed (Slavik et al., 2004). Point source disturbances on Arctic lakes from municipal water management have also demonstrated long-term changes to the systems. However, with removal of the stress, the ecosystem will begin to recover with time as shown by Douglas and Smol (2000), but the resulting species assemblages are either slower to respond to the removal of stress or a new community may permanently displace the predisturbance community.

Given the sensitivity of the tundra biome (both terrestrial and aquatic ecosystems) and its apparent slow recovery to disturbances (natural and anthropogenic), the application and wider use of practices to remediate, rehabilitate, and even restore anthropogenic degraded Arctic ecosystems are required. More importantly, further study needs to be conducted on novel techniques for restoration projects in the Arctic that take into account the complexity and fragility of Arctic ecosystems. Ecological engineering is one area of study that could be more proactively applied in the Arctic tundra biome.

Ecological engineering, although still a rather new field, has grown substantially over the past decade (Mitsch, 2012). The works of Odum (1983), Straskraba (1993), Jorgensen and Mitsch (1989), Mitsch and Jorgensen (2004), Jones (2012), and Mitsch (2012) chronicle the growth and application of ecological engineering as a field. Although there are some slight deviations in approach and understanding of its meaning as described by Jones (2012), the overarching goal remains consistent. Ecologists and engineering practitioners must be able to combine the disciplines of ecology and engineering to create or restore ecosystems that are sustainable and benefit both humans and nature. Despite the growth of the field, there remains very little evidence of the application of ecological engineering in the Arctic.

Mitsch (2012) best describes some of the basic concepts of ecological engineering: He suggests that (i) projects should focus on an ecosystem's self-design; (ii) the practitioner should focus on system thinking (or examining the ecosystem as a whole); and (iii) ecological engineering should be used as an acid test for ecological theory.

In this review, we use the definition and goals of ecological engineering presented by Mitsch and Jorgensen (2004) to guide our discussion and identify future applications of ecological engineering in Arctic environments. Mitsch and Jorgensen (2004) describe ecological engineering as an applied science that involves both the creation and restoration of sustainable ecosystems for the benefit of nature and humans. More simply stated, the applied science of ecological engineering should use biological and ecological techniques to reclaim degraded landscapes, remediate contaminated systems with the goal of restoring the system to a previous state, or rehabilitate the structure and function of an ecosystem toward a sustainable trajectory. However, in many cases, the latter is the predominant solution, especially in Arctic tundra conditions (as well as other biomes) where restoring a system to its previous state is difficult because of numerous socioeconomic and environmental factors that pose daunting challenges.

In this chapter, we review the current applications of ecological engineering principles in the Canadian Arctic, with anecdotes from Alaska and other Arctic localities. We also identify and describe the many barriers to recovery of degraded Arctic systems and how conscious efforts using principles of ecological engineering that could be used as best practices may potentially overcome some of the barriers to rehabilitating or remediating large-scale degraded sites in remote localities of the Arctic. In doing so, we will describe (i) the research conducted using ecological engineering methods used in the Arctic to date; (ii) their application by practitioners; and (iii) the logistical barriers to conducting ecological engineering projects in the Arctic.

20.2 REVIEW OF APPLICATION OF ECOLOGICAL ENGINEERING IN THE ARCTIC: 1970s TO PRESENT

The use of ecological engineering principles in the Arctic is evident in a number of different disciplines, including remediation of contaminated soils, rehabilitated roads, and even in the treatment of municipal wastewater through wetland technologies. A number of "how-to" articles are present in the peer-reviewed literature that identify key steps for rehabilitation and restoration of tundra ecosystems (Jorgenson and Joyce, 1994; Firlotte and Staniforth, 1995). Jorgenson and Joyce (1994) described several strategies to rehabilitate (revegetation and creation systems) disturbed land by oil development in Arctic Alaska.

20.2.1 **Revegetation of disturbed tundra**

Revegetation of denuded Arctic landscapes has been a particular interest of researchers since the 1970s. Many studies have focused on recent disturbances caused by the oil and gas industry. The works of McKendrick and team have been particularly dominant (McKendrick, 1987, 1991; Elliott and McKendrick, 1995; Streever et al., 2003). Many of the studies have noted that revegetation of sites damaged by development is possible; however, as Streever et al. (2003) suggest, it is important to properly define the performance standard.

The main concern of many of the strategies includes the length of time for reestablishment of these sites, and also the establishment of less diverse and/or weedy grasses. Forbes and Jefferies (1999) concluded in their review of the revegetation of disturbed Arctic sites that rates of reestablishment on sites are very slow. Of particular issue are gravel sites, where moisture, nutrients, and organic matter are very low, making establishment of native vegetation difficult. McKendrick (1991) indicated that for revegetation to predisturbance vegetation diversity levels required 20 years. These early studies have also found a number of species that have proven to be very effective at becoming established in disturbed gravel sites. Other assisted revegetation trials in Arctic region have primarily involved nutrient and organic matter amendments to the soil (Firlotte and Staniforth, 1995; Handa and Jefferies, 2000; Kidd et al., 2006).

Common soil amendments to enhance organic matter development has been the planting of grass strips to stimulate organic matter accumulation (Runolfsson, 1987), or the direct application of some form of fertilizer or nutrient (Kidd et al., 2006). In the sub-Arctic Russian town of Monchegorsk, sewage sludge has even been used to successfully amend soil denuded by metal pollution from a nearby smelter (Gorbacheva et al., 2009).

Direct and indirect seeding has also been extensively used to promote revegetation of denuded Arctic sites. Many of the early studies involved the direct seeding of exotic species which were thought to provide initial vegetation cover to reduce erosion and improve the rate of natural revegetation (Webber and Ives, 1978; Densmore and Holmes, 1987). However, the positive effects of using exotic species were outweighed by the negative impacts (Forbes, 1992; Forbes and Jefferies, 1999). Indirect methods of seeding denuded locations have also been demonstrated by increasing seed rain from adjacent undisturbed sites through fertilization with phosphorus and potassium, resulting in increased flowering density (Chapin and Chapin, 1980). More recently, studies have been undertaken to determine the application potential of certain Arctic species for restoration efforts as described in Hagen (2002). Finally, Shirazi et al. (1998) conducted restoration trial plots by manipulating the thermal regime for planted native Arctic species.

Despite some notable successes, Forbes and Jefferies (1999) advised caution and the need for more long-term experimental trials using various restoration techniques. One area that was identified as lacking in the literature is the study of soil manipulation to improve microtopographic heterogeneity. The microtopography of the

Arctic tundra is complex and greatly influences moisture regimes, organic matter accretion/decretion, gas exchange, nutrient availability, carbon dynamics, thermal conditions of the soil, and therefore the biotic communities that reside (Chapin et al., 1979; Andrus et al., 1983; Ohlson and Dahlberg, 1991; Christensen et al., 2000; Zona et al., 2011). Reestablishment of plant communities in the Arctic is difficult (Forbes and Jefferies, 1999), and monoculture stands of weedy grasses as discussed above have been a common result.

This strategy has been successfully used in the establishment of microstructures as part of the restoration or rehabilitation plans that are extensively employed in systems other than the tundra, including peatlands (Ferland and Rochefort, 1997; Price et al., 1998), forests (Simmons et al., 2011), created wetlands (Moser et al., 2007), and prairie restoration (Biederman and Whisenant, 2011), among others. Our review only identified one peer-reviewed article on assessing microtopography and revegetation following anthropogenic disturbances in sub-, low-, and high Arctic regions. This particular article by Campbell and Bergeron (2012) only discusses natural revegetation of winter road cuts and does not deal with assisted revegetation. Methods should be tested to mimic the microtopographic heterogeneity created by frost heaves and other environmental forces.

20.2.2 Soil remediation

There are a number of contaminated sites throughout the Arctic, ranging from decommissioned military sites along the Distant Early Warning (DEW) line, oil extraction, pipelines in the sub-Arctic, to commonly decommissioned energy fuel stores in remote communities. Since the 1990s, anthropogenic contaminants, in particular polychlorinated biphenyls (PCBs), have been under intense scientific investigation in the Canadian Arctic. PCBs have spanned various abiotic and biotic environments (Braune et al., 1999; Muir et al., 1999; Braune et al., 2010; Mueller et al., 2011), and the influence of PCB contamination is seen at some distance away from the source (Bright et al., 1995).

The DEW line has received particular attention due to the extent of contamination from the decommissioned military sites (Poland et al., 2001; Stow et al., 2005). Stow et al. (2005) reported that 18 of 21 of the DEW line sites undergoing remediation were contaminated with PCBs. Poland et al. (2001) describe the remediation of a number of PCB-contaminated sites in the Canadian Arctic and the remediation options available at the time. They suggested bioremediation through the use of aerobic/anaerobic bacteria, white rot fungus, and genetic engineering. However, none of the bioremediation options were found to be feasible. In one instance, they packaged and shipped soils to southern Canada for disposal. The application of bioremediation techniques on contaminated soils in the Arctic since this time has become more prevalent in ecological engineering technology presented in the peer-reviewed literature (McCarthy et al., 2004; Paudyn et al., 2008; Sanscartier et al., 2009; Yang et al., 2009; Kalinovich et al., 2012; Ferrera-Rodriguez et al., 2013). Permeable reactive barriers (PRB) have been used to treat polychlorinated biphenyl (PCB)–contaminated soil in the high Arctic. Kalinovich et al. (2012) found that they were able to successfully trap PCB-contaminated soil with a PRB system that was modified for high Arctic conditions.

Bioremediation of soils contaminated by petroleum-based hydrocarbons has also been utilized with some success in the Arctic. However, similar to the application of PRB, the application of psychrotolerant microorganisms to mineralize petroleum-based hydrocarbons in Arctic soil has also only become popular in the past decade. Mohn and Stewart (2000) first analyzed feasibility bioremediation of hydrocarbon-contaminated soils in the 1990s. A scattered number of studies have been conducted investigating the utilization of landfarming, biopiles, and the application of various microorganism inoculants to successfully remediate soil (Mohn et al., 2001; McCarthy et al., 2004; Filler et al., 2006; Paudyn et al., 2008). One particular site that has received significant attention concerning the application of landfarming techniques has been the Fairbanks International Airport crash-fire-rescue training facility (Braley, 1993; Reynolds et al., 1998). Researchers observed that landfarming with the addition of nutrients and plants Paudyn et al. (2008) could be successfully used to remediate diesel-contaminated soils at the decommissioned military base on Resolution Island, Nunavut. They observed that landfarming, with the addition of fertilizer and aeration through soil mixing, biostimulated the biotic community and produced the best results for remediation of the soil. The use of only a single technique did little to remediate the soil.

Bioaugmentation and biostimulation of biopiles and landfarming have become popular ways to improve the degradation rate of soil contaminants. A study in Barrow, Alaska, found that amending the soil with phosphorus and nitrogen, along with regular tillage with heavy equipment, allowed for remediation of moderately contaminated soil (McCarthy et al., 2004). Experimentation with thermal enhancement of bioremediation techniques has also been accomplished through both active and passive heating (Filler et al., 2006). Filler et al. (2001) found that at a Prudhoe Bay bioremediation site, enhancing the thermal environment provided superior degradation of diesel-hydrocarbon-contaminated soil.

Similar to all other remediation techniques being employed in the Arctic, bioremediation of contaminated soil is also subject to the constraints of low temperature. Initial results by Mohn et al. (1997) found that temperature severely limited removal of PCBs in contaminated soil through the use of native Arctic microflora. As discussed previously, the rate of recovery from contamination of PCBs and hydrocarbons is thought to be slower in Arctic environments than in warmer climates. Other factors that influence the successful bioremediation of PCB- and hydrocarbon-contaminated soil include increases in soil moisture, nutrient availability, and porosity (soil texture). Soil moisture is of particular importance for bioremediation because hydrocarbon-degrading aerobic microorganisms used in landfarming and biopiles require adequate levels of soil moisture without filling too much pore space.

Like all projects in the Canadian Arctic, environmental factors are not the only limiting factors preventing the study of these systems. Socioeconomics is also key factor inhibiting remediation projects. McCarthy et al. (2004) suggest that landfarming is an economical and technically simple method of remediation for remote sites. A life cycle assessment of two remediation alternatives for diesel-contaminated soil in Iqaluit, Nunavut, demonstrated that landfarming was superior to shipping contaminated soil to southern Canada for disposal in a landfill (Bolton, 2012). Filler et al. (2006) also suggested that landfarming is a cost–effective remediation strategy for cold climate contaminated sites. However, in even more remote localities, where resources from larger Arctic communities such as Iqaluit are not available, the long-term nature of remediation projects may not be economically feasible.

20.2.3 **Wetlands for wastewater treatment and rehabilitation**

The use of tundra to treat domestic wastewater in the Canadian Arctic is the most common management strategy for remote communities and resource exploration/ extraction camps (Yates et al., 2012a). Economically, the tundra treatment wetlands at present are the most feasible solution, given the socioeconomic, sociocultural, and climatic regimes of the region (Johonson, 2010). Similar needs have been demonstrated in Greenland (Jenssens, 2011). The performance of these systems during the ice-free period as noted by Yates et al. (2012b) is very good and indicates the potential for the extensive use of phytotechnologies in the region. However, many of these systems are largely unmanaged, with very limited regulatory monitoring. Kadlec and Johnson (2008) modeled and engineered a tundra wetland complex for municipal wastewater treatment in Cambridge Bay, Nunavut, in order to augment flow, increase residency time, and improve resulting effluent. In Chapter 22, of this book, Yates et al. tested a pilot-scale horizontal subsurface flow constructed wetland system in Baker Lake, Nunavut, with varying success. They found that the activation of the biofilm after winter and overloading of the system and cool soil temperatures resulted in poor performance of the system. This pilot study only scratched the surface of determining feasibility of constructed ecological systems in the Arctic.

Only several other studies testing alternative wastewater treatment systems have been identified in the Arctic, including the Riznyk et al. (1993) study of peat leachmounds in sub-Arctic Alaska, which demonstrated some success. However, widespread application of ecological engineered systems remains minimal.

With the rapidly growing population of many Arctic communities and expanding resource development activities, more studies are needed to not only optimize performance of natural systems to accommodate the demand of wastewater management, but also seek alternative methods to manage wastewater.

Wastewater is only one of many wastes being largely managed through the assimilative capacity of the Arctic tundra. Given the presence of modern wastewaster and waste in the Arctic, ranging from mining effluents, chemical waste dumps, to more inert discharges of graywater (Lockhart et al., 1992; MacDonald et al., 2000; Jenssens, 2011), a greater array of treatment options are needed, such as hybrid constructed wetland systems that can be engineered to effectively treat a variety of waste types.

In light of the presence of contaminants in the Arctic, it is surprising that wetland creation in Arctic regions is minimal, at least in the peer-reviewed literature. Created wetlands have been successfully adopted not only for the treatment of wastewater but also leachate and remediation, minewater, industrial wastes, and reclamation (Mitsch and Jorgensen, 2004; Kadlec and Wallace, 2009). The creation of wetlands in the Arctic given the hydrologic characteristics is thought to be quite feasible (Handa and Jefferies, 2000). Despite this, there have been only a few studies conducted on wetland and lake creation in the Arctic. Much of the work has revolved around creating systems following gravel extraction for mining or oil extraction activities and revegetation after disturbance (Jorgenson et al., 1992; Handa and Jefferies, 2000; Jorgenson et al., 2004).

20.3 BARRIERS TO THE APPLICATION ECOLOGICAL ENGINEERING IN THE ARCTIC

Conducting research, undertaking design projects, and implementing these treatment options are complex and challenging endeavors within the Arctic region. The obvious obstacle is the environment itself; however, the greatest challenges to conducting work in the Arctic are less visible. Most resource extraction projects are located in remote locations, and the majority of communities, at least in the Canadian Arctic, have fly-in access only. The remote localities of many communities result in socioeconomic tensions and a lack of resources necessary to undertake small-scale research trials, let alone large-scale remediation or rehabilitation projects. Therefore, communities in Arctic Canada need to rely heavily on federal government assistance to undertake projects (Ritter, 2007; Johnson, 2010).

The remoteness of many sites makes the long-term study of contamination, remediation, and restoration of sites difficult. Many of the decommissioned DEW line sites are hundreds of kilometers away from communities, and many resource extraction sites are also located at some distance from communities or are also fly-in access only. The result is that monitoring many of these locations is problematic not only logistically, but also economically.

But more specifically, regulatory monitoring and reporting of wetlands used for municipal wastewater treatment has been largely nonexistent (Yates et al., 2012a). And long-term monitoring of rehabilitation and revegetation projects has only been conducted in a few cases (Forbes, 1992; Streever et al., 2003; Jorgenson et al., 2004). Only a few papers from peer-reviewed and gray literature sources emerge to describe best practices for mining reclamation, at least in the Canadian Arctic (Bowman and Baker, 1998; Hoos and Williams, 1999). As a result, it is difficult for managers to determine which strategies are most successful.

As with ecological engineering in any other region, climate change will have significant impact on how we approach the restoration, rehabilitation, and remediation of many sites. Climate change is expected to have significant impacts on the Arctic on communities of microorganisms, vegetation, avian species, mammals, and marine life, among others (Cornelissen et al., 2001; Ford et al., 2006; Moore and Huntington, 2008; Kirk et al., 2011; Deslippe et al., 2012). The challenge with ecological engineering will be determining how to restore or remediate sites so that they will be resilient with impending climatic changes, especially in the Arctic.

Despite the many barriers, there are clear opportunities for wider application of ecological engineering in the Arctic. Ecological engineering has been identified to provide cost–effective means to remediate contaminated soils, stabilize eroding substrate, and treat wastewater, among many others, where conventional engineering or ecology will not solve the problem alone.

20.4 MOVING RESEARCH TO APPLICATION

In the complex socioeconomic environment of many Arctic communities, with specific reference to Canadian and Alaskan communities, ecological engineering projects should focus on the development of technology that adopts minimalistic resource demands and solves some of the larger issues in the Arctic communities. In the recent International Polar Year program, some of the major issues that face remote Arctic communities were studied. Parlee and Furgal (2012) identified a number of major issues facing the Canadian Arctic. Primary concerns for Arctic communities as population growth increases will be the availability of drinking water (White et al., 1995; White et al., 2007), economical electricity production (Hamilton et al., 2012), food insecurity (Huet et al., 2012), and the introduction of even more anthropogenic contaminants (Filler et al., 2006). Climate change in particular, with melting ice and stability of the landscape, imposes increased risk for land users, and further resource development may cause more environmental stress than produce offsets (Parlee and Furgal, 2012). In light of impending growth, more ecological engineering technologies need to be investigated to determine their feasibility in the Arctic environment, and those already proved feasible need to be broadly applied.

Researchers need to not only continue finding feasible methods to amend disturbed Arctic environments but also target ways to integrate best practices of ecological engineering techniques into the regulatory framework. In the late 1990s, Bowman and Baker identified significant shortcomings in the mine reclamation policy structure of the Northwest Territories. In more recent literature, the topic of reclamation only receives little acknowledgment and often suggests its difficulty in the undertaking (Couch, 2002).

20.5 CONCLUSIONS

In the past four decades, concepts of ecological engineering have been applied in the Arctic with varying degrees of success. Climate is a significant barrier to the success of many of the ecological engineering projects that have been undertaken. However, logistics and economics of conducting research in the Arctic have been the single greatest hurdle to determine the long-term feasibility and successful application of the ecological engineering in Arctic. Despite the barriers, ecological engineering has shown to be an excellent candidate to provide adaptive solutions to the complex ecological and engineering problems presented in the Arctic. Research should focus on long-term application, creation of system physiology, continued use of native species (vegetation and microorganisms), as well as integrating ecological engineering into best management practices for remediation, revegetation, treatment of wastes, and restoration of Arctic sites.

References

Andrus, R., Wagner, D., Titus, J., 1983. Vertical zonation of Sphagnum mosses along hummock-hollow gradients. Can. J. Bot. Rev. Can. Bot. 61, 3128–3139.

Benstead, J.P., Deegan, L.A., Peterson, B.J., Huryn, A.D., Bowden, W.B., Suberkropp, K., Buzby, K.M., Green, A.C., Vacca, J.A., 2005. Responses of a beaded Arctic stream to short-term N and P fertilisation. Freshw. Biol. 50, 277–290.

Benstead, J.P., Green, A.C., Deegan, L.A., Peterson, B.J., Slavik, K., Bowden, W.B., Hershey, A.E., 2007. Recovery of three arctic stream reaches from experimental nutrient enrichment. Freshw. Biol. 52, 1077–1089.

Biederman, L.A., Whisenant, S.G., 2011. Using mounds to create microtopography alters plant community development early in restoration. Restor. Ecol. 19, 53–61.

Bolton, M., 2012. Comparing Two Remediation Alternative for Diesel-Contaminated Soil in the Arctic Using Life Cycle Assessment. Master's Dissertation. Queen's University, Ontario, Canada.

Bowman, B., Baker, D., 1998. Mine Reclamation Planning in the Canadian North. Canadian Arctic Resources Committee. http://www.carc.org/pdfs/NMPWorkingPaper1BowmanandBaker.pdf.

Braley, W.A., 1993. The Fairbanks International Airport experimental bioremediation project preliminary report. Fairbanks: Alaska Department of Transportation and Public Facilities.

Braune, B., Muir, D., DeMarch, B., Gamberg, M., Poole, K., Currie, R., et al., 1999. Spatial and temporal trends of contaminants in Canadian arctic freshwater and terrestrial ecosystems: a review. Sci. Total Environ. 230, 145–207.

Braune, B.M., Mallory, M.L., Butt, C.M., Mabury, S.A., Muir, D.C.G., 2010. Persistent halogenated organic contaminants and mercury in northern fulmars (fulmarus glacialis) from the canadian. arctic. Environ. Pollut. 158, 3513–3519.

Bright, D., Dushenko, W., Grundy, S., Reimer, K., 1995. Evidence for short-range transport of polychlorinated-biphenyls in the Canadian Arctic using congener signatures of pcbs in soils. Sci. Total Environ. 251–263.

Campbell, D., Bergeron, J., 2012. Natural revegetation of winter roads on peat lands in the Hudson Bay Lowland, Canada. Arct. Antarct. Alp. Res. 44, 155–163.

Chapin, D.M., Bledsoe, C.S. (Eds.), 1992. Nitrogen Fixation in Arctic Plant Communities. Academic Press, Inc., San Diego.

Chapin, F., Chapin, M., 1980. Revegetation of an Arctic disturbed site by native Tundra species. J. Appl. Ecol. 17, 449–456.

Chapin, F., Vancleve, K., Chapin, M., 1979. Soil-temperature and nutrient cycling in the Tussock growth form of Eriophorum-Vaginatum. J. Ecol. 67, 169–189.

Chernov, Y., 2002. Arctic biota: taxonomic diversity. Zool. ZH. 81, 1411–1431.

Christensen, T., Friborg, T., Sommerkorn, M., Kaplan, J., Illeris, L., Soegaard, H., Nordstroem, C., Jonasson, S., 2000. Trace gas exchange in a high-arctic valley 1. Variations in CO2 and CH4 flux between tundra vegetation types. Glob. Biogeochem. Cycles 14, 701–713.

Cornelissen, J.H.C., Callaghan, T.V., Alatalo, J.M., Michelsen, A., Graglia, E., Hartley, A.E., Hik, D.S., Hobbie, S.E., Press, M.C., Robinson, C.H., Henry, G.H.R., Shaver, G.R., Phoenix, G.K., Jones, D.G., Jonasson, S., Chapin, F.S., Molau, U., Neill, C., Lee, J.A., Melillo, J.M., Sveinbjornsson, B., Aerts, R., 2001. Global change and arctic ecosystems: is lichen decline a function of increases in vascular plant biomass? J. Ecol. 89, 984–994.

Couch, W.J., 2002. Strategic resolution of policy, environmental and socio-economic impacts in Canadian Arctic diamond mining: BHP's NWT diamond project. Impact Assess. Project Appraisal 20, 265–278.

Cramer, W., 1997. Modeling the possible impact of climate change on broad-scale vegetation structure: examples from northern Europe. In: Oechel, W.C., Callaghan, T., Gilmanov, T., Holten, J.I., Maxwell, B., Molau, U., Sveinbjörnsson, B. (Eds.), Global Change and Arctic Terrestrial Ecosystems. Springer-Verlag, New York, NY, pp. 312–329.

Densmore, R., Holmes, K., 1987. Assisted revegetation in Denali-National-Park, Alaska, USA. Arct. Alp. Res. 19, 544–548.

Deslippe, J.R., Hartmann, M., Simard, S.W., Mohn, W.W., 2012. Long-term warming alters the composition of Arctic soil microbial communities. FEMS Microbiol. Ecol. 82, 303–315.

Douglas, M.S.V., Smol, J.P., 2000. Eutrophication and recovery in the High Arctic: Meretta Lake (Cornwallis Island, Nunavut, Canada) revisited. Hydrobiologia 431, 193–204.

Elliott, C., McKendrick, J., 1995. Revegetated Coal Surface Mines in Alaska—Do They Meet the Forage Needs of Native Big Game? Issues and Technology in the Management of Impacted Wildlife – Proceedings 6th Symposium, 85–92.

Ferland, C., Rochefort, L., 1997. Restoration techniques for Sphagnum-dominated peatlands. Can. J. Bot.-Rev. Can. Bot. 75, 1110–1118.

Ferrera-Rodriguez, O., Greer, C.W., Juck, D., Consaul, L.L., Martinez-Romero, E., Whyte, L.G., 2013. Hydrocarbon-degrading potential of microbial communities from Arctic plants. J. Appl. Microbiol. 114, 71–83.

Filler, D., Lindstrom, J., Braddock, J., Johnson, R., Nickalaski, R., 2001. Integral biopile components for successful bioremediation in the arctic. Cold Reg. Sci. Technol. 32, 143–156.

Filler, D., Reynolds, C., Snape, I., Daugulis, A., Barnes, D., Williams, P., 2006. Advances in engineered remediation for use in the Arctic and Antarctica. Polar Rec. 42, 111–120.

Firlotte, N., Staniforth, R., 1995. Strategies for revegetation of disturbed gravel areas in climate stressed sub-Arctic environments with special reference to Churchill, Manitoba, Canada—a Literature-Review. Clim. Res. 5, 49–52.

Forbes, B., 1992. Tundra disturbance studies.1. Long-term effects of vehicles on species richness and biomass. Environ. Conserv. 19, 48–58.

Forbes, B.C., Jefferies, R.L., 1999. Revegetation of disturbed arctic sites: constraints and applications. Biol. Conserv. 88, 15–24.

Forbes, B., Ebersole, J., Strandberg, B., 2001. Anthropogenic disturbance and patch dynamics in circumpolar arctic ecosystems. Conserv. Biol. 15, 954–969.

Ford, J., Smit, B., Wandel, J., 2006. Vulnerability to climate change in the Arctic: a case study from Arctic Bay, Canada. Glob. Environ. Change 16, 145–160.

Gorbacheva, T.T., Kikuchi, R., Gorbachev, P.A., 2009. Evaluation of extractable elements in artificial substratum made from sewage sludge: approach to remediation of degraded land in the Arctic. Land Degradation Dev. 20, 119–128.

Hagen, D., 2002. Propagation of native Arctic and alpine species with a restoration potential. Polar Res. 21, 37–47.

Hamilton, L.C., White, D.M., Lammers, R.B., Myerchin, G., 2012. Population, climate, and electricity use in the Arctic integrated analysis of Alaska community data. Popul. Environ. 33, 269–283.

Handa, I., Jefferies, R., 2000. Assisted revegetation trials in degraded salt-marshes. J. Appl. Ecol. 37, 944–958.

Hobbie, S.E., 2007. Arctic ecology. In: Pugnaire, F.I., Valladares, F. (Eds.), Functional Plant Ecology, second ed. CRC Press, New York, pp. 369–388.

Hoos, R., Williams, W., 1999. Environmental Management at BHP's Ekati Diamond Mine in the Western Arctic. A A Balkema Publishers, Netherlands, LEIDEN; SCHIPHOLWEG 107C, PO BOX 447, 2316 XC LEIDEN.

Houghton, R.A., Skole, D.L., 1990. Carbon. In: Turner, B.L., Clark, W.C., Kates, R.W., Richards, J.F., Matthews, J.T., Meyer, W.B. (Eds.), The Earth as Transformed by Human Action. Cambridge University Press, Cambridge, UK, pp. 393–408.

Huet, C., Rosol, R., Egeland, G.M., 2012. The prevalence of food insecurity is high and the diet quality poor in inuit communities. J. Nutr. 142, 541–547.

Jenssens, P., 2011. Wastewater treatment in cold/arctic climate with a focus on small scale and onsite systems, 28th Alaska Health Summit, January 10-13.

Johnson, K., 2010. The social context of wastewater management in remote communities. Environ. Sci. Eng. Mag. Summer, 28–30.

Jones, C.G., 2012. Grand challenges for the future of ecological engineering. Ecol. Eng. 45, 80–84.

Jorgensen, S.E., Mitsch, W.J., 1989. Ecological Engineering: An Introduction to Ecotechnology. Wiley, New York.

Jorgenson, M., Joyce, M., 1994. 6 Strategies for rehabilitating land disturbed by oil development in Arctic Alaska. Arctic 47, 374–390.

Jorgenson, M., Cater, T., Jacobs, L., Joyce, M., 1992. Wetland Creation and Revegetation on Overburden in Arctic Alaska. 2nd International Symposium – Mining In the Arctic, 265–278.

Jorgenson, M.T., Kidd, J.G., Carter, T.C., Bishop, S., Racine, C.H., 2004. Long-term evaluation of methods for rehabilitation of lands disturbed by industrial development in the Arctic. 31.

Kadlec, R.H., Johnson, K., 2008. Cambridge Bay, Nunavut, wetland planning study. Journal of the Northern Territories Water and Waste Association. Fall, 30–33.

Kadlec, R.H., Wallace, S., 2009. Treatment Wetlands, second ed. CRC Press, Boca Raton, FL.

Kalinovich, I.K., Rutter, A., Rowe, R.K., Poland, J.S., 2012. Design and application of surface PRBs for PCB remediation in the Canadian Arctic. J. Environ. Manage. 101, 124–133.

Kidd, J.G., Streever, B., Jorgenson, M.T., 2006. Site characteristics and plant community development following partial gravel removal in an arctic oilfield. Arct. Antarct. Alp. Res. 38, 384–393.

Kirk, J.L., Muir, D.C.M., Antoniades, D., Douglas, M.S.V., Evans, M.S., Jackson, T.A., Kling, H., Lamoureux, S., Lim, D.S.S., Pienitz, R., Smol, J.P., Stewart, K., Wang, X., Yang, F., 2011. Climate change and mercury accumulation in Canadian high and subarctic lakes. Environ. Sci. Technol. 45, 964–970.

Kumpula, T., Pajunen, A., Kaarlejarvi, E., Forbes, B.C., Stammler, F., 2011. Land use and land cover change in Arctic Russia: ecological and social implications of industrial development. Global Environ. Change Human Policy Dimens. 21, 550–562.

Lockhart, W., Wagemann, R., Tracey, B., Sutherland, D., Thomas, D., 1992. Presence and implications of chemical contaminants in the fresh-waters of the Canadian Arctic. Sci. Total Environ. 122, 165–243.

MacDonald, R., Barrie, L., Bidleman, T., Diamond, M., Gregor, D., Semkin, R., Strachan, W., Li, Y., Wania, F., Alaee, M., Alexeeva, L., Backus, S., Bailey, R., Bewers, J., Gobeil, C., Halsall, C., Harner, T., Hoff, J., Jantunen, L., Lockhart, W., Mackay, D., Muir, D., Pudykiewicz, J., Reimer, K., Smith, J., Stern, G., Schroeder, W., Wagemann, R., Yunker, M., 2000. Contaminants in the Canadian Arctic: 5 years of progress in understanding sources, occurrence and pathways. Sci. Total Environ. 254, 93–234.

McCarthy, K., Walker, L., Vigoren, L., Bartel, J., 2004. Remediation of spilled petroleum hydrocarbons by in situ landfarming at an arctic site. Cold Reg. Sci. Technol. 40, 31–39.

Mcgraw, J., Shaver, G., 1982. Seedling density and seedling survival in Alaskan cotton grass Tussock Tundra. Holarct. Ecol. 5, 212–217.

McKendrick, J., 1987. Plant succession on disturbed sites, North Slope, Alaska, USA. Arct. Alp. Res. 19, 554–565.

McKendrick, J., 1991. Colonizing tundra plants to vegetate abondoned gravel pads in arctic Alaska. Adv. Ecol. 1, 209–223.

Michelutti, N., Hermanson, M.H., Smol, J.P., Dillon, P.J., Douglas, M.S.V., 2007. Delayed response of diatom assemblages to sewage inputs in an Arctic lake. Aquat. Sci. 69, 523–533.

Mitsch, W.J., 2012. What is ecological engineering? Ecol. Eng. 45, 5–12.

Mitsch, W.J., Jorgensen, S.E., 2004. Ecological Engineering and Ecosystem Restoration. John Wiley & Sons, New Jersey.

Mohn, W., Stewart, G., 2000. Limiting factors for hydrocarbon biodegradation at low temperature in arctic soils. Soil Biol. Biochem. 32, 1161–1172.

Mohn, W., Westerberg, K., Cullen, W., Reimer, K., 1997. Aerobic biodegradation of biphenyl and polychlorinated biphenyls by arctic soil microorganisms. Appl. Environ. Microbiol. 63, 3378–3384.

Mohn, W., Radziminski, C., Fortin, M., Reimer, K., 2001. On site bioremediation of hydrocarbon-contaminated arctic tundra soils in inoculated biopiles. Appl. Microbiol. Biotechnol. 57, 242–247.

Moore, S.E., Huntington, H.P., 2008. Arctic mammals and climate change: impacts and resilience. Ecol. Appl. 18, S165.

Moser, K., Ahn, C., Noe, G., 2007. Characterization of microtopography and its influence on vegetation patterns in created wetlands. Wetlands 27, 1081–1097.

Mueller, C.E., De Silva, A.O., Small, J., Williamson, M., Wang, X., Morris, A., et al., 2011. Biomagnification of perfluorinated compounds in a remote terrestrial food chain: Lichen-caribou-wolf. Environ. Sci. Technol. 45, 8665–8673.

Muir, D., Braune, B., DeMarch, B., Norstrom, R., Wagemann, R., Lockhart, L., et al., 1999. Spatial and temporal trends and effects of contaminants in the canadian arctic marine ecosystem: a review. Sci. Total Environ. 230, 83–144.

Nadelhoffer, K.J., Giblin, A.E., Shaver, G.R., Laundre, J.A., 1991. Effects of temperature and substrate quality on element mineralization in six Arctic soils. Ecology 72, 242–253.

Odum, H.T., 1983. Systems Ecology: An Introduction. Wiley, New York, Toronto.

Ohlson, M., Dahlberg, B., 1991. Rate of peat increment in Hummock and Lawn communities on Swedish mires during the last 150 years. Oikos 61, 369–378.

Parlee, B., Furgal, C., 2012. Well-being and environmental change in the arctic: a synthesis of selected research from Canada's International Polar Year program. Clim. Chang. 115, 13–34.

Paudyn, K., Rutter, A., Rowe, R.K., Poland, J.S., 2008. Remediation of hydrocarbon contaminated soils in the Canadian Arctic by landfarming. Cold Reg. Sci. Technol. 53, 102–114.

Poland, J., Mitchell, S., Rutter, A., 2001. Remediation of former military bases in the Canadian arctic. Cold Regions Sci. Tech. 32, 93–105.

Price, J., Rochefort, L., Quinty, F., 1998. Energy and moisture considerations on cutover peatlands: surface microtopography, mulch cover and Sphagnum regeneration. Ecol. Eng. 10, 293–312.

Reynolds, C.M., Braley, W.A., Travis, M.D., Perry, L.B., Iskandar, I.K., 1998. Bioremediation of hydrocarbon-contaminated soils and groundwater in northern climates. US Army Corps of Engineers, Cold Regions Research and Engineering Laboratory.

Reynolds, J., Tapper, S., 1996. Control of mammalian predators in game management and conservation. Mammal Rev. 26, 127–155.

Ritter, T.L., 2007. Sharing environmental health practice in the North American Arctic: a focus on water and wastewater service. J. Environ. Health 69, 50–55.

Riznyk, R.Z., Rockwell, J., Reid, L.C., Reid, S.L., 1993. Peat leachmount treatment of residential waste-water in sub-Arctic Alaska. Water Air Soil Pollut. 69, 165–177.

Runolfsson, S., 1987. Land reclamation on Iceland. Arct. Alp. Res. 19 (4), 514–517.

Sanscartier, D., Laing, T., Reimer, K., Zeeb, B., 2009. Bioremediation of weathered petroleum hydrocarbon soil contamination in the Canadian High Arctic: laboratory and field studies. Chemosphere 77, 1121–1126.

Shirazi, M., Haggerty, P., Hendricks, C., Reporter, M., 1998. The role of thermal regime in tundra plant community restoration. Restor. Ecol. 6, 111–117.

Simmons, M.E., Wu, X.B., Whisenant, S.G., 2011. Plant and soil responses to created microtopography and soil treatments in bottom land hardwood forest restoration. Restor. Ecol. 19, 136–146.

Slavik, K., Peterson, B., Deegan, L., Bowden, W., Hershey, A., Hobbie, J., 2004. Long-term responses of the Kuparuk river ecosystem to phosphorus fertilization. Ecology 85, 939–954.

Stow, J., Sova, J., Reimer, K., 2005. The relative influence of distant and local (DEW-line) PCB sources in the Canadian Arctic. Sci. Total Environ. 342, 107–118.

Straskraba, M., 1993. Ecotechnology as a new means for environmental-management. Ecol. Eng. 2, 311–331.

Streever, W., McKendrick, J., Fanter, L., Anderson, S., Kidd, J., Portier, K., 2003. Evaluation of percent cover requirements for revegetation of disturbed sites on Alaska's North Slope. Arctic 56, 234–248.

Walker, D., Walker, M., 1991. History and pattern of disturbance in Alaskan Arctic terrestrial ecosystems—A hierarchical approach to analyzing landscape change. J. Appl. Ecol. 28, 244–276.

Webber, P., Ives VES, J., 1978. Damage and recovery of Tundra vegetation. Environ. Conserv. 5, 171–182.

White, K.D., Byrd, L.A., Robertson, S.C., ODriscoll, J.P., King, T., 1995. Evaluation of peat biofilters for onsite sewage management. J. Environ. Health 58, 11–17.

White, D.M., Gerlach, S.C., Loring, P., Tidwell, A.C., Chambers, M.C., 2007. Food and water security in a changing arctic climate. Environ. Res. Lett. 2, 045018.

Yang, S., Jin, H., Wei, Z., He, R., Ji, Y., Li, X., Yu, S., 2009. Bioremediation of oil spills in cold environments: a review. Pedosphere 19, 371–381.

Yates, C.N., Wootton, B., Jørgensen, S.E., Murphy, S.D., 2012a. Use of wetlands for wastewater treatment in Arctic regions. In: Jorgensen, S.E. (Ed.), Encyclopedia of Environmental Management. Taylor Francis, New York.

Yates, C.N., Wootton, B.C., Murphy, S.D., 2012b. Performance assessment of arctic tundra municipal wastewater treatment wetlands through an arctic summer. Ecological Engineering 44, 160–173.

Zona, D., Lipson, D.A., Zulueta, R.C., Oberbauer, S.F., Oechel, W.C., 2011. Microtopographic controls on ecosystem functioning in the Arctic Coastal Plain. J. Geophys. Res. Biogeosci. 116, G00I08.

Exploratory Performance Testing of a Pilot Scale HSSF Wetland in the Canadian Arctic

Colin N. Yates[a],*, Brent C. Wootton[b], Stephen D. Murphy[a], Sven Erik Jørgensen[c]

[a]*Faculty of Environment, Waterloo, Ontario, Canada*
[b]*Centre for Alternative Wastewater Treatment, Fleming College, Lindsay, Ontario, Canada*
[c]*Professor Emeritus, Environmental Chemistry, University of Copenhagen, Copenhagen, Denmark*
*Corresponding author: e-mail address: cyates@uwaterloo.ca; sd2murph@uwaterloo.ca

21.1 INTRODUCTION

In the context of ecological engineering and restoration (Mitsch and Jørgensen, 2004), constructed wetlands (CWs) have become a popular low-cost, high-efficiency technology for the treatment of many different types of wastewater (Campbell and Ogden, 1999; Kadlec, 2009) and have been applied widely around the world, including tropical, temperate, and cold temperate environments (Greenway and Simpson, 1996; Wittgren and Maehlum, 1997; Wallace et al., 2001). However, what Spieles and Mitsch (2000) stated is still true today—their long-term effectiveness and sustainability requires study. This is especially true in the Arctic regions of Canada where they have yet to be experimentally tested.

In the cold temperate regions of North America, continental Europe, and Scandinavia, performance of CWs has been well documented (Wittgren and Maehlum, 1997). Free water surface wetlands, horizontal subsurface flow (HSSF), and vertical flow (VF) have all been adopted throughout these regions. Mander and Jenssen (2003) describe these treatment wetlands as facing two main operating challenges in cold climates: (1) failure of system hydraulics, due to a change in viscosity or a freezing of the wastewater and (2) the low temperatures leading to inadequate purification.

With respect to temperature in cold climate environments, removal of the chemical oxygen demand (COD) and biological oxygen demand (BOD) has been shown to be uninfluenced down to 5 °C. Greenway and Woolley (1999) and Vymazal (2002) have shown that organic matter removal in wastewater through anaerobic and aerobic bacteria can remain active to 5 °C. However, prolonged temperatures below 5 °C have many limitations for treatment of wastewater in wetlands: environmental variables that may indirectly or directly affect performance include freezing (ice), reduction in microbial community biomass, plant dynamics, and the mineralization of organics. Resulting heat loss in temperate environments generally occurs in late winter (Kadlec, 2009), whereas in an Arctic environment, this would be expected to occur much more rapidly. Even though substantial attention has been given to finding effective measures to limit the effect of temperature on CWs systems, very little is known about these technologies when employed in regions where mean annual temperature is well below 0 °C.

Natural tundra wetland systems have been extensively used for the treatment of wastewater in remote communities of the Canadian Arctic. Wootton et al. (2008) noted that 11 such wetland treatment systems are currently being used in Nunavut. These wetlands are often used to polish continuous discharge from lagoons and facultative lakes, as well as decanted lagoon wastewater, and to treat raw wastewater. Kadlec and Johnson (2008) described a natural wetland in Cambridge Bay, Nunavut, that was augmented or engineered to enhance treatment through the use of berms to channel wastewater and increase residency time. Preliminary observations from tundra treatment wetlands in the Canadian Arctic showed that during the summer months (July to mid-September) treatment of wastewater is high. Yates et al. (2012) observed 94%, 67%, 52%, 92%, 99%, 99.99%, and 99.99% removal of $cBOD_5$, COD, TSS, total phosphorus (TP), NH_3—N, *Escherichia coli* and total coliforms, respectively, for a natural wetland system in Chesterfield Inlet, Nunavut. Treatment during the winter months using treatment wetlands is not feasible due to the climate. However, the successful use of natural wetlands in the Canadian far north suggests that CWs may be a viable alternative technology for remote Arctic communities during summer months. CWs have also been shown to be an economical and a resource-conservative technology appropriate for developing countries, rural areas, and in other small communities in cold temperate climates that have limited ability for large capital investments (Kivaisi, 2001; Werker et al., 2002; Wallace and Knight, 2006). We identified Arctic communities as localities that may potentially benefit from the technology and places to test CWs effectiveness in an extreme cold climate environment. Our objectives were (i) to conduct exploratory studies on the treatment efficacy of a small scale pilot HSSF-CW during the short Arctic summer (a first in the Canadian Arctic) and (ii) to determine how a CWs system would respond in one of the most extreme cold climate wastewater treatment environments.

21.2 METHODS

The Hamlet of Baker Lake (64°N, 96°W) is the only inland community of Nunavut. Baker Lake has average summer (June–August) temperatures between 5 and 12 °C, and mid-winter (December–March) −27 to −32.3 °C. The yearly average temperature for the community is −11.8 °C (Environment Canada, 2010). The landscape is dominated by low granite ridges and a glacial till moraine with underlying mineral soils. The current municipal wastewater treatment facility is composed of a small detention pond ($\sim$60 m^2) that drains overland through a sedge wetland and into a series of small natural lakes with riparian wetland complexes between. Currently, the community discharges 167 m³/day (167,000 L/day) into the holding pond.

The HSSF system tested in this study consists of four in-line cells, with a total treatment area of 15 m², designed by fourth author, Jorgensen (Table 21.1). The cells were built with recycled insulated fiberglass holding tanks and connected with 0.025-m diameter polyvinyl piping. The influent piping feeding the CW was installed through the berm side of the pretreatment holding pond and sunk below the surface. Piping was shallow buried to minimize late and early season freezing.

Local screened aggregate with a porosity of 0.40 was used as the bed media. Perforated sampling ports were installed in the media at the influent and effluent of each wetland cell. Each of the cells were planted with approximately 10 (dependent on plug size) *Carex aquatilis* (Stans) and *Poa glauca* (Vahl) plugs, two species that are indigenous to the adjacent natural treatment wetland. These species were selected because they have been commonly found in areas of high wastewater loading, are known to be nitrophilic, and demonstrate phenotypic plasticity (Aiken, 2007). Additional plugs were planted in 2009 to increase vegetation cover in the cells. In 2008, wastewater was fed through the system to establish the plant community and biofilm. Wastewater flow (m³/day) was measured with a collection tank, which was emptied daily. We sampled the system in the summer of 2009 (June 21–August 10) and again in 2010 (June 21–August 13) corresponding with the frost-free season in the community. Samples were collected from the holding tank and from the effluent of the system three times per week. In 2009, the system was fed minimally pretreated wastewater from the community of Baker Lake, and in 2010, the wastewater was diluted to reduce the organic load. In 2010, the organic load was reduced from 66 kg BOD ha^{-1}day^{-1} to approximately 23 kg BOD ha^{-1}day^{-1}. We maintained an average theoretical hydraulic residency time (HRT) of ~9 d for both years. Longer residency times of 8–14 d have been shown to be more affective in temperatures below 15 °C (Akratos and Tsihrintzis, 2007). Through both summer field seasons, we sampled for COD, cBOD$_5$, total suspended solids (TSS), *E. coli*, total coliforms,

Table 21.1 Pilot Constructed Wetland Dimensions in Baker Lake

Cell #	Length (m)	Width (m)	Area (m²)	Depth of Water (m)	Depth of Gravel (m)	Total Saturated Water Volume (m³)	Water Only Volume (0.4 Porosity) (m³)
1	2.26	1.98	4.47	0.33	0.36	1.48	0.59
2	2.16	1.73	3.74	0.37	0.51	1.38	0.55
3	2.16	1.73	3.74	0.3	0.51	1.12	0.46
4	2.13	1.55	3.30	0.38	0.46	1.25	0.50
Total			15.25	$\bar{x}=0.35$	$\bar{x}=0.46$	5.25	2.10

TP, ammonia $(NH_3 - N)$, and temperature. Additional parameters of nitrate $(NO_3^- - N)$, and phosphate $(PO_4^{3-} - P)$ were more extensively monitored in 2010. Parameters were analyzed according to standard methods for water and wastewater (Eaton and Franson, 2005).

We calculated expected effluent concentrations using the first-order kinetic model (P-k-C*) in order to compare observed effluent values for cBOD$_5$ and TSS (Equations 21.2 and 21.3). We recalculated the rate constant for the P-k-C* model at 10 °C using the van't Hoff–Arrhenius equation as described in Crites and Tchobanoglous (1998) (Equation 21.1):

$$\frac{d(\ln k)}{dT} = E/RT^2 \tag{21.1}$$

The P-k-C* model is described in Campbell and Ogden (1999) as:

$$As = \frac{Q(\ln Co - \ln Ce)}{k_t * d * n} \tag{21.2}$$

The k_t value for the P-k-C* model was determined by using a k_{10} value of 1.0; the θ-factor used was 1.14. A high θ-factor was deemed appropriate for extreme temperature cases as determined for a Minnesota HSSF wetland with a temperature range from 1–17 °C, as outlined in Kadlec (2009).

The equation for TSS removal also described in Campbell and Ogden (1999) as:

$$TSS_{eff} = TSS_{inf} * (0.0158 + 0.0011 * HLR) \tag{21.3}$$

21.3 RESULTS/DISCUSSION

cBOD$_5$ in the Baker Lake holding cell was observed to be an average of 421 mg/L for the summer of 2009 (Table 21.2). In 2010, the diluted wastewater that was fed to the system was maintained at an average cBOD$_5$ concentration of 164 mg/L.

Average removal of wastewater constituents was observed to be greatest during last week of July, 2009 (Table 21.2). Performance of the wetland would be expected to be highest during this time in Baker Lake CW because this would correspond with the season's highest average mean daily air temperature (11.4 °C in July) (Environment Canada, 2010). The observed average temperature of wetland effluent was 11.8 °C, while the influent wastewater temperature exhibited a summer time average of 17.1 °C.

In 2009, we observed promising performance in the HSSF system, despite the high organic load from minimally pretreated wastewater. The percent removal of cBOD$_5$ and COD averaged approximately 25–32% change in concentration, respectively. Organic solids and total solids removal were observed to be 33% and 53%, respectively. Pathogen parameters of *E. coli* and total coliforms also showed

Table 21.2 The Average Weekly Loading and Performance of Pilot System from July to Mid-August 2009

Week	Parameter	Flow (L/day)	Loading (kg/ha/day)	Influent Concentration	Obs. Effluent Concentration	Expected Effluent	% Change
1	COD	387	253	998	773		23
	cBOD$_5$		118	464	212	66	54
	DO		0.15	0.6	1.5		−60
	TSS		43	168	36	11	79
	VSS		13	52	20		62
	E. coli		–	6.61	5.78		0.83[a]
	Total coliforms		–	8.83	6.93		1.90[a]
	NH$_3$		28	110	62		44
	TP		3.8	15	15		0
	Temp. (°C)		–	19.1	12.3		36
2	COD	383	158	628	476		24
	cBOD$_5$		97	384	380	53	1
	DO		–	ND	4.4		ND
	TSS		47	189	56	11	70
	VSS		37	149	31		79
	E. coli		–	8.45	5.48		2.97[a]
	Total coliforms		–	9.29	6.61		2.68[a]
	NH$_3^+$		20	80	96		−20
	TP		3.4	13.4	14.5		−8
	Temp. (°C)		–	18.2	12.9		29
3	COD	240	150	952	567		40
	cBOD$_5$		78	493	434	12	12
	DO		0.09	0.4	3.5		−89
	TSS		22	141	17	10	88

Continued

Table 21.2 The Average Weekly Loading and Performance of Pilot System from July to Mid-August 2009—cont'd

Week	Parameter	Flow (L/day)	Loading (kg/ha/day)	Influent Concentration	Obs. Effluent Concentration	Expected Effluent	% Change
	VSS		20	128	16		88
	E. coli		–	6.60	5.47		1.13[a]
	Total coliforms		–	8.21	6.79		1.42[a]
	NH$_3$		17	110	101		8
	TP		2.7	17	17.9		–5
	Temp. (°C)		–	15	9.5		37
4	COD	167	90	824	565		31
	cBOD$_5$		36	333	256	4	23
	DO		0.03	0.3	4.8		–94
	TSS		12	108	23	10	79
	VSS		11	98	21		65
	E. coli		–	6.60	5.47		1.13[a]
	Total coliforms		–	8.11	6.51		1.60[a]
	NH$_3^+$		12	110	107		3
	TP		1.6	14.4	14.6		–2
	Temp. (°C)		–	16.9	12.4		26
5	COD	101	63	956	454		53
	cBOD$_5$		30	460	145	0.3	69
	DO		0.02	0.3	1.8		–83
	TSS		13	200	37	11	82
	VSS		12	185	31		83
	E. coli		–	6.38	5.48		0.90[a]
	Total coliforms		–	8.36	6.37		1.99[a]
	NH$_3^+$		4.0	60	80		–33
	TP		1.1	17	12.1		29
	Temp. (°C)		–	17.3	12.5		28

6						
COD	149	77	792	691		13
cBOD$_5$		38	393	459	2	−17
DO		0.02	0.16	3.2	11	−95
TSS		21	211	39		82
VSS		19	193	34		82
E. coli		–	6.60	5.48		1.12[a]
Total coliforms		–	8.84	6.38		2.46[a]
NH$_3^+$		6.8	70	92		−31
TP		ND	ND	13.1		ND
Temp. (°C)		–	16.3	10.9		33
Avg.						
COD	238	134	858	588		32
cBOD$_5$		66	421	314	17	25
DO		0.062	0.4	3.2		−88
TSS		26	169	80	11	53
VSS		20	134	90		33
E. coli		–	7.70	5.54		2.16[a]
Total coliforms		–	8.80	6.65		2.15[a]
NH$_3^+$		14	90	90		0
TP		2.4	15.1	14.34		5
Temp. (°C)		–	17.1	11.8		31

All parameters in mg/L unless otherwise stated. ND indicates an absence of data during the week of study.
[a]Log units. E. coli and total coliforms in log_{10} CFU/100 mL.

promising changes as concentrations were reduced by a few logs. The only parameters to change minimally were NH_3—N and TP, as would be expected in a HSSF system fed with minimally pretreated wastewater in cold climates.

In response to our findings in 2009, we replicated the experiment in 2010 but reduced the organic loading rate by diluting the wastewater entering the system, believing the overall percent removal could be improved. This was done to accommodate for the low BOD removal values from 2009. Average loading for 2010 was calculated as 23 kg BOD $ha^{-1}day^{-1}$ compared to 66 kg BOD $ha^{-1}day^{-1}$ in 2009. Despite the reduction in loading in 2010, we observed no observable decrease in concentration of wastewater parameters from the influent to the effluent of the system as we had expected. In most samples, the concentrations of in BOD, COD, TP, and TSS exiting the wetland had increased, not decreased as anticipated (Table 21.3). Only pathogens, *E. coli* and total coliforms, were observed to decrease in concentration in the "treated" effluent (Table 21.3). Also concentrations of $NH_3 - N$ were observed to increase in the wetland effluent, despite large increases of $NO_3^- - N$.

Again, we would like to recognize the limitations of the short duration of the study periods but seek to describe our findings as purely exploratory as we test CWs systems in some of the most extreme cold climate operating conditions. We present the following discussion with this understanding, while exploring the influence of Arctic climatic conditions on treatment performance.

Table 21.3 The Average Flow, Loading, and Performance of Pilot System in 2010

Parameter	Flow (L/day)	Loading (kg/ha/day)	Influent Concentration	Obs. Effluent Concentration	Expected Effluent	% Change
COD	210	35.8	260	369		−30
cBOD$_5$		22.7	164	200	4.5	−18
DO		0.94	6.8	1.9		−72
TSS		3	21.7	25	9	−13
VSS		0.68	4.9	1.6		68
E. coli		–	4.51	3.95		0.56[a]
Total coliforms		–	6.91	6.40		0.51[a]
NH$_3$		0.66	4.8	25.5		−431
TP		0.3	2.2	5.5		−60
NO$_3^-$		0.1	0.51	0.56		−9
PO$_4^{3-}$		0.5	3.8	10.6		−64
Temp. (°C)		–	12.7	10.8		15

All values presented in mg/L unless otherwise stated.
[a]*Log units. E. coli and total coliforms in log_{10} CFU/100 mL.*

In an Arctic environment, temperature is the greatest factor indirectly limiting the treatment of wastewater by negatively impinging on a number of process functions important for the mineralization of organic matter and nutrient cycling (Pugnaire and Valladares, 2007). In natural Arctic environments, microbial communities are at their lowest population levels during the summer months. This is because of a lack of available nutrients, after microbial communities have used much of the available C and N in the soil in the early spring (Edwards and Jefferies, 2010). In a nutrient- and carbon-enriched environment such as a treatment wetland, N and C should not be the limiting factors of microbial growth; therefore, temperature and oxygen are the most likely causes of reduced decomposition of organic matter and other wastewater contaminants in a treatment wetland environment.

As reported in the results, we observed low organic matter removal in our treatment system in 2009. HSSF systems generally have lower organic removal than other systems such as VF CWs because oxygen is limiting in HSSF but is required for the mineralization of organic material through microbial respiration (Gajewska and Obarska-Pempkowiak, 2009). Experts initially hypothesized that vegetation in HSSF systems would oxygenate the rhizosphere, providing aerobic conditions necessary for nitrification and to allow the mineralization of organic matter (Brix, 1997). However, later studies consistently revealed that the actual transfer of oxygen from plant roots was insufficient for mineralization of organic matter, and especially to nitrify $NH_3 - N$ (Tanner, 2001; Tanner and Kadlec, 2003; Kouki et al., 2009).

Therefore, again the importance of temperature in the breakdown of organic matter in cold climate HSSF systems comes to the forefront. Gikas et al. (2007) observed that temperature did not strongly impede organic matter mineralization down to 5 °C. But organic matter mineralization rates are much lower in Arctic environments than in temperate regions (Sullivan et al., 2008). Therefore, organic matter accumulation in consistently colder temperatures is much greater, explaining the low observed COD and $cBOD_5$ removal rates from the CWs trials of 2009. This is particularly important to subsurface flow systems, where organic matter accumulation can lead to clogging and saturation of the bed media (Wallace and Knight, 2006). Also in 2009, we observed TSS removal in the pilot systems up to 80%, indicating that sedimentation and filtration was occurring in the system. Expected removal for TSS was calculated to be approximately 95% as predicted from the Campbell and Ogden (1999) equation (see Equation 21.3). The high remaining BOD values would indicate that that decomposition of the organic material was not occurring, even for the smaller than expected removals of suspended matter observed in the system. Despite changing the k value in the P-k-C* model to represent Arctic summer temperature conditions, expected BOD levels were not representative of observed BOD levels in the effluent of the pilot system.

After two seasons of operation, the HSSF system appeared to fail in the second year of operation, producing increases in concentration of parameters in the wetland effluent. We hypothesize that there were a number of factors contributing to the poor treatment. First, a major factor was likely the overloading of the system in 2009, which led to a saturation of the bed media with organic matter. The increased concentrations of water quality parameters COD, cBOD$_5$, and TSS in the effluent in 2010 point toward this hypothesis, because dissolved organic and particulate matter may have been resuspended during treatment thus contributing to the higher values exiting the wetland system. The higher effluent values observed in 2010 suggests that the wetland system was actually contributing organic matter and nutrients to the waste stream as it traveled through the wetland. We calculated that from the 2009 field season approximately 1 kg of TSS remained in system, of which 0.5 kg was in the form of volatile solids. Kadlec (2009) and Knowles et al. (2011) both suggest that in HSSF wetlands, water velocity is not enough to produce a shear force capable of causing resuspension or disassociation of particles. Rather than picking up particles, surfacing of water on top the media would likely be the result of complete clogging of pore space with solids (Maloszewski et al., 2006). Because we did not observe surfacing of wastewater in the system, it would suggest that significant clogging of the pore space was not occurring. This observation suggests that particle retention on the media surfaces was poor because of oversaturation, poor electrostatic interaction between the particles and the bed media, and/or a difference in ionic strength of incoming wastewater. Hermansson (1999) states that adhesion of particles to a surface is dependent on the media, bulk fluid, and charge on the particle. If attachment was poor, Knowles et al. (2011) suggest that in SSF wetlands, particles could be released back into solution by peptization. This would result in the release of any number of particles back in solution, including phosphorus, solids, and dissolved organics, which we observed in 2010.

Another reason for the poor performance in 2010 may be related to a low rate of mineralization caused primarily by low temperatures and an anaerobic environment, which ultimately led to incomplete mineralization of organic matter and an accumulation of this organic fraction within the system. The additional load was observed in 2010 as unrespired forms of organic C, N, and P remaining in the media accounting for the elevated cBOD$_5$, TP, and NH$_3$—N in the wetland effluent. These results suggest that the system's primary treatment mechanism was sedimentation, rather than the decomposition and transformation of organics and nutrients as initially thought may be occurring after the 2009 trial. When comparing these findings to decomposition and mineralization of nutrients in Arctic tundra soils, it would be expected that decomposition of organic matter and respiration of C would be very slow, especially in an anaerobic system (Sullivan et al., 2008), because waterlogging, cold temperatures, and soil quality can work to stabilize C, P, and N in the soil in Arctic environments. Furthermore, buried soil organic matter has been shown to have significantly reduced mineralization rates in Arctic soils (Kaiser et al., 2007). Despite a rest period following the 2009 sampling when the system was turned off during the freezing months, accumulated organic material in the system was still not mineralized, as suggested could happen in southern conditions (Platzer and Mauch, 1997). This process could explain why in 2010 we observed increases in PO$_4$$^{3-}$ − P concentration through each consecutive cell despite the decrease in oxygen and yet no decrease in BOD or COD. The only evidence we observed that refutes this hypothesis was the mean decrease in concentration of volatile suspended solids (VSS) in 2010. This result conflicts with the increase in TSS, cBOD$_5$, and COD in 2010. The mean increase in TSS would suggest a mineral fraction was responsible for the increase in concentration, rather than released particulate organics, whereas the increase in cBOD$_5$ and COD suggest dissolved organics were in excess in the system.

An analysis of the soil media extracted from the cells following the completion of the studies provides further indication that the bed was saturated and releasing excess nutrients in 2010. Analysis of nutrients showed soil solution concentrations orders of magnitude in greater concentration than influent water (Table 21.4).

Higher concentration of nutrients in the soil solution of the wetland bed media would allow for dissociation from the soil solution into the lower concentrated pore water, resulting in the higher concentrations of effluent observed. For example, the mean concentration of $NH_3 - N$ in the wetland effluent from 2010 (Table 21.3) is very similar in concentration to concentrations observed in the soil solution of the bed media (Table 21.4).

Finally, the presence of elevated $NO_3^- - N$ in the wetland effluent in 2010 may be explained by poor vertical mixing of water through the soil media, resulting in vertical stratification and preferential flows of varying temperatures and dissolved oxygen concentration. Warmer soil temperatures and higher dissolved oxygen in the upper layers of the soil media would result in an optimal environment for the nitrification of $NH_3 - N$. Although Kadlec (2009) suggest vertical stratification caused by temperature happens infrequently in HSSF systems, we believe it should not be discounted in an environment where soil temperatures are consistently below 5 °C. Also, porous media have been shown to be prone to stratification due to differences in electrical conductivity, and gravel media often prevent adequate mixing. Kadlec et al. (2003) observed such an event in an HSSF located in the cold temperate climate of northern Minnesota, where differences in conductivity of varying water sources fed into system resulted in a large vertical stratification and less treatment occurring in the deeper flow paths. Similarly, we observed high conductivity wastewater fed into the Baker Lake pilot system in 2009, on average 1210 μS, and in 2010, we observed conductivity of influent to be on average of 245 μS from the diluted wastewater. Had we observed concentrations of wastewater at different depths in the system as Kadlec et al. (2003), we may have also recorded less treatment with greater depth.

Table 21.4 Soil Solution Concentrations Observed from Each Treatment Cell Following Study Period

Parameter	Cell 1	Cell 2	Cell 3	Cell 4	Mean
TP	1100	1010	964	1090	1041
PO_4^{3-}	1.8	3.9	7.2	4	4.2
NH_3^+	22.7	26.4	33.9	33.9	29.2
All concentrations were recorded in mg/L.					

None of presented causes for the poor 2010 performance are likely sufficient in themselves to be solely responsible. In fact, we believe all explanations provided could have, in part, either independently or interrelationally contributed to the poorer performance. For example, the low conductivity wastewater fed into the system in 2010 could have favored disassociation of ionic particles remaining in the system from the 2009 trials, leading to elevated nutrients exiting the wetland, while also causing vertical stratification within media. We suggest that further studies should be undertaken to expand our understanding of CWs in extreme cold climate environments, specifically investigating optimal depth of media and monitoring temperature at different depths, as well as studies to examine different organic loading and its corresponding mineralization rates in the Canadian Arctic.

21.4 CONCLUSIONS

Despite promising performance results of the pilot HSSF system in 2009, the pilot HSSF failed in 2010, even with a lower organic load to the system. Several factors potentially led to the systems failure. High organic loading prior to biofilm and plant establishment and high organic loading during the first year of study saturated the system with organics. This, coupled with the use of predominately anaerobic technology in an extreme cold climate environment, would cause mineralization of organics to be very slow. The result was an additional organic load being exerted on the system during 2010 study, which was observed in the increased concentrations of BOD in the effluent. Also, the use of dilute wastewater could have created vertical stratification of the pore water and/or providing a low electrical conductivity environment causing dissociation of weakly adhered particles to the media. This may explain the greater concentrations of N and P ions in the 2010 effluent compared to influent.

We suggest further studies examining the influence of soil depth on subsurface treatment, as well as continuing to investigate appropriate organic loading for constructed in extreme cold climates. In environments such as the Canadian Arctic, deeper substrate mediums in subsurface flow systems may not be appropriate given the short frost-free period.

Acknowledgments

We would like to thank the Government of Canada Programme for the International Polar Year and Nasivvik Centre for Inuit Health and Changing Environments for funding this project. Also a special acknowledgment for their assistance is extended to the Hamlet of Baker Lake, the Nunavut Community Government Services, and the staff of Baker Lake's Jonah Amitnaaq Secondary School. We are also very appreciative of the hard work of all the staff at the Centre for Alternative Wastewater Treatment at Fleming College.

References

Aiken, S.G., 2007. Flora of the Canadian Arctic Archipelago. NRC Research Press: Canadian Museum of Nature, Ottawa.

Akratos, C.S., Tsihrintzis, V.A., 2007. Effect of temperature, HRT, vegetation and porous media on removal efficiency of pilot-scale horizontal subsurface flow constructed wetlands. Ecol. Eng. 29, 173–191.

Brix, H., 1997. Do macrophytes play a role in constructed treatment wetlands? Water Sci. Technol. 35 (5), 11–17.

Campbell, C.S., Ogden, M.H., 1999. Constructed Wetlands in the Sustainable Landscape. Wiley, New York.

Crites, R., Tchobanoglous, G., 1998. Small and Decentralized Wastewater Management Systems. WCB/McGraw-Hill, Boston.

Eaton, A.D., Franson, M.A.H., 2005. Standard Methods for the Examination of Water and Wastewater. American Public Health Association, Washington, DC.

Edwards, K.A., Jefferies, R.L., 2010. Nitrogen uptake by Carex aquatilis during the winter-spring transition in a low Arctic wet meadow. J. Ecol. 98 (4), 737–744.

Environment Canada, 2010. Canadian Climate Normals. In: http://www.climate. weatheroffice.gc.ca/climate_normals/index_e.html. Accessed June 10, 2010.

Gajewska, M., Obarska-Pempkowiak, H., 2009. 20 years of experience of hybrid constructed wetlands exploitation in Poland. Rocz. Ochr. Sr. 11, 875–888.

Gikas, G.D., Akratos, C.S., Tsihrintzis, V.A., 2007. Performance monitoring of a vertical flow constructed wetland treating municipal wastewater. Global Nest J. 9 (3), 277–285.

Greenway, M., Simpson, J.S., 1996. Artificial wetlands for municipal wastewater treatment and water re-use in tropical and arid Australia. In: Recycling the Resource—Proceedings of the Second International Conference on Ecological Engineering for Wastewater Treatment, 5–6, pp. 367–369.

Greenway, M., Woolley, A., 1999. Constructed wetlands in Queensland: performance efficiency and nutrient bioaccumulation. Ecol. Eng. 12 (1–2), 39–55.

Hermansson, M., 1999. The DLVO theory in microbial adhesion. Colloids Surf. B Biointerfaces 14 (1–4), 105–119.

Kadlec, R.H., Johnson, K., 2008. Cambridge Bay, Nunavut, wetland planning study. J. Northern Territories Water Waste Assoc. Fall, 30–33.

Kadlec, R.H., 2009. Treatment Wetlands. CRC Press, Boca Raton.

Kadlec, R.H., Axler, R., McCarthy, B., Henneck, J., 2003. Subsurface treatment wetlands in the cold climate of Minnesota. In: Mander, Ü., Jenssen, P.D. (Eds.), Constructed Wetlands for Wastewater Treatment in Cold Climates. WIT Press, Boston, pp. 19–52.

Kaiser, C., Meyer, H., Biasi, C., Rusalimova, O., Barsukov, P., Richter, A., 2007. Conservation of soil organic matter through cryoturbation in arctic soils in Siberia. J. Geophys. Res. Biogeosci. 112 (G2), G02017.

Kivaisi, A.K., 2001. The potential for constructed wetlands for wastewater treatment and reuse in developing countries: a review. Ecol. Eng. 16 (4), 545–560.

Knowles, P., Dotro, G., Nivala, J., García, J., 2011. Clogging in subsurface-flow treatment wetlands: occurrence and contributing factors. Ecol. Eng. 37 (2), 99–112.

Kouki, S., M'hiri, F., Saidi, N., Belaid, S., Hassen, A., 2009. Performances of a constructed wetland treating domestic wastewaters during a macrophytes life cycle. Desalination 246 (1–3), 452–467.

Maloszewski, P., Wachniew, P., Czuprynski, P., 2006. Study of hydraulic parameters in heterogeneous gravel beds: constructed wetland in Nowa Slupia (Poland). J. Hydrol. 331 (3–4), 630–642.

Mander, Ü., Jenssen, P.D., 2003. Constructed Wetlands for Wastewater Treatment in Cold Climates. WIT Press, Southampton.

Mitsch, W.J., Jørgensen, S.E., 2004. Ecological Engineering and Ecosystem Restoration. John Wiley and Sons, New Jersey.

Platzer, C., Mauch, K., 1997. Soil clogging in vertical flow reed beds—mechanisms, parameters, consequences and solutions? Water Sci. Technol. 35 (5), 175–181.

Pugnaire, F.I., Valladares, F., 2007. Functional plant ecology. CRC Press, Boca Raton.

Spieles, D., Mitsch, W., 2000. The effects of season and hydrologic and chemical loading on nitrate retention in constructed wetlands: a comparison of low- and high-nutrient riverine systems. Ecol. Eng. 14 (1–2), 77–91.

Sullivan, P.F., Arens, S.J.T., Chimner, R.A., Welker, J.M., 2008. Temperature and microtopography interact to control carbon cycling in a high arctic fen. Ecosystems 11 (1), 61–76.

Tanner, C.C., 2001. Plants as ecosystem engineers in subsurface-flow treatment wetlands. Water Sci. Technol. 44 (11–12), 9–17.

Tanner, C.C., Kadlec, R.H., 2003. Oxygen flux implications of observed nitrogen removal rates in subsurface-flow treatment wetlands. Water Sci. Technol. 48 (5), 191–198.

Vymazal, J., 2002. The use of sub-surface constructed wetlands for wastewater treatment in the Czech Republic: 10 years experience. Ecol. Eng. 18 (5), 633–646.

Wallace, S., Knight, R.L., 2006. Small-Scale Constructed Wetland Treatment Systems; Feasibility, Design Criteria and OandM Requirements. WERF, UK.

Wallace, S., Parkin, G., Cross, C., 2001. Cold climate wetlands: design and performance. Water Sci. Technol. 44 (11–12), 259–265.

Werker, A.G., Dougherty, J.M., McHenry, J.L., Van Loon, W.A., 2002. Treatment variability for wetland wastewater treatment design in cold climates. Ecol. Eng. 19 (1), 1–11.

Wittgren, H.B., Maehlum, T., 1997. Wastewater treatment wetlands in cold climates. Water Sci. Technol. 35 (5), 45–53.

Wootton, B., Durkalec, A., and Ashley, S., 2008. Canadian Council of Ministers of the Environment Draft Canada-wide Strategy for the Management of Municipal Wastewater Effluent: Nunavut Regional Impact Analysis. Inuit Tapiriit Kanatami.

Yates, C.N., Wootton, B.C., Murphy, S.D., 2012. Performance assessment of arctic tundra municipal wastewater treatment wetlands through an arctic summer. Ecol. Eng. 44, 160–173.

Practical Aspects, Logistical Challenges, and Regulatory Considerations for Modeling and Managing Treatment Wetlands in the Canadian Arctic

22

Colin N. Yates[a],*, Gordon C. Balch[b], Brent C. Wootton[b], Sven Erik Jørgensen[c]

[a]*Faculty of Environment, Waterloo, Ontario, Canada*
[b]*Centre for Alternative Wastewater Treatment, Fleming College, Lindsay, Ontario, Canada*
[c]*Professor Emeritus, Environmental Chemistry, University of Copenhagen, Copenhagen, Denmark*
**Corresponding author: e-mail address: cyates@uwaterloo.ca*

22.1 INTRODUCTION

Communities in the Canadian Arctic and sub-Arctic are remote, sparsely populated, and resource dependent (Chabot and Duhaime, 1998; Parlee and Furgal, 2012). Many communities are not connected to transportation corridors to southern Canada and are therefore dependent upon self-supported infrastructure to deliver community services, such as wastewater treatment, solid waste disposal, and drinking water (Warren et al., 2005; Ritter, 2007). However, many of the communities lack economic resources, access to trained personnel, and infrastructure to deliver services as would be expected for municipalities commonly found in Canada (Johnson, 2010). Further, extreme climatic conditions make infrastructure commonly used for wastewater treatment in temperate developed regions of Canada impractical, especially those used for the treatment of wastewater.

Yates et al. (2012a) used the term *extreme cold climate wastewater treatment* to describe the treatment of municipal wastewater in the Canadian Arctic, because the mean annual ambient air temperatures of many Canadian Arctic communities are well below 0 °C. In addition, environmental extremes include treatment systems that overtop permafrost, as well as periods of 24 h of daylight or darkness. The influence of all these environmental factors on the treatment of wastewater is still largely unknown.

In the past, small Arctic communities in Canada, as well as other remote Arctic communities in Alaska and Greenland, relied upon dilution of small volumes of wastewater by discharging them directly into natural sources of receiving water

Developments in Environmental Modelling, Volume 26, ISSN 0167-8892, http://dx.doi.org/10.1016/B978-0-444-63249-4.00024-5

(Ritter, 2007; Jenssen, 2011; Yates et al., 2012a). However, with rapid population growth and continuous exploration for natural resource development, demand for monitoring of existing systems, the construction of treatment systems has become warranted. Many communities in the Canadian Arctic have relied on direct discharge into tundra wetlands (Wootton and Yates, 2010). Regulatory monitoring of many systems, specifically in the Canadian Arctic, has been very limited, and those that have been monitored have been poorly documented.

In this chapter, we review and discuss technical elements for the application of beneficial practices of management, design, and regulation of wetland treatment systems in the Arctic, with specific reference to systems in the Canadian Arctic, and excerpts from a few case studies conducted in Alaska and Greenland. We also make reference to current understanding of wastewater treatment in Arctic conditions and the known performance of Arctic tundra treatment wetlands.

22.1.1 Arctic context

The Arctic tundra is one of the world's most extreme biomes, with winter temperatures that can dip below $-50\,°C$. On occasion, temperatures will rise above $20\,°C$ in some locations during the summer; however, most summer averages remain around $10\,°C$. Precipitation is also characteristically low, 15–25 cm/year. The tundra is primarily characterized by permafrost, which limits plant growth and reproduction. As a result, the tundra is one of the lowest biomass-producing biomes, with low biodiversity. The mineralization of organic matter and the development of soil are slow. Biotic processes are strongly influenced directly or indirectly by temperature and light.

Due to the extreme climatic conditions, the Arctic tundra has seen minimal human activity until the past several decades, which have coincided with increased exploration for natural resources and the rapid growth of many communities. In the past three decades, the Canadian Arctic has seen particularly rapid growth and development. Many Arctic communities have undergone a rapid growth in population, as well as growing corporate interest in resource development and the construction of their associated camps. Despite the rapid growth, many communities remain remote, isolated from transportation corridors, and struggling with various socioeconomic barriers as well as a maturing of their political legitimacy (Chabot and Duhaime, 1998).

Still, at present, the Canadian Arctic remains one of the most sparsely populated regions in the world. Many of the communities have populations under 1000 people, yet have population growth expectations much higher than the Canadian average. Currently, communities in the tundra portions of Canada have a population of approximately 40,000 in 50 communities spread throughout the region (Statistics Canada, 2006). In boreal portions of Arctic Canada, another 36 municipal systems would be included. However, the majority of the boreal communities are connected by physical transportation corridors and therefore have access to a different suite of resources. However, this does not imply that management and treatment of wastewater in these communities is not a challenge.

Despite their small size, all of the communities in the tundra region are required to support the provision of their own municipal services for waste collection and disposal, drinking water treatment and distribution, as well as compliance monitoring reporting. With respect to the treatment of wastewater, almost all communities haul wastewater to lagoons, facultative lakes, or holding ponds that continuously discharge onto the tundra. A few examples of communities that have sewer or utilidor systems include Rankin Inlet and Iqaluit in the territory of Nunavut and Inuvik in the Northwest Territories, all of which lie above 62°N latitude (Figure 22.1). A number of dated reports exist providing descriptions of the systems in the Nunavut Territory, Northwest Territories, Northern Quebec, and Labrador (Johnson and Wilson, 1999; Sikumiut Environmental Management Ltd, 2008; Wootton et al., 2008a,b,c). However, a number of the systems have been upgraded since these reports were released.

22.2 LOGISTICAL CHALLENGES

Given the remote nature of communities in the Canadian Arctic, there are a number of logistical challenges that confront the management and treatment of wastewater in the region. Surprisingly, adequate locations for the siting of wastewater management facilities are at a premium in many Arctic communities despite the vast surrounding landscape. The majority of Arctic communities lie along coastlines, which are often rocky, barren, and have an absence of material suitable for the construction of lagoon berms. As a result, many communities rely on (i) either facultative lakes or natural depressions on the landscape that can be easily bermed to temporarily hold the wastewater and (ii) tundra wetlands that receive the primarily treated effluent from the berms and in turn provide additional treatment before final discharge, which is often to flow into some receiving body of water (often Hudson Bay or the Arctic Ocean). The physical distance between the management zone and the community is also an important consideration in the siting of these facilities. Harsh long winters make navigation of routes to disposal areas problematic, especially during frequent ground blizzard conditions that can last for days (Grainge, 1969).

Many of the logistical challenges regarding wastewater management faced by Arctic communities at the time of their formation in the early 1960s remain. Some early literature on wastewater management in the Arctic describes the challenges for Arctic and sub-Arctic wastewater management (see Dawson and Grainge, 1969; Grainge, 1969; Tilsworth and Smith, 1984). Grainge (1969) initially suggested that utilidor networks (heated piped systems) for Arctic communities were superior to haulage systems, because of lower maintenance cost, lower chance of contamination, and a reduced chance of accidental spillage. However, Grainge also clearly described the logistical challenges with respect to designing piped systems in Arctic communities because of permafrost, topography, poor town layout, and the high construction costs.

The complex nature of Arctic communities has meant that the logistical challenges of disposal and management have largely remained unchanged to this day.

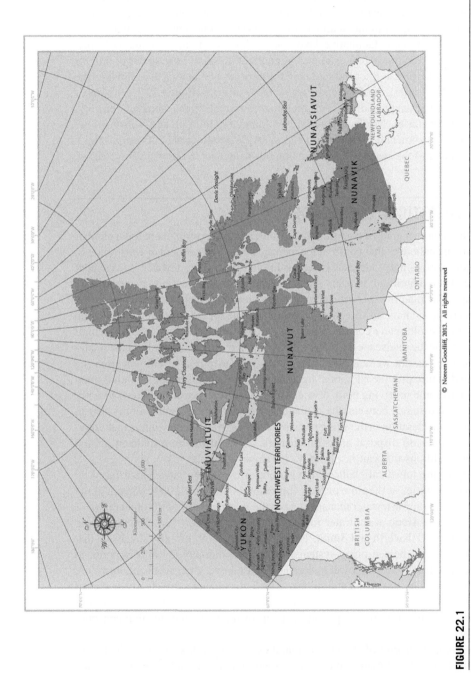

FIGURE 22.1

Remote communities in the Canadian Arctic. (For color version of this figure, the reader is referred to the online version of this chapter.)

Map created by Noreen Goodliff.

Haulage systems remain the most common form of collection, despite periods of no service because of service repairs to vehicles or inclement weather. Because of the way wastewater is managed, it is difficult to quantify volumes of wastewater discharged into the treatment system. Best estimates are generated by knowing the volume of potable water delivered to each building, which is reported by municipalities under the territorial water-taking license agreements (Government of Nunavut, 2002).

A major logistical issue for all Arctic communities is access to trained personnel required for the management and operation of municipal infrastructure, including wastewater facilities. This administrative challenge can be found at multiple levels government; often expressed as a lack of resources for hiring, and even more so, a lack of resources for training and retaining qualified personnel (Johnson, 2010). Constant shuffling of staff results in poor management of records and ultimately an erosion of the community's knowledge-base (Johnson, 2008). This has become an endemic problem for the majority of Canadian Arctic communities.

The lack of accredited laboratories close to Arctic communities creates a logistical hurdle for compliance testing required for wastewater facilities. The analyses of many wastewater parameters are time sensitive and require completion within defined time lines. Sample shipments often require more than the 24 h to reach the closest laboratory for analysis. Therefore, when samples are collected for regulatory purposes, the quality of the sample arriving at the laboratory is often poor.

22.2.1 The wastewater system
22.2.1.1 Handling

Wastewater streams in Arctic communities are often quite homogenous. In most communities, it only contains blackwater and gray water. In some communities, gray water is discharged directly onto the land from the residence. Wastewater from commercial sources is typically limited to a few service providers in the community, generally consisting of a hotel, grocery, and hardware stores. Industrial wastewater is not a contributing factor in the majority of Arctic communities. Occasionally, communities will have process/packaging plants for fish or other locally harvested foodstuffs. Airports and fueling depots provide the only other source of industrial wastewaters. However, must of the waste generated in these systems is not incorporated into the municipal waste stream, as these systems are often diffused across the landscape.

As described earlier, wastewater is managed at the source most commonly by haulage systems. Residences and commercial buildings utilize temporary insulated holding tanks located just outside or under raised buildings. Tanks are pumped out and sewage hauled to the disposal area via pumper trucks. Drinking water is also distributed via tanker trucks. As a result, water use in these communities is significantly less than the Canadian average. The average for many communities in Nunavut is approximately 90 L/day/per, but some communities have usage averages below 70 L/day/per (Wootton and Yates, 2010). In these communities, low usage is often the result of logistical issues described earlier, particularly, the breakdown of chlorination plants or pumper trucks. The current system also accounts for unsanitary conditions present at various periods throughout Arctic as a result of low water usage and poor management (Warren et al., 2005; Ritter, 2007).

22.2.1.2 Wastewater treatment

As discussed by Yates et al. (2012a) and Gunnarsdottir et al. (2013), treatment systems in the Arctic are generally simple or nonexistent. Gunnarsdottir et al. (2013) found in Greenland that all wastewater is directly discharged into the ocean. Although direct discharge to the ocean is not the predominant method of wastewater management in Arctic Canada, treatment is still relatively rudimentary. Only a few examples exist in the Canadian Arctic where mechanical systems are used, and these systems have largely done so unsuccessfully (Yates et al., 2012a). Facultative lagoons and facultative lakes have been the most prevalent treatment systems across the Canadian Arctic. In most cases, the lagoons are unlined systems constructed from various local sources of coarse- and fine-grade aggregate material. Little research has been conducted on the performance of lagoon systems in the Canadian Arctic, but what has been done suggests that treatment via lagoons can be quite variable (Miyamoto and Heinke, 1979; Heinke et al., 1991; Johnson and Wilson, 1999). Given the absence of performance validation data, a precautionary approach to their efficacy should be taken given the environmental conditions of the Arctic.

Many of the lagoons and holding cells continuously exfiltrate into the receiving environment, although some are also annually decanted. Exfiltrate and decanted wastewater is then often polished by the adjacent tundra, which in many cases is or has become a wet-sedge tundra landscape because of the long-term discharge of primary treated wastewater into the environment. These tundra wetland systems have become identified as important components of the treatment train for wastewater in remote Arctic communities, which includes both lagoon and wetland. Until recently, many communities (e.g., Chesterfield Inlet, NU, and Repulse Bay, NU) relied solely on the wetland complex to treat wastewater discharged directly into the wetland with little to no primary treatment. Despite the differences and logistical challenges of the Arctic region, management practices and system designs used here are primarily fashioned using approaches originally designed for use in temperate regions.

The reason tundra wetland systems have been recognized as a treatment method is simple; they are a cost–effective, easily managed method for treatment of wastewater in Arctic communities (Johnson, 2010; Yates et al., 2012a). However, like lagoons, the performance of these systems remains largely unstudied. Yates et al. (2012b) observed that the performance of treatment wetlands in several Arctic communities during the ice-free period met performance standards set for temperate regions of Canada (Table 22.1).

However, the actual mechanisms of treatment in these systems can still only largely be speculated. As previously discussed, the logistics of studying these

Table 22.1 Average Concentration of Influent and Influent in Six Treatment Wetlands Located in Nunavut Territory, Canada

	Influent Concentration				Effluent Concentration				% Change
	Mean	Standard Deviation	Max	Min	Mean	Standard Deviation	Max	Min	
TC (cfu/100 mL)	99,700,000	186,000,000	969,600,000	2000	2640	10,100	79,400	11	99.99
E. coli (cfu/100 mL)	5,620,000	11,600,000	68,500,000	600	220	655	4510	3	99.99
cBOD$_5$ (mg/L)	233	223	1022	7	16	23	174	0	93
COD (mg/L)	341	215	1058	96	60	51	198	1	83
TSS (mg/L)	133	285	1775	2	14	19	84	0	90
NH$_3$—N (mg/L)	46.8	31.5	142.8	3.2	2.1	4.1	17.1	0.0	96
TP (mg/L)	7.7	3.6	14.8	1.3	0.7	0.8	3.6	0.0	91
DO	2.9	3.5	12.3	0.2	10.1	1.7	15.7	4.2	71

Averages were taken from weekly samples between mid-June and the end of September. Volume discharge ranged from 36 to 235 m^3/day.

systems is complex, and the diffuse nature of natural wetlands in general makes understanding the mechanistic elements a complicated and long-term effort. A characterization study by Yates et al. (2012a) described the treatment wetland in Paulatuk, Northwest Territories, in Canada. Wastewater entering this wetland received primary treatment from a large facultative lake. As a result, concentration of primary treated wastewater entering the wetland was observed to be low; 3 mg/L for NH_3—N and 40 mg/L for $cBOD_5$. As shown in Figure 22.2, NH_3—N was observed to quickly dissipate in the wetland, achieving background levels of 1 µg/L prior to discharging into the Arctic Ocean.

However, not all wetlands in Arctic communities are as easily defined as the example of Paulatuk. Previously unpublished work by the authors conducting a characterization study in Ulukhaktok (Holman), Northwest Territories, Canada, found this wetland to be a hydrologically closed system with no distinguishable flow path or outflow (Figure 22.3).

Piezometer and hydraulic head tests revealed movement of wastewater through the system was very slow. From calculating the hydraulic conductivity of 0.02 m/d and infield observation, we speculated that much of the flow was sheet flow. The wetland area in Ulukhaktok is slightly concave, with no distinguishable point

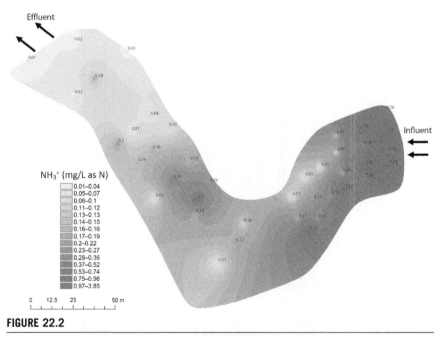

FIGURE 22.2

Concentration gradient of NH_3—N in Paulatuk, NWT, treatment wetland (as shown in Yates et al., 2012a). (For color version of this figure, the reader is referred to the online version of this chapter.)

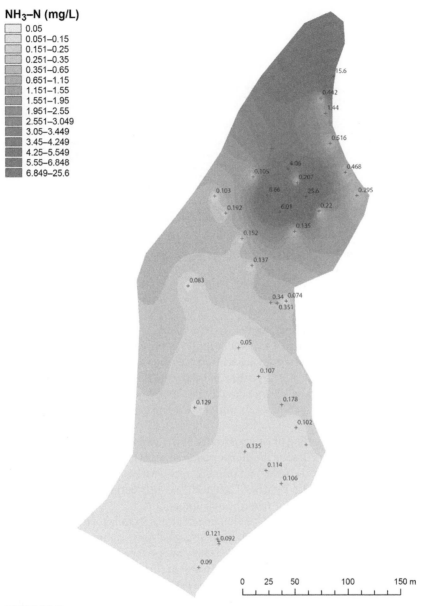

NH₃–N (mg/L)

	0.05
	0.051–0.15
	0.151–0.25
	0.251–0.35
	0.351–0.65
	0.651–1.15
	1.151–1.55
	1.551–1.95
	1.951–2.55
	2.551–3.049
	3.05–3.449
	3.45–4.249
	4.25–5.549
	5.55–6.848
	6.849–25.6

FIGURE 22.3

Concentration gradient of NH₃—N at the Ulukhaktok treatment wetland. (For color version of this figure, the reader is referred to the online version of this chapter.)

FIGURE 22.4

Ulukhaktok treatment wetland. (For color version of this figure, the reader is referred to the online version of this chapter.)

of discharge (Figure 22.4), making this system unique from all we had previously characterized. It was speculated that water loss was primarily via evaporation.

In comparison, other systems, such as Repulse Bay, are composed entirely of a series of natural water bodies as described and shown in the aerial imagery presented in Yates et al. (2012b) and discharge directly into a body of water. Many discharge points are also diffuse, which for regulatory reasons makes the point of control difficult to define.

The overall understanding of natural treatment wetlands of Arctic systems is limited (Yates et al., 2012b), despite the global wealth of knowledge regarding the use of constructed wetlands for the treatment of municipal wastewater (see Tanner, 2001; Kadlec and Wallace, 2009; Vymazal, 2009, 2011). Our current understanding of Arctic systems generated from our research throughout the Canadian Arctic has shown that wetlands that received continuous exfiltrates from lagoons or facultative lakes, despite cold ambient air temperature, permafrost soils, minimal soil depth, and a growing season of as little as 2 months, achieved or exceeded performance standards set for temperate regions of Canada. However, there are a number of critical questions that need to be answered as we try to determine appropriate performance standards for wetland treatment systems in the Canadian Arctic: (i) what impact does the cold climate have on the microbial community and how does this impact treatment performance? (ii) Can the wetlands provide adequate treatment during the spring freshet at a time when the loading rates may be high due to a backlog of winter effluent that requires treatment? (iii) What percentage of treatment is resulting from physiochemical/biological processes inherent within the wetlands and what percentage results from dilution alone? (iv) Are the existing wetland systems oversized or undersized and can they accommodate future growth and the resulting increases in wastewater effluent volumes?

22.2.2 Current regulatory framework in the Canadian Arctic

As described earlier, performance standards for wastewater effluents are currently in transition within Canada because the federal government is developing national performance standards for municipal wastewater effluents. Presently, individual territories, through the use of territorial water boards, maintain their own performance standards. Territorial water boards operate at arm's length from the territorial government. The performance standards are generally community specific. Further, each territory and its respective communities are also responsible for monitoring their treatment systems. In 2009, the Canadian Council of Ministers of the Environment released the final draft of the *Canada-wide Strategy for the Management of Municipal Wastewater Effluent*, which details regulatory changes to be implemented through the Canadian *Fisheries Act*. The intent of the strategy is to ensure there are no deleterious effects to the water bodies receiving the treated effluent, particularly with regard to fish health and/or fish habitat. This strategy has identified specific national performance standards for effluents of Canadian wastewater treatment facilities at 25 mg/L for the parameters cBOD and total suspended solids (TSS), 1.25 mg/L for unionized ammonia expressed as NH_3—N at 15 ± 1 °C, and a standard of 0.02 mg/L of total residual chlorine (Canadian Council of Ministers of the Environment, 2009). However, at this time, standards have not been set for Canada's Far North because of logistical and environmental issues. The Federal Government recognizes that environmental and socioeconomic conditions in portions of Canada's Far North (Nunavut, Northwest Territories, and regions located north of the 54th parallel in Quebec, Newfoundland, and Labrador) are unique, and therefore should have separate performance standards for municipal wastewater effluent. Given the absence of performance data and knowledge on Arctic systems, the Canadian Council of Ministers of the Environment granted a 5-year research period to determine appropriate performance standards.

The difficulty of regulating Arctic treatment systems has been previously discussed in this chapter in terms of logistical constraints imposed upon managing and regulating wetland in the Arctic, but wetlands themselves in any environment are difficult to regulate (Kadlec and Tilton, 1979). Natural wetlands are by nature open and diffuse systems, often with poorly defined boundaries, flow patterns, and permeable boarders (Kadlec, 2009). These conditions present challenges for wastewater regulators who require well-defined points of control. Further discussion is needed to address this concern if wetlands are to become more formally recognized as part of the wastewater treatment chain. There is concern that if wetlands become part of the formalized treatment chain and are expected to meet prescribed performance standards, then how are performance standards best monitored or enforced and in particular if performance standards are exceeded, how is corrective action to be taken when no point of control exists?

22.3 BEST PRACTICES

22.3.1 Sampling

Our research in the Arctic has led to the use of a number of techniques to measure performance of wetlands used in Arctic communities. Because there are no formal points of control, or control structures at outlets in so many of the systems, sampling is difficult. In diffuse and dynamic systems like wetlands, flow patterns change on an annual basis (Mitsch and Gosselink, 1986); therefore, establishing formal sampling points is not realistic. As described earlier, many lagoons discharge through continuous exfiltration. We have employed lysimeters to collect water passing just below the soil surface with some success. However, with this method the use of TSSs as a regulatory parameter is problematic, because the mineral fraction of TSS is substantially elevated. In some cases, exfiltrates emerge on the surface and are easily sampled with a grab sample. However, there may be numerous exfiltration points along a lagoon berm. We suggest that a number of sample points be placed along the berm edge, and if necessary sampling of both surface water and subsurface water be done. In cases where wastewater is discharged directly into the wetland, samples should be obtained directly from the haulage truck. Also, multiple samples should be taken from the truck to ensure heterogeneity of the sample. We suggest a similar approach for monitoring final post or discharge point of wetlands, because the point of discharge may be as difficult to identify as the point of influence.

The timing of sample collection is important to ensure key treatment periods are captured in an attempt to better understand treatment processes within these Arctic tundra wetlands. Arctic temperature and environmental conditions change rapidly with seasonal swings. Because of this, it is crucial that wetlands are sampled at various times throughout the year. This is particularly true for those communities where the winter holding capacity for raw wastewater is minimal, and the spring freshet may facilitate the rapid release of untreated wastewater. Capturing this event would therefore be important to understanding the efficacy of the system throughout different seasons of the year. At minimum, sampling should take place three times within the year. It is recognized that accessing these communities more frequently may not be feasible given the limited availability of resources (both economic and personnel) to do so. Samples should be conducted during thaw (to capture any freshet event—if logistically possible), the active summer period of the wetland, and just before freeze up. These shoulder periods are important, as they will represent the period of poorest performance in the wetland.

22.3.2 Management/infrastructure

Management and timing of management in the Arctic is especially critical to performance. The authors have witnessed management strategies that follow southern management protocols. For example, the decanting of lagoons in the high Arctic in late September into already frozen wetland systems is an example of a southern practice that is not appropriate for northern applications. In temperate locations, the decanting of lagoons in spring during high stream flow and in the fall is a long recommended management strategy (Tchobanoglous, 1979; Crites and Tchobanoglous, 1998). However, as identified above, the performance of lagoons in Arctic conditions has not been definitively shown to meet performance standards developed for temperature regimes. It is thought that Arctic lagoons can only reduce suspended solids, with only minimal nutrient removal occurring Arctic. As early as 40 years ago, evidence was presented to suggest that spring and fall decants in Arctic conditions should only be continued with great caution (Smith and Cameron, 1975). Our evidence suggests that decants into adjacent wetlands should occur in mid-summer when temperatures are the highest and plant growth is still occurring to absorb nutrients. Also, it may be advisable to perform a number of decants to reduce the hydraulic and organic load on the receiving wetland. However, this would need further testing and validation.

Various approaches have also been adopted in Arctic systems to increase residency time, especially in wetlands. Makeshift berms have been used in a number of communities to redirect preferential flows, increase residency time, and therefore contact time for microbial activity. These activities would achieve the longer residency times that have been recommended for wastewater treatment within Arctic wastewater (Tilsworth and Smith, 1984; Prince et al., 1995). Retention berms have been used to slow flows in two known locations: Cambridge Bay and Arviat, Nunavut. However, the effectiveness of these berms to increase residency and therefore treatment is not well documented.

22.3.3 Modeling

Standardized assessment methods and predictive tools are needed for wetlands. There are currently a variety of approaches that can be applied to predict the future capacity of constructed wetlands. These become particularly difficult when attempts are made to apply them to natural wetlands where the hydrology and physical dimensions of the natural wetlands are not fully known. The best approaches are likely those that incorporate site-specific performance data into the model as an attempt to calibrate the model to an individual wetland. The SubWet 2.0 model has been modified for use within natural wetlands of northern Canada and has the capacity to be calibrated to site-specific conditions of individual wetlands. Many of the other currently available predictive tools are based on either (i) first-order kinetic models for which aerial rate coefficients have often been developed for constructed wetlands situated in warmer regions resulting in low predictive precision for natural tundra wetlands or (ii) are more complicated two-dimensional or three-dimensional mathematical models that require a high sophistication of input parameters, which are generally unknown for natural wetlands. These models are also more costly and require specialized expertise to operate.

The only other evidence for modeling of wetlands in the Arctic was work by Kadlec and Johnson (2008). Kadlec and Johnson (2008) used first-order kinetic models to predict the performance of a treatment wetland in Cambridge Bay, Nunavut, 50 years into the future. They used very low-rate coefficients for biological processes, specifically nitrogen (actual values were not published). Based on their models, they expected that nitrogen and phosphorus would see minimal removal; 10–9 and 2.5–2.1 mg/L change NH_3—N and TP, respectively, from the influent to effluence of the wetland.

Much more study is required to determine ideal rate coefficients for Arctic environments, as well as other variables, such as evaporation, plant transpiration, and nutrient uptake by plants. A wealth of knowledge has been generated by plant ecologists who have been conducting work on vegetation communities for almost half a century (see Chapin and Bloom, 1976; Chapin et al., 1979, 1993; Schimel and Chapin, 1996; Hobbie, 2007; Edwards and Jefferies, 2010). Future work needs to adapt the knowledge of Arctic ecologists to wastewater treatment through the use of wetlands and realign management strategies to correspond with this understanding.

22.4 RECOMMENDATIONS

Despite 50 years of wastewater treatment in the Arctic, we still have little knowledge to guide our management decisions in the region, specifically those systems that utilize wetlands for treatment. Here, we highlight a few major gaps in our current understanding of the science.

22.4.1 Data gaps/science needs

More research attention is therefore needed to answer key questions required for the enhancement of ecological modeling and the development of management tools required by regulators to ensure protection for the fragile Arctic environment. Some of the data gaps currently existing around the treatment and management of wastewater include the following.

(i) *Spring freshet*: A limited amount of information generated from these studies suggests that the level of treatment may vary seasonally and in particular during the spring freshet when subsurface soils are still frozen and the wastewater that has accumulated over the wintertime on top of the wetland surface begins to melt. At this point, we do not have a good understanding of early season variability, including which wetlands may be overwhelmed by high organic loadings and which wetlands have the capacity to assimilate high springtime loadings. Similarly, we do not know the effects of dilution from spring meltwater. Dilution from meltwater may mitigate concentration-based effects.

(ii) *Year-to-year variability*: Monitoring treatment performance data collected in wetlands is limited and as such it is rare to have comparable data from 1 year to the next. As a result, little is known regarding how treatment efficiencies may vary and what factors influence this variability.

(iii) *Hydrology*: Site-specific information regarding subsurface and surface flow is generally lacking for most wetland sites. The volume of wastewater entering the wetland can be estimated from the volume of waste hauled to the site; however, it becomes difficult to determine flow volumes exfiltrating from the lagoon berm and just as difficult to determine how much of this flow travels overland and what portion travels subsurface. Likewise, it is difficult to determine the volume of new water entering the wetland either via surface or subsurface flow and how this might influence wastewater strength through dilution. Knowing this information can help significantly in both interpreting the results and predicting how the wetland would perform under different organic loading regimes.

22.4.2 Tundra wetland as part of the treatment chain

Natural wetlands were never specifically engineered to perform as a wastewater treatment option. Because of this, there has not been formal recognition or designation that would set these lands apart as part of the treatment chain. Future discussions should focus on the need and/or merit of formally recognizing these lands in land use planning documents. This will ensure that they have special designation as part of the treatment train so that performance standards would be enforced at the site of discharge from the wetland, not within the wetland.

22.5 CONCLUSIONS

As Arctic communities continue to grow rapidly and resource development expands through the coming decades, the demand for sound technologies to meet the need will only become more prevalent. Treatment wetlands provide one end-of-the-pipe solution. Yet, we need to develop a deeper understanding of their mechanisms in order to better manage them and the Arctic wastewater treatment chain as a whole. However, other onsite wastewater management technologies should also be explored.

References

Canadian Council of Ministers of the Environment, 2009. Canada-wide strategy for the management of municipal wastewater effluent. Whitehorse, Yukon.

Chabot, M., Duhaime, G., 1998. Land-use planning and participation: the case of Inuit public housing (Nunavik, Canada). Habitat Int. 22, 429–447.

Chapin, F.S., Bloom, A., 1976. Phosphate absorption—adaptation of tundra graminoids to a low-temperature, low phosphorus environment. Oikos 27, 111–121.

Chapin, F., Vancleve, K., Chapin, M., 1979. Soil-temperature and nutrient cycling in the tussock growth form of eriophorum-vaginatum. J. Ecol. 67, 169–189.

Chapin, F.S., Moilanen, L., Kielland, K., 1993. Preferential use of organic nitrogen for growth by a nonmycorrhizal arctic sedge. Nature 361, 150–153.

Crites, R., Tchobanoglous, G., 1998. Small and Decentralized Wastewater Management Systems. WCB/McGraw-Hill, Boston.

Dawson, R.N., Grainge, J.W., 1969. Proposed design criteria for waste-water lagoons in arctic and sub-arctic regions. J. Water Pollut. Control Federation 41, 237–246.

Edwards, K.A., Jefferies, R.L., 2010. Nitrogen uptake by carex aquatilis during the winter-spring transition in a low arctic wet meadow. J. Ecol. 98, 737–744.

Government of Nunavut, 2002. Nunavut Waters and Nunavut Surface Rights Tribunal Act. S.C. c. 10. Accessed 622 at www.canlii.org/ca/sta/n-28.8/whole.html.

Grainge, J.W., 1969. Arctic heated pipe water and waste water systems. Water Res. 3, 47–64.

Gunnarsdottir, R., Jenssen, P.D., Jensen, P.E., Villumsen, A., Kallenborn, R., 2013. A review of wastewater handling in the arctic with special reference to pharmaceuticals and personal care products (PPCPs) and microbial pollution. Ecol. Eng. 50, 76–85.

Heinke, G.W., Smith, D.W., Finch, G.R., 1991. Guidelines for the planning and design of waste-water lagoon systems in cold climates. Can. J. Civ. Eng. 18, 556–567.

Hobbie, S.E., 2007. Arctic ecology. In: Pugnaire, F.I., Valladares, F. (Eds.), Functional Plant Ecology, second ed. CRC Press, New York, pp. 369–388.

Jenssen, P., 2011. Wastewater treatment in cold/arctic climate with a focus on small scale and onsite systems. In: 28th Alaska Health Summit, Anchorage, Alaska.

Johnson, K., 2008. Inuit Position Paper Regarding the CCME Canada-Wide Strategy for the Management of Municipal Wastewater Effluent and Environment Canada's Proposed Regulatory Framework for Wastewater. Inuit Tapiriit Kanatami, Ottawa, ON.

Johnson, K., 2010. The social context of wastewater management in remote communities. Environ. Sci. Eng. Mag, Summer, 28–30.

Johnson, K., Wilson, A., 1999. Sewage treatment systems in communities and camps of the Northwest Territories and Nunavut Territory. In: Cold Regions Specialty Conference of the Canadian Society for Civil Engineering.

Kadlec, R.H., 2009. Wastewater treatment at the Houghton lake wetland: hydrology and water quality. Ecol. Eng. 35, 1287–1311.

Kadlec, R.H., Johnson, K., 2008. Cambridge bay, nunavut, wetland planning study. J. Northern Territories Water Waste Assoc. Fall, 30–33.

Kadlec, R., Tilton, D., 1979. Use of fresh-water wetlands as a tertiary wastewater-treatment alternative. CRC Crit. Rev. Environ. Control 9, 185–212.

Kadlec, R.H., Wallace, S., 2009. Treatment Wetlands, second ed. CRC Press, Boca Raton, FL.

Mitsch, W.J., Gosselink, J.G., 1986. Wetlands. Van Nostrand Reinhold Co., New York.

Miyamoto, H.K., Heinke, G.W., 1979. Performance evaluation of an arctic sewage lagoon. Can. J. Civ. Eng. 6, 324–328.

Parlee, B., Furgal, C., 2012. Well-being and environmental change in the arctic: a synthesis of selected research from canada's international polar year program. Clim. Chang. 115, 13–34.

Prince, D.S., Smith, D.W., Stanley, S.J., 1995. Intermittent-discharge lagoons for use in cold regions. J. Cold Reg. Eng. 9, 183–194.

Ritter, T.L., 2007. Sharing environmental health practice in the North American Arctic: a focus on water and wastewater service. J. Environ. Health 69, 50–55.

Schimel, J.P., Chapin, F.S., 1996. Tundra plant uptake of amino acid and NH4 + nitrogen in situ: plants complete well for amino acid N. Ecology 77, 2142–2147.

Sikumiut Environmental Management Ltd, 2008. Canadian Council of Ministers of the Environment Draft Canada Wide Strategy for the Management of Municipal Wastewater Effluent: Nunatsiavut Regional Impact Analysis. Inuit Tapariit Kanatami.

Smith, D.W., Cameron, J.J., 1975. Wastewater treatment and disposal alternatives in northern regionsReport No. IWR-62. In: Proceedings, Environmental Standards for northern Regions, Institute of Water Resources. University of Alaska, Fairbanks, AK, pp. 357–389.

Statistics Canada. Aboriginal Peoples in Canada in 2006: Inuit, Métis and First Nations, 2006 Census. Catalogue no. 97-558-XIE.

Tanner, C.C., 2001. Plants as ecosystem engineers in subsurface-flow treatment wetlands. Water Sci. Technol. 44, 9–17.

Tchobanoglous, G., 1979. Wastewater Engineering: Treatment, Disposal, Reuse, second ed. McGraw-Hill, New York; Montreal.

Tilsworth, T., Smith, D.W., 1984. Cold climate facultative lagoons. Can. J. Civ. Eng. 11, 542–555.

Vymazal, J., 2009. The use constructed wetlands with horizontal sub-surface flow for various types of wastewater. Ecol. Eng. 35, 1–17.

Vymazal, J., 2011. Constructed wetlands for wastewater treatment: Five decades of experience. Environ. Sci. Technol. 45, 61–69.

Warren, J.A., Berner, J.E., Curtis, T., 2005. Climate change and human health: infrastructure impacts to small remote communities in the north. J. Circumpolar Health 64, 487–497.

Wootton, B., Yates, C.N., 2010. Wetlands: simple and effective wastewater treatment for the north. Meridian, Canadian Polar Commission Newsletter. Fall/Winter.

Wootton, B., Durkalec, A., Ashley, S., 2008a. Canadian Council of Ministers of the Environment Draft Canada-Wide Strategy for the Management of Municipal Wastewater Effluent: Nunavut Regional Impact Analysis. Inuit Tapiriit Kanatami. https://www.itk.ca/publication/inuit-position-management-municipal-wastewater.

Wootton, B., Durkalec, A., Ashley, S., 2008b. Canadian Council of Ministers of the Environment Draft Canada Wide Strategy for the Management of Municipal Wastewater Effluent: Nunavik Regional Impact Analysis. Inuit Tapiriit Kanatami. https://www.itk.ca/publication/inuit-position-management-municipal-wastewater.

Wootton, B., Durkalec, A., Ashley, S., 2008c. Canadian Council of Ministers of the Environment Draft Canada-Wide Strategy for the Management of Municipal Wastewater Effluent: Inuvialuit Settlement Region Impact Analysis. Inuit Tapiriit Kanatami. https://www.itk.ca/publication/inuit-position-management-municipal-wastewater.

Yates, C.N., Wootton, B., Jørgensen, S.E., Murphy, S.D., 2012a. Use of wetlands for wastewater treatment in arctic regions. In: Jorgensen, S.E. (Ed.), Encyclopedia of Environmental Management, first ed. Taylor Francis, New York.

Yates, C.N., Wootton, B.C., Murphy, S.D., 2012b. Performance assessment of arctic tundra municipal wastewater treatment wetlands through an arctic summer. Ecol. Eng. 44, 160–173.

Modeling of Municipal Wastewater Treatment in a System Consisting of Waste Stabilization Ponds, Constructed Wetlands and Fish Ponds in Tanzania

23

T.A. Irene[a],*, L. Yohana[a], M. Senzia[b], M. Mbogo[a], T.S.A. Mbwette[a]

[a]*Department of Environmental Studies, The Open University of Tanzania, Dar es Salaam, Tanzania*
[b]*Civil Engineering, Arusha Technical College, Arusha, Tanzania*
*Corresponding author: e-mail address: irene.tarimo@out.ac.tz

23.1 INTRODUCTION

The chapter present results from a study conducted in Tanzania (Irene, 2013) on a continuing effort to develop management tools for the environmental pollution control. The effort led to development of a dynamic model accounting for treatment processing of nitrogen transformation and removal via an integrated wastewater treatment plant (IWTP) consisting of system of waste stabilization ponds (WSPs), horizontal subsurface flow constructed wetlands (HSSFCW) and fish ponds in series. The study is limited to nitrogen transformation, removal and re-use/recycling based on a wastewater treatment system located in Mabogini Moshi, Kilimanjaro Region, in the northern part of Tanzania.

The wastewater treatment system in Mabogini was constructed to treat municipal wastewater of Moshi Municipality. It was designed to receive an average daily flow of $4500 \, \text{m}^3/\text{d}$ and currently it is operated at an average daily flow of $3600-3800 \, \text{m}^3$. The system comprises of nine (09) cells of full-scale WSPs, a sludge drying bed, a pilot scale of HSSFCW and a pilot scale of fish pond (Figure 23.1).

However, in the study only the maturation pond (MTP2), HSSFCW and fish pond were considered. The effluent from the outlet system is used for aquaculture and agriculture for income generation and food purposes for a group of 56 beneficiaries.

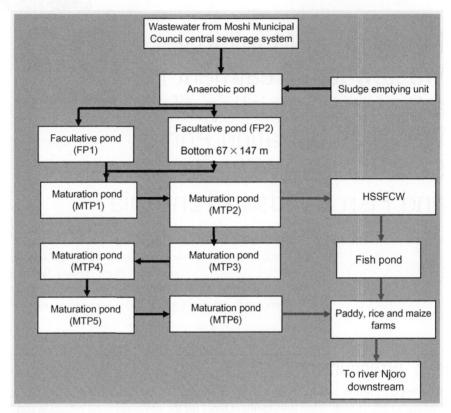

FIGURE 23.1

Wastewater treatment system at Mabogini-Moshi. (Note that the blue color indicates the targeted study area.). (For interpretation of the references to color in this figure legend, the reader is referred to the online version of this chapter.)

23.2 PREVIOUS EFFORTS IN MODELING OF WASTEWATER TREATMENT

Previously, studies on the modeling effort for wastewater treatment are mainly based on a single treatment system such as WSPs, dynamic roughing filter (DRF), HSSFCW and fish ponds (Fritz *et al.,* 1979; Pano and Middlebrooks, 1982; Kayombo[†], 2001; Kaseva *et al.,* 2002; Senzia *et al.,* 2002, 2004; Kimwaga, 2004; Mayo and Bigambo, 2005; Yohana, 2009).

In their book Kadlec and Wallace (2009) presented a number of models in HSSFCW and free water surface flow constructed wetlands (FWSFCW), including

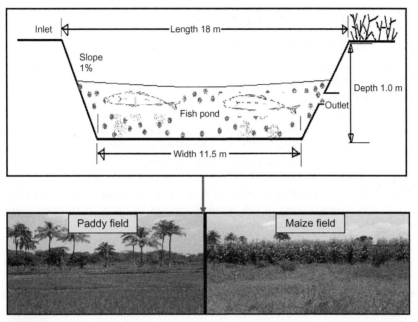

FIGURE 23.2

Paddy and maize fields irrigated with the treated wastewater outlet from fish pond in Mabogini-Moshi IWTP. (For the color version of this figure, the reader is referred to the online version of this chapter.)

information from McBride and Tanner (2000), Wynn and Liehr (2001), Senzia *et al.* (2002), Langergraber (2001), Liu and Dahab (2004), Dorge (1994), Jørgensen (1994), Kadlec and Knight (1996), Martin and Reddy (1997), Gerke *et al.* (2001), Jørgensen and Bendoricchio (2001), Jørgensen and Fath (2011) and Howell *et al.* (2005).

This study adds value to increasing attention on the resource-oriented sanitation concept in which reuse of wastewater is predominantly practiced, especially in suburban centers of municipalities in Tanzania and elsewhere (USEPA, 2000; Mbwette *et al.*, 2001; Metcalf and Eddy, 2003; Senzia *et al.*, 2009; Yohana, 2009; Valero *et al.*, 2010). Integrated wastewater treatment systems are able to provide multiuse activities for effluents, including aquaculture and agriculture (Figure 23.2). The World Health Organization (WHO) has provided guidelines on the reuse of excreta, wastewater and greywater (WHO, 2006).

Modeling, in addition to being used for design purposes, also aids in providing information on the characteristics of expected outlets from the system and thus safeguards human and other users and the downstream environment.

23.3 SAMPLING AND MODEL DEVELOPMENT

23.3.1 Sampling and analysis

Grab samples were collected on a daily basis for 3 months, April, August and October 2010, at the inlet and outlet of the maturation pond (MTP2). A similar sampling procedure was used for HSSFCW and fish pond. The samples were analyzed in a water quality laboratory for biological oxygen demand (BOD_5), total Kjeldahl nitrogen (TKN), organic nitrogen (Org-N), ammonia nitrogen (NH_3-N), nitrate nitrogen (NO_3-N), nitrite nitrogen (NO_2-N), chlorophyll-a, nitrogen in the sediments (N-Sedim), nitrogen in the constructed wetland plants (N-Plants), nitrogen in fish (N-Fish), total suspended solids (TSS), turbidity and the fecal coliforms (FC) using Standard Methods of the Examination of Water and Wastewater Analysis (APHA, 2005). In situ measurements of pH, temperature and DO were performed using a multiparameter spectrophotometer (model 156, 2001).

23.3.2 Model state variables and processes/flows

The model considers state variables of nitrogen in the system that contains water, substrate/gravel, plants and fish. In maturation ponds the model has three state variables, namely, organic nitrogen (Org-N), ammonia nitrogen (NH_3-N) and nitrate nitrogen (NO_3-N). The processes included are mineralization, nitrification, denitrification, nutrient uptakes by microorganisms and settling of organic matters.

In HSSFCW, the model has four state variables: organic nitrogen (Org-N), ammonia nitrogen (NH_3-N), nitrate nitrogen (NO_3-N) and nitrogen in the plants (N-Plants). The processes are mineralization, nitrification, denitrification, nutrient uptake by plants, accumulation/accretion of organic matter and decaying /decomposition of plants.

In the fish pond, the model has four state variables: organic nitrogen (Org-N), ammonia nitrogen (NH_3-N) and nitrate nitrogen (NO_3-N), nitrogen in planktons (N-Planktons), nitrogen in sediments (N-sediments) and nitrogen in fish (N-Fish). The processes are mineralization, nitrification, denitrification, nutrient uptake by microorganisms, uptake of plankton by fish, sedimentation, accumulation/accretion of organic matter, excretion and regeneration. Figure 23.3 shows the conceptual diagram in the integrated system of maturation pond, HSSFCW and fish pond.

23.4 MATHEMATICAL EQUATIONS

23.4.1 WSP (maturation pond)

The differential equations for the state variables are presented in Equations (23.1)–(23.12):

$$\frac{d(\text{Org-N})}{dt} = \frac{Q_i}{V}(\text{Org-N})_i - \frac{Q_e}{V}(\text{Org-N}) - r_m - r_s + r_1 + r_2 \qquad (23.1)$$

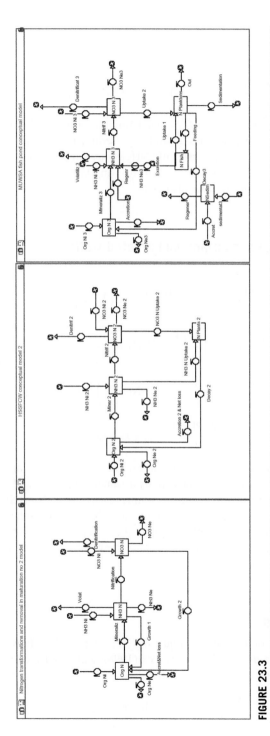

FIGURE 23.3

Nitrogen conceptual diagram of integrated system of maturation pond, HSSFCW and fish pond drawn by STELLA II software (STELLA® 9.1.4). (For the color version of this figure, the reader is referred to the online version of this chapter.)

$$\frac{d(NH_3\text{-}N)}{dt} = \frac{Q_i}{V}(NH_3\text{-}N)_i - \frac{Q_e}{V}(NH_3\text{-}N) - r_1 + r_m + r_v \qquad (23.2)$$

$$\frac{d(NO_3\text{-}N)}{dt} = \frac{Q_i}{V}(NO_3\text{-}N)_i - \frac{Q_e}{V}(NO_3\text{-}N) - r_n + r_2 + r_d \qquad (23.3)$$

where Q_i = inlet flow rate (m^3/d); Q_e = outlet flow rate (m^3/d); V = volume of the pond (m^3); r_m = mineralization rate (mg/l/d); r_s = sedimentation rate; r_n = nitrification rate (mg/l d); r_1 = uptake rate of NH$_3$-N by microorganisms (mg/l/d); r_2 = uptake rate of NO$_3$-N by microorganisms (mg/l/d); r_v = volatilization rate (mg/l/d) and r_d = denitrification rate (mg/l/d).

23.4.2 Horizontal subsurface constructed wetlands

$$\frac{d(Org\text{-}N)}{dt} = \left(\frac{Q_i}{A}Org\text{-}N_i\right) - \left(\frac{Ks \cdot S}{L}Org\text{-}N\right) - r_m - r_a + r_{dc} \qquad (23.4)$$

$$\frac{d(NH_3\text{-}N)}{dt} = \left(\frac{Q_i}{A}NH_3\text{-}N_i\right) - \left(\frac{Ks \cdot S}{L}NH_3\text{-}N\right) - r_n - r_1 - r_m + r_r \qquad (23.5)$$

$$\frac{d(NO_3\text{-}N)}{dt} = \left(\frac{Q_i}{A}NO_3\text{-}N_i\right) - \left(\frac{Ks \cdot S}{L}NO_3\text{-}N\right) + r_n - r_2 - r_{dn} \qquad (23.6)$$

$$\frac{d(N\text{-}Plant)}{dt} = r_1 + r_2 - r_{dc} \qquad (23.7)$$

where Q_i = inlet flow rate (250 m^3/d), A = area of the HSSFCW (972 m^2), L = length of the bed (54 m), Ks is hydraulic conductivity assumed (86.4 m/d) and S = the slope of the bed (1%). The processes included r_n = nitrification rate (gN/m^2 d), r_{dn} = denitrifying rate (gN/m^2 d) and r_m = mineralization rate (gN/m^2 d). The other features are r_1 = uptake rate of (NH$_3$-N) by the plants (gN/m^2 d), r_2 = uptake rate of (NO$_3$-N) by the plants, r_{dc} = decaying rate (gN/m^2 d), r_a = accretion rate of (Org-N) in the sediments (gN/m^2 d), N-Plant = nitrogen in the plants.

23.4.3 Fish pond

The following sets of differential Equations (23.8)–(23.12) follow the studies done by Lorenzen *et al.* (1997), Hargreaves (1998), Jamu and Piedrahita (2000), Burford and Lorenzen (2004) and Yohana (2009).

$$\frac{dC_{Org\text{-}N}}{dt} = (1-p)A_t - r_w C_{Org\text{-}N} \qquad (23.8)$$

$$\frac{dC_{NH_3-N}}{dt} = (pNt + rM_{N\text{-Sed}}) - (n + v + r_W) - r_g c C_{Chl} \left(\frac{C_{NH_3-N}}{C_{NH_3-N} + C_{NO_3-N}} \right) \quad (23.9)$$

$$\frac{dC_{NO_3-N}}{dt} = (nC_{NH_3-N} - r_W C_{NO_3-N}) - r_g c C_{Chl} \left(\frac{C_{NO_3-N}}{C_{NO_3-N} + C_{NH_3-N}} \right) \quad (23.10)$$

$$\frac{dC_{Chl}}{dt} = r_g C_{Chl} - (r_W + s) C_{Chl} \quad (23.11)$$

$$\frac{dM_{N\text{-Sed}}}{dt} = scC_{Chl} - rM_{N\text{-Sed}} \quad (23.12)$$

where dC is change in concentration; dt is change of time; p is the proportion of nitrogen entering the fish pond with the outlet and the remainder entering as organic nitrogen; A_t is total nitrogen inflow (mg/l d); r is regeneration rate of ammonia nitrogen in the pond/d; pNt is the proportion of nitrogen that undergo nitrification at certain time; $M_{N\text{-Sed}}$ is mass of nitrogen (mg/l) in the sediments; n is nitrification rate/d; r_g is the planktons growth rate/d; v is volatilization rate of ammonia/d; r_W is water exchange rate/d; c is nitrogen/chlorophyll ratio of phytoplankton; Org-N is organic nitrogen concentration dissolved in the fish pond water (mg/l); C_{NH_3-N} is ammonia nitrogen concentration (mg/l); C_{NO_3-N} is nitrate nitrogen concentration (mg/l); C_{Chl} is chlorophyll-a concentration; s is sedimentation/accretion rate of planktons per day.

23.5 MODEL SIMULATIONS AND OUTPUT

Modeling of nitrogen inputs, transformation and removal in the IWTP was done by using STELLA II Software (STELLA II® 9.1.4). The data were processed by using fourth-order Runge-Kutta approximations incorporated in the software.

23.5.1 Waste stabilization pond (maturation pond)

The simulated and measured/observed value of Org-N, NH$_3$-N and NO$_3$-N in WSP together with their correlation are as shown Figure 23.4a and c.

From Figure 23.4a–c, there is good agreement (R^2) between observed value and simulated value, which indicates that the model developed can predict nitrogen transformation and removal in the maturation pond.

23.5.2 Nitrogen mass balance in WSP

Figure 23.5 shows mass balance in the WSP. Accretion and net loss of organic nitrogen to the sediment was the major removal of nitrogen from the pond, followed by denitrification. The two processes account for 40.94% (22.37 kg/d) of the removal of inflow nitrogen (57.066 kg/d) to the WSP. The major process of nitrogen transformation in WSP maturation pond was found to be uptake of ammonia nitrogen by microorganisms (22.07 kg/d).

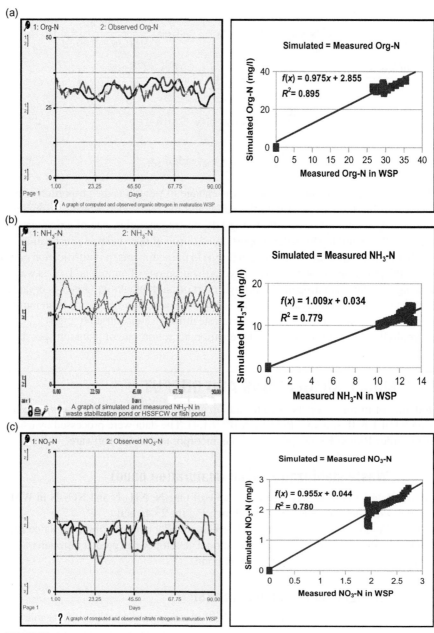

FIGURE 23.4

(a) Simulated and measured Org-N in WSP (with $R^2 = 0.895$). (b) Simulated and measured NH$_3$-N in WSP (with $R^2 = 0.779$). (c) Simulated and measured NO$_3$-N in WSP (with $R^2 = 0.780$). (For the color version of this figure, the reader is referred to the online version of this chapter.)

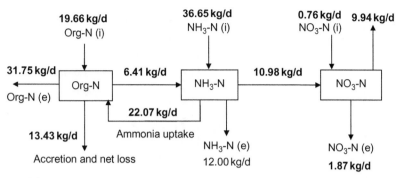

FIGURE 23.5

Nitrogen mass balance in waste stabilization pond.

A similar study carried on primary facultative WSPs (Senzia *et al.*, 2002) indicated that sedimentation was also the major route in permanently removing nitrogen from the pond.

23.5.3 Horizontal subsurface flow constructed wetlands

The simulated and measured/observed value of state variables together with their correlation are as shown Figure 23.6a–c.

There is good agreement between measured and observed values (R^2), indicating that the model developed can be used in predicting nitrogen transformation and removal in HSSFCW.

23.5.4 Mass balance in the HSSFCW

Figure 23.7 shows mass balance in HSSFCW. The numbers in the brackets represent the following: (1) organic nitrogen flowing to the HSSFCW, (2) ammonia nitrogen flowing to the HSSFCW, (3) nitrate nitrogen flowing to the HSSFCW, (4) denitrification of NO_3-N to $N_2(g)$, (5) Org-N flowing out of the HSSFCW, (6) mineralization of Org-N to NH_3-N, (7) nitrification of NH_3-N to NO_3-N, (8) nitrate nitrogen flowing out of the HSSFCW, (9) Org-N lost to the sediments by accretion, (10) decaying of organic matter returned to Org-N, (11) NH_3-N flowing out of the HSSFCW, (12) NH_3-N uptake by the *Phragmites mauritianus* plants for their growth and (13) NO_3-N uptake by the *Phragmites mauritianus* plants.

From the mass balance in Figure 23.7, accretion of organic nitrogen, nitrification/denitrification and plant uptake were the major mechanisms for nitrogen removal. On average, accretion, denitrification and plant uptake accounted for 31.6%, 5.5% and 1.6%, respectively, of the removal of the nitrogen flowing to the wetlands

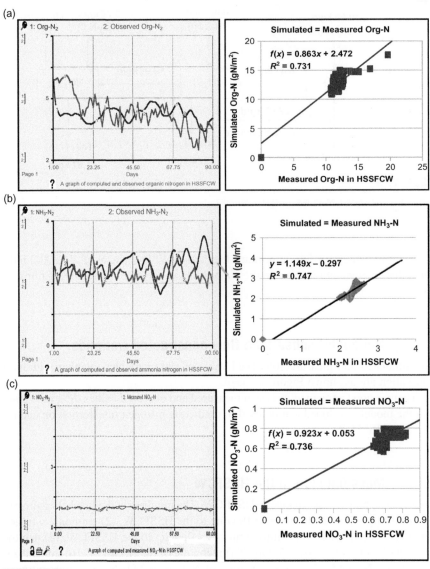

FIGURE 23.6

(a) Simulated and measured Org-N in HSSFCW (with $R^2 = 0.731$). (b) Simulated and measured NH_3-N in HSSFCW in (with $R^2 = 0.530$). (c) Simulated and measured NO_3-N in HSSFCW ($R^2 = 0.736$). (For the color version of this figure, the reader is referred to the online version of this chapter.)

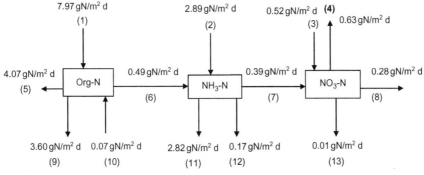

FIGURE 23.7

Nitrogen mass balance in constructed wetlands.

(4154 g/m^2 y). However due to the fact that 0.07 gN/m^2 d of the nitrogen in the wetlands was transformed through decay process back to the organic nitrogen, only 1584 gN/m^2 y (38.1%) of the inflow nitrogen can permanently be removed from the wetlands if harvesting of plants will take place. Kadlec and Wallace (2009) reported a total removal of 253 gN/m^2 y (47.7%) for a constructed wetland system loading of 530 gN/m^2 y.

Mineralization of organic nitrogen to ammonia nitrogen is a major route for nitrogen transformation, accounting for 0.49 gN/m^2 d (178.9 gN/m^2 y), which is 4.3% of inflow nitrogen.

23.5.5 Fish pond

Figure 23.8a–c give the results of the measured and simulated nitrogen in the fish pond.

23.5.6 Mass balance in the fish pond

Figure 23.9 shows the nitrogen mass balance in the fish pond. The numbers in the brackets represent: (1)organic nitrogen flowing to the fish pond, (2) ammonia nitrogen flowing to the fish pond, (3) nitrate nitrogen flowing to the fish pond, (4) denitrification of NO$_3$-N to N$_2$(g), (5) Org-N flowing out of the fish pond, (6) mineralization of Org-N to NH$_3$-N, (7) nitrification of NH$_3$-N to NO$_3$-N, (8) nitrate nitrogen flowing out of the fish pond, (9) Org-N lost to the sediments by accretion/settling, (10) decaying organic matter returns to Org-N, (11) regeneration of NH$_3$-N from the sediments back to water column (12) NH$_3$-N flowing out of the fish pond, (13) excretion of NH$_3$-N from the fish back to water column, (14) uptake *of* NH$_3$-N by the phytoplanktons for their growth and (15) NO$_3$-N uptake by the phytoplankton for their growth.

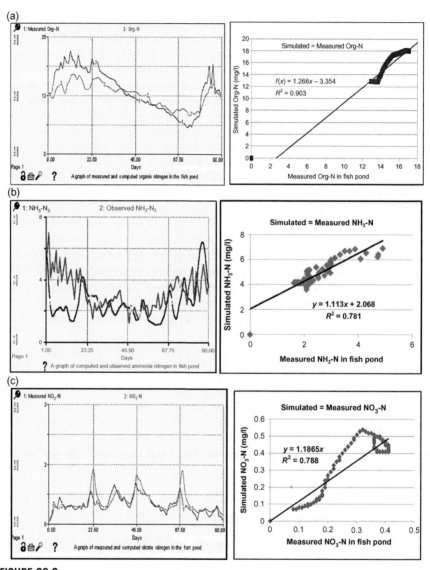

FIGURE 23.8

(a) Simulated and measured Org-N in the fish pond ($R^2 = 0.983$). (b) Simulated and measured values of NH$_3$-N in the fish pond ($R^2 = 0.713$). (c) Simulated and measured values of NO$_3$-N in the fish pond ($R^2 = 0.788$). (For the color version of this figure, the reader is referred to the online version of this chapter.)

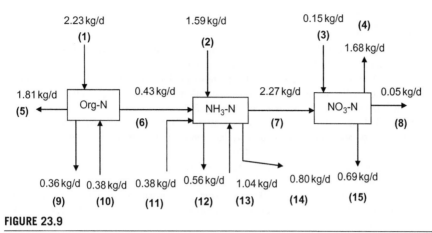

FIGURE 23.9

Nitrogen mass balance in the fish pond.

From Figure 23.9, the major nitrogen transformation route in the fish pond is nitrification of NH_3-N to NO_3-N, which accounts for 828.6 Kg/y. Denitrification is responsible for 42.3% removal of nitrogen in the fish pond.

23.6 CONCLUSIONS AND RECOMMENDATIONS
23.6.1 Conclusions

A dynamic model developed for nitrogen transformation, removal and re-use in an IWTP consisting of system of WSPs, HSSFCW and fish ponds in series indicated that:

(a) This model can simulate the transformation, removal and re-use mechanisms that are normally taking place in wastewater treatment system and recycling.
(b) The major route for the nitrogen removal is sedimentation for both the WSP and the constructed wetlands while denitrification is the major removal mechanism in fish pond.
(c) Of the 20,828 kg/y loading to the WSP, 4901 kg/y is removed through sedimentation and 3628 kg/y is removed by denitrification.
(d) Out of the 4154 g/m² y total loading to the constructed wetlands, 1584 gN/m² y (38.1%) can be permanently removed from the wetlands if harvesting of plants will take place.
(e) It can be used as management tool for water bodies downstream of the system.
(f) It can be used by engineers and policymakers for design purposes, especially in tropical climates around the world.

References

APHA, 2005. Standard Methods for the Examination of Water and Wastewater, 24th ed. American Public Health Association, Washington, DC.

Burford, M.A., Lorenzen, K., 2004. Modelling nitrogen dynamics in intensive shrimp ponds: the role of sediment remineralization. Aquaculture 229, 129–145, Available online at www.sciencedirect.com (accessed on 4th July 2011).

Dorge, J., 1994. Modelling Nitrogen transformation in freshwater wetlands, estimating nitrogen retention and removal in natural wetlands in relation to their hydrology and nutrient loadings. Ecol. Model. 75–76, 409–420.

Fritz, J.J., Middleton, A.C., Meredith, D.D., 1979. Dynamic process modelling of wastewater stabilization ponds. J. WPCF 51 (11), 2724–2743.

Gerke, S., Baker, L.A., Xu, Y., 2001. Nitrogen transformations in a wetland receiving lagoon effluent: sequential model and implications for water reuse. Water Res. 35 (16), 3857–3866.

Hargreaves, J.A., 1998. Nitrogen Biogeochemistry of aquaculture ponds. Aquaculture 166, 181–212.

Howell, C.J., Crohn, D.M., Omary, M., 2005. Simulating nutrient cycling and removal through treatment wetlands in arid/semiarid environments. Ecol. Eng. 25 (1), 25–40.

Irene, A.T., 2013. Modelling Nitrogen Transformation, Removal and Re-use in an Integrated Wastewater Treatment Plant (IWTP). PhD Thesis, The Open University of Tanzania, Faculty of Science, Technology and Environmental Studies, Department of Environmental Studies, p. 200, Unpublished.

Jamu, D.M., Piedrahita, R.H., 2000. An organic matter and Nitrogen dynamics model for the ecological analysis of integrated aquaculture/agriculture systems: I. Model development and calibration. Environ. Model Softw. 17, 571–582, Available on www.elsevier.com/locate/envsoft (accessed 20th September 2011).

Jørgensen, S.E., 1994. A general model of nitrogen removal by wetlands. In: Mitsch, W.J. (Eds.), Global Wetlands: Old World and New. Elserier Science, Amsterdam.

Jørgensen, S.E., Bendoricchio, G., 2001. Fundamentals of Ecological Modelling. Elsevier Science, B.V., The Netherlands.

Jørgensen, S.E., Fath, B.D., 2011. Fundamentals of Ecological Modelling: Application in Environmental Management and Research, fourth ed. Elsevier B.V., Denmark.

Kadlec, R.H., Knight, R., 1996. Treatment Wetlands. Lewis Publishers, Boca Raton, FL, pp. 373–440.

Kadlec, R.H., Wallace, S.D., 2009. Treatment Wetlands, second ed. Taylor and Francis Group, LLC, Boca Raton, FL.

Kaseva, M.E., Mbwette, T.S.A., Katima, J.H.Y., 2002. Domestic sewage treatment in a pilot plant composed of septic tank and a constructed wetland system—a case study in University College of Lands and Architectural Studies (UCLAS) and University of Dar, Prospective College of Engineering and Technology, Tanzania. In: Proceedings of the 8th International Conference on Wetland Systems for Water Pollution Control, vol. 1, pp. 367–379.

Kayombo[†], S., 2001. Development of a Holistic Ecological Model for Design of Facultative Waste Stabilization Ponds in Tropical Climates. Royal Danish School of Pharmacy, Institute for Analytical and Pharmacetial Chemistry, University of Copenhagen, Denmark, PhD Thesis.

Kimwaga, R.J., 2004. Modelling of Coupled Dynamic Roughing Filters and Horizontal Subsurface Flow Constructed Wetlands to Tertiary Treatment of Waste Stabilization Pond Outlets. Engineering at the University of Dar es Salaam, pp. 46–53, A Thesis submitted in fulfillment for the degree of Doctor of Philosophy (PhD).

Langergraber, G., 2001. Development of a Simulation Tool for Subsurface Flow Constructed Wetlands. IWGA SIG University of Vienna, Austria, PhD Thesis.

Liu, W., Dahab, M.F., 2004. Nitrogen transformation modelling in subsurface flow constructed wetlands. In: Lienard, A., Burnett, H. (Eds.), Proceedings of the 9th International Conference on Wetlands System for Water Pollution Control, 26–30 September 2004. Association Scientifique et Technique pour l'EAU et l'Environment (ASTEE), Cemargef and IWA, Avignon, France, pp. 405–412.

Lorenzen, K., Struve, J., Cowan, V.J., 1997. Impact of farming intensity and water management on nitrogen dynamics in intensive pond culture: a mathematical model applied to thai commercial shrimp farms. Aquaculture Res. 28, 493–507.

Martin, J.F., Reddy, K.R., 1997. Interaction and spatial distribution of wetland nitrogen processes. Ecol. modell. 105, 1–21.

Mayo, A.W., Bigambo, T., 2005. Nitrogen transformation in horizontal subsurface flow constructed wetlands I: model development. Phys. Chem. Earth 30 (11–16), 658–667. Available online 5 October, 2005 at info@ajol.info & ScienceDirect (accessed 12.10.11 & 21.02.14).

Mbwette, T.S.A., Katima, J.H.Y., Jorgensen, S.E., 2001. Application of wetland systems and waste stabilization ponds in water pollution control. In: Mbwette, T.S.A. et al., (Eds.), 2001 WSP Project, Dar es Salaam. IKR, Faculty of Engineering, University of Dar es Salaam, Tanzania, pp. 1–17.

McBride, G.B., Tanner, C.C., 2000. Modelling biofilm nitrogen transformations in constructed wetlands mesocosms with fluctuating water levels. Ecol. Eng. 14 (1–2), 93–196.

Metcalf and Eddy, 2003. Wastewater Engineering: Treatment, Disposal and Re-use. Mc Graw-Hill, New York, NY.

Pano, A., Middlebrooks, E.J., 1982. Ammonia nitrogen removal in facultative wastewater stabilization ponds. J. Water Pollut. Control Fed. 54, 344–351.

Senzia, M.A., Mayo, A.W., Mbwette, T.S.A., Katima, J.H.Y., Jørgensen, S.E., 2002. Modelling nitrogen transformation and removal in primary facultative ponds. Ecol. Model. 154, 207–215. University of Dar es Salam, pp. 2.

Senzia, A.M., Mashauri, D.A., Mayo, A.W., 2004. Modelling nitrogen transformation in horizontal subsurface flow constructed wetlands planted with Phragmites mauritianus. J. Civil Eng. Res. Pract. 1 (2), 1–15, Available at info@ajol.info (accessed on 22nd September 2011).

Senzia, M.A., Marwa, J., Einsk, J., Kimwaga, R., Kimaro, T., Mashauri, D., 2009. Health risks of irrigation with treated urban wastewater. In: Shaw, D. (Eds.), Proceedings of the 34th WEDC on Water, Sanitation and Hygiene. Sustainable Development and Multisectoral Approaches. International Conference. UN Conference Centre, Addis Ababa, Ethiopia. WEDC, Loughborough.

STELLA® v 9.1.4; Copyright © 1985–2010, isee systems inc. (purchased July 2011).

USEPA, 2000. Constructed Wetlands Treatment of Municipal Wastewaters. U.S. EPA 625/R99/010, Cincinnati, Ohio, USA.

Valero, C.M.A., Mara, D.D., Newton, R.J., 2010. Nitrogen removal in maturation waste stabilisation ponds via biological uptake and sedimentation of dead biomass. Water Sci. Technol. 61 (4), 1027–1034.

World Health Organization, 2006. Guidelines for the safe use of wastewater, excreta and greywater. Wastewater Use in Agriculture. World Health Organization, Geneva, Switzarland, p. 176.

Wynn, T.M., Liehr, S.K., 2001. Development of a constructed sub-surface flow wetland simulation model. Ecol. Eng. 16 (4), 519–536.

Yohana, L., 2009. The Potential Re-Use of Treated Wastewater from a Horizontal Subsurface Flow Constructed Wetland for Aquaculture production: Modelling of Nitrogen Dynamics and removal in aquaculture pond (PhD Thesis). Water Resources Engineering (WRE) Department at University of Dar es Salaam, Tanzania.

A Novel Subsurface Upflow Wetland with the Aid of Biosorption-Activated Media for Nutrient Removal

24

Ni-Bin Chang*, Martin P. Wanielista, Zhemin Xuan

Department of Civil, Environmental, and Construction, University of Central Florida,
Orlando, FL, USA
*Corresponding author: e-mail address: nchang@ucf.edu

24.1 INTRODUCTION

The septic tank is normally an underground, watertight container made of concrete, fiberglass, or other durable material that provides primary wastewater treatment (settling of solids). It is connected to the standard drainfield constructed with a series of parallel, underground, perforated pipes that allow septic tank effluent to percolate into the surrounding soil in the vadose (unsaturated) zone where it is expected that most of the residual nutrients may be assimilated. Several types of effluent distribution are applicable in standard drainfield systems. These include gravity systems, low-pressure dosed systems, and drip irrigation systems. Some systems require an additional pump. Through various physical, chemical, and biological processes, most bacteria, viruses, and nutrients in wastewater are expected to be consumed or filtered as the wastewater passes through the soil. After treatment, the effluent enters the vadose zone and ultimately a groundwater aquifer acts as a receiving water body. When properly constructed and maintained, the septic system can provide years of safe, reliable, cost-effective service.

Due to widespread concerns about the impacts of onsite sewage treatment and disposal systems (OSTDS) on ground and surface waters, scientists, engineers, and manufacturers in the wastewater treatment industry have developed a wide range of alternative active and passive technologies designed to address increasing hydraulic loads, energy-saving requirements, and improved removal of nutrients and pathogens from onsite wastewater treatment. These alternative systems require increased testing to verify system performance; pollutant transport and fate; resultant environmental impacts; and an integration of the planning, design, siting, installation, maintenance, and management functions. Cost effectiveness, system reliability, and proper management become the major concerns in their use. In general, passive

Developments in Environmental Modelling, Volume 26, ISSN 0167-8892, http://dx.doi.org/10.1016/B978-0-444-63249-4.00026-9

technologies (those without more than one pump) might be advantageous due to their cost effectiveness, system reliability, and low maintenance requirement. This triggers an acute need to perform a thorough technology assessment, screening, and prioritization. Given the above issues with conventional and performance-based OSTDSs, a new generation of passive onsite wastewater treatment technologies with nutrient removal capacity is much needed to effectively remove nutrients and better protect public health and our ground and surface waters in a cost-effective manner. Reactive media are effluent materials from a septic tank or pretreatment device that pass through prior to reaching the groundwater. These reactive media may include, but are not limited to, soil, sawdust, zeolites, tire crumb, vegetative removal, sulfur, spodosols, or other media.

For some years, wetlands have been playing an important role in water conservation, climate regulation, soil erosion control, flood storage for use in drought, and environment purification. Based on the same principles for wastewater purification using natural wetlands, man-made constructed wetlands with effective management can improve effluent water quality. Constructed wetlands can be generally divided into two main types: free water surface flow (SF) wetlands and subsurface flow (SSF) wetlands (US EPA, 1993). Both wetland systems remove nitrogen in the water through a variety of mechanisms, including biological, physical, and chemical reactions (US EPA, 1993). The biological functions, such as ammonification, nitrification–denitrification, and plant uptake, under appropriate conditions are regarded as crucial for nitrogen (N) removal (US EPA, 1993). Precipitation of a particular form of phosphorus (P) is the main path for phosphorus removal. Microbial absorption and accumulation also play a role (US EPA, 1993).

Constructed wetlands with different designs and hydraulic residence times (HRT) have been implemented and successfully operated for wastewater treatment in many parts of the world (Surrency, 1993; Steer et al., 2002; Mbuligwe, 2005; Tee et al., 2009), yet the design for nutrient removal is not always successful (Mann and Bavor, 1993; Billore et al., 1999). This chapter presents an innovative design and implementation of a subsurface upflow wetland (SUW) in a field-scale study. The biosorption-activated media (BAM) are used before the standard drainfield design, and the SUW is used to replace the conventional drainfield. The SUW must have a seepage area for the effluent from the SUW if reuse of the water is not planned. Hence, this new type of constructed wetland system, which was deemed an integral part of the septic tank system for domestic wastewater treatment in this study, differs from others in that (1) a suite of selected recipes of BAM in concert with native plant species was used for wastewater treatment, and (2) the hydraulic flow through the system is in an upflow direction, in which the discharge is below the surface of the wetland (Chang et al., 2011a). The objectives of this research concentrate on the following critical questions that have not been fully answered in the literature:

(1) What are effective treatment media for removing nutrients from septic tank effluent?

(2) What are the underlying processes of such treatment media and their associated functions, effectiveness, and longevity?

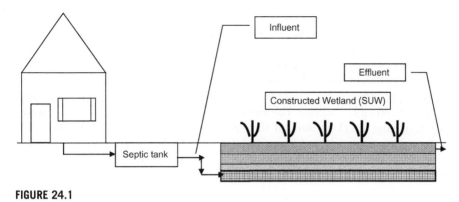

FIGURE 24.1

Schematic of an SUW-based OSTDS.

(3) What insights are available on how such systems have been designed, installed, maintained, controlled, and replaced that may be applicable to onsite sewage treatment?

(4) What comparative basis can be used when different sorption media are used in passive treatment processes and are compared against other treatment trains, such as the use of a conventional drainfield?

In our study, a novel SUW system filled with BAM (e.g., mixes of recycled and natural materials) along with selected plant species was tested as a substitute for the conventional drainfield in a septic tank system. Four parallel SUWs (i.e., three with plants vs. one without plants) were built to handle 454 L day^{-1} (120 gallons day^{-1}, GPD) of septic wastewater flow with a 7-day HRT. The objectives of this study were thus to assess (1) the performance of an SUW to further reduce nutrients in the effluent of a septic tank while maintaining the usual functions of reducing biochemical oxygen demand (BOD) and removing total suspended solid (TSS) and harmful bacteria such as *Escherichia coli* (*E. coli*), and (2) the overall role of the planted vegetation. The installation of the SUW must follow pretreatment from a septic tank (Figure 24.1). The effluent of the SUW may be routed to final disposal or may be reused according to the regulatory requirements of the site location.

24.2 REGULATION AND POLICY

24.2.1 Current regulation of water quality and OSTDS standards

Current regulation of water quality and OSTDS standards might give rise to more insight about where the technology frontiers are located. The Florida Department of Environmental Protection (DEP) is charged with implementing the requirements of the Federal Clean Water Act and the Florida Water Pollution Control Act as set forth in Chapter 403, Florida Statutes. DEP has established by rule a water body classification system and the supporting surface water quality standards, which are designed to protect the beneficial uses set forth in the water body classes. With respect to nutrients, DEP has adopted a narrative nutrient criterion that states that nutrient levels shall not create an imbalance of flora and fauna. DEP currently is working on numeric nutrient criteria and has established water body specific ones with the adoption of Total Maximum Daily Loads (TMDLs) for those water bodies impaired by nutrients. For example, the TMDL for Wekiwa Springs is a monthly average of 286 μg L^{-1} nitrate. The Florida Department of Health (DOH) is charged with regulating OSTDSs through their authority in Chapter 381, F.S., and their implementing regulations in Chapter 64E-6, F.A.C. DOH's mission is the protection of public health, not water quality, and it uses the drinking water standard of 10 mg L^{-1} nitrate as its goal (Chapter 64E-6, F.A.C.).

24.2.2 **NSF 245 standards**

National Sanitation Foundation and the American National Standards Institute (NSF/ANSI) Standard 245, published in 2007, was developed for residential wastewater treatment systems designed to provide for nitrogen reduction. The evaluation involves 6 months of performance testing, incorporating stress tests to simulate wash day, working parents, power outages, and vacation conditions. The standard is set up to evaluate systems having rated capacities between 400 and 1500 gallons day^{-1}. Technologies testing against Standard 245 must either be Standard 40 certified (ANSI-40) or be evaluated against Standard 40 at the same time (NSF, 2009). The NSF 245/ANSI-40 influent concentration standards for testing are:

- BOD$_5$: 100–300 mg L^{-1}
- TSS: 100–350 mg L^{-1}
- TKN: 35–70 mg L^{-1} as N
- Alkalinity: greater than 175 mg L^{-1} as CaCO$_3$ (alkalinity may be adjusted if inadequate)
- Temperature: 10–30 °C
- pH: 6.5–9 SU

Environmental Technology Verification (ETV) protocols are developed for specific technology areas and serve as templates for developing test plans for the evaluation of individual technologies at specific locations. The ETV protocols for suggested average influent requirements are (NSF, 2009):

- CBOD$_5$: 100–450 mg L^{-1}
- TSS: 100–500 mg L^{-1}
- TKN: 25–70 mg L^{-1}
- Total P: 3–20 mg L^{-1}
- Alkalinity: greater than 60 mg L^{-1}
- Temperature: 10–30 °C

The NSF Standard 245 would allow chemical addition to adjust influent's alkalinity using sodium bicarbonate. Throughout the testing, samples are collected during design loading periods and evaluated against the pass/fail requirements. NSF states that an OSTDS must meet the following effluent concentrations averaged over the course of the testing period in order to meet Standard 245 (NSF, 2009):

- $CBOD_5$: 25 mg L^{-1}
- TSS: 30 mg L^{-1}
- TN: less than 50% of average of all influent TN samples
- pH: 6.0–9.0 SU.

24.3 BIOSORPTION-ACTIVATED MEDIA

Soil augmentation with sorption media mixes results in improvements in nutrient removal of current treatment technologies used for stormwater management, wastewater treatment, landfill leachate treatment, groundwater remediation, and treatment of drinking water (Chang et al., 2010a,b). The use of these sorption media in the engineered processes and natural systems may remove not only the nutrients, but also some other pollutants, such as heavy metals, pathogens, pesticides, and toxins (TCE, PAH, etc.,). Sorption is important for phosphorus removal because of the adsorption, absorption, and precipitation effects. With such functionality, a biofilm can be formed on the surface of soil or media particles to allow microbes to assimilate nitrogen species, although nitrogen cannot be removed by sorption directly. It is indicative that sorption provides an amenable environment for subsequent nitrification and denitrification.

The importance of developing specific wetland porous media instead of conventional soil, sand, and gravel to gain better pollutant removal capacity has been widely recognized (Chang et al., 2011a). Mann and Bavor (1993) represented the pioneer trial during the early period when laboratory-scale phosphorus adsorption was compared between regional gravels and alternative adsorptive media including industrial slag and ash by-products. Mann and Bavor (1993) reported the maximum adsorption capacity of regional gravels as 25.8–47.5 μg P g^{-1}, blast furnace slag as 160–420 μg P g^{-1}, and fly ash as 260 μg P g^{-1}, which warranted further research involving the inclusion of industrial waste substrata (Chang et al., 2011a). Later efforts demonstrated that removal of ammonia, nitrite, nitrate, and phosphorus can be achieved collectively or independently by using media including sawdust (Kim et al., 2000; Gan et al., 2004; Schipper et al., 2005), tire crumb (Shin et al., 1999; Lisi et al., 2004; Smith et al., 2008), sand (Birch et al., 2005; Hsieh and Davis, 2005; Seelsaen et al., 2006), clay (Gálvez et al., 2003; Gisvold et al., 2000), zeolite (Li, 2003; Birch et al., 2005; Seelsaen et al., 2006), sulfur (Zhang, 2002; Ray et al., 2006), and limestone (Kim et al., 2000; Zhang, 2002), which are deemed as multifunctional materials applicable to natural systems and built environments that can improve both physicochemical and microbiological processes (Chang et al., 2010a,b). Green sorption medium consisting of recycled and natural materials was shown to provide a favorable environment for nutrient removal (Xuan et al., 2009).

Coombes and Collett (1995) used crushed basalt and limestone chippings in their horizontal flow *Phragmites australis* wetland and found that ammoniacal nitrogen in the effluent averaged <2 mg L^{-1}. Pant et al. (2001) found that Fonthill sand performed better in removing P from wastewater in a comparison among three types of media (Lockport dolomite, Queenston shale, and Fonthill sand). Vohla et al. (2007) tested a designed oil-shale ash derived from oil-shale combustion for P retention. Integration between different wetland species and green sorption media has not yet been examined to explore the most cost-effective and sustainable solution. Park and Polprasert (2008) investigated P removal using an integrated constructed wetland system packed with oyster shells as adsorption and filtration media and found a removal efficiency of 7% of N and 98.3% of P. Tee et al. (2009) reported a better performance of planted constructed wetlands with gravel and raw rice husk-based media for phenol and N removal compared with unplanted ones.

The adsorption, absorption, ion exchange, and precipitation processes are intertwined with the overall physicochemical process when removing nutrients via BAM with different ingredients. Pollutants removed by the adsorption process in green sorption media (i.e., BAM) may subsequently desorb. If there are organic sources in the environment, hydrolysis converts particulate organic N to soluble organic N, and ammonification in turn releases ammonia into the water bodies. Ammonia may be sorbed by clay in bioretention filters filled with sorption media.

Overall, the removal of the nutrient could be enhanced in BAM because its pore structure can retain nitrogen and phosphorus through sorptive processes, making nitrogen available to the microorganisms and plant uptake as well as immobilizing phosphorus to a maximum degree. Select recipes of various BAM for the removal of nutrients in dealing with a variety of inflow conditions were tested in a laboratory setting using columns filled with different BAM associated with varying temperature regimes (Xuan et al., 2010a; Chang et al., 2011b). The ingredients of BAM tested in previous studies include tire crumb, organics, sand, and expanded clay in mixes with limestone (US Patent 8153005 B1, 2011; US Patent 8252182 B1, 2012).

For example, a BAM layer composed of 50% citrus grove sand, 15% tire crumbs, 15% sawdust, and 20% limestone was used to improve nutrient removal in a prototype testing. As HRT increased, the percent removal increased, but at a decreasing rate (Figure 24.2); these results were used to establish an HRT of 7 days. Based on a nitrogen influent concentration of 50 mg L^{-1} with no plant coverage, the effluent concentration at 7 days HRT and under a stress condition would produce an effluent nitrogen concentration of 10 mg L^{-1}. After a long period of trial and error, the final recipe adopted in this study was (1) Pollution Control Media (P Media): 50% citrus grove sand, 15% tire crumbs, 15% sawdust, and 20% limestone; and (2) Growth Media (G Media): 75% expanded clay, 10% vermiculite, and 15% peat moss (all percentages are by volume). Greater removal is expected with plant coverage.

24.4 FIELD-SCALE STUDY

24.4.1 Process design

An SUW receiving septic effluent from BPW Scholarship House (a 15-person dormitory at the University of Central Florida) processed 454 m^3 (120 gallons)

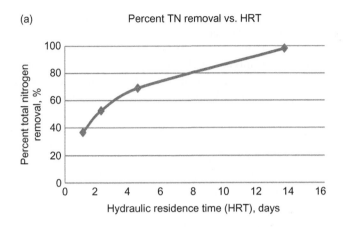

(a)

Percent TN removal vs. HRT

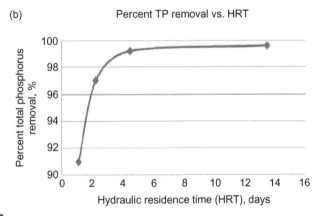

(b)

Percent TP removal vs. HRT

FIGURE 24.2

Effectiveness of nitrogen removal as a function of HRT with no plant coverage. (For the color version of this figure, the reader is referred to the online version of this chapter.)

of influent per day in this wastewater treatment study (Chang et al., 2011a). The septic tank that received the influent before the SUW system has a capacity of 1000 GPD, providing 2–3 days HRT. A gravel-filled gravity distribution system included a header pipe, even distribution box, distribution pipe, flow meter, and four wetland cells packed with special green sorption media (recycled and natural materials) (Chang et al., 2011a). Within the full-scale field study, a new set of green sorption media was used for both nutrient and pathogen removal in the SUW (Chang et al., 2011a). Because a high-porosity gravel was used as the substrate at the bottom of each cell, an innovative upflow (i.e., outlet of SUW is higher than inlet) design was introduced to induce a uniform upflow hydraulic pattern and an amenable nitrification–denitrification environment as well as avoid clogging and flooding, which overcomes the main disadvantage of conventional SSF wetlands.

In addition, this design could result in maximal reduction of the effect of stormwater. In a conventional SSF wetland (i.e., inlet is higher than outlet), the stormwater moves downward, permeates the whole cell, and finally drains out of the bottom of the cell with wastewater in rainy days. For our SUW, the water level in the cell was kept as high as the outlet, and part of the stormwater drained directly through the higher outlet instead of mixing with the wastewater beneath (Chang et al., 2011a); therefore, less stormwater could reach the sampling ports, which allowed us to make a more accurate evaluation of our designed media performance (Chang et al., 2011a).

Through various physical, chemical, and biological processes, most bacteria and viruses in wastewater, as well as nutrients, were consumed and intercepted as the wastewater effluent traveled upward through the sand layer (i.e., aerobic layer at the bottom) and pollution control layer (i.e., anaerobic layer in the middle) to the growth media layer (Chang et al., 2011a). Collectively, the gravel layer beneath the sand layer and the plant species inserted in the growth media in this unique SUW might promote removal of pathogens, nitrogen, and phosphorus via nitrification, denitrification, adsorption, absorption, ion exchange, filtration, and precipitation.

To test this possibility, three kinds of plant species were individually planted in three cells to compare with the parallel control cell with no plant species (see Figure 24.3) (Chang et al., 2011a). To measure the full-pipe water flow we used a DIGI-FLOW™ F-1000RT paddlewheel meter, a battery-powered digital flow meter (Chang et al., 2011a) that measures a total as well as an instantaneous flow measurement. This meter was installed at every entrance and exit of the four wetland cells to record the inflow and the outflow and to determine the mass balance of the wetland system. A totalizer was used to measure the daily total flow of the wetlands (Chang et al., 2011a).

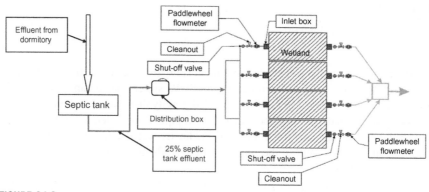

FIGURE 24.3

Configuration of septic tank followed by four-cell wetlands for treating 25% of the septic effluent. For comparison, the remaining 75% was sent to other wastewater treatment facilities in the same pilot plant (Chang et al., 2011a). (For the color version of this figure, the reader is referred to the online version of this chapter.)

The four parallel 1.52 m wide × 3.05 m long × 0.91 m deep ($5 \times 10 \times 3$ ft^3, respectively) cells in this test bed (Figure 24.4) each contained an impermeable liner, a gravel substrate, fabric interlayer, sand, P media, G media, and selected plants. The main function of the G media layer was to support the root zone and to speed the maturation process of the treatment system. The 30.48-cm (12 in.) P media layer was used to remove nutrients and TSS and reduce BOD. A 15.24-cm (6-in.) sand layer was then added beneath the P media to improve the removal of pathogens and TSS. The 30.48-cm (12 in.) thick gravel substrate created additional pore space, allowing water to spread across the bottom of wetland more freely while maintaining a certain flow rate.

The purpose of the separation fabric liner on the top of the gravel layer was to keep the sand stable above the gravel layer. Once the gravel layer was fully saturated, the water level would gradually rise, passing through the sand and P medium layer up to the outlet. Chowdhury et al. (2008) provided an overview of flow pattern in a SUW by conducting a bromide tracer study. They found that, in a bottom inlet and top outlet condition, a gravel layer added at bottom caused the flow to be mostly vertical, which provided strong evidence for our hydraulic pattern hypothesis. In each wetland, two customized oxygenators (PVC pipe wrapped with fabric at bottom) were inserted on both sides of the inlet into the gravel layer to enhance nitrification at the bottom of the wetland cells and therefore fulfill the design configuration for the SUW (Chang et al., 2011a).

The samplers (airstone connected with rubber tubing to the surface) were installed at the interface between different layers with three depths: B, M, and T (Chang et al., 2011a). Horizontally, the samplers in the four wetland cells were 33%, 67%, and 100% (positions 1–3, respectively) along the length of the wetland (Figure 24.4) (Chang et al., 2011a) at each interface. B series ports provided a mixture of the bottom three samples; M series ports a mixture of the middle three samples; and T ports a mixture of the top three samples (Chang et al., 2011a).

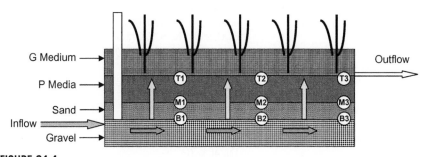

FIGURE 24.4

Locations of the sample points in the wetland cross-section (Xuan et al., 2010a,b). *Note:* B, M, and T series ports are installed for water sample collection.

24.4.2 Selection of plant species

As components of wetland systems, plant species have an irreplaceable function in pollutant purification. In subsurface wetland systems, the biofilm around the plant rhizosphere provides a potential attachment site for bacteria to achieve the nitrification and the aerobic degradation of soluble organics (Chang et al., 2011a). Based on the characteristics of oxygen transmission, an aerobic–anoxic–anaerobic state was seen around the rhizosphere, the equivalent of a series of parallel anaerobic–anoxic–oxic (A^2O) processing units, which is a commonly used nutrient removal process in wastewater treatment (Chang et al., 2011a). Aerobic areas near the root zone were conducive to nitrification, and anaerobic areas away from the roots were conducive to denitrification, both of which might perform the final clean-up of residual nitrogen in the septic effluent (Chang et al., 2011a). Nitrate would thus be effectively removed by denitrification in rhizospheric zones (Chang et al., 2011a).

Total N and P (TN and TP) can be removed if the plants are harvested routinely. Seidel's work (1955) is known as the first attempt to use the wetland vegetation to remove various pollutants from wastewater. Since then, researchers have continued to test different vegetation in natural soil substrates to improve pollutant removal efficiencies in SF and SSF (Tables 24.1 and 24.2, respectively) (Chang et al., 2011a).

Only *Phragmites australis* (in case 1b and 1f in SF; Tables 24.1 and 24.2) showed a positive result with respect to the removal of nutrients (about 90% TN removal); however, *P. australis* is a typical kind of emergent vegetation that is unsuitable for planting in our SUW because emergent vegetations are typically planted in free water surface wetlands (Chang et al., 2011a). Under the criteria for screening plant species described in the previous study (Xuan et al., 2009), three kinds of native

Table 24.1 SF Wetland Performance Based on Different Kinds of Vegetation (Chang et al., 2011a)

SF	Plant	Removal Efficiency	Reference
1a	*Typha latifolia, Phragmites australis, Sparganium erectum*	80% COD, 83% BOD, 45% TN, 47% TP	Cadelli et al. (1998)
1b	*Phragmites australis*	98% SS, 87% COD, 96% BOD, 91% TN, 60% OrthoP	Cadelli et al. (1998)
1c	*Phragmites australis, Scirpus lacustris*	68% COD, 83% BOD, 26% TN, 2% Ortho P	Cadelli et al. (1998)
1d	*Lemna* sp.	96% SS, 75% COD, 90% BOD, 43% TN, 47% TP	Cadelli et al. (1998)
1e	*Lemna* sp.	98% SS, 96% COD, 94% BOD, 49% TN, 49% TP	Cadelli et al. (1998)
1f	*Phragmites australis*	87% COD, 97% BOD, 89% TN, 46% TP	Cadelli et al. (1998)

Note: *Surface flow wetland (SF), suspended solid (SS), biochemical oxygen demand (BOD), chemical oxygen demand (COD), total phosphorus (TP), total nitrogen (TN).*

Table 24.2 SSF Wetland Performance Based on Different Kinds of Vegetation (Chang et al., 2011a)

SSF	Plant	Removal Efficiency	Reference
1	*Phragmites*	90% COD, 96% BOD, 92% SS, 63% TP, 36% TN	Haberl et al. (1998)
2	*Scirpus cyperinus, Typha latifolia*	73.4% NH_4^+-N, 67.5% TKN	Huang et al. (2000)
3a	*Typha latifolia, T. angustofolia, Scirpus taebormontanii*	92% BOD, 87% TSS, 99.6% Fecal, 41% TN, 50% TP	Henneck et al. (2001)
3b	*Typha* sp.	82% BOD, 86% TSS, 92.4% Fecal, 51% TN, 59% TP	Henneck et al. (2001)
3c	*Typha latifolia*	83% BOD, 81% TSS, 99.9% Fecal, 54% TN, 97% TP	Henneck et al. (2001)
4a	*Phragmites mau Ritianus*	25.2% NH_4^+-N, 56.3% COD, 57% TC, 68% FC	Kaseva (2004)
4b	*Typha latifolia*	23% NO_2-N, 23% NH_4^+-N, 60.7% COD, 60% TC, 72% FC	Kaseva (2004)
5a	*Cyperus papyrus*	75.3% NH_4^+-N, 83.2% TRP	Kyambadde et al. (2004)
5b	*Miscanthidium violaceum*	61.5% NH_4^+-N, 48.4% TRP	Kyambadde et al. (2004)
6	*Phragmites australis*	30% of TP , 50% Denitrification	Brix and Arias (2005)
7	*Phragmites* and *Typha*	27% TKN, 19% NH_4^+-N, 4% Nitrite	Keffala and Ghrabi (2005)
8a	*Juncus effusus L.*	54% NH_4^+-N, 55% TN, 95% TP	Xuan et al. (2009)
8b	*Panicum hemitomon*	88% NH_4^+-N, 85% TN, 94% TP	Xuan et al. (2009)
8c	*Zizaniopsis miliacea*	78% NH_4^+-N, 79% TN, 95% TP	Xuan et al. (2009)

Note: Subsurface wetland (SSF), ammonium-nitrogen (NH_4^+–N), ammonium (NH_4^+), nitrite (NO_2^-), total reactive phosphorus (TRP), total Kjeldahl nitrogen (TKN), nitrite-nitrogen (NO_2–N), fecal coliform (FC), total carbon (TC), total suspended solid (TSS), biochemical oxygen demand (BOD), chemical oxygen demand (COD), total phosphorus (TP), total nitrogen (TN).

vegetation with the same volume and net price, canna (*Canna flaccida*), blue flag (*Iris versicolor L.*), and bulrush (*Juncus effusus L.*) (Figure 24.5) were ultimately selected and evenly planted in wetland cells 1, 2, and 3, respectively (Chang et al., 2011a). Seedlings of these three kinds of plant were purchased from a local nursery and planted 2 months before the experiment period with a density of 7–8 plants per m^2 (Chang et al., 2011a). Wetland cell 4 is the control case with the same layered green sorption media but without any plantings (Chang et al., 2011a).

FIGURE 24.5

Plant species selected in the study: (a) canna, (b) blue flag, (c) bulrush (Chang et al., 2011a). (For the color version of this figure, the reader is referred to the online version of this chapter.)

24.4.3 Sampling and analysis

A 24-h composite sample (a representative sample combined by the multiple samples at regular intervals) was taken at every sampling port in proportion to the actual flow load during that 24 h period. Fecal coliform and *E. coli* samples were collected in a 100-mL sterile polyethylene flask, which was immediately sealed, labeled, and delivered to the external certified laboratory for analysis within the same day. A clean polyethylene jug was used to store each sample for analysis of other parameters. Once the samples were taken, the containers were stored in a chilled cooler (4 °C) until the 24-h composite samples were completed. Samples that required appropriate preservatives were processed according to a quality assurance/quality control (QA/QC) protocol; 200 mL of each sample was filtered through a 0.45-μm filter, half of which was held at <pH 2.0. Each sample was delivered to the external certified laboratory in an ice chest within the same day to ensure the integrity of the samples.

A mass balance analysis was conducted based on the measurements during a 1-month study period. TN in the media was measured as the amount being sorbed in media during the wastewater treatment process. N_2 discharge into the air was calculated as the residual term in the TN budget. By the same token, the TN and TP in influent and effluent were calculated as the product of the water flux and nutrient concentration. Porosity in different layers was taken into account to finally summarize the amount of TN and TP in pore water. The amount of phosphorus released from the plants can be ignored since no fading leaves fell in the given period. The amount of uptake and release by microbes can be balanced out by the dynamic equilibrium between uptake and release by microbes. Due to technical complexity, precipitation in the septic wetland was calculated as the residual term of the phosphorus budget. Ultimately, the nitrogen and phosphorus budget can be simplified by the following equations:

$$\text{In} = \text{Out} + \text{Uptake by plant} + \text{Sorption in media} + \text{Storage in wastewater}$$
$$+ \text{N}_2 \text{ discharge by denitrification} \qquad (24.1)$$

$$\text{In} = \text{Out} + \text{Uptake by plant} + \text{Sorption in media} + \text{Storage in wastewater}$$
$$+ \text{Precipitation} \qquad (24.2)$$

At the beginning and the end of the study period, samples of P media and G media were randomly collected from each wetland cell and mixed in a 1-L (quart-sized) resealable plastic bag to form a composite sample. Similarly, samples of leaf and root were collected in a random 30×30 cm^2 area and analyzed to measure nutrient uptake by plant tissue.

The water quality in the wetland system was monitored weekly from September 2–30, 2009. Using grab-sample analysis, dissolved oxygen (DO), pH, and temperature were measured on site by an HACH HQ40d field case. In addition, ammonia-nitrogen (NH_3-N), nitrite-nitrogen (NO_2-N), nitrate-nitrogen (NO_3-N), organic nitrogen–nitrogen (ON-N), TN, organic phosphorus (OP-P), soluble reactive phosphorus–phosphorus (SRP-P), and TP were measured by a certified lab (Table 24.3).

All media and plant samples were collected on September 2 and 30, 2009, as representative samples of the initial and final stages of the experimental period. Media samples were delivered to a soil laboratory at Pennsylvania State University (PSU) for analysis. Plant samples were sent to an agricultural lab located in Orange City, Florida. For real-time polymerase chain reaction (PCR) analysis at University of Central Florida (UCF), each media sample from different wetlands was collected into a 1.5-mL microcentrifuge tube and held at -20 °C until use.

Table 24.3 Outline of Analysis Methods

Parameter	Analytical Method	Testing Location
pH	USEPA 150.1	On site
NH_3-N	EPA 350.1	Certified lab
NO_2-N	EPA 353.2	Certified lab
NO_3-N	EPA 353.2	Certified lab
ON-N	EPA 350.2	Certified lab
SRP-P	EPA 365.3	Certified lab
TP	Alkaline persulfate digestion	Certified lab
DO	Manufacturer manual	On site
Temperature	Manufacturer manual	On site
Uptake by plant	N: Kjeldahl digestion	Agricultural lab
	P: wet digestion with nitric and perchloric acids	
Nutrient Sorption media	DTPA Saturated Media Extract Method	PSU lab
Quantity of denitrifiers	Real-time PCR	UCF lab

A detailed procedure of the DNA extraction and real-time PCR analysis was provided in a companion study (Xuan et al., 2009). The real-time PCR was applied to gain insight into the denitrifier activity across the green sorption media given that the substrate and enzyme are not the limiting factors in the treated effluent. Both P and G media samples from each wetland were collected into 1.5 mL microcentrifuge tubes and held at $-20\,°C$ until use. DNA from the sample was extracted in duplicate using a SoilMaster® DNA Extraction Kit (EPICENTRE), and 50 mg of sample instead of 100 mg (default value in manufacturer's instructions) was weighed into the 1.5-mL microcentrifuge tube to decrease the effects of enzymatic inhibitors. Finally, the 300-µL extracted DNA template was prepared.

Real-time PCR quantification was performed by the Stepone® (Applied Biosystems) PCR instrument to amplify *nirK* gene from the denitrifiers with a pair of primers, nirK876 (5′-ATYGGCGGVCAYGGCGA-3′) and nirK1040 (5′-GCCTCGATCAGRTTRTGGTT-3′) (Braker et al., 1998). The PCR mixture was prepared using 12.5 µL of the SYBR Green PCR Master Mix kit, 10 µM of each primer, standard DNA or extracted DNA from samples, and diethylpyrocarbonate-treated water to complete a 25-µL total volume. The protocols for *nirK* real-time PCR were 120 s at 50 °C, 900 s at 95 °C, followed by four operational cycles: 15 s at 95 °C for denaturation, 30 s at 63 °C for annealing, 30 s at 72 °C for extension, and 30 s at 80 °C for a final data acquisition step. However, the annealing temperature in cycle 3 above (i.e., 30 s at 63 °C) was progressively decreased by 1°C increments down to 58 °C via six touchdowns associated with each cycle to form a total of 24 runs. After these 24 runs, a final cycle with an annealing temperature of 58 °C was repeated 40 × (Henry et al., 2004).

24.4.4 Performance-based comparisons between wetland cells

The average wastewater loadings were measured before and after the four SUW test wetland cells were operational (Tables 24.4 and 24.5). A summary of removal efficiency data for all tests provides a comparative study across all four wetland cells (Table 24.6) and shows that the canna-based SUW cell (cell #1) performed best. For normal warm weather conditions, the nutrient removal efficiency of SUW planted with canna is reported in reference publications as 97.1% and 98.3% for TN and TP, respectively (Chang et al., 2011a).

Canna in SUW cell 1 is known as a seasonal plant that withers in freezing temperatures. Winter 2009–2010, reported as one of the coldest winters since records began, was the cold-weather test period designed to answer the following questions: (1) Will canna continue to function without its foliage? and (2) What is the nutrient removal efficiency of SUW planted with canna in cold weather? Samples collected from the effluent of the canna SUW compared with the control cell during the cold period show that nitrogen removal in the canna cell was 87.4% TN compared with the 41.0% TN in the control cell. Most nitrogen removal occurs at the root zone, and the root zones were marginally affected by the colder air temperatures.

Table 24.4 Average Waste Loadings in Influent Were Measured Before the SUW ($n=5$) (Chang et al., 2011a)

Ammonia-N (g day^{-1})	Nitrate-N (mg day^{-1})	TKN (g day^{-1})	Influent Average Loading Rate			
			TN (g day^{-1})	SRP (g day^{-1})	TP (g day^{-1})	E. coli (cfu/day)
25.4±1.0	1.4±0.8	31.6±2.2	31.6±2.2	1.8±0.3	2.5±0.2	$2.18 \times 10^9 \pm 2.39 \times 10^9$

Note: Nitrate-nitrogen (NO_3^+-N), soluble reactive phosphorus (SRP), total Kjeldahl nitrogen (TKN), total phosphorus (TP), total nitrogen (TN).

Table 24.5 Average Waste Loadings in Effluent Were Measured After the SUW ($n=5$) (Chang et al., 2011a)

SUW	Ammonia-N (g day^{-1})	Nitrate-N (mg day^{-1})	TKN (g day^{-1})	Average Loading Rate			
				TN (g day^{-1})	SRP (mg day^{-1})	TP (mg day^{-1})	E. coli (cfu/day)
1 (Canna)	0.08±0.06	0.8±0.6	0.23±0.06	0.23±0.06	2.1±0.9	10.9±4.0	<1134
2 (Blue Flag)	0.13±0.02	0.4±0.1	0.54±0.48	0.54±0.48	4.4±5.0	27.4±25.8	$2.0 \times 10^4 \pm 3.2 \times 10^4$
3 (Bulrush)	2.4±2.1	0.3±0.1	2.9±2.1	2.9±2.1	1.6±0.7	12.6±4.2	$7.2 \times 10^4 \pm 1.1 \times 10^5$
4 (Control)	5.9±3.0	0.4±0.1	7.1±2.2	7.1±2.2	3.1±3.6	90.7±45.7	$2.8 \times 10^3 \pm 1.7 \times 10^3$

Note: Ammonium-nitrogen (NH_4^+-N), nitrate-nitrogen (NO_3^+-N), soluble reactive phosphorus (SRP), total Kjeldahl nitrogen (TKN), total phosphorus (TP), total nitrogen (TN).

Table 24.6 Summary of Performance Test Data Expressed as Removal Effectiveness ($n=5$) (Chang et al., 2011a)

Parameter	Removal Efficiency (%) Based on Concentration			
	Wetland Cell 1 (Canna)	Wetland Cell 2 (Blue Flag)	Wetland Cell 3 (Bulrush)	Wetland Cell 4 (Control)
Fecal Coliform	99.98	97.06	99.76	99.74
E. coli	100.00	99.94	99.80	100.00
BOD_5	89.4	89.6	85.2	75.2
$CBOD_5$	87.6	86.5	81.1	79.5
NH_3-N	98.6	97.9	62.0	27.6
TN	97.1	93.0	62.5	10.5
SRP	99.5	99.0	99.6	99.3
TP	98.3	95.7	98.0	85.7

Note: Ammonium-nitrogen (NH_4^+–N), soluble reactive phosphorus (SRP), total phosphorus (TP), total nitrogen (TN), 5-day biochemical oxygen demand (BOD_5), 5-day carbonaceous biochemical oxygen demand ($CBOD_5$).

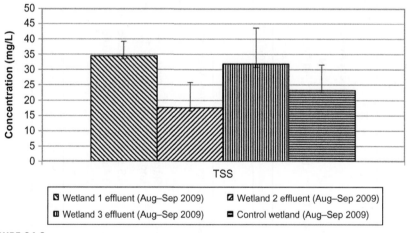

FIGURE 24.6

TSS effluent concentrations from SUW ($n=5$) (Chang et al., 2011a).

24.4.5 SUW effluent performance concentrations

TSS and 5-day carbonaceous biological oxygen demand ($CBOD_5$) concentrations of the SUW effluents (Figures 24.6 and 24.7) indicated that TSS concentrations were near 30 mg L^{-1}, with an average of all four cells below 30 mg L^{-1}, the NSF 245 requirement for effluent TSS, and $CBOD_5$ concentrations averaged <5 mg L^{-1}

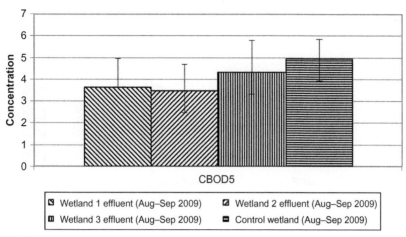

FIGURE 24.7

CBOD$_5$ effluent concentrations from SUW ($n=5$) (Chang et al., 2011a).

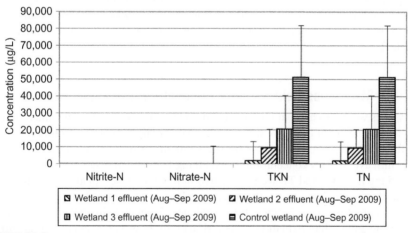

FIGURE 24.8

Nitrogen effluent concentration from SUW ($n=5$) (Chang et al., 2011a).

(NSF 245 requirement 25 mg L^{-1}). Effluent concentrations for nitrogen and phosphorus-species (Figures 24.8 and 24.9) show that, overall, SUW cells 1 and 2 performed better in removing nitrogen, with nitrate, nitrite, and TN concentrations than the cell 3 and the control cell. TN in SUW cells 1 and 2 was <1 mg L^{-1} (Figure 24.8), and nitrogen concentrations in the effluent of SUW cells 1 and 2 were below the U.S. Environmental Protection Agency (EPA) maximum contaminant level of nitrate (10 mg L^{-1}) and nitrite (1 mg L^{-1}) for drinking water. The maximum TN concentration from SUW cell 1 was 2.578 mg L^{-1}. With this performance-based

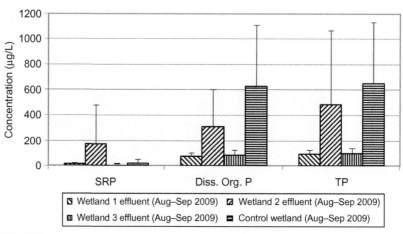

FIGURE 24.9

Phosphorus effluent concentration from SUW ($n=5$) (Chang et al., 2011a).

comparison, the canna wetland cell is deemed the best; therefore, a summary of the mean, maximum, and minimum effluent concentrations was compiled for all water quality parameters associated with the operation of the SUW planted with canna (Table 24.7).

24.4.6 SUW mass balance with removal performance data

Flow rates were measured at both inlet and outlet of the SUW; thus, the mass of nutrients into and out of the SUW was calculated by multiplying the concentrations and the flow rate. The mass balance is for the 1-month period with rainfall of 9.65 mm (3.80 in.), average TN concentration of 0.80 mg L^{-1}, and average TP concentration of 0.25 mg L^{-1}. TN and TP were measured in the pore water, plant tissue, and soil media; thus, the unknowns are the reactive values of precipitate for the phosphorus balance and nitrogen gas for the nitrogen balance. It is uncertain how nitrogen is lost (anaerobic ammonium oxidation or denitrification); however, a significant quantity of denitrifying bacteria was present in all SUW cells (Figure 24.10) (Chang et al., 2011a).

Ultimately, the nitrogen and phosphorus budget can be simplified by Equations (24.1) and (24.2) (Chang et al., 2011a). The mass balance for nitrogen and phosphorus (Figures 24.11 and 24.12, respectively) illustrate that mass removals are high for both nitrogen and phosphorus, as expected, because of the concentration changes and evapotranspiration occurring within the SUW. Mass balance of water over the four wetland cells (Figure 24.13) shows that in four sets of flow measurements over time, the average discharge flow of wetland cells 1, 2, 3, and 4 were about 76%, 82%, 81%, and 84%, respectively, lower than the inflow. A simulated rainfall below 50 mm (20 in.) per hour indicates rain flowing into the SUW, and above that level

Table 24.7 Summary of the Mean, Maximum, and Minimum Effluent Concentrations for the Canna-Planted Wetland ($n=5$) in which the bold face values of TN and TP are the major concern in this study (Chang et al., 2011a)

	SUW Wetland 1 Effluent (Aug–Sep 2009)			
	Average	**Min**	**Max**	**Median**
Alkalinity (mg L^{-1})	378.8	277.0	465.0	38.01
TSS (mg L^{-1})	34.6	29.0	42.0	34.0
BOD$_5$ (mg L^{-1})	5.22	2.9	8.7	3.7
CBOD$_5$ (mg L^{-1})	3.6	2.3	5.5	3.1
Ammonia-N (μg L^{-1})	859.6	304.0	1437.0	972.0
NO$_X$-N[a] (μg L^{-1})	7.8	4.0	16.0	5.0
Nitrite-N (μg L^{-1})	1.6	1.0	3.0	1.0
Nitrate-N (μg L^{-1})	6.2	2.0	14.0	4.0
Org. N (μg L^{-1})	1097.4	337.0	2030.0	1139.0
TKN (μg L^{-1})	1957.0	1536.0	2576.0	1711.0
TN (μg L^{-1})	**1964.2**	**1540.0**	**2578.0**	**1727.0**
SRP (μg L^{-1})	18.0	11.0	27.0	17.0
Org. P (μg L^{-1})	78.0	38.0	101.0	79.0
TP (μg L^{-1})	**96.0**	**51.0**	**125.0**	**96.0**
Fecal (cfu 100 mL^{-1})	657.0	1.0	3000.0	20.0
E. coli (cfu 100 mL^{-1})	6.8	1.0	30.0	1.0

[a]*[NO$_X$-N]=[Nitrite-N]+[Nitrate-N].*

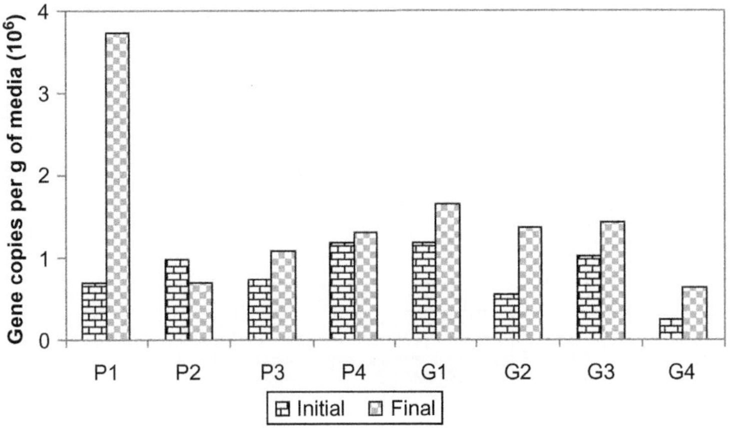

FIGURE 24.10

Quantity of denitrifying bacteria (P1=P media in wetland 1, G1=G media in wetland 1) (Chang et al., 2011a). (For the color version of this figure, the reader is referred to the online version of this chapter.)

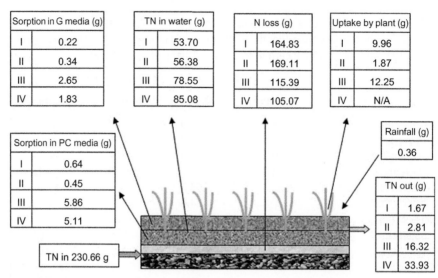

FIGURE 24.11

Mass balance for nitrogen in four SUW cells (September 2–30, 2009) (Chang et al., 2011a). *Note*: Roman numerals I, II, III, and IV represent wetland cells 1, 2, 3, and 4, respectively. (For the color version of this figure, the reader is referred to the online version of this chapter.)

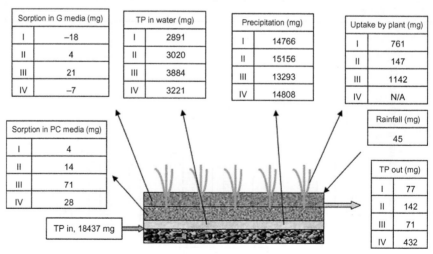

FIGURE 24.12

Mass balance for phosphorus in four SUW cells (September 2–30, 2009) (Chang et al., 2011a). *Note*: Roman numerals I, II, III, and IV represent wetland cells 1, 2, 3, and 4, respectively. (For the color version of this figure, the reader is referred to the online version of this chapter.)

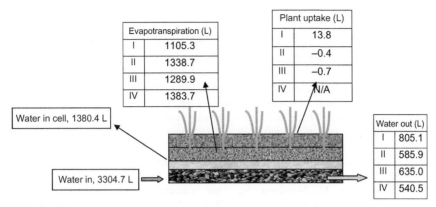

FIGURE 24.13

Mass balance of water in four wetland cells. *Note*: Roman numerals I, II, III, and IV represent wetland cells 1, 2, 3, and 4, respectively. (For the color version of this figure, the reader is referred to the online version of this chapter.)

indicates rain flowing away from the SUW by gravity without any interaction with the treated wastewater (Xuan et al., 2010b).

24.5 CONCLUSIONS

Passive nutrient removal OSTDSs are expected to be preferred choices for future generations. Pairing their excellent nutrient reduction and their energy-efficient operation with low-cost maintenance should stimulate demand for them in the future marketplace. The substantial ability of the SUW to remove nutrients was confirmed by this study. Findings indicate that wetland cell 1, which was planted with canna, achieved a removal efficiency of 97.1% and 98.3% for TN and TP, respectively. Canna would be a highly competitive SUW plant candidate in terms of esthetics and nutrient control year-round. Denitrification was successful as evidenced by the more than threefold increase in denitrifiers in the P media of wetland cell 1. The field campaign of this pilot study has eventually proven SUW to (1) be effective in nutrient reduction with the use of BAM and (2) maintain operating reliability. Yet the coupling mechanism between hydrodynamics, geochemical interactions, and microbiological activities in such an SUW remains unclear. A mechanistic model to address the system dynamics is expected to be the ultimate tool to answer this scientific question in the next chapter.

Acknowledgments

The authors are grateful for the financial support provided by an Urban Nonpoint Source Research Grant from the Bureau of Watershed Restoration, Florida Department of Environmental Protection.

References

Billore, S.K., Sharmaj, K., Dass, P., Nelson, R., 1999. Horizontal subsurface flow gravel bed constructed wetland with Phragmites karka in Central India. Water Sci. Technol. 40 (3), 163–171.

Birch, G.F., Fazeli, M.S., Matthai, C., 2005. Efficiency of infiltration basin in removing contaminants from urban storm water, Environ. Monit. Assess. 101, 23–38.

Braker, G., Fesefeldt, A., Witzel, K.P., 1998. Development of PCR primer systems for amplification of nitrite reductase genes (nirK and nirS) to detect denitrifying bacteria in environmental samples. Appl. Environ. Microbiol. 64, 3769–3775.

Brix, H., Arias, C.A., 2005. The use of vertical flow constructed wetlands for on-site treatment of domestic wastewater: new Danish guidelines. Ecol. Eng. 25, 491–500.

Cadelli, D., Radoux, M., Nemcova, M., 1998. Constructed wetlands in Belgium. In: Vymazal, H. Brix, Cooper, P.F., Green, M.B., Haberl, R. (Eds.), Constructed Wetland for Wastewater Treatment in Europe. Backhuys publishers, Leiden, The Netherlands, pp. 77–93.

Chang, N.B., Hossain, F., Wanielista, M., 2010a. Use of filter media for nutrient removal in natural systems and built environments (I): previous trends and perspectives. Environ. Eng. Sci. 27 (9), 689–706.

Chang, N.B., Wanielista, M., Makkeasorn, A., 2010b. Use of filter media for nutrient removal in natural systems and built environments (II): design challenges and application potentials. Environ. Eng. Sci. 27 (9), 707–720.

Chang, N.B., Xuan, Z., Daranpob, A., Wanielista, M., 2011a. A subsurface upflow wetland system for nutrient and pathogen removal in on-site sewage treatment and disposal systems. Environ. Eng. Sci. 28 (1), 11–24.

Chang, N.B., Wanielista, M., Henderson, D., 2011b. Temperature effects on functionalized filter media for nutrient removal in stormwater treatment. Environ. Prog. Sustain. Energy 30 (3), 309–317.

Chowdhury, R., Apul, D., Dwyer, D., 2008. Preliminary studies for designing a wetland for arsenic treatment. In: Groundwater: Modelling, Management and Contamination. Nova Science Publishers, Inc., pp. 151–166.

Coombes, C., Collett, P.J., 1995. Use of constructed wetland to protect bathing water quality. Water Sci. Technol. 32 (3), 149–158.

Gálvez, J.M., Gómez, M.A., Hontoria, E., López, J.G., 2003. Influence of hydraulic loading and air flow rate on urban wastewater nitrogen removal with a submerged fixed-file reactor. J. Hazard. Mater. B101, 219–229.

Gan, Q., Allen, S.J., Matthews, R., 2004. Activation of waste MDF sawdust charcoal and its reactive dye adsorption characteristics. Waste Manag. 24, 841–848.

Gisvold, B., Ødegaard, H., Føllesdal, M., 2000. Enhanced removal of ammonium by combined nitrification/ adsorption in expanded clay aggregate filters. Water Sci. Technol. 41 (4–5), 409–416.

Haberl, R., Perfler, R., Laber, J., Grabher, D., 1998. Constructed wetland for wastewater treatment in Europe. In: Brix, H., Cooper, P.F., Green, M.B., Haberl, R. (Eds.), Vymazal. Backhuys publishers, Leiden, The Netherlands, pp. 67–76.

Henneck, J., Axler, R., McCarthy, B., Geerts, S.M., Christophenson, S.H., Anderson, J., Crosby, J., 2001. Onsite treatment of septic tank effluent in Minnesota using SSF constructed wetland: performance, costs and maintenance, on-site wastewater treatment.

In: Proc. of the 9th National Symposium on Individual and Small Community Sewage Systems, ASAE, St. Joseph, MI.

Henry, S., Baudion, E., Lopez-Gutierrez, J.C., Martin-Laurent, F., Brauman, A., Philippot, L., 2004. Quantification of denitrifying bacteria in soils by nirK gene targeted real-time PCR. J. Microbiol. Methods 59, 327–335.

Hsieh, C.H., Davis, A.P., 2005. Evaluation and optimization of bioretention media for treatment of urban storm water runoff. J. Environ. Eng. 131 (11), 1521–1531.

Huang, J., Reneau, R.B., Hagedorn, C., 2000. Nitrogen removal in constructed wetlands employed to treat domestic wastewater. Water Res. 34 (9), 2582–2588.

Kaseva, M.E., 2004. Performance of a sub-surface flow constructed wetland in polishing pre-treated wastewater- a tropical case study. Water Res. 38, 681–687.

Keffala, C., Ghrabi, A., 2005. Nitrogen and bacterial removal in constructed wetlands treating domestic waste water. Desalination 185, 383–389.

Kim, H., Seagren, E.A., Davis, A.P., 2000. Engineering bioretention for removal of nitrate from stormwater runoff. In: WEFTEC 2000 Conference Proceedings on CDROM Research Symposium, Nitrogen Removal, Session 19, Anaheim, CA, October, 2000.

Kyambadde, J., Kansiime, F., Gumaelius, L., Dalhammar, G., 2004. A comparative study of Cyperus Papyrus and Miscanthidium Violaceum based constructed wetlands for wastewater treatment in a tropical climate. Water Res. 38, 475–485.

Li, Z., 2003. Use of surfactant-modified zeolite as fertilizer carriers to control nitrate release. Microporous Mesoporous Mater. 61 (1), 181–188.

Lisi, R.D., Park, J.K., Stier, J.C., 2004. Mitigation nutrient leaching with a sub-surface drainage layer of granulated tires. Waste Manag. 24, 831–839.

Mann, R.A., Bavor, H.J., 1993. Phosphorus removal in constructed wetlands using gravel and industrial waste substrata. Water Sci. Technol. 27 (1), 107–113.

Mbuligwe, S.E., 2005. Applicability of a septic tank/engineered wetland coupled system in the treatment and recycling of wastewater. Environ. Manag. 35 (1), 99–108.

National Sanitation Foundation, 2009. Standard 245: Nitrogen Reduction, ANSI/NSF Standard 245.

Pant, H.K., Reddy, K.R., Lemon, E., 2001. Phosphorus retention capacity of root bed media of sub-surface flow constructed wetlands. Ecol. Eng. 17 (4), 345–355.

Park, W.H., Polprasert, C., 2008. Roles of oyster shells in an integrated constructed wetland system designed for P removal. Ecol. Eng. 34, 50–56.

Ray, A.B., Selvakumar, A., Tafuri, A.N., 2006. Removal of selected pollutants from aqueous media by hardwood mulch. J. Hazard. Mater. B136, 213–218.

Schipper, L.A., Barkle, G.F., Vukovic, M.V., 2005. Maximum rates of nitrate removal in a denitrification wall. J. Environ. Qual. 34, 1270–1276.

Seelsaen, N., McLaughlan, R., Moore, S., Ball, J., Stuetz, R., 2006. Pollutants removal efficiency of alternative filtration media in storm water treatment. Water Sci. Technol. 54 (6–7), 299–305.

Seidel, K., 1955. Die Flechtbinse, Scirpus lacustris L. Oekologie, Morphologie und Entwicklung, ihre Stellung bei den Völkern und ihre wirtschaftliche Bedeutung. Schweizerbart'sche Verlagsbuchhandlung, Stuttgart, Germany, pp. 37–52.

Shin, H.S., Yoo, K.S., Park, J.K., 1999. Removal of polychlorinated phenols in sequential anaerobic-aerobic biofilm reactors packed with tire chips. Water Environ. Res. 71 (3), 363–367.

Smith, D.P., Otis, R., Flint, M., 2008. Florida Passive Nitrogen Removal Study (Final Report). Florida Department of Health, Tallahassee, FL.

Steer, D., Fraser, L., Boddy, J., Seibert, B., 2002. Efficiency of small constructed wetlands for subsurface treatment of single-family domestic effluent. Ecol. Eng. 18, 429–440.

Surrency, D., 1993. Evaluation of aquatic plants for constructed wetlands. In: Moshiri, G.A. (Ed.), Constructed Wetlands for Water Quality Improvement. Lewis, Boca Raton, FL, pp. 349–357.

Tee, H.C., Seng, C.E., Noor, A., Lim, P.E., 2009. Performance comparison of constructed wetlands with gravel- and rice husk-based media for phenol and nitrogen removal. Sci. Total Environ. 407, 3563–3571.

US EPA, 1993. Subsurface Flow Constructed Wetlands for Wastewater Treatment. EPA 832-R-93-008 Washington D.C., USA.

US Patent 8153005 B1, 2011 (Martin P. Wanielista, Ni-Bin Chang): Retention/Detention Pond Stormwater Treatment System, issued on August 23, 2011.

US Patent 8252182 B1, 2012 (Ni-Bin Chang, Martin P. Wanielista): A Subsurface Upflow Wetland System for Nutrient and Pathogen Removal in Wastewater Treatment Systems, issued on August 28, 2012.

Vohla, C., Alas, R., Nurk, K., Baatz, S., Mander, U., 2007. Dynamics of phosphorus, nitrogen and carbon removal in a horizontal subsurface flow constructed wetland. Sci. Total Environ. 380 (1–3), 66–74.

Xuan, Z., Chang, N.B., Makkeasorn, A., Wanielista, M., 2009. Initial test of a subsurface constructed wetland with green sorption media for nutrient removal in on-site wastewater treatment systems. Water Qual. Expo. Health 1 (3), 159–169.

Xuan, Z., Chang, N.B., Wanielista, M., Hossain, F., 2010a. Laboratory-scale characterization of the green sorption medium for wastewater treatment to improve nutrient removal. Environ. Eng. Sci. 27 (4), 301–312.

Xuan, Z., Chang, N.B., Daranpob, A., Wanielista, M.P., 2010b. Modeling the subsurface upflow wetlands (SUW) systems for wastewater effluent treatment. Environ. Eng. Sci. 27 (10), 879–888.

Zhang, T.C., 2002. Nitrate removal in sulfur: limestone pond reactor. J. Environ. Eng. 128 (1), 73–84.

Tracer-based System Dynamic Modeling for Designing a Subsurface Upflow Wetland for Nutrient Removal

25

Ni-Bin Chang*, Martin P. Wanielista, Zhemin Xuan

Department of Civil, Environmental, and Construction, University of Central Florida,
Orlando, FL, USA
*Corresponding author: e-mail address: nchang@ucf.edu

25.1 INTRODUCTION

The constructed wetland, an effective small-scale wastewater treatment system with low energy, maintenance requirements, and operational costs has been widely used to treat various kinds of wastewater throughout the world (Vymazal and Kröpfelová, 2009). While the constructed wetlands showed their remarkable removal efficiency of organic matter, nutrients, pathogens, and other substances, the stricter water quality standard motivated many more advanced studies with regard to higher commercial, esthetic, habitat, and sustainable value. Constructed wetlands have been popular in the ecological engineering regime; yet modeling the physical, chemical, and biological processes within these wetlands is a long-standing challenge in the past decades.

The constructed wetlands can be divided into three main types according to the different hydrologic modes: free water surface (FWS) wetlands, horizontal subsurface flow (HSSF OR SSF) wetlands, and vertical flow (VF) wetlands (Kadlec and Wallace, 2008). FWS wetlands include emergent vegetation, soil, or medium to support the emergent vegetation, and a water surface above the substrate. In the HSSF, the wastewater is fed in the inlet and passes the filter medium under the surface until it reaches the outlet zone through the subsurface pathways. VF constructed wetlands generally consist of a gravel layer at the bottom topped with a sand layer. When intermittently feeding with a large batch, the wastewater percolates vertically until it reaches a drainage network. With a variety of wetland systems being applied successfully at the field scale, the designed models of constructed wetland with systems dynamics characteristics have gradually gained growing attention during the past decades.

Developments in Environmental Modelling, Volume 26, ISSN 0167-8892, http://dx.doi.org/10.1016/B978-0-444-63249-4.00027-0

As promising sustainable infrastructure alternatives for wastewater treatment, constructed wetlands can remove nutrients with the aid of complex interactions among water, substrate, plants, and microorganisms through a variety of physical, chemical, and biological mechanisms. Physical functions are mainly filtration by the substrate layer, plant-root zone, and sedimentation; chemical reactions consist mainly of chemical precipitation, adsorption, ion exchange, and oxidation–reduction (REDOX) reactions; and biochemical reactions mainly refer to microorganism degrading and pollutant removal in aerobic, anoxic, and anaerobic conditions (Xuan et al., 2010). The performance of constructed wetlands relies on many factors intertwined with these various processes. To be in concert with our field-scale pilot testing of a new-generation SUW system, this paper highlights an advancement of modeling the SUW system with a layer-structured compartmental simulation model.

Within the constructed wetland, the transition of nitrogen from one phase to another is commonly referred to as the nitrogen cycle. Ammonia combines with organic materials to create ammonium (NH_4^+). In the presence of ammonia-oxidizing bacteria (AOB) and nitrite-oxidizing bacteria (NOB), ammonium is converted to nitrite (NO_2^-) and further to nitrate (NO_3^-). These two reactions are collectively called nitrification. Denitrification, conversely, performed by denitrifying community, is an anaerobic respiration process using nitrate as a final electron acceptor and results in stepwise microbiological reduction of nitrate, nitrite, nitric oxide (NO), nitrous oxide (N_2O) to nitrogen gas (N_2) (Koike and Hattori, 1978). Nitrate removal rates are directly influenced by the slow-growing bacteria that govern nitrification and denitrification.

The objective of this study is to perform integrated tracer-based system dynamics modeling for simulation analyses of nutrient removal in the lined media filter. For the identification of hydraulic or flow patterns in the media filter, a tracer study was conducted to determine the direction and velocity of water movement in the media filter. Due to the advantages of low detection limits, zero natural background, low relative cost, and easy onsite analysis, Rhodamine WT was selected as the water tracing dye to determine the hydraulic pattern and hydraulic retention time of the media filter. It investigates the coupling mechanism between hydrodynamics, geochemical interactions, and microbiological activities in an SUW system. First, a tracer study, the most direct approach to obtain the information of internal hydraulics in constructed wetlands, was applied to determine the direction and velocity of water movement and to track the overall hydraulic pathways aiding in the system dynamic modeling analysis (Chang et al., 2012). An ideal tracer should follow the same path as the water and should have the following characteristics: easy detection, inexpensive analysis procedure, low toxicity, high solubility, and low background in the system tested. To further understand the system features and dynamic rates, and thus the reliability of the SUW design, a layer-structured compartmental simulation model was developed. This systems dynamic model used STELLA® graphical representations to illustrate the essential biochemical mechanisms of the nitrification and denitrification processes based on the holistic physical upflow patterns within a sorption media-based SUW system, as introduced in the previous chapter (Xuan et al.,

2010; Chang et al., 2011). Such a systems dynamic model may be used repetitively to design future SUW systems at different locales once the inflow conditions and required effluent standards are modified.

25.2 SITE DESCRIPTION

In this study, a subsurface upflow constructed wetland (SUW) system receiving septic effluent from a student dormitory handled 454 L (120 gallons) per day influent for a wastewater treatment study using the green sorption media along with plant species. The wastewater was intermittently pumped into the constructed wetland by about 15.12 L (4 gallons) at a time. It had been shown that green sorption medium consisting of recycled and natural materials provide a favorable environment for nutrient removal (Xuan et al., 2009). There were four parallel cells in this test bed each with the dimensions of 1.52 m wide × 3.05 m long × 0.91 m deep (5 ft wide × 10 ft long × 3 ft deep). Each of four cells consisted of an impermeable liner, a gravel substrate layer, a sand layer, a green sorption media layer, a growth media layer and selected plants; a gravel-filled gravity distribution system including header pipe, distribution pipe, collection pipe, flow meter, and a planted bed of special green sorption media with an underdrain collection system. Two PVC pipes with fabric filter at bottom were inserted into the gravel layer close to the inlet of each cell to introduce more oxygen for promoting nitrification. Three sets of plant species were tested against the control case where there was no plant species (Xuan et al., 2009).

The main criteria for choosing plants in this study included: (1) native—long-term survival or minimal environmental problems; (2) perennial—ability to function all year long and don't need to replant after harvesting; (3) good rooting system—help nitrification; (4) high yield—evaporate more water since yield and water use are closely correlated (i.e., this criteria is flexible); (5) high protein content—plant will take more nitrogen since protein content is correlated with nitrogen content; and (6) ability to tolerate trimming or grazing—biomass is harvested and consequently nutrients are removed. Three kinds of herbaceous perennial plants including canna (*Canna Flaccida*), blue flag (*Iris versicolor L.*), and bulrush (*Juncus effusus L.*) were eventually selected due to the features of the local availability, biomass production, nutrient content, and the similar size. Seedlings of those three kinds of plant were purchased from local nursery and planted 2 months before the experiment period. Since the wetland plants had been acclimated and taken shape during the experiment, the plant growing rate was treated as a constant for simplification.

The research team at UCF decided to follow six criteria to screen those possible sorption media to support both pollution control and plant growth: (1) the relevance of nitrification or denitrification process or both; (2) the hydraulic permeability; (3) the cost level; (4) the removal efficiency as evidenced in the literature with regard to adsorption, precipitation, and filtration capacity; (5) the availability in Florida; and (6) additional environmental benefits. All of the four treatment units (cells) in our wetland system were filled from bottom to top with gravel, sand, "Pollution Control

Media" denoted as PC medium hereafter and "Expanded Clay Growth Media" denoted as G medium hereafter. The 15.24 (6-in.) sand layer had the main function of removing the pathogen from the septic effluent and worked as a buffer layer between gravel and PC media layers. A 30.48-cm (12-in.) layer of PC media (50% citrus grove sand, 15% tire crumb, 15% sawdust, and 20% limestone) was used to remove most of the nutrients, TSS, and BOD, at the depth of 30.48 cm (12 in.) beneath the G media layer. At the top place, a 30.48-cm (12-in.) growth media layer (75% expanded clay, 10% vermiculite, and 15% peat moss) was used to support the root zone. Since nitrification is considered to be the primary rate-limiting step for nitrogen removal unless the wastewaters are pre-nitrified or the oxygen can be diffused more efficiently into the upper layer of the root zone via some specific growth media, the expanded clay growth media was used to ensure vibrant plant growth and efficient oxygen diffusion. Once the wastewater fully saturated the gravel layer, the water level would rise up gradually, passing through the sand and PC medium layer up to the outlet (see Figure 25.1). Chowdhury et al. (2008) reported a bromide tracer study in a similar SUW. They found that a gravel layer added at the bottom caused the flow to be mostly in the vertical direction, which provided strong evidence for our hydraulic pattern hypothesis. The samplers were installed at the interface between different layers with three depths. Horizontally, the samplers in the four wetland cells were located at 33%, 67%, and 100% along the length of the wetland. Sample identities (IDs) here were defined for following discussion as below: (1) "port B": mixture of bottom three samples, (2) "port M": mixture of middle three samples, (3) "port T": mixture of top three samples (Figure 25.2).

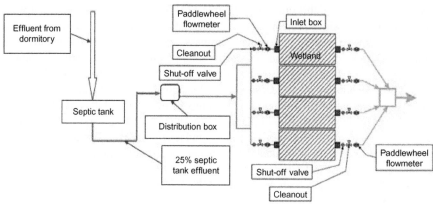

FIGURE 25.1

Configuration of a septic tank followed by a four-cell wetland system including shut-off valve, cleanout, and flow meter. (For the color version of this figure, the reader is referred to the online version of this chapter.)

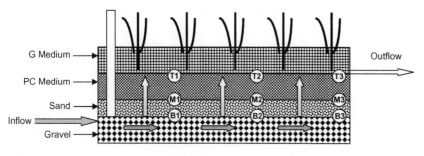

FIGURE 25.2

Locations of the sample points in the wetland cross-section (Xuan et al., 2010). *Note*: B, M, and T series ports are installed for water sample collection.

Three kinds of plant species were individually planted in three cells (1–3) to compare with the parallel control cell with no plant species (cell 4) (Figure 25.1) (Chang et al., 2011). Using the criteria for screening plant species as described in Xuan et al. (2009), we selected three species of native vegetation with similar plant density and cost: canna (*Canna Flaccida*; wetland cell 1), blue flag (*Iris versicolor L.*; wetland cell 2), and bulrush (*Juncus effusus L.*; wetland cell 3). We chose wetland cell 1 for demonstration and measured the water quality data used to calibrate the model each layer of the SUW. The data collection points were identified as the sand/gravel interface, the PC media interface, and the growth media (G media)/PC media interface, which was the location of the discharge. B series ports provided a mixture of the bottom three samples; M series ports a mixture of the middle three samples; and T ports a mixture of the top three samples (Chang et al., 2011). Each data point is a composite of at least three measurements during a day. The water quality measurements, along with the species of nutrients, were determined for each measuring point using canna planted in cell 1 as an example (Table 25.1).

25.3 THE TRACER STUDY

25.3.1 Tracer study

There are three most popular choices for a tracer: isotope (Kadlec et al., 2005; Ronkanen and Kløve, 2007, 2008), ions, and dyes. The isotope technology has high accuracy but is expensive. Ionic compounds, especially bromide, have been widely used as a groundwater tracer (Harman et al., 1996; Wang et al., 2008). Małoszewski et al. (2006) used instantaneously injected bromide to evaluate of hydraulic characteristics of a duckweed pond in Mniów, Poland. Yet, ionic tracers rely on less reliable measuring probes. Dyes have advantages of low detection limits, near zero natural background, and low relative cost. One of the most popular dyes is rhodamine WT (RWT) (Lin et al., 2003; Dierberg and DeBusk, 2005; Giraldi et al., 2009).

Table 25.1 Water Quality Measurements in SUW Layers (Canna) (Xuan et al., 2010)

	Alk. (mg L^{-1})		TSS (mg L^{-1})		BOD$_5$ (mg L^{-1})		CBOD$_5$ (mg L^{-1})	
	Ave.	St. Dev.	Ave.	St. Dev.	Ave.	St. Dev.	Ave.	St. Dev.
Sand/gravel ($n=4$)	386.8	85.2	37.5	3.1	14.3	4.2	11.5	4.3
PC media/sand ($n=4$)	462.3	74.6	51.0	18.0	12.5	4.5	11.2	4.1
G media/PC media ($n=12$)	472.5	100.2	38.4	7.5	4.8	1.5	3.7	0.7

	pH		Temp (°C)		Cond. (µs cm^{-1})	
	Ave.	St. Dev.	Ave.	St. Dev.	Ave.	St. Dev.
Sand/gravel ($n=12$)	7.21	0.38	30.52	2.41	1414	477
PC media/sand ($n=12$)	7.11	0.32	30.65	2.67	1363	490
G media/PC media ($n=36$)	7.12	0.36	30.53	2.54	1273	280

	Org. N-N (µg L^{-1})		NH$_3$-N (µg L^{-1})		NO$_x$-N (µg L^{-1})		TN (µg L^{-1})	
	Ave.	St. Dev.	Ave.	St. Dev.	Ave.	St. Dev.	Ave.	St. Dev.
Sand/gravel ($n=4$)	7826.8	7786.8	20,583.0	6966.1	14.3	17.8	28,422.8	11,842.1
PC media/sand ($n=4$)	5241.5	3823.8	14,049.5	9603.6	10.5	11.0	19,299.5	10,094.3
G media/PC media ($n=12$)	1470.5	615.5	1164.3	897.7	6.8	3.5	2639.7	1010.6

	OP-P (µg L^{-1})		Org. P-P (µg L^{-1})		TP (µg L^{-1})	
	Ave.	St. Dev.	Ave.	St. Dev.	Ave.	St. Dev.
Sand/gravel ($n=4$)	16.3	8.5	375.8	218.7	392.0	222.8
PC media/sand ($n=4$)	16.5	8.1	132.5	130.2	149.0	126.6
G media/PC media ($n=12$)	19.0	7.4	94.5	54.4	113.5	54.6

Based on its advantages of low detection limits, zero natural background, low cost, and easy operation, RWT was selected to determine the hydraulic patterns and hydraulic residence time (HRT) of the SUW system (Xuan et al., 2010). RWT, also known as Acid Red #388, is a synthetic red to pink dye with brilliant fluorescent qualities, a molecular formula of $C_{29}H_{29}N_2O_5ClNa_2$, and a CAS Number of 37299-86-8. It is often used as a tracer in water to determine the rate and direction of flow and transport. In our study, the RWT liquid (20% solution) was purchased from Keystone Aniline Corporation. 70 mL of 2×10^7 PPB (1.4 g active ingredient) Rhodamine WT solution was added into the pipe before the inlet of the media filter. Five sets of data were collected and measured in April 2010. The tracer with 25% of the designed water loading was dosed into the media filter. The 3D distribution of tracer in the media filter was plotted by Voxler® (Golden software). The tracer HRT was then calculated following Headley and Kadlec's practical guide (Headley and Kadlec, 2007).

Following the RWT injection in July 2010, 10 sets of 50 mL water samples from nine sampling ports were collected using a peristaltic pump and then measured with an Aquafluor™ (Turner Designs 998-0851) handheld fluorometer. The linear detection range for the RWT is 0.4–300 ppb (active ingredient). Because RWT fluorescence is susceptible to photolysis and sensitive to temperature, samples were collected in glass bottles and kept in the dark prior to analysis. In addition, the solution with known concentration was analyzed on site for calibration prior to the sample measurement (Xuan et al., 2010). The minimal root mean square error divided by the peak height of the distribution was evaluated to determine the goodness of fit for the tracer study. Following tracer analysis, the RWT distribution was demonstrated by the three-dimensional (3D) data visualization software package, Voxler® (Golden Software, Inc.) (Chang et al., 2012).

25.3.2 Hydraulic pathways of SUW

The distribution of tracer in the SUW was plotted by Voxler®, a robust program that can display the data in a variety of formats: 3D volrender, isosurfaces, contours, 3D slices, orthographic and oblique images, scatter plots, stream lines, and vector plots. A profile view of the tracer distribution in wetland during an 18-day period (Figure 25.3) shows 10 small images representing a series of detailed flow sequences of water with dye illumination (Chang et al., 2012).

On each small image, the inlet is on the bottom left and the outlet is on the upper right. Day 2 small image indicates that the tracer flowed with water throughout the bottom layer and moved upward along two vertical edges of the wetland (i.e., front and back edge) within 2 days. The blue color in the middle shows that the tracer had not yet become homogeneous in that region, which means there might be some "hydraulic retardation" with time in the middle of the wetland. Because most of the tracer at the top layer came from the bottom rather than horizontal movement, the overall observation confirmed our concept of the hydraulic "upflow" design. By day 3, the tracer gradually faded at the inlet side and continued to rise at the

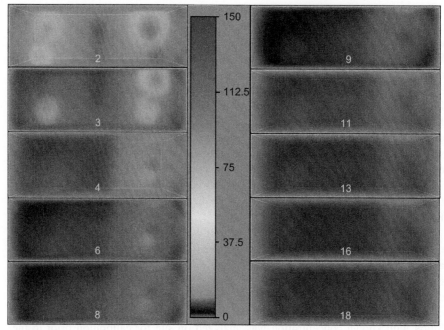

FIGURE 25.3

Profile view of tracer distribution in the wetland. Left five images: days 2, 3, 4, 6, and 8; right five images: days 9, 11, 13, 16, and 18. On each small image, the inlet is on the bottom left and the outlet is on the upper right; vertical scale is concentration (ppb) (Chang et al., 2012). (For the color version of this figure, the reader is referred to the online version of this chapter.)

outlet side. From day 4 to day 8, there was a rising progress of tracer in the middle, then the peak of the tracer moved out of the outlet from day 9 to day 11. Finally, the remaining tracer flowed gently out of the wetland from day 11 to day 18. Globally, the spatiotemporal tracer distribution provides strong support for the upflow hypothesis in biosystem engineering design of a layer-structured compartmental simulation model, discussed in the next section.

25.4 MODELING THE SUW SYSTEM

25.4.1 Conceptual model

The satisfactory nutrient and pathogen removal efficiency and upflow pattern have been fully proven in the previous chapter (see Chapter 24). To be in concert with our field-scale pilot testing of a new-generation SUW system, the following text highlights an advancement of modeling the SUW system with a layer-structured compartmental simulation model. This is the first wetland model of its kind in the world to address the complexity between plant nutrient uptake and media sorption. Such a system dynamics model using STELLA® as a means for a graphical formulation was applied to illustrate the essential mechanism of the nitrification and denitrification processes within a sorption media-based SUW system, which can be recognized as one of the major passive onsite wastewater treatment technologies in this decade.

In the context of biochemical mechanisms, we considered five main nitrogen transformations in the SUW to be consistent with transformation kinetics and as representative of the SUW (Kadlec and Wallace, 2008).

a. Organic nitrogen to ammonium nitrogen (ammonification or mineralization): Organic nitrogen cannot be utilized directly by plants but is gradually transformed to NH_4^+ by heterotrophic microorganisms:

$$NH_2CONH_2 + H_2O \rightarrow 2NH_3 + CO_2, \tag{25.1}$$

$$NH_3 + H_2O \leftrightarrow NH_4^+ + OH^- \tag{25.2}$$

b. Ammonium nitrogen to nitrate nitrogen (nitrification): In aerobic oxidized condition, ammonium transforms to NO_3^- through the process of nitrification in two steps by AOB:

$$2NH_4^+ + 3O_2 \rightarrow 2NO_2^- + 2H_2O + 4H^+, \tag{25.3}$$

and by nitrite-oxidizing bacteria (NOB):

$$2NO_2^- + O_2 \rightarrow 2NO_3^- \tag{25.4}$$

When there is adequate oxygen available, nitrification can also occur in the oxidized rhizosphere of plants.

c. Nitrate nitrogen to gaseous nitrogen (denitrification). Denitrifiers use the oxygen from NO_3^- instead of O_2 to convert NO_3^- to nitrogen oxide and N_2:

$$NO_3^- + 0.833CH_3OH \rightarrow 0.5N_2 + 0.833CO_2 + 1.167H_2O + OH^-, \tag{25.5}$$

$$NO_3^- + 0.208C_6H_{12}O_6 + \rightarrow 0.5N_2 + 1.25CO_2 + 0.75H_2O + OH^-. \tag{25.6}$$

d. Nitrate or ammonium nitrogen to organic nitrogen (assimilation or immobilization): Immobilization can be considered as the reverse reaction of mineralization. Inorganic nitrogen (NO_3^- and NH_4^+) is converted to organic nitrogen by microbes and used by plants, which roughly is counted as plant uptake in the model.

e. Biomass nitrogen to organic nitrogen (decomposition): because the plant grew well and had no residue in late summer, this part of nitrogen transformation can be ignored.

The term *Residence Time Distribution (RTD)*, characterizing chemical reactors, was first proposed by Danckwerts (1953). The term was oftentimes used to discuss the type of mixing in constructed wetland (Werner and Kadlec, 1996). The RTD function is generally measured by injecting an impulse of tracer and measuring the tracer concentration as a function of time at interior wetland points as well as the outlet. Many wetland systems were modeled as a number, N, of stirred tank reactors in series. In fitting experimental data, the form for the Tanks-In-Series model (TIS) is given by (Kadlec and Wallace, 2008).

$$\text{RTD} = \frac{N}{(N-1)!}\left(N\frac{t}{\tau}\right)^{N-1}\exp\left(-N\frac{t}{\tau}\right) \tag{25.7}$$

Equation (25.7) can be considered as single continuous stirred tank reactor (CSTR) when $N=1$ and the plug flow reactor (PFR), $N=\infty$.

In the earlier stage of designing the treatment model, the constructed wetlands were considered as a "black box." Scientists focused on the influent and effluent concentration and fit the result with the designed linear or power equation to build the relationship between them. The regressions in Table 25.2 were derived for wetland nitrogen and phosphorus removal of SSF wetlands (Kadlec and Knight, 1996). This kind of model oversimplified the constructed wetland treatment system, which has an extremely complicated physical, chemical, and biological process. Not only the influent concentration and HRT but also hydrodynamic conditions, such as wetland dimension, porosity, and conductivity of media can affect the removal efficiency of pollutants of concern. Gradually, the first-order kinetics equation or Monod type equations were widely accepted and applied to replace the regression method. Kadlec and Knight (1996) had summarized the nitrogen removal equations of modeling the constructed wetland (Table 25.3).

Table 25.2 Regression Models for Wetland Nitrogen and Phosphorus Removal

Parameters	Regression equations
Organic N	$C_2 = 0.1C_1 + 1.0$
Ammonium N	$C_2 = 0.46C_1 + 3.3$
Nitrate N	$C_2 = 0.62C_1$
Total Kjeldahl nitrogen (TKN)	$C_2 = 0.752C_1^{0.821}q^{0.076}$
Total nitrogen(TN)	$C_2 = 0.46C_1 + 0.124q + 2.6$
Phosphorus	$C_2 = 0.51C_1^{1.10}$

Note: C_2, concentration at the outlet; C_1, concentration at the inlet; q, hydraulic loading rate.

Table 25.3 Parameter and Corresponding Formula of Modeling the Constructed Wetland (Kadlec and Knight, 1996)

Parameter	Formula
Nitrifier growth rate (u_{NITR})	$u_{NITR} = 172e^{0.098(T-15)}[1 - 0.833(7.2 - pH)]\left(\frac{C_{AN}}{1+C_{AN}}\right)\left(\frac{C_{DO}}{1.3+C_{DO}}\right)$
Denitrifier growth rate (u_{DENITR})	$u_{DENITR} = u_{DENITR_{max}}\left[\left(\frac{C_{NN}}{K_{DENITR}+C_{NN}}\right)\left(\frac{C_{ORGC}}{K_{ORGC}+C_{ORGC}}\right)\right]$
Outlet concentration of ammonium nitrogen (C_{AN})	$C_{AN} = (C_{ANi})e^{-k_{AN}/(Q/A)} + \left(\frac{k_{ON}}{k_{AN}-k_{ON}}\right)\left(C_{ONi} - C_{ON}^{*}\right)$ $\left(e^{-k_{ON}/(Q/A)} - e^{-k_{AN}/(Q/A)}\right)$

Note: $K_{DENITRN}$, denitrification half-saturation constant, $mg\,L^{-1}$; K_{ORGC}, organic nitrogen half-saturation constant, $mg\,L^{-1}$; k_{ON}, k_{AN}, first-order organic nitrogen, ammonium loss rate, $g\,m^{-2}\,yr^{-1}$; C_{AN}, C_{DO}, C_{NN}, C_{ORGC}, C_{ON}, concentration of ammonium, dissolved oxygen, nitrite + nitrate, organic carbon, organic nitrogen, $mg\,L^{-1}$; C_{ANi}, C_{ONi}, inlet concentration of ammonium nitrogen, organic nitrogen, $mg\,L^{-1}$; C_{ON}^{*}, background concentration of organic nitrogen, $mg\,L^{-1}$; Q/A, hydraulic loading rate.

Tunçsiper et al. (2006) simulated removal efficiencies of nitrogenous pollutants in SSF and FWS constructed wetland systems. Two types of the models (first-order plug flow and multiple regressions) were used to evaluate the system performances. Nitrification, denitrification, and ammonification rate constant values in SSF and FWS systems were $0.898\,d^{-1}$, $0.486\,d^{-1}$, and $0.986\,d^{-1}$, and $0.541\,d^{-1}$, $0.502\,d^{-1}$, and $0.908\,d^{-1}$, respectively. They found that the first-order plug flow model clearly estimated slightly higher or lower values than observed when compared with the other models. Jou et al. (2008) tried a constructed wetland for restoring a creek. The ecological treatment system removed 64.0% of suspended solids (SS), 43.0% of biochemical oxygen demand (BOD), and 11.0% of ammonia nitrogen. A first-order biokinetic model was used to estimate the reductions of BOD and nitrogenous biochemical oxygen demand (NBOD). They reported that the first-order biokinetic model appeared useful for estimating BOD and NBOD reductions in a constructed wetland. However, the fatal limitation of the first-order kinetics is that the constructed wetland system is required to keep the same flow rate, concentration, and ideal plug flow. To make the dynamic modeling of the constructed wetland processes more acceptable and flexible, Pastor et al. (2003) proposed the design optimization of constructed wetland for wastewater treatment by combining a first principal model and an artificial neural network (ANN), which had a main advantage for better representing highly nonlinear multi-input/multi-output system. Tomenko et al. (2007) compared multiple regression analysis and two artificial neural networks (ANN)-multilayer perceptron (MLP) and radial basis function network in terms of their accuracy and efficiency when applied to prediction of the BOD concentration at effluent and intermediate points of subsurface flow constructed wetlands. The dataset was normalized and transformed using principal component analysis to increase the efficiency of the modeling. Artificial neural networks models were eventually cross-validated to find optimal network architectures and values of training algorithm parameters.

The models mentioned above just provide a limited understanding of specific items, which were even separately analyzed. The mechanistic approach for modeling constructed wetland systems has been highly regarded by people who prefer mystery of the whole wetland treatment process. Wynn and Liehr (2001) developed a mechanistic compartmental simulation model, which included six linked submodels: the carbon cycle, the nitrogen cycle, a water balance, an oxygen balance, autotrophic bacteria growth, and heterotrophic bacteria growth. Darcy's law was used to describe the flow through the media. The wetland was regarded as either a single continuously stirred tank reactor (CSTR) or a series of CSTRs instead of plug flow reactors, which was considered to be a better reactor model for simulating non-ideal plug flow. Monod kinetics was utilized to describe microbial growth rate. Transformations, such as nitrification and denitrification, were then linked directly to microbial growth. In general, except for the oxygen, the result of effluent BOD, organic nitrogen, ammonium, and nitrate concentration fit the model well.

Langergraber (2001) presented a multicomponent reactive transport module CW2D to model the biochemical transformation and degradation processes in SSF CWs. The mathematical structure of CW2D was based on that of the ASMs (Henze et al., 2000). The CW2D consisted of 12 components, nine processes, and 46 parameters. The HYDRUS-2D was incorporated by using Richards' equation to describe the variably saturated water flow conditions. Water uptake by plant roots was accounted as a sink term in the flow equation. The components considered ammonium, nitrite, nitrate, and nitrogen gas; dissolved oxygen; organic matter; inorganic phosphorus; heterotropic and two species of autotrophic microooranisms. The rates of the biochemical elimination and transformation processes were described by using Monod-type of equation. Recently, Giraldi et al. (2010) developed a mathematical model (FITOVERT) to analyze the hydrodynamics of a one-dimensional VF CW under three different saturation conditions: complete saturation, partial saturation, and complete drainage by dosing rhodamine WT in steady-state conditions. Richards' equation was used for modeling the variably saturated conditions, while van Genuchten-Mualem function was used to describe the relationships between pressure head, hydraulic conductivity, and water content. In particular, the porosity reduction due to bacteria growth and accumulation of particulate component (i.e., clogging process) can be simulated by FITOVERT.

25.4.2 Model construction

Assume that each media layer is a continuously stirred treatment reactor (CSTR). Based on the above understanding, the conceptual model for nitrogen removal of SUW is shown in Figure 25.4. Each media layer was assumed to be a Completely Stirred Tank Reactor (CSTR) based on the tracer test described earlier. According

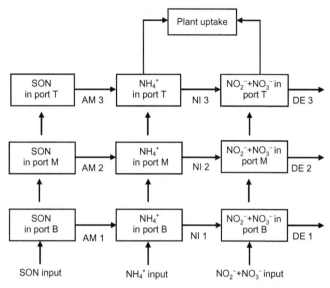

FIGURE 25.4

General conceptual model of nitrogen removal in SUW (Chang et al., 2012). *Note*: SON, Soluble Organic Nitrogen; AM, ammonification; NI, nitrification; DE, denitrification; B, Bottom layer; M, Middle layer; T, Top layer.

Table 25.4 Description of Symbols in Stock and Flow Diagram (Chang et al., 2012)

Symbol	Description
"sand ON"	ON (μg day^{-1}) in sand layer
"sand NH$_4$"	NH$_4$ (μg day^{-1}) in sand layer
"sand NO$_2$ and NO$_3$"	NO$_2$+NO$_3$ (μg day^{-1}) in sand layer
"sand AM"	Ammonification (μg day^{-1}) in sand layer
"sand NI"	Nitrification (μg day^{-1}) in sand layer
"sand DE"	Denitrification (μg day^{-1}) in sand layer
"sand to PC ON"	ON (μg day^{-1}) transfer from sand layer to PC layer
"sand to PC NH$_4$"	NH$_4$ (μg day^{-1}) transfer from sand layer to PC layer
"sand to PC NO$_2$ and NO$_3$"	NO$_2$+NO$_3$ (μg day^{-1}) transfer from sand layer to PC layer
"r_a sand"	Ammonification rate (day^{-1}) in sand layer
"r_n sand"	Nitrification rate (day^{-1}) in sand layer
"r_d sand"	Denitrification rate (day^{-1}) in sand layer

to the water quality database collected in the field study in previous chapter and the assumption that each layer is a separate reactor (CSTR), we used a STELLA$^{\circledR}$ simulation program to create a conceptual model for nitrogen removal in an SUW (Figure 25.4; Table 25.4) and compared it to a common stock and flow diagram

for nitrogen removal (Figure 25.4). The modeling structure follows the layered structure for nitrogen removal.

The empirical formula in Equation (25.8) is valid for water temperatures between
about 5 and 30 °C (the final expression of nitrification rate was eventually reorga-
nized as Equation 25.8). The model calibration began with adjusting the ammonification rate (i.e., the nutrient source, organic nitrogen (ON), in sand layer) to
minimize discrepancies between modeled and measured values.

$$r_n = \frac{u_N}{Y_N} C_T C_{pH} \left(\frac{C_{DO}}{1.3 + C_{DO}} \right) C_{AN} \tag{25.8}$$

$$C_T = \begin{cases} e^{0.098(T-15)}, & \text{for } T < 30°C; \\ e^{0.098(30-15)}, & \text{for } T \geq 30°C; \end{cases} \quad C_{pH} = \begin{cases} 1 - 0.833(7.0 - pH), & \text{for } pH < 7.0; \\ 1, & \text{for } pH \geq 7.0 \end{cases}$$

25.4.3 Model equations

Equations (25.7)–(25.9) were used to predict organic nitrogen (ON), ammonia
(NH_4), and the sum of nitrite and nitrate ($NO_2 + NO_3$), respectively. The unit form,
$\mu g \, L^{-1} day^{-1}$, was used for all flows and stocks. Only plant uptake is a real and ultimate stock (Figure 25.4). The remaining nine stocks have their own outflow to
reach a steady-state condition; thus, the value in stock can be represented as the
"instantaneous concentration" in a unit volume or a point (i.e., sampling port).
We assumed that the upflow rate decreased linearly due to the evapotranspiration
and plant uptake with the increase of the water level. V was considered as the effective volume (product of volume and porosity) of each layer where water flows.
The $NO_2 + NO_3$ concentrations in all layers were so low that the $NO_2 + NO_3$ uptake
by plant was negligible. We assumed a constant rate of biomass production for
simplification as expressed in Equations (25.9)–(25.11):

$$dON/dt = \frac{Q_{in}}{V_{in}} ON_{in} - \frac{Q_{out}}{V_{out}} ON_{out} - r_a, \tag{25.9}$$

$$dNH_4/dt = \frac{Q_{in}}{V_{in}} NH_{4_{in}} - \frac{Q_{out}}{V_{out}} NH_{4_{out}} + r_a - r_n - r_p \text{ (only in G media layer)}, \tag{25.10}$$

and

$$d(NO_2 + NO_3)/dt = \frac{Q_{in}}{V_{in}} (NO_2 + NO_3)_{in} - \frac{Q_{out}}{V_{out}} (NO_2 + NO_3)_{out} + r_n - r_d. \tag{25.11}$$

The remaining parameters must be measured or assumed so they can be determined
holistically via the model calibration stage. The stock and flow diagram of nitrogen
removal in SUW using the STELLA® simulation program (Figure 25.5) demonstrates that the modeling structure follows the layered structure for nitrogen removal.

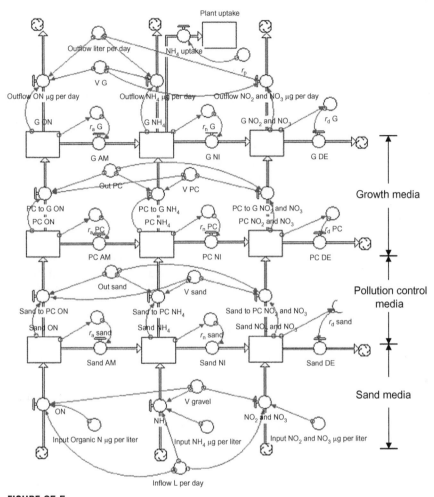

FIGURE 25.5

Flow diagram of nitrogen removal model (Xuan et al., 2010). *Note*: The symbols used in above are defined in Table 25.2. (For the color version of this figure, the reader is referred to the online version of this chapter.)

25.4.4 Model calibration and validation

For model calibration, the concentration and flow data from SUW cell 1 were selected to develop the system dynamics model with the aid of first and second datasets. A third dataset for average rate of plant growth was used to validate the model. The average results of three sample runs along with the measured hydraulics properties (Table 25.5) were used to calibrate the SUW system dynamic model for estimating nitrogen removal; Runge-Kutta 4 was used as the integration method. Nitrification has a wide optimum pH range from 7.0 to 9.0 (Sajuni et al., 2010);

Table 25.5 Hydraulics Values Used in SUW Model (Xuan et al., 2010)

Parameters	Description	Values
Q_{in}	Inflow rate	113.4 L day^{-1}
Q_{sand}	Flow rate out of sand layer	93 L day^{-1}
Q_{PC}	Flow rate out of PC media layer	52 L day^{-1}
Q_{out}	Outflow rate	31.5 L day^{-1}
Φg	Porosity of gravel	0.34
Φs	Porosity of sand	0.43
Φ_{PC}	Porosity of PC media	0.42
Φ_G	Porosity of G media	0.50

Table 25.6 Rate Equations of Ammonification, Nitrification, and Denitrification in Model (Xuan et al., 2010)

	Rate Equations	Unit	In Sand Layer	In PC Media Layer	In G Media Layer
k_a	$r_a = k_a C_{ON}$	day^{-1}	0.08	0.42	0.28
$\dfrac{u_N}{Y_N}$	$r_n = \dfrac{u_N}{Y_N} C_T C_{pH} \left(\dfrac{C_{DO}}{1.3 + C_{DO}} \right) C_{AN}$	day^{-1}	0.12	0.18	0.37
DO	$r_n = \dfrac{u_N}{Y_N} C_T C_{pH} \left(\dfrac{C_{DO}}{1.3 + C_{DO}} \right) C_{AN}$	mg L^{-1}	3.41	3.39	2.51
pH	$r_n = \dfrac{u_N}{Y_N} C_T C_{pH} \left(\dfrac{C_{DO}}{1.3 + C_{DO}} \right) C_{AN}$	N/A	7.02	7.00	7.01
T	$r_n = \dfrac{u_N}{Y_N} C_T C_{pH} \left(\dfrac{C_{DO}}{1.3 + C_{DO}} \right) C_{AN}$	°C	29.94	30.08	29.69
K_{20d}	$r_d = K_{20d} \theta_d^{(T-20)} C_{NN}$	day^{-1}	180	235	80
r_p	$r_p = iNPgp$	day^{-1}	N/A	N/A	140

pH below 7.0 adversely effects on ammonia oxidation (Lin and Lee, 2001). The empirical formula is valid for water temperatures between about 5 and 30 °C (Equation 25.8).

The model calibration began by adjusting the ammonification rate (i.e., the nutrient source, ON, in sand layer) to minimize the discrepancies between modeled and measured values. The model calibration then moved along the direction of nutrient transport (i.e., from bottom to top) and nitrogen transformation (i.e., from left to right) in relation to all three parameters of interest (Figure 25.3), which included the rates of ammonification, nitrification, and denitrification. Their final values were determined based on assigned parameter values in an effort for model calibration (Tables 25.6 and 25.7). After these iterations of model calibration, the final agreement between the measured and simulated values of ON, NH$_4$, and the sum of

Table 25.7 Rate of Ammonification, Nitrification, and Denitrification in Model (Xuan et al., 2010)

	r_a (mg L^{-1}day^{-1})	r_n (mg L^{-1}day^{-1})	r_d (mg L^{-1}day^{-1})
Sand layer	0.63	7.71	6.37
PC layer	2.20	7.84	6.18
G layer	0.41	1.24	1.35

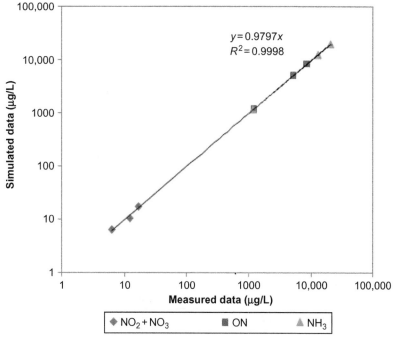

FIGURE 25.6

Correlation between the measured and simulated values in model (Xuan et al., 2010). (For the color version of this figure, the reader is referred to the online version of this chapter.)

$NO_2 + NO_3$ were quantified by the slope of the regression line of 0.9791 and the R^2 value of 0.9998 (Figure 25.6), which supports the model calibration.

The experimental data for third run was used for model validation. Table 25.8 lists the measured environmental values of the third run. The correlation between the measured and simulated values is shown in Figure 25.7. The slope of the regression line was 0.9532 and correlation (R^2) was about 0.9644, which shows the model validation, corroborating previous calibrated data shown in Table 25.7. The values of sum of nitrite and nitrate ($NO_2 + NO_3$) led to a slightly lower R^2 value. The extremely low concentration, which is close to the lower detection limit, might increase the deviation. The ammonification rate constant (k_a) in PC media increased up to fivefold compared with

Table 25.8 Temperature, pH, and Dissolved Oxygen Value Used in Model Validation (the Third Run)

	DO (mg L^{-1})	pH (Unitless)	Temperature (°C)
Sand layer	3.02	7.77	32.23
PC layer	2.68	7.40	32.37
G layer	2.73	7.44	33.04

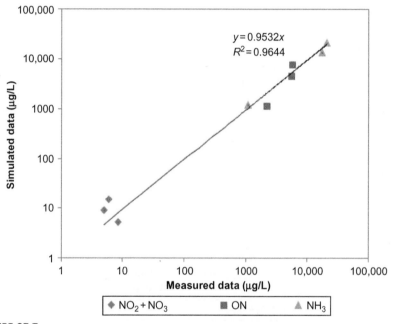

FIGURE 25.7

Correlation between the measured and simulated values in model (Xuan et al., 2010). (For the color version of this figure, the reader is referred to the online version of this chapter.)

that in sand layer. The denitrification rate constant in PC media was 30% more than that in sand layer and 3 × as much as in G media. May et al. (1990) found that most nitrifiers were associated with roots rather than the gravel layer. Similarly, we found higher specific yield of nitrification rate (u_N/Y_N) in G media, which is the root zone layer.

Because the field-collected water quality data were obtained with a flow rate of only 114 L day^{-1} (30 gallons day^{-1}), the flow rate for the SUW was increased in the model up to the average of 456 L day^{-1} (120 gallons day^{-1}). The inflow concentration for all three forms of nitrogen was kept the same as during the calibration and verification modeling: 14.0 mg L^{-1} of ON, 55.1 mg L^{-1} of NH$_4$, and 7.0 µg L^{-1} as the sum of NO$_2$+NO$_3$. The results indicated that an HRT of 6–7 days in the SUW was needed to maintain nitrogen levels below 10 mg L^{-1}.

Overall, model calibration and validation received excellent R-squared values of 0.9998 and 0.9644, respectively (Xuan et al., 2010). This close agreement with the measured data confirms that the developed system dynamics model provides a reliable tool for designing this particular type of constructed wetland (Xuan et al., 2010), as well as validates the operation over time.

25.5 SENSITIVITY ANALYSIS

The exceptional ability of wetlands for nutrients removal in our study has been fully confirmed. However, the wetland 1 just treated the wastewater with the loading of 113.4 L day^{-1} (30 gallons day^{-1}), which is smaller than the amount of wastewater produced from most common family. People might wonder how the SUW work under higher loading to fully meet the requirement of household wastewater treatment. In such a case, the superiority of the dynamic simulation model is manifested. A new wastewater loading number input and a gentle press on "run" button relieves all the effort to manually increase the wastewater loading into wetland and collect the water samples for analyses. Keeping the inflow concentration for all three forms of nitrogen: 14.0 mg L^{-1} of organic nitrogen (ON), 55.1 mg L^{-1} of ammonium (NH$_4$), and 7.0 µg L^{-1} of the sum of nitrite and nitrate (NO$_2$+NO$_3$), 378 L day^{-1} (100 gallons day^{-1}), 576 L day^{-1} (200 gallons per day), 1134 L day^{-1} (300 gallons day^{-1}), and 1512 L day^{-1} (400 gallons day^{-1}) were input as the inflow rate into the model interface, all the parameters were kept the same as used in model calibration. The concentration of organic nitrogen (ON), ammonium (NH$_4$), and the sum of nitrite and nitrate (NO$_2$+NO$_3$) from the outlet were shown as follows (Figure 25.8).

With the flow rate of 378 L day^{-1}, three forms of nitrogen keep increasing with the time. With the increase up to fourfold wastewater loading, the concentrations of NH$_4$ and NO$_2$+NO$_3$ increased with almost the same ratio. The ON concentration had a less increase after triple loading. With the loading of 1,512 L day^{-1}, the concentrations of NH$_4$, NO$_2$+NO$_3$, and ON were less than 42 mg L^{-1}, 250 µg L^{-1}, and 16 mg L^{-1}, respectively. The NO$_2$+NO$_3$ concentration was still far beyond the maximum contaminant level (MCL) drinking water standard. With the wastewater loading increase, we can obviously see that the concentrations of nitrogen reach a stable level after the 2-day treatment. That is to say, the dimension of wetland had been overdesigned due to the remarkable nitrogen removal of the media. Half of original dimension is more than enough. The complexity of nitrification rate has significant influence on the model accuracy. Further sensitivity analyses, especially for the nitrification rate, may certainly help us understand the mechanism according to the nitrogen removal leading to modify the model up to a more sophisticated level in the future. Temperature (T), pH, and dissolved oxygen (DO) are all the variables of the nitrification rate equation. Certain ranges of these three parameters were introduced to examine how they individually work on the nitrification rate.

As Table 25.9 shows, the nitrification rate is hardly affected by temperature. Instead, DO and pH value are critical for the nitrification. The lower level of DO resulted in an enlarged range of variation of nitrification rate presumably because of the Monod-style expression. The G media layer had an extreme low DO value, 1.3 mg L^{-1}, which might explain the 31.18% decrease of the nitrification rate. Slightly acidic wastewater with pH as 6.67 also might produce a decrease of 27.49% in the nitrification rate.

Recently, two more nitrogen transformations, ANAMMOX (anaerobic ammonia oxidation) and nitrate-ammonification (conversion of ammonia to nitrate under anaerobic conditions), have been studied in the CWs (Dong and Sun, 2007). Both transformations might have contributed the high nitrogen removal efficiency in our study. However, the extent of these reactions in CWs is far from certain. There is still a lack of information about these processes in CWs and their role in treatment process

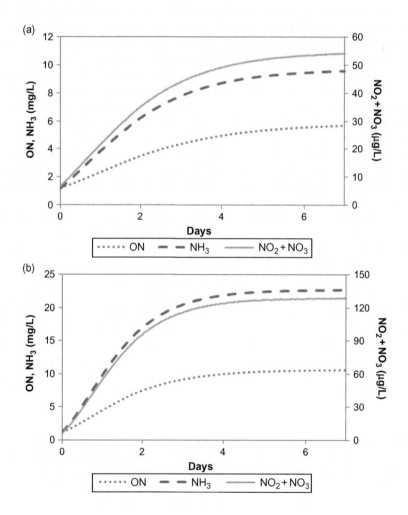

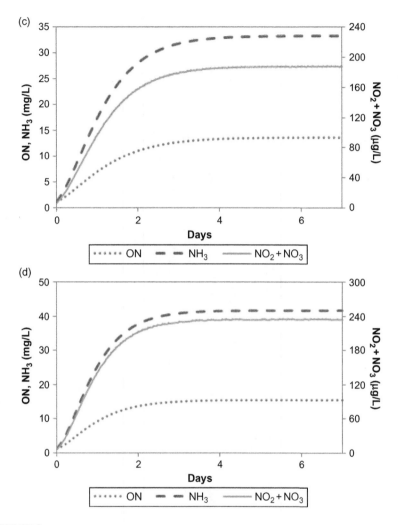

FIGURE 25.8

Effluent quality of different wastewater loading: (a) 378 L day^{-1} (100 GPD), (b) 756 L day^{-1} (200 GPD), (c) 1134 L day^{-1} (300 GPD), and (d) 1512 L day^{-1} (400 GPD) (Xuan et al., 2010). (For the color version of this figure, the reader is referred to the online version of this chapter.)

(Vymazal, 2007). Thus, we temporarily count those effects as an integral part of generalized nitrification/denitrification in our model if they do exist. Even they can be confirmed, our system dynamic model will still be useful and applicable after just adding one set of transformation rate to respond to these two more nitrogen transformations.

Table 25.9 Min and Max Value of Temperature, pH, and Dissolved Oxygen in Our Case and How Much Percentage They Correspondingly Influence the Nitrification Rate Compared with the Average Value

| | DO (mg L^{-1}) | | pH (Unitless) | | Temperature (°C) | |
	MIN	MAX	MIN	MAX	MIN	MAX
Sand layer	2.87 (⁻5.16%)	4.46 (⁺6.70%)	6.86 (⁻11.66%)	7.46 (⁺0.00%)	26.1 (⁻0.35%)	33.2 (⁺0.01%)
PC layer	2.24 (⁻9.69%)	4.56 (⁺7.11%)	6.81 (⁻15.83%)	7.35 (⁺0.00%)	25.5 (⁻0.36%)	33.6 (⁺0.00%)
G layer	1.3 (⁻31.18%)	3.77 (⁺2.35%)	6.67 (⁻27.49%)	7.4 (⁺0.00%)	26.3 (⁻0.28%)	33.1 (⁺0.03%)

⁺, increase; ⁻, decrease (Xuan et al., 2010).

Table 25.10 Spatial Distribution of Nitrogen Concentration, μg L^{-1} (Chang et al., 2012)

	NH$_3$-N	NO$_2$-N + NO$_3$-N	ON-N	TN
Septic effluent	56,061.8	7.5	13,670.3	69,739.5
Port B	27,159.0	7.0	5325.3	32,489.0
Port M	16,653.3	8.5	5069.8	21,730.0
Port T1	839.8	5.0	3684.8	4526.8
Port T2	1169.8	16.8	1497.0	2681.3
Port T3	1185.0	8.8	3607.0	4799.3

Note: TN, total nitrogen, ON-N, organic nitrogen; NH$_3$-N, ammonia nitrogen; port B, the composite of three bottom samples; port M, the composite of three middle samples.

25.6 FINAL REMARKS

In our study, composite samples were collected across ports M and B (Figure 25.2), assuming that water quality would not vary much in these two layers. Because of the root zone heterogeneity, sampling across three ports at the top layer between G and PC media was carried out separately. Water quality analysis spatially across these three layers and ports (Table 25.10) showed a systematic trend of nitrification and denitrification layer by layer from the bottom to the top layers.

The nutrient removal in the SUW reveals the affinity between transport observations and fate phenomena. In concert with the upflow pattern, the concentration of TN, ON-N, and NH$_3$-N decreased layer by layer. In particular, ON-N and TN were lower in the middle top ports (i.e., T2) due to the delayed flow, evidenced by the hydraulic retardation effect in the tracer study (Figure 25.1). Therefore, a deeper SUW is recommended in the future to reach a longer HRT and achieve a higher ON-N removal efficiency. A more sufficient aeration via the use of more oxygenators could enhance nitrification and improve NH$_3$-N removal, however. Such a synergistic modification could further promote overall performance (Chang et al., 2012).

25.7 CONCLUSIONS

A tracer study was carried out to confirm upflow hydraulic patterns in concert with a transport model to assess the nutrient removal efficiencies across different locations of the SUW system.

The system dynamic model was confirmed as a convenient tool to describe the coupling between hydrodynamics, geochemical interactions, and microbiological activities in this SUW system. The predicted and measured concentrations of the ON, NH$_4$, and NO$_2$ + NO$_3$ in the SUW were in close agreement. The ammonification rate constant (k_a) in designed media (PC and G media) increased up to fivefold compared with that in sand layer. The denitrification rate constant in PC media is 30% more than that in sand layer and $3\times$ as much as in G media. Higher specific yield of nitrification rate (u_N/Y_N) was found in G media, the root zone layer. Planting vegetation with deeper roots, which may reach the bottom of PC media, is recommended to further increase DO, nitrification rate, and NH_4 uptake during the initial period. The modified nitrification rate expression formula was used in model validation and proved to be feasible.

Acknowledgments

The authors are grateful for the financial support provided by an Urban Nonpoint Source Research Grant from the Bureau of Watershed Restoration, Florida Department of Environmental Protection.

References

Chang, N.B., Xuan, Z., Daranpob, A., Wanielista, M.P., 2011. A subsurface upflow wetland system for nutrient and pathogen removal in on-site sewage treatment and disposal systems. Environ. Eng. Sci. 28 (1), 11–24.

Chang, N.B., Xuan, Z., Wanielista, M.P., 2012. A tracer study for addressing the interactions between hydraulic retention time and transport processes in a subsurface wetland system for nutrient removal. Bioproc. Biosyst. Eng. 35 (3), 399–406.

Chowdhury, R., Apul, D., Dwyer, D., 2008. Preliminary studies for designing a wetland for arsenic treatment. Groundwater: Modelling, Management and Contamination. Nova Science Publishers, Inc., Hauppauge NY, pp. 151–166.

Danckwerts, P.V., 1953. Continuous flow systems: distribution of residence times. Chem. Eng. Sci. 2 (1), 1–13.

Dierberg, F.E., DeBusk, T.A., 2005. An evaluation of two tracers in surface-flow wetlands: rhodamine-WT and lithium. Wetlands 25 (1), 8–25.

Dong, Z., Sun, T., 2007. A potential new process for improving nitrogen removal in constructed wetlands—promoting coexistence of partial-nitrification and ANAMMOX. Ecol. Eng. 31 (2), 69–78.

Giraldi, D., De'Michieli Vitturi, M., Zaramella, M., Marion, A., Iannelli, R., 2009. Hydrodynamics of vertical subsurface flow constructed wetlands: tracer tests with rhodamine WT and numerical modeling. Ecol. Eng. 35, 265–273.

Giraldi, D., de Michieli Vitturi, M., Iannelli, R., 2010. FITOVERT: a dynamic numerical model of subsurface vertical flow constructed wetlands. Environ. Model. Softw. 25 (5), 633–640.

Harman, J., Robertson, W.D., Cherry, J.A., Zanini, L., 1996. Impacts on a sand aquifer from an old septic system: nitrate and phosphate. Ground Water 34, 1105–1114.

Headley, T.R., Kadlec, R.H., 2007. Conducting hydraulic tracer studies of constructed wetlands: a practical guide. Ecohydrol. Hydrobiol. 7, 269–282.

Henze, M., Gujer, W., Mino, T., van Loosdrecht, M., 2000. Activated sludge models ASM1, ASM2, ASM2d and ASM3: Scientific and technical report No 9. IWA Publishing, London, UK.

Jou, C.U., Chen, S.W., Kao, C.M., Lee, C.L., 2008. Assessing the efficiency of a constructed wetland using a first-order biokinetic model. Wetlands 28 (1), 215–219.

Kadlec, R.H., Knight, R., 1996. Treatment Wetlands. CRC Press, Boca Raton, FL.

Kadlec, R.H., Wallace, S.D., 2008. Treatment Wetlands, second ed. CRC Press, Boca Raton, FL.

Kadlec, R.H., Tanner, C.C., Hally, V.M., Gibbs, M.M., 2005. Nitrogen spiraling in subsurface-flow constructed wetlands: implications for treatment response. Ecol. Eng. 25, 365–381.

Koike, I., Hattori, A., 1978. Denitrification and ammonia formation in anaerobic coastal sediment. Appl. Environ. Microbiol. 35 (2), 278–282.

Langergraber, G., 2001. Development of a Simulation Tool for Subsurface Flow Constructed Wetlands. Wiener Mitteilungen, vol. 169, p. 207.

Lin, S.D., Lee, C.C., 2001. Water and Wastewater Calculations Manual. McGraw-Hill, New York, NY.

Lin, A.Y.C., Debroux, J.F., Cunningham, J.A., Reinhard, M., 2003. Comparison of rhodamine WT and bromide in the determination of hydraulic characteristics of constructed wetlands. Ecol. Eng. 20, 75–88.

Małoszewski, P., Wachniew, P., Czupryński, P., 2006. Hydraulic characteristics of a wastewater treatment pond evaluated through tracer test and multi-flow mathematical approach. Pol. J. Environ. Stud. 15 (1), 105–110.

May, E., Butler, J.E., Ford, M.G., Ashworth, R., Williams, J.B., Bahgat, M.M.M., 1990. Chemical and microbiological processes in gravel-bed hydroponic (GBH) systems for sewage treatment. In: Cooper, P.F., Findlater, B.C. (Eds.), Constructed Wetlands in Water Pollution Control. Pergamon Press, Oxford, UK, pp. 33–40.

Pastor, R., Benqlilou, C., Paz, D., Cardenas, G., Espuna, A., Puigjaner, L., 2003. Design optimisation of constructed wetlands for wastewater treatment. Resour. Conserv. Recycl. 37 (3), 193–204.

Ronkanen, A.K., Kløve, B., 2007. Use of stabile isotopes and tracers to detect preferential flow patterns in a peatland treating municipal wastewater. J. Hydrol. 347, 418–429.

Ronkanen, A.K., Kløve, B., 2008. Hydraulics and flow modelling of water treatment wetlands constructed on peatlands in Northern Finland. Water Res. 42, 3826–3836.

Sajuni, N.R., Ahmad, A.L., Vadivelu, V.M., 2010. Effect of filter media characteristics, pH and temperature on the ammonia removal in the wastewater. J. App. Sci. 10, 1146–1150.

Tomenko, V., Ahmed, S., Popov, V., 2007. Modelling constructed wetland treatment system performance. Ecol. Model. 205, 355–364.

Tunçsiper, B., Ayaz, S.C., Akça, L., 2006. Modelling and evaluation of nitrogen removal performance in subsurface flow and free water surface constructed wetlands. Water Sci. Technol. 53 (12), 111–120.

Vymazal, J., 2007. Removal of nutrients in various types of constructed wetlands. Sci. Total Environ. 380, 48–65.

Vymazal, J., Kröpfelová, L., 2009. Removal of nitrogen in constructed wetlands with horizontal sub-surfaceflow: a review. Wetlands 29 (4), 1114–1124.

Wang, B., Jin, M., Nimmo, J.R., Yang, L., Wang, W., 2008. Estimating groundwater recharge in Hebei Plain, China under varying land use practices using tritium and bromide tracers. J. Hydrol. 356, 209–222.

Werner, T.M., Kadlec, R.H., 1996. Application of residence time distribution to stormwater treatment system. Ecol. Eng. 7, 213–234.

Wynn, T.M., Liehr, S.K., 2001. Development of a constructed subsurface flow wetland simulation model. Ecol. Eng. 16, 519–536.

Xuan, Z., Chang, N.B., Makkeasorn, A., Wanielista, M., 2009. Initial test of a subsurface constructed wetland with green sorption media for nutrient removal in on-site wastewater treatment systems. Water Qual. Expo. Health 1 (3), 159–169.

Xuan, Z., Chang, N.B., Daranpob, A., Wanielista, M.P., 2010. Modeling the subsurface upflow wetlands (SUW) systems for wastewater effluent treatment. Environ. Eng. Sci. 27 (10), 879–888.

Index

Note: Page numbers followed by *f* indicate figures and *t* indicate tables.

Printed and bound by CPI Group (UK) Ltd, Croydon, CR0 4YY

08/05/2025

01864826-0003